AF370783

Feeding Strategies to Improve Sustainability and Welfare in Animal Production

Feeding Strategies to Improve Sustainability and Welfare in Animal Production

Editors

Fulvia Bovera
Giovanni Piccolo

MDPI • Basel • Beijing • Wuhan • Barcelona • Belgrade • Manchester • Tokyo • Cluj • Tianjin

Editors
Fulvia Bovera
Università degli Studi di Napoli Federico II
Italy

Giovanni Piccolo
Università degli Studi di Napoli Federico II
Italy

Editorial Office
MDPI
St. Alban-Anlage 66
4052 Basel, Switzerland

This is a reprint of articles from the Special Issue published online in the open access journal *Animals* (ISSN 2076-2615) (available at: https://www.mdpi.com/journal/animals/special_issues/Feeding_strategies_to_improve_sustainability_and_welfare_in_animal_production).

For citation purposes, cite each article independently as indicated on the article page online and as indicated below:

LastName, A.A.; LastName, B.B.; LastName, C.C. Article Title. *Journal Name* **Year**, *Volume Number*, Page Range.

ISBN 978-3-03943-252-3 (Hbk)
ISBN 978-3-03943-253-0 (PDF)

Contents

About the Editors

Fulvia Bovera Ph.D., DVM. She graduated in Veterinary Medicine, summa cum laude, in 1997. In 2001, she obtained her Ph.D. in Animal Production. She has been researcher since 2005 and associate professor since 2017 at the Department of Veterinary Medicine and Animal Production, University of Napoli Federico II. She is Section Editor of Italian Journal of Animal Science a Taylor & Francis open access journal (IF 2019: 1.805; Rank Q1). She is author/co-author of 120 Scopus indexed experimental papers (h-index: 25).
Main research topics:
- The use of insect meals in poultry and fish nutrition;
- Alternatives to antibiotics in rabbit production;
- Improvement of honey bee production.

Giovanni Piccolo Ph.D., He graduated in Animal Production, summa cum laude, in 2000. In 2004, he obtained his Ph.D. in Animal Production. He has been researcher since 2005 and associate professor since 2020 at the Department of Veterinary Medicine and Animal Production, University of Napoli Federico II. From 2012 to 2014, he was scientific coordinator of a project on the breeding of blackspot sea bream (Pagellus bogaraveo), an innovative fish species for Italian aquaculture, funded by the Puglia Region as part of the EFF-The European Fisheries Fund 2007–2013 (382,000 €). In 2014 and 2017, he won two Transnational Access (TNA) grants in the AQUAEXCEL and AQUAXCEL 2020 programs as project scientific coordinator. He is author/co-author of 69 Scopus indexed experimental papers (h-index: 20)
Main research topics:
- The use of innovative protein sources in fish and poultry nutrition;
- The use of nutraceuticals in fish;
- Breeding of new fish species.

Preface to "Feeding Strategies to Improve Sustainability and Welfare in Animal Production"

Most accredited FAO statistics predict that in 30 years, the world's population will have reached 9 billion people. In order to satisfy the nutritional needs of humans, the demand for raw materials, especially protein sources, will increase. It has been estimated that by 2050, the production of meat will have increased by 50%, while the demand for fish, milk, and eggs will have grown by 75%. An increase in animal products requires an increase in farmed animals, and this will be accompanied by a significant intensification in livestock farming (higher animal densities and production units, more concentrated feed, pharmaceuticals, and vaccinations, etc.). A large number of animals, farmed in relatively small areas, will result in a larger demand for protein and energy sources on which to feed them and in the deposition of large amounts of excreta, containing nitrogen, phosphorus, organic matter, and fecal microbes, in the water, with a consequent contamination of water systems globally, which will include surface water eutrophication and groundwater nitrate enrichment. Thus, the livestock sector is an important user of natural resources and has a great influence on air, soil, and water quality, the global climate, and biodiversity maintenance. Our research proposes innovative ideas to control the environmental damage through the management of animal nutrition. At the same time, the perception of animals as sentient beings capable of feeling emotions, like joy and pain, will increase in prevalence in the future. Thus, it will be increasingly important to adopt nutritional strategies and breeding techniques capable of increasing animal welfare and at the same time to reduce the use of pharmacological treatments in full respect of the environment, animal health, and food safety.

Fulvia Bovera, Giovanni Piccolo
Editors

Article

Microbial and Fungal Phytases Can Affect Growth Performance, Nutrient Digestibility and Blood Profile of Broilers Fed Different Levels of Non-Phytic Phosphorous

Youssef A. Attia [1,*], Fulvia Bovera [2,*], Francesco Iannaccone [2], Mohammed A. Al-Harthi [1], Abdulaziz A. Alaqil [3], Hassan S. Zeweil [4] and Ali E. Mansour [5]

[1] Arid Land Agriculture Department, Faculty of Meteorology, Environment and Arid Land Agriculture, King Abdulaziz University, P.O. Box 80208, Jeddah 21589, Saudi Arabia; malharthi@kau.edu.sa

[2] Department of Veterinary Medicine and Animal Production, University of Napoli Federico II, via F. Delpino 1, 80137 Napoli, Italy; francesco.iannaccone@unina.it

[3] Department of Animal and Fish Production, King Faisal University, Al-Hufof, Al-Hassa 31982, Saudi Arabia; aalaqil@yahoo.com.

[4] Department of Animal and Fish Production, Faculty of Agriculture, Saba Basha, Alexandria University, Alexandria 21527, Egypt; hszeweil@yahoo.com

[5] Ministry of agriculture, Animal Production section, Behiri Governorate 22951, Egypt; alierfatmansour@gmail.com

* Correspondence: yaattia@kau.edu.sa (Y.A.A.); bovera@unina.it (F.B.); Tel.: +00966568575961 (Y.A.A.); +390812536497 (F.B.)

Received: 28 February 2020; Accepted: 25 March 2020; Published: 30 March 2020

Simple Summary: To reduce the environmental pollution is a must to preserve the health of the world. The environmental impact of poultry farming is receiving an increasing attention due to several emissions among these is phosphorus. This element is in general present in the commercial diets of broilers or laying hens in an amount exceeding the real needing of the animals, and, therefore, a great amount of phosphorus ends in the excreta. Thus, optimizing the amount of phosphorous in the diets of poultry could partially alleviate the environmental impact of these farms.

Abstract: A total of 420 day old chicks were divided into seven groups (5 replicates of 12 chicks/group) fed isoproteic and isoenergetic diets. The control group was fed diets containing 0.50%, 0.45% and 0.40% of non-phytic phosphorous (nPP) in starter (1–35), grower (37–56) and finisher (57–64 d) periods, respectively. The three intermediate nPP (IntnPP) groups were fed diets with 0.40%, 0.35% and 0.30% nPP according to the growth period and were submitted to three dietary treatments: unsupplemented; supplemented with 500 FTU/kg diet of an *Aspergillus niger* phytase (IntnPP_fp) and supplemented with 500 FTU/kg diet of an *Escherichia coli* phytase (IntnPP_bp). The three low nPP groups fed diets contained 0.30%, 0.25% and 0.20% nPP and were submitted to the same dietary treatments than IntnPP to obtain LnPP, LnPP_fp and LnPP_bp groups. IntnPP and LnPP groups had lower body weight gain and feed, crude protein (CP) and metabolizable energy (ME) intake ($p < 0.05$) than the control. Feed conversion ratio of IntnPP was more favorable ($p < 0.01$) than the LnPP group. CP and ME conversion ratios worsened ($p < 0.01$) in IntnPP and LnPP groups in comparison to the control. The nPP conversion ratio improved ($p < 0.01$) from the control to the LnPP group. Fungal phytase reduced ($p < 0.05$) feed, CP, ME and nPP intake than the bacterial one. IntnPP and LnPP diets had a lower digestibility of CP ($p < 0.01$) and CF ($p = 0.01$) than the control. IntnPP and LnPP groups showed a higher ($p < 0.05$) economic efficiency than the control. Blood total protein was the lowest ($p < 0.05$) in the LnPP group, the control group showed the lowest ($p < 0.05$) level of albumin and IntnPP group had the lowest ($p < 0.01$) globulin level. The use of bacterial phytase increased ($p < 0.01$) total protein and globulin and decreased ($p < 0.05$) the plasma cholesterol in comparison to fungal phytase. Decreasing nPP levels in colored slow-growing broilers diet negatively affects growth performance and the use of phytase can partly alleviate these negative effects, but the efficiency of different enzyme sources (bacterial or fungal) was tied to the dietary nPP levels.

Keywords: phytase; Sasso broiler; non-phytic phosphorous; productive performance; blood profiles

---

1. Introduction

Phosphorus (P) is an essential nutrient for plants and animals and is critically important in the production of poultry. Phosphorus is well represented in the vegetables ingredients commonly used in poultry diet preparation (as soybean or corn), however 50%–80% of total phosphate in plant seeds is stored as phytate P [1]. Phosphorous phytate is poorly available to intestine absorption of monogastric animals and can also reduce the digestibility of other nutrients as well as the performance of animals owing to its antinutritional effect [2,3]. It is possible to add a further amount of non-phytic P to the diet, but this allows an increase of the amount of P released, especially from the waste of intensive poultry farms. It is well known that excess of P in the water led to its eutrophication and the subsequent enrichment of surface water by plant nutrients is a form of pollution [4]. This problem has led to numerous studies to reduce the P amount in the waste from the poultry industry. Several studies indicated great differences in the non-phytate phosphorus (nPP, inorganic) requirement of broilers compared to the data presented in the 1994 by the National Research Council (NRC). Waldroup et al. [5] showed that the nPP requirement for the starter phase of broilers ranges from 0.37% to 0.39%. Angel et al. [6,7] fixed the nPP requirement of the broiler between 0.32% and 0.28% in the growing period (18–32 d), 0.24% and 0.19% in the finisher period (32–42 d) and 0.16% and 0.11% in the withdrawal period (42–49 d). Dhandu and Angel [8] reported a requirement of 0.20% nPP for the finisher period and 0.16% nPP in the withdrawal period of broiler. All these nPP requirements are substantially lower than that reported by the NRC [9]. These data suggested that it is possible to reduce the P concentration in the diet and this can reduce the amount of P in the manure without negative effects on poultry performance. The addition of phytase to the diet can help to release P from phytate and thus reduce the amount of non-phytic P supplemented to the diets. Phytase is the most common enzyme used in the feed for monogastric animals: it can reduce the antinutritional effect of phytate and improve the digestibility of phosphorous, calcium, amino acids and energy, as well as reduce the negative impact of inorganic P excretion to the environment [10]. Phytase are phosphatases hydrolyzing one or more phosphate group. Depending on the position of the phosphate group on the myo-inositol ring, which they firstly hydrolyze, they belong to one of two subclasses, 3-phytase and 6-phytase [11]. Ghazalah et al. [12] and Selim [13] indicated that phytase supplementation to the diet of fast-growing chickens can replace around 0.1% of dietary P; however, the data are inconsistent due to type of diet, dietary composition, type and age of chicks, type and dose of phytase. So, there are many areas of research that need to be addressed such as the effect of type of phytase and the efficacy of bacterial (phytase-6) vs. fungal phytase (phytase-3) on phosphorus utilization, due to the exist contradictory results [6,14–18].

The aim of our research was to study the effect of microbial or fungal phytase on growth performance, nutrient digestibility, carcass and meat quality and blood profile of colored slow-growing Sasso broilers fed diets with different supplementation of non-phytic phosphorous.

2. Materials and Methods

2.1. Chickens, Experimental Design and Husbandry

The department committee of Animal and Poultry Production accepts all the procedures done in the current study. These procedures suggest minimal stress to the animal to ensure rights and welfare by eliminating harm or suffering to animals according to the official decrees of the Ministry of Agriculture in Egypt regarding animal welfare (Decree No. 27 (1967) that enforces the humane treatment of animals generally).

A total of 420 unsexed day-old colored broilers (Sasso strain) were homogeneously divided into 7 groups (5 replicates of 12 chicks each). Chicks were housed in 35 floor pens (1.0 m × 1.0 m/pen) furnished with rice hulls as a litter. Along the experimental period, chicks were fed three different diets, according to the animal age: starter (1–35 d of age); grower (36–56) and finisher diet (57–64). The experimental period was listed from 1 to 64 d of age. The control group fed diets supplemented with 0.50%, 0.45% and 0.40% of non-phytic phosphorus (nPP), respectively for each period. The three groups fed intermediate nPP (IntnPP) diets supplemented with 0.40%, 0.35% and 0.30% nPP in the starter, grower and finisher periods, respectively were submitted to three dietary treatments: unsupplemented, supplemented with 500 FTU of a fungal phytase (IntnPP_fp, BASF Germany, n-3 *Aspergillus niger* phytase)/kg diet and supplemented with 500 FTU of a bacterial phytase (IntPP_bp, Phyzyme® XP, n-6 *Escherichia coli* phytase, Danisco Animal Nutrition)/kg diet. The three groups fed low nPP (LnPP) diets were supplemented with 0.30%, 0.25% and 0.20% nPP in the starter, grower and finisher periods, respectively and were submitted to the same dietary treatments than IntnPP groups to obtain LnPP, Lnpp_fp and LnPP_bp fed groups.

The experimental diets (Table 1) were formulated according to NRC [9]. Washed building sand was used at small amount (0.14%–0.50%) to keep nutrients profiles of different diets similar as far as possible as, otherwise the change in feedstuffs contents may affect nutrients profiles.

Table 1. Ingredients profiles and nutrient compositions (as the fed basis) of the experimental diets fed during the starter, grower and finisher diets.

Items	Starter			Grower			Finisher		
	0.50 nPP	0.45 nPP	0.35 nPP	0.45 nPP	0.35 nPP	0.25 nPP	0.40 nPP	0.30 nPP	0.20 nPP
Ingredients									
Yellow corn	58.30	58.35	58.35	62.65	62.65	62.70	63.00	63.00	63.55
Soybean meal	32.00	32.00	32.00	27.50	27.60	27.50	28.25	28.25	28.20
Fish meal 65 % CP	3.00	3.00	3.00	3.00	3.00	3.00	-	-	-
Limestone	1.00	1.35	1.73	0.90	1.25	1.60	0.85	1.20	1.55
Dicalcium Phosphate	1.85	1.25	0.65	1.60	1.00	0.35	1.75	1.10	0.50
Vit + Min premix [1]	0.30	0.30	0.30	0.30	0.30	0.30	0.30	0.30	0.30
NaCl	0.30	0.30	0.30	0.30	0.30	0.30	0.30	0.30	0.30
Methionine	0.15	0.15	0.15	0.15	0.15	0.15	0.15	0.15	0.15
Lysine	0.10	0.10	0.10	0.10	0.10	0.10	0.10	0.10	0.10
Vegetable oils	3.00	3.00	3.00	3.50	3.50	3.50	5.30	5.30	5.30
Washed building sand	0.00	0.20	0.42	0.00	0.15	0.50	0.00	0.30	0.14
Chemical-nutritional characteristics									
Dry matter, % [2]	89.61	89.83	89.67	89.61	89.83	89.67	89.57	89.72	89.66
CP, % [2]	21.03	21.04	21.04	19.49	19.42	19.40	17.80	17.83	17.78
ME, MJ/kg [3]	12.33	12.33	12.33	12.65	12.66	12.65	12.93	12.93	12.97
SAA, % [3]	0.85	0.85	0.85	0.80	0.80	0.80	0.74	0.74	0.74
Lysine, % [3]	1.24	1.24	1.24	1.13	1.13	1.13	1.00	1.00	1.00
Calcium, % [3]	1.00	1.00	1.00	0.90	0.90	0.90	0.80	0.80	0.80
Av. P, % [2]	0.47	0.38	0.26	0.41	0.33	0.23	0.37	0.28	0.17
Ether extract, % [2]	5.47	5.51	5.45	6.21	6.18	6.22	7.63	7.68	7.59
Crude fiber, % [2]	3.47	3.47	3.49	3.31	3.36	3.28	3.33	3.36	3.29
Ash, % [2]	9.24	9.24	9.70	9.30	9.41	9.88	9.18	9.50	9.37
NFE, % [2]	50.4	50.57	49.99	51.3	51.46	50.89	51.63	51.35	51.63

[1] Vit+Min mixture provides per kilogram of the diet: vitamin A (retinyl acetate) 24 mg, vitamin E (dl-α-tocopheryl acetate) 20 mg, menadione 2.3 mg, Vitamin D3 (cholecalciferol) 0.05mg, riboflavin 5.5 mg, calcium pantothenate 12 mg, nicotinic acid 50 mg, choline chloride 600 mg, vitamin B12 10 *g, vitamin B6 3 mg, thiamine 3 mg, folic acid 1 mg, d biotin 0.50 mg. Trace mineral (milligrams per kilogram of diet): Mn 80 Zn 60, Fe 35, Cu 8, Se 0.60. [2] Analyzed values. [3] Calculated values. nPP: non phytic phosphorous; CP: crude protein; ME: metabolizable energy; SAA: sulphate amino-acids; Av. P: available phosphorous; NFE: nitrogen free extract

Feed samples of the starter, grower and finisher diets were chemically analyzed for dry matter, crude protein, ether extract, crude fiber and ash according to the Association of Official Agricultural Chemists (AOAC) [19] official methods. The nitrogen free extract was determined by subtracting the sum of all fractions mentioned above from the dry matter (DM).

Mash form diets and clean water were offered ad libitum. Birds were illuminated with 24 h light cycle during the first 3 days of age and then with 23:1 light-dark cycle, according to Attia et al. [20]. The vaccinations and medical care were conducted according to veterinary indications. Chicks were farmed under similar environmental, managerial, and hygienic condition and breed according to Hendrix Genetic indication for the Sasso strain (available at: https://www.hendrix-genetics.com/en/news/new-brand-promise-sasso/).

2.2. Data Collection

Birds were weighed (g) in the morning before offering feeds at the beginning (day 1) and at the end of the trial (day 64) to calculate body weight gain for each replicate. Each group was provided daily with enough pre-weighed feed of its corresponding diet. The remainder and scattered feed as well as the consumed feed were weekly calculated for each replicate to calculate feed intake in the periods 1–64 days of age. The consumption of CP, ME and nPP was calculated by multiplying the CP, ME and nPP contents of the experimental diets by its corresponding feed consumption during the entire experimental period.

Conversion indexes of feed (feed conversion ratio (FCR), g feed/g gain), crude protein (crude protein conversion ratio, CPCR, g protein/g gain) and ME (metabolizable energy conversion ratio, MECR, MJ/g gain) of the diets were calculated during the experimental period. Utilization of P as the nPP conversion ratio was calculated as g of nPP consumption required to produce 1000 g body weight gain. Mortality rate was calculated as the percentage of birds dead in each treatment during the entire experimental period.

At 64 days of age, 6 broilers per treatment as three replicates of two males each were used to measure the nutrient digestibility of the diets according to the total collection method described by Attia et al. [20]. The broilers were fasted for 24 h, thus fed on their experimental diets for 120 h; feed intake and excreta were measured during 72 h. The excreta, collected for each replicate, sprayed with 4% boric acid to capture the ammonia in the form of ammonium borate. The excreta were cleaned from feathers and feed, weighed and dried in a forced air oven at 70 °C until constant weight. Samples were finely ground and placed in screw-top glass jars until analyses.

Fecal nitrogen was separated from urine in the excreta samples according to Jakobsen et al. [21]. Total nitrogen, fecal nitrogen, fat, crude fiber and ash contents were determined in the excrement and feeds according to AOAC [19] procedures (dry matter, method number 934.01; crude protein, method number 954.01; ether extract, method number 920.39; crude fiber, method number 954.18 and ash, method number 942.05) and expressed on a dry matter basis. Phosphorus concentrations were determined utilizing flame spectrophotometric techniques using the method of Haraguchi and Fuwa [22].

The apparent digestibility of the nutrients (dry matter, crude protein, fat and fiber) and the apparent retention of ash were calculated by dividing the daily amount retained (g/d) by the amount intake (g/d). The daily amount retained is equal to the amount intake (% nutrient in feed × g of feed consumed) minus that lost in the excreta (% nutrient in excreta × g of excreta voided). Metabolic nitrogen was equal to excreta nitrogen − fecal nitrogen.

At 64 d of age, six birds (3 males and 3 females) were randomly collected from each treatment, weighed after fasting overnight, slaughtered, their feathers were plucked and the inedible parts (head, legs and inedible viscera) were taken aside. Thus, the remaining carcass (dressed carcasses) was weighed and expressed as a percentage of the body weight. The internal organs (liver, pancreas and spleen) were weighed and expressed as percentage of live body weight. Abdominal fat (located in the abdominal cavity, surrounding the intestines and around the heart) was separated, weighed and expressed as a percentage of the live body weight.

A sample of 50% breast muscle + 50% thigh muscle was weighed and kept in an electric drying oven at 70 °C until constant weight. The dried meat was ground to pass through a sieve (1 mm^2) and then carefully mixed. The air-dried samples were kept into tight glass container for the subsequent chemical analysis [19]. Protein content of meat was calculated as 100 − moisture − ash − fat. The ability of meat to hold water (WHC), meat tenderness, pH, color intensity and drip loss were measured according to Bovera et al. [23].

The right tibia was removed from the carcass of 6 birds per group, cleaned from tissues, set in hexane for 48 h to remove fat and dry in an oven until constant weight. Their length (mm), width (mm), weight (g) as well as ash, Ca and P contents were determined. Length and width of tibia were determined using a Vernier caliber. The width of tibia was measured at three points in the middle and at the end of both sides. Ca and P contents were determined after ashing at 600 °C, according to the method of AOAC [24].

Blood samples were collected at 64 d of age from 6 broilers per group in heparinized tubes. Plasma was separated by centrifugation at 3000 rpm for 15 minutes and stored at −20 °C until analysis. Concentrations of plasma Ca, inorganic P and alkaline phosphatase were determined according to Attia et al. [25]. All biochemical traits of blood plasma (total protein, albumin, alaninine aminotransferase (ALT), aspartate aminotransferase (AST), total lipids and cholesterol) were determined using commercial diagnosing kits (Diamond Diagnostics Company, Egypt) as reported by Attia et al. [26–28]. Globulin concentration was calculated as the difference total protein − albumin.

Economic evaluation for all experimental diets was made. Economic efficiency was calculated as described by Attia et al. [29].

2.3. Statistical Analysis

Data were analyzed using the GLM procedure of Statistical Analysis Software (SAS) version 6.11 (SAS Institute Inc.: Cary, NC, USA) [30] by one-way design as the phytases was added to two nPP levels only resulted in an unbalanced experimental deign as the recommended phosphorus levels usually not supplemented with phytases. In addition, the use of factorial analyses in this case can generate main effects that were confounded by the interactions effect. The effect of nPP was compared between the unsupplemented levels and phytases within different nPP levels. Mean difference at $p \leq 0.05$ was tested using the Student Newman–Keuls test. The replicate was the experimental unit. Data in percentage were transformed to log 10 before running the analyses of variance.

3. Results

No deaths were observed in the groups along the trial and all the animals were healthy.

Data on the broilers' growth performance and feed efficiency in the entire experimental period (1–64 d) are reported in Table 2. The control and IntnPP_bp groups showed higher ($p < 0.01$) body weight gain than the other groups, except for LnPP_fp. The LnPP group showed also lower ($p < 0.01$) body weight gain (BWG) than LnPP_fp group. Feed, crude protein and energy intakes were lower ($p < 0.01$) in LnPP and LnPP_fp than the other groups. nPP intake was higher ($p < 0.01$) in the control, followed by all the IntnPP and then all the LnPP groups. The feed conversion ratio of LnPP group was higher ($p < 0.01$) than the control, IntnPP_bp and LnPP_fp groups. Crude protein and metabolizable energy conversion ratios of LnPP were higher ($p < 0.01$) than the control, IntnPP_bp and _fp and LnPP_fp. IntnPP_bp and LnPP_fp groups showed lower crude protein and metabolizable energy conversion ratios ($p < 0.01$) than the control and LnPP groups.

Table 2. Growth performance, feed, crude protein, energy intakes and their conversion during days 1-64 of age of Sasso chickens fed diets containing different non-phytate phosphorus levels with or without two types of commercial phytases.

Groups	BWG (g)	FI (g)	CP Intake (g)	ME intake (MJ)	nPP intake (g)	FCR	CPCR	MECR
Control	1769 [a]	4025 [a]	794 [a]	50.78 [a]	17.17 [a]	2.36 [b]	0.449 [b]	6.86 [b]
IntnPP	1552 [bc]	3704 [a]	732 [a]	46.76 [a]	13.16 [b]	2.39 [ab]	0.472 [abc]	7.20 [abc]
IntnPP_bp	1668 [a]	3706 [a]	733 [a]	46.83 [a]	13.19 [b]	2.22 [b]	0.439 [c]	6.72 [c]
IntnPP_fp	1598 [bc]	3709 [a]	731 [a]	46.69 [a]	13.10 [b]	2.32 [ab]	0.457 [bc]	6.98 [bc]
LnPP	1478 [c]	3675 [b]	726 [b]	46.37 [b]	9.37 [c]	2.49 [a]	0.491 [a]	7.50 [a]
LnPP_bp	1586 [bc]	3738 [a]	742 [a]	47.35 [a]	9.59 [c]	2.36 [ab]	0.468 [abc]	7.14 [abc]
LnPP_fp	1646 [ab]	3689 [b]	725 [b]	46.34 [b]	9.37 [c]	2.24 [b]	0.441 [c]	6.73 [c]
RMSE	204.2	42.7	8.56	2.15	0.376	0.088	0.0173	0.267
p value	0.01	0.01	0.05	0.04	0.0001	0.001	0.001	0.001

[abc] means within a column under the same treatment with different superscripts are significantly different, NS= not significant, RMSE= root mean square error. BWG = body weight gain; FI= feed intake, CP= crude protein, ME= metabolizable energy, nPP= Non-phytate phosphorus, FCR= feed conversion ratio, CPPR= crude protein conversion ratio, MECR= metabolizable energy conversion ratio; IntnPP: intermediate nPP group; LnPP: low nPP group; bp: bacterial phytase; fp: fungal phytase.

Table 3 shows the effect of dietary treatments on nutrient digestibility, ash retention and fate of nitrogen in broilers. Both IntnPP and LnPP diets had a lower digestibility of CP ($p < 0.01$) and CF ($p = 0.01$) and a lower ($p < 0.01$) ash retention. Excreta and metabolic nitrogen were higher in IntnPP and LnPP diets in comparison to the control. The effect of the treatments revealed that the use of both phytases improved CP digestibility only when supplemented to IntnPP diets. CF digestibility improved due to the addition of both phytases to IntnPP diets, while fungal phytase supplementation only improved CF digestibility in the LnPP groups. The excreta nitrogen in IntnPP groups decreased ($p < 0.01$) due to use bacterial phytase in comparison to the unsupplemented group while within the LnPP groups the bacterial phytase increased the percentage of excreta nitrogen in comparison to fungal phytase. Fecal nitrogen in the IntnPP and LnPP groups was similar to that of the control and lowered ($p < 0.01$) due to the use of both types of phytase.

Table 3. Apparent digestibility, ash retention and excreta and fecal nitrogen of 64-day-old Sasso chickens fed diets containing different non-phytate phosphorus levels with or without two types of commercial phytases.

Groups	Apparent Digestibility, %				Ash Retention, %	Excreta Nitrogen, %	Fecal Nitrogen, %
	DM	CP	CF	EE			
Control	80.6	77.6 [a]	30.7 [a]	78.6	31.4 [a]	4.99 [b]	2.53 [a]
IntnPP	81.0	75.3 [b]	28.8 [b]	77.4	30.4 [b]	5.32 [a]	2.54 [a]
IntnPP_bp	80.7	77.6 [a]	31.7 [a]	77.9	33.5 [a]	4.05 [b]	2.33 [b]
IntnPP_fp	81.2	77.5 [a]	31.4 [a]	78.7	33.1 [a]	5.00 [ab]	2.30 [b]
LnPP	80.4	75.3 [b]	28.6 [b]	77.7	30.9 [b]	5.29 [ab]	2.50 [a]
LnPP_bp	82.2	75.5 [b]	29.0 [ab]	79.6	34.1 [a]	5.42 [a]	2.33 [b]
LnPP_fp	81.8	76.7 [ab]	31.6 [a]	78.0	33.7 [a]	5.04 [b]	2.27 [b]
RMSE	2.05	1.27	1.69	2.19	1.24	0.854	0.883
p value	NS	0.007	0.01	NS	0.0001	0.001	0.007

[ab] means within a column under the same treatment with different superscripts are significantly different, NS= not significant, RMSE= root mean square error; IntnPP: intermediate nPP group; LnPP: low nPP group; bp: bacterial phytase; fp: fungal phytase; DM = dry matter; CP = crude protein; CF = crude fiber; EE = ether extract.

Table 4 shows carcass traits and economic indexes of broilers as affected by dietary treatments. LnPP diets had a lower ($p = 0.01$) liver percentage than the other groups. Liver percentage, reduced due to use of fungal phytase in IntnPP groups only.

Table 4. Carcass characteristics of 64-day-old Sasso chickens fed diets containing different non-phytate phosphorus levels with or without two types of commercial phytases.

Groups	Dressing %	Liver %	Pancreas %	Spleen %	Abdominal Fat %	Economic Efficiency %
Control	70.8	2.40 [a]	0.200	0.165	1.93	41.2
IntnPP	69.7	2.47 [a]	0.186	0.144	2.52	38.7
IntnPP_bp	69.9	2.43 [a]	0.240	0.183	2.36	45.5
IntnPP_fp	70.6	2.20 [b]	0.193	0.181	2.41	47.8
LnPP	70.4	2.22 [b]	0.217	0.162	1.66	42.4
LnPP_bp	71.4	2.30 [ab]	0.199	0.132	1.54	47.8
LnPP_fp	70.5	2.14 [b]	0.222	0.154	3.00	46.1
RMSE	14.7	0.43	0.045	0.056	1.17	0.553
p value	NS	0.01	NS	NS	NS	NS

[ab] means within a column under the same treatment with different superscripts are significantly different, NS= not significant, RMSE= root mean square error; IntnPP: intermediate nPP group; LnPP: low nPP group; bp: bacterial phytase; fp: fungal phytase.

Tibia characteristics were presented in Table 5. Reducing nPP level in the diets progressively reduced ($p < 0.05$) tibia weight and phosphorus. LnPP diets had a lower ($p < 0.05$) tibia diameter and ash percentage than the other groups, while both IntnPP and LnPP diets reduced ($p = 0.01$) Ca percentage in tibia in comparison to the control. An effect of the treatments was observed for almost all the parameter in the Table except for tibia length. No differences were observed within IntnPP groups for tibia weight, but within LnnPP groups there is a progressive increase of tibia weight from unsupplemented to bacterial and fungal phytase groups with fungal phytase showed stronger effect. Tibia diameter was decreased due to supplementation of fungal phytase in comparison to the other groups in IntnPP diets while both phytases increased tibia diameter in LnPP diets in comparison to the unsupplemented group. Tibia ash was unaffected by dietary treatments in IntnPP groups while both phytases increased ($p < 0.05$) this criterion in comparison to the unsupplemented group when broilers fed LnPP diets. Regarding tibia calcium, no differences were observed among LnPP groups while in IntnPP groups both phytases increased tibia calcium percentage in comparison to the unsupplemented group. Tibia P was unaffected by dietary treatments in IntnPP diets, while in LnPP diets the addition of bacterial phytase increased P percentage in comparison to the unsupplemented group.

Table 5. Tibia characteristics of 64-day-old Sasso chickens fed diets containing different non-phytate phosphorus levels with or without two types of commercial phytases.

Groups	Tibia Length (mm)	Tibia Weight (g)	Tibia Diameter (mm)	Tibia Ash (%)	Tibia Calcium (%)	Tibia Phosphorus (%)
Control	11.0	7.59 [a]	3.59 [a]	45.1 [a]	37.3 [a]	14.40 [a]
IntnPP	10.8	7.21 [ab]	3.40 [a]	44.8 [a]	34.7 [b]	13.59 [ab]
IntnPP_bp	10.7	7.22 [ab]	3.50 [a]	45.1 [a]	37.1 [a]	15.42 [a]
IntnPP_fp	10.6	7.03 [b]	3.12 [b]	45.0 [a]	37.4 [a]	14.43 [ab]
LnPP	10.6	6.20 [c]	3.18 [b]	43.8 [b]	35.3 [b]	13.15 [b]
LnPP_bp	10.9	7.04 [b]	3.51 [a]	44.8 [a]	36.5 [ab]	14.42 [a]
LnPP_fp	11.0	7.58 [a]	3.52 [a]	45.0 [a]	36.6 [ab]	14.36 [ab]
RMSE	0.739	1.57	0.407	1.01	0.888	1.08
p value	NS	0.04	0.04	0.03	0.01	0.02

[abc] means within a row under the column treatment with different superscripts are significantly different, NS= not significant, RMSE= root mean square error, IntnPP: intermediate nPP group; LnPP: low nPP group; bp: bacterial phytase; fp: fungal phytase.

No effects of dietary treatments were observed for chemical and physical characteristics of meat (Table 6).

Table 6. Chemical composition and physical characteristics of meat of 64-day-old Sasso chickens fed diets containing different non-phytate phosphorus levels with or without two types of commercial phytases.

Groups	Dry Matter %	Protein %	Lipid %	Ash %	pH	Color Intensity	Tenderness cm^2/0.3 g	WHC cm^2/0.3 g
Control	27.0	75.1	18.9	4.90	6.71	0.270	2.98	5.82
IntnPP	27.3	74.4	19.5	4.80	6.74	0.288	2.91	5.74
IntnPP_bp	26.4	74.4	19.6	4.60	6.75	0.295	2.89	5.70
IntnPP_fp	26.8	75.5	18.6	4.60	6.70	0.257	2.93	5.94
LnPP	26.6	75.2	18.3	4.80	6.69	0.270	2.90	5.61
LnPP_bp	26.5	74.7	18.6	5.30	6.70	0.288	2.98	5.74
LnPP_fp	26.2	74.9	18.9	4.80	6.63	0.275	2.91	5.65
RMSE	0.916	1.14	6.02	0.537	0.132	0.034	0.148	0.323
p value	NS	NS	NS	NS	NS	NS	NS	NS

NS= not significant, RMSE= root mean square error, pH=hydrogen power, WHC= water holding capacity; IntnPP: intermediate nPP group; LnPP: low nPP group; bp: bacterial phytase; fp: fungal phytase.

Table 7 shows the effect of dietary treatments on blood criteria. Total protein was the lowest ($p < 0.05$) in LnnP group, the control group showed the lowest ($p < 0.05$) level of albumin and IntnPP group had the lowest ($p < 0.01$) globulin level in blood. Cholesterol was reduced ($p < 0.05$) in IntnPP than the other two groups; the control group showed the highest ($p < 0.05$) level of phosphorus.

Table 7. Metabolic profiles of 64-day-old Sasso chickens fed diets containing different non-phytate phosphorus levels with or without two types of commercial phytases.

Groups	TP, g/dL	Alb, g/dL	Glo, g/dL	TL, mg/dL	Chol, mg/dL	Calcium mg/dL	Phosphorus mg/dL	ALP U/L	AST U/L	ALT U/L
Control	4.15 [a]	1.11 [b]	3.00 [a]	631	111.8 [a]	10.9	6.70 [a]	52.6	10.5	5.24
IntnPP	4.05 [a]	1.55 [a]	1.88 [c]	692	83.7 [b]	9.8 [b]	6.04 [b]	49.2	10.5	5.33
IntnPP_bp	4.36 [a]	1.43 [b]	3.93 [a]	694	97.8 [b]	12.3 [a]	6.68 [a]	54.1	10.4	5.57
IntnPP_fp	4.29 [a]	1.57 [a]	2.73 [b]	634	105.0 [ab]	12.4 [a]	6.63 [a]	50.5	10.7	5.37
LnPP	3.31 [b]	1.67 [a]	2.38 [b]	691	117.2 [a]	10.0 [b]	6.10 [b]	50.9	10.7	5.34
LnPP_bp	4.50 [a]	1.69 [a]	2.84 [b]	689	87.2 [b]	12.2 [a]	6.30 [ab]	51.5	11.0	5.57
LnPP_fp	3.86 [ab]	1.48 [b]	2.36 [b]	687	119.1 [a]	12.0 [a]	6.64 [a]	50.0	10.3	5.04
RMSE	0.771	0.133	0.775	53.2	17.84	4.76	0.167	3.74	3.99	0.675
p value	0.04	0.01	0.04	NS	0.01	0.03	0.01	NS	NS	NS

[ab] means within a column under the same treatment with different superscripts are significantly different, NS= not significant, RMSE= root mean square error, TP: total protein, Alb: albumin; Glob: globulin; TL: total lipids; Chol: cholesterol; ALP=alkaline phosphatase; AST= asparate amino transferase; ALT= alainine amino transferase; IntnPP: intermediate nPP group; LnPP: low nPP group; bp: bacterial phytase; fp: fungal phytase.

The use of bacterial phytase increased the total protein in comparison to the unsupplemented group when LnPP albumin was detected due to the use of bacterial phytase in IntnPP diets and of fungal phytase in LnPP diets. Bacterial phytase was able to decrease plasma cholesterol in comparison to the other groups when a broiler was fed LnPP diets while no differences were observed among IntnPP groups. Both phytases increased ($p < 0.05$) the phosphorus level in comparison to the unsupplemented group when broilers were fed IntnPP diets while with LnPP diets the addition of fungal phytase increased phosphorus level in comparison to the unsupplemented group.

4. Discussion

The decrease of broiler performance due to the reduction of nPP levels in the diets without enzyme supplementation, confirmed the inability of birds to utilize the phytic sources of P, and also confirmed that 0.50, 0.45 and 0.40 nPP levels in the starter, grower and finisher diet (without phytase), respectively were adequate to maintain growth performance.

The worsening of broiler growth performance as weight gain, feed and nutrient intakes are in line with findings by other authors [31–33]. According to Karimi et al. [34] the FCR worsened when nPP levels were under 30 mg/kg considering the average value of the three experimental diets. However, the yield of protein and energy was decreased already at the intermediate level of nPP. This result can be ascribed to the combination of low protein and energy intake and a low digestibility of crude protein and fiber found at the intermediate and low nPP levels. Our results disagree with the findings of Karimi et al. [34] who concluded that feeding 0.40% followed by 0.35 % nPP diets during starter and grower periods respectively results in sufficient feed intake of broilers, with such levels actually being in excess of that required for maintaining FCR or mortality rate comparable to the controls. In fact, in our study the reduction of feed intake at the intermediate level of nnP is important and tied to a severe reduction of body weight gain of broilers (−12.27% and −16.50% than the control, respectively for IntnPP and LnPP groups). This discrepancy is probably tied to the age of animals used in our trial. In fact, as a slow-growth genetic type (colored broilers), the slaughter ages of our broilers was 64 days more than double than the slaughter age of broilers used by Karimi et al. [34] in their trial. It is well known that phosphorus requirements are tied to animal age and probably the amounts of nnP at the intermediate and low levels are not sufficient to allow a sufficient storage of the phosphorus in the broilers body, specifically in the bones, with negative effects on animal growth. This is confirmed by a reduction of circulating phosphorus, tibia weight and tibia Ca and P contents observed in our trial with both IntnPP and LnPP diets. In this regard, we must consider that bone density and strength were positively correlated to calcium and phosphorus content in the tibia [35] and that bone characteristics are important in broilers also as indicators of animal health and welfare. The balance between Ca and P in poultry diets is of the utmost importance, particularly in starter diets; Ca: P should be approximately 2:1 [36,37]. In our trial the Ca:P ratio based on calculated and analyzed values of phosphorus in the starter diets was 2:1 vs. 2: 0.94 in the control and 2:0.8 vs. 2:0.76 and 2:0.6 vs. 2:0.54, respectively for IntnPP and LnPP groups.

The plasma protein showed differences between the different fraction among nPP levels but, very interesting, there is also a difference in albumin to globulin ratio, which is the lowest (0.37) for the control group, while IntnPP and LnPP groups showed similar values (0.82 and 0.70, respectively): high globulin levels and low albumin/globulin ratios indicates better disease resistance and immune response of birds [38].

The results indicate that the effect of type of phytase was correlated with nPP concentrations showing that the differences between the two types of phytase were maximized when the amount of nPP increase in the broiler diet. It is well established that bacterial phytase have a higher efficacy than fungal phytase due to its resistance to low pH an digestive protease [10], so when the amount of nPP in the diet decreases, the bacteria phytase would show a higher efficiency in broiler performance improvement. However, the opportunity to observe improvements in phytase activity is influenced by the level of nPP also in another way: as the nPP level in the diet decreases the efficiency of its utilization increases. In our trial, the phosphorus retention reaches high values and thus the potential to show an improvement in efficiency activity of phytase is greatly reduced. This agrees to other studies in broilers [39,40] and in turkeys [41], which showed that digested P per unit of phytase decreases as the ratio available or non-phytate P/phytase decreases.

This double mechanism of action could be responsible of the results obtained in our trial. As an example, feed, protein and metabolizable energy intakes were not different among unsupplemented, bacterial or fungal phytase groups when nPP level was intermediate, while at the low nPP level bacterial phytase increase all the three criteria when compared to unsupplemented or fungal phytase groups.

The decrease of CP digestibility due to lower nPP availability is in line with the findings of Xue et al. [42], who observed that, like to pigs, also in poultry there is a linear interrelationship between N and P digestion, quantified by the authors as 10:1. This relationship between protein and P could be linked in various aspects. For example, both muscle and bone developments occur concurrently in growing animals, and both require an adequate amount of AA and P; but also a protein deficiency may affect gut development and therefore the absorption and utilization of P in the small intestine [42]. In general, the researches on this topic focused on the effect of diet protein level modification on P absorption in the developing animals, but as there is a linear correlation, it is also possible that deficiencies of P can affect protein digestion, as shown in our research. A further confirmation is that the use of phytase has a positive impact on CP digestion. As known, phytase has a positive effect on protein digestibility as phytate can bind an amount of free amino acids, such as lysine [43]. In our trial, fungal phytase at the low nPP level was more effective than bacterial phytase on protein availability for metabolic purposes. The increase of feed and nutrient intake in broiler fed bacterial phytase at low nPP levels could be tied to low bioavailability. In this regard, the differences in CP digestibility among low nPP groups were not significant but the group supplemented with fungal phytase had a CP digestibility not different from the groups supplemented with both phytases in broiler fed intermediate nPP levels. This trend to decrease nutrient digestibility as nPP decreases, with beneficial effects due to the use of phytase was observed also for crude fiber and could be ascribed to a higher efficiency of fungal than bacterial phytase. This point needs further study to correlate the phytase activity to different metabolic condition of broilers. On the other hand, our results found a further confirmation in the improvement of nutrient conversion indexes in the fungal phytase low nPP group when compared to the unsupplemented one. However, when low-nPP diets were fed to the broiler, the addition of both phytase had no effects on nPP intake or the conversion ratio.

It is not easy to understand the decrease of the blood cholesterol level in the IntnPP unsupplemented group, but it is very interesting to observe that in broilers that were fed an LnPP diet supplemented with bacterial phytase, the levels of blood cholesterol decreased in comparison to the control and the other LnPP group, suggesting that bacterial phytase had a greater potential than fungal phytase in lowering cholesterol. Several studies [44–46] reported that dietary phytate decreased the concentrations of serum total and low-density lipoprotein (LDL) cholesterol in mice or rats. Jariwalla et al. [47] suggested that phytate decreased serum cholesterol by affecting the zinc: copper ratio, as phytate is a strong chelator of divalent cations. Perhaps more likely, the ratio of these cations to each other may affect cholesterol absorption or metabolism. Thus, the addition of phytase can act increasing the cholesterol levels in blood as shown by Liu et al. [48].

The fungal phytase also showed a higher activity than bacterial when low-nPP diets were fed to the broiler as the levels of P in plasma were not different between the two type of phytases at both intermediate and low-nPP diets but, with low-nPP levels fungal phytase was able to increase plasma P levels in comparison to the unsupplemented group. However, at the same nPP levels in the diet, the percentage of P in tibia was not different between phytase levels but bacterial phytase had a higher percentage than the control.

5. Conclusions

Decreasing non phytic phosphorous levels in the colored slow-growing broilers diet negatively affects the growth performance and the digestibility of some nutrients (crude protein and fiber). The use of phytase can partly alleviate these negative effects but the efficiency of different enzyme sources (bacterial or fungal) was tied to the nPP levels in the diet. When the amount of non-phytic phosphorous was decreased, the efficiency of P utilization increased and thus in some cases was difficult to appreciate the differences reported in the literature on enzyme efficiency due to the different source and, in other cases, the fungal phytase seemed to show a higher efficiency in comparison to the bacterial phytase. Further investigations need to understand well the interaction between non phytic phosphorus in the broiler diet and efficiency of bacterial or fungal phytase.

Author Contributions: Conceptualization, Y.A.A. and F.B.; methodology, A.E.M.; data collection, and formal analysis, M.A.A.-H. and A.A.A.; ; statistical anaylses, A.A.A.; investigation, H.S.Z.; resources, F.I.; writing—original draft preparation, Y.A.A., M.A.A.-H., and F.B.; writing—review and editing, F.I.; visualization, Y.A.A.; supervision All authors have read and agreed to the published version of the manuscript.

Acknowledgments: In this section you can acknowledge any support given which is not covered by the author contribution or funding sections. This may include administrative and technical support, or donations in kind (e.g., materials used for experiments).

Conflicts of Interest: The authors declare no conflict of interest.

References

1. Kumar, V.; Sinha, A.H.; Makkar, H.P.S.; De Boeck, G.; Becker, K. Phytate and phytase in fish nutrition. *J. Anim. Physiol. Anim. Nutr.* **2012**, *96*, 335–364. [CrossRef]
2. Woyengo, T.A.; Nyachoti, C.M. Review: Anti-nutritional effects of phytic acid in diets for pigs and poultry: Current knowledge and directions for future research. *Can. J. Anim. Sci.* **2013**, *93*, 9–21. [CrossRef]
3. Morgan, N.K.; Walk, C.L.; Bedford, M.R.; Scholey, D.V.; Burton, E.J. Effect of feeding broilers diets differing in susceptible phytate content. *Anim. Nutr.* **2016**, *2*, 33–39. [CrossRef]
4. Withers, P.J.; Davidson, I.A.; Foy, R.H. The contribution of agricultural phosphorus to eutrophication. In *The Fertilizer Society*; Proc. No. 365; Greenhill House: Thorpe Wood, Petersborough, UK, 1995.
5. Waldroup, P.W.; Kersey, J.H.; Saleh, E.A.; Fritts, C.A.; Yan, F.; Stilborn, H.L.; Crum, R.C., Jr.; Raboy, V. Nonphytate phosphorus requirement and phosphorus excretion of broiler chicks fed diets composed of normal or high available phosphate corn with and without microbial phytase. *Poult. Sci.* **2000**, *79*, 1451–1459. [CrossRef]
6. Angel, R.; Applegate, T.J.; Christman, M. Effect of dietary non-phytate phosphorous (nPP) on performance and bone measurements in broilers fed on a four-phase feeding system. *Poult. Sci.* **2000**, *79*, 21–22.
7. Angel, R.; Applegate, T.J.; Christman, M.; Mitchell, A.D. Effect of dietary non-phytate phosphorous (nPP) on broiler performance and bone measurements in the starter and grower phase. *Poult. Sci.* **2000**, *79* (Suppl. 1), 22.
8. Dhandu, A.S.; Angel, R. Broiler non-phytinphosphorus requirement in the finisher and withdrawal phases of four phases feeding program. *Poult. Sci.* **2003**, *82*, 1257–1265. [CrossRef]
9. NRC National Research Council. *Nutrient Requirements of Poultry*, 9th ed.; National Academy Press: Washington, DC, USA, 1994.
10. Dersjant-Li, Y.; Awati, A.; Schulze, H.; Partridge, G. Phytase in non-ruminant animal nutrition: A critical review on phytase activities in the gastrointestinal tract and influencing factors. *J. Sci. Food Agric.* **2015**, *30*, 878–896. [CrossRef]
11. Żyła, K.; Mika, M.; Stodolak, B.; Wikiera, A.; Koreleski, J.; Świątkiewicz, S. Towards complete dephosphorylation and total conversion of phytases in poultry feeds. *Poult. Sci.* **2004**, *83*, 1175–1186. [CrossRef]
12. Ghazalah, A.A.; Abd EI-Gawad, A.H.; Soliman, M.S.; Youssef, A.W. Effect of enzyme preparation on performance of broilers fed corn-soybean meal based diets. *Egypt. Poult. Sci. J.* **2005**, *25*, 295–316.
13. Selim, W. Effect of Two Sources of Microbial Phytase on Productive Performance Meat Quality and Dietary Protein and Energy Utilization by Sasso Chickens. Master's Thesis, Faculty of Agriculture-Damanhour, Alexandria University, Alexandria, Egypt, 2007.
14. Valaja, J.; Perttila, S.; Partanen, K.; Kiiskinen, T. Effect of two microbial phytase preparations on phosphorus utilisation in broilers fed maize-soybean meal based diets. *Agric. Food Sci.* **2001**, *10*, 19–26. [CrossRef]
15. Camden, B.J.; Morel, P.C.H.; Thomas, D.V.; Ravindran, V.; Bedford, M.R. Effectiveness of exogenous microbial phytase in improving the bioavailabilies of phosphorus and other nutrients in maize-soybean diets for broilers. *Anim. Sci.* **2001**, *73*, 289–297. [CrossRef]
16. Augspurger, N.R.; Webel, D.M.; Lei, X.G.; Baker, D.H. Efficacy of an E. coli phytase expressed in yeast for releasing phytate-bound phosphorus in young chicks and pigs. *J. Anim. Sci.* **2003**, *81*, 474–483. [CrossRef]
17. Roberson, K.D.; Kalbfleisch, J.L.; Pan, W.; Charbeneau, R.A. Effect of corn distiller's dried grains with solubles at various levels on performance of laying hens and egg yolk color. *Int. J. Poult. Sci.* **2005**, *4*, 44–51.
18. Attia, Y.A.; Addeo, N.F.; Al-Hamid, A.A.; Bovera, F. Effects of phytase supplementation to diets with or without zinc addition on growth performance and zinc utilization of white pekin ducks. *Animals* **2019**, *9*, 280. [CrossRef]
19. Association of Official Analytical Chemists, AOAC. *Official Methods of Analysis of the Association of Official Analytical Chemists*, 18th ed.; AOAC: Washington, DC, USA, 2004.

20. Attia, Y.A.; El-Tahawy, W.S.; Abd El-Hamid, A.E.; Hassan, S.S.; Nizza, A.; El-Kelaway, M.I. Effect of phytase with or without multienzyme supplementation on performance and nutrient digestibility of young broiler chicks fed mash or crumble diets. *Ital. J. Anim. Sci.* **2012**, *11*, 303–308. [CrossRef]

21. Jakobsen, P.E.; Gertov, K.; Nilsen, S.H. Frdjelighed frogmed fierbrae. "Digestibility trails with poultry". Bereting fra for sogslabortoriet. *Kabenhaven* **1960**, *56*, 1–34.

22. Haraguchi, H.; Fuwa, K. Determination of phosphorus by molecular absorption flame spectrometry using the phosphorus monoxide band. *Anal. Chem.* **1976**, *48*, 784–786. [CrossRef]

23. Bovera, F.; Lestingi, A.; Iannaccone, F.; Tateo, A.; Nizza, A. Use of dietary mannanoligosaccharides during rabbit fattening period: Effects on growth performance, feed nutrient digestibility, carcass traits, and meat quality. *J. Anim. Sci.* **2012**, *90*, 3858–3866. [CrossRef]

24. Association of Analytical Communities (AOAC) International. *Official Methods of Analysis*, 15th ed.; AOAC International: Washington, DC, USA, 1990.

25. Attia, Y.A.; Qota, M.A.; Bovera, F.; Tag El-Din, A.E.; Mansour, S.A. Effect of amount and source of manganese and/or phytase supplementation on productive and reproductive performance and some physiological traits of dual-purpose cross-bred hens in the tropics. *Br. Poult. Sci.* **2010**, *51*, 235–245. [CrossRef]

26. Attia, Y.A.; Abd El-Hamid, A.E.; Bovera, F.; El-Sayed, M.I.M. Reproductive performance of rabbit does submitted to an oral glucose supplementation. *Animal* **2009**, *3*, 1401–1407. [CrossRef] [PubMed]

27. Attia, Y.A.; Al-Hanoun, A.; Bovera, F. Effect of different levels of bee pollen on performance and blood profile of New Zealand White bucks and growth performance of their offspring during summer and winter months. *J. Anim. Physiol. Anim. Nutr.* **2011**, *95*, 17–26. [CrossRef]

28. Attia, Y.A.; Al-Hanoun, A.; Tag El-Din, A.E.; Bovera, F.; Shewika, E. Effect of bee pollen levels on productive, reproductive and blood traits of NZW rabbits. *J. Anim. Physiol. Anim. Nutr.* **2011**, *95*, 294–303. [CrossRef]

29. Attia, Y.A.; Hamed, R.S.; Abd El-Hamid, A.E.; Al-Harthi, M.A.; Shahba, H.A.; Bovera, F. Performance, blood profile, carcass and meat traits and tissue morphology in growing rabbits fed mannanoligosaccharides and zinc-bacitracin continuously or intermittently. *Anim. Sci. Pap. Rep.* **2015**, *33*, 85–101.

30. SAS Institute. *SAS/STAT User's Guide Statistics*; SAS Institute Inc.: Cary, NC, USA, 2002.

31. Singh, P.K.; Khatta, V.K.; Thakur, R.S.; Dey, S.; Sangwan, M.L. Effects of Phytase Supplementation on the Performance of Broiler Chickens Fed Maize and Wheat Based Diets with Different Levels of Non-phytate Phosphorus. *Asian-Aust. J. Anim. Sci.* **2003**, *16*, 1642–1649. [CrossRef]

32. Karimi, A. Responses of broiler chicks to non-phytate phosphorus levels and phytase supplementation. *Int. J. Poult. Sci.* **2006**, *5*, 251–254.

33. Wilkinson, S.J.; Bedford, M.R.; Cowieson, A.J. Effect of dietary nonphytate phosphorus and calcium concentration on calcium appetite of broiler chicks. *Poult. Sci.* **2014**, *93*, 1695–1703. [CrossRef]

34. Karimi, A.; Bedford, M.R.; Sadeghi, G.; Ghobadi, Z. Influence of Dietary Non-phytate Phosphorous Levels and Phytase Supplementation on the Performance and Bone Characteristics of Broilers. *Braz. J. Poult. Sci.* **2011**, *1*, 43–51. [CrossRef]

35. Barreiro, F.R.; Sagula, A.L.; Junqueira, O.M.; Pereira, G.T.; Baraldi-Artoni, S.M. Densitometric and biochemical values of broiler tibias at different ages. *Poult. Sci.* **2009**, *88*, 2644–2648. [CrossRef]

36. Whitehead, C.C. Nutrition and poultry welfare. *World's Poult. Sci. J.* **2001**, *58*, 349–356. [CrossRef]

37. Fleming, R.H. Nutritional factors affecting poultry bone health. *Proc. Nutr. Soc.* **2008**, *67*, 177–183. [CrossRef]

38. Bovera, F.; Loponte, R.; Pero, M.E.; Cutrignelli, M.I.; Calabrò, S.; Musco, N.; Vassalotti, G.; Panettieri, V.; Lombardi, P.; Piccolo, G.; et al. Laying performance, blood profiles, nutrient digestibility and inner organs traits of hens fed an insect meal from Hermetia illucens larvae. *Res. Vet. Sci.* **2018**, *120*, 86–93. [CrossRef]

39. Dembow, D.M.; Ravindran, V.; Kornegay, T. Improving phosphorus availability in soybean meal for broilers by supplemental phytase. *Poult. Sci.* **1995**, *74*, 1831–1842. [CrossRef]

40. Kornegay, E.; Denbow, D.; Ravindran, V. Responses of broilers to graded levels of Natuphos phytase added to corn soybean meal based diets containing three levels of non-phytate phosphorus. *Br. J. Nutr.* **1996**, *75*, 839–852. [CrossRef]

41. Ravindran, V.; Bryden, W.; Kornegay, E. Phytates: Occurrence, bioavailability and implications in poultry nutrition. *Poult. Avian Biol. Rev.* **1995**, *6*, 125–143.

42. Xue, P.C.; Ajuwon, K.M.; Adeola, O. Phosphorus and nitrogen utilization responses of broiler chickens to dietary crude protein and phosphorus levels. *Poult. Sci.* **2016**, *95*, 2615–2623. [CrossRef]

43. Ravindran, V.; Selle, P.H.; Bryden, W.L. Effect of phytase supplementation individually and in combination with glycanase on the nutritive value of wheat and barley. *Poult. Sci.* **1999**, *78*, 1588–1595. [CrossRef]

44. Lee, D.H.; Choi, S.U.; Hwang, Y.H. Culture Conditions and Characterisations of a new phytase-producing fungal isolate, Aspergillus sp. L117. *Mycobiology* **2005**, *33*, 223–229. [CrossRef]

45. Lee, J.; Yunjaie, C.; Peter-Changwhan, L.; Seungha, K.; Jinduck, B.; Jaiesoon, C. Recombinant production of Penicillium oxalicum PJ3 phytase in Pichia pastoris. *World J. Microbiol. Biotechnol.* **2007**, *23*, 443–446. [CrossRef]

46. Szkudelski, T. Phytic acid-induced metabolic changes in the rat. *J. Anim. Physiol. Anim. Nutr.* **2005**, *89*, 397–402. [CrossRef]

47. Jarivalla, R.J.; Sabin, R.; Lawson, S.; Herman, Z.S. Lowering of serum cholesterol and tryglicerides and modulation of divalent cations by dietary phytase. *J. Appl. Nutr.* **1990**, *42*, 18–28.

48. Liu, N.; Ru, Y.J.; Li, F.D. Effect of dietary phytate and phytase on metabolic change of blood and intestinal mucosa in chickens. *J. Anim. Physiol. Anim. Nutr.* **2010**, *94*, 368–374. [CrossRef]

Article

Multiple Amino Acid Supplementations to Low-Protein Diets: Effect on Performance, Carcass Yield, Meat Quality and Nitrogen Excretion of Finishing Broilers under Hot Climate Conditions

Youssef A. Attia [1,*], Fulvia Bovera [2,*], Jinquan Wang [3], Mohammed A. Al-Harthi [1] and Woo Kyun Kim [3,*]

[1]   Arid Land Agriculture Department, Faculty of Meteorology, Environment and Arid Land Agriculture, King Abdulaziz University, Jeddah 21589, Saudi Arabia; malharthi@kau.edu.sa
[2]   Department of Veterinary Medicine and Animal Production, University of Napoli Federico II, via F. Delpino1, 80137 Napoli, Italy
[3]   Department of Poultry science, University of Georgia, Athens, GA 30602, USA; jq.wang@uga.edu
*   Correspondence: yaattia@kau.edu.sa (Y.A.A.); bovera@unina.it (F.B.); wkkim@uga.edu (W.K.K.)

Received: 26 March 2020; Accepted: 12 May 2020; Published: 3 June 2020

Simple Summary: Crude protein is an essential nutrient in poultry feed. Reducing the use of crude protein not only reduces the feed cost, but also minimizes pollution during poultry production. Thus, finding the minimum protein requirement in broiler diet without compromising broiler growth is the objective in the present study. Supplementing essential amino acids including methionine and lysine to the low-protein diet showed comparable growth performance and carcass yield to the regular protein diet. Thus, reducing the crude protein level is possible if the essential amino acid balance is adequate for broiler growth.

Abstract: The objective of this study was to evaluate the effect of low-protein diets with amino acid supplementation on growth performance, carcass yield, meat quality and nitrogen excretion of broilers raised under hot climate conditions during the finisher period. In trial 1, broilers from 28 to 49 days of age were fed 18% crude protein (CP) as a positive control or 15% CP supplemented with (1) DL-methionine (Met) + L-lysine (Lys), (2) Met + Lys + L-Arginine (Arg), or (3) Met + Lys + L-Valine (Val). In trial 2, broilers from 30 to 45 days of age, were fed an 18% CP diet as a positive control or 15% CP supplemented with Met, Lys, Arg, Val, L-Isoleucine (Ile) or combination with glycine (Gly) and/or urea as nitrogen sources: (1) Met + Lys, (2) Met + Lys + Arg, (3) Met + Lys + Val, (4) Met + Lys + Ile, (5) Met + Lys + Arg +Val + Ile + Gly, and (6) Met+ Lys + Arg + Val + Ile + Gly + urea. Protein use was improved by feeding low-protein amino acid-supplemented diets as compared to the high-protein diet. Feeding 15% crude protein diet supplemented with only methionine and lysine had no negative effects on carcass yield, CP, total lipids and moisture% of breast meat while decreasing nitrogen excretion by 21%.

Keywords: broilers; amino acid; low crude protein; carcass yield; nitrogen excretion; hot climate

---

1. Introduction

In hot climates, the growth of broilers is usually slow, as high temperature adversely affects feed intake, and high-protein diets, used to sustain animal growth, may impair broiler tolerance to heat stress due to the high heat increment induced by protein metabolism [1,2]. Decreasing dietary protein levels of broiler diets could be a good strategy, but the amino acid requirements for broiler chicks must be satisfied by supplementing synthetic amino acids to the basal diets. In this way it is possible to achieve several goals: (1) minimize the excess of amino acid input and thus nitrogen pollution; (2) improve the bird tolerance to hot climate; (3) obtain satisfactory growth performance; all this can increase the profit for the poultry industry [3–5].

Environmental pollution due to nitrogen excretion is an important issue in poultry houses and results in negative impacts on the health of workers and birds, soil, and ground water [4,5]. Thus, reducing dietary protein levels with proper amino acid supplementation has been of great interest [2,4,6–14]. It is well recognized that methionine (Met) is typically the first limiting amino acid in practical broiler diets, whereas lysine (Lys) is the second in broiler diets. However, this depends on the ingredient composition of the diets [4,15,16]. There are several commercially available synthetic amino acids, such as methionine, lysine, threonine (Thr), tryptophan (Trp), and arginine (Arg) [4,12,17] that can be used to fulfill amino acid requirements in low-protein diets.

Fancher and Jensen [6], feeding female broilers on low-protein, corn/soybean meal diets supplemented with synthetic amino acids from 3 to 6 weeks of age, found worse performance in comparison to a higher-protein finisher diet. Liu et al. [18] reported that a decrease in protein level from 22.5% to 20% impaired body weight gain from 1 to 21 days of age. Along the same line, Holsheimer and Janssen [19] reported that Thr, Trp and Arg were limiting in diets containing 17% CP supplemented with Met and Lys when compared to 20% CP diet during 3–7 weeks of age. They also indicated that 0.77% Thr and 0.22% Trp were enough in finisher broiler diets during 3–7 weeks of age. Han et al. [7] observed that broiler chicks fed low-protein diets (16% CP) supplemented with Met, Lys, Arg, valine (Val), and glutamic acid during 3-6 weeks of age had high body weight and better feed conversion ratio (FCR) than those fed a 20% CP diet while maintaining similar body fat content. Moreover, Laudadio et al. [8] indicated that female broiler chicks fed 17% CP diet supplemented with Met plus cysteine (Cys) showed higher body weight than chickens fed 20% CP during 3–7 weeks of age. Lipstein et al. [20] and Lecercq et al. [21] reported that low-protein diets supplemented with Met and Lys prevented increased carcass fat deposition during the finisher period. However, severely low crude protein diets may increase abdominal fat deposition in carcass [22,23]. A 3% reduction in CP compromised body weight and feed conversion during 21–42 day of age but the supplementation of Val+Ile (isoleucine)+Arg+Gly was able to restore both [24].

Broiler chicks consume the highest proportion (~60%) of their feed during the finisher period, when they are more susceptible to heat stress because of their bigger body size. However, few studies have been performed to evaluate the effects of low-protein diets supplemented with multiple key synthetic amino acids on growth performance and carcass yield, under hot climate during finisher period. Thus, the objective of this study, organized in two different trials (Trial 1 and 2) was to evaluate the effect of low-protein diets supplemented with different levels of various essential amino acids on body weight gain, feed conversion ratio, nitrogen excretion, carcass yield, and inner organ weight of broilers during the finisher period and farmed under hot climate conditions.

2. Materials and Methods

2.1. Animal Husbandry:

Broiler chicks were reared in battery-brooders (35 cm length × 25 cm width × 30 cm height) with 23:1 light-dark cycle at day 1, gradually reduced to 18:6 light-dark cycle. Birds were fed from tube feeders and drank from automatic nipple drinkers. Mash diets and water were offered ad libitum throughout the trials. Birds were vaccinated against Newcastle disease virus with Hatchner (B1) at 7 days old and Lasota at 20, 30 and 45 days of age. The average indoor ambient temperature and relative humidity were (mean ± standard deviation): 34 ± 6 °C and 54 ± 9% in Trial 1 and 35 ± 5 °C and 57 ± 11% in Trial 2, respectively. The experimental protocol was approved by the Animal and Poultry Production Scientific and Ethics Committee of Damanhour University. The care and handling of the animals were performed so as to maintain their rights, ensure their welfare, and cause minimal stress, according to International Guidelines for research involving animals (Directive2010/63/EU).

2.2. Experimental Design and Diets

Trial 1 comprised 7 groups (each with 25 Hubbard male broilers, 5 replicates of 5 birds) which had the same feed ingredients and only differed in the protein source and the type of amino acid included. Two different CP diets were formulated based on the protein source: vegetable diets with corn soybean meal (groups 2, 3 and 4) or animal protein diets with fish and meat meals (groups 5, 6 and 7). From 28 to 49 days of age, the groups were submitted to the following dietary treatment: 18% CP (group 1); 15% CP supplemented with Met and Lys (group 2 and group 5); 15% CP supplemented with Met, Lys and Arg (group 3 and group 6); 15% CP supplemented with Met, Lys, Arg and Val (group 4 and group 7). In trial 1, the amounts of Met and Lys supplemented were 0.20% and 0.28% for vegetable diets (group 2, 3 and 4), and 0.19% and 0.26% for animal diets (groups 5, 6 and 7), respectively. The supplemented amounts of Arg, Val and Ile were 0.17%, 0.15% and 0.0% for the vegetable protein diet (group 2, 3 and 4), and 0.22%, 0.15% and 0.05% for the animal protein diet (groups 5, 6 and 7), respectively.

Trial 2 comprised 7 groups (25 male Hubbard broilers, 5 replicates of 5 birds per group) which had the same feed ingredients and only differed in the type of amino acid included. From 30 to 45 days of age, the groups were submitted to the following dietary treatments: group 1 as positive control with 18% CP; 15% CP supplemented with Met at 0.19%, Lys at 0.25% (group 2); 15% CP supplemented with Met at 0.19%, Lys at 0.25% and Arg 0.21% (group 3); 15% CP supplemented with Met at 0.19%, Lys at 0.25% and Val at 0.10% (group 4); 15% CP supplemented with Met at 0.19%, Lys at 0.25% and Ile (group 5); 15% CP supplemented with Met at 0.19%, Lys at 0.25%, Arg at 0.21% Val at 0.10%, Ile and Gly as an amino nitrogen source (group 6); and 15% CP supplemented with Met at 0.19%, Lys at 0.25%, Arg at 0.21% Val at 0.10%, and Ile (as a nitrogen source to equalize nitrogen content between the positive control diet and the 15% CP diet; group 7).

The diets (Trial 1 and 2) were formulated based on Nutrient Requirements of Poultry from National Research Council (NRC) [25] tabulated values for feedstuffs (Table 1). Met and Lys levels of the low protein 15% CP diets were equal to those of the positive control diet. Amino acids were supplemented and analyzed by Evonik (Degussa-Hüls AG, Frankfurt am Main, Germany). Amino acids were supplemented to meet NRC amino acid requirements [25] for broiler chicks.

Table 1. Composition, and calculated and analyzed nutrients of diets in trial 1 and 2.

Ingredients, %	Trial 1			Trial 2	
	18%	15% Animal	15% Plant	18%	15%
Yellow corn	71.63	73.00	73.00	66.86	72.40
Soybean meal	21.00	15.00	20.00	24.71	16.90
Fish meal (72% CP herring)	2.00	1.73	0.00	2.00	2.00
Meat meal	2.00	1.73	0.00	0.00	0.00
Soybean oil	0.50	1.45	1.70	1.65	1.56
Bone meal	1.40	1.68	2.38	1.70	1.88
Lime stone	0.82	0.72	0.63	0.75	0.72
Vit + Min premix [1]	0.25	0.25	0.25	0.25	0.25
NaCl	0.30	0.30	0.30	0.30	0.30
DL-methionine	0.10	0.19	0.20	0.11	0.19
L-lysine	0.00	0.26	0.28	0.06	0.25
Sand	0.00	3.69	1.26	1.61	3.55
Total	100.0	100.0	100.0	100.0	100.0

Table 1. *Cont.*

Ingredients, %	Trial 1			Trial 2	
	18%	**15% Animal**	**15% Plant**	**18%**	**15%**
Calculated values, %					
ME kcal/ kg	3008	3000	3041	3000	3004
Tryptophan	0.23	0.18	0.19	0.24	0.19
Ca	1.01	1.01	1.01	0.91	0.91
Available P	0.41	0.41	0.41	0.37	0.37
Determined values [2], %					
Crude protein	19.04	15.54	15.18	18.63	15.23
Methionine	0.30	0.25	0.24	0.31	0.26
TSAA	0.92	0.70	0.68	0.59	0.51
Lysine	0.92	0.70	0.72	0.96	0.74
Arg	1.02	0.84	0.81	1.03	0.86
Val	0.89	0.69	0.71	0.82	0.71
Ile	0.71	0.62	0.60	0.73	0.59

[1] Premix provides per kg of diet: vitamin A, 8000 international units (IU); vitamin E, 9 mg; menadione (as menadione sodium bisulfite), 150 IU; Vit. D3, 1,000 ICU; riboflavin, 4.0 mg; Ca pantothenate, 10 mg; nicotinic acid, 12 mg; choline chloride, 300 mg; vitamin B12,2 mg; vitamin B6, 1.2 mg; thiamine (as thiamine mononitrate), 2.0 mg; folic acid, 40 mg; d-biotin, 0.05 mg. Trace minerals (mg per kg of diet): Mn, 75; Zn, 40; Fe, 40; Cu, 3; Se, 0.15; iodine, 0.8 and 500 mg an antioxidant. [2] Calculated based on the analyzed amino acids of feed ingredients except for tryptophan, which was not determined.

2.3. Growth Performance and Carcass Yield

Chicks were weighed at the beginning, at 28 days of age in Trial 1 and at 30 days in Trial 2, and at the end at 49 days of age for Trial 1 and 45 days of age for Trial 2. Average initial body weights in Trial 1 and 2 were (mean ± standard deviation) 1,008 ± 10 g and 1,164 ± 12 g, respectively. In both trials there were no differences among groups for initial body weight and the Body Weight Gain was calculated as the difference between final and initial body weight. Additionally, for both trials, the amounts of administered and residual feed were weighed on the same day as the weighing of the birds, and total feed and protein intakes were calculated as the difference between administered and residual feed. Feed and protein conversion ratios were calculated by dividing feed and protein intake by body weight gain. In Trial 2, at 40 days of age, represented samples (500 g) of excreta (n = 5 per treatment) as one sample per replicate were collected and cleaned from feathers, feeds and scabies, and then moisture and nitrogen contents were determined according to the Association of Official Agricultural Chemists (AOAC) [26]. A total of 35 broilers at the end of Trials 1 and 2 were slaughtered for carcass evaluation. In Trial 1, moisture, crude protein, ether extract, and ash of the skinless boneless breast meat (pectorals major) from the slaughtered birds were further determined in pooled triplicate samples according to AOAC [26]. The crude protein, ether extract, and ash content were divided by dry weight for breast meat chemical characteristics.

2.4. Statistical Analysis:

Data from each trial were analyzed using the general linear model procedure of Statistical Analysis Software [27] using a one-way analysis of variance. The experimental unit was the replicate. Mean differences were tested using student Newman Keuls Test [27] to evaluate the differences among means at $P \leq 0.05$.

3. Results

In general, broilers in both trials showed signs of heat stress, including panting, lying down, and flapping their wings to dissipate heat, but differences in behavior responses were not quantitative among different treatments (data not shown).

In Trials 1 and 2, during the whole finishing period, there were no significant differences among groups for body weight, weight gain, feed intake and feed conversion ratio (Tables 2 and 3).

Table 2. Growth performance from 28 to 49 day of age (Trial 1).

Item	Groups							SEM [1]	P Value
	1	**2**	**3**	**4**	**5**	**6**	**7**		
Body weight at 28 d, g	1009	1006	1011	1005	1002	1018	1006	15.3	0.99
Body weight at 49 d, g	1964	1940	1970	1938	1947	1970	1853	28.7	0.88
Body weight gain, g	955	934	959	933	945	952	947	21.1	0.78
Feed intake, g	2212	2223	2257	2263	2223	2226	2226	12.7	0.15
Protein intake, g	409.2 [a]	333.4 [b]	338.5 [b]	339.4 [b]	333.5 [b]	333.9 [b]	333.9 [b]	1.96	0.01
Feed conversion ratio, g/g	2.32	2.38	2.35	2.43	2.35	2.34	2.35	0.11	0.74
Protein Conversion Ratio, g/g	0.428 [a]	0.357 [b]	0.353 [b]	0.364 [b]	0.353 [b]	0.351 [b]	0.353 [b]	0.009	0.01

[a,b]: Means within a row not sharing a common a superscript differ significantly $P \leq 0.05$, based on Duncan's test;
[1] SEM, standard error of the mean.

Table 3. Growth performance from 30 to 45 day of age (Trial 2).

Item	Groups							SEM	P Value
	1	**2**	**3**	**4**	**5**	**6**	**7**		
Body weight 30 d, g	1155	1176	1167	1153	1171	1160	1166	25.1	0.99
Body weight 45 d, g	1959	1977	1992	1988	1981	1993	1971	37.2	0.99
Body weight gain,	804	801	824	835	810	834	806	25.8	0.95
Feed intake, g	1717	1756	1763	1782	1796	1763	1753	23.4	0.38
Protein intake, g	309 [a]	263 [b]	264 [c]	267 [b]	269 [b]	265 [b]	263 [b]	3.05	0.01
Feed conversion ratio, g/g	2.14	2.19	2.14	2.13	2.22	2.11	2.17	0085	0.98
Protein conversion ratio, g/g	0.384 [a]	0.329 [b]	0.321 [b]	0.320 [b]	0.333 [b]	0.317 [b]	0.326 [b]	0.014	0.01
Excreta nitrogen, %	4.03	3.18	3.25	3.47	3.79	3.71	3.90	0.54	0.17
Excreta dry matter, %	21.0	20.9	21.8	21.1	21.6	22.2	22.7	1.6	0.46

[a–c] Means within a row not sharing a common a superscript differ significantly $P \leq 0.05$, based on Duncan's test.

However, in both trials the 15% CP diet groups with amino acid supplementation showed lower ($P < 0.05$) protein intake and protein conversion ratio (PCR) compared to group 1 (Tables 2 and 3).

In Trial 1, there were no differences among groups for carcass yield, body organs and breast meat chemical compositions (Table 4). In Trial 2 (Table 5), the supplemented groups 2, 3 and 4 showed a significantly higher ($P < 0.05$) gizzard percentage compared to groups 1 and 7. Groups 2, 3 and 7 showed a lower ($P < 0.05$) spleen percentage compared to group 1 (Table 5). However, there were no differences among the other carcass traits.

Table 4. Carcass yield, body organs percentage and breast meat chemical composition (Trial 1).

Item	Groups							SEM	P Value
	1	**2**	**3**	**4**	**5**	**6**	**7**		
	Carcass yield								
Dressing [1], %	62.2	61.9	62.2	62.0	63.2	63.4	62.5	0.97	0.17
Breast + wings, %	23.4	23.3	22.5	24.4	23.6	22.4	22.1	0.77	0.40
Thigh + legs, %	21.7	22.3	22.0	21.6	21.6	23.6	22.6	0.67	0.40
Abdominal fat, %	1.71	1.82	1.83	1.81	1.80	2.19	1.93	0.24	0.36
Liver, %	2.42	2.67	2.24	2.18	2.14	2.31	2.29	0.44	0.10
Heart, %	0.64	0.65	0.64	0.69	0.62	0.65	0.65	0.041	0.93
Pancreas, %	0.23	0.25	0.22	0.21	0.24	0.24	0.23	0.012	0.43
Spleen, %	0.131	0.125	0.108	0.161	0.125	0.134	0.149	0.013	0.14
	Chemical characteristics of breast meat								
Moisture, %	74.2	74.4	73.8	75.1	74.9	74.6	74.9	2.42	0.58
Crude protein, %	82.7	81.4	81.4	82.0	81.9	82.0	82.0	1.73	0.74
Ether extract, %	13.1	13.4	13.8	13.0	13.2	13.6	13.5	0.74	0.68
Ash, %	4.2	5.2	4.8	5.0	4.9	4.4	4.5	0.12	0.39

[1] Head, giblets, feet and eviscerate were not included.

Table 5. Carcass yield of broiler chickens (Trial 2).

Item	Groups							SEM	P Value
	1	2	3	4	5	6	7		
Dressing [1], %	62.4	63.7	61.8	60.8	61.2	63.2	60.0	0.96	0.11
Breast + wings, %	30.5	31.1	30.2	29.2	30.4	32.4	29.6	0.71	0.07
Thigh + legs, %	32.9	34.8	32.1	32.1	31.7	31.8	31.4	0.89	0.16
Abdominal fat, %	1.71	1.80	2.10	1.84	1.45	1.33	1.62	0.25	0.39
Liver, %	1.84	1.95	2.22	2.09	2.02	1.93	1.99	0.13	0.49
Heart, %	0.44	0.46	0.51	0.47	0.42	0.44	0.45	0..03	0.43
Gizzard, %	1.54 [b]	2.00 [a]	1.90 [a]	1.88 [a]	1.83 [a,b]	1.76 [a,b]	1.53 [b]	0.10	<0.01
Pancreas, %	0.21	0.18	0.20	0.19	0.20	0.18	0.19	0.015	0.69
Spleen, %	0.129 [a]	0.083 [c]	0.087 [b,c]	0.124 [a,b]	0.115 [a–c]	0.121 [a,b]	0.088 [b,c]	0.012	0.03

[a–c] Means within a row not sharing a common a superscript differ significantly $P \leq 0.05$, based on Duncan's test;
[1] Head, giblets, feet and eviscerate were not included.

4. Discussion

4.1. Growth Performance, Feed/Protein Conversion Ratio and Nitrogen Excretion of Broiler Chicks

Our previous studies showed that broiler growth was low at high temperature ($34 \pm 6 – 35 \pm 5$ °C and 54 ± 9 and 57 ± 11% Relative Humidity) [1,2]. In the present study, the results from Trials 1 and 2 indicated that reducing the CP level from 18% to 15% with the supplementation of amino acids under hot climates and during the finisher period reduced the overall protein intake of broilers without detriment to growth performance. In both trials, the low final body weight compared to Attia et al. [28] could be attributed to the high ambient temperature and humidity (34 and 35 °C and 54% and 57%, respectively). Heat stress in tropical and subtropical regions not only causes economic effects in the form of retarded growth performance, but also induces animal health and welfare concerns [29]. The hot temperatures during the daytime retarded the growth of 7-week-old broilers under 2 kg in both trials. Extreme environmental conditions should be avoided in general through ventilation or reducing stock density; however, hot climate conditions are unavoidable in tropical areas. The results in the current study indicate that satisfactory broiler growth and feed conversion could be achieved with a reduction by 16.6% (from 18% to 15%) of the crude protein content of the diet in finishing broilers raised under hot climate conditions, but the low-protein diet must be supplemented with adequate amounts of Met and Lys, and the same results were also obtained with only vegetable or partially animal protein sources in the diet.

Several studies have suggested that reducing protein levels in diets with adequate supplementation of indispensable amino acids will not compromise broiler growth performance during the finisher period [10–14]. Recent research indicates that in broilers from 1 to 35 days of age, the CP of the diets can be reduced from 210 to 180 g/kg without detriment to broiler performance, but further reduction from 180 to 165 g/kg of diet CP worsened FCR, although the level of essential amino acids was maintained [30]. Additionally, van Harn et al. [31] observed that, in broilers from 1 to 35 days of age, during the finisher period, the crude protein of the diets could be reduced by 2.2–2.3 percentage points (from 19.8%) without negative effects on bird performance if the essential amino acids were correctly balanced, but further decrease could penalize some carcass traits. Chrystal et al. [30] attributed these results to a possible deficiency of non-essential amino acids such as glycine and serine at the lowest level of diet CP. As animal age increases in the present trial (28–49 and 30–45 day of age for Trials 1 and 2, respectively), the requirements of birds for non-essential amino acids could be satisfied even while reducing the protein content of the diet to 15%. The retarded growth performance under hot climate also contributed to the change in requirements. However, the current results could not distinguish the confounding effects of heat stress and reduction of protein. It is interesting to observe that, under hot climate conditions, reducing the CP level to 15% with key amino acid supplementation can maintain the same growth performance of finishing broilers as 18% CP.

Arginine or Valine supplementation in 15% CP animal/vegetable diets over Met and Lys did not show any additive impact in growth performance. In the literature, the availabilies of Val and Arg in fish meal were low (67.3% and 62.6%, respectively); thus, supplementation of Val and Arg was necessary to maintain optimum growth performance in the fish meal-containing diet [32,33]. Wide variations in amino acid digestibility have been reported in animal source proteins, particularly in terms of batch-to-batch differences [34]. The differences in responses to Arg or Val addition may be due to the difference in dietary protein sources, inclusion levels and amino acid requirements for broiler males and females [35–38]. It has been suggested that the protein and/or amino acid levels of the diets, strain, and the age of the birds may affect the response to the level of amino acid addition [7,39]. The supplementation of Met, Lys, Thr, Arg and Val was found to improve the growth of broilers [39], and the supplementation of Met, Lys, Arg, Val and Thr plus glutamic acid to 16% CP diet during 3–6 weeks of age also increased growth performance [36].

In the 14% CP soybean meal diet, the limiting amino acids were Met, Lys, Thr and Val [40], whereas Met, Lys, Thr, Arg, Val and Ile were limited in the 12% CP diet [41]. It has also been reported a low-protein diet (17%) with balanced Lys, Met, Thr and Trp levels had the same performance as a 19% CP diet during 22 to 56 days of age [42,43]. However, the present research indicates that 15% CP diet supplemented with Met + Lys showed no significant differences in growth and FCR compared to the positive control diet. Therefore, our trial suggests that limiting amino acids are related to CP percentage, environmental temperature and ingredient composition (availability of amino acids) of the experimental diets.

In the present study, soybean meal, a high-quality protein ingredient [34], has been used as primary protein source. The lack of response to Ile in Trial 2 indicated that this amino acid is not limiting in 15% CP diet under heat stress and/or there might be the antagonism between Ile and leucine, because both are branched-chain amino acids [17,44,45]. This inconsistency among many studies may be attributed to diet composition, amino acid content and availability in ingredients, age of birds, and environment conditions (temperature and humidity).

Glycine as a source of amino nitrogen or urea, as a nitrogenous source had no additive influence on the growth of broilers in Trial 2, indicating that the 15% CP diet supplemented with Met and Lys furnishes adequate amounts of dispensable amino acids for broiler growth during 30–45 days of age under a hot climate. Awad et al. [46] reported that the addition of glycine did not compensate the depressed growth performance in broilers under hot climate. The lack of responses to glycine as a nonessential amino acid on improving broiler performance is in contrast to the finding of Dean et al. [47]. Supplementation of glycine enhanced growth performance of broilers fed a low protein amino acid-supplemented diet because low-protein diets are low in nitrogen sources for making non-essential amino acids and other essential nitrogen compounds in the body [47]. Urea in non-ruminant animals had no nutrient values [48,49]. Thus, further studies are needed to evaluate the amino acid requirements under different conditions involving ingredients and rearing conditions.

The present results indicated that the diet containing 15% of CP supplemented with Met and Lys may satisfy all amino acid requirements for optimal broiler growth under hot climate. Some studies have shown that broilers eat more feed to satisfy their protein or amino acid needs under hot climate [9,50,51]. The feed conversion ratios in this study are similar to other studies showing that there were no significant differences in feed intake and efficiency of feed use among birds fed a low protein and a high-protein diets [8,51–53].

The present results indicated that protein intake was high, and protein/gain ratio was poorer in the 18% CP group than the groups fed low-protein amino acid-supplemented diets. Aletor et al. [50] indicated that a lower dietary protein level improves the protein use due to the lower protein turnover rate. There was a numerical 21.1% reduction in nitrogen excreted when 15% CP supplemented with Met and Lys in the present study. These results are in agreement with several studies reporting that low-protein amino acid-supplemented diets had no negative effect on broiler performance, while reducing nitrogen excretion by 33.6% [24,43,52,54,55]. The reduction of nitrogen excretion also influences ammonia volatilization in poultry houses: supplementing amino acids in low-protein diets reduced ammonia volatilization by 45% in poultry [4,5,56].

In general, due to the limits of the experimental conditions, it is not clear that the amino acid requirement change was triggered by heat stress or a decrease on performance or both; thus, further research is needed.

4.2. Carcass Yield, Body Organs and Chemical Composition of Breast Meat

Carcass yield results in the current study indicate that Met and Lys supplementation in 15% CP diet provides sufficient amount of amino acids for maintaining optimum growth and development of body organs compared to 18% or 18.5% CP diets under hot climate. Alleman and Leclercq [51] reported that reducing CP content from 20% to 16% during 23–44 days of age had no negative effect on breast muscle percentage. However, several researchers have concluded that lowering the protein level in diet, maintaining appropriate levels of essential amino acids (Arg, Ile, Val and Trp) allowed normal protein gain, but increased fat deposition in both fat and lean lines of broilers, revealing that protein or lipid deposition was potentially controlled independently [8,57,58].

In agreement with the present results, Aletor et al. [50] found that decreasing dietary CP percentage while satisfying the amino acid requirements of broilers had no detrimental effect on relative weights of breast or thigh, which is similar to the present findings. However, in contrast to the present results, there were significant increases in the relative weights of liver and abdominal fat which may indicate higher lipogenetic activity in the body. Other studies have shown a significant increase in abdominal fat deposition when broilers or ducks were fed low-CP diets [22,23,59]. However, according to the current study, under hot climates, broilers may use amino acid and protein from low-CP diets better than those under the normal temperature conditions.

The gizzard and spleen percentage differences were unexpected in the current study. Further studies are necessary to explain this change. The protein level in finisher diets for broiler chickens could be reduced from 18% to 15% without adverse effect on broiler growth performance during 28–49 days of age under hot climates if Met and Lys are supplemented to meet requirements while decreasing nitrogen excretion by 21%. Therefore, protein use and economic efficiency in birds fed 15% CP diet supplemented with Met and Lys were improved because dietary protein input was lower compared to the high-protein control diet. Furthermore, the non-significant influences on carcass yield, abdominal fats, and total lipids of meat indicated that these traits are not compromised feeding the low-protein diet supplemented with Met and Lys.

5. Conclusions

Under hot climate conditions and during the finisher period, feeding a low-protein diet fortified with essential amino acids improved protein use and deceased nitrogen excretion without negative effect on performance and carcass yield of broiler. Thus, in the above-mentioned environmental conditions, the protein level of the finishing broiler diets can be reduced from 18% to 15% with an adequate supplementation of only methionine and lysine.

Author Contributions: Conceptualization, Y.A.A. and F.B.; methodology and data collection, and formal analysis, M.A.A.-H.; writing—review and editing, W.K.K. and J.W.; resources, writing—original draft preparation, review and proofreading, Y.A.A., M.A.A.-H. and F.B. All authors have read and agreed to the published version of the manuscript.

Funding: This research received no external funding.

Acknowledgments: This article was funded by the Deanship of Scientific Research (DSR), King Abdulaziz University, Jeddah. Therefore, the authors appreciate the technical and financial support of the DSR.

Conflicts of Interest: The authors declare no conflict of interest.

References

1. Attia, Y.A.; Böhmer, B.M.; Roth-Maier, D.A. Responses of broiler chicks raised under constant relatively high ambient temperature to enzymes, amino acid supplementations, or diet density. *Archive Für Geflügelkunde* **2006**, *70*, 80–91.
2. Attia, Y.A.; Hassan, R.A.; Tag El-Din, A.E.; Abou-Shehema, B.M. Effect of ascorbic acid or increasing metabolizable energy level with or without supplementation of some essential amino acids on productive and physiological traits of slow-growing chicks exposed to chronic heat stress. *J. Anim. Physiol. Anim. Nutr.* **2011**, *95*, 744–755. [CrossRef] [PubMed]
3. Dozier, W.A.; Kidd, M.T.; Corzo, A. Dietary amino acid responses of broiler chickens. *J. Appl. Poult. Res.* **2008**, *17*, 157–167. [CrossRef]
4. Kim, J.H.; Patterson, P.H.; Kim, W.K. Impact of dietary crude protein, synthetic amino acid and keto acid formulation on nitrogen excretion. *Int. J. Poult. Sci.* **2014**, *13*, 429–436. [CrossRef]
5. Chalova, V.I.; Kim, J.H.; Patterson, P.H.; Ricke, S.C.; Kim, W.K. Reduction of nitrogen excretion and emissions from poultry: A review for conventional poultry. *World's Poult. Sci. J.* **2016**, *72*, 509–520. [CrossRef]
6. Fancher, B.I.; Jensen, L.S. Dietary protein level and essential amino acid content: Influence upon female broiler performance during the grower period. *Poult. Sci.* **1989**, *68*, 897–908. [CrossRef]
7. Han, Y.; Suzuki, H.; Parsons, C.M.; Baker, H. Amino acid fortification of a low-protein corn and soybean meal diet for chicks. *Poult. Sci.* **1992**, *71*, 1168–1178. [CrossRef]
8. Laudadio, V.; Dambrosio, A.; Normanno, G.; Khan, R.U.; Naz, S.; Rowghani, E.; Tufarelli, V.; Bunchasak, C.U.; Tanaka, S.K.; Ohtani, S.; et al. The effect of supplementing methionine plus cystine to a low-protein diet on the growth performance and fat accumulation of growing broiler chicks. *Asian-Australas. J. Anim. Sci.* **1997**, *10*, 185–191.
9. Alleman, F.; Michel, J.; Chagneau, A.M.; Leclercq, B. The effects of dietary protein independent of essential amino acids on growth and body composition in genetically lean and fat chickens. *Br. Poult. Sci.* **2000**, *41*, 214–218. [CrossRef]
10. Aletor, V.A.; Hamid, I.I.; Niess, E.; Pfeffer, E. Low-protein amino acids-supplemented diets in broiler-chickens: Effects on performance, carcass characteristics whole-body–composition and efficiencies of nutrient utilization. *J. Sci. Food Agric.* **2000**, *80*, 547–554. [CrossRef]
11. Aletor, V.A.; Roth, F.X.; Paulicks, B.R.; Roth-Maier, D.A. Growth, body-fat deposition, nitrogen excretion and efficiencies of nutrients utilization in broiler-chicks fed low-protein diets supplemented with amino acids, conjugated linoleic acid or an -glucosidase inhibitor. *Archive Geflügelk* **2001**, *66*, 21–30.
12. Laudadio, V.; Dambrosio, A.; Normanno, G.; Khan, R.U.; Naz, S.; Rowghani, E.; Tufarelli, V. Effect of reducing dietary protein level on performance responses and some microbiological aspects of broiler chickens under summer environmental conditions. *Avian Biol. Res.* **2012**, *5*, 88–92. [CrossRef]
13. McGill, E.; Kamyab, A.; Firman, J.D. Low crude protein corn and soybean meal diets with amino acid supplementation for broilers in the starter period. 1. Effects of feeding 15% crude protein. *Int. J. Poult. Sci.* **2012**, *11*, 161–165.
14. McGill, E.; Kamyab, A.; Firman, J.D. Low crude protein corn and soybean meal diets with amino acid supplementation for broilers in the starter period. 2. Effects of feeding 13% crude protein. *Int. J. Poult. Sci.* **2012**, *11*, 166–171.
15. Scott, M.L.; Nesheim, M.C.; Young, R.J. *Nutrition of the Chicken*, 3rd ed.; M. L. Scott and Associates: Ithaca, NY, USA, 1982.
16. Fernandez, S.R.; Aoyagi, S.; Han, Y.; Parsons, C.; Baker, D.H. Limiting order of amino acids in corn and soybean meal for growth of the chick. *Poult. Sci.* **1994**, *73*, 1887–1896. [CrossRef]
17. Baker, D.M. Problems and pitfalls in animal experiments designed to establish dietary requirements for essential nutrients. *J. Nutr.* **1986**, *116*, 2339–2349. [CrossRef]
18. Liu, N.; Wang, J.Q.; Gu, K.T.; Deng, Q.Q.; Wang, J.P. Effects of dietary protein levels and multienzyme supplementation on growth performance and markers of gut health of broilers fed a miscellaneous meal based diet. *Anim. Feed Sci. Technol.* **2017**, *234*, 110–117. [CrossRef]
19. Holsheimer, J.P.; Janssen, W.M.M.A. Limiting amino acids in low protein maize-soybean meal diets fed to broiler chicks from 3 to 7 weeks of age. *Br. Poult. Sci.* **1991**, *32*, 151–158. [CrossRef]

20. Lipstein, B.; Bornstein, S.; Bartov, I. The replacement of some of the soybean meal by the first limiting amino acids in practical broiler diets. 3. Effects of protein concentration and amino acid supplementation in broiler finisher diets on fat deposition on the carcass. *Br. Poult. Sci.* **1975**, *16*, 627–635. [CrossRef]

21. Leclercq, B.; Chagneau, A.M.; Cochard, T.; Khoury, J. Comparative responses of genetically lean and fat chickens to lysine, arginine and non-essential amino acid supply. I. Growth and body composition. *Br. Poult. Sci.* **1994**, *35*, 687–696. [CrossRef]

22. Sterling, K.G.; Costa, E.F.; Henry, M.H.; Pesti, G.M.; Bakalli, R.I. Responses of broiler chickens to cottonseed and soybean meal-based diets at several protein levels. *Poult. Sci.* **2002**, *81*, 217–226. [CrossRef] [PubMed]

23. Xie, M.; Jiang, Y.; Tang, J.; Wen, Z.G.; Zhang, Q.; Huang, W.; Hou, S.S. Effects of low-protein diets on growth performance and carcass yield of growing White Pekin ducks. *Poult. Sci.* **2017**, *96*, 1370–1375. [CrossRef] [PubMed]

24. Ospina-Rojas, I.C.; Murakami, A.E.; Duarte, C.R.A.; Eyng, C.; Oliveira, A.L.; Janeiro, V. Valine, isoleucine, arginine and glycine supplementation of low-protein diets for broiler chickens during the starter and grower phases. *Br. Poult. Sci.* **2014**, *55*, 766–773. [CrossRef] [PubMed]

25. National Research Council (NRC). *Nutrient Requirements of Poultry*, 9th ed.; National Academy of Sciences, National Research Council: Washington, DC, USA, 1994.

26. Association Official Analytical Chemistry (AOAC). *Official Methods of Analysis*, 19th ed.; Association Official Analytical Chemistry: Washington, DC, USA, 2007.

27. SAS Institute. *SAS®User's Guide for Personal Computer*; SAS Institute Inc.: Cary, NC, USA, 2007.

28. Attia, Y.A.; Al-Harthi, M.A.; Hassan, S.S. Turmeric (Curcuma longa Linn.) as a phytogenic growth promoter alternative for antibiotic and comparable to mannan oligosaccharides for broiler chicks. *Rev. Mex. Cienc. Pecu.* **2017**, *8*, 11–21. [CrossRef]

29. Quinteiro-Filho, W.M.; Ribeiro, A.; Ferraz-de-Paula, V.; Pinheiro, M.L.; Sakai, M.; Sá, L.R.M.D.; Palermo-Neto, J. Heat stress impairs performance parameters, induces intestinal injury, and decreases macrophage activity in broiler chickens. *Poult. Sci.* **2010**, *89*, 1905–1914. [CrossRef]

30. Chrystal, P.V.; Moss, A.F.; Khoddami, A.; Naranjo, V.D.; Selle, P.H.; Liu, S.Y. Impacts of reduced-crude protein diets on key parameters in male broiler chickens offered maize-based diets. *Poult. Sci.* **2020**, *99*, 505–516. [CrossRef]

31. Van Harn, J.; Dijkslag, M.A.; van Krimpen, M.M. Effect of low protein diets supplemented with free amino acids on growth performance, slaughter yield, litter quality, and footpad lesions of male broilers. *Poult. Sci.* **2019**, *98*, 4868–4877. [CrossRef]

32. Smith, R.E. Assessment of the availability of the amino acid in fish meal, soybean meal and feather meal by chick growth assay. *Poult. Sci.* **1968**, *47*, 1624. [CrossRef]

33. Miller, D.; Kifer, R.R. Factors affecting protein evaluation of fish meal by chick bioassay. *Poult. Sci.* **1970**, *49*, 999–1004. [CrossRef]

34. Ravindran, V.; Hew, L.; Ravindran, G.; Bryden, W. Apparent ileal digestibility of amino acids in dietary ingredients for broiler chickens. *Anim. Sci.* **2005**, *81*, 85–97. [CrossRef]

35. Han, Y.; Baker, D.H. Effects of sex, heat stress, body weight and genetic strain on the lysine requirement of broiler chicks. *Poult. Sci.* **1993**, *72*, 701–708. [CrossRef] [PubMed]

36. Han, Y.; Baker, D.H. Lysine requirement of fast and slow growing broiler chicks. *Poult. Sci.* **1991**, *70*, 2108–2114. [CrossRef] [PubMed]

37. Aggoor, F.A.M.; El-Naggar, N.M.; Mehrez, A.Z.; Attia, Y.A.; Qota, E.M.A. Effect of different dietary protein and energy levels during roaster period on: 1-Performance and economic evaluation of broiler chicks. *Egypt. Poult. Sci.* **1997**, *17*, 81–105.

38. Costa, M.J.; Da Zaragoza-Santacruz, S.; Frost, T.J.; Halley, J.; Pesti, G.M. Straight-run vs. sex separate rearing for 2 broiler genetic lines Part 1: Live production parameters, carcass yield, and feeding behavior. *Poult. Sci.* **2017**, *96*, 2641–2661. [CrossRef]

39. Waguespack, A.M.; Powell, S.; Bidner, T.D.; Payne, R.L.; Southern, L.L. Effect of incremental levels of L-lysine and determination of the limiting amino acids in low crude protein corn-soybean meal diets for broilers. *Poult. Sci.* **2009**, *88*, 1216–1226. [CrossRef]

40. Warnick, R.E.; Anderson, J.O. Limiting essential amino acids in soybean meal for growing chickens and the effects of heat upon availability of the essential amino acids. *Poult. Sci.* **1968**, *47*, 281–287. [CrossRef]

41. Schwartz, R.W.; Bray, D.J. Limiting amino acids in 40:60 and 15:85 blends of corn: Soybean protein for the chicks. *Poult. Sci.* **1975**, *54*, 1814.

42. Uzu, G. Limit of reduction of the protein level in broiler feeds. *Poult. Sci.* **1982**, *61*, 1557–1558.

43. Shao, D.; Shen, Y.; Zhao, X.; Wang, Q.; Hu, Y.; Shi, S.; Tong, H. Low-protein diets with balanced amino acids reduce nitrogen excretion and foot pad dermatitis without affecting the growth performance and meat quality of free-range yellow broilers. *Ital. J. Anim. Sci.* **2018**, *17*, 698–705. [CrossRef]

44. Jensen, L.S.; Mendonca, C.X. Amino acid nutrition of broilers during the grower period. In *Georgia Nutrition Conference for the Feed Industry*; University of Georgia: Athens, GA, USA, 1988; pp. 76–83.

45. D'Mello, J.P.E. (Ed.) Amino acid imbalances, antagonisms and toxicities. In *Amino Acid in Farm Animal Nutrition*; CAB International: Wallingford, UK, 1994; pp. 63–98.

46. Awad, E.A.; Zulkifli, I.; Soleimani, A.F.; Loh, T.C. Individual non-essential amino acids fortification of a low-protein diet for broilers under hot and humid tropical climate. *Poult. Sci.* **2014**, *94*, 2772–2777. [CrossRef]

47. Dean, D.W.; Bidner, T.D.; Southern, L.L. Glycine supplementation to low protein, amino acid-supplemented diets supports optimal performance of broiler chicks. *Poult. Sci.* **2006**, *85*, 288–296. [CrossRef] [PubMed]

48. El-Boushy, A.R. Broiler growth response from practical low-protein diets supplemented with urea and diammonium hydrogen phosphate. *Neth. J. Agric. Sci.* **1980**, *28*, 147–155.

49. Kagan, A.; Balloun, S.L. Urea and aspartic acid supplementation of low-protein broiler diets. *Br. Poult. Sci.* **1976**, *17*, 403–413. [CrossRef] [PubMed]

50. Aletor, V.A.; Eder, K.; Becker, K.; Paulicks, B.R.; Roth, F.X.; Roth-Maier, D.A. The effects of conjugated linoleic acids or an alpha-glucosidase inhibitor on tissue lipid concentrations and fatty acid composition of broiler chicks fed a low-protein diet. *Poult. Sci.* **2003**, *82*, 796–804. [CrossRef]

51. Alleman, F.; Leclercq, B. Effect of dietary protein and environmental temperature on growth performance and water intake of male broiler chickens. *Poult. Sci.* **1997**, *38*, 607–610. [CrossRef]

52. Jacob, J.; Blair, R.; Ibrahim, S.; Scott, T.; Newberry, R. Using reduced protein diets to minimize nitrogen excretion of broilers. *Poult. Sci.* **1995**, *74* (Suppl. 1), 127.

53. Nahashon, S.M.; Bartlett, J.R.; Smith, E.J. Responses to dietary crude protein and energy levels by cross of chickens involving White Plymouth Rock. *Poult. Sci.* **1995**, *74* (Suppl. 1), 207.

54. Hegedüs, M. Dietary factors influencing protein utilization: A review. *Acta Vet. Hung.* **1992**, *40*, 133–143.

55. Moran, E.T.; Bushong, R.D.; Bilgili, S.F. Reducing dietary crude protein for broilers while satisfying amino acid requirements by least-cost formulation: Live performance, litter composition, and yield of fast-food carcass cute of six weeks. *Poult. Sci.* **1992**, *71*, 1687–1694. [CrossRef]

56. Keshavarz, K.; Austic, R.E. The use of low-protein, low-phosphorus, amino acid- and phytase-supplemented diets on laying hens performance and nitrogen and phosphorus excretion. *Poult. Sci.* **2004**, *83*, 75–83. [CrossRef]

57. Leclercq, B. The influence of dietary protein content on the performance of genetically lean or fat growing chickens. *Br. Poult. Sci.* **1983**, *24*, 581–587. [CrossRef] [PubMed]

58. Cahaner, A.; Leenstra, F. Effects of high temperature on growth and efficiency of male and female broilers from lines selected for high weight gain, favorable Feed conversion, and high or low fat content. *Poult. Sci.* **1992**, *71*, 1237–1250. [CrossRef] [PubMed]

59. Smith, E.R.; Pesti, G.M.; Bakalli, R.I.; Ware, G.O.; Menten, J.F.M. Further studies on the influence of genotype and dietary protein on the performance of broilers. *Poult. Sci.* **1998**, *77*, 1678–1687. [CrossRef] [PubMed]

Article

Honey Bee Pollen in Meagre (*Argyrosomus regius*) Juvenile Diets: Effects on Growth, Diet Digestibility, Intestinal Traits, and Biochemical Markers Related to Health and Stress

Valentina Panettieri [1], Stavros Chatzifotis [2], Concetta Maria Messina [3], Ike Olivotto [4], Simona Manuguerra [3], Basilio Randazzo [4], Andrea Ariano [1], Fulvia Bovera [1,*], Andrea Santulli [3,5], Lorella Severino [1] and Giovanni Piccolo [1]

[1] Department of Veterinary Medicine and Animal Production, University of Napoli Federico II, Via F. Delpino 1, 80137 Napoli, Italy; valentina.panettieri@unina.it (V.P.); andrea.ariano@unina.it (A.A.); lorella.severino@unina.it (L.S.); giovanni.piccolo@unina.it (G.P.)

[2] Institute of Marine Biology, Biotechnology and Aquaculture, Hellenic Centre for Marine Research, Gournes Pediados P.O. Box 2214,71003 Heraklion, Crete, Greece; stavros@hcmr.gr

[3] DiSTeM, Marine Biochemistry and Ecotoxicology Laboratory, University of Palermo, Via G. Barlotta 4, 91100 Trapani, Italy; concetta.messina@unipa.it (C.M.M.); simona.manuguerra@unipa.it (S.M.); andrea.santulli@unipa.it (A.S.)

[4] Department of sea science, University Polytechnic of Marche, via Brecce Bianche, 60100 Ancona, Italy; i.olivotto@staff.univpm.it (I.O.); b.randazzo@staff.univpm.it (B.R.)

[5] Consorzio Universitario della Provincia di Trapani, Institute of Marine Biology, Via G. Barlotta 4, 91100 Trapani, Italy

* Correspondence: bovera@unina.it; Tel.: +39-081-253-6497

Received: 23 December 2019; Accepted: 29 January 2020; Published: 31 January 2020

Simple Summary: Recently, several studies have focused on the use of nutraceuticals and honey bee products to improve the welfare and sustainability of animal husbandry. Honey bee pollen is rich in bioactive substances, presenting a strong antioxidant activity with possible positive effects on growth performance and non-specific immune responses in reared fish. Despite its favorable characteristics, the addition of honey bee pollen to a meagre (*Argyrosomus regius*) diet in our trial resulted in a reduction of growth performances and diet digestibility, histological alterations of intestinal morphology, and high levels of biomolecular stress markers, probably due to its complex ultrastructure, which is indigestible for monogastric animals. These negative effects could be overcome by using bioactive component extraction methods and thus eliminating the indigestible fractions. Our results confirmed the general assumption that it should always be considered necessary to test nutraceutical additives of natural origin in a given species in order to verify the effective positive action and exclude any negative repercussions on animal health.

Abstract: This research aimed to evaluate the effects of the inclusion of honey bee pollen (HBP) in meagre (*Argyrosoumus regius*) juveniles' diets on growth performance, diet digestibility, intestinal morphology, and immunohistochemistry. Furthermore, stress-related molecular markers and biochemical blood profile of fish were evaluated, together with mineral trace and toxic element concentration in pollen and diets. Specimens of meagre (360) of 3.34 ± 0.14 g initial body weight, were randomly allocated to twelve 500 L circular tanks (30 fish per tank). Four diets were formulated: a control diet and three experimental diets with 1%, 2.5%, and 4% of HBP inclusion. All the growth parameters and crude protein and ether extract digestibility coefficients were negatively linearly affected by increased HBP inclusion ($p < 0.0001$). Histology of medium intestine showed slight signs of alterations in group HPB1 and HPB2.5 compared to control. Fish from HBP4 group showed severe alterations at the intestinal mucosa level. Immunohistochemical detection of TNF-α in the medium intestine showed the presence of TNF-α+ cells in the lamina propria, which resulted in accordance with the increased level of the TNF-α protein detected by immunoblotting in the liver. This stress situation was confirmed by the increased hepatic level of HSP70 ($p < 0.05$) in fish fed the HBP4 diet and by the linear decrease of total serum protein levels in HBP-containing diets ($p < 0.0001$). These negative effects can be related to the ultrastructure of the bee pollen grain walls, which make the bioactive substances unavailable and can irritate the intestine of a carnivorous fish such as meagre.

Keywords: meagre; honey bee pollen; growth trial; digestibility trial; TNF-α; HSP70; intestinal immunohistochemistry; toxic elements; trace elements; total serum protein

---

1. Introduction

In the past 20 years, researchers have focused their attention on several natural molecules to be used as possible antioxidant therapeutic and preventive agents [1].

A growing amount of scientific evidence is demonstrating that supplementing diets with natural compounds that act as protective factors or immunostimulants can improve growth performance, product quality, fish health, and physiological response to stress situations and diseases [2–6].

Bee-derived products have been widely investigated for their positive characteristics, such as their strong antioxidant activity, their positive effects on non-specific immune responses, and their capacity to increase growth performance in various animal species [7–9]. Research has been carried out to test the inclusion of these nutraceutical products in aquafeeds because of their antioxidant potential to prevent or treat aquatic animal diseases and to increase their performance [8,10].

Honey bee pollen (HBP) consists of the male generative cells gathered by honeybees from flower stamens and the anthers of flowers, which are collected by foraging bees and carried to the hives, where pollen agglutinates with bee secretions and the addition of nectar [11,12]. The consumption of HBP has increased in the recent years because it has been considered a healthy and therapeutic product because of its nutritional properties, source of proteins, lipids, vitamins, minerals, amino acids, carotenoids, flavonoids, phenolic compounds, antioxidants, carotenes, and xanthophylls [13,14]. Phenolic acid, flavonoids, and tannins act as potent antioxidants and protective agents and their free radical scavenging activity encourages their application in the biomedical field [15,16].

The introduction of HBP in freshwater fish diets has led to improved growth and immune status [17], increase in the number of phagocytic cells, (i.e., neutrophils and monocytes), and reduction of the mortality caused by *Aeromonas hydrophila* in Nile tilapia (*Oreochromis niloticus*) [8].

Despite these favourable characteristics, no research has been carried out to evaluate the effects of the inclusion of HBP in diets for marine fish species. The present research intended to fill this gap, evaluating growth performance, diet digestibility, intestinal morphology and immunohistochemistry, stress-related molecular markers, and biochemical blood profile in meagre (*Argyrosoumus regius*) juveniles fed diets containing increasing levels of commercial HBP. In addition, the research aimed to contribute to the knowledge of the mineral trace element concentration in honey bee pollen and in fish diets.

2. Materials and Methods

A chestnut honey bee pollen (HBP) purchased from a local organic farm located in the city of Naples (Napoli, Italy) was chosen for the research. Both growth and digestibility trials were conducted at the Institute of Marine Biology, Biotechnology and Aquaculture (IMBBC) of the Hellenic Centre for Marine Research (Crete, Greece). The experimental protocol was designed according to the guidelines of European Directive (2010/63/EU) on the protection of animals used for scientific purposes. The digestibility trial was approved by the experts of ASSEMBLE Plus project (Contract number: SR020220189047). The growth trial procedures and the blood sampling were approved by ethics experts of the Trans National Access (TNA) selection panel of the AQUAEXCEL 2020 (project number: AE040069).

2.1. Growth Trial

2.1.1. Fish and Experimental Conditions

Meagre juveniles were obtained from the IMBBC hatchery. After a 2 week period of acclimation to the experimental conditions, 360 fish of 3.34 ± 0.14 g initial body weight were individually weighed under moderate anaesthesia conditions (anaesthetic: MS222-Tricain Methanesulphonate at 50 mg/L dosage) and randomly allocated to twelve 500 L circular tanks (30 fish per tank) supplied by flow-through borehole aerated seawater (renewal 200% per hour).

The trial was conducted from July to October 2018 under constant temperature (19.0 ± 0.5 °C), salinity (36‰), and dissolved oxygen (DO 8 ± 0.5 ppm) and a dark:light cycle of 12:12 h.

2.1.2. Fish Diets

To determine the effects of the inclusion of HBP, four isonitrogenous (53% crude protein (CP) on as fed basis), isolipidic (15% ether extract (EE) on as fed basis), and isocaloric diets were formulated to meet the meagre nutrient and energy requirements according to Chatzifotis et al. [18]: a control diet with no addition of HBP (HBP0) and three experimental diets in which HBP was included at 1% (HBP1), 2.5% (HBP2.5), and 4% (HBP4) on as fed basis.

The experimental feeds were prepared at the IMBBC laboratory. All dietary ingredients were finely ground and thoroughly mixed; water was then blended into the mixture to obtain an appropriate consistency for pelleting using a 2.5 mm die meat grinder. After pelleting, the diets were dried in a ventilated oven at 40 °C for 24 h and stored in plastic bags until use.

Chemical composition of the experimental diets was determined as follows: dry matter (DM), ash, CP, EE, and crude fiber (CF) (procedure numbers 934.01, 942.05, 954.01, 920.39, and 962.09, respectively) according to AOAC [19]. Before starting the analysis, the pellets were finely ground using a cutting mill. The ingredients and approximate composition of the diets are reported in Table 1 and the approximate composition of the bee pollen is reported in Table 2.

Table 1. Ingredients and chemical composition of experimental diets for meagre.

Ingredients (g/kg)	HBP0	HBP1	HBP2.5	HBP4
Fish meal	440	438	435	430
Soybean meal	240	238	235	230
Corn gluten meal	100	98	96	95
Wheat flour	100	98	93	90
Fish oil	100	98	96	95
Honey bee pollen	0	10	25	40
Mineral mix	10	10	10	10
Vitamin mix	10	10	10	10
Chemical Composition				
Dry matter (%)	90.0	92.7	90.6	93.0
Ash (%)	9.4	9.8	9.0	9.4
Crude protein (%)	53.0	53.6	52.5	52.5
Ether extract (%)	15.0	14.9	15.0	14.8
Crude fiber (%)	7.7	7.7	7.7	7.5
Na (%)	0.086	0.075	0.058	0.075
Mg (%)	0.130	0.117	0.131	0.125
K (ppb)	381.65	351.92	381.93	377.90
Ca (ppb)	870.39	696.49	456.83	562.36

Table 2. Chemical composition of honey bee pollen.

	DM (%)	Ash (%)	CP (%)	EE (%)	CF (%)	Na (%)	Mg (%)	K (%)	Ca (%)
HBP	85.02	3.79	23.18	7.82	19.23	0.067	0.062	0.019	0.061

Abbreviations: HBP: honey bee pollen; DM: dry matter; CP: crude protein; EE: ether extract; CF: crude fiber.

The experimental diets were randomly assigned to triplicate groups of 30 fish.

Each diet was administered three times per day (09.00 am; 12.00 pm, and 4.00 pm) to apparent satiety (until the first pellet was refused), 7 days per week. Any not-ingested pellet was recovered, dried, and weighed. The exact quantity of feed distributed in each tank was recorded daily. The trial lasted 88 days.

2.1.3. Trace Elements

The trace elements contained in the honey bee pollen and in the diets were also determined. Before analysis, the samples (0.5 ± 0.02 g) were placed in a Teflon vessel with 5.0 mL of 65% HNO_3 and 2.0 mL of 30% H_2O_2 (Romil UpA). The vessel was sealed and placed in a microwave digestion system (Milestone, Bergamo, Italy). Microwave-assisted digestion was performed with a mineralization program for 15 min at 200 °C. The vessel was then cooled at 30 °C, the digestion mixture was transferred into a 50.0 mL flask, and the final volume was obtained by adding Milli-Q water [20]. Trace element concentrations were determined by inductively coupled plasma optical emission spectroscopy (ICP-OES) technique using an Optima 2100 DV instrument (PerkinElmer Inc., Wellesley, MA, USA) coupled with a CETAC U5000AT (CETAC Technologies, Omaha, NE, USA). The calibration curve and two blanks were run during each set of analyses, to check the purity of the chemicals. A reference material (CRM DORM-4, National Research Council of Canada (NRC-CNRC), Ottawa (Ontario), Canada) was also included for quality control. All the values of the reference materials were within the certified limits.

The instrumental detection limits are expressed as wet weight (w.w.) and were determined following the protocol described by Perkin Elmer ICP, application study number 57 [21].

2.1.4. Growth Performance

At the end of the trial, fish were starved for 1 day, lightly anesthetized (MS222-Tricain methanesulphonate at 50 mg/L dosage), and group weighed.

The following growth performance indexes and protein efficiency ratios (PER) were calculated according to the following formulas [22]:

- Weight gain (WG%) = 100 × ((FBW, final body weight (g)—IBW, initial body weight (g))/initial live weight (g))
- Daily intake rate (DIR, %/day) = 100 × ((feed intake (g)/mean weight (g))/days)
- Specific growth rate (SGR, %/day) = ((lnFBW − lnIBW)/number of feeding days) × 100
- Feed conversion ratio (FCR) = (total feed supplied (g)/weight gain (g))
- Protein efficiency ratio (PER) = (weight gain (g)/total protein fed (g))

2.2. Somatic Traits

At the end of the feeding trial, 45 fish per treatment (15 fish per replicate) were randomly chosen, weighed, and sacrificed by over-anaesthesia (MS222-Tricain methanesulphonate at 250 mg/L dosage) and dissected. Liver and gut were weighed to determine hepatosomatic (HSI) and viscerosomatic (VSI) indexes as described in Piccolo et al. [23]. Liver, muscle, and intestine of four fish for each tank were sampled.

2.3. Digestibility Trial

Dry matter, protein, and lipid digestibility of the tested diets was evaluated in a separate trial. One hundred and eighty meagre of 23.53 ± 2.16 g initial weight obtained from the IMBBC hatchery were distributed in 12 circular fiberglass tanks (3 tanks per treatment; 15 fish per tank) of 270 L equipped with a settling column. The water and environmental conditions were the same as described for the growth trial.

Diets were prepared at the IMBBC laboratory following the same procedure described above and using the same inclusion levels of HBP (1%, 2.5%, 4%). The apparent digestibility coefficients were measured using the indirect acid-insoluble ash (AIA) method; for this reason, 1% celite® (Fluka, St. Gallen, Switzerland) was added to the diets used in the growth trial as an inert marker.

Fish were fed the experimental diets ad libitum three times a day for 4 weeks. Feces accumulated in the settling column was collected daily before the morning meal, centrifuged at 7000 rpm for 10 min, and stored at −20 °C until analysis.

Thirty minutes after the last feeding in the afternoon, tanks were cleaned to remove excess of feces, and then the settling column was placed.

Feeds and feces were analyzed according to AOAC [19]. The AIA contents of feeds and feces were determined according to Vogtmann et al. [24].

The apparent digestibility coefficients of dry matter (ADC DM), crude protein (ADC CP), and ether extract (ADC EE) were calculated following Palmegiano et al. [25].

2.4. Histology

For the histological analysis, segments of medium intestine were isolated from growth trial fish and immediately fixed in modified buffered Karnovsk fixative for at least 24 h. After fixation, samples were dehydrated in graded alcohol solutions, cleared in xylene, and embedded in solid paraffin. Cross-sections of 5 µm, cut with a LEICA microtome (Leica Microsystems Srl, Buccinasco (MI), Italy), were stained with Mayer's haematoxylin and eosin Y and examined under a Zeiss Axio Imager.A2 microscope (Carl Zeiss Microscopy GmbH, Jena, Germany); images were acquired by means of a combined colour digital camera Axiocam 503 (Carl Zeiss Microscopy GmbH, Jena, Germany). For the morphometric evaluation of intestinal fold height, and mucous cells abundance, nine fish for each treatment (three for each tank) were considered. Three sections for each fish were observed at intervals of about 200 µm in order to avoid repetitions in morphometric measurements of folds and quantification of mucous cells. Intestinal folds were measured from the apex to the base, excluding the underlying connective layer and values have been expressed by mean and standard deviation (SD).

2.5. Immunohistochemistry

For the immunodetection of TNF-α in the medium intestine, nine fish for each treatment (three for each tank) were used and sections (5 µm) were placed on gelatinized slides (0.5% fish gelatin, 0.05% chromopotassium sulfate) in order to prevent section detachment during the reaction. Sections were deparaffinized and rehydrated through serial graded ethanol solutions. Endogenous peroxidase activity was blocked by treating sections with 0.3% hydrogen peroxide for 10 min at room temperature. To prevent aspecific antibody binding, slides were rinsed in 0.01 M phosphate-buffered saline (PBS), pH 7.4, for 15 min and blocked using 5% BSA in Tris for 20 min. After rinsing in PBS for 15 min, sections were incubated overnight at 4 °C with anti-TNF-α rabbit polyclonal primary antibody specific for zebrafish (ANASPEC). After rinsing the slides with PBS, the sections were incubated with the secondary fluorescent antibody (Goat Anti-Rabbit IgG H&L, Alexa Fluor 488 No. ab150077; Abcam, dilution 1:400) at room temperature for 1 h and 30 min. Following incubation, the slides were mounted with Fluoreshield Mounding Medium with DAPI (ab104139) for nucleus staining. The sections were observed under a Zeiss Axio Imager M2 microscope and images were captured with a high-resolution camera Zeiss Axiocam 105 color (Carl Zeiss Microscopy GmbH, Jena, Germany). Negative controls were obtained by incubation without the primary antibody. For the relative quantification of TNF-α+ cells, three sections from three different fish for each treatment were analyzed. Observations were performed counting TNF-α+-labeled cells in three undamaged folds for each section. An arbitrary unit was adopted on the mean of the count: 0–20 = +, 20–50 = ++, 50–>100 = +++.

2.6. Extraction of Total Proteins and Detection of Hepatic Biomolecular Markers by Immunoblotting

For total protein extraction, aliquots of lyophilized livers were homogenized on ice (1:3 w/v) with 100 mM HEPES pH 7.4, 10 mM EDTA, 10 mM EGTA, and 4 M NaCl 23.37 g / 100 mL, in the presence of 10 µL protease inhibitor mix (NaF 42 mg/mL, Aprotinine, Na_3PO_4 183.9 mg/mL, Leupeptine 4×). The homogenate was centrifuged at 6000 rpm at 4 °C for 15 min and the supernatant was stored on ice for protein quantification according to Lowry et al. [26]. Biomolecular markers in liver were analyzed by immunoblotting.

Equivalent amounts of proteins (30 µg) diluted with Laemmli buffer (1970) (Sigma-Aldrich, St. Louis, MO, USA) were loaded, after protein denaturation for 5 min at 90 °C, on pre-cast gel for SDS–polyacrylamide electrophoresis (SDS-PAGE) (Bio-Rad, Hercules, CA, USA) and blotted using a Trans Blot Turbo Transfer System (Bio-Rad, Hercules, CA, USA). The correct amount of protein loaded was confirmed by red Ponceau staining. Filters were used for protein detection by primary antibodies (AbI) specifics for TNF-α (Ab polyclonal from rabbit) and HSP70 (Ab monoclonal from rabbit). (Sigma-Aldrich, Dorset, UK; Santa Cruz, CA, USA). The primary antibodies were diluted in buffer at the concentrations suggested by the manufactures for each AbI. In relation to the origin of the AbI, the appropriate secondary antibodies were used (anti-mouse or anti-rabbit secondary); as means of detection, the secondary antibodies were conjugated with horseradish peroxidase (GAR/M-HRP Bio-Rad, Hercules, CA, USA) or alkaline phosphatase. The signals originated by immunoreaction were detected using enhanced chemo-luminescent (ECL) reagents (Clarity Western ECL Blotting Substrate, Bio-Rad, Hercules, CA, USA) and a BCIP/NBT substrate system (Bio-Rad). Images were obtained, photographed, and digitalized with Chemi Doc XRS (Bio-Rad, Hercules, CA, USA), and further analyzed with Image Lab software (Bio-Rad, Hercules, CA, USA). The results were expressed as fold increase of each treatment in relation to the respective control; the images shown are representative of three separate experiments, for which the mean quantification is reported in each figure, together with the significance of the differences ($p < 0.05$).

2.7. Assessment of Biochemical Parameters in Blood Samples

Biochemical measurements were performed on blood samples from 2-month-old fish of the species *Argyrosomus regius*, within a weight 20–25 g. The purpose of the test was to evaluate the effect of bee pollen enrichment in accordance with the requirements of the AquaExcel2020 program.

On the day of the sampling, fish were euthanized, and their weight and length was measured. Immediately, blood was collected from the caudal vessel via heparinized syringes. Hematocrit was measured, using capillary tubes and a specialized centrifuge to determine the percentage of red blood cells by volume. Additionally, hemoglobin was determined using a colorimetric assay kit (Spinreact, Girona, Spain).

Following these procedures, blood samples were centrifuged at 2000× *g* for 10 min and the resulting plasma was stored at −20 °C until further analysis. Plasma samples were used to assess the following biochemical parameters: cholesterol, triglycerides, albumin, total proteins, glucose, creatinine, sodium, potassium, calcium, phosphorus, lactate, glutamic oxaloacetic transaminase (GOT), and glutamic pyruvic transaminase (GTP).

More specifically, plasma cholesterol (CO/PAP, Biosis, Greece), triglycerides (GPO/PAP, Biosis, Athens, Greece), albumin (BCG, Biosis, Greece), total serum proteins (Biuret, Biosis, Greece), glucose (GOD/PAP, Biosis, Greece), calcium (Arsenazo III, Biosis, Greece), phosphorus (UV, Biosis, Greece), lactate (LO-POD, Spinreact, Girona, Spain), and creatinine (Cayman Chemical Company, Ann Arbor, MI, USA) concentrations were assessed using commercial enzymatic colorimetric kits. Moreover,

GOT-AST (IFCC/LIQUID, Biosis, Greece), GPT-ALT (IFCC/LIQUID, Biosis, Greece) concentrations were estimated using enzymatic kinetic assays. Finally, a flame-photometer was used to determine plasma sodium and potassium concentrations.

2.8. Statistical Analysis

The data were tested for normal distribution and equal variances before analysis (SAS, 2000, SAS Institute Inc., Cary, NC, USA). All the data were analyzed by one-way ANOVA, using the GLM procedure of SAS (2000), according to the model

$$Yij = m + Di + eij, \tag{1}$$

For growth and digestibility trials, the experimental unit was the tank and each value was obtained as the average of the 30 fish in the tank. For somatic indexes, the experimental unit was the individual fish (15 fish for each tank). Finally, for the blood parameters, the unit was derived from the average of four fish per tank.

In addition, the mean comparison was performed using orthogonal contrast analysis. The examined components were linear and quadratic (SAS, 2000, SAS Institute Inc., Cary, NC, USA). $P <$ 0.05 was considered the threshold for statistical significance.

3. Results

3.1. Trace Elements

Table 3 shows the content of toxic and essential trace elements in experimental feeds and honey bee pollen.

Table 3. Toxic and essential trace elements in experimental feeds and honey bee pollen.

Trace Elements (ppb)	HBP0	HBP1	HBP2.5	HBP4	Honey Bee Pollen
Fe	499.27	490.53	511.96	512.95	134.63
Cu	40.17	38.13	44.73	38.86	6.16
Zn	186.06	178.50	193.95	191.28	29.77
Co	0.15	0.15	0.15	0.15	0.02
Al	62.98	46.01	56.82	66.21	58.13
Se	0.55	0.53	0.54	0.57	1.85
Mn	259.04	262.06	257.79	268.78	31,73
Pb	0.22	0.23	0.23	0.24	0.16
Cd	0.12	0.10	0.11	0.11	0.03
Cr	0.81	0.79	0.83	0.82	0.30
Ni	1.92	1.81	1.90	1.88	0.34
As	2.45	2.23	2.41	2.35	0.59

3.2. Growth Performance

In Table 4, the growth parameters measured during the trial are reported.

Table 4. Growth performance of meagre fed experimental diets.

	HBP0	HBP1	HBP2.5	HBP4	*p*-Values			
					RMSE	Linear [1]	Quadratic [1]	Cubic [1]
Initial BW (g)	3.35	3.34	3.34	3.34				
Final BW (g)	26.78	23.29	22.69	21.36	0.66	<0.0001	<0.0001	0.0659
FCR	0.96	1.04	1.18	1.23	0.06	0.0007	0.0003	0.3186
SGR	2.37	2.21	2.18	2.11	0.08	0.0044	0.0062	0.4158
Intake‰ ABW/d	16.48	17.23	19.08	19.43	0.98	0.0061	0.0025	0.336
WG%	705.3	606.9	579.7	539.7	52.9	0.0050	0.0059	0.5564
PER	1.96	1.80	1.61	1.55	0.10	0.0008	0.0004	0.5619

Abbreviations: Initial BW: initial body weight; Final BW: final body weight; FCR: feed conversion ratio; SGR: specific growth rate; Intake‰ ABW/d: intake per kg adjusted body weight; WG%: weight gain percentage per initial body weight; PER: protein efficiency ratio. [1] Contrast analysis.

All the growth parameters were significantly influenced by the diet. Daily feed intake increased as bee pollen inclusion in the diet increased ($p < 0.01$), while final weight ($p < 0.0001$), FCR ($p < 0.001$), SGR ($p < 0.01$), PER ($p < 0.001$), and WG %ABW ($p < 0.01$) were negatively linearly affected by bee pollen inclusion in the diet. The growth parameters reached the worst values in the HBP4 diet.

3.3. Somatic Indexes, Slaughter Traits

In Table 5, the viscerosomatic indexes and the slaughter traits are shown.

Table 5. Somatic indexes of meagre fed experimental diets.

	HBP0	HBP1	HBP2.5	HBP4		*p*-Values		
					RMSE	Linear [1]	Quadratic [1]	Cubic [1]
Final Weight	37.4	33.9	36.8	33.9	4.41	0.0584	0.316	0.037
VSI (%)	4.37	4.27	4.20	4.11	0.70	0.3299	0.333	0.9185
HIS (%)	1.62	1.69	1.67	1.72	0.36	0.4245	0.517	0.6901
K	1.17	1.05	1.06	1.06	0.13	0.0397	0.078	0.3916
Gut length %BW	37.50	45.64	40.09	44.05	7.46	0.0413	0.282	0.0206

Final weight: final body weight; HIS: hepatosomatic index; VSI: viscerosomatic index; K: condiction factor; [1] Contrast analysis.

No significant differences in somatic indexes emerged among the fish fed the experimental diets, except for the gut length, which showed a linear increase with increasing honey bee pollen inclusion in the diet.

3.4. Digestibility Trial

The estimated ADC of DM, CP, and EE of the four diets are reported in Table 6.

Table 6. Digestibility trial results.

	HBP0	HBP1	HBP2.5	HBP4		*p*-Values		
					RMSE	Linear [1]	Quadratic [1]	Cubic [1]
ADC DM	64.24	74.28	36.03	45.63	0.82	<0.0001	<0.0001	0.0001
ADC CP	90.26	91.36	78.71	78.56	0.76	<0.0001	<0.0001	0,0003
ADC EE	93.45	92.15	85.725	85.91	0.91	<0.0001	<0.0001	<0.0001

ADC DM: dry matter apparent digestibility coefficient; ADC CP: crude protein apparent digestibility coefficient; ADCEE: ether extract apparent digestibility coefficient. [1] Contrast analysis.

ADC DM linearly decreased as the HBP content in the diets increased. Furthermore, linear and quadratic components indicated a gradual reduction of crude protein and ether extract digestibility coefficients going from the control to HBP4 diet. Nevertheless, the ADCs of HBP1 diet resulted higher than those shown by HBP2.5 and HBP4 groups, as indicated by the quadratic component of the variance ($p < 0.001$).

3.5. Histology

Histology of the medium intestine showed slight signs of alterations in group HPB1 and HPB2.5 when compared with the HBP0 group. No significant differences were shown by the morphometric evaluation of intestinal folds among experimental groups (Table 7).

Table 7. Morphometric evaluation of mucosal folds.

	MF Height (μm)
HBP0	462.3 ± 18.7
HBP1	487.8 ± 23.5
HBP2.5	453.1 ± 24.8
HBP4	476.6 ± 46.4

MF height: mucosal fold height. Data are expressed as mean and SD.

A significant linear increase in mucous cells in response to pollen inclusion was observed (Figure 1).

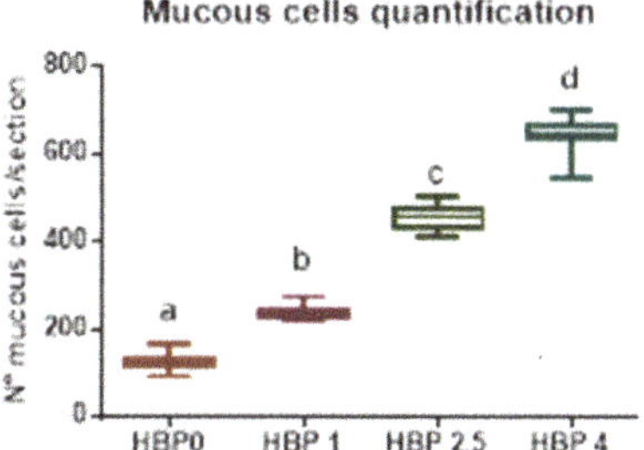

Figure 1. Abundance of mucous cell in medium intestine. The number of mucous cells is reported as mean of the observations performed on three transversal intestinal sections from individual fish (n = 9) from the different experimental groups. Different letters indicate significant differences among groups ($p < 0.05$).

The most frequently encountered alterations consisted of leucocyte infiltration at the level of the lamina propria and thickening of the submucosa (Figure 2g). Intestine samples from HBP4 group showed severe alteration at the level of the mucosa, even preserving epithelial integrity. In this group, a substantial increase in the number of mucous cells was observed (Figures 1 and 2). Moreover, melano macrophage like, melanin-containing cells were often observed at the base of the epithelial layer and infiltrating lamina propria and submucosa (Figure 2h).

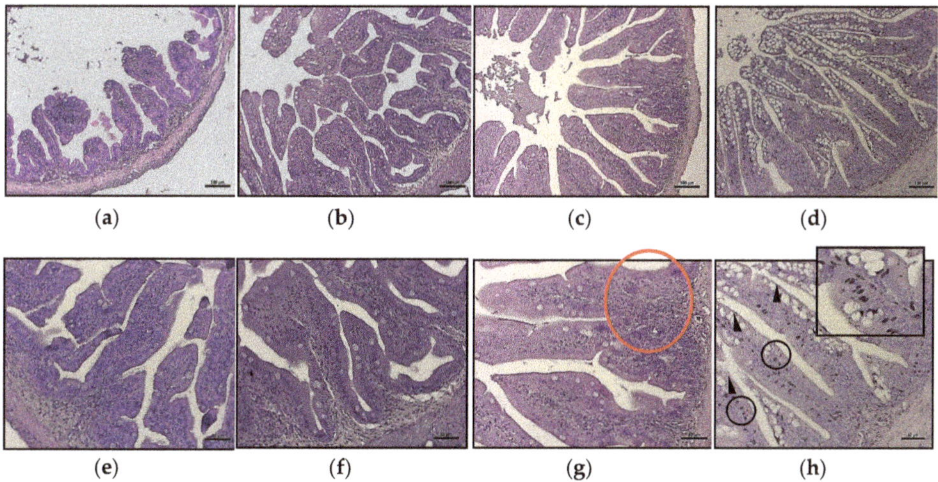

Figure 2. Intestinal histology of meagre fed experimental diets after 88 days of trial. Histology of *A. regius* medium intestine from the different experimental groups: (**a,e**) normal histology from HBP0 group; (**b,f**) HBP1 group did not show signs of pathological alteration; (**c,g**) in the HBP2.5 group, it was possible to detect inflammatory influx in some portions of the intestinal submucosa (red circle); (**d,h**) the HBP4 group showed a severe degree of mucipar hyperplasia of the mucosa (arrowhead) with an abundant presence of melano macrophages at the base of the enterocytes (black circle and box). Scale bar: a, b, c, d = 200 μm; e, f, g, h = 50 μm.

3.6. Immunohistochemistry

Immunohistochemical detection of TNF-α in medium intestine showed the presence of TNF-α+ cells in the lamina propria and submucosa of bee-pollen-treated fish. While a weak reaction was recorded in the epithelial layer of intestinal folds in all treated groups, a moderate influx of TNF-α+ cells was recorded in particular in groups HBP2.5 and HBP4 (Figure 3). In particular, round macrophage/neutrophil-like TNF-α+ cells were observed in the submucosa and lamina propria of HBP1, HBP2.5, and HBP4 groups. The relative abundance evaluation of TNF-α+ cells infiltrating submucosa and lamina propria showed low values in HBP1, while HBP2.5 and HBP4 showed high values with no significant differences between the two groups (Table 8) at the base of the epithelial layer and infiltrating lamina propria and submucosa.

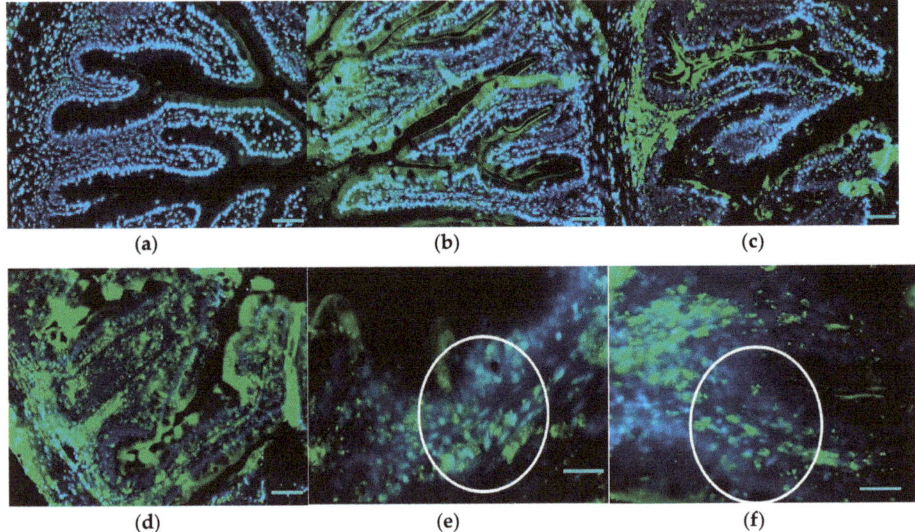

Figure 3. Immunohistochemical detection of TNF-α in medium intestine of the different experimental groups (**a–d**). High magnification images showing TNF-α+ cells (white circle) in the submucosa and in the lamina propria of intestine from HBP2.5 (**e**) and HBP4 (**f**) groups. Positive reactions are shown in green. Nuclei were stained with DAPI (blue). Scale bar: a,b,c,d = 50 μm; e,f = 20 μm.

Table 8. Relative density of TNF-α+ cells in medium intestine.

CTRL	HBP1	HBP2.5	HBP4
+	+	+++	+++

3.7. Hepatic Biomolecular Markers

The evaluation of the molecular markers related to immunostimulation and stress resistance showed that the levels of the protein TNF-α presented a significant increase only in fish fed the diet with the highest level of HBP inclusion (HBP4) ($p < 0.05$), while HSP70 resulted increased both in HBP2.5 and HBP4 groups compared to the control group (Figure 4).

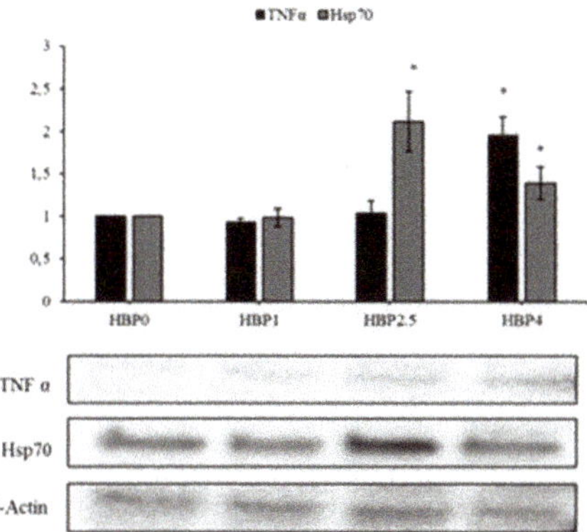

Figure 4. Immunoblot of TNFα, HSP70, evaluated on samples of the liver of *A. regius* fed on three different diets containing increasing levels of commercial HBP (HBP1, HHBP2.5, HBP4) and in standard condition (HBP0). β-Actin was used as the internal control. The images are representative of at least three separate experiments. The relative protein quantification is represented in the graphic (* $p < 0.05$ vs. HBP0).

3.8. Assessment of Blood Samples

The results obtained from the blood analysis are shown in Table 9.

Table 9. Biochemical assessment of blood samples.

	HBP0	HBP1	HBP2.5	HBP4	*p*-Value			
					RMSE	Linear [1]	Quadratic [1]	Cubic [1]
HCT (%)	28.67	26.00	27.00	26.67	3.19	0.46	0.62	0.56
HGB (g/dL)	3.83	3.80	3.76	3.56	0.40	0.43	0.45	0.88
Glucose (mmol/L)	4.71	5.21	4.92	5.96	1.20	0.24	0.34	0.51
TSP (g/dL)	10.20	8.15	6.53	3.73	0.98	<0.0001	<0.0001	0.58
Albumin (mg/dL)	1.61	1.91	1.38	1.64	0.60	0.94	0.60	0.33
Globulin (mg/dL)	8.56	5.72	5.14	2.07	1.27	0.0004	0.0007	0.2331
A/G	0.19	0.43	0.36	0.83	0.24	0.01	0.03	0.27
Creatinine (μmol/L)	80.00	132.50	94.67	104.33	16.51	0.16	0.76	0.03
GOT (UI/L)	360.00	345.00	355.67	243.33	67.7	0.11	0.16	0.48
GPT (U/L)	335.50	298.50	273.33	386.67	49.45	0.30	0.48	0.41
Triglyceride (mmol/L)	13.83	16.43	15.10	17.10	2.45	0.14	0.28	0.28
Cholesterol (mg/dL)	6.63	7.63	9.55	6.90	0.91	0.73	0.19	0.07
P (mg/dL)	7.57	9.50	9.77	11.90	4.63	0.29	0.32	0.78
Ca (mg/dL)	27.07	30.87	22.53	33.47	7.31	0.32	0.75	0.14
Na (mmol/L)	210.33	179.67	206.67	187.00	24.35	0.27	0.67	0.14
K (mmol/L)	2.70	2.17	2.07	2.53	0.73	0.79	0.75	0.95
Lactate (mmol/L)	3.17	4.15	2.75	3.3	0.73	0.83	0.45	0.11

HCT: hematocrit; HGB: hemoglobin; TSP: total serum protein; A/G: albumin to globulin ratio; GOT: glutamic oxaloacetic transaminase; GTP: glutamic pyruvic transaminase. statistically different. [1] Contrast analysis.

This table highlights the differences among the groups. Total serum protein, globulin, and albumin to globulin ratio linearly and quadratically decreased as the HBP content in the diets increased.

4. Discussion

The trial results showed that honey bee pollen induced negative effects on meagre growth performance and diet digestibility. These data were not in line with the current, although limited bibliography, in which bee pollen had notable beneficial effects in other fish species, as described by El-Asely for Nile tilapia and Choobkar for rainbow trout [8,10], even if it must be remarked that these latter authors used a bee pollen alcoholic extract.

The bad results obtained with the in vivo results can be explained by looking at the other performed investigations. The intestinal histology evidenced pathological alteration of the mucosa morphology, showing a severe mucipar hyperplasia and an abundant presence of melano macrophage-like cells at the base of the enterocytes. These alterations became more evident as the inclusion of bee pollen in the diets increased, where, besides the inflammatory influx, the migration and infiltration of melanin-containing cells indicated a further, extreme index of an immune response. The presence of TNF-α cells, resembling in shape neutrophil cells or macrophages, in fish fed different HBP inclusion corroborated the histological results. Tumor necrosis factor-alpha is an important pro-inflammatory cytokine playing an important role in cell survival through the activation, proliferation, and differentiation of macrophages. It is released by several immune cells during infection and tissue damage and its inhibition increases susceptibility to disease and reduces capacity to resolve infection [27–30]. TNF-α isoforms have been identified in various fish species [27,28,31], showing high levels of conservative regions with mammals and a constitutive expression in healthy fish tissues [28]. Inflammation is a highly energy-demanding process and it has been widely shown that intestinal inflammation correlated to anti-nutritional factors in fish diet is able to affect growth in fish [32,33].

Regarding the variation of the liver biomolecular markers related to general stress response and immunostimulation, in this case the increase of TNF-α could also be ascribed to a significant production of mediators of inflammation, acting as a strategy of protection [34]. The positive modulation of TNF-α has been observed in specimens of *Cyprinus carpio* treated, during vaccination, with *Aloe vera*, and also on immunity response in poultry, mice, and humans [35]. In that case, the effects were attributed to some poorly defined compounds of the plant [35].

Members of the HSP70 family are widely used as biomarkers of environmental stress in ecological and toxicological studies in fish [36–38]. The functions of the different HSP70 family members depend on their cellular localization, acting both as chaperons and immunomodulators [38], with the general aim of maintaining protein homeostasis [39] in normal and stressful conditions. HSP70 can be induced in response to thermal stress, hypoxia, oxidative stress, ultraviolet radiation, nutrient deprivation, osmotic pressure, heavy metals, chemical agents, microbial infections, and inflammation [37–41].

The levels of HSP70 protein increased in fish fed with the highest doses of bee pollen inclusion (HBP2.5 and HBP4) (Figure 4), indicating a stress condition that could be related to metabolic or oxidative stress. In fact, this protein is normally present at a low level in organisms with a good balance of antioxidants and over-expressed in a situation of oxidative stress [42–44]. Our previous study on the inclusion of dehydrated lemon peel in the diet for seabream showed a reduction of the HSP70 levels in fish fed on increased levels of bioactive compounds, due to the antioxidant composition of this resources, which is able to guarantee protection against oxidative stress [2]. Accordingly, Di Giancamillo et al. [45] described a decrease of HSP70 in pigs fed a diet supplemented with a natural verbascoside extract, compared with pig fed a high-fat diet. These authors suggested a role of HSP70 in modulating hepatic oxidative stress [45].

The main evidence that emerged from the blood analyses was the linear reduction of the total serum protein levels, concomitant with the increase of HBP inclusion in diets. TSP is a non-destructive parameter that is robust, easy to measure everywhere, and cheap, representing a suitable way of monitoring the overall welfare of fish by its regular increase [46]. It has been reported that this parameter tends to decrease in fish in response to various stress conditions such as hyperoxia, hypercapnia, stocking density, transfer to another tank, or nodavirus injection [46]. Many explanations have been suggested to explain TSP decrease, such as changes in blood volume and plasma hydration, alterations in permeability of environmental barriers, tissue destruction, malabsorption, and tissue injury by parasites or other pathogens [47]. These results confirm what emerged in the intestinal histology and immunohistochemistry and in hepatic-stress-related biomolecular markers.

Given that the fish of the different groups were subjected to the same environmental conditions, and received isoproteic and isoenergetic diets that varied only according to the level of pollen inclusion, our results can be explained through a negative action of the pollen in terms of intestinal inflammation.

Concerning the mineral content, the macro elements and the trace element contents in our HBP were in line with the values reported by Taha [48]. Compared to the maximum levels (MLs) of heavy metals set by the EU Commission (Directive (EC) No 2002/32 [49]), Cd, Pb, and as levels in honey bee pollen were lower than the maximum values established for feedstuff and feed materials, indicating that the risk of exposure to heavy metals deriving from consumption of the honey bee pollen used in this trial was relatively low and in compliance with EU regulations. However, in all the diets used in the trial, the levels of potentially toxic elements resulted lower than the MLs stated by EU regulation establishing the following MLs of heavy metals content in mg/kg (ppm) relative to a feed with a moisture content of 12%: the Cd MLs in the feed materials and complete feed are 2.0 and 0.5 mg/kg, respectively; the Pb MLs in the feed materials and complete feed are 10.0 and 5.0 mg/kg, respectively; the As MLs in the feed materials and complete feed are both 2 mg/kg. These elements, therefore, may not have been important in determining the observed deterioration of the growth performance and health status of the intestine in such fish species.

To deeply comprehend the reason for such negative results, it is necessary to dwell on the microscopic structure of bee pollen. Honey bee pollen is composed of a multitude of microscopic particles (6–200 μm in diameter) with variable shape, usually spherical or oval. The pollen cell walls consist of a series of stratified concentric layers [50]. As Roulstone described, the outermost layer of the pollen wall is the pollenkitt [51], a semi-solid coating comprised primarily of neutral lipids, hydrocarbons, terpenoids, and carotenoid pigments [52]. Inside the pollenkitt is the exine, an often intricately ridged matrix of the complex carbohydrate sporopollenin. Sporopollenin is a compound that provides chemical resistance to bee pollen and preserves the compounds within it [53]. The chemical analysis of sporopollenin has been limited due to the difficulty of obtaining large quantities of exine without affecting or modifying the structure of this molecule. Atkin et al. [53] recently found that this molecule has an empirical formula $C_{90}H_{144}O_{27}$. Nowadays, it is suggested that the sporopollenin consists of a series of biopolymers, homologous to the compounds cutin, suberin, and lignin.

The exine greatly resists decay and digestion, but is commonly perforated by one-to-several pores or slits (germination pores) that lead to the inner wall layer, known as the intine. The intine, composed primarily of cellulose and pectin, is also very resistant, and forms the final barrier to the nutrient-rich cytoplasm. From reports based on literature, it is known that the exine is rich in antioxidant compounds, while the intine has the rest of the nutritional and bioactive compounds [54]. Thus, any animal consuming pollen consumes pollenkitt nutrients through external probing of pollen grains or ingestion of pollen grains, but must penetrate or dismantle two resistant pollen wall layers in order to access cytoplasmic nutrients [51]. As the pollen has such complex ultrastructure, it is not certain that the beneficial substances contained in it are available for monogastric animals. Previous studies have shown that the exine and intine are indigestible by the human digestive tract [55]. Furthermore, administering pollen suspended in water to mice by means of gastric intubation resulted in grains being found in the feces [56]. The artificial digestion at various pH values and after various incubation times appeared unaffected after acid treatment (corresponding to gastric digestion), with the exception of the digestion of substances located outside the walls. Partial digestion of grains occurred only during alkaline treatment in the presence of pancreatic enzymes [57]; grains may be partially emptied if the enzymes are able to penetrate the exine pores, and it depends also on the thickness of the intin layer.

Taking into account the difficulties of monogastric animals in digesting bee pollen grains and the similarity of pollen walls with indigestible fibrous substances such as suberin and lignin, it can be assumed that bee pollen had a pro-inflammatory action on the intestinal mucosa of meagre, a fish with carnivorous feeding habits [19,58]. In contrast, the bee pollen structure may not represent a problem to herbivorous fish species such as the Nile tilapia, with an enzymatic profile more suitable for the digestion of fibrous substances and in which positive effects of honey bee pollen addition to the diet have been reported [8].

5. Conclusions

In conclusion, the results of the present trial showed that the addition of honey bee pollen to juvenile meagre diets not only did not have the positive effects described in the literature for other fish species, but it even had negative effects on both growth performance and diet digestibility. Moreover, the inflammatory state of the intestine progressively worsened as the level of HBP inclusion increased. These effects could be ascribed to the ultrastructure of the bee pollen grain walls (exine and intine), which made the bioactive substances unavailable for the fish, while the intestinal inflammation could have been due to the bee pollen grains' chemical composition rich in lignin- and suberin-like substances, which can irritate the intestine of monogastric animals such as carnivorous fish. These negative HBP effects could be overcome by using an extraction method able to concentrate the bioactive substances and eliminate the indigestible fractions. As a general conclusion, it is worth highlighting that it should always be considered necessary to test nutraceutical additives of natural origin in each species in order to verify the effective positive action and exclude any negative repercussions on animal health.

Author Contributions: V.P., G.P., F.B., contributed to the conception of the experimental design; V.P. and S.C., conducted the digestibility and the growth trials and performed the samples collection; S.C. performed the blood analyses (at "Biology—Physiology of Marine Organisms", University of Crete, Prof. M. Pavlidis); C.M.M., S.M., and A.S., performed the biomolecular stress related markers analyses; A.A. and L.S., performed the trace and toxic elements analyses; I.O. and B.R., performed the intestinal histology and immunohistochemistry analyses; F.B., C.M.M., and I.O., performed the statistical analysis; V.P., C.M.M., I.O., S.C., S.M., B.R., F.B., A.A., G.P., supported the acquisition and interpretation of data and contributed in writing the article. All authors have read and agreed to the published version of the manuscript.

Funding: The digestibility trial received funding by the European Union's Horizon 2020 research and innovation program under grant agreement No 730984, "ASSEMBLE Plus project" (Contract number: SR020220189047). The growth trial the blood analyses were funded by the European Union's Horizon 2020 research and innovation program under the TNA program (Project ID AE040069) within "AQUAEXCEL2020" project, grant agreement 652831.

Conflicts of Interest: The authors declare no conflict of interest.

References

1. Ferreira, A.; Proença, C.; Serralheiro, M.L.M.; Araújo, M.E.M. The in vitro screening for acetylcholinesterase inhibition and antioxidant activity of medicinal plants from Portugal. *J. Ethnopharmacol.* **2006**, *108*, 31–37. [CrossRef]

2. García Beltrán, J.M.; Espinosa, C.; Guardiola, F.A.; Manuguerra, S.; Santulli, A.; Messina, C.M.; Esteban, M.Á. Effects of dietary dehydrated lemon peel on some biochemical markers related to general metabolism, welfare and stress in gilthead seabream (*Sparus aurata* L.). *Aquac. Res.* **2019**, *50*, 3181–3191. [CrossRef]

3. Guardiola, F.A.; Bahi, A.; Messina, C.M.; Mahdhi, A.; Santulli, A.; Arena, R.C.; Bakhrouf, A.; Esteban, M.A. Quality and antioxidant response of gilthead seabream (*Sparus aurata* L.) to dietary supplements of fenugreek (*Trigonella foenum graecum*) alone or combined with probiotic strains. *Fish Shellfish Immun.* **2017**, *63*, 277–284. [CrossRef] [PubMed]

4. Bahi, A.; Guardiola, F.A.; Messina, C.M.; Mahdhi, A.; Cerezuela, R.; Santulli, A.; Bakhrouf, A.; Esteban, M.A. Effects of dietary administration of fenugreek seeds, alone or in combination with probiotics, on growth performance parameters, humoral immune response and gene expression of gilthead seabream (*Sparus aurata* L.). *Fish Shellfish Immun.* **2017**, *60*, 50–58. [CrossRef] [PubMed]

5. Bulfon, C.; Volpatti, D.; Galeotti, M. Current research on the use of plant-derived products in farmed fish. *Aquac. Res.* **2015**, *46*, 513–551. [CrossRef]

6. Messina, M.; Piccolo, G.; Tulli, F.; Messina, C.M.; Cardinaletti, G.; Tibaldi, E. Lipid composition and metabolism of European sea bass (*Dicentrarchus labrax* L.) fed diets containing wheat gluten and legumes meals as substitutes for fish meal. *Aquaculture* **2013**, *376*, 6–14. [CrossRef]

7. Babaei, S.; Shaban, R.; Torshizi, K.M.A.; Gholam, T.H.; Miran, K.S.N. Effect of Honey, royal jelly and bee pollen on performance, immune system and blood parameters in Japanese quail. *J. Anim. Prod.* **2016**, *17*, 311–320.

8.   El-Asely, A.M.; Abbass, A.A.; Austin, B. Honey bee pollen improves growth, immunity and protection of Nile tilapia (*Oreochromis niloticus*) against infection with *Aeromonas hydrophila*. *Fish Shellfish Immun.* **2014**, *40*, 500–506. [CrossRef]

9.   Attia, Y.A.; Al-Hanoun, A.; Tag El-Din, A.E.; Bovera, F.; Shewika, Y.E. Effect of bee pollen levels on productive, reproductive and blood traits of NZW rabbits. *J. Anim. Physiol. Anim. Nutr.* **2010**, *95*, 294–303. [CrossRef]

10.   Choobkar, N.; Kakoolaki, S.; Mohammadi, F.; Rezaeimanesh, M. The effect of dietary propolis and pollen extracts on growth performance and haematological responses of rainbow trout (*Onchorhynchus mykiss*). *J. Aquat. Anim. Health* **2017**, *3*, 16–25. [CrossRef]

11.   LeBlanc, B.W.; Davis, O.K.; Boue, S.; DeLucca, A.; Deeby, T. Antioxidant activity of Sonoran Desert bee pollen. *Food Chem.* **2009**, *115*, 1299–1305. [CrossRef]

12.   Thorp, R.W. The collection of pollen by bees. In *Pollen and Pollination*; Dafni, A., Hesse, M., Pacini, E., Eds.; Springer: Vienna, Austria, 2000; pp. 211–223. [CrossRef]

13.   Campos, M.G.R.; Bogdanov, S.; de Almeida-Muradian, L.B.; Szczesna, T.; Mancebo, Y.; Frigerio, C.; Ferreira, F. Pollen composition and standardisation of analytical methods. *J. Apic. Res.* **2008**, *47*, 154–161. [CrossRef]

14.   Jean-Prost, P.; Le Conte, Y. *Apicultura: Conocimiento de la Abeja. Manejo de la Colmena*, 4th ed.; Ediciones Paraninfo: Madrid, Spain, 2007.

15.   Attia, Y.A.; Al-Hamid, A.A.; Ibrahim, M.S.; Al-Harthi, M.A.; Bovera, F.; Elnaggar, A.S. Productive performance, biochemical and hematological traits of broiler chickens supplemented with propolis, bee pollen, and mannan oligosaccharides continuously or intermittently. *Livest. Sci.* **2014**, *164*, 87–95. [CrossRef]

16.   Mărghitaş, L.A.; Stanciu, O.G.; Dezmirean, D.S.; Bobiş, O.; Popescu, O.; Bogdanov, S.; Campos, M. In vitro antioxidant capacity of honeybee-collected pollen of selected floral origin harvested from Romania. *Food Chem.* **2009**, *115*, 878–883. [CrossRef]

17.   Abbass, A.A.; El-Asely, A.M.; Kandiel, M.M. Effects of Dietary Propolis and Pollen on Growth Performance, Fecundity and Some Hematological Parameters of Oreochromis niloticus. *Turk. J. Fish. Aquat. Sci.* **2012**, *12*, 851–859. [CrossRef]

18.   Chatzifotis, S.; Panagiotidou, M.; Papaioannou, N.; Pavlidis, M.; Nengas, I.; Mylonas, C. Effect of dietary lipid levels on growth, feed utilization, body composition and serum metabolites of meagre (*Argyrosomus regius*) juveniles. *Aquaculture* **2010**, *307*, 65–70. [CrossRef]

19.   AOAC. *Official Methods of Analysis*, 18th ed.; Association of Official Analytical Chemists: Arlington, VA, USA, 2004; Available online: https://www.aoac.org/ (accessed on 30 January 2020).

20.   Ariano, A.; Marrone, R.; Andreini, R.; Smaldone, G.; Velotto, S.; Montagnaro, S.; Anastasio, A.; Severino, L. Metal concentration in muscle and digestive gland of common octopus (*Octopus vulgaris*) from two coastal site in Southern Tyrrhenian Sea (Italy). *Molecules* **2019**, *24*, 2401. [CrossRef]

21.   Barnard, T.W.; Crockett, M.I.; Ivaldi, J.C.; Lundberg, P.L. Design and evaluation of an echelle grating optical system for ICP-OES. *Anal. Chem.* **1993**, *65*, 1225–1230. [CrossRef]

22.   Piccolo, G.; Iaconisi, V.; Marono, S.; Gasco, L.; Loponte, R.; Nizza, S.; Bovera, F.; Parisi, G. Effect of *Tenebrio molitor* larvae meal on growth performance, in vivo nutrients digestibility, somatic and marketable indexes of gilthead sea bream (*Sparus aurata*). *Anim. Feed Sci. Technol.* **2017**, *226*, 12–20. [CrossRef]

23.   Piccolo, G.; Centoducati, G.; Bovera, F.; Marrone, R.; Nizza, A. Effects of Mannan Oligosaccharide and Inulin on Sharpsnout Seabream (*Diplodus Puntazzo*) in the Context of Partial Fish Meal Substitution by Soybean Meal. *Ital. J. Anim. Sci.* **2013**, *12*, 132–138. [CrossRef]

24.   Vogtmann, H.; Pfirter, H.P.; Prabucki, A.L. A new method of determining metabolisability of energy and digestibility of fatty acids in broiler diets. *Br. Poult. Sci.* **1975**, *16*, 531–534. [CrossRef]

25.   Palmegiano, G.B.; Dapra, F.; Forneris, G.; Gai, F.; Gasco, L.; Guo, K.; Peiretti, P.G.; Sicuro, B.; Zoccarato, I. Rice protein concentrate meal as a potential ingredient in practical diets for rainbow trout (*Oncorhynchus mykiss*). *Aquaculture* **2006**, *258*, 357–367. [CrossRef]

26.   Lowry, O.H.; Rosebrough, N.J.; Farr, A.L.; Randall, R.J. Protein measurement with the Folin phenol reagent. *J. Biol. Chem.* **1951**, *193*, 265–275.

27.   Hong, S.; Li, R.; Xu, Q.; Secombes, C.J.; Wang, T. Two Types of TNF-α Exist in Teleost Fish: Phylogeny, Expression, and Bioactivity Analysis of Type-II TNF-α3 in Rainbow Trout *Oncorhynchus mykiss*. *J. Immunol.* **2013**, *15*, 5959–5972. [CrossRef]

28. Roca, F.J.; Mulero, I.; López-Muñoz, A.; Sepulcre, M.P.; Renshaw, S.A.; Meseguer, J.; Mulero, V. Evolution of the Inflammatory Response in Vertebrates: Fish TNF-α Is a Powerful Activator of Endothelial Cells but Hardly Activates Phagocytes. *J. Immunol.* **2008**, *181*, 5071–5081. [CrossRef]

29. Johnston, B.L.; Conly, J.M. Tumour necrosis factor inhibitors and infection: What is there to know for infectious diseases physicians? *Can. J. Infect. Dis. Med. Microbiol.* **2006**, *17*, 209–212. [CrossRef]

30. Piccolo, G.; Bovera, F.; Lombardi, P.; Mastellone, V.; Nizza, S.; Di Meo, C.; Marono, S.; Nizza, A. Effect of *Lactobacillus plantarum* on growth performance and hematological traits of European sea bass (*Dicentrarchus labrax*). *Aquac. Int.* **2015**, *23*, 1025–1032. [CrossRef]

31. García-Castillo, J.; Pelegrín, P.; Mulero, V.; Meseguer, J. Molecular cloning and expression analysis of tumor necrosis factor α from a marine fish reveal its constitutive expression and ubiquitous nature. *Immunogenetics* **2002**, *54*, 200–207. [CrossRef]

32. Wang, Y.; Wang, L.; Zhang, C.; Song, K. Effects of substituting fishmeal with soybean meal on growth performance and intestinal morphology in orange-spotted grouper (*Epinephelus coioides*). *Aquac. Rep.* **2017**, *5*, 52–57. [CrossRef]

33. Zhong, J.; Feng, L.; Jiang, W.; Wu, P.; Liu, Y.; Jiang, J.; Kuang, S.; Tang, L.; Zhou, X. Phytic acid disrupted intestinal immune status and suppressed growth performance in on-growing grass carp (*Ctenopharyngodon idella*). *Fish Shellfish Immunol.* **2019**, *92*, 536–551. [CrossRef]

34. Chang, H.R.; Arsenijevic, D.; Pechère, J.C.; Piguet, P.F.; Mensi, N.; Girardier, L.; Dulloo, A.G. Dietary supplementation with fish oil enhances in vivo synthesis of tumor necrosis factor. *Immunol. Lett.* **1992**, *34*, 13–17. [CrossRef]

35. Abdy, E.; Alishahi, M.; Tabandeh, M.R.; Ghorbanpoor, M.; Jafari, H. Comparative effect of Freund's adjuvant and *Aloe vera* L. gel on the expression of TNF-α and IL-1β in vaccinated *Cyprinus carpio* L. with *Aeromonas hydrophila* C. bacterin. *Aquac. Res.* **2016**, *47*, 2543–2552. [CrossRef]

36. Girilal, M.; Krishnakumar, V.; Poornima, P.; Fayaz, A.M.; Kalaichelvan, P.T. A comparative study on biologically and chemically synthesized silver nanoparticles induced Heat Shock Proteins on fresh water fish *Oreochromis niloticus*. *Chemosphere* **2015**, *139*, 461–468. [CrossRef]

37. Metzger, D.C.H.; Hemmer-Hansen, J.; Schulte, P.M. Conserved structure and expression of *hsp70* paralogs in teleost fishes. *Comp. Biochem. Physiol Part D Genomics Proteomics* **2016**, *18*, 10–20. [CrossRef]

38. Radons, J. The human HSP70 family of chaperones: Where do we stand? *Cell Stress Chaperones* **2016**, *21*, 379–404. [CrossRef]

39. Zuo, D.; Subjeck, J.; Wang, X.Y. Unfolding the role of large heat shock proteins: New insights and therapeutic implications. *Front. Immunol.* **2016**, *7*, 75–90. [CrossRef]

40. Han, Y.L.; Yang, W.X.; Long, L.L.; Sheng, Z.; Zhou, Y.; Zhao, Y.Q.; Wang, Y.F.; Zhu, J.Q. Molecular cloning, expression pattern, and chemical analysis of heat shock protein 70 (HSP70) in the mudskipper *Boleophthalmus pectinirostris*: Evidence for its role in regulating spermatogenesis. *Gene* **2016**, *575*, 331–338. [CrossRef]

41. Teigen, L.E.; Orczewska, J.I.; McLaughlin, J.; O'Brien, K.M. Cold acclimation increases levels of some heat shock protein and sirtuin isoforms in threespine stickleback. *Comp. Biochem. Physiol. Part A Mol. Integr. Physiol.* **2015**, *188*, 139–147. [CrossRef]

42. Messina, C.M.; Pizzo, F.; Santulli, A.; Bušelić, I.; Boban, M.; Orhanović, S.; Mladineo, I. *Anisakis pegreffii* (Nematoda: Anisakidae) products modulate oxidative stress and apoptosis-related biomarkers in human cell lines. *Parasit. Vectors* **2016**, *9*, 1–10. [CrossRef]

43. Messina, C.M.; Faggio, C.; Laudicella, V.A.; Sanfilippo, M.; Trischitta, F.; Santulli, A. Effect of sodium dodecyl sulfate (SDS) on stress response in the Mediterranean mussel (*Mytilus Galloprovincialis*): Regulatory volume decrease (Rvd) and modulation of biochemical markers related to oxidative stress. *Aquat. Toxicol.* **2014**, *157*, 94–100. [CrossRef]

44. Chopra, A.S.; Kuratnik, A.; Scocchera, E.W.; Wright, D.L.; Giardina, C. Identification of novel compounds that enhance colon cancer cell sensitivity to inflammatory apoptotic ligands. *Cancer Biol. Ther.* **2013**, *14*, 436–449. [CrossRef]

45. Di Giancamillo, A.; Rossi, R.; Pastorelli, G.; Deponti, D.; Carollo, V.; Casamassima, D.; Domeneghini, C.; Corino, C. The effects of dietary verbascoside on blood and liver oxidative stress status induced by a high *n*-6 polyunsaturated fatty acids diet in piglets. *J. Anim. Sci.* **2015**, *93*, 2849–2859. [CrossRef] [PubMed]

46. Coeurdacier, J.L.; Dutto, G.; Gasset, E.; Blancheton, J.P. Is total serum protein a good indicator for welfare in reared sea bass (*Dicentrarchus labrax*)? *Aquat. Living Resour.* **2011**, *24*, 121–127. [CrossRef]

47. Steinhagen, D.; Oesterreich, B.; Korting, W. Carp coccidiosis: Clinical and hematological observations of carp infected with *Goussia carpelli*. *Dis. Aquat. Org.* **1997**, *30*, 137–143. [CrossRef]

48. Taha, E.A. Chemical composition and amounts of mineral elements in honey bee-collected pollen in relation to botanic origin. *J. Apic. Sci.* **2015**, *59*, 75–81. [CrossRef]

49. Directive (EC). No 2002/32 of the European Parliament and of the Council of 7 May 2002 on Undesirable Substances in Animal Feed (OJ L 140, 30.5.2002, p. 10). Available online: https://eur-lex.europa.eu/legal-content/EN/TXT/?uri=CELEX%3A32002L0032 (accessed on 30 January 2020).

50. Domínguez, C.M.; Bermudez, J.C.; de Cuenca, M.C. Valorization alternatives of Colombian bee pollen for its use as food resource. *Vitae* **2014**, *21*, 237–247.

51. T'ai, H.R.; Cane, J.H. Pollen nutritional content and digestibility for animals. In *Plant Systematics and Evolution*; Springer: Vienna, Austria, 1999; pp. 187–209.

52. Dobson, H.E.M. Survey of pollen and pollenkitt lipids-chemical cues to flower visitors? *Am. J. Bot.* **1988**, *75*, 170–182. [CrossRef]

53. Atkin, S.L.; Barrier, S.; Cui, Z.; Fletcher, P.D.I.; Mackenzie, G.; Panel, V.; Sol, V.; Zhang, X. UV and visible light screening by individual sporopollenin exines derived from *Lycopodium clavatum* (club moss) and *Ambrosia trifida* (giant ragweed). *J. Photochem. Photobiol. B Biol.* **2011**, *102*, 209–217. [CrossRef] [PubMed]

54. Mackenzie, G.; Beckett, S.; Atkin, S.; Taboada, A.D. Pollen and Spore Shells. In *Microencapsulation in the Food Industry*, 1st ed.; Gaonkar, A.G., Vasisht, N., Khare, A.R., Sobel, R., Eds.; Academic Press: Cambridge, MA, USA, 2014; pp. 283–297. [CrossRef]

55. Franchi, G.; Corti, P.; Pompella, A. Microspectrophotometric evaluation of digestibility of pollen grains. *Plant Foods Hum. Nutr.* **1997**, *50*, 115–126. [CrossRef]

56. Franchi, G.G. Research on pollen digestibility. *Atti. Soc. Tosc. Sci. Nat.* **1987**, *B 94*, 43–52.

57. Franchi, G.G.; Franchi, G.; Corti, P. *Indagini Preliminari Sulla Cessione del Contenuto del Polline Nell'apparato Digerente*; Summaries of papers presented at 3° Congresso Nazionale della Società Italiana di Farrnacognosia; Università degli Studi: Sassari, Italy, 1985.

58. Chatzifotis, S.; Panagiotidou, M.; Divanach, P. Effect of protein and lipid dietary levels on growth of juveniles meagre (*Argyrosomus regius*). *Aquac. Int.* **2012**, *307*, 65–70. [CrossRef]

Article

Quality of Eggs and Albumen Technological Properties as Affected by *Hermetia Illucens* Larvae Meal in Hens' Diet and Hen Age

Giulia Secci [1], Fulvia Bovera [2], Giuliana Parisi [1,*] and Giuseppe Moniello [3]

[1] Department of Agriculture, Food, Environment and Forestry, University of Firenze, via delle Cascine 5, 50144 Firenze, Italy; giulia.secci@unifi.it

[2] Department of Veterinary Medicine and Animal Production, University of Napoli Federico II, via Federico Delpino 1, 80137 Napoli, Italy; fulvia.bovera@unina.it

[3] Department of Veterinary Medicine, University of Sassari, via Vienna 2, 07100 Sassari, Italy; moniello@uniss.it

* Correspondence: giuliana.parisi@unifi.it; Tel.: +39-055-2755590

Received: 29 October 2019; Accepted: 30 December 2019; Published: 3 January 2020

Simple Summary: Concurrently with the actual challenges in the poultry industry, we aimed to assess the changes induced by the inclusion of an alternative protein source (partially defatted *Hermetia illucens* larva meal, HI) at two different levels as well as hen age on the overall egg quality and deepened their effect of albumen technological properties. This study could provide useful information for the egg supply chain in order to optimize egg utilization, as a whole or as egg products, thus preventing food losses or undesirable wastes. Specifically, based on the obtained results, the eggs laid by hens fed the highest inclusion of HI would be directed towards egg product industry due to their reduced eggshell percentage and thickness which could increase their fragility. Contrariwise, due to the impaired albumen technological properties, as foaming, found in the egg laid by 27–35 wk-old hens, they could be preferentially sold as a whole.

Abstract: The impact on the egg quality and the albumen technological properties were here evaluated as affected by diet and hen age (A) of 162 Hy-line Brown hens. Three isoproteic and isoenergetic diets were formulated respecting the requirements for Hy-line hens: the control diet (C) based on conventional protein sources, and other two where vegetable proteins were substituted at 25% (HI25) and 50% (HI50) by partially defatted *Hermetia illucens* larva meal (HI). Ten eggs collected from each group at the hen ages of 20, 27, and 35 weeks were evaluated. The eggshell percentage and thickness were significantly reduced in the HI50 eggs (11.93% and 476 μm, respectively) compared to the C (12.34%, 542 μm) and HI25 (12.54%, 516 μm). The aging lowered ($p = 0.05$) the protein and increased ($p < 0.001$) water contents of the eggs. Although the foam capacity of the HI50 albumen was halved than the C group ($p < 0.05$), it was unaffected by the aging. Additionally, this did not impair the volume and the textural properties of a batter (angel cake) in which it was included. On the opposite, the textural characteristics of the cake made by the oldest hens (i.e., 35 wk-old) were compromised. In conclusion, the diet and hen age differently affected egg quality and its technological properties, which could be positive to obtain eggs to destine directly to the market or to the egg industry.

Keywords: black soldier fly; foam capacity; angel cake; texture; week of deposition

---

1. Introduction

Standing on the data proposed by the Food and Agriculture Organization in 2018 [1], eggs are the second-fastest growing industry in the world, with more than 50% growth forecast in the next 2 decades. Europe production in 2019 amounts to 7.5 million tonnes of eggs thanks to the 400 million laying hens kept throughout the European Union (27 Member States) [2]. Italy is the fourth producer with approximately 817,000 tons of eggs consumed as a whole or destined to the egg industry. Indeed, despite eggs play a very important role in human nutrition as a precious source of proteins, essential amino acids, lipid and several trace elements, the industry is continuously asking for egg products, such as liquid, powdered, concentrated whole egg or its separated main components, yolk and albumen. For instance, the egg allowance directed toward processing chains has been estimated at 32% of the whole Italian production [3]. This interest is linked to the unique functional properties of eggs, such as foaming, gelling and emulsification. The foaming capacity belongs to the egg white protein and it is defined as "the ability to rapidly adsorb on the air-liquid interface during whipping or bubbling and by its ability to form a cohesive viscoelastic film by way of intermolecular interactions" [4]. Thanks to these characteristics, albumen is widely utilized as an ingredient in bakery products, like bread, cakes and meringues, ice creams and several other processed foods.

The increase in egg and egg-derived product demand is however dependent on the laying hen farming, a sector that is affected by some critical aspects, such as feed supplying. As one of the main concerns of the last decade, the unsustainability of feed, especially soybean meal, is quite debated. Indeed, despite being an important source of essential amino acids, soybean production for feed directly competes with human nutrition and is largely associated with water pollution and exploitation of the lands [5].

A big opportunity for companies looking at more sustainable protein sources has been recently identifying in insect meals. Although processed animal proteins (PAPs) are still banned as livestock feeds in Europe, the European Commission voted in 2017 to introduce seven insect species, processed as meal or oil, into fish feeds and pet food [6]. That is why the stakeholders are confident that these protein sources would be allowed in other feeds in the coming years. As the interest in this alternative ingredient has grown, researchers are checking the feasibility of partial or complete substitution of the conventional protein sources with insect meal in poultry feed. Among the other species, *Hermetia illucens* meal (HI) has deserved major attention for laying hen diets. Indeed, in addition to the other advantages related to the rearing process of this species of insect, HI is a valuable protein source (40–44 g/100 g of dry matter, DM) and it provides calcium and phosphorous (50 to 80 g/kg dry matter, 6 to 15 g/kg dry matter, respectively), which are fundamental for laying nutrition [7]. HI has been proved to be a feasible substitute for vegetable protein sources in the hen diet when considering animal growth and performance [8–10] and egg quality [11–13]. However, the authors focused on the egg deposition, quality characteristics as the albumen height and the yolk color and its chemical composition, more than the albumen technological properties, which could be of interest to the egg industry. Thus, this study aimed to test the effect of partial substitution of the conventional protein sources with a partially defatted *Hermetia illucens* larva meal on the overall egg quality and on the technological properties of the albumen from the eggs produced by Hy-line Brown laying hens and collected at three different hen ages.

2. Materials and Methods

2.1. In Vivo Trial

All the hens were treated in accordance with the Directive of the European Parliament of the Council on the Protection of Animals Used for Scientific Purpose and in agreement with the Institutional Animal Care and Use Committee of the University of Naples Federico II, D.lgs n. 26 04/03/2014. All the experiments involving hens were approved by the Bioethical Committee of the University of Naples Federico II, under the number of protocol 2017/0017676.

The HI tested in this study was purchased from Hermetia Deutschland GmbH & Co KG (Baruth/Mark, Germany), chemically characterized as reported in Bovera et al. [14] and utilized to formulate the experimental diets, presented as follows. A control diet was formulated based on the conventional protein sources (corn-soybean based diet, C) while the other two were based on two different substitution levels of the conventional protein source with the partially defatted *Hermetia illucens* larva meal (HI, ether extract equal to 8.34% as fed). Specifically, the 25 and 50% of the dietary proteins were substituted by HI in the H25 and HI50 groups, respectively. The diets, whose ingredients and chemical composition are shown in Table 1, were formulated in order to result isoproteic and isoenergetic and to satisfy the Hy-line Brown requirements, as specified in Bovera et al. [14]. A total of 162 sixteen weeks-old Hy-line Brown hens (average live weight 1.41 kg ± 0.13) were housed in a commercial private laying hens farm (total flocks housed: 40,000 hens), located in Sardinia (Italy), for 20 weeks. The hens were equally divided into three feeding groups. Each group was housed in modified cages (18 hens/cage, 800 cm^2/hen, nine replicates of six hens each) adjacent to a nest box for egg deposition. The trial started after four weeks of adaptation, when hens were 20 weeks old. The mashed feeds were administrated *ad libitum* to the hens through a feeding belt, manually loaded. A dark: light cycle of 9:15 h was set. More details concerning the farming conditions as well as the data on feed intake, feed conversion ratio, weight gains and laying percentage can be retrieved in Bovera et al. [14]. The collection of the eggs for the purposes of the present trial occurred at three different hen ages (A), i.e., namely 20-wk-old (A20), 27-wk-old (A27), and 35-wk-old (A35). Specifically, the eggs were collected every day, then the overall weekly production per each dietary treatment was pooled. Ten undamaged eggs were selected from each pool and destined to the present trial, hence stored at −80 °C and moved to the Department of Agriculture, Food, Environment and Forestry (University of Firenze, Italy) where they were subjected to the scheduled analyses.

2.2. Egg Quality and Albumen Foaming Properties

After an over-night thawing in a refrigerated room (4 °C), the eggs were weighed prior to being broken and to separate each component (eggshell, albumen, and yolk) one from each other. The weights of the albumen and yolk were recorded and utilized for the calculation of their relative percentage. Concerning the eggshells, they were weighed after the removal of the inner membrane and the drying of the internal surface with common kitchen paper. After this, the thickness of the eggshell was measured by a manual caliper (Salmoiraghi, Milan, Italy) on three points (equator, round and apex). After this, the whole eggshell was grounded prior to ash quantification [15]. The color of the yolk was measured on three points by a Konica Minolta colorimeter CR-200 (Chiyoda, Japan) and the data were expressed as lightness (L*), redness (a*) and yellowness (b*) indexes [16].

The following analyses were conducted one-by-one on each albumen, without pooling the ones belonging to the same treatments. The chalazae were removed from each albumen prior to gently homogenate it by stirring for 15 min (speed set n. 4 of the magnetic stirrer, Type Age, Velp Scientific, Usmate Velate, Italy) in order to avoid foam formation and water losses, then the pH was determined using a SevenGo pH meter (Mettler-Toledo, Columbus, OH, USA) at room temperature. Ten mL of each albumen were allotted to the evaluation of egg white technological properties, as foam ability and foam stability, following the methodology recently proposed [17]. Briefly, the albumen was whipped for 1 min at 20 °C in a 50 mL plastic cylinder by a high-speed mixer (IKA Werke T25, Staufen, Germany) operating at 9500 rpm and the volume of the foam was recorded 30 s after whipping (V0) and holding 30 min (V30). The data were inserted into the following formula for the determination of the foam capacity (FC, %) and the foam stability (FS, %) [18]:

$$FC = (V0/Valb) \times 100 \tag{1}$$

$$FS = (V30/V0) \times 100 \tag{2}$$

where Valb is the initial volume of the albumen.

The other 20 g of the egg white from each sample were destined to the angel cake preparation, as detailed in the next section. The remaining albumen (around 8 g) was freeze-dried prior to being subject to water and crude protein content quantification [15].

Table 1. Ingredients (g/kg), chemical composition (g/100 g as fed), and metabolizable energy (MJ/kg) of the experimental diets C (control diet based on maize grain and soybean meal), HI25 (*Hermetia illucens* larva meal substituted the 25% of dietary vegetable protein) and HI50 (*Hermetia illucens* larva meal substituted the 50% of dietary vegetable protein).

	Diets		
	C	HI25	HI50
Ingredients			
Maize grain	605.5	597.5	630.5
Soybean meal	265	200	95
Insect meal	-	73	145
CaCO$_3$ grains	80	80	80
Vegetable oil	10	10	10
MinVit *	10	10	10
Methionine	2.5	2.5	2.5
Monocalcium phosphate	5	5	5
Celite	20	20	20
Salt	2	2	2
Chemical Composition			
Dry matter [1]	91.53	91.39	91.62
Crude protein [1]	16.45	16.32	17.03
Ether extract [1]	3.17	3.61	4.06
Metabolizable Energy [2]	11.85	11.90	11.89
Ash	14.00	14.21	14.13
Ca	4.63	4.75	4.51
Total P	0.68	0.69	0.65
Lysine [2]	0.86	0.97	1.00
Methionine [2]	0.63	0.58	0.61
Methionine + Cysteine	0.84	0.80	0.82
Isoleucine	0.72	0.73	0.75
Threonine	0.63	0.64	0.65
Tryptophan	0.18	0.15	0.16
Valine	0.93	1.07	1.14

[1] Determined according to AOAC (2004). [2] Calculated according to NRC (1994).* Provided per kilogram: vitamin A (retinyl acetate), 20,000 IU; vitamin D3 (cholecalciferol), 6000 IU; vitamin E (dl-α-tocopheryl acetate), 80 IU; vitamin B1(thiamine monophosphate), 3 mg; vitamin B2 (riboflavin), 12 mg; vitamin B6 (pyridoxine hydrochloride), 8 mg; vitamin B12 (cyanocobalamin), 0.04 mg;, vitamin K3 (menadione), 4.8 mg; vitamin H (d biotin), 0.2 mg; vitamin PP (nicotinic acid), 48 mg; folic acid, 2 mg; calcium pantothenate, 20 mg; manganous oxide, 200 mg; ferrous carbonate, 80 mg; cupric sulphate pentahydrate, 20 mg; zinc oxide, 120 mg; basic carbonate monohydrate, 0.4 mg; anhydrous calcium iodate, 2 mg; sodium selenite, 0.4 mg; choline chloride, 800 mg; 4–6-phytase, 1800 FYT; D.L. methionine, 2600 mg; canthaxanthin, 8 mg.

2.3. Angel Cake Preparation and Characterisation

Twenty g of egg white, 14 g of white sugar (Italia Zuccheri, Minerbio, Italy), 8 g of wheat flour "00" (Il Molino Chiavazza, Casalgrasso, Italy), and 2 g of rice starch (Pedon S.p.A., Molvena, Italy) were utilized for the production of the angel cake (one for each sample), which is a common batter utilized to test the foaming capacity of the protein ingredients. Albumen and sugar were whipped together for 60 s with a hand mixer equipped with a wire whisk attachment (Kenwood HDP408WH Triblade Mixer, Kenwood, Woking, UK), set at its maximum speed level. All the other dry ingredients were manually incorporated into the foam by beating on high until stiff peaks form (approximately 40 s). An aliquot corresponding to 31.711 ± 1.072 g of batter was spooned into one cavity of an ungreased silicone pan (mold size: 23.5 × 21.2 × 2.5 cm, total 6 cavities, cavity size: 8.0 × 5.5 × 2.5 cm, Shen zhen shi jun yang ke ji you xian gong si, Guangdong, China). The batters were baked for 15 min in a pre-heated static oven (180 °C, Mod. S370EB, Smeg, Guastalla, Italy), then the angel cakes were left to cool at room temperature for 1 h. Some physical parameters such as the weight and the maximum height of each cake were recorded, then baking loss (BL, %) was calculated as proposed [19]:

$$BL = (WCk\text{-}Wb) \times 100/WCb \tag{3}$$

where WCk was the weight of the angel cake, Wb was the weight of the batter and WCb was the water content of the batter obtained following the AOAC method [15]. Both angel cake dough and the baked cakes were analyzed for water and protein contents [15].

The color and texture profile analyses were assessed. Specifically, the cakes were cut vertically in the middle. Once opened, the color values were determined with a Konica Minolta colorimeter (Chiyoda, Japan) [16] on two points of the two halves. Finally, 4 × 4 cm pieces were cut from the two halves of each cake, the crust was removed and then both pieces were analyzed, using a Zwick Roell® texturometer (model KAF-TC 0901279, Zwick GmbH & Co., Ulm, Germany) equipped with a 50 N load cell. The texture profiles analysis was determined through a double compression cycle (crosshead speed: 1 mm s^{-1}) until the 60% deformation of the initial height, as previously done [19]. The hardness (N), cohesiveness (the area of work during the second compression divided by the area of work during the first compression), springiness (the distance of the detected height during the second compression divided by the original compression distance) and chewiness (Hardness x Cohesiveness x Springiness) were calculated from the force–time diagram (Test-Xpert2 software version 3.0, Zwick GmbH & Co., Ulm, Germany) [20].

2.4. Statistical Analysis

The data collected were processed by a 2-way ANOVA using the PROC GLM of SAS/STAT Software, Version 9 [21] by using Diet (D), Hen Age (A) and their interaction (D × A) as fixed factors. The comparison among the means was achieved by conducting Tukey's test [21] and the significance was set at $p < 0.05$. The means are accompanied by the Root Mean Square Error (RMSE) for characterizing the variability of the groups. Each egg within the feeding group and hen age was considered as a biological replicate.

3. Results

The egg, eggshell and yolk weights significantly increased due to the inclusion of HI in the feed for laying hens, irrespective of the insect inclusion level utilized (Table 2). Despite this, while looking at the percentage of the main components of the egg, only the eggshell resulted significantly diminished in the HI50 eggs compared to the C and HI25 groups. Accordingly, the thickness of the eggshell was reduced while increasing HI in the feed.

Table 2. Egg weight (g), egg components weight (g) and percentage (% on egg weight) and eggshell thickness (μm) as affected by the diet (C, HI25, HI50) and hen ages (20, 27, and 35 weeks).

Item	Diet, D			Hen Age, A			D	A	D × A	RMSE
	C	HI25	HI50	A20	A27	A35				
Egg Weight, g	57.56 [b]	61.10 [a]	60.82 [a]	60.79	59.36	59.43	*	ns	ns	4.413
Eggshell, g	7.10 [b]	7.65 [a]	7.24 [a,b]	7.26	7.60	7.14	*	ns	ns	0.643
Yolk, g	12.93 [b]	14.30 [a]	14.34 [a]	13.92	13.46	14.19	***	ns	***	1.298
Albumen, g	37.53	39.15	38.47	39.62	37.43	38.10	ns	ns	ns	4.091
Eggshell, %	12.34 [a]	12.54 [a]	11.93 [b]	11.96 [Y]	12.83 [X]	12.02 [Y]	*	***	ns	0.741
Yolk, %	22.49	23.44	23.59	22.92	22.71	23.89	ns	ns	ns	1.896
Albumen, %	65.18	64.02	63.21	65.12	63.19	64.10	ns	ns	ns	4.016
Eggshell Thickness, μm	542 [a]	516 [a,b]	476 [b]	499	540	494	*	ns	ns	72.249

RMSE: Root Means Square Error. a, b are significant different means among diets. X, Y are significant different means among hen ages. ns: not significant ($p > 0.05$), * $p < 0.05$, *** $p < 0.001$.

Furthermore, the eggshell percentage varied ($p < 0.05$) along with the hen age, showing the highest value at A27. These modifications in the eggshell characteristics cannot be attributed to the different ash content since it resulted unaffected by both D and A and equal to 76.4 ± 2.5 g/100 g of eggshell. Since L*, a*, and b* values resulted significantly affected by the interaction D × A at $p < 0.026$, 0.0001, and 0.049, respectively, we summarised the means obtained for this interaction in Table 3. No difference in the lightness of yolks was found as affected by the diet in the A20 group (Table 3). Conversely, the H50 eggs laid at A27 and A35 showed L* value significantly different from the C ones. The major differences were in the a* value, found significantly higher in the C yolks than in the H25 and H50 ones at A20 and A27. A greater decrease of the a* value was recorded in the C group increasing the time of laid, whereas the H50 raised its maximum a* value in the yolk from the eggs collected at A35. The b* index appeared scarcely affected by the diet, indeed the H50 yolks were significantly less yellow than the other two groups only in the A20 group. Furthermore, the same group significantly increased the b* index from the A27.

Table 3. Color parameter (lightness, L*, redness, a*, and yellowness, b*, indexes) values of the yolk from hens fed three different diets (C, HI25, HI50) and collected at three different hen ages (20, 27, and 35 weeks).

Colour Parameter	Diet	Hen Age (Weeks)		
		A20	A27	A35
L*	C	65.804 [ab]	64.686 [b,Y]	67.865 [a,XY]
	H25	66.665	68.553 [X]	68.671 [X]
	H50	68.858 [a]	68.723 [a,X]	65.537 [b,Y]
a*	C	6.514 [ab,X]	8.338 [a,X]	2.874 [b,X]
	H25	1.353 [ab,Y]	2.395 [a,Y]	−1.747 [b,Y]
	H50	−0.437 [b,Y]	−2.422 [b,Z]	5.268 [a,X]
b*	C	31.907 [X]	32.428	32.558
	H25	33.463 [X]	34.551	31.020
	H50	20.458 [b,Y]	28.391 [a]	33.300 [a]

[a,b]: different superscript letters are significantly different means within the row. [X,Y,Z]: different superscript letters are significantly different means within the column.

Table 4 shows the results of the technological properties and chemical composition analyses of the albumen. The FC was decreased by the presence of HI in the diet of laying hens, whereas the FS ad the pH value, as well as the water and protein contents were unaffected. A deep effect of the A can be observed for pH and chemical components of the albumen. For instance, a significant decrease in the pH value ($p < 0.01$) was found after the A27. The albumen of the eggs collected at the A35 showed a significantly ($p = 0.05$) lower protein content than those collected at the A20, while their water content increased with the aging.

Table 4. Technological properties (foam capacity and foam volume stability as %) and chemical composition (g/100 g of albumen) of the albumen from the eggs laid by hens fed three different diets (C, HI25, HI50) and collected at three different hen ages (20, 27, and 35 weeks).

Item	Diet, D			Hen Age, A			D	A	D × A	RMSE
	C	HI25	HI50	A20	A27	A35				
pH	8.12	8.08	8.05	8.19 [X]	8.04 [Y]	8.03 [Y]	Ns	**	ns	0.143
Foam Capacity	43.33 [a]	29.44 [b]	20.74 [b]	34.82 [X]	39.44 [X]	19.26 [Y]	***	***	*	14.036
Foam Stability	59.20	52.55	63.05	52.58	57.34	64.89	Ns	ns	ns	19.331
Water Content	87.30	87.47	87.11	87.39 [XY]	85.44 [Y]	89.05 [X]	Ns	***	ns	3.003
Crude Protein Content	11.20	11.03	11.39	11.11 [X]	13.06 [X]	9.45 [Y]	Ns	***	ns	3.003

RMSE: Root Means Square Error. [a,b] are significant different means among diets. [X,Y] are significant different means among hen ages. ns: not significant ($p > 0.05$). * $p < 0.05$, ** $p < 0.01$, *** $p < 0.001$.

Data related to the significant interaction between D and A emerged for the FC are reported in Figure 1. Although the H50 albumens had a halved ($p < 0.05$) FC than the C ones, they kept this property constant while increasing the hen age, while both the C and H25 groups reduced their FC between the A20 and the A35.

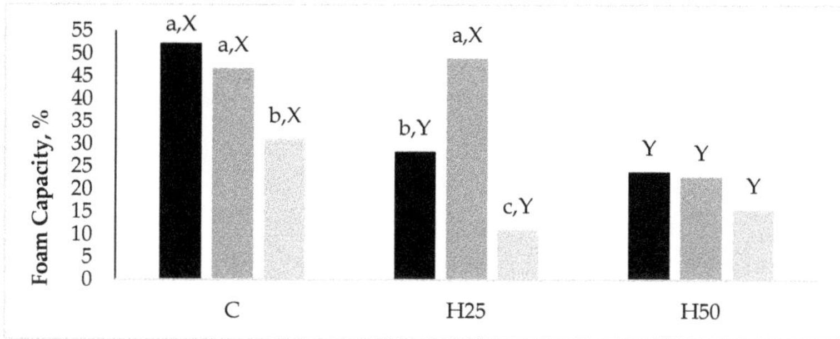

Figure 1. Effect of the interaction Diet x Hen Age on the foam capacity (%) of albumen from eggs laid by hens fed three experimental diets (C, H25, and H50) at three different hen ages (A20, black bars; A27, dark grey bars; A35, grey bars). a, b, c indicate significantly different means among the hen ages, within the diets. X, Y indicate significantly different means among the diets, within the hen ages.

The FC of the egg white was even determined by the analyses of physical parameters conducted on the angel cakes (Table 5). The textural properties of the angel cake were not affected by the dietary inclusion of *Hermetia illucens*, while A20 induced a significant reduction of the chewiness.

Table 5. Physical and chemical characteristics of the angel cakes obtained by the albumen from the eggs laid by hens fed three different diets (C, HI25, HI50) and collected at three different hen ages (20, 27, and 35 weeks).

Item	Diet, D			Hen Age, A			D	A	D × A	RMSE
	C	HI25	HI50	A20	A27	A35				
Height, mm	20.25 [a]	15.95 [b]	18.65 [a]	17.58	19.31	17.97	***	ns	*	2.39
Baking Loss, %	43.28 [a]	44.67 [a]	35.19 [b]	41.28 [X]	45.23 [X]	36.63 [Y]	***	***	*	7.871
Colour Parameters										
L*	87.956	87.653	88.360	88.257	87.690	88.023	ns	ns	ns	1.509
a*	−2.705	−2.742	−2.644	−2.669	−2.650	−2.772	ns	ns	ns	0.630
b*	13.494 [a]	13.335 [a]	11.936 [b]	13.034	13.024	12.707	***	ns	***	0.847
Texture										
Hardness, N	6.39	5.52	5.95	5.90	6.85	5.13	ns	ns	ns	3.088
Cohesiveness	0.71	0.72	0.70	0.71	0.70	0.72	ns	ns	ns	0.027
Springiness	8.92	9.22	8.67	9.94 [X]	7.89 [Y]	8.98 [X]	ns	***	ns	1.565
Chewiness	36.63	34.23	34.29	40.64 [X]	36.42 [X]	28.01 [Y]	ns	***	ns	11.121

RMSE: Root Means Square Error. a, b are significant different means among diets. X, Y are significant different means among hen ages. ns: not significant ($p > 0.05$). * $p < 0.05$, *** $p < 0.001$.

A significant interaction (D × A) emerged for height, b* value and baking loss of the samples (Table 6). The height of the angel cakes produced from the C egg whites was unaffected by the A, whereas the angel cakes obtained by the H25 eggs collected during the A35 significantly lost their height compared to those produced from the H25 eggs of the A20. While looking at the difference among the dietary treatments, the angel cake from the C eggs was higher than the H50 only considering the A20. In addition, the H25 egg whites collected at the A27 and A35 significantly decreased the height of the angel cake compared both to the C and H50 groups. Concerning the baking loss, the age did not affect the water retention ability of the angel cake made with the C albumen, while the highest inclusion of HI in the diet increased the ability of the batter to retain water, especially at the A35. The b* value of the angel cake obtained with the C albumens was unaffected by the age, while the yellowness was significantly higher using the albumen from the H25 eggs collected at the A35 than that collected at the A20. Conversely, the H50 angel cake was slight but significantly discoloured when prepared with the egg whites from the A27 and A35. Regarding the differences among the diets, the angel cakes produced with the H25 albumens collected at the A20 had a lower ($p < 0.001$) b* value than those of the C and H50 groups. In contrast, the H50 cakes prepared with the eggs collected at the A27 and A35 showed a lower ($p < 0.001$) b* index than the C and HI25.

Table 6. Results of the interaction between the treatments (Diet × Hen Age) for angel cake characteristics.

Item	Diet	Hen Age (Weeks)		
		A20	A27	A35
Height, mm	C	19.45 [X]	21.41 [X]	19.88 [X]
	H25	17.11 [a,XY]	16.49 [ab,Y]	14.26 [b,Y]
	H50	16.17 [b,Y]	20.03 [a,X]	19.77 [a,X]
Baking Loss, %	C	44.97	42.26 [Y]	42.60 [X]
	H25	40.86 [b]	54.30 [a,X]	38.84 [b,X]
	H50	38.01 [a]	39.12 [a,Y]	28.44 [b,Y]
b* Value	C	13.59 [X]	13.78 [X]	13.113 [X]
	H25	12.584 [b,Y]	13.49 [ab,X]	13.932 [a,X]
	H50	12.929 [a,XY]	11.803 [b,Y]	11.077 [b,Y]

[a,b] are significant different means within the row. [X, Y] are significant different means within the column.

Finally, Table 7 depicts that the batters prepared with the egg whites collected at the A27 had a slight but significantly lower water content than those from the egg whites collected at the other two considered ages.

Table 7. Chemical composition (g/100 g fresh sample) of the angel cake as raw and baked batter.

Item	Diet, D			Hen Age, A			D	A	D × A	RMSE
	C	HI25	HI50	A20	A27	A35				
Angel Cake Batter										
Water	40.75	40.68	40.52	40.79 [X]	39.74 [Y]	41.42 [X]	ns	***	ns	1.254
Crude protein	8.11	8.07	8.36	8.38	8.22	7.95	ns	ns	ns	0.744
Angel Cake Baked										
Water	28.61	29.13	27.72	29.68 [X]	26.08 [Y]	29.70 [X]	ns	***	ns	2.736
Crude protein	9.77	9.65	10.16	9.96	10.08	9.54	ns	ns	ns	0.963

RMSE: Root Means Square Error. [a,b] are significant different means among diets. [X,Y] are significant different means among hen ages. ns: not significant ($p > 0.05$), *** $p < 0.001$.

4. Discussion

The main factors affecting the egg weight are the dietary metabolized energy [22] and the size of the yolk that, in turn, is influenced by the body weight of the hens [23]. Since the administered diets in the present trial were isoenergetic, it seems clear to attribute the major changes in HI egg weights to the increase in the size of their yolks. Nevertheless, this modification cannot be attributable to the different body weight of the hens, resulted unaffected by the dietary intervention as well as the feeding intake and the feed conversion ratio [14]. Thus, the diet could have influenced the egg and yolk weights. Leeson and Summers [23] suggested that egg weight is very sensitive to methionine and total sulphur amino acid levels, which however were balanced in all the three administered diets.

Our hypothesis is that the significantly ($p = 0.032$) augmented length of the jejunum found in the HI25 and HI50 hens, being respectively 4.68 and 4.64 (as percentage of the live weight), compared to the C one (3.45% live weight), [14] could have promoted the absorption of the amino acids, hence resulting in a higher assumption of these nutrients. This possible explanation strongly needs the support of other comprehensive studies, since the effect of HI meal on hen egg weight is still scarcely investigated and conflicting in results [13,24]. The increased egg weight found with the HI meal inclusion in the hens' diet may require an increase in eggshell weight because of a rising in calcium deposition. Notwithstanding, authors underlined that egg-laying birds have a limited amount of calcium available to produce the shell, approximately 2.0–2.5 g Ca^{2+}, irrespective of the egg size weight [25]. Hence, the production of heavier eggs in the HI50 group might explain the significant reduction in shell percentage and even in its thickness. Since the eggshell weight and thickness are physical variables correlated with the egg strength, resistance to physical and pathogenic challenges from laying to the transportation and selling phase [26], our results should be considered when evaluating the suitability to include the HI meal in the laying hens' feed.

The egg internal quality, determined by several parameters (as albumen height and Haugh Unit), is even associated with the yolk quality, whose color is a fundamental characteristic [27]. The pigments contained in feeds mainly derive by plant ingredients, however, Secci et al. [13] underlined that even the HI larva meal may contain carotenoids and tocopherols. These pigments have a strong affinity for non-polar molecules, such as lipid, and they generally absorb wavelengths ranging from 400 to 550 nm, coloring as yellow, orange, or red. Thus, carotenoids can be easily accumulated in the egg yolk, which is rich in fat, producing the major modification of the yolk redness index, as we found.

Although the egg production and quality as affected by *H. illucens* in laying hen diets have been recently examined [11–13,24], the focus areas were the egg deposition and quality characteristics as the albumen height, the yolk color and its chemical composition. Nonetheless, the albumen has relevant importance for the egg industry mainly because of its unique functional properties, such as foaming that promote its extensive use as an ingredient in several processed food. In the present study, the diet did not affect the egg white pH and protein content, thus supporting previous findings [13]. The protein concentration as well as pH and cooking temperature, in addition to the physical-chemical properties of proteins, are well-established factors affecting the foaming ability [28]. Although in the present study no significant difference among the diets was found for pH and protein values, the foam ability of the HI25 and HI50 egg whites was decreased compared to that of the C ones. Since the temperature was controlled during the experiment, we hypothesize that the HI diets could modify the concentration, or the proportion of the single protein fractions contained in the egg white. As previously noted [4], the albumen foaming properties are the result of the interaction among the different proteins, such as globulins, ovalbumin, ovotransferrin, lysozyme, ovomucoid and ovomucin. Each fraction has its own ability to form a voluminous foam (foam capacity), to maintain it (foam stability) and to increase the volume of the batter containing it (i.e., angel cake) which differs from the overall foaming ability of the whole albumen [4]. For instance, globulins have the highest foaming index, around 4.71 cm^3/g min, more than 7 times higher than that of the ovalbumin (0.59), while ovomucin and ovomucoid show no foaming ability [4]. Nevertheless, Mine [4] reviewed that these last two proteins contributed to the final volume of the angel cakes made with them, whereas the ovalbumin produced cakes with a volume comparable to the one obtained by globulins, being 308 and 330 cm^3, respectively [4]. Standing on the mentioned literature, the possible role of the diet on protein matrix composition should be investigated to support or not our results about the technological properties of the egg white.

The albumen chemical composition varied while increasing the hen age, in line with previous findings [29] which proposed that the increase in hen age and weight led to a reduction in the crude protein content of the albumen. Changes in pH and protein content are consistent with the reduction of the foam capacity while using eggs from the 35th week of age. Instead, the maximum height raised by the angel cake varied due to the interaction D × WDA, thus supporting the previous hypothesis on the induced modification of proteins.

Baking loss is an important technological characteristic of a batter and plays a key role both for the quality of the final baked products and their shelf life. Indeed, the water content can promote microbiological growth during the storage, inducing a loss of the shelf life length. Considering that proteins may affect the water retention during cooking, the high affinity to the water of the white from eggs laid by hens fed the HI deserves other studies.

It has been previously demonstrated that the HI inclusion in the diet affects the yolk color values of the eggs [11–13], probably because of the pigments contained in the insect meal [13]. Standing on our knowledge, this is the first time that a difference in color emerged in a baked product made by egg white derived by hens fed with HI and, for this reason, little comparisons are possible.

The textural attributes of food are strictly related to its quality and consumers' acceptance. In the present study, the HI meal presence in the diet did not affect the textural properties of the angel cake, similarly to other authors who found that different dietary types of protein (i.e., soybean meal, cottonseed protein, double-zero rapeseed meal, individually or in combination with equal crude protein) administered to the Jinghong laying hens did not affect the cooked yolk hardness and springiness [30]. Furthermore, the deep reduction in the chewiness that occurred between the eggs laid at 20 and 35 weeks of hens' age agrees with the finite effect of the substitution of egg white by vegetable protein on angel cake chewiness [19]. Since chewiness can be enhanced due to a reinforcement of the protein entanglement in the networks [19], it is possible that variations in the egg white protein-gluten-starch connections occurred with hen aging, although more studies are necessary to define what kind of interaction is eventually developed.

5. Conclusions

The challenge for the foreseeable future will be to increase the sustainability of poultry production, including the egg supply chain. If one of the steps could be the substitution of the conventional protein sources, i.e., soybean meal, with the alternative ingredients, as insects in the case of their approval by European legislation, another point could be how to optimize the use of poultry products. In this regard, the present study provides the first information on how *infra-vitam* factors, such as diet and hens' age, could affect egg components and albumen technological properties, thus suggesting a differential use of the eggs based on hens' farming. Here, the use of eggs laid by hens fed the HI50 diet could be directed towards to egg product industry instead of selling as whole eggs because of their reduced eggshell percentage and thickness. Contrariwise, the eggs laid by hens from 27th to 35th weeks of age could be sold as whole because of their impaired albumen technological properties and consequently the chewiness of the baked cakes, which could be negatively perceived by the consumers.

Author Contributions: Conceptualization, F.B., G.M., and G.P.; methodology, G.S.; formal analysis, G.S.; data curation, G.P.; writing-original draft preparation, G.S.; writing-review and editing, F.B., G.M., G.S, and G.P.; supervision, G.P. All authors have read and agreed to the published version of the manuscript.

Funding: This research received no external funding.

Acknowledgments: This work was supported by the contribution of the Fondo di Ateneo per la Ricerca 2019, University of Sassari (Responsible: G. Moniello) and by the Fondo di Ateneo per la Ricerca 2019, University of Florence (Responsible: Giuliana Parisi).

Conflicts of Interest: The authors declare no conflict of interest.

References

1. FAO. Global Forum on Food Security and Nutrition. Report of Activity n. 154; 2018. Available online: http://www.fao.org/3/CA3569EN/ca3569en.pdf (accessed on 7 October 2019).
2. Committee for the Common Organization of the Agricultural Markets. EU Market Situation for Eggs. Available online: http://circabc.europa.eu (accessed on 5 December 2019).
3. Unaitalia. Annata Avicola. 2017. Available online: http://www.unaitalia.com/mercato/annata-avicola (accessed on 5 December 2019).

4. Mine, Y. Recent advances in the understanding of egg white protein functionality. *Trends Food Sci. Tech.* **1995**, *6*, 225–232. [CrossRef]

5. FAO. *The State of Food and Agriculture*; FAO: Rome, Italy, 2009.

6. European Commission. Commission Regulation (EU) 2017/893 of 24 May 2017. *O.J. L.* **2017**, *L138*, 92–116.

7. Makkar, H.P.; Tran, G.; Heuzé, V.; Ankers, P. State-of-the-art on use of insects as animal feed. *Anim. Feed Sci. Technol.* **2014**, *197*, 1–33. [CrossRef]

8. Cutrignelli, M.I.; Messina, M.; Tulli, F.; Randazzo, B.; Olivotto, I.; Gasco, L.; Loponte, R.; Bovera, F. Evaluation of an insect meal of the Black Soldier Fly (*Hermetia illucens*) as soybean substitute: Intestinal morphometry, enzymatic and microbial activity in laying hens. *Res. Vet. Sci.* **2018**, *117*, 209–215. [CrossRef] [PubMed]

9. Marono, S.; Loponte, R.; Lombardi, P.; Vassalotti, G.; Pero, M.E.; Russo, F.; Gasco, L.; Parisi, G.; Piccolo, G.; Nizza, S.; et al. Productive performance and blood profiles of laying hens fed *Hermetia illucens* larvae meal as total replacement of soybean meal from 24 to 45 weeks of age. *Poult. Sci.* **2017**, *96*, 1783–1790. [CrossRef]

10. Maurer, V.; Holinger, M.; Amsler, Z.; Früh, B.; Wohlfahrt, J.; Stamer, A.; Leiber, F. Replacement of soybean cake by *Hermetia illucens* meal in diets for layers. *J. Insects Food Feed* **2016**, *2*, 83–90. [CrossRef]

11. Kawasaki, K.; Hashimoto, Y.; Hori, A.; Kawasaki, T.; Hirayasu, H.; Iwase, S.; Hashizume, A.; Ido, A.M.; Miura, C.; Miura, T.; et al. Evaluation of black soldier fly (*Hermetia illucens*) larvae and pre-pupae raised on household organic waste, as potential ingredients for poultry feed. *Animals* **2019**, *9*, 98. [CrossRef]

12. Mwaniki, Z.; Neijat, M.; Kiarie, E. Egg production and quality responses of adding up to 7.5% defatted black soldier fly larvae meal in a corn-soybean meal diet fed to Shaver White Leghorns from wk 19 to 27 of age. *Poult. Sci.* **2018**, *97*, 2829–2835. [CrossRef]

13. Secci, G.; Bovera, F.; Nizza, S.; Baronti, N.; Gasco, L.; Conte, G.; Serra, A.; Bonelli, A.; Parisi, G. Quality of eggs from Lohmann Brown Classic laying hens fed black soldier fly meal as substitute for soya bean. *Animal* **2018**, *12*, 2191–2197. [CrossRef]

14. Bovera, F.; Loponte, R.; Pero, M.E.; Cutrignelli, M.I.; Calabrò, S.; Musco, N.; Panettieri, V.; Lombardi, P.; Piccolo, G.; Di Meo, C.; et al. Laying performance, blood profiles, nutrient digestibility and inner organs traits of hens fed an insect meal from *Hermetia illucens* larvae. *Res. Vet. Sci.* **2018**, *120*, 86–93. [CrossRef]

15. AOAC. *Official Methods of Analysis of AOAC International*; Association of Official Analysis Chemists International: Arlington, TX, USA, 2012.

16. CIE (Commission Internationale de l'Éclairage). *Colorimetry. Publication n. 15*; Bureau Central de la CIE: Wien, Austria, 2004.

17. Fernández-Martín, F.; Pérez-Mateos, M.; Dadashi, S.; Gómez-Guillén, C.M.; Sanz, P.D. Impact of magnetic assisted freezing in the physicochemical and functional properties of egg components. Part 1: Egg white. *Innov. Food Sci. Emerg.* **2017**, *44*, 131–138. [CrossRef]

18. Lomakina, K.; Míková, K. A study of the factors affecting the foaming properties of egg white—A review. *Czech J. Food Sci.* **2006**, *24*, 110–118. [CrossRef]

19. Jarpa-Parra, M.; Wong, L.; Wismer, W.; Temelli, F.; Han, J.; Huang, W.; Eckhart, E.; Tian, Z.; Shi, K.; Sun, T.; et al. Quality characteristics of angel food cake and muffin using lentil protein as egg/milk replacer. *Int. J. Food Sci. Technol.* **2017**, *52*, 1604–1613. [CrossRef]

20. Bourne, M.C. Texture profile analysis. *Food Technol.* **1978**, *32*, 62–66.

21. Statistical Analyses System (SAS). *SAS/STAT Software*; Version 9; SAS Institute Inc.: Cary, NC, USA, 2000.

22. Al-Qazzaz, M.F.A.; Ismail, D.; Akit, H.; Idris, L.H. Effect of using insect larvae meal as a complete protein source on quality and productivity characteristics of laying hens. *Rev. Bras. Zootecn.* **2016**, *45*, 518–523. [CrossRef]

23. Leeson, S.; Summers, J.D. Feeding program for laying hens. In *Commercial Poultry Nutrition*; Nottingham University Press: Nottingham, UK, 2009; pp. 193–214.

24. Ruhnke, I.; Normant, C.; Campbell, D.L.M.; Iqbal, Z.; Lee, C.; Hinch, G.N.; Roberts, J. Impact of on-range choice feeding with black soldier fly larvae (*Hermetia illucens*) on flock performance, egg quality, and range use of free-range laying hens. *Anim. Nutr.* **2018**, *4*, 452–460. [CrossRef]

25. Jonchère, V.; Brionne, A.; Gautron, J.; Nys, Y. Identification of uterine ion transporters for mineralisation precursors of the avian eggshell. *BMC Physiol.* **2012**, *12*, 1–17. [CrossRef]

26. Ketta, M.; Tůamová, E. Eggshell structure, measurements, and quality-affecting factors in laying hens: A review. *Czech J. Anim. Sci.* **2016**, *61*, 299–309. [CrossRef]

27. Roberts, J.R. Factors affecting egg internal quality and egg shell quality in laying hens. *J. Poult. Sci.* **2004**, *41*, 161–177. [CrossRef]
28. Alleoni, A.C.C. Albumen protein and functional properties of gelation and foaming. *Sci. Agric.* **2006**, *63*, 291–298. [CrossRef]
29. Anyaegbu, B.C.; Onunkwo, D.N.; Irebuisi, D. Evaluation of hen age, body weight and egg weight on percent egg component and the internal composition of Harco hen eggs. *Sci. Afr. J. Sci. Issues Res. Essays* **2016**, *4*, 943–945.
30. Wang, X.C.; Zhang, H.J.; Wu, S.G.; Yue, H.Y.; Wang, J.; Li, J.; Qi, G.H. Dietary protein sources affect internal quality of raw and cooked shell eggs under refrigerated conditions. *Asian-Australas. J. Anim. Sci.* **2015**, *28*, 1641–1648. [CrossRef] [PubMed]

Article

Effects of Fennel Seed Powder Supplementation on Growth Performance, Carcass Characteristics, Meat Quality, and Economic Efficiency of Broilers under Thermoneutral and Chronic Heat Stress Conditions

Ahmed A. AL-Sagan [1], Shady Khalil [2], Elsayed O. S. Hussein [3] and Youssef A. Attia [4,*]

[1] King Abdulaziz City for Science & Technology, P.O. Box 6086, Riyadh 11442, Saudi Arabia; abdeen@kacst.edu.sa
[2] AlzChem Trostberg GmbH, 83308 Trostberg, Germany; shady.khalil@alzchem.com
[3] Department of Animal Production, College of Food and Agriculture Sciences, King Saud University, Riyadh 11451, Saudi Arabia; shessin@ksu.edu.sa
[4] Arid Land Agriculture Department, Faculty of Meteorology, Environment, and Arid Land Agriculture, King Abdulaziz University, Jeddah 21589, Saudi Arabia
* Correspondence: yaattia@kau.edu.sa

Received: 30 November 2019; Accepted: 23 January 2020; Published: 26 January 2020

Simple Summary: Nowadays, great attention has been given to phytogenic products as a growth enhancer due to their safety and eco friendly influences on animal nutrition. Thus, this investigation looked at the use of fennel seed powder at 0, 1.6, and 3.2% as a dietary additive on the performance, carcass traits, meat quality, and efficiency of the production of broiler chickens raised under thermoneutral and chronic heat stress conditions. In essence, 3.2% fennel seed powder in the diet of broilers enhanced the growth rate under chronic heat stress and decreased breast meat redness and temperature, suggesting that 3.2% fennel seed could be used as an agent for enhancing the broiler's tolerance during chronic heat stress (CHS) condition from 19 to 41 days of age.

Abstract: Nowadays, phytogenic products have received great attention as a growth promoter due to their safety and environmentally friendly effect as a replacement for classical growth promoters such as antibiotics in animal nutrition. Thus, this research seeks the possibility of using fennel seed powder (FSP) as a dietary additive from 19 to 41 days of age on productive performance, carcass traits, meat quality, and production efficiency of broiler chickens raised under thermoneutral and chronic heat stress conditions. Thus, 216 one-day-old Ross-308 broiler chicks were divided into two equal groups. The first group was placed in an independent temperature-controlled room at 23 ± 2 °C. The broiler chicks from the second group were placed in a heat-stressed room and exposed to chronic heat stress conditions (32 ± 2 °C) for seven hours per day from 8 a.m. to 3 p.m. The experimental design was 2 × 3 factorial including two environmental temperatures (thermoneutral vs chronic heat stress) and three experimental diets that contained 0, 1.6, and 3.2% FSP. The chickens were randomly assigned to 18-floor pens per room temperature, representing six replicates per treatment and six birds per replicate. The results showed that dietary fennel seed powder during days 19–41 of age enhanced the growth rate of broiler chickens and improved breast meat redness and reduced temperature under chronic heat stress. In conclusion, 3.2% of fennel seed powder could be used as an agent for enhancing the broiler's tolerance during chronic heat stress condition from 19 to 41 days of age. Moreover, it is necessary to study in further detail the nitrite and nitrate contents in FSP and their impacts on muscle redness (a*) as well as muscle temperature.

Keywords: broilers; heat stress; fennel seed; performance; carcass characteristics; meat quality

1. Introduction

In hot regions, the poultry industry is significantly affected by high ambient temperatures, especially when combined with high relative air moisture contents (humidity) [1–3]. A further increase of 0.6–2.5 °C is expected to occur due to global warming over the next 50 years [4]. Optimum temperatures for broiler performance are 34–32, 32–28, 28–26, 26–24, 18–24, and 18–24 °C for the first, second, third, fourth, fifth, and sixth weeks of age, respectively [5]. Therefore, it is necessary to reevaluate poultry management to minimize heat stress.

Signs of heat stress begin to appear when the environmental temperature increases above 30 °C [6]. High temperatures can cause serious physiological disturbances [7]. Since birds do not have sweat glands, they evaporate their excess sweat through panting [8]. It has been estimated that approximately 0.5 kcals of energy are lost per gram of water evaporated during panting [9]. Urinary output is another mechanism that accompanies the excreta, with an increase in water intake to reduce body heat. This deviation in normal physiology reduces broiler productivity and leads to economic losses [9]. Generally, heat stress is considered as an inducer of oxidative stress [3,10,11] and can negatively affect animal performance and meat quality [12,13]. Moreover, nutritional intervention, along with proper management, is required to relieve the negative effect of heat stress [1]. Nutritional management can include feed restriction and dietary antioxidant supplementation [14,15] as well as the maintenance of electrolyte balance [16].

Recently, there has been great interest in phytochemicals, which are effective stress-alleviating agents [17,18]. The use of phytochemicals from medicinal plants in poultry diets is more related to product quality and consumer health than to production efficiency [19]. Fennel (Foeniculum vulgare, Apiaceae) seeds contain health-promoting volatile essential oil compounds such as fenchone, anethole, myrcene, limonene, chavicol, cineole, anisic aldehyde, and pinene as well as amino acids, phenolic compounds, and flavonoids [20]. These seeds also have apoptotic, hepatoprotective, antithrombotic, antiviral, antimicrobial, antispasmodic, anti-inflammatory, antimutagenic, antipyretic, and antinociceptive properties [14,20]. These active ingredients are also known to have digestive, anti-flatulent, and carminative properties [21]. This medicinal plant is a potential source of natural phytochemicals, especially antioxidants [22]. Furthermore, it can be used as a therapeutic agent to alleviate heat stress and stress-related diseases [15]. Fennel seeds contain high levels of nitrites and nitrates (65 to 376.7 mg/100 g), which are known to play crucial roles in maintaining vascular and digestive functions [23]. In mammals, nitrates are reduced to nitrites and eventually to nitric oxide (NO), which plays an essential role in the maintenance of the cardiovascular system [23]. In mammals and poultry, NO plays an important role in the regulation of various physiological functions that are mediated by the hypothalamus such as thermoregulation, fever, water balance, and cardiovascular regulation [24]. It has been found that inducible nitric oxide synthase (iNOS) increases in cold stress, heat stress, and inflammatory response [25] and as a response to lipopolysaccharide injection [26]. Since NO is produced from arginine metabolism [27,28], it can be speculated that fennel seed powder (FSP) may supply some NO, helping to reduce arginine metabolism for NO production [29].

Therefore, the objective of this study was to evaluate the potential of different levels of fennel seed powder supplemented to broiler diet as a physiological stimulator for the growth and development of chicks through its impacts on production performance traits, carcass characteristics, digestive organs, and meat quality under thermoneutral and heat stress conditions.

2. Materials and Methods

2.1. Animals and Treatments

King Abdulaziz City for Science and Technology, Riyadh, Saudi Arabia approved the experimental procedures. It recommends animal rights, welfare, and minimal stress and did not cause any harm or suffering to animals according to the Royal Decree M59 in 14/9/1431H.

Two hundred and sixteen one-day-old Ross-308 broiler chicks were purchased from the Al-Wadi company (Saudi Arabia) and used in this experiment. The chicks were housed in thermostatically controlled rooms. They were fed with a commercial starter mash based on corn and soybean ration (crude protein (CP), 23.2%; metabolizable energy (ME), 12.97 MJ/kg) during day 1 to 18 of age, and were managed according to the Ross-308 broiler guide for husbandry and health care practice.

On day 19, the broilers were divided into two equal groups. The first group was placed in an independent temperature-controlled room at 23 ± 2 °C. The broiler chicks from the second group were placed in a heat-stressed room and exposed to chronic heat stress conditions (32 ± 2 °C) for seven hours per day from 8:00 a.m. to 3:00 p.m. The indoor relative humidity and temperature under thermoneutral conditions during the experimental period for the broiler was 51.2% and 23.1 °C, 57% and 23.9 °C, and 53% and 24.7 °C during days 19–26, 27–33, and 34–41 of age, respectively. Under chronic heat stress conditions, the indoor relative humidity and temperature was 44.9% and 31.5 °C, 47.8% and 31.8 °C, and 41.8% and 31.9 °C during days 19–26, 27–33, and 34–41 of age, respectively. The experimental diets were formulated according to [30], as shown in Table 1. The nitrate contents were calculated using the minimum average reported by [23].

Table 1. Ingredients and nutrient composition of finishers (days 19 to 41), diets as fed basis.

Ingredients and Composition, g/kg	Fennel, g/kg		
	0	16	32
Corn	621	611	602
Soybean meal, 48% CP	287	283	277
Gluten Meal	30.0	30.0	30.0
Fennel seed powder	0.0	16.0	32.0
Palm oil	24.4	22.8	21.8
Dicalcium phosphate	17.5	17.1	16.8
Limestone	7.0	6.8	6.8
Dl-methionine	0.6	0.7	0.8
Lysine-HCl	1.9	2.0	2.2
L-threonine	0.5	0.6	0.7
Premix [1]	5.0	5.0	5.0
Salt	4.6	4.5	4.4
Choline chloride, 60%	0.5	0.5	0.5
Calculated composition			
ME, MJ/kg [2]	13.18	13.18	13.18
Dry Matter, g/kg [3]	894	894	894
Crude protein, g/kg [2]	205	204	204
Digestible methionine, g/kg [2]	5.18	5.23	5.28
Digestible lysine, g/kg [2]	11.0	11.0	11.0
Digestible methionine + cysteine, g/kg [2]	8.4	8.4	8.4
Digestible threonine, g/kg	7.2	7.2	7.2
Digestible arginine, g/kg	11.6	11.4	11.2
Nitrate, mg/kg	19.5	54.6	89.7
Nitrite, mg/kg	NA	NA	NA
Ether extract, g/kg [3]	46.2	44.3	42.9
Crude fiber, g/kg [3]	51.6	52.4	53.8
Calcium, g/kg [2]	8.5	8.5	8.5
Available Phosphorus, g/kg [2]	4.2	4.2	4.2
Magnesium, g/kg	1.59	1.59	1.60
Potassium, g/kg	7.93	8.24	8.54
Na (Sodium), g/kg	2.0	2.0	2.0
Cl (Chloride), g/kg	3.78	3.73	3.69
Dietary electrolyte balance, mEq/kg	183	193	201

[1] Vitamin-mineral premix contains the following per kg: vitamin A, 2,400,000 IU; vitamin D, 1,000,000 IU; vitamin E, 16,000 IU; vitamin K, 800 mg; vitamin B1, 600 mg; vitamin B2, 1600 mg; vitamin B6, 1000 mg; vitamin B12, 6 mg; niacin, 8000 mg; folic acid, 400 mg; pantothenic acid, 3000 mg; biotin 40 mg; antioxidant, 3000 mg; cobalt, 80 mg; copper, 2000 mg; iodine, 400; iron, 1200 mg; manganese, 18,000 mg; selenium, 60 mg, and zinc, 14,000 mg; DL-methionine, 32 g; lysine, 25 g. NA, data not available in addition.

The experimental design was 2 × 3 factorial including two environmental temperatures (thermoneutral vs. chronic heat stress), and three experimental diets that contained 0, 1.6, and 3.2% FSP. The two rooms were equipped with ultrasonic humidifiers to supply constant relative humidity to each room. The chickens were randomly assigned to 18-floor pens per room temperature, representing six replicates per treatment. Each floor pen (0.75 × 0.75 m) had six birds. Fennel seeds were purchased from the local market (imported from India) and ground into a fine powder using an electric mill. The FSP was added to the experimental diet in a mash form at the top. To assure that the diets were consumed, we offered a per weighted amount of feed daily according to the expected feed intake of each pen. Before, next meal, we collect the residual and mixed well with the new amount. From our observation, birds consumed almost all feeds and excessed 97% daily. The provision of feed and water was ad libitum throughout the experimental period. The chicks were exposed to continuous light from 1–6 days of age, then to 23:1 light-dark cycle throughout the experimental period. Feeding on experimental diets was initiated at 19 days of age.

2.2. Measurements

2.2.1. Production Performance Measurements

Live body weight (LBW), feed intake (FI) as fed basis, and feed conversion ratio (FCR) were measured for each floor pen every week. Then, the data were averaged for each pen to calculate the FI, body weight gain (BWG), and the FCR for the entire experimental period (19–41 days of age), where the BWG and FCR were calculated as follows:

BWG = Final body weight at 41 days of age-initial body weight at 19 days of age.

FCR = Amount of feed consumed (g)/live weight of chicks (g).

Nitrate intake was calculated using the feed intake data and the average concentration (220.85 mg/100 g) of nitrate in fennel seeds [23] and the available literature data.

Dietary electrolyte balance = Na concentration × 10000/ molecular weights of Na+K concentration × 10000/molecular weights of K- chloride × 10000/ molecular weights Cl.

Nitrate intake = Amount of feed consumed × concentration of nitrates in feedstuffs.

European production index and economic efficiency were determined, according to [18].

2.2.2. Carcass and Meat Characteristics

On day 41, six chicks were randomly taken from each treatment, one per replicate to represent all treatment replicates, individually weighed, and slaughtered. The chicks were scalded at 54 °C for 2 min and then de-feathered, and finally, their heads were removed. After de-feathering, breast muscles and abdominal fat were immediately removed from the hot carcass and weighed. The internal organs were removed in a precise anatomical manner from the beginning of the esophagus to the end of the outlet [31].

The intestines were cleaned and weighed using a 0.1 g sensitive balance, and the data were expressed as a percentage of carcass weight. The clean carcass weight (dressing yield) was expressed as a percentage of live body weight. The dressing percentage was determined by dividing the clean carcass weight by the live body weight. Abdominal fat, leg weight, breast weight, intestinal weight, gizzard weight, liver weight, and heart weight were determined as a percentage of carcass weight, as indicated by [32].

2.2.3. pH, Temperature, and Meat Color

The pH of the boneless breast fillets was measured at 15 min post-mortem, using an electronic TPX-90i pH meter (Toko Chemical Laboratories Co. Ltd., Tokyo, Japan) with a needle-type electrode (CE201S-SR; Toko Chemical Laboratories Co. Ltd., Tokyo, Japan). Each sample was measured at three points, and their mean value was recorded. The breast muscle temperature was determined 15 min after slaughter, with a portable digital thermocouple (EcoScan Series, Temp JKT, Eutech Instruments,

Vernon Hills, IL, USA). Then, the CIE color parameters (L* (lightness), a* (redness), and b* (yellowness) values) were determined at three points of the breast muscle sample surface on the dorsal side using a CR-400 colorimeter (Konica Minolta Sensing Inc., Osaka, Japan).

2.2.4. Economic Efficiency

Economic efficiency was calculated according to [33]. The assessment was carried out using the income-cost divided by the cost and multiplied by 100. The income is equal to the selling price of body weight at the end of the experiment. The cost included the cost of day-old chicks, veterinary care and other husbandry costs, and cost of feeding. The difference between the income and cost is the net revenue, which is divided by total cost and multiplied by 100 to obtain the economic efficiency.

2.3. Statistical Analysis

The general linear model procedure of the SAS Software (Cary, NC, USA) [34] was used for the analysis of the obtained data. A 3 × 2 factorial experiment with six replicates was used in a completely randomized block design. The replicate was the experimental unit. Percentage data were transformed to arcsine before the analysis. Differences in treatment means were compared using the Bonferroni Test at $p < 0.05$.

3. Results

3.1. Growth Performance

Table 2 shows the effects of the inclusion of FSP in Ross-308 broiler diets and heat stress on BWG, FI, FCR, survival rate, and production index. The interaction effect (Temp × FSP) was observed on the BWG and production index. Under normal temperatures, no differences were noticed among the groups. However, under heat stress, the highest BWG and production index were recorded in the 3.2% FSP ($p < 0.05$). These results reflect the positive effects of feeding the chicks with a diet containing 3.2% FSP on the BWG and production index under heat stress conditions.

Table 2. Broiler body weight gain, feed intake, and feed conversion ratio following supplementation with fennel seed powder under thermoneutral or heat stress conditions.

Temp	Diet	BWG (g/bird)	FI (g)	FCR (g Feed/g Weight Gain)	Survival Rate (%)	European Production Index
Normal	Control	2149 [a]	3226	1.50	100	341 [a]
	FSP (1.6%)	2092 [a]	3137	1.50	100	332 [a]
	FSP (3.2%)	2141 [a]	3052	1.43	100	358 [a]
High	Control	1740 [c]	2832	1.63	100	255 [c]
	FSP (1.6%)	1751 [bc]	2854	1.63	100	256 [bc]
	FSP (3.2%)	1821 [b]	2992	1.64	97.2	264 [b]
SEM		31.2	29.9	0.017	0	11.6
ANOVA						
Temp * FSP		0.050	0.072	0.170	0.863	0.043
Temp		0.001	0.001	0.001	0.932	0.001
FSP		0.012	0.885	0.415	0.912	0.013

[a–c] Means with varying superscripts differ significantly ($p < 0.05$). BWG, body weight gain; FI, feed intake; FCR, feed conversion ratio. [a–c] Means with varying superscripts differ significantly ($p < 0.05$) BWG, body weight gain; FI, feed intake; FCR, feed conversion ratio.

Heat stress significantly decreased feed intake and impaired FCR compared to the thermoneutral group. However, the effect of FSP and the interaction between fennel and heat stress were not significant.

3.2. Carcass Characteristics

Table 3 presents the carcass characteristics affected by the inclusion of FSP in broiler diets under heat stress conditions. The results indicate that the percentages of dressing, abdominal fat, leg, gizzard, and liver were not significantly different among the heat stress groups and/or the dietary FSP groups. However, significant differences were observed in the percentages of breast muscle due to FSP diets, and in the intestine of broiler chicks due to the interaction effects (Table 3). The highest percentage of the intestine was observed in broilers fed with diets tested under thermoneutral conditions. Groups on CHS had significantly decreased heart percentage compared to the thermoneutral groups. The heart was significantly smaller in broilers fed the 1.6% fennel diet compared to the unsupplemented control. The breast was significantly greater in broilers fed the 1.6% fennel diet compared to the unsupplemented control.

Table 3. Carcass characteristics affected by supplementation with fennel seed powder under thermoneutral or heat stress conditions.

Temp.	Diet	Dressing (%)	Fat (%)	Leg (%)	Breast (%)	Intestine (%)	Gizzard (%)	Heart (%)	Liver (%)
	Control	69.7	2.49	30.9	36.6	5.92 [b]	3.32	0.92	2.72
Normal	FSP (1.6%)	69.0	2.09	28.7	39.4	8.17 [a]	3.34	0.84	2.91
	FSP (3.2%)	68.2	2.22	29.1	39.2	9.14 [a]	3.43	0.95	2.99
	Control	68.8	1.99	29.7	36.6	8.08 [ab]	3.75	0.83	2.93
High	FSP (1.6%)	68.7	1.98	29.8	38.5	7.70 [ab]	3.56	0.65	2.75
	FSP (3.2%)	69.2	2.27	28.5	38.0	7.67 [ab]	3.50	0.78	2.87
SEM		0.29	0.081	0.03	0.39	0.241	0.082	0.021	0.072
ANOVA									
Temp* FSP		0.385	0.361	0.173	0.791	0.003	0.652	0.640	0.510
Temp		0.894	0.248	0.658	0.355	0.856	0.136	0.001	0.867
FSP		0.713	0.489	0.069	0.030	0.03	0.902	0.021	0.801

[a,b] Means with varying superscripts differ significantly ($p < 0.05$).

3.3. Meat Quality

The results of nitrate intake, pH, and L*, a*, and b* values as well as temperature are shown in Table 4. Nitrate intake was significantly affected by temperature and levels of FSP. Nitrate intake increased with increasing FSP within each temperature, and significantly decreased due to exposure to high temperature. CHS had a significant impact on broiler meat quality. Breast muscle pH was neither affected by CHS nor by dietary supplementation of FSP ($p > 0.05$). Breast muscle color of broilers subjected to heat stress, as indicated by lightness (L* values), redness (a* values), and yellowness (b* values), had variable responses compared to those under normal temperature conditions. An interaction effect was observed in the redness (a*) of breast muscles ($p < 0.05$). A significant difference between dietary treatments were found under normal temperature conditions (1.6 vs. 3.2 FSP), but non-significant differences under heat stress.

The dietary supplementation of FSP under both environmental conditions did not affect the yellowness (b* values) and lightness (L* values) of breast meat, but the effect of temperature on the b* and L* values was significant. The value of b* significantly decreased with CHS, while the L* value increased (Table 4).

Breast muscle temperatures were significantly influenced by heat stress and/or dietary FSP supplementation. The highest value (29.67 °C) was observed in broilers fed with the control diet under CHS. Moreover, broilers supplemented with 1.6 and 3.2% FSP had significantly lowered breast temperatures than the unsupplemented control.

Table 4. Broiler meat quality affected by supplementation with fennel seed powder under thermoneutral or heat stress conditions.

Temp	Diet	Nitrate Intake, mg/Chick	a*	b*	L*	Temp	pH
Normal	Control	62.8 [e]	4.04 [ab]	4.51	45.4	29.5 [a]	6.69
	FSP (1.6%)	171.4 [c]	5.31 [a]	4.31	47.2	28.2 [b]	6.74
	FSP (3.2%)	276.1 [a]	3.47 [bc]	4.88	45.6	27.9 [b]	6.71
High	Control	56.2 [f]	3.28 [bc]	3.49	47.1	29.7 [a]	6.73
	FSP (1.6%)	156.3 [d]	2.56 [bc]	2.89	47.9	29.3 [b]	6.61
	FSP (3.2%)	265.7 [b]	2.23 [c]	2.59	48.0	29.1 [b]	6.65
SEM		1.61	0.213	0.172	0.37	0.16	0.021
ANOVA							
Temp * FSP		0.115	0.029	0.098	0.612	0.007	0.509
Temp		0.005	0.0001	<0.0001	0.030	0.009	0.618
FSP		0.0001	0.016	0.596	0.311	0.139	0.932

[a–f] Means with varying superscripts differ significantly ($p < 0.05$).

3.4. Economic Efficiency

Table 5 shows the impact of temperature, funnel supplementation, and the interaction between them on economic efficiency. As expected, CHS had significant negative effects on the economic efficiency traits of broilers. On the other hand, funnel fortification had significant impacts on the total revenue, net revenue, and economic efficiency, showing a positive effect of 3.2% FSP. The interaction effect was significant for most of the traits, except for economic efficiency. The results indicate that feeding 3.2% FSP under CHS increased the feeding cost and total cost compared to other FSP levels. Total and net revenue was significantly decreased due to feeding 1.6% FSP under thermoneutral compared to other FSP levels. In addition, feeding 3.2% FSP under CHS increased the total and net revenue compared to 1.6% FSP.

Table 5. Economic efficiency of boilers affected by supplementation with fennel seed powder under thermoneutral or heat stress conditions.

Temp	Diet	Feeding Cost ($)	Total Cost ($)	Total Revenue ($)	Net Revenue ($)	Economic Efficiency (%)
Normal	Control	1.12 [a]	1.92 [a]	3.06 [a]	1.14 [a]	59.6
	FSP (1.6%)	1.10 [a]	1.90 [a]	2.98 [b]	1.081 [b]	57.0
	FSP (3.2%)	1.09 [a]	1.89 [a]	3.05 [a]	1.17 [a]	61.8
High	Control	1.00 [c]	1.79 [c]	2.48 [e]	0.681 [d]	37.9
	FSP (1.6%)	1.00 [c]	1.79 [c]	2.50 [d]	0.696 [d]	38.6
	FSP (3.2%)	1.05 [b]	1.85 [b]	2.60 [c]	0.749 [c]	40.6
SEM		0.032	0.030	0.056	0.0431	2.11
ANOVA						
Temp*FSP		0.022	0.022	0.001	0.023	0.134
Temp		0.001	0.001	0.001	0.001	0.001
FSP		0.270	0.270	0.001	0.002	0.013

[a–e] Means with varying superscripts differ significantly ($p < 0.05$). BWG, body weight gain; FI, feed intake; FCR, feed conversion ratio.

4. Discussion

The main effect of the temperature on broiler performance was significant and confirmed the success of heat stress induction. The adaptation to heat stress markedly influences broiler performance, production index, and economic efficiency [3,13,35]. Broilers at market weight generate approximately 5–10 kcals of energy per hour as a normal physiological process [9]. Under heat stress, broilers dissipate heat from their body to the surrounding environment, assuming that it is at a lower temperature than the body (41 °C) [9]. As body temperature rises above 41.5 °C, broilers are considered to be under heat stress, and their performance is more likely to be adversely affected. However, mortality usually occurs when the temperature is above 42 °C [9].

To lose the extra heat gained from the environment, birds must reduce the FI to decrease metabolic heat production from feed, and thus, their BWG decreases [16]. FSP intake promotes BWG and FCR. Likewise, dietary FSP supplementation significantly increased the growth of quail chicks [16]. In the present study, the best BWG, FCR, breast muscle, and liver percentage were obtained when broilers under normal condition were supplemented with 3.2% of FSP. It is worth noting that FI significantly decreased under CHS conditions among all treatments when compared to the control. The improved performance of broilers on FSP diets may be due to enhanced digestibility and enriched antioxidants status. Fennel also has strong antiviral, antimicrobial, and anti-inflammatory effects, which may improve gut health and eliminate pathogens. It was reported that the FSP is a rich source of essential oil (anethole, fenchone, methyl chavicol, limonene, phellandrene, camphene, pinine, anisic acid, and palmitic, oleic, linoleic, and petroselenic acids, volatile compounds, flavonoids, phenolics [19–21].

Our results showed significant alterations in broiler performance, production, and economic efficiency under CHS. This finding is in line with those reported by [13,16]. Dietary supplementation with 3.2% FSP improved broiler growth rate by 4.7% under CHS, while, under optimum temperatures, FSP-supplemented groups displayed similar growth, feed intake, FCR, production index, and economic efficiency. Moreover, our findings were in agreement with those of [20,36]. However, diets with fennel powder of 1, 2, and 3 g/kg resulted in significant improvements in broiler growth and FCR, while feed intake was unaffected [28]. Heat stress may cause a physiological dysfunction that triggers the body to utilize nutrients to synthesize critical proteins instead of using them for growth, which allows broilers to decrease against the oxidative damage of heat stress [2,3,37].

The results obtained on the carcass characteristics in the experimental conditions did not show differences in these characteristics, except for heart percentage, which was affected by heat stress. Supplementation of 1.6% and 3.2% FSP seemed to improve breast percentage. Treatments with FSP, particularly at 1.6%, tended to decrease heart percentages. The results of carcass characteristics in this study were in line with those obtained by [38], who reported that the addition of FSP at 1, 2, and 3 g/kg to the diet resulted in insignificant differences for all carcass characteristics. Similar results were also reported by [20]. The results indicate that FSP had no adverse effects on carcass traits. Moreover, phytogenic plants and their essential oils are used as preservation approaches to enhance the sensory attributes and prolong the shelf life of animal products [19,21,36,39].

The results showed that meat quality parameters such as meat color, which is an important visual criterion, were significantly affected by heat stress. The color of meat is correlated with myoglobin and hemoglobin concentrations and their status [40,41]. The current results are in agreement with the results reported by [3,42]. The highest value of a* was reported in the 1.6% FSP group under normal temperatures. However, it was only similar to the control group under the same conditions. It is worth noting that the a* value is extremely important to consumers.

Furthermore, a high redness (a*) value yields an undercooked appearance. The a* value can be affected by bird age, stress before slaughter, and consumption of dietary nitrates [43]. Nitrates have been reported to be at high concentrations (65 to 376.7 mg/100 g) in FSP [23,44–46], which may explain the obtained results and have both beneficial and adverse effects [47]. The change in a* value parallels the change in nitrate intake, which increases with increasing FSP concentration and decreased under CHS condition. The nitrate intakes ranged from 56.2 in the control groups to 276 mg/ bird during 19–41 days of age, with an average daily intake of nitrate ranging from 2.44 to 12 mg/day. The toxic level of nitrate for chickens ranges from 450–900 ppm and is 658 ppm for nitrite [30]. Thus, the intake of nitrate was far from the toxic level. The concentration of nitrate in animal feedstuffs is considered safe at 4, 44, and 22 mg/kg in soybean meal, oat grain, and maize grain, respectively [48,49]. Thus, nitrite poisoning is rare in poultry fed cereal/grain/oilseeds meal based diets [50–52].

Additionally, NO was known to react with a myoglobin increased a* value [27]. Nevertheless, the latter may act as pro- or anti-oxidant, according to its concentration. With higher NO concentration, it may rapidly react with superoxide radical forming proxy nitrite [47], which oxidizes myoglobin, resulting in low a* value [43]. On the other hand, low a* values may indicate more oxidized myoglobin in birds subjected to heat stress, as reported by [53].

Heat stress increases breast meat temperatures, which may accelerate oxidative rancidity and increase the peroxidation biomarker. Similarly, CHS was found to increase body temperature in a study conducted by [54]. However, in our study, the FSP-supplemented groups showed significant reductions in breast muscle temperature under both normal and heat stress conditions, revealing a beneficial effect of FSP. This finding, along with the increase in BWG in broilers under CHS and fed 3.2% FSP diet, can be explained on the basis that FSP contains NO, which increases blood flow to the body surface and upper respiratory tract to dissipate body temperature, as reported by [53], suggesting that FSP may be used as a CHS alleviating agent.

In the present investigation, no differences were observed in the pH of meat among different temperatures and/or fennel groups. The pH is one of the most important physical traits for the qualitative profile of meat and is commonly utilized as an assessment of sensory qualities of the technological properties of meat. Meat pH is related to the water-holding capacity [40,55], which is correlated positively with the texture, juiciness, and flavor of meat [56].

5. Conclusions

According to the findings of this study, the dietary inclusion of FSP, especially at 3.2%, resulted in a beneficial impact on the growth performance and carcass quality of broilers under heat stress conditions. Furthermore, CHS significantly decreased the production index and economic efficiency, but FSP did not affect both criteria.

Author Contributions: A.A.A.-S. conducted the experimental design, fieldwork, and first draft of the manuscript; S.K. carried out the data collection and contribution to the experimental set-up, E.O.S.H., lab work, and reading the manuscript; and Y.A.A. undertook the statistical analyses and proofreading of the manuscript. All authors have read and agreed to the published version of the manuscript.

Funding: This article was funded by the Deanship of Scientific Research (DSR), King Abdulaziz University, Jeddah. Therefore, the authors appreciate the technical and financial support of the DSR.

Acknowledgments: The authors express their sincere gratitude and appreciation to King Abdulaziz City for Science and Technology (KACST) for allowing the use of its facilities to conduct this study.

Conflicts of Interest: The authors declare no conflict of interest.

References

1. Suganya, T.; Senthilkumar, S.; Deepa, K.; Amutha, R. Nutritional management to alleviate heat stress in broilers. *Int. J. Sci. Environ. Technol.* **2015**, *4*, 661–666.
2. Attia, Y.A.; Hassan, S.S. Broiler tolerance to heat stress at various dietary protein/energy levels. *Europ. Poult. Sci.* **2017**, *81*. [CrossRef]
3. Attia, Y.A.; Al-Harthi, M.A.; Elnaggar, A.S. Productive, physiological and immunological responses of two broiler strains fed different dietary regimens and exposed to heat stress. *Ita, J. Anim. Sci.* **2018**, *17*, 686–697. [CrossRef]
4. Yahav, S. Alleviating heat stress in domestic fowl: Different strategies. *World's Poult. Sci. J.* **2009**, *65*, 719–732. [CrossRef]
5. Cassuce, D.C.; Tinôco, I.; de, F.F.; Baêta, F.C.; Zolnier, S.; Cecon, P.R.; Vieira, M.; de, F.A. Thermal comfort temperature update for broiler chickens up to 21 days of age. *Eng. Agrícola Jaboticabal* **2013**, *33*, 28–36. [CrossRef]

6.   Yardibi, H.; Gülhan, T. The effects of vitamin E on the antioxidant system, egg production, and egg quality in heat-stressed laying hens. Turk. *J. Vet. Anim. Sci.* **2008**, *32*, 319–325.

7.   Whitehead, C.C.; Keller, T. An update on ascorbic acid in poultry. *World's Poult. Sci. J.* **2003**, *59*, 161–184. [CrossRef]

8.   Vandana, G.D.; Sejian, V. Towards identifying climate resilient poultry birds. *JDVAR* **2018**, *7*, 84–85.

9.   Leeson, S.; Summers, J.D. *Commercial Poultry Nutrition*, 3rd ed.; Nottingham University Press: Nottingham, UK, 2009; pp. 261–262.

10. Lin, H.; Eddy, D.; Johan, B. Acute heat stress induces oxidative stress in broiler chickens. *Comp. Biochem. Physiol. A Mol. Integr. Physiol.* **2006**, *144*, 11–17. [CrossRef]

11. Akbarian, A.; Joris, M.; Jeroen, D.; Maryam, M.; Abolghasem, G.; Stefaan, D.S. Association 0 -between heat stress and oxidative stress in poultry; mitochondrial dysfunction and dietary interventions with phytochemicals. *J. Anim. Sci. Biotechnol.* **2016**, *7*, 37. [CrossRef]

12. Akşit, M.; Yalçin, S.; Özkan, S.; Metin, K.; Özdemir, D. Effects of temperature during rearing and crating on stress parameters and meat quality of broilers. *Poult. Sci.* **2006**, *85*, 1867–1874. [CrossRef] [PubMed]

13. Attia, Y.A.; Hassan, R.A.; Tag El-Din, A.E. Abou- Shehema, B.M. Effect of ascorbic acid or increasing metabolizable energy level with or without supplementation of some essential amino acids on productive and physiological traits of slow-growing chicks exposed to chronic heat stress. *J. Anim. Physiol. Anim. Nutr.* **2011**, *95*, 744–755. [CrossRef] [PubMed]

14. Guimarães, R.; Barros, L.; Carvalho, A.M.; Ferreira, I.C.F.R. Infusions and decoctions of mixed herbs used in folk medicine: Synergism in antioxidant potential. *Phytotherapy Res.* **2011**, *25*, 1209–1214. [CrossRef]

15. Koppula, S.; Kumar, H. Foeniculum vulgare Mill (Umbelliferae) attenuates stress and improves memory in Wister rats. *Trop. J. Pharm. Res.* **2013**, *12*, 553–558. [CrossRef]

16. Gous, R.M.; Morris, T.R. Nutritional interventions in alleviating the effects of high temperatures in broiler production. *Proc. Nutr. Soc.* **2005**, *61*, 463–475. [CrossRef]

17. Sahin, K.; Orhan, C.; Smith, M.O.; Sahin, N. Molecular targets of dietary phytochemicals for the alleviation of heat stress in poultry. *World's Poult. Sci. J.* **2013**, *69*, 113–123. [CrossRef]

18. Attia, Y.A.; Bakhashwain, A.A.; Bertu, N.K. Utilisation of thyme powder (Thyme Vulgaris, L.) as a growth promoter alternative to antibiotics for broiler chickens raised in a hot climate. *Europ. Poult. Sci.* **2018**, *82*. [CrossRef]

19. Gharaghani, H.; Shariatmadari, F.; Torshizi, M.A. Effect of fennel (Foeniculum vulgare Mill.) used as a feed additive on the egg quality of laying hens under heat stress. *Rev. Bras. Ciênc. Avícola* **2015**, *17*, 199–207. [CrossRef]

20. El-Deek, A.A.; Al-Harthi, M.A.; Attia, Y.A.; Hannfy, M.M. Effect of anise (Pimpinella anisum), ginger (Zingiber officinale roscoe) and fennel (Foeniculum vulgare) and their mixture on performance of broilers. *Arch. Geflugelkunde* **2003**, *67*, 92–96.

21. Badgujar, S.B.; Patel, V.V.; Bandivdekar, A.H. Foeniculum vulgare Mill: A Review of Its Botany, Phytochemistry, Pharmacology, Contemporary Application, and Toxicology. *BioMed Res. Int.* **2014**, *2014*, 842674. [CrossRef]

22. Rather, M.A.; Dar, B.A.; Sofi, S.N.; Bhat, B.A.; Qurishi, M.A. *Foeniculum vulgare mill*: A comprehensive review of its traditional use, phytochemistry, pharmacology, and safety. *Arab. J. Chem.* **2016**, *9*, 1574–1583. [CrossRef]

23. Swaminathan, A.; Sridhara, S.R.; Sinha, S.; Nagarajan, S.; Balaguru, U.M.; Siamwala, J.H.; Rajendran, S.; Saran, U.; Chatterjee, S. Nitrites derived from Foeniculum vulgare (fennel) seeds promotes vascular functions. *J. Food Sci.* **2012**, *77*, 273–279. [CrossRef] [PubMed]

24. Dunai, V.; Tzschentke, B. Impact of environmental thermal stimulation on activation of hypothalamic neuronal nitric oxide synthase during the prenatal ontogenesis in Muscovy ducks. *The Sci. World, J.* **2012**, *2012*, 416936. [CrossRef]

25. Zeng, T.; Li, J.J.; Wang, D.Q.; Li, G.Q.; Wang, G.L.; Lu, L.Z. Effects of heat stress on antioxidant defense system, inflammatory injury, and heat shock proteins of Muscovy and Pekin ducks: Evidence for differential thermal sensitivities. *Cell Stress Chaperones* **2014**, *19*, 895–901. [CrossRef] [PubMed]

26. Dantonio, V.; Batalhão, M.E.; Fernandes, M.H.; Komegae, E.N.; Buqui, G.A.; Lopes, N.P.; Bícego, K.C. Nitric oxide and fever: Immune-to-brain signaling vs. thermogenesis in chicks. *Am. J. Physiol. Regul. Integr. Comp. Physiol.* **2016**, *310*(10), R896–R905. [CrossRef] [PubMed]

27. Zhang, W.; Al-Hijazeen, M.; Himali, S.; Eun, J.L.; Dong, U.A. Breast meat quality of broiler chickens can be affected by managing the level of nitric oxide. *Poult. Sci.* **2013**, *92*, 3044–3049. [CrossRef]

28. Zhang, Z.Y.; Jia, G.Q.; Zuo, J.J.; Zhang, Y.; Lei, J.; Ren, L.; Feng, D.Y. Effects of constant and cyclic heat stress on muscle metabolism and meat quality of broiler breast fillet and thigh meat. *Poult. Sci.* **2012**, *91*, 2931–2937. [CrossRef]

29. Assreuy, J.; Cunha, F.Q.; Liew, F.Y.; Moncada, S. Feedback inhibition of nitric oxide synthase activity by nitric oxide. *Br. J. Pharmacol.* **1993**, *108*, 833–837. [CrossRef]

30. National Research Council. *Nutrient requirements of poultry*, 9th ed.; National Academy Press: Washington, DC, USA, 1991.

31. Fletcher, D.L. Poultry meat color in poultry meat science. *Poult. Sci. Symp. Ser.* **1999**, *25*, 159–175.

32. AL-Fayad, H.A.; Naji, S.A.; Abid, N.N. *Poultry products technology*, 2nd ed.; Ministry of Higher Education and Scientific Research, College of Agriculture, University of Baghdad: Baghdad, Iraq, 2011.

33. Attia, Y.A.; Hamed, R.S.; Abd El-Hamid, A.E.; Shahba, H.A.; Bovera, F. Effect of inulin and mannan-oligosaccharides compared with zinc-bacitracin on growing performance, nutrient digestibility and hematological profiles of growing rabbits. *Anim. Prod. Sci.* **2015**, *55*, 80–86. [CrossRef]

34. SAS Institute. *SAS/STAT User's Guide (Version 9.1)*; SAS Inst. Inc.: Cary, NC, USA, 2004.

35. Hassan, A.; Abd Elazeem, M.H.; Hussein, M.M.; Osman, M.M.; Abd El-wahed, Z.H. Effect of chronic heat stress on broiler chicks performance and immune system. *SCVMJ* **2007**, *12*, 55–68.

36. Gharaghani, H.; Shariatmadari, M.; Karimi Torshizi, A. Comparison of oxidative quality of meat of chickens fed corn or wheat based diets with fennel (Foeniculum vulgare mil.), antibiotic and probiotic as feed additive, under different storage conditions. *Arch. Geflugelkunde* **2013**, *77*, 199–205.

37. Butcher, G.D.; Miles, R.D. *Interrelationship of Nutrition and Immunity*; University of Florida Cooperative Extension Service, Institute of Food and Agricultural Sciences, EDIS: Gainesville, FL, USA, 2002; Volume VM139, pp. 1–8.

38. Mohammed, A.A.; Abbas, R.J. The effect of using fennel seeds (Foeniculum vulgare L.) on productive performance of broiler chickens. *Int. J. Poult. Sci.* **2009**, *8*, 642–644. [CrossRef]

39. Bozkurt, M.; Küçükyılmaz, K.; Pamukçu, M.; Çabuk, M.; Alçiçek, A.; Çatlı, U. Long-term effects of dietary supplementation with an essential oil mixture on the growth and laying performance of two layer strains. *Ita. J. Anim. Sci.* **2012**, *11*, 23–28. [CrossRef]

40. Brewer, S. Irradiation effects on meat color – a review. *Meat Sci.* **2004**, *68*, 1–17. [CrossRef]

41. Attia, Y.A.; El-Tahawy, W.S.; Abd El-Hamid, A.E.; Nizza, A.; El-Kelway, M.I.; Al-Harthi, M.A.; Bovera, F. Effect of feed form, pellet diameter and enzymes supplementation on carcass characteristics, meat quality, blood plasma constituents and stress indicators of broilers. *Arch. Tierz.* **2014**, *57*, 1–14. [CrossRef]

42. Yalçin, S.; Önenç, A.; Özkan, S.; Güler, H.C.; Siegel, P.B. Meat quality of heat-stressed broilers: Effects of thermal conditioning at pre- and postnatal stages. In *Book of Abstract. XVIIth European Symposium On the Quality of Poultry Meat, Doorwert, The Netherlands, 23–26 May 2005*; WPSA: Beekbergen, The Netherlands, 2005; p. 265.

43. Smith, D.P.; Northcutt, J.K. Red discoloration of fully cooked chicken products. *J. Appl. Poult. Res.* **2003**, *12*, 515–521. [CrossRef]

44. Koudela, M.; Petříková, K. Nutritional compositions and yield of sweet fennel cultivars - Foeniculum vulgare Mill. ssp. vulgare var. azoricum (Mill.). *Thell Hort. Sci.* **2008**, *35*, 1–6.

45. Bukhari, H.; Shehzad, A.; Saeed, K.; Sadiq, B.M.; Tanveer, S.; Iftikhar, T. Compositional Profiling of Fennel Seed. *Pak. J. Food Sci.* **2014**, *24*, 132–136.

46. Hord Norman, G.; Tang, Y.; Bryan Nathan, S. Food Sources of Nitrates and Nitrites: The Physiologic Context for Potential Health Benefits. *Am. J. Clin. Nut.* **2009**, *90*, 1–10. [CrossRef]

47. Møller Jens, K.S.; Skibsted Leif, H. Myoglobins - The link between discoloration and lipid oxidation in muscle and meat. *Quimica Nova* **2006**, *29*, 1270–1278.

48. Cockburn, A.; Brambilla, G.; Fernandez, M.L.; Arcella, D.; Bordajandi, L.R.; Cottrill, B.; van Peteghem, C.; Dorne, J.L. Nitrite in feed: From Animal health to human health. *Toxicol. Appl. Pharmacol.* **2013**, *270*, 209–217. [CrossRef] [PubMed]

49. Crowley, J.W. Effects of nitrate on livestock. American Society of Agricultural Engineers Paper Number 80-20026. Cited by Undersander, D., Combs, D., Shaver, R.,Thomas, D. Available online: http://www.uwex.edu/ces/forage/pubs/nitrate.htm1985 (accessed on 20 March 2017).

50. Attia, Y.A.; Abd El Hamid, A.E.; Ismaiel, A.M.; de Oliveira Maria, C.; Al-Harthi, M.A.; El- Naggar Asmaa, S.H.; Simon, G.A. Nitrate detoxification using antioxidants and probiotics in the water of rabbits. *Revista Colombiana De Ciencias Pecuarias (RCCP)* **2018**, *31*, 130–138. [CrossRef]
51. Attia, Y.A.; El Hamid, E.A.; Ismaiel, A.M.; El-Nagar, A. The detoxication of nitrate by two antioxidants or a probiotic and the effects on blood and seminal plasma profiles and reproductive function of NZW rabbit bucks. *Animal* **2013**, *7*, 591–601. [CrossRef]
52. Ologhobo, A.D.; Adegede, H.I.; Maduagiwu, E.N. Occurrence of nitrate, nitrite and volatile nitrosamines in certain feedstuffs and animal products. *Nutr Health.* **1996**, *11*, 109–114. [CrossRef] [PubMed]
53. Zhu, W.; Jiang, W.; Wu, L.Y. Dietary L-Arginine supplement alleviates hepatic heat stress and improves feed conversion ratio of Pekin ducks exposed to high environmental temperature. *J. Anim. Physiol. Anim. Nutr.* **2014**, *98*, 1124–1131. [CrossRef]
54. Vinoth, A.; Thirunalasundari, T.; Shanmugam, M.; Rajkumar, U. Effect of early age thermal conditioning on expression of heat shock proteins in liver tissue and biochemical stress indicators in colored broiler chicken. *Eur. J. Exp. Biol.* **2016**, *6*, 53–63.
55. Jung, S.; Choe, J.H.; Kim, B.; Yun, H.Z.; Kruk, A.; Jo, C. Effect of dietary mixture of gallic acid and linoleic acid on antioxidative potential and quality of chest meat from broilers. *Meat Sci.* **2010**, *86*, 520–526. [CrossRef]
56. Abdullah, A.Y.; Musallam, H.S. Effect of different levels of energy on carcass composition and meat quality of male black goat kids. *Livest Sci.* **2007**, *107*, 70–80. [CrossRef]

Article

Influence of Different Time and Frequency of Multienzyme Application on the Efficiency of Broiler Chicken Rearing and Some Selected Metabolic Indicators

Youssef A. Attia [1,*], **Mohammed A. Al-Harthi** [1,*] and **Ali S. El-Shafey** [2]

[1] Arid Land Agriculture Department, Faculty of Meteorology, Environment and Arid Land Agriculture, King Abdulaziz University, Jeddah 21589, Saudi Arabia

[2] Animal and Poultry Production Department, Faculty of Agriculture, Damanhour University, Damanhour 22516, Egypt; alielshafey2015@gmail.com

* Correspondence: yaattia@kau.edu.sa (Y.A.A.); malharthi@kau.edu.sa (M.A.A.-H.)

Received: 25 February 2020; Accepted: 6 March 2020; Published: 8 March 2020

Simple Summary: Enzymes are a useful, valuable and economic tool to improve feed utilization and thus animal performance. Costs of enzymes supplementation and time and application frequency of enzymes have received little attention other than dose and type of enzymes. This study showed that multienzymes offered intermittently during both early and late growth periods up to market age of broilers enhanced productive performance and economic profits and can substitute the daily administration with a considerable lowering of the supplementation cost.

Abstract: This study looks at the influence of time and/or frequency of multienzymes application on productivity, carcass characteristics, metabolic profile, and red blood cell characteristics of broiler chickens. Two hundred and eighty, one-day-old Arbor Acres broiler male chicks were randomly distributed into seven treatment groups. Each group consisted of eight replicates of five unsexed birds. The same basal diet was fed in a crumble form to all experimental groups: group one was the unsupplemented control that did not receive multienzymes supplementation. Additionally, multienzymes in water were supplemented in six groups in a factorial arrangement, including three times of application (starter time only which included days 1–21 of age, grower time only which included days 22–37 of age, and starter and grower time which included days 1–37 of age) and two application frequencies (continuously or intermittently). In the continuous application, the multienzymes were added to water over 24 h in a day, while in the intermittent frequency multienzymes were added to water for one day followed by a day off according to the time of application. Regardless of time and frequency of application, enzymes supplementation significantly increased growth rate, feed intake, European Production Index (EPI), protein digestibility, serum albumin, and high-density lipoprotein (HDL). Intermittent multienzymes application during days 1–21 of age or days 22–37 of age resulted in significantly greater growth, better feed conversion rate (FCR), and higher EPI of broilers during the whole rearing period than those under continuous multienzymes during different growth periods. Besides, intermittent multienzymes addition during days 1–37 of age improved FCR of broiler chicks compared to constant application. The intermittent addition of multienzymes during days 1–21 of age or 22–37 days of age and days 1–37 of age caused a significant increase in dry matter (DM) digestibility than the continuous application. The intermittent addition of multienzymes during days 1–21 of age significantly increased the digestibility of crude protein (CP), ether extract (EE), and crude fiber (CF) compared to continuous application. A similar trend was shown in the digestibility of CP and EE due to intermittent use during days 22–37 of age. Intermittent enzymes addition significantly increased high density lipoprotein (HDL) of groups receiving enzymes during days 22–37 of age compared to continuous application of enzymes. In conclusion, the use of multienzymes intermittently during days 1–21 of age and 22–37 days of age significantly increased growth, improved FCR, and raised EPI. Intermittent use can replace continuous multienzyme applications which can save 68.6% of the cost, even though further research is need from the cost-saving edge.

Keywords: broilers; application time and way; digestibility; carcass traits; metabolic profiles

---

1. Introduction

The digestive tract of chickens develops over time and is nearly mature by the fourth week of age, hence the digestibility of nutrients is weak and feed utilization is inadequate during this early age, resulting in increased costs of feeding and environmental pollution [1–4]. Nutrient digestibility and use of feeds were shown to improve over time and by enzyme supplementations [5,6]. Supplementation with enzymes in broiler diets increased the activity of the digestive enzymes [7,8] and improved the endogenous enzyme production, thus increasing the quantity of nutrients available for absorption [8–10]. Nonetheless, multienzymes impacts dietary composition, enzymes type, concentration and profiles, and strain and age of birds [11–14]. It is clear from the literature that the efficacy of enzymes decreased with the increasing age of chickens, and thus it is correlated with maturation of gut in terms of capacity, endogenous enzyme secretion, and ecology [3,12,13]. It was found that enzymes supplementation had more pronounced effects on broiler chickens than laying hens [2,14] and during the early stage of growth than later ones. The enzyme supplementation is effective for the total feeding cost because costs can be cut considerably if the efficiency of multienzymes can be determined according to the time of feeding [15]. Even so, until now and to the best of our knowledge, no studies that have addressed the relationship between time of growth and time and frequency of multienzymes administration on chickens' performance, nutrient digestibility, carcass traits, metabolic parameters, and blood constituents are available in the literature. Hence, the hypothesis of this research assumes that multienzymes improve the growth performance of broilers, and the intermittent application during the early growth period may replace the continuous use without adverse effects on growth performance, carcass traits and metabolic profits while enhancing economic profits.

2. Materials and Methods

2.1. Animals and Dietary Treatments

Two hundred and eighty, one-day-old Arbor Acres broiler chicks were randomly distributed into seven treatment groups. Each group consisted of eight replicates of five male chicks each. The same basal diet was offered in crumble form to all experimental groups. The group one (unsupplemented control group) was fed an un-supplemented diet. Further, there were six multienzyme-supplemented groups in a factorial arrangement that included three groups supplemented during different times of growth by two frequencies of administration. The times of growth were: the starter time only (days 1–21 of age), the grower time only (days 22–37 of age), and the starter plus grower time (days 1–37 of age). The administration of the multienzymes within each time of application were continuous or intermittent. In the continuous frequency, the multienzymes were added to the water over 24 h in a day. In the intermittent frequency, the multienzymes were added to the water for one day, followed by a day off. The diets in this experiment were formulated to meet the requirements of broiler chickens [16].

The multienzymes (Galzym®produced by Tex Biosciences (P) limited, Ashok Nagar, Chennai, India) used were a combination of a group of exogenous and fibrolytic enzymes consisting of: cellulase (100,000,000 units), xylanase (1,500,000 units), lipase (6500 units), alpha amylase (25,0000 units), protease (40,0000 units) and Pectinase (30,000 units). The recommended dose of enzymes was 0.333 mL/1 L water. The composition of the experimental diets is presented in Table 1.

Table 1. Ingredients and chemical-nutritional composition of the experimental diets (as feed basis) during different experimental periods.

Ingredients (g/kg)	Diets	
	Starter	Grower-Finisher
Maize	502.3	508.1
Rye	0	50
Soybean meal, 44% crude protein	328	244
Dicalcium phosphate	18.00	16.00
Limestone	10.00	10.00
NaCl	3.00	4.50
Full fat soybean meal	100	130
Vit+Min premix [1]	3.00	3.00
L-Lysine	1.00	1.90
DL-Methionine	2.00	2.50
Vegetable oil	22.70	20.00
Celite [TM]	10.0	10.0
Calculated and determined composition, (g/kg)		
Dry matter [2]	891	896
Crude protein, % [2]	223	205
Metabolisable energy (MJ/kg) [3]	12.50	12.64
Crude fat,% [2]	58	61
Crude fibre,% [2]	34.1	33.3
Nitrogen free extract [2]	513.4	532
Calcium [3]	8.72	8.13
Available phosphorus [3]	4.17	3.88
Methionine [3]	5.42	5.68
Methionine+cystine [3]	9.10	9.05
Lysine [3]	13.08	12.41
Ash [2]	62.5	64.7

[1] Vit+Min mix. provides per kilogram of the diets: Vit. A, 12,000 internation unit (IU), vit. E (dl-α-tocopheryl acetate) 20 mg, menadione 2.3 mg, Vit. D_3, 2200 international chick unit (ICU), riboflavin 5.5 mg, calcium pantothenate 12 mg, nicotinic acid 50 mg, Choline 250 mg, vit. B_{12} 10 µg, vit. B_6 3 mg, thiamine 3 mg, folic acid 1 mg, d-biotin 0.05 mg. Trace mineral (mg/kg of diets): Mn 80 Zn 60, Fe 35, Cu 8, Selenium 0.1 mg. [2] Analyzed values. [3] Calculated values.

All the experimental procedures were approved by Deanship of Scientific Research King Abdulaziz University, under proposal number G 170-155-1440H. The treatment of animals was according to the Royal Decree number M59 in 14/9/1431H, which recommends animal rights, welfare, and minimal stress and prevents any harm or suffering.

2.2. Animal Housing and Management

Broilers were housed in batteries of five chickens per cage ($30 \times 35 \times 45$ cm). Chicks were offered free access to feed and water during the experimental period. The housing temperature was 33 °C during the 1st week and declined gradually by 2 °C each week and was kept constant at 25 °C until slaughter. The average minimum and maximum outdoor and indoor relative humidity and temperature were 25.7 and 38.6%, 48.5 and 54.2%, 28.7 and 37.5 °C, and 27.8 and 34.6 °C, respectively. The light-dark cycle schedule was 23:1 h daily.

2.3. Experimental Measurements

Broilers in each replicate were weighed (g) at day 1, 21, and 37 of age, and the body weight gain (BWG) (g/chick) was calculated. Feed intake was recorded for each replicate (g/chick), and thereby FCR (g feed/g gain), and survival rate (SR, 100 − mortality rate) were calculated during days 1–21, 22–37, and 1–37 of age. Water intake was not measured due to a lack of precise research facility. European production index was for the index of economic evaluation, as reported by [12].

Apparent digestibility for crude protein (CP), dry matter (DM), ether extract (EE), crude fiber (CF), nitrogen-free extract (NFE) and ash during different times of growth using various methods of application was determined using Celite^{TM} (Celite Corp., Lompar, CA, USA) as a marker according to Dourado et al. [17]. The total collection method of excreta was followed for 5 days in each application time during days 15–20, 25–30, and 32–36 of age, respectively. In each time (1–21), (22–37) and (1–37) excreta and feed samples were constituted, and samples of each (25%) were collected, cleaned from feed residuals, scale and feathers, and then the DM, CP, EE, CF and ash of feeds and excreta were assayed according to [18] and expressed on dry matter basis. The NFE of feed and excreta was estimated by differences between the dry matter and sum of CP, EE, CF, and ash. The method adopted by [17] was used for analyses of acid-insoluble ash (AIA). The AIA recovery was estimated as g execrated/g intake × 100 using dry matter basis of excreta voided and feed intake data. It was found to range from 94% to 109%.

At 37 days of age, eight male broiler chicks per treatment group were slaughtered after 8 h fasting according to the Islamic method using a sharp knife and cutting into the jugular vein, carotid artery and windpipe, processed, and the weight of the carcass and inner organs was measured and estimated as a percentage of live body weight.

At 37 days of age, eight blood samples of 5 mL each per treatment group were collected and equally divided into two clean tubes with or without heparin. The tubes without heparin were centrifuged at 1500× g for 10 min at 4 °C, and then serum was separated and stored at −18 °C until analysis. The serum profiles were determined using commercial diagnostic kits (Diamond Diagnostics Company, Cairo, Egypt). Total protein (g/dl) [19], albumin (g/dl) [20], and globulin (g/dl) [21] were estimated. In addition, triglycerides [22], total cholesterol [23], high-density lipoprotein (HDL) [24]; and low-density lipoprotein (LDL) [25] were appraised.

Red blood cells [26], hemoglobin [27], packed cells volume (PCV) [28], mean cell volume (MCV), mean cell hemoglobin (MCH), and mean cell hemoglobin concentration (MCHC) [29] were manually measured.

2.4. Statistical Analysis

The statistical analysis software (SAS) was done using a completely randomized design and data were subjected to a two-way ANOVA procedure based on only time of supplementation and administration frequency without the unsupplemented control [30]. The statistical model included the effects of the time of multienzyme application (starter, grower-finisher and whole period), frequency of application (continuously vs. intermittently) and their interactions according to the subsequent model:

$$Yijk = \mu + Di + AMj + (D \times AM) + eijk \tag{1}$$

where: Y = the dependent variables; μ = general mean; D = time effect; AM = administration frequency effect; (D × AM) = interaction between the two factors; e = random error. Orthogonal contrast was done to test the effect of enzyme treatment regardless of time and frequency of application against the unsupplemented control. Before analysis, all percentages were subjected to logarithmic transformation (log10x+1) to normalize data distribution. The mean differences were determined using student Newman Keuls [30]. When p values found to be between 0.05 and 0.1 were considered as a tendency and thus indicate in the results "tend to be higher/lower."

3. Results

The effect of different times of multienzymes application in water continuously or intermittently on the production performance of broiler chickens is summarized in Table 2.

The contrast analyses showed that multienzymes supplementation significantly increased the growth of broilers during days 1–21 of age and days 1–37 of age compared to the unsupplemented control.

Both application time and application frequency had significant effects on the growth of broilers during 1–21, 22–37, and 1–37 days of age, but these effects were confounded by the significant interaction between the time and the frequency of application.

The interaction between application time and application frequency of multienzymes was significant for 1–21, 22–37, and days 1–37 of age. It was found that intermittent use during days 1–21 of age significantly improved the rate of growth of broilers during periods 1–21, 22–37, and 1–37 days of age compared with other continuous or intermittent groups. Moreover, intermittent enzyme application during days 22–37 of age substantially increased the growth of broilers during periods 1–37 days of age compared to continuous enzyme supplementation during periods 1–21 and 22–37 days of age, and permanent and intermittent use during periods 1–37 days of age. Enzyme application either as constant or intermittent during days 1–37 of age showed a similar growth rate during all tested periods.

The contrast analyses indicated that feed intake was significantly higher for chickens fed enzyme-supplemented diets compared to the un-supplemented ones during periods 1–21 and 1–37 days of age.

Feed intake during periods 1–21, 22–37, and 1–37 days of age was not significantly affected by the time of application and the interaction between time and frequency of enzyme application. However, feed intake during 22–37 and 1–37 days of age was significantly higher for broilers fed diets supplemented with enzymes continuously compared to those on the intermittent addition.

The contrast analyses for FCR showed no significant difference between the enzyme treatments and the un-supplemented group. The time of application significantly affected FCR during days 1–37 of age and application frequency influenced FCR during different tested periods, but these effects were confounded by the interaction between the time and frequency of application during days 1–21 and 1–37 of age.

In general, intermittent application during days 1–21 and 22–37 of age significantly improved FCR of broilers compared to continuous supplementation during periods 1–21 and 1–37 days of age. Besides, intermittent enzyme application during 1–37 days of age significantly improved FCR for broilers during days 1–37 compared to constant use.

There was no significant effect of time and/or frequency of enzyme application on the survival rate of broilers during days 1–37 of age.

The contrast analyses showed that enzyme application increased EPI compared to the unsupplemented control. The time and/or frequency of enzyme application significantly affected EPI. The intermittent enzyme application during days 1–21 of age significantly improved EPI of broilers compared to other groups. In addition, intermittent enzyme application during days 22–37 of age significantly enhanced EPI of broilers compared with other groups on continuous or intermittent frequency, except for intermittent frequency during days 1–21 of age. The intermittent application during 1–21 days of age had a greater effect on EPI than the same use during 22–37 days of age.

Results concerning the influence of the time and/or application frequency on nutrients digestibility of broilers are shown in Table 3. The contrast analyses demonstrated that enzyme supplementations significantly increased only crude protein digestibility compared to unsupplemented control, and had no effect on other digestibility traits.

Time and frequency of enzyme application had a significant effect on the digestibility of DM, CP, EE, and CF, but these factors were confounded by the interaction between the time and frequency of enzyme application. The interaction between application time and/or application frequency had a significant impact on most nutrient digestibility with the exception of the digestibility of NFE ($p = 0.834$) and ash ($p = 0.078$).

The interaction effect indicates that the intermittent use of enzymes during 1–21, 22–37, and days 1–37 of age improved the DM digestibility compared with continuous applications.

Table 2. Effect of multienzymes supplemented in water continuously or intermittently during the starter, grower, and the whole time on growth performance of broiler chicks.

Treatments/Period	Body Weight Gain (g)			Feed Intake (g)			Feed Conversion Rate (g/g)			Survival Rate (%)	Production Index
	1–21 d	22–37 d	1–37 d	1–21 d	22–37 d	1–37 d	1–21 d	22–37 d	1–37 d		
						Effect of enzyme treatments vs unsupplemented control (contrast)					
Control	573 [b]	1113	1686 [b]	907 [b]	2119	3026 [b]	1.59	1.91	1.80	97.5	246 [b]
Enzymes	674 [a]	1139	1813 [a]	1031 [a]	2093	3125 [a]	1.53	1.84	1.72	99.6	286 [a]
						Effect of administration time					
1–21	692 [a]	1158 [a]	1850 [a]	1031	2066	3096	1.50	1.79	1.68 [b]	100	301 [a]
22–37	676 [b]	1156 [a]	1831 [a]	1020	2106	3126	1.51	1.82	1.71 [ab]	98.8	287 [b]
1–37	654 [c]	1102 [b]	1755 [b]	1043	2108	3152	1.57	1.90	1.77 [a]	100	269 [c]
						Effect of administration frequency					
Continuous	658 [b]	1117 [b]	1775 [b]	1035	2128 [a]	3164 [a]	1.58 [a]	1.91 [a]	1.78 [a]	99.2	267 [b]
Intermittent	690 [a]	1160 [a]	1850 [a]	1027	2058 [b]	3086 [b]	1.47 [b]	1.77 [b]	1.66 [b]	100	304 [a]
						Interaction between time and frequency of application					
1–21 Con	653 [c]	1093 [d]	1746 [d]	1047	2107	3154	1.60 [a]	1.93	1.81 [a]	100	261 [c]
1–21 Int	731 [a]	1222 [a]	1953 [a]	1014	2024	3038	1.39 [e]	1.66	1.56 [d]	100	340 [a]
22–37 Con	667 [bc]	1143 [bc]	1810 [c]	1027	2143	3170	1.54 [c]	1.88	1.75 [b]	97.5	273 [c]
22–37 Int	685 [b]	1167 [b]	1852 [b]	1012	2069	3082	1.48 [d]	1.77	1.66 [c]	100	300 [b]
1–37 Con	652 [c]	1114 [cd]	1766 [d]	1031	2135	3167	1.58 [ab]	1.92	1.79 [a]	100	267 [c]
1–37 Int	654 [c]	1089 [d]	1743 [d]	1054	2081	3136	1.55 [bc]	1.88	1.75 [b]	100	270 [c]
RMSE	32.9	41.3	37.8	52.4	84.9	88.5	0.071	0.109	0.067	5.21	13.5
						p value					
Control vs. enzymes	0.002	0.609	0.011	0.001	0.975	0.012	0.620	0.672	0.750	0.465	0.001
Time	0.048	0.009	0.001	0.616	0.462	0.391	0.088	0.112	0.018	0.552	0.001
Frequency	0.012	0.009	0.001	0.679	0.034	0.024	0.001	0.002	0.001	0.693	0.001
Interaction	0.041	0.001	0.001	0.492	0.927	0.562	0.017	0.066	0.005	0.348	0.001

D = day; Con = Continuous; Int = Intermittent; RMSE = Root mean square error; a–d, means with different superscripts in the same column in similar treatment groups are significantly different.

Table 3. Effect of multienzymes supplemented in water continuously or intermittently on apparent nutrient digestibility of broiler chicks.

Treatments/Period	Apparent Nutrient Digestibility, %					
	Dry Matter	Crude Protein	Ether Extract	Crude Fiber	Nitrogen Free Extract	Ash
Unsupplemented control vs. enzymes treatment (contrast)						
Control	83.2	74.4 [b]	85.2	27.3	76.9	34.8
Enzymes	85.6	79.5 [a]	88.4	29.7	77.6	37.3
Effect of administration time						
1–21	86.8 [a]	80.5 [a]	89.6 [a]	30.2 [a]	78.0	37.9
22–37	85.0 [b]	79.5 [b]	87.9 [b]	29.8 [a]	77.3	37.0
1–37	84.9 [b]	78.5 [c]	87.6 [b]	28.9 [b]	77.3	36.9
Effect of administration frequency						
Continuous	84.2 [b]	78.5 [b]	87.2 [b]	28.8 [b]	77.1 [b]	36.8 [b]
Intermittent	86.9 [a]	80.5 [a]	89.5 [a]	30.5 [a]	78.0 [a]	37.8 [a]
Interaction between time and frequency of application						
1–21 Con	83.9 [c]	78.5 [c]	86.9 [d]	28.3 [e]	76.8	36.7 [b]
1–21 Int	89.7 [a]	82.5 [a]	92.3 [a]	32.1 [a]	77.9	39.2 [a]
22–37 Con	84.3 [c]	78.7 [c]	87.1 [cd]	29.5 [bc]	77.4	36.6 [b]
22–37 Int	85.5 [b]	80.2 [b]	88.7 [b]	30.1 [b]	78.6	37.1 [b]
1–37 Con	84.3 [c]	78.2 [c]	87.6 [cd]	28.6 [de]	77.1	36.9 [b]
1–37 Int	85.7 [b]	78.8 [c]	87.7 [c]	29.3 [cd]	77.5	37.1 [b]
RMSE	2.61	1.76	2.74	1.52	2.42	1.74
p value						
Control vs. enzymes	0.732	0.001	0.554	0.866	0.399	0.271
Time	0.045	0.002	0.049	0.036	0.565	0.116
Fequency	0.001	0.001	0.001	0.001	0.157	0.027
Interaction	0.011	0.011	0.011	0.002	0.834	0.078

Con = Continuous; Int = Intermittent; RMSE = Root means square error; a–e, means with different superscripts in the same column in similar treatment groups are significantly different.

The intermittent application of multienzymes during days 1–21 of age improved the digestibility of CP, EE, and CF compared to the other groups. The intermittent application of enzymes during days 22–37 of age enhanced the digestibility of CP and EE compared with the continuous and intermittent use during days 1–37 of age. The intermittent application during 1–21 days of age increased the digestibility of DM, CP, EE, and CF more than intermittent use during 22–37 days of age.

There were no significant differences between continuous and intermittent enzyme applications during days 1–37 of age on the digestibility of CP, EE, and CF.

Regardless of the time of application, intermittent supplementation increased ash digestibility compared to continuous supplementation.

The carcass characteristics and the inner organs of the broilers influenced by time of multienzymes application and/or the way of addition are shown in Table 4. The contrast analyses showed no significant effect of multienzymes compared to the unsupplemented control on all carcass traits and inner body organs.

Dressing, abdominal fat, gizzard, proventriculus, intestinal tract, liver, and heart were not significantly affected by the time and frequency, or by the interaction between time and frequency of application of enzymes.

The pancreas percentage was affected ($p = 0.051$) by the time of application, showing a trend of high value with enzymes supplemented during days 22–37 of age compared to the other periods. The effect of the application frequency approached significant ($p = 0.092$) for the heart percentage, showing the increase was proportional due to intermittent use.

The biochemical indices of blood serum of the broilers affected by the time of application and/or the application frequency are shown in Table 5.

The contrast analyses showed that enzyme treatments significantly increased the albumin, albumin/globulin ratio, total lipids, HDL, and HDL/LDL ratio in blood serum compared to the unsupplemented control, but decreased serum globulin. Differences between the enzymes' treatment and the unsupplemented control approached significant for total protein ($p < 0.054$).

There was no significant impact of time of application and/or frequency of use on blood serum proteins (total protein, albumin, globulin, albumin/globulin ratio). On the other hand, plasma total lipids were significantly affected by the application time, showing that total lipids were lower when enzymes were applied during days 22–37 of age than during the other times. The effect of time of application approached significant ($p < 0.063$) for serum triglycerides, with the highest being the 1–21 days group.

There was a significant interaction between application time and application frequency for serum HDL, LDL, and LDL/HDL ratio. Enzyme application intermittently during days 1–21 of age significantly decreased serum LDL compared to other groups, except for the continuous use during days 1–37 of age. On the other hand, the continuous offer of multienzymes during days 1–21 and 1–37 of age significantly increased the HDL and HDL/LDL ratio compared to the intermittent application, but the contrary was observed during days 22–37 of age.

Characteristics of red blood cells of broiler chickens are reported in Table 6. There were no significant effects due to enzyme treatments compared to the unsupplemented control when contrast analyses were considered for all red blood parameters.

The application time, application frequency, and the interaction between time and frequency of enzyme application did not significantly affect red blood cell traits, except for a significant increase in MCHC ($p < 0.039$) due to the continuous use compared to intermittent addition.

Table 4. Effect of multienzymes supplemented in water continuously or intermittently during the starter, grower, and the whole times on carcass characteristics and inner body organs of 37-day-old broiler chicks.

Treatments/Period	Dressing and Inner Organs,%							
	Dressing	Abdominal Fat	Gizzard	Proventriculus	Intestinal	Liver	Pancreas	Heart
Unsupplemented control vs enzymes treatment (contrast)								
Control	74.7	0.983	1.23	0.333	4.83	2.23	0.235	0.418
Enzymes	71.7	1.12	1.17	0.387	5.08	2.04	0.190	0.457
Effect of administration time								
1–21	74.1	1.078	1.16	0.379	5.12	2.14	0.186	0.434
22–37	70.5	1.170	1.21	0.388	5.15	2.10	0.213	0.476
1–37	70.3	1.140	1.12	0.394	4.96	1.88	0.171	0.461
Effect of administration frequency								
Continuous	71.6	1.199	1.15	0.4_1	5.04	2.07	0.193	0.439
Intermittent	71.7	1.054	1.18	0.363	5.11	2.01	0.187	0.475
Interaction between time and frequency of application								
1–21 Con	73.1	1.058	1.07	0.402	5.06	2.18	0.182	0.414
1–21 Int	75.1	1.098	1.25	0.356	5.18	2.10	0.190	0.454
22–37 Con	71.9	1.190	1.20	0.372	5.08	2.10	0.230	0.454
22–37 Int	68.7	1.150	1.22	0.404	5.22	2.10	0.196	0.498
1–37 Con	69.7	1.350	1.17	0.458	4.98	1.92	0.168	0.450
1–37 Int	71.3	0.877	1.07	0.330	4.94	1.84	0.174	0.472
RMSE	4.41	0.329	0.169	0.078	1.06	0.292	0.036	0.055
p value								
Control vs. enzymes	0.233	0.174	0.333	0.268	0.906	0.172	0.553	0.311
Time	0.122	0.821	0.510	0.911	0.911	0.122	0.051	0.244
Frequency	0.938	0.222	0.625	0.110	0.851	0.622	0.624	0.092
Interaction	0.341	0.219	0.184	0.093	0.979	0.941	0.369	0.893

Con = Continuous; Int = Intermittent; RMSE = Root means square error.

Table 5. Effect of multienzymes supplemented in water continuously or intermittently during the starter, grower, and the whole times on blood biochemical constituents of 37-day-old broiler chicks.

Treatments/Period	Total Protein, g/dL	Albumin, g/dL	Globulin, g/dL	A/G Ratio	Total Lipids, mg/dL	Triglycerides, mmol/L	Cholesterol, mmol/L	HDL, mmol/L	LDL, mol/L	LDL/HDL Ratio
					Unsupplemented control vs. enzyme treatments (contrast)					
Control	6.38	2.92 [b]	3.46 [a]	0.844 [b]	7.4 [b]	2.07	5.56	0.936 [b]	2.23	0.421 [b]
Enzymes	6.21	2.98 [a]	3.23 [b]	0.931 [a]	9.3 [a]	1.94	5.73	0.978 [a]	2.09	0.466 [a]
					Effect of administration time					
1–21	6.24	3.03	3.21	0.953	9.7 [a]	1.95	5.73	0.993 [a]	2.12	0.469
22–37	6.17	3.03	3.15	0.975	8.5 [b]	1.92	5.73	0.996[a]	2.09	0.475
1–37	6.21	2.87	3.34	0.864	9.7 [a]	1.94	5.73	0.944[b]	2.07	0.456
					Effect of administration frequency					
Continuous	6.18	2.99	3.18	0.951	9.6	1.93	5.66[b]	0.998[a]	2.12[a]	0.471
Intermittent	6.23	2.96	3.28	0.910	9.0	1.94	5.79[a]	0.957[b]	2.07[b]	0.461
					Interaction between time and frequency of application					
1–21 Con	6.26	3.10	3.16	0.984	9.4	1.94	5.66	1.071 [a]	2.21[a]	0.485 [a]
1–21 Int	6.22	2.96	3.16	0.923	10.0	1.97	5.81	0.915 [bc]	2.02[c]	0.453 [bc]
22–37 Con	6.10	3.08	3.02	1.036	9.4	1.92	5.68	0.946 [bc]	2.09[b]	0.452 [bc]
22–37 Int	6.24	2.98	3.28	0.915	7.6	1.92	5.78	1.045 [a]	2.10[b]	0.497 [a]
1–37 Con	6.18	2.80	3.38	0.834	10.0	1.92	5.66	0.978 [b]	2.05[bc]	0.476 [b]
1–37 Int	6.24	2.94	3.30	0.893	9.4	1.95	5.80	0.910 [c]	2.10[b]	0.434 [c]
RMSE	0.146	0.206	0.251	0.131	1.13	0.030	0.062	0.047	0.051	0.024
					p value					
Control vs. enzymes	0.054	0.048	0.009	0.025	0.001	0.731	0.896	0.005	0.351	0.029
Time	0.689	0.156	0.249	0.150	0.039	0.063	0.995	0.035	0.219	0.197
Frequency	0.327	0.661	0.318	0.395	0.159	0.188	0.001	0.023	0.025	0.281
Interaction	0.399	0.279	0.333	0.311	0.081	0.384	0.605	0.001	0.001	0.001

Con = Continuous; Int = Intermittent; HDL = High density lipoprotein; LDL = Low density lipoprotein; RMSE = Root mean square error; a–c, means with different superscripts in the same column in similar treatment groups are significantly different.

Table 6. Effect of multienzymes supplemented in water continuously or intermittently during the starter, grower and the complete times on red blood cell characteristics in broiler chicks.

Treatments/Period	Red Blood Cell Characteristics, %					
	RBC's (106/mm^3)	Hemoglobin (g/dL)	PCV (mL/1 mL)	MCV, μm^3/RBC	MHC, pg	MCHC, %
Unsupplemented control vs enzymes (contrast)						
Control	1.52	10.5	32.3	217	70.9	32.9
Enzymes	1.33	11.2	33.2	250	84.0	33.6
Effect of administration time						
1–21	1.34	11.4	33.5	252	84.8	33.7
22–35	1.31	11.1	33.0	253	84.91	33.6
1–35	1.35	11.1	33.1	246	82.47	33.5
Effect of administration frequency						
Continuous	1.31	11.3	33.2	253	86.4	34.1
Intermittent	1.35	11.1	33.2	246	81.6	33.1
Interaction between time and frequency of application						
1–21 Con	1.34	11.6	33.8	253	86.7	34.3
1–21 Int	1.34	11.1	33.2	250	82.9	33.1
22–35 Con	1.26	11.0	32.4	257	87.3	33.9
22–35 Int	1.36	11.2	33.6	248	82.5	33.3
1–35 Con	1.34	11.4	33.4	250	85.4	34.1
1–35 Int	1.36	10.8	32.8	242	79.6	32.9
RMSE	0.089	0.87	1.89	21.1	7.87	1.26
p value						
Control vs. enzymes	0.991	0.718	0.741	0.848	0.799	0.862
Time	0..589	0.838	0.824	0.756	0.742	0.948
Frequency	0.232	0.302	0.993	0.370	0.108	0.039
Interaction	0.429	0.501	0.482	0.916	0.959	0.825

Con = Continuous; Int = Intermittent; PCV = Packed cell volume; MCV = Mean cell volume; MCH = Mean hemogloblin concentration; MCHC = Mean cell hemogloblin concentration; RMSE = Root means square error.

4. Discussion

It is interesting to report that enzyme supplementation regardless of time and frequency of application enhanced growth performance and protein digestibility compared to unsupplemented controls. The increase in the growth of enzyme-supplemented groups was connected with increasing feed intake and digestibility of crude protein. These results are similar to the findings by [8,12,31–34]. The effect of the enzyme cocktail was attributed to improved digestibility and increased protein and energy availability for growth [32–40].

The intermittent enzymes application during days 1–21 of age resulted in better growth, improved FCR, and improved EPI compared to the continuous administration along the same period, but also to the constant or intermittent use during the other tested times (22–37 and days 1–37 of age). The positive effects of offering enzymes during the early growth phase (1–21 days of age) may be due to the supportive influence of exogenous enzymes on gut development and functions [3,13,15], and thus enhancing the digestibility of nutrients during the first stage of growth [1,10]. The gut of young chicks is under development and maturation in terms of enzyme secretion, capacity, and ecology during day 1–28 of age [1,5].

The results demonstrated that intermittent administration was adequate, which can save considerable cost (68.6%) of enzyme application. The high growth and better FCR of broilers on the intermittent use condition are tied to the increased digestibility of DM, CP, EE, and CF. Similar results concerning the positive impact of enzymes on nutrient digestibility were recently cited [6,8,13,31].

However, continuous use of enzymes throughout the experimental period (1–37 days of age) neither improved growth performance nor the digestibility of nutrients, and this may be due to the negative feedback mechanism of exogenous enzymes on pancreatic enzyme secretion [12,15].

It was found that the application of multienzymes during the period from 1 to 21 days of age yielded stronger effects than the use during the later ages and the whole period of growth on BWG and FCR than those supplemented during 22–37 or and days 1–37 of age. This may be because the digestive tract is under development up to four weeks of age. The enzyme applications during days 1–37 of age yield the least response. The continuous use during days 22–37 of age produced better growth during days 1–37 of age and FCR during days 1–21 and 1–37 of age than other permanent groups. The relationship between the time of application and growth performance of broilers needs, therefore, further research.

The positive impact of enzymes on animal productivity reported herein is in concert with those observed by [7,14,32]. These enhancements were attributed to the increasing activity of digestive enzymes [7,13], as estimated by indirect measurements, such as growth performance, or directly by digestibility of nutrients and eliminating the adverse effects of anti-nutritional factors [6,33,34]. Besides, multienzymes were found to increase the use of energy in sorghum and maize-soybean meal-based diets due to enhancing cereal cell walls and starch digestion [3,10]. Enzyme application may improve beneficial intestinal microbiota favoring *Lactobacillus* [14,31] and reduce environmental pollution [6,15] due to increased digestibility of nutrients (protein, fat and fiber, and minerals), while decreasing nutrients available in the hindgut for growth of harmful microbiota and increasing animal productive performance [9,13,15]. However, multienzymes' influence depends on enzyme type and concentrations, diet profile, form [4,6,31,35], and stage of growth [11,15].

Based on the present results, an enzyme blend supplemented in water during a different time of growth by various application frequencies at the recommended level did not affect most of the carcass and organ traits of broiler chickens. However, the pancreas percentage was increased due to enzyme supplementations during days 22–37 of age, suggesting a negative feedback mechanism of exogenous enzymes on the pancreas function [2,36]. There was also a decrease in proventriculus percentage when enzymes were added during the early time (1–21 days of age) and the growing time (22–37 days of age). In general, the absence of effects of enzymes on most of carcass and organ traits is constant with the findings by [7,37–39]. These authors indicated that application with enzymes did not affect carcass parameters and body organs of broilers and Japanese quail except for liver percent that was reduced with multienzymes application. Likewise, carcass traits were not influenced by dietary enzyme addition to broiler chicken diets [29,40,41].

In general, with few exceptions, our results showed that there were no significant effects from the application time and/or application frequency on the blood serum biochemistry and hematology. In other studies, dietary enzyme addition did not affect serum protein concentrations (alpha 1-, alpha 2-, beta and gamma-globulins and albumin) and albumin/globulin ratios [42–45].

The most pronounced effects found herein were a decrease in plasma cholesterol due to the continuous addition of enzymes and an increase in HDL with the intermittent applications during days 22–37 of age. There was also a decrease in LDL when enzymes were supplemented intermittently during days 1–21 of age. These beneficial effects due to multienzyme application on lipid metabolites require further investigation. However, enzymes had no significant effect on concentrations of triglyceride and cholesterol compared with animals fed on the same diet without the addition of enzymes [46]. Similar to the current results, different blood cholesterols and HDL/LDL ratios were increased in the unsupplemented control groups than enzyme-supplemented groups, showing the positive influence of enzymes on blood cholesterol [7,41,47].

5. Conclusions

Enzymes positively affect growth rate, feed intake, EPI, protein digestibility, serum albumin, total lipids, and HDL compared to the control group, irrespective of time and frequency of enzyme application. The application of multienzymes intermittently during 1–21 days of age and 22–35 days of age for broilers significantly increased growth, improved FCR, and enhanced EPI. This can replace continuous multienzymes applications while saving 68.6% of the cost, even though further research is needed to explore the beneficial economic impact of of time and frequency of enzyme application.

Author Contributions: Y.A.A., M.A.A.-H., A.S.E.-S. designed the experiment, and A.S.E.-S. carried the fieldwork and LAB analyses. Y.A.A. ran the statistical analyses. M.A.A.-H. prepared the first draft. All authors contributed to the final preparation and revision of the manuscript. All authors have read and agreed to the published version of the manuscript.

Funding: This project was funded by the Deanship of Scientific Research (DSR), King Abdulaziz University, Jeddah, Saudi Arabia, under grant no."G-170-155-1440 H". The authors, therefore, acknowledge with thanks DSR for technical and financial support.

Conflicts of Interest: The authors declare no conflict of interest.

References

1. Noy, Y.; Sklan, D. Digestion and absorption in the young chick. *Poult. Sci.* **1995**, *74*, 366–373. [CrossRef] [PubMed]
2. Attia, Y.A.; Tag El-Din, A.E.; Zeweil, H.S.; Hussein, A.S.; Qota, E.M.; Arafat, M.A. The effect of application of enzyme on laying and reproductive performance in Japanese Quail hens fed nigella seed meal. *J. Poult. Sci.* **2008**, *45*, 110–115. [CrossRef]
3. Attia, Y.A.; El-Tahawy, W.S.; Abd El-Hamid, A.E.; Hassan, S.S.; Nizza, A.; El-Kelaway, M.I. Effect of phytase with or without multienzyme application on performance and nutrient digestibility of young broiler chicks fed mash or crumble diets. *Ital. J Anim. Sci.* **2012**, *11*, 303–308.
4. Fafiolu, A.O.; Oduguwa, O.O.; Jegede, A.V.; Tukura, C.C.; Olarotimi, I.D.; Teniola, A.A.; Alabi, J.O. Assessment of enzyme application on growth performance and apparent nutrient digestibility in diets containing undecorticated sunflower seed meal in layer chicks. *Poult. Sci.* **2015**, *94*, 1917–1922. [CrossRef]
5. Lee, S.A.; Apajalahti, J.; Vienola, K.; Gonzalez-Ortiz, G.; Fontes, C.M.G.A.; Bedford, M.R. Age and dietary xylanase application affects ileal sugar residues and short chain fatty acid concentration in the ileum and caecum of broiler chickens. *J. Anim. Feed Sci. Technol.* **2017**, *234*, 29–42. [CrossRef]
6. Attia, Y.A. The use of enzymes in poultry diets trends and approaches, a minimum Review. *Acta Sci. Nutr. Health* **2019**, *3*, 95–97.
7. Alagawany, M.; Attia, A.I.; Ibrahim, Z.A.; Mahmoud, R.A.; El-Sayed, S.A. The effectiveness of dietary sunflower meal and exogenous enzyme on growth, digestive enzymes, carcass traits, and blood chemistry of broilers. *Environ. Sci. Pol. Res.* **2017**, *24*, 12319–12327. [CrossRef]
8. Angel, C.R.; Saylor, W.; Vieira, S.L.; Ward, N. Effects of a monocomponent protease on performance and protein utilization in 7- to 22-day-old broiler chickens. *Poult. Sci.* **2011**, *90*, 2281–2286. [CrossRef]
9. Cozannet, P.; Kidd, M.T.; Neto, R.M.; Geraert, P.A. Next-generation non-starch polysaccharide-degrading, multi-carbohydrase complex rich in xylanase and arabinofuranosidase to enhance broiler feed digestibility. *Poult. Sci.* **2017**, *96*, 2743–2750. [CrossRef]
10. Williams, M.P.; O'Neil, H.V.M.; York, T.; Lee, J.T. Effects of nutrient variability in corn and xylanase inclusion on broiler performance, nutrient utilisation, and volatile fatty acid profiles. *J. Appl. Anim. Nutr.* **2018**, *6*, e1. [CrossRef]
11. Hong, D.; Burrows, H.; Adeola, O. Addition of enzyme to starter and grower diets for ducks. *Poult. Sci.* **2002**, *81*, 842–1849. [CrossRef]
12. Attia, Y.A.; El-Tahawy, W.S.; Abd El-Hamid, A.E.; Nizza, A.; Bovera, F.; Al-Harthi, M.A.; El-Kelway, M.I. Effect of feed form, pellet diameter and enzymes application on growth performance and nutrient digestibility of broiler during days 21–37 of age. *Arch. Tierzucht* **2014**, *57*, 1–11.
13. Aftab, U.; Bedford, M.R. The use of NSP enzymes in poultry nutrition: Myths and realities. *Worlds Poult. Sci. J.* **2018**, *74*, 277–286. [CrossRef]

14. Al-Harthi, M.A.; Attia, Y.A.; Al-sagan, A.; Elgandy, M.F. The effects of autoclaving or/and multi-enzymes complex application on performance, egg quality and profitability of laying hens fed whole Prosopis juliflora pods meal diet. *Eur. Poult. Sci.* **2018**, *82*. [CrossRef]

15. Choct, M. Enzymes for the feed industry: Past, present and future. *World's Poult. Sci. J.* **2006**, *62*, 5–16. [CrossRef]

16. National Research Council, NRC. *Nutrient Requirement of Poultry*; National Academy Press: Washington, DC, USA, 1994.

17. Dourado, L.R.B.; Siqueira, J.S.; Sakomura, N.K.; Pinheiro, S.R.F.; Marcato, S.M.; Fernandes, J.B.K.; Silva, J.H.V. Poultry feed metabolizable energy determination using total or partial excreta collection methods. *Rev. Bras. Cienc. Avic.* **2010**, *12*, 129–132. [CrossRef]

18. Association of Official Analytical Chemists, AOAC. *Official Methods of Analysis*, 18th ed.; Association of Official Analytical Chemists: Washington, DC, USA, 2004.

19. Henry, R.; Cannon, D.; Winkelman, J. *Clinical Chemistry, Principles and Techniques*, 2nd ed.; Harper and Row: New York, NY, USA, 1974.

20. Doumas, B.T.; Watson, W.A.; Biggs, H.G. Albumin standards and the measurement of serum albumin with bromcresol green. *Clin. Chim. Acta* **1971**, *31*, 21–30. [CrossRef]

21. Coles, E.H. *Veterinary Clinical Pathology*, 1st ed.; W.B. Saunder: Philadelphia, PA, USA, 1974.

22. Fossati, P.; Prencipe, L. Serum triglycerides determined colorimetrically with an enzyme that produces hydrogen peroxide. *Clin. Chem.* **1982**, *28*, 2077–2080. [CrossRef]

23. Stein, E.A. Quantitative enzymatic colorimetric determination of total cholesterol in serum or plasma. In *Textbook of Clinical Chemistry*; Tietz, N.W., Ed.; WB. Saunders: Philadelphia, PA, USA, 1986; pp. 879–886.

24. Lopez-Virella, M.F.; Stone, S.; Eills, S.; Collwel, J.A. Determination of HDL-cholesterol using enzymatic method. *Clin. Chem.* **1977**, *23*, 882–884.

25. Friedewald, W.T.; Levy, R.I.; Frederickson, D.S. Estimation of the concentration of low-density lipoprotein cholesterol in plasma without use of the preparative ultracentrifuge. *Clin. Chem.* **1972**, *18*, 499–502. [CrossRef] [PubMed]

26. Helper, O.E. *Manual of Clinical Laboratory Methods*, 4th ed.; Charles C Thomas Publishing Ltd.: Springfield, IL, USA, 1966.

27. Tietz, N.W. *Fundamentals of Clinical Chemistry*, 2nd ed.; Saunders Company Publishing: Philadelphia, PA, USA, 1982.

28. Wintrobe, M.M. *Clinical Hematology*; Lea & Febiger Publishing: Philadelphia, PA, USA, 1963.

29. Attia, Y.A.; El-Tahawy, W.S.; Abd El-Hamid, A.E.; Nizza, A.; El-Kelway, M.I.; Al-Harthi, M.A.; Bovera, F. Effect of feed form, pellet diameter and enzymes application on carcass characteristics, meat quality, blood plasma constituents and stress indicators of broilers. *Arch. Tierzucht* **2014**, *57*, 1–14.

30. SAS Institute. *SAS/STAT User's Guide Statistics*; SAS Institute INC: Cary, NC, USA, 2009.

31. Borda-Molina, D.; Zubar, T.; Siegert, W.; Camarinha-Silva, A.; Feuerstein, D.; Rodehutscord, M. Effects of protease and phytase supplements on small intestinal microbiota and amino acid digestibility in broiler chickens. *Poult. Sci.* **2019**, *98*, 2906–2918. [CrossRef]

32. Zeng, Q.; Huang, X.; Luo, Y.; Ding, X.; Bai, S.; Wang, J.; Xuan, Y.; Su, Z.; Liu, Y.; Zhang, K. Effects of a multi-enzyme complex on growth performance, nutrient utilization and bone mineralization of meat duck. *J. Anim. Sci. Biotechnol.* **2015**, *6*, 12. [CrossRef]

33. Kocher, A.; Hower, J.M.; Moran, C.A. A dualenzyme product containing protease in broiler diet: Efficacy and tolerance. *J. App. Anim. Nut.* **2015**, *3*, 1–14.

34. Abdel-Hafeez, H.M.; Saleh, E.S.E.; Tawfeek, S.S.; Youssef, I.M.I.; Abdel-Daim, A.S.A. Utilization of potato peels and sugar beet pulp with and without enzyme application in broiler chicken diets: Effects on performance, serum biochemical indices and carcass traits. *J. Anim. Phys. Anim. Nutr.* **2016**, *102*, 56–66. [CrossRef]

35. Nourmohammadi, R.; Hosseini, S.M.; Farhangfar, H.; Bashtani, M. Effect of citric acid and microbial phytase enzyme on ileal digestibility of some nutrients in broiler chicks fed corn-soybean meal diets. *Ital. J. Anim. Sci.* **2012**, *11*, 36–40. [CrossRef]

36. Attia, Y.A.; Qota, E.M.A.; Aggoor, F.A.M.; Kies, A.K. Value of rice bran, its maximal utilization and upgrading by phytase and other enzymes and diet- formulation based on available amino acid for broiler chicks. *Arch. Geflügelk.* **2003**, *67*, 157–166.

37. Mushtaq, T.; Sarwara, M.; Ahmad, G.; Mirza, M.A.; Ahmad, T.; Mushtaq, M.M.H.; Kamran, Z. Influence of sunflower meal based diets supplemented with exogenous enzyme and digestible lysine on performance, digestibility and carcass response of broiler chickens. *Anim. Feed Sci. Technol.* **2009**, *149*, 275–286. [CrossRef]

38. De Araujo, W.; Albino, L.F.T.; Rostagno, H.S.; Hannas, M.I.; Pessoa, G.B.S.; Messias, R.K.G.; Lelis, G.R.; Ribeiro, V., Jr. Sunflower meal and enzyme application of the diet of 21-to 42-d-old broilers. *Braz. J. Poult. Sci.* **2014**, *16*, 17–24. [CrossRef]

39. Rabie, M.H.; Abo El-Maaty, H.M.A. Growth performance of Japanese quail as affected by dietary protein level and enzyme application. *Asian J. Anim. Vet. Adv.* **2015**, *10*, 74–85. [CrossRef]

40. Dalólio, F.S.; Moreira, J.; Albino, L.F.T.; Vaz, D.P.; Pinheiro, S.R.F.; Valadares, L.R.; Pires, A.V. Exogenous enzymes in diets for broilers. *Rev. Bras. Saúde Prod. Anim.* **2016**, *17*, 149–161.

41. Al-Harthi, M.A. The effect of olive cake, with or without enzymes application, on growth performance, carcass characteristics, lymphoid organs and lipid metabolism of broiler chickens. *Rev. Bras. Cienc. Avic.* **2017**, *19*, 83–90. [CrossRef]

42. Mehri, M.; Adibmoradi, M.; Samie, A.; Shivazad, M. Effects of Mannanase on broiler performance, gut morphology and immune system. *Afr. J. Biotech.* **2010**, *9*, 6221–6228.

43. Gheisari, A.A.; Zavareh, M.S.; Toghyani, M.; Bahadoran, R. Application of incremental program, an effective way to optimize dilatory inclusion rate of guar meal in broiler chicks. *Livest. Sci.* **2011**, *140*, 117–123. [CrossRef]

44. El-Serwy, A.; Shoeib, M.S.; Fathey, I.A. Performance of broiler chicks fed mash or pelleted diets containing corn-with-cobs meal with or without enzyme application. *J. Anim. Poult. Prod.* **2012**, *3*, 137–155.

45. Dinani, O.P.; Tyagi, P.K.; Tyagi, P.K.; Mandal, A.B.; Jaiswal, S.K. Effect of feeding fermented guar meal vis-à-vis toasted guar meal with or without enzyme application on immune response, caeca micro flora status and blood biochemical parameters of broiler quails international. *J. Pure Appl. Biosci.* **2017**, *5*, 624–630. [CrossRef]

46. El-Katcha, M.I.; Soltan, M.A.; El-Kaney, H.F.; Karwarie, E.R. Growth performance, blood parameters, immune response and carcass traits of broiler chicks fed on graded levels of wheat instead of corn without or with enzyme application. *Alex. J. Vet. Sci.* **2014**, *40*, 95–111.

47. Zarfar, A. Effect of hemicell enzyme on the performance, growth parameter, some blood factors and ileal digestibility of broiler chickens fed corn/soybean-based diets. *J. Cell Anim. Biol.* **2013**, *7*, 5–91.

Article

Effect of Supplementation with Trimethylglycine (Betaine) and/or Vitamins on Semen Quality, Fertility, Antioxidant Status, DNA Repair and Welfare of Roosters Exposed to Chronic Heat Stress

Youssef A. Attia [1,2,*], Asmaa Sh. El-Naggar [2], Bahaa M. Abou-Shehema [3] and Ahmed A. Abdella [3]

[1] Arid Land Agriculture Department, Faculty of Meteorology, Environment and Arid Land Agriculture, King Abdulaziz University, Jeddah 21589, Saudi Arabia
[2] Animal and Poultry Production Department, Faculty of Agriculture, Damanhour University, Damanhour 22713, Egypt
[3] Department of Poultry Nutrition, Animal production Research Institute, Agriculture Research Center, Ministry of Agriculture and Land Reclamation, Alexandria 21917, Egypt
* Correspondence: yaattia@kau.edu.sa or attia0103753095@gmail.com

Received: 18 June 2019; Accepted: 2 August 2019; Published: 12 August 2019

Simple Summary: Semen, reproductive traits, and the welfare of males are negatively affected by environmental stressors. Stress-alleviating agents, such as vitamins and osmoregulators, may improve semen quality, seminal and blood plasma constituents, antioxidants' status, and the welfare of roosters exposed to chronic heat stress (CHS). It has been shown that betaine (Bet) may be a useful tool for improving the reproductive traits of roosters exposed to CHS, and may have comparable effects to vitamin C and/or E, thus improving the breeding strategy.

Abstract: In this study, we investigated the influence of betaine (Bet, 1000 mg/kg), with or without vitamin C (VC, 200 mg/kg ascorbic acid) and/or vitamin E (VE, 150 mg/kg α-tocopherol acetate) on semen quality, seminal and blood plasma constituents, antioxidants' status, DNA repair, and the welfare of chronic heat stress (CHS)-exposed roosters. A total of 54 roosters were divided into six groups of nine replicates. One group was kept under thermoneutral conditions, whereas the other five were kept under CHS. One of the five groups served as an unsupplemented CHS group, and was fed with a basal diet. The other four CHS groups were supplemented with Bet, Bet + VC, Bet + VE, and Bet + VC + VE, respectively. Our data indicate that supplementation with Bet, Bet + VC, Bet + VE, and Bet + VC + VE, resulted in complete recovery of the CHS effect on sperm concentration and livability, semen pH, and fertility compared to the thermoneutral group. Seminal plasma total antioxidant capacity (TAC) was significantly ($p < 0.05$) increased with Bet, with or without vitamins, compared to the thermoneutral and CHS groups. Urea and blood plasma malondialdehyde (MDA) were totally recovered with Bet, with or without vitamin treatments. Both the jejunum and ileum DNA were partially recovered following Bet, with or without vitamin supplementation. In conclusion, Bet, at 1000 mg/kg feed, may be a useful agent for increasing semen quality, fertility, welfare, and to improve the breeding strategy of breeder males in hot climates.

Keywords: betaine; vitamin C; vitamin E; heat stress; roosters; semen quality; blood constituents

---

1. Introduction

Oxidative stress plays a key role in sperm function and motility, cell quality, and fertility, since lipid peroxidation increases under chronic heat stress (CHS), particularly when the temperature exceeds 27 °C [1]. Heat stress negatively influences testis functions [2,3], causes DNA damage [2,4], lowers the quality of semen [5], and reduces animal wellbeing [2]. In addition, HS causes abnormal mitotic division during spermatogenesis and produces abnormal spermatozoa that may not complete the fertilization process [6].

Avian spermatozoa are rich in polyunsaturated fatty acids (PUFA) and are, therefore, subjective to reactive oxygen species (ROS), causing male infertility [7]. Therefore, antioxidants are essential for avoiding male infertility by limiting free radical production [1,8]. Betaine (Bet) is a trimethylglycine produced by choline oxidation [9], and it is implicated in methionine and choline sparing [10,11], in fat distribution, and immune responses [12]. Bet has also been found to improve tolerance to stress [9,13], but studies on the reproduction of roosters remain limited. However, a study conducted in boars is available [14], which indicates that dietary 0.63 and 1.26% Bet elevated the amount of sperm compared to the control animals. Adding 200 mg/kg of Bet to the diet decreased the mortality of heat-stressed chickens; enhanced their immunity and health status [15]; and increased their antioxidant status, such as glutathione peroxidase (GPx) and superoxide dismutase (SOD), while decreasing malondialdhyde (MDA) in blood serum [16] and in breast muscle tissue [17].

Vitamin C (VC, ascorbic acid) has antioxidant activity that protects animals under stress conditions [18], and animals can synthesize VC under normal conditions. However, adding VC improves performance, immunological status, and welfare of the animals exposed to HS [8,18]. Chickens require VC to metabolize minerals and amino acids, as well as for the activity of leukocytes and 1,25-dihydroxy vitamin D, collagen, the regulation of body temperature, corticosterone secretion, adrenaline biosynthesis [1], and testosterone synthesis [19]. Heat stress impairs ascorbic acid absorption and increases the need for VC [20]. Moreover, the impact of VC on semen quality is quite inconsistent [6] and dose-specific [1]. This may be due to the heat destruction of VC [21].

Alpha-tocopherol (vitamin E, VE) is a natural antioxidant and an excellent biological chain-breaking antioxidant that is involved in tissue protection from oxidative damage and ROS [22]. A VE diet that exceeds 15 mg/kg (100–150 mg/kg) [23] was found to boost the semen quality and fertilizing ability of roosters [24–26] and boars [27]. Furthermore, a 400 mg/day dose of VE was recommended for the treatment of male infertility in men [28]. Therefore, dietary supplementation with VE is essential for compacting the adverse effects of HS on the lipid peroxidation of spermatozoa cell membranes [22], and it enhances plasma cells, lymphocytes, and macrophages against oxidative damage [29]. The influence of HS and the combined effect of different antioxidant agents working by different modes of action to relieve the negative impact of CHS are scarce in the available literature and particularly in the nutrition of roosters. Currently, no studies address the effects of Bet alone or in combination with antioxidant vitamins on semen quality, reproductive and breeding strategy of males to elucidate their synergetic effect on modes of action. Hoehler and Marquardt [30] indicated that the in vivo antioxidant influence of VE might be greater than that of VC. Supplementation with Bet, VC, or a combination of both, resulted in significantly increased total protein and globulin of laying hens exposed to HS, while serum triglyceride, total cholesterol, HDL-cholesterol, and glucose were significantly decreased in comparison to the control group [31]. In addition, Attia et al. [15] found that Bet and VC have similar effects on production traits, metabolism, blood constituents, and the wellbeing of laying hens exposed to chronic heat stress (CHS).

Therefore, the aim of this study was to test the influence of the HS and Bet (Bet + VC, Bet + VE, and Bet + VC + VE) on the quality of semen, seminal and blood plasma metabolites, reproductive performance, the wellbeing, and the DNA of intestinal segments of roosters.

2. Materials and Methods

2.1. Animals and Treatments

The protocol for this experiment was approved by the scientific committee of the Animal Production Research Institute under the registration code no: 9-2-4-3-10-1. The committee recommended that care and handling of the animals maintains their rights, welfare, with minimal stress, according to International Guidelines for research involving animals (Directive2010/63/EU).

A total of 54, 32-week-old male Mandarah (a dual-purpose breed) chickens, with a similar initial body weight were distributed randomly among six treatment groups of nine males during weeks 32–52 of age. Each male served as a replicate, and was individually housed in galvanized wire cages in batteries with standard dimensions ($30 \times 50 \times 60$ cm) in an environmentally controlled lightproof house (close system; controlled for temperature, humidity, and light). Each cage was provided with a manual feeder and nipple waters. Chickens were offered free access to mash diets and water throughout the experimental period. The experimental diets' chemical composition was done according to [32].

The health care, housing condition, heat stress protocol, indoor and outdoor temperature, relative humidity (RH), and light schedule were similar to those reported in [15]. The roosters were reared (indoor) either at an optimum temperature of 22–24 °C, with RH of 45–55%, serving as the thermoneutral group (positive control). They were fed with a basal diet (Table 1) or under CHS (38 ± 1 °C; 55–65% RH) for three successive days a week, from 11.00 a.m. to 15.00 p.m., and returned to the thermoneutral condition thereafter. Roosters under CHS were divided into four groups: roosters kept under CHS and fed with a basal diet without additional Bet, Bet + VC, Bet + VE, and Bet + VC + VE serving as the CHS negative control. The Bet group, roosters kept under CHS and fed with a basal diet supplemented with 1000 mg/kg of Bet (natural Betafin® S4 contain 93% dry Bet, Danisco Animal Nutrition, Marlborough, UK). The Bet +VC group, roosters kept under CHS and fed a basal diet supplemented with 1000 mg/kg betaine + 200 mg/kg ascorbic acid (L-ascorbic acid; a heat stabilized product, Hoffmann-La Roche, Basel, Switzerland). The Bet+VE group, roosters kept under CHS and fed a basal diet supplemented with 1000 mg/kg betaine + 150 mg/kg α-tocopherol acetate (VE; α-tocopherol acetate, Hoffmann-La Roche, Switzerland). The Bet+VC+VE group, roosters kept under CHS and fed with a basal diet supplemented with 1000 mg/kg betaine + with 200 mg/kg ascorbic acid + 150 mg/kg α-tocopherol acetate (VC + VE).

Table 1. Ingredients and composition of the experimental diet (as fed basis).

Ingredients and Composition, g/kg	Amount
Yellow corn, ground	663.3
Soybean meal, 48% crude protein	242.0
Wheat bran	65.0
Limestone	10.0
Dicalcium phosphate	13.2
Vit + Min Premix [1]	2.5
NaCl	2.5
DL–methionine	1.5
Calculated and determined composition	
Metabolizable energy, MJ/kg [2]	11.98
Dry matter, g/kg [3]	917.3
CP, g/kg[3]	179.6
Ether extract, g/kg [3]	28.5
Crude fibre, g/kg [3]	4.78
Methionine, g/kg [2]	4.1
Methionine + Cysteine, TSAA, g/kg [2]	6.70
Lysine, g/kg [2]	8.8
Calcium, g/kg [2]	1.10
Available P, g/kg [2]	3.94

[1] Vit + Min mixture provides per kilogram of diet: vitamin A, 12,000 IU; vitamin E, 10 IU; menadione, 3 mg; vitamin D3, 2200 ICU; riboflavin, 10 mg; Ca pantothenate, 10 mg; nicotinic acid, 20 mg; choline chloride, 500 mg; vitamin B12, 10 µg; vitamin B6, 1.5 mg; vitamin B1, 2.2 mg; folic acid, 1 mg; and biotin, 50 µg. Trace minerals (milligrams per kilogram of diet): Mn, 55; Zn, 50; Fe, 30; Cu, 10; Se, 0.10; antioxidants, 3 mg. [2] Calculated according to the National Research council (NRC) [23] tabulated values for feedstuffs.[3] Determined values based on Reference [32].

2.2. Data Collection

Roosters were individually weighted at 32 and 52 weeks of age, body weight gain was calculated from the difference between the initial and final body weight. Daily feed intake (g/bird) and mortality were recorded for each replicate.

Semen was collected weekly from all roosters after 10 weeks of treatments at week 42 of age, and maintained for another 10 weeks to determine the semen's quality as outlined by Attia et al. [33]. Semen collection was performed under the procedure of abdominal massage. The internal temperature of the semen at the time of collection was kept in the range of 41–44 °C using a water bath (at 37 °C). Semen samples were then transferred immediately after collection to the laboratory to determine spermatozoa quality. Moreover, special attention was given to protect semen from cold shocks and direct light. Throughout the course of semen collection, time, place of collection, and collector were kept constant.

At weeks 44, 48, and 52 of age, the roosters' fertility was evaluated. Semen samples were artificially collected by abdominal massage and used for the artificial insemination of hens. Semen was used after 1:1 dilution using a 0.9% saline solution as a diluent [34]. The semen of each male was used to inseminate 10 hens. Each hen was inseminated with 0.5 mL of semen over two successive days. After two days of insemination, the eggs were collected for ten days, and stored at room temperature (22–24 °C with 45–55% RH), incubated (37.6°C, 55% RH), and hatched (36.8°C, 65% RH) in an automatic incubator. Fertility was calculated by dividing the number of fertile eggs by the total eggs set.

At 52 weeks of age, blood samples (5 mL) were withdrawn from the wing vein after each treatment. Blood samples ($n = 5$) after each treatment were collected heparin (an anticoagulant agent) tubes in the morning from the overnight-fasted roosters. Blood plasma and seminal plasma were obtained by blood and semen centrifugation at 1500× g for 20 min, and kept at −20 °C until the analysis was performed [33].

Plasma and seminal plasma metabolites, seminal and blood plasma total antioxidant capacity (TAC), and malondialdhyde (MDA) were determined using a diagnostic kit and following the manufacture's recommendation (Diamond diagnostics, 23 EL-Montazah St. Heliopolis, Cairo, Egypt, http://www.diamonddiagnostics.com). Blood plasma creatinine was measured using special kits delivered from N.S. BIOTEC (http://www.nsbiotec.com). The aspartate aminotransferase (AST) and alanine aminotransferase (ALT) as (U/L) in the blood and seminal plasma were determined using commercial kits produced by the Pasteur Lab (http://www.pasteurvetlab.com). Blood plasma alkaline phosphatase was measured according to the method by Yan et al. [35]. Seminal plasma α-, β-, and γ-globulin were determined using commercial ELISA according to the method by Bianchi et al. [36].

Blood hematological parameters, such as hemoglobin (Hgb; %), were determined according to the method by Tietz [37]. Red blood cells (RBCs) were counted on a bright line hemocytometer using a light microscope at 400× magnification according to the methods by Helper [38], and Hawkeye and Dennett [39]. Packed cell volume (PCV; %) was measured according to the method by Wintrobe [40]. The mean cell volume (MCV), the mean cell hemoglobin (MCH), and the mean cell hemoglobin concentration (MCHC), were estimated as absolute values as reported by Attia et al. [34]. The phagocytic activity and index were determined according to the method by Kawahara [41]. White blood cells (WBCs) were assessed according to the methods by Helper [38], and Dennett [39], using a light microscope at 100× magnification. The blood film was prepared according to the method by Lucky [42] to determine different leucocytes.

High molecular weight DNA was extracted from intestinal parts (jejunum and ileum) according to the method by Sambrook et al. [43], with some modification according to Abdel-Fattah [44]. The DNA concentration was estimated from the optical density (O.D.) reading of a UV spectrophotometer at a 260 nm wave length (1.0 O.D. = 50 µg DNA/mL of solution), according to the method by Charles [45].

2.3. Statistical Analysis

Data were tested using the GLM procedure published by SAS® (SAS Institute, Cary, NC, USA) [46], using one-way ANOVA according to the following model: $yij = \mu + \tau j + \varepsilon ij$, where μ = the general mean, τj = the effect of treatment, and εij = the experimental error. A $p \leq 0.05$ value of significance for the student Newman Klaus test was used for testing mean differences among the experimental groups. Prior to the analyses, all the percentages were subjected to logarithmic transformation to normalize data distribution.

3. Results

Overall, neither diseases symptoms nor mortality occurred throughout the experimental period.

The initial roosters' body weights (BW) were not significantly ($p > 0.001$) different, while their BW changes and feed intake significantly decreased with the increase in the level of CHS, as indicated in Table 2. Body weight changes were completely resorted with high supplementations, except for Bet + VC, which caused partial recovery. All Bet-supplemented groups produced similar partial recovery in feed intake.

Table 2 indicates that the group on CHS without antioxidant addition significantly decreased semen's physical characteristics. The ejaculate volume, sperm concentrate, concentrate/ejaculate, sperm motility %, sperm livability %, total live sperm/ejaculate, semen quality factor, and fertility for CHS groups were significantly decreased ($p > 0.001$) by 23.2, 23.6, 40.8, 13.1, 9.3, 46.1, 46.0, and 17.9%, respectively, compared to the thermoneutral control group. However, sperm mortality (%) and pH significantly increased ($p > 0.001$) by 62.8 and 3.7%, respectively.

The antioxidants, either individually or combined, induced complete recovery in sperm concentration, livability, pH, and fertility. Moreover, supplementation of Bet + VC + VE restored ejaculation volume, concentration/ejaculate, sperm motility, total live sperm/ejaculate, and the semen's quality factor to the thermoneutral group, but differences among these groups and Bet or Bet + VE groups were not significant.

Groups of CHS roosters without antioxidants had significantly impaired total protein, globulin, AST, ALT, and TAC in seminal plasma compared to the thermoneutral group (Table 2). Antioxidants significantly ($p > 0.001$) restored total protein, globulin, AST, ALT, TAC, and MDA, indicating no significant difference from the thermoneutral group. When Bet + VC + VE were added, this caused an increase in γ-globulin compared to the group on Bet alone. Antioxidant groups showed increased TAC in comparison to both the thermoneutral and CHS groups. The different treatments did not cause any significant effects on plasma albumin, albumin/globulin ratio α- and β-globulin, and AST/ALT ratio.

Results in Table 3 show that the CHS group without antioxidants had significantly impaired RBCs, Hgb, PCV %, pH, PA, and PI compared to the thermoneutral group, the values decreased by 17.8, 20.6, 19.0, 2.8,2 and 28.9%, respectively. When antioxidants were added, a complete recovery of all hematological traits occurred except for PI, which was recovered only when the three additives were combined. Furthermore, PA was partially recovered similarly due to the different antioxidant supplementations. There were no significant effects from the different treatments on MCV, MCH, and MCHC.

White blood cell counts (−14.5%), lymphocytes (−9.4%), heterophile (+15.4%), H/L ratio as the welfare index (+22.9%), were negatively affected under the CHS exposure without antioxidants condition compared to the thermoneutral group. The addition of Bet + VC, Bet + VE, and Bet + VC + VE completely recovered the negative effects of HS on the WBC parameters and index of wellbeing (H/L). Monocyte, basophil, and eosinophil percentages were insignificantly affected by the different treatments.

Results in Table 4 show that CHS caused a significant impairment in plasma glucose, protein profiles except for plasma globulin, lipid profiles, and the liver function index, renal function index except for plasma urea/creatinine ratio, TAC, and MDA.

Table 2. Effects of dietary betaine (Bet), with or without vitamin C (VC) and vitamin E (VE) supplementation, on semen quality in Mandarah rooster chickens reared under heat stress condition.

Parameters	Control (+)	Heat Stress Treatments					p-Value	SEM
		Control (−)	+ Bet	+ Bet + VC	+ Bet + VE	+ Bet + VC + VE		
Roosters performance								
Initial BW (32 wk), g	2142	2129	2115	2126	2128	2129	37.0	0.989
BW gain, g	513 [a]	401 [c]	490 [a]	452 [b]	486 [a]	497 [a]	11.4	0.001
Feed intake, g/bird/d	136 [a]	123 [c]	131 [b]	131 [b]	129 [b]	130 [b]	0.846	0.001
Semen quality characteristics								
Ejaculate volume (mL)	0.561 [a]	0.431 [d]	0.528 [b,c]	0.523 [c]	0.539 [a,b,c]	0.549 [a,b]	0.009	0.001
Concentrate/mL ($\times 10^9$ sperm)	2.76 [a]	2.11 [b]	2.65 [a]	2.62 [a]	2.66 [a]	2.70 [a]	0.054	0.001
Concentrate/ejaculate ($\times 10^9$ sperm)	1.55 [a]	0.917 [d]	1.40 [b,c]	1.38 [c]	1.43 [b,c]	1.49 [a,b]	0.039	0.001
Sperm motility (%)	90.8 [a]	78.9 [c]	88.7 [a,b]	88.4 [b]	89.1 [a,b]	89.9 [a,b]	0.830	0.001
Sperm livability (%)	87.2 [a]	79.1 [b]	86.5 [a]	86.6 [a]	86.9 [a]	86.7 [a]	0.502	0.001
Total live sperm/ejaculate ($\times 10^9$ sperm)	1.35 [a]	0.728 [d]	1.22 [b,c]	1.19 [c]	1.25 [b,c]	1.29 [a,b]	0.034	0.001
Semen Ph	7.28 [b]	7.55 [a]	7.37 [b]	7.29 [b]	7.33 [b]	7.30 [b]	0.032	0.001
Semen quality factor	1349 [a]	728 [d]	1216 [b,c]	1193 [c]	1249 [b,c]	1289 [a,b]	34.3	0.001
Fertility	96.5 [a]	79.2 [b]	95.6 [a]	95.6 [a]	95.7 [a]	96.8 [a]	5.64	0.001
Seminal plasma constituents								
Total protein, (g/dL)	6.13 [a]	5.48 [b]	5.83 [a,b]	5.80 [a,b]	5.73 [a,b]	5.83 [a,b]	0.170	0.037
Albumin, (g/dL)	2.30	2.28	2.43	2.30	2.10	2.10	0.115	0.072
Globulin, (g/dL)	3.83 [a]	3.20 [b]	3.40 [a,b]	3.50 [a,b]	3.63 [a,b]	3.73 [a]	0.155	0.009
A/G ratio	0.604	0.714	0.717	0.659	0.582	0.567	0.050	0.022
α–globulin, (g/dL)	1.63	1.48	1.68	1.68	1.63	1.40	0.099	0.059
β–globulin, (g/dL)	1.20	1.08	1.20	1.18	1.13	1.18	0.077	0.546
γ–globulin, (g/dL)	1.00 [a,b]	0.64 [a,b]	0.52 [b]	0.62 [a,b]	0.87 [a,b]	1.15 [a]	0.194	0.035
AST, U/L	41.3 [b]	51.8 [a]	44.3 [b]	43.8 [b]	43.8 [b]	42.8 [b]	1.21	0.001
ALT, U/L	15.0 [b]	19.8 [a]	16.8 [b]	16.0 [b]	16.3 [b]	15.8 [b]	0.898	0.001
AST/ALT	2.77	2.65	2.66	2.76	2.68	2.75	0.170	0.967
TAC, Mmol/dL	409 [b]	324 [c]	430 [a]	433 [a]	438 [a]	439 [a]	4.09	0.001
MDA, Mmol/dL	0.925 [a,b]	1.10 [a]	0.800 [b]	0.725 [b]	0.700 [b]	0.675 [b]	0.089	0.001

BW = body weight; A/G ratio = albumin/globulin ratio; AST = aspatate amino transferase; ALT = alanine amino transferase; TAC = total antioxidant capacity; malondialdehyde (MDA); SEM = standard error of the mean. [a–c] means with different superscripts in the same column in similar treatment groups are significantly different.

Table 3. Effects of dietary betaine (Bet), with or without vitamin C (VC) and vitamin E (VE) supplementation, on hematological parameters, some immunological traits in Mandarah rooster chickens reared under heat stress condition.

Parameters	Control (+)	Heat Stress Treatments					SEM	*p*-Value
		Control (−)	+ Bet	+ Bet + VC	+ Bet + VE	+ Bet + VC + VE		
Hematological parameters and some immunological traits.								
RBCs, ×10^6/mm^3	1.52 [a]	1.25 [b]	1.47 [a]	1.43 [a]	1.40 [a]	1.45 [a]	0.067	0.001
Hgb, g/dL	10.7 [a]	8.50 [b]	10.2 [a]	10.2 [a]	9.83 [a]	10.5 [a]	0.585	0.012
PCV, %	32.7 [a]	26.5 [b]	29.7 [a,b]	29.5 [a,b]	30.7 [a,b]	30.3 [a,b]	1.54	0.015
MCV, µm^3/red blood cell	216	214	203	206	220	212	15.4	0.892
MCH, pg/Dl	70.6	68.6	69.9	72.1	70.6	73.2	6.35	0.984
MCHC, %	32.7	32.2	34.4	34.9	32.1	34.8	2.20	0.604
Blood pH	7.54 [b]	7.76 [a]	7.56 [b]	7.62 [b]	7.62 [b]	7.55 [b]	0.030	0.001
PA, %	20.0 [a]	15.8 [c]	17.3 [b]	17.7 [b]	18.0 [b]	18.0 [b]	0.679	0.001
PI, %	1.73 [a]	1.23 [d]	1.47 [bc]	1.57 [a,b,c]	1.43 [c]	1.65 [a,b]	0.076	0.001
White blood cell parameters								
WBCs, ×10^3/mm^3	26.3 [a]	22.5 [c]	24.3 [b]	24.7 [a,b]	24.7 [a,b]	26.0 [a,b]	0.622	0.001
Lymphocyte, %	47.8 [a]	43.3 [c]	45.3 [b]	46.2 [a,b]	46.2 [a,b]	47.2 [a]	0.660	0.001
Monocyte, %	6.83	7.50	8.17	7.33	7.33	8.00	0.615	0.325
Basophil, %	0.500	0.500	0.167	0.333	0.333	0.667	0.285	0.265
Eosinophil, %	9.50	9.33	9.67	9.33	9.33	10.00	0.505	0.779
Heterophile, %	35.3 [b]	39.3 [a]	36.7 [b]	36.8 [b]	36.8 [b]	34.2 [b]	0.980	0.001
H/L ratio	0.739 [b,c]	0.909 [a]	0.810 [b]	0.797 [b,c]	0.797 [b,c]	0.725 [c]	0.028	0.001

RBC = red blood cells; PCV = packed-cell volume; Hgb = hemoglobin; MCV = mean corpuscular volume; MCH = mean corpuscular hemoglobin; MCHC= mean corpuscular hemoglobin concentration; PA = phagocytic activity; PI = phagocytic index; SEM = standard error of the mean. [a–c] Mears with different superscripts in the same column in similar treatment groups are significantly different.

Table 4. Effects of dietary betaine (Bet), with or without vitamin C (VC) and vitamin E (VE) supplementation, on some blood biochemical constituents in Mandarah rooster chickens reared under heat stress condition.

Parameters	Control (+)	Heat Stress Treatments					SEM	p-Value
		Control (−)	+ Bet	+ Bet + VC	+ Bet + VE	+ Bet + VC + VE		
Glucose, (mg/dL)	228 [a]	210 [b]	223 [a]	223 [a]	219 [a]	222 [a]	2.93	0.001
Total protein, (g/dL)	5.92 [a]	5.13 [c]	5.62 [a,b]	5.47 [b,c]	5.38 [b,c]	5.59 [a,b]	0.140	0.001
Albumin, (g/dL)	2.64 [a]	1.82 [d]	2.30 [b]	2.21 [b,c]	2.13 [c]	2.31 [b]	0.054	0.001
Globulin, (g/dL)	3.28	3.32	3.32	3.26	3.26	3.28	0.173	0.998
A/G ratio	0.803 [a]	0.550 [c]	0.708 [a,b]	0.680 [a,b]	0.653 [b,c]	0.710 [a,b]	0.049	0.003
Total lipids, g/dL	4.42 [b]	5.54 [a]	4.74 [b]	4.51 [b]	4.56 [b]	4.59 [b]	0.226	0.001
Triglycerides, mg/dL	150 [c]	175 [a]	165 [b]	159 [b]	161 [b]	158 [b]	2.09	0.001
Cholesterol, mg/dL	136 [b]	154 [a]	144 [b]	141 [b]	138 [b]	139 [b]	2.71	0.001
AST, U/L	40.1 [b]	61.5 [a]	42.6 [b]	42.4 [b]	41.8 [b]	41.5 [b]	1.10	0.001
ALT, U/L	17.3 [b]	21.9 [a]	18.0 [b]	17.7 [b]	18.0 [b]	17.6 [b]	0.276	0.001
AST/ALT	2.32 [b]	2.81 [a]	2.36 [b]	2.39 [b]	2.32 [b]	2.36 [b]	0.065	0.001
ALP, U/l	171 [d]	192 [a]	181 [b]	180 [b]	176 [c]	173 [d]	1.42	0.001
Creatinine, (mg/dL)	3.19 [c]	3.30 [a]	3.26 [b]	3.24 [b,c]	3.22 [b,c]	3.20 [c]	0.021	0.001
Urea, (mg/dL)	3.40 [b]	3.58 [a]	3.43 [b]	3.45 [b]	3.43 [b]	3.41 [b]	0.040	0.001
Urea/creatinine ratio	1.07	1.08	1.05	1.06	1.07	1.07	0.012	0.450
TAC, Mmol/dL	431 [d]	374 [e]	437 [c]	456 [b]	460 [a,b]	461 [a]	1.92	0.001
MDA, Mmol/dL	0.915 [b]	1.27 [a]	0.898 [b]	0.843 [c]	0.818 [c]	0.817 [c]	0.024	0.001
DNAJ mg/g tissue	27.0 [a]	22.8 [d]	25.6 [b]	24.6 [c]	25.3 [b]	24.8 [c]	0.206	0.001
DNAI mg/g tissue	25.2 [b]	20.2 [e]	26.2 [a]	23.3 [d]	26.5 [a]	24.0 [c]	0.204	0.001

A/G ratio = albumin/globulin ratio; ALP = alkaline phosphatase; AST = aspratate amino transferase; ALT = alanine amino transferase; TAC = total antioxidant capacity; malondialdehyde (MDA). DNAJ = jejunum DNA; DNAI = ileum DNA. SEM = standard error of the mean. [a–e] Means with different superscripts in the same column in similar treatment groups are significantly different.

The addition of Bet alone or in combination with vitamins caused similar complete recovery in plasma glucose, the A/G ratio, total lipids, cholesterol, AST, ALT, the AST/ALT ratio, urea, and TAC. Instead, Bet and Bet + VC + VE caused a complete recovery in plasma total protein, and the combination of Bet + VC, Bet + VE, and Bet + VC + VE groups resulted in a complete recovery in plasma creatinine. The combination of Bet + VE and Bet + VC + VE completely recovered TAC, and had stronger effect than Bet alone. Furthermore, the three agents had a stronger effect compared to Bet + VC. Bet treatments caused a complete recovery in MDA, and the combined agents resulted in a stronger effect compared to Bet alone.

Antioxidant supplementation showed a partial recovery in plasma albumin with Bet and Bet + VC + VE, and resulted in a stronger recovery compared to the Bet + VE group. In addition, plasma triglyceride was partially recovered similarly due to the different antioxidants. Plasma alkaline phosphatase was partially recovered by the antioxidants. However, Bet + VE and Bet + VC + VE produced stronger effects compared to Bet and Bet + VC. Furthermore, the combination of the three agents resulted in a stronger effect compared to Bet + VE. The combination of Bet + VC, Bet + VE, and Bet + VC + VE showed a synergetic effect on TAC and MDA, which surpassed the effect of Bet and the thermoneutral groups.

Jejunum and ileum DNA concentrations were significantly affected, negatively by HS and positively by antioxidants (Table 4). The CHS treatment without antioxidants significantly decreased jejunum and ileum DNA concentrations by 10.2% and 19.8%, respectively, compared to the thermoneutral group. Antioxidants significantly increased DNA concentrations in the jejunum and ileum compared to the CHS group. With Bet and Bet + VE surpassing the other antioxidant groups to result in complete recovery of ileum DNA only.

4. Discussion

Heat stress negatively affects both animal and human welfare. The use of Bet with or without vitamin fortifications to relief the adverse influence of HS is an essential tool in human and animal nutrition. In this study, we demonstrate that HS negatively affects fertility and semen quality, and that Bet supplementation alone showed promising relieving effects. The negative effects of CHS caused an increase in the H/L ratio in roosters exposed to HS, suggesting their low [1,6]. Similarly, H/L appears to be a more reliable indicator for determining stress and welfare in poultry [47], which has negative effects on immunity and disease resistance [18,48]. The reduction in TAC and elevation in MDA confirmed the adverse effects of CHS on antioxidant conditions [49,50]. In addition, blood biochemistry and immunity indices were negatively affected herein by HS. More specifically, we observed a decrease in seminal and blood plasma total protein, and blood plasma albumin (nonspecific immune protein), γ-globulin (innate immunity), PA and PI (nonspecific immunity), WBCs and lymphocyte (cell mediated immunity), and DNA of the intestinal segments. In addition, RBCs, Hgb concentration, and PCV were adversely affected due to CHS, showing low animal welfare and health. This could be attributed to the decline in RBCs, which reduces oxygen uptake, resulting in less metabolic heat loss. In accordance with the present results, studies described in References [51] and [52] show that HS decreases Hgb and PCV while increasing blood pH. This was in agreement with the low availability of essential nutrients for DNA synthesis, functions, and repairs [53,54] and increased water intake [48].

Our study shows a decrease in the semen quality and fertility of roosters exposed to HS, and this was associated with the reduction in this groups feed intake. A HS > 31 °C is known to depress rooster sperm motility, viability, and fertilization potential [6]. This could be attributed to the decrease in sperm motility and the number of spermatozoa stored in the sperm host gland of hens [1]. In addition, HS can have negative effects on testosterone, causing hypertrophy and weakening of Leydig cell function [55]. The sperm's damaged DNA can determine abnormal spermatozoa, which could cause low male fertility and subsequently fewer surviving embryos [4,56,57]. Even if the sperm can fertilize eggs normally, embryos that have received an injured paternal genome could die, causing poor performance [58].

The biochemical and immunological change in blood and seminal plasma was reflected in lower semen quality and reproductive efficiency, showing low breeding strategy under CHS.

In the present study, supplementing Bet alone caused a complete recovery in BW, sperm concentration, motility, livability of sperm, semen pH, fertility, seminal plasma total protein, globulin, seminal plasma AST and ALT, RBCs, Hgb, PCV, blood pH, heterophile and H/L ratio, blood plasma glucose, total protein, A/G ratio, total lipids, cholesterol, AST, ALT, urea, and MDA of roosters exposed to HS. Instead, Bet showed partial recovery in feed intake (96.3%), ejaculate volume (94.1%), sperm concentrate/ejaculate (92.7%), total sperm/ejaculate (90.4%), semen quality factor (87.2%), seminal plasma MDA (86.5%), PA (86.5%), PI (85%), WBCs (93.2%), lymphocyte (94.8%), blood plasma albumin (87.1%), triglycerides (−10%), alkaline phosphatase (−5.8%), and creatinine (−2.2%). Bet also restored the H/L ratio to the level of the thermoneutral group, providing supporting evidence for improving welfare. These data indicate that the effect of Bet on HS-exposed animals depends on the investigated traits. In addition, Bet helps in sustaining the metabolic function when cells are under osmotic pressure, and Bet acts as a methyl donor group in protein metabolism [15]. Bet increases protein deposition in broilers [18] and ducks [9], and reduces blood urea-N by up to 47% [59], and the H/L ratio [15,48,60]. Furthermore, Bet shows an inverse relationship with obesity criteria in males [11,61].

The positive effect of Bet on reproduction and semen quality was further emphasized by the elevated feed intake, by up to 6.1% compared to the CHS group. Similarly, Singh et al. [62] found that Bet at 1.3 and 2 g/kg diet significantly increased feed intake of broiler chickens under thermal stress. In the literature, dietary Bet of 0.63 and 1.26% for boars increased sperm concentration by 6% and 13%, respectively, and total sperm output in comparison to the control group [14]. The complete recovery in fertility was validated by th increased motility, concentration, pH, and livability of the sperm in the antioxidant-fortified groups, with sperm motility being the most important indicator [1,6].

The positive effect of Bet on TAC (+32.7%) and MDA (−27.3%) in seminal plasma, while the corresponding values were +16.8 and −29.3% in blood plasma of Bet-fortified groups, suggesting that Bet is a potential antioxidant. In addition, Bet increased RBCs (17.6%), Hgb (20%), PCV (12.1%), pH (2.5%), PA (9.5%), PI (19.5%), WBCs (8%), lymphocyte (4.6%), and the H/L ratio (10.9%), demonstrating the positive impact of Bet on the health status. Similarly, Hossein and Hossein [16] reported that the H/L ratio, total IgG, and antibody titers against SRBCs were similar for the Bet-fortified group and thermoneutral group. Bet fortification decreased Urea-N levels in blood of pigs by 47%. Bet at 200 mg/kg increased survival %, immunity, and health in heat exposed chickens [15,59].

Furthermore, Bet decreased AST leakage from sperm, hepatocytes, and plasma ALP, and improved renal function (urea and creatinine). The positive effect of Bet on seminal plasma protein, globulin and total protein, blood plasma albumin, globulin and the albumin/globulin ratio indicate the boosting impact of Bet on liver function by increasing protein synthesis as a methyl group donor [10], in addition to decreasing AST and ALT [18].

Lipid metabolites (total lipids, triglycerides, cholesterol) were also improved by Bet fortification. Furthermore, MDA significantly decreased due to Bet supplementation showing the antioxidant effects of Bet. In the literature, Bet decreases meat lipid, increases the yield of breast meat [63], and improves Hgb and PCV [18]. Bet also has a sparing effect on methionine and choline [10], and thus increases choline for biosynthesis of very low-density lipoproteins, which prevent the lipid deposition via increased lipid removal from the liver [64], and regulates cholesterol in chicken [65].

It is interesting to report that Bet restored DNA in the ilium and caused a partial recovery in the jejunum DNA [66]. Bet, as a methyl donor group, increases the methionine and/or cysteine for glutathione synthesis that protects the cell from ROS and reaction metabolites, and boosts the synesthesia of DNA [17,67]. In addition, Bet is an osmolyte that stabilize proteins, cell membranes, organelles, and cells under stress [13,15]. Furthermore, recent studies by Gholami et al. [68], Nosrati et al. [69], and dos Santos et al. [70] indicate that Bet improves immunity, plasma biochemistry, and plasma osmolality of broiler chickens, suggesting its multibeneficial effects on many body functions.

In general, VE + Bet did not exceed the effect of Bet alone, even the combination of the three agents did not surpass the effect of Bet alone or Bet + VE. Body weight changes, ejaculate volume, concentration per ejaculate, total live sperm/ejaculate, and the semen quality factor of roosters supplemented with Bet + VC + VE surpassed only those of Bet + VC. Bet + VC + VE had an additive effect on seminal γ-globulin that surpassed other supplements, showing a synergetic effect over the Bet group in the current study. A similar effect was observed in H/L when the combination of the three agents exceeded the effect of Bet alone. In addition, Bet + VE or Bet + VC + VE improved plasma ALP better than Bet alone or Bet + VC, showing a synergetic effect. TAC and MDA improved due to VC and VC + VE addition compared to Bet, demonstrating an additional impact over Bet. Bet alone or Bet + VE resulted in the greatest recovery in jejunum and ileum DNA, suggesting that Bet alone is adequate for recovering the negative impact of HS on jejunum and ilium DNA. The drawback effect of Bet + VC compared to Bet alone and other combination was further validated by a drawback in the jejunum and ilium DNA recovery of VC-supplemented groups. This may adversely affect intestinal absorption capacity.

The synergetic effect of VE supplementation over Bet increased lymphocytes (cell mediated immunity), and decreased plasma albumin (nonspecific immune protein) and plasma alkaline phosphatase (index of liver function and bone mineralization), showing that a combination of Bet + VE might be beneficial for antioxidant status, and thus the animals' immunity. The synergetic effect shown in blood biochemistry could be elucidated by the antioxidant influence of VE as confirmed herein by the increase in TAC and decrease in MDA [8]. Vitamin E is known to prevent oxidation of vitamin A and lipids. Furthermore, by increasing oxygen levels, it supports muscle growth and increases oxygen assimilation by RBCs [7]. In addition, VE acts as an essential part of the antioxidant system to prevent peroxidation of PUFA in the sperm's membrane [71], reducing the adverse influence of corticosterone induced by stress [7,22]. VE protects cells such as macrophages, plasma cells, and lymphocytes against oxidative damage and increases the proliferation and functions of immune cells, thereby boosting animal welfare. Therefore, a 250 mg/kg VE is optimum for partly alleviating the negative influences of HS [72]. In this study, VE at 150 mg/kg feed caused a significant reduction in MDA concentrations and improved sperm motility, confirming the protective influences of VE on semen quality. This result supports VE use for treating male fertility dysfunction [25,28]. In addition, ALP, AST, and ALT were significantly decreased in males receiving 150 IU VE compared to the control roosters (15 IU VE), suggesting that 150 IU VE may be valuable for semen quality [25].

A synergetic effect of Bet + VC + VE induced a decrease in blood plasma ALP, creatinine, and MDA, and increased TAC over Bet. However, the three agents combined did not surpass the influence of Bet + VE alone. The combination of the tested agents did not exceed the effect of Bet alone on DNA repair in the intestinal segments, with Bet causing the highest and Bet + VC the lowest recovery, respectively. Jacob [73] quoted that VC enriches VE antioxidant activity by reducing tocopheroxyl radicals to their active form of VE or by sparing VE availability. In addition, Hoehler and Marquardt [30] indicated that the in vivo antioxidant influence of VE might be greater than that of VC. VE increased antibody titer and Bet + VC significantly elevated serum total protein and globulin of hens exposed to HS compared to controls [74], while decreasing serum glucose, triglyceride, HDL, and cholesterol [31]. However, Bet + VC had a similar impact on production and metabolic profiles of laying hens exposed to CHS [15].

5. Conclusions

Betaine at 1000 mg/kg feed for roosters had comparable effects to the combination of 1000 mg Bet with 200 mg VC and/or 150 mg E per kg diet. Therefore, Bet at 1000 mg/kg feed may be an adequate agent for compacting CHS, given the improvement of semen quality, fertility, physiological, antioxidant status, wellbeing, and intestinal DNA damage of breeder roosters, suggesting that Bet is a valuable tool for enhancing the breeding strategy of roosters in hot regions.

Author Contributions: A.S.E.-N. and Y.A.A., designed the experiment; B.M.A.-S., A.A.A., and A.S.E.-N., conducted the field work and lab analyses. All the authors contributed to the preparation of the manuscript; Y.A.A., revised and approved for the manuscript for final submission.

Funding: This article was funded by the Deanship of Scientific Research (DSR), King Abdulaziz University, Jeddah. The authors, therefore, acknowledge with thanks DSR's technical and financial support.

Conflicts of Interest: The authors declare no conflict of interest.

References

1. Khan, R.U.; Naz, S.; Nikousefat, Z.; Selvaggi, M.; Laudadio, V.; Tufarelli, V. Effect of ascorbic acid on heat-stressed poultry. *World's Poult. Sci. J.* **2012**, *68*, 477–489. [CrossRef]

2. Ayo, J.O.; Obidi, J.A.; Rekwot, P.I. Effects of heat stress on the well–being, fertility, and hatchability of chickens in the northern guinea savannah zone of Nigeria: A review. *ISRN Vet. Sci.* **2011**. [CrossRef] [PubMed]

3. Chen, Z.; Zhang, J.R.; Zhou, Y.W.; Liang, C.; Jiang, Y.Y. Effect of heat stress on the pituitary and testicular development of Wenchang chicks. *Arch. Anim. Breed.* **2015**, *58*, 373–378. [CrossRef]

4. Pérez-Crespo, M.; Pintado, B.; Gutiérrez-Adán, A. Scrotal heat stress effects on sperm viability, sperm DNA integrity, and the offspring sex ratio in mice. *Mol. Reprod. Dev.* **2008**, *75*, 40–47. [CrossRef] [PubMed]

5. Türk, G.; Çeribaşı, A.O.; Şimşek, Ü.G.; Çeribaşı, S.; Güvenç, M.; Kaya, Ş.Ö.; Çiftçi, M.; Sönmez, M.; Yüce, A.; Bayrakdar, A.; et al. Dietary rosemary oil alleviates heat stress–induced structural and functional damage through lipid peroxidation in the testes of growing Japanese quail. *Anim. Reprod. Sci.* **2016**, *164*, 133–143. [CrossRef] [PubMed]

6. McDaniel, C.D.; Hood, J.E.; Parker, H.M. An attempt at alleviating heat stress infertility in male broiler breeder chickens with dietary ascorbic acid. *Int. J. Poult. Sci.* **2004**, *3*, 593–602.

7. Surai, P.F.; Kochish, I.I.; Fisinin, V.I. Antioxidant systems in poultry biology: Nutritional modulation of vital genes. *Eur. Poult. Sci.* **2017**, *81*. [CrossRef]

8. Khan, R.U. Antioxidants and Poultry semen quality. *World's Poult. Sci. J.* **2011**, *67*, 297–308. [CrossRef]

9. Wang, Y.Z.; Xu, Z.R.; Feng, G. The effect of betaine and DL–Methionine on growth performance and cass characteristics in meat ducks. *Anim. Feed Sci. Technol.* **2004**, *116*, 151–159. [CrossRef]

10. Hassan, R.A.; Attia, Y.A.; El–Ganzory, E.H. Growth, carcass quality, and blood serum constituents of slow growth chicks as affected by betaine Additions to diets containing 1. Different levels of choline. *Int. J. Poult. Sci.* **2005**, *4*, 840–850.

11. Gao, X.; Randell, E.; Zhou, H.; Sun, G. Higher serum choline and betaine levels are associated with better body composition in male but not female population. *PLoS ONE* **2018**, *13*, e0193114. [CrossRef] [PubMed]

12. Attia, Y.A.; Hassan, R.A.; Shehatta, M.H.; Abd El-Hady, S.B. Growth, carcass quality and blood serum constituents of slow growth chicks as affected by betaine additions to diets containing 2. Different levels of methionine. *Int. J. Poult. Sci.* **2005**, *11*, 856–865.

13. Ahmed Mervat, M.N.; Zienhom, S.H.; Ismail Ahmed, A.; Abdel-Wareth, A. Application of betaine as feed additives in poultry nutrition-a review. *J. Exp. Appl. Anim. Sci.* **2018**, *2*, 266–272. [CrossRef]

14. Cabezón, F.A.; Stewart, K.R.; Schinckel, A.P.; Barnes, W.; Boyd, R.D.; Wilcock, P.; Woodliff, J. Effect of natural betaine on estimates of semen quality in mature AI boars during summer heat stress. *Anim. Reprod. Sci.* **2016**, *170*, 25–37. [CrossRef] [PubMed]

15. Attia, Y.A.; Abd-El-Hamid, A.E.; Abedalla, A.A.; Berika, M.A.; Al-Harthi, M.A.; Kucuk, O.; Sahin, K.; Abou–Shehema, B.M. Laying performance, digestibility and plasma hormones in laying hens exposed to chronic heat stress as affected by betaine, vitamin C, and/or vitamin E supplementation. *Springerplus* **2016**, *5*, 1619. [CrossRef] [PubMed]

16. Hossein, A.S.; Hossein, A.G. Alleviation of chronic heat stress in broilers by dietary supplementation of betaine and turmeric rhizome powder: Dynamics of performance, leukocyte profile, humoral immunity, and antioxidant status. *Trop. Anim. Health Prod.* **2016**, *48*, 181–188.

17. Alirezaei, M.; Jelodar, G.; Ghayemi, Z. Antioxidant defense of betaine against oxidative stress induced by ethanol in the rat testes. *Int. J. Pept. Res. Ther.* **2012**, *18*, 239–247. [CrossRef]

18. Attia, Y.A.; Hassan, R.A.; Qota, E.M. Recovery from adverse effects of heat stress on slow-growing chicks in the tropics 1: Effect of ascorbic acid and different levels of betaine. *Trop. Anim. Health Prod.* **2009**, *41*, 807–818. [CrossRef] [PubMed]

19. McDowell, L.R. *Vitamins in Animal Nutrition*; Academic Press Inc.: San Diego, CA, USA; New York, NY, USA; Boston, MA, USA; London, UK; Sydney, Australia; Toyko, Japan; Toronto, ON, Canada, 1989.

20. Klasing, K.C. Nutritional modulation of resistance to infectious diseases. *Poult. Sci.* **1998**, *77*, 1119–1125. [CrossRef]

21. Mckee, J.S.; Harrison, P.C. Effects of supplemental ascorbic acid on the performance of broiler breeders following exposure to an elevated environmental temperature. *Poult. Sci.* **1995**, *74*, 1029–1038. [CrossRef]

22. Metwally, M.A. Efficacy of different dietary levels of antioxidants vitamin C, E and their combination on the performance of heat-stressed laying hens. *Egypt. Poult. Sci. J.* **2005**, *25*, 613–636.

23. National Research Council (NRC). *Nutrient Requirement of Poultry*; National Academy Press: Washington, DC, USA, 1994.

24. Surai, P.F. Vitamin E in avian reproduction. *Poult. Avi. Biol. Rev.* **1999**, *10*, 1–60.

25. Biswas, A.; Mohan, J.; Sastry, K.V.H.; Tyagi, J.S. Effect of dietary vitamin E on the cloacal gland, foam and semen characteristics of male Japanese quail. *Theriogenology* **2007**, *67*, 259–263. [CrossRef] [PubMed]

26. Ebied, T.A. Vitamin E and organic selenium enhances the antioxidative status and quality of chicken semen under high ambient temperature. *Br. Poult. Sci.* **2012**, *53*, 708–714. [CrossRef] [PubMed]

27. Echeverria, S.; Santos, R.; Centurion, F.; Ake, R.; Alfaro, M.; Rodriguez, J. Effects of dietary selenium and vitamin E on semen quality and sperm morphology of young boars during warm and fresh season. *J. Anim. Vet. Adv.* **2009**, *8*, 2311–2317.

28. Keskes-Ammar, L.; Feki-Chakroun, N.; Rebai, T.; Sahnoun, Z.; Ghozzi, H.; Hammami, S.; Zghal, K.; Fki, H.; Damak, J.; Bahloul, A. Sperm oxidative stress and the effect of an oral vitamin E and selenium supplement on semen quality in infertile men. *J. Reprod. Syst.* **2003**, *49*, 89–94.

29. Traber, M.G.; Atkinson, J. Vitamin E, antioxidant and nothing more. *Free Radic. Biol. Med.* **2007**, *43*, 4–15. [CrossRef]

30. Hoehler, D.; Marquardt, R.R. Influence of Vitamin E and C on the toxic effects of ochratoxin A and T–2 Toxin in chicks. *Poult. Sci.* **1996**, *75*, 1508–1515. [CrossRef]

31. Ezzat, W.; Shoeib, M.S.; Mousa, S.M.M.; Bealish, A.M.A.; Ibrahiem, Z.A. Impact of betaine, vitamin C and folic acid supplementation to the diet on productive and reproductive performance of Matrouh poultry strain under Egyptian summer condition. *Egypt. Poult. Sci. J.* **2011**, *31*, 521–537.

32. Association of official analytical chemists. *Official Methods of Analysis*; AOAC: Washington, DC, USA, 2004.

33. Attia, Y.A.; Abd El-Hamid, E.A.; El-Hanoun, A.M.; Al-Harthi, M.A.; Mansour, G.M.; Abdella, M.M. Responses of the fertility, semen quality, blood constituents, immunity and antioxidant status of rabbit bucks to type and magnetizing of water. *Ann. Anim. Sci.* **2015**, *15*, 387–407. [CrossRef]

34. Bootwalla, S.M.; Miles, R.D. Development of diluents for domestic fowl semen. *World's Poult. Sci. J.* **1992**, *48*, 121–128. [CrossRef]

35. Yan, F.; Kersey, J.H.; Fritts, C.A.; Waldroup, P.W.; Stilborn, H.L.; Crumm, R.C., Jr.; Rice, D.W. DGKC: Recommendations of the German Society for Clinical Chemistry, Z. Klin. *Chim. Clin. Biochem.* **1972**, *10*, 182.

36. Bianchi, A.T.J.; Moonen–Leusen, H.W.M.; van der Heijden, P.J.; Bokhout, B.A. The use of a double antibody sandwich ELISA and monoclonal antibodies for the assessment of porcine IgM, IgG, and IgA concentrations. *Vet. Immunol. Immunopathol.* **1995**, *44*, 309–317. [CrossRef]

37. Tietz, N.W. *Fundamental of Clinical Chemistry 3rd Edition*; W B Saunders Co.: Philadelphia, PA, USA, 1982.

38. Helper, O.E. *Manual of Clinical Laboratory Methods*; Thomas Spring Field: Illinois, IL, USA, 1966.

39. Hawkeye, C.M.; Dennett, T.B. *A Color Atlas of Comparative Veterinary Hematology*; Wolf publishing Limited: London, UK, 1989.

40. Wintrobe, M.M. *Clinical Hematology*, 5th ed.; Lea and Febiger: Philadelphia, PA, USA, 1965.

41. Kawahara, E.; Ueda, T.; Nomura, S. In vitro phagocytic activity of white spotted shark cells after injection with Aermonassalmonicida extracellular products. *GyobokenkyuJpn* **1991**, *26*, 213–214.

42. Lucky, Z. *Methods for the Diagnosis of Fish Diseases*; Ameruno Publishing Co. PVT, LTd.: New Delhi, India; New York, NY, USA, 1977.

43. Sambrook, J.; Fristsch, E.F.; Maniatis, M. *Molecular Cloning A Laboratory Manual*, 2nd ed.; Cold Spring Harber Laboratory: New York, NY, USA, 1989.

44. Abdel-Fattah, A.F.; Mohamed, E.H.; Ramadan, G. Effect of thymus extract on immunologic reactively of chicken vaccinated with infectious bursal disease virus. *J. Vet. Med. Sci.* **1995**, *61*, 811–817. [CrossRef]

45. Charles, P. *Isolation of Deoxyribonucleic Acid: Uptake of Informative Moleculas by Living Cells*; Ledouix, L., Ed.; North Holland Publishing Company: Amesterdam, The Netherlands; London, UK, 1970.

46. *Statistical Analyses Software, SAS, Institute: SAS User's Guide*; SAS, Institute: Cary, NC, USA, 2007.

47. Lentfer, T.L.; Pendl, H.; Gebhardt-Henrich, S.G.; Fröhlich, E.K.; Von Borell, E. H/L ratio as a measurement of stress in laying hens-methodology and reliability. *Br. Poult Sci.* **2015**, *56*, 157–163. [CrossRef]

48. Attia, Y.A.; Hassan, R.A.; Tag El-Din, A.E.; Abou-Shehema, B.M. Effect of ascorbic acid or increasing metabolizable energy level with or without supplementation of some essential amino acids on productive and physiological traits of slow-growing chicks exposed to chronic heat stress. *J. Anim. Phys. Anim. Nutr.* **2011**, *95*, 744–755. [CrossRef]

49. Sahin, K.; Sahin, N.; Yaralioglu, S. Effects of vitamin C and vitamin E on lipid peroxidation, blood serum metabolites, and mineral concentrations of laying hens reared at high ambient temperature. *Biol. Trace Elem. Res.* **2002**, *85*, 35–45. [CrossRef]

50. Zeng, T.; Li, J.J.; Wang, D.Q.; Li, G.Q.; Wang, G.L.; Lu, L.Z. Effects of heat stress on antioxidant defense system, inflammatory injury, and heat shock proteins of Muscovy and Pekin ducks: Evidence for differential thermal sensitivities. *Cell Stress Chaperones* **2014**, *19*, 895–901. [CrossRef]

51. Toyomizu, M.; Tokuda, M.; Mujahid, A.; Akiba, Y. Progressive alteration to core temperature, respiration and blood acid-base balance in broiler chickens exposed to acute heat stress. *J. Poult. Sci.* **2005**, *42*, 110–118. [CrossRef]

52. Daghir, N.J. *Poultry Production in Hot Climates*, 2nd ed.; CAB International: Wallinford, Oxfordshire, UK, 2008; p. 387.

53. Habashy, W.S.; Milfort, M.C.; Adomako, K.; Attia, Y.A.; Rekaya, R.; Aggrey, S.E. Effect of heat stress on amino acid digestibility and transporters in meat–type chickens. *Poult. Sci.* **2017**, *96*, 312–319. [CrossRef]

54. Habashy, W.S.; Milfort, M.C.; Fuller, A.L.; Attia, Y.A.; Rekaya, R.; Aggrey, S.E. Effect of heat stress on protein utilization and nutrient transporters in meat–type chickens. *Int. J. Biometeorol.* **2017**. [CrossRef]

55. Stone, B.A.; Seamark, R.F. Effects of acute and chronic testicular hyperthermia on levels of testosterone and corticosteroids in plasma of boars. *Anim. Reprod. Sci.* **1984**, *7*, 391–403. [CrossRef]

56. Peña, S., Jr.; Gummow, B.; Parker, A.J.; Paris, D.B.B.P. Revisiting summer infertility in the pig: Could heat stress–induced sperm DNA damage negatively affect early embryo development? *Anim. Reprod. Sci.* **2016**, *57*, 1975–1983. [CrossRef]

57. Zheng, W.W.; Song, G.; Wang, Q.L.; Liu, S.W.; Zhu, X.L.; Deng, S.M.; Zhong, A.; Tan, Y.M.; Tan, Y. Sperm DNA damage has a negative effect on early embryonic development following in vitro fertilization. *Asian J. Androl.* **2018**, *20*, 75–79. [CrossRef]

58. Sutovsky, P.; Aarabi, M.; Miranda-Vizuete, A.; Oko, R. Negative biomarker based male fertility evaluation: Sperm phenotypes associated with molecular-level anomalies. *Asian J. Androl.* **2015**, *174*, 554–560. [CrossRef]

59. Eklund, M.; Bauer, E.; Wamatu, J.; Mosenthinm, R. Potential nutritional and physiological functions of betaine in livestock. *Nutr. Res. Rev.* **2005**, *18*, 31–48. [CrossRef]

60. Attia, Y.A.; Böhmer Barbara, M.; Roth-Maier, D.A. Responses of broiler chicks raised under constant relatively high ambient temperature to enzymes, amino acid supplementations, or diet density. *Arch. Geflügelk.* **2006**, *70*, 80–91.

61. Attia, Y.A.; Burke, W.H.; Yamani, Y.A.; Jensen, L.S. Energy allotments and performance of broiler breeders 1. Males. *Poult. Sci.* **1995**, *74*, 247–260. [CrossRef]

62. Singh, A.K.; Ghosh, T.K.; Creswell, D.C.; Haldar, S. Effects of supplementation of betaine hydrochloride on physiological performances of broilers exposed to thermal stress. *Open Access Anim. Phys.* **2015**, *7*, 111–120.

63. Metzler-Zebeli, B.U.; Eklund, M.; Mosenthin, R. Impact of osmoregulatory and methyl donor functions of betaine on intestinal health and performance in poultry. *World's Poult. Sci. J.* **2009**, *65*, 419–442. [CrossRef]

64. Yao, Z.; Vance, D. Head group specificity in the requirement of phosphatidylcholine biosynthesis for very low density lipoprotein secretion from cultured hepatocytes. *J. Biol. Chem.* **1989**, *264*, 11373–11380.

65. Yun, H.; Sun, Q.; Li, X.; Wang, M.; Cai, D.; Li, X.; Zhao, R. In Ovo injection of betaine affects hepatic cholesterol metabolism through epigenetic gene regulation in newly hatched chicks. *PLoS ONE* **2015**, *10*, e0122643.

66. Zabrodina, V.V.; Shreder, O.V.; Shreder, E.D.; Durnev, A.D. Effect of afobazole and betaine on cognitive disorders in the offspring of rats with streptozotocin–induced diabetes and their relationship with DNA damage. *Bull. Exp. Biol. Med.* **2016**, *161*, 359–366. [CrossRef]

67. Zhang, M.; Hong, Z.; Li, H.; Lai, F.; Li, X.F.; Tang, Y.; Tian, M.; Hui, W. Antioxidant mechanism of betaine without free radicals scavenging ability. *J Agric. Food Chem.* **2016**, *64*, 7921–7930. [CrossRef]
68. Gholami, J.; Qotbi, A.A.A.; Seidavi, A.R.; Meluzzi, A.; Tavaniello, S.; Maiorano, G. Effects of in ovo administration of betaine and choline on atchability results, growth and carcass characteristics and immune response of broiler chickens. *Ital. J. Anim. Sci.* **2015**, *14*, 187–192. [CrossRef]
69. Nosrati, M.; Javandel, F.; Camacho, L.M.; Khusro, A.; Cipriano, M.; Seidavi, A.R.; Salem, A.Z.M. The effects of antibiotic, probiotic, organic acid, vitamin C, and Echinacea purpurea extract on performance, carcass characteristics, blood chemistry, microbiota, and immunity of broiler chickens. *J. Appl. Poult. Res.* **2017**, *26*, 295–306.
70. Dos Santos, T.T.; Baal, S.C.S.; Lee, S.A.; e Silva, F.R.O.; Scheraiber, M.; da Silva, A.V.F. Influence of dietary fibre and betaine on mucus production and digesta and plasma osmolality of broiler chicks from hatch to 14 days of age. *Livest. Sci.* **2019**, *220*, 67–73. [CrossRef]
71. Tran, L.V.; Malla, B.A.; Kumar, S.A.K. Polyunsaturated fatty acids in male ruminant reproduction—A Review. *Asian Aust. J. Anim. Sci.* **2017**, *30*, 622–637. [CrossRef]
72. Bollinger-Lee, S.; Williams, P.E.; Whitehead, C.C. Optimal dietary concentration of vitamin E for alleviating the effect of heat stress on egg production in laying hens. *Br. Poult. Sci.* **1999**, *40*, 102–107. [CrossRef]
73. Jacob, R.A. The integrated antioxidant system. *Nutr. Res.* **1995**, *15*, 755–766. [CrossRef]
74. Mohiti-Asli, M.; Shariatmadari, F.; Lotfollahian, H. The influence of dietary vitamin E and selenium on egg production parameters, serum and yolk cholesterol and antibody response of laying hen exposed to high environmental temperature. *Arch. Geflügelk.* **2010**, *74*, 43–50.

Article

Protein and Amino Acid Content in Four Brands of Commercial Table Eggs in Retail Markets in Relation to Human Requirements

Youssef A. Attia [1],*, Mohammed A. Al-Harthi [1], Mohamed A. Korish [1] and Mohamed H. Shiboob [2]

[1] Arid Land Agriculture Department, Faculty of Meteorology, Environment, and Arid Land Agriculture, King Abdulaziz University, Jeddah 21589, Saudi Arabia; malharthi@kau.edu.sa (M.A.A.-H.); mmkorish@yahoo.com (M.A.K.)

[2] Environmental Department, Faculty of Meteorology, Environment and Arid land Agriculture, King Abdulaziz University, P.O. Box 80208, Jeddah 21589, Saudi Arabia; mshiboob@kau.edu.sa

*   Correspondence: yaattia@kau.edu.sa

Received: 4 February 2020; Accepted: 24 February 2020; Published: 1 March 2020

Simple Summary: At present, great attention has been paid to the nutritional values of animal products to resilient humans against pathogens, boost their immunity, and cure diseases. Thus, this research investigated the nutritional value of four sources of commercial table eggs in the retail market in Jeddah, KSA, with the possible presence of raw protein, amino acid content, and protein quality indicators for different parts of eggs. The examined eggs showed a different percentage of essential and non-essential amino acids and antioxidant amino acids, suggesting a potential for enriching the nutritional values and prolonging the shelf life of the eggs by various nutritional strategic ways to enhance the antioxidant amino acids and the essential amino acid profile in eggs.

Abstract: Considering the common believe that all eggs in the retail market are nutritionally similar, four different commercial sources of eggs (A, B, C, and D) available in a retail market were collected to investigate the crude protein and amino acid content, as well as the protein quality in the whole edible part of eggs (albumen + yolk), egg albumen, and egg yolk, separately. Five egg samples per source were collected four times during the experimental period, which resulted in a total number of 20 samples that were pooled to finally present five samples per source of eggs. The results show that crude protein in albumen was significantly higher in A and B than that of C and D, but the difference was found among edible parts of eggs such as yolk > whole edible part > albumen. Essential amino acids (arginine, histidine, isoleucine, lysine, methionine, methionine + cysteine, phenylalanine, phenylalanine + tyrosine, threonine, and valine) of eggs significantly differed according to the source of eggs, but eggs from different sources could provide from 17.4–26.7% of recommended daily allowance (RDA) of amino acids for adults. Essential amino acids (EAAs) were higher ($p \leq 0.05$) in eggs from sources A and B than in source D, while source C exhibited intermediate values. Source B had greater ($p \leq 0.05$) non-essential amino acids (NEAAs) than did sources C and D in whole edible egg, while source A displayed intermediate values. The phenylalanine + tyrosine, histidine, and lysine were the 1st, 2nd, and 3rd limiting amino acids in all sources of eggs. In conclusion, the investigated eggs showed different EAAs/NEAAs ratio and antioxidant amino acids, indicating a potential for enhancing nutritional values and extending the shelf life of eggs by different nutritional additions.

Keywords: egg sources; amino acids ratios; antioxidant; recommended daily allowance

---

1. Introduction

Eggs are a portion of essential food for human consumption and contain most of the necessary nutrients, based on the daily need [1,2]. Eggs have all the crucial nutrients for life and are a valuable source of protein\amino acids, antioxidants, and bioactive components [3,4]. Furthermore, there has been considerable progress in the egg production industry, including the improvement of the genetic makeup of layers and production capacity, egg quality, and layers management [5,6].

Protein and amino acids are primary components of eggs and play critical roles in egg consumption and nutrition as they present the main part of the muscle, body function, hormones, enzymes and body fluids [7,8]. The total amino acids (TAAs) in eggs were 10.0 and 10.1 mg/g in the dry yolk of eggs in corn- and wheat-based diets, respectively [4]. According to the same authors, lysine was found to be 929, 1182 and 760 mg/100 g in fresh whole eggs, yolk, and albumen, respectively; for methionine, the corresponding values were 400, 375, and 396 [9]. The essential amino acids (EAAs) on dry yolk accounted for about 46–47%, while aromatic amino acids represented about 11% of the TAAs; arginine was the most copious amino acid (about 1.5 mg/g) in all samples, followed by glutamic acid (1.2 mg/g) and lysine (0.9 mg/g). The authors added that yolk samples content of the TAAs decreased by different cooking procedures, except for the boiled yolk of hens that were fed a corn-based diet, and tyrosine and tryptophan were noticed to be the key providers to the antioxidant properties of eggs.

Eggs were also found to be a rich source of antioxidants [4,9] with the amino acid cysteine having the potential as an antioxidant amino acid, and the combination of strong, weak, and non-antioxidant amino acids increase the antioxidant capacity [9,10]. Besides, the amino acid content of eggs was found to be affected by poultry breeds [5] and species [6], preparation method [11,12] and components (whole, albumen, and yolk) of an egg [13]. The literature review indicates that there is room to improve the nutritional values of eggs by breeds and dietary and management practices due to the differences found in amino acids and antioxidant profiles of table eggs [1,5,6]. However, data for amino acids profile and amino acids ratio of eggs in the retail market are scarce, and there is a common belief that all eggs are nutritionally equal. The present research aims to investigate the crude protein and amino acid content and to evaluate the protein and amino acid ratio in commercial table eggs from four sources in the retail market in terms of recommended daily allowance (RDA) for adults.

2. Materials and Methods

2.1. Sample Collection

The Deanship of Scientific Research, King Abdulaziz University Saudi Arabia, approved the experimental procedures under protocol no. D-156-155-1438. It recommends animal rights, welfare, and minimal stress and did not cause any harm or suffering to animals, according to the Royal Decree M59 on 14/9/1431H.

Four commercial table egg sources, named A, B, C, and D in the retail market from white eggshell hybrids laying hens, originated from Single Comb White Leghorn commercially available in Saudi Arabia. The hens had 40–60 weeks of age. The hens were housed in cages in environmental control houses, and fed commercial layer diets contained 17% crude protein (CP), 11.60 MJ/kg, 3.5% Ca, and 0.35% none-phytate phosphorus kcal, in mash form and offered free access to feed and water and eliminated with 14:10 light-dark cycle. The medical care, vaccination, and husbandry practice were as suggested by the primary breeders and carried under the supervision of a veterinarian.

Fresh eggs of grade A classes, stored at 5 °C, were gathered randomly to represent various sources of eggs in retail markets in Jeddah city, Saudi Arabia. The eggs were clean, of normal eggshell index, and of medium size, 55–60 g, of grade A. Thirty eggs were collected from A, B, C, D sources four times during February, March, April, and May 2017. The eggs of each source per time were broken opened, and five samples of each albumen, yolk, and albumen + yolk were randomly collected. Thus, there were 20 egg samples per time per egg part. After the last egg collection in May 2017, at 60 weeks of age, the samples (20 samples) of each egg source per part were pooled over time to finally present five samples per source of eggs per egg part albumen, yolk, and albumen + yolk, and thus used for protein and amino acids analyses.

2.2. Measurements

2.2.1. Determination of Crude Protein

The protein content of eggs (yolk, albumen and yolk with albumen) was determined following method number 954.01, the principle of the Kjeldahl method [14]. One gram of sample was digested using 15 mL sulphuric acid (96%), utilizing an electrically heated block digester. The resultant digest was made alkaline by dilution with 50 mL of 40% sodium hydroxide. Thereafter the diluted sample was rapidly steam distilled for ammonia into 25 mL 4% boric acid. The sample was then manually titrated with 0.2 N hydrochloric acid. The protein content was estimated using 6.25 as a conversion factor of nitrogen to crude protein. Triplicate samples were analyzed, and the results are expressed as g/100 g dry matter basis of a sample [14].

2.2.2. Determination of Amino Acids

The egg samples were dried using method number 934.01, and defatted using method number 920.39, as per reference [14]. A part of each sample, 0.2 g was hydrolyzed with 6 N HCl (10 mL) in a sealed tube and then heated in an oven at 100 °C for 24 h [15]. For the analysis of methionine and cysteine, samples were oxidized by performic acid before hydrolysis [16]. The resulting solution was brought to 25 mL with de-ionized water. After filtration, 5 mL of hydrolysate was evaporated until free from HCl. Then, the residue was dissolved in a diluting citrate buffer. The amino acids were measured using an Automatic Amino Acid Analyzer model AAA400 (Ingos Ltd., K Nouzovu 2090, 14316 Prague 4, Czech). The column was filled with Resin material and Ninhydrin reagent. The separation of amino acids depends on using different gradient pH buffers. Acid hydrolysis of samples was carried out following [15,17]. Tryptophan was determined, as reported by [18] and modified by [19].

2.2.3. Predicted Protein Quality and Amino Acids Ratios

The predicted protein efficiency ratio (P-PER) of the different sources of eggs was estimated from their amino acid content, according to [20]:

$$\text{P-PER} = -0.468 + 0.454 \,(\text{Leucine}) - 0.105 \,(\text{Tyrosine})$$

2.2.4. Amino Acids Ratios of Different Egg Sources

The quality of the amino acids was determined by estimating the ratio of amino acids in the egg samples compared with the requirements stated as a ratio [21]. The Amino Acid Score (AAS) was then predicted by applying the formula in reference [22].

The essential amino acid to TAAs ratio and the cystine to sulfur amino acids ratio were, thus calculated. The EAAs/non-essential amino acids (NEAAs) ratio was also calculated. The total of the aromatic amino acids was also estimated, and the ratio of aromatic amino acids to TAAs was calculated.

Tyrosine + tryptophan to TAAs and total aromatic amino acids were also calculated as indices of the antioxidant property of eggs [4].

2.3. Statistical Analysis

The Data were evaluated using the one-way ANOVA of SAS software program [23] according to the following model:

$$Y_{ij} = \mu + A_i + e_{ij},$$

where: μ = general mean; A_i: influence of the source of the egg; e_{ij}: random error. The same model was used for the evaluation of amino acid patterns of different parts of the eggs.

The percentage of data were normalized by applying transformations to arcsin before running the analysis. Means differences were compared using the Student-Newman Keuls test ($p \leq 0.05$).

3. Results and Discussion

3.1. Growth Performance Crude Protein and Amino Acids Pattern of Whole Egg

The content of the crude protein and amino acids in the whole edible part of the egg is presented in Table 1. Crude protein of whole eggs was not significantly different among different egg sources. The results are in agreement with those recently published by Secci et al. [24,25] Arginine, lysine, threonine, and proline were significantly greater in eggs from sources C and D than in those of A, B, but phenylalanine + tyrosine (aromatic amino acids along with histidine and tryptophan) showed the contrary trend. However, the difference between sources A and B in these amino acids was not significant. Tyrosine is conditional NEAAs and an important for photosynthesis and signal transduction processes [26]. Phenylketonuria is a genetic disease in which the tyrosine becomes EAAs because tyrosine cannot be synthesized from phenylalanine [26]. The EAAs/NEAAs were significantly greater for sources A, B, and then source D, and the latter was greater than source C.

Table 1. Crude protein and amino acid patterns of whole edible parts of eggs of different sources in the retail market.

Crude Protein (g/100 g)/Amino Acid (mg/100 g)	Source of Eggs				Statistical Analysis	
	A	B	C	D	RMSE	*p*-Value
Crude protein	14.03	14.02	13.77	13.88	0.395	0.721
Arginine	769.0 [b]	693.3 [b]	915.3 [a]	882.3 [a]	61.65	0.001
Histidine	296.8 [bc]	279.0 [c]	315.8 [ab]	331.8 [a]	12.94	0.001
Isoleucine	698.3 [b]	720.0 [a]	641.8 [c]	613.3 [d]	10.82	0.001
Leucine	1100.0	1081.3	1128.0	1057.8	36.19	0.097
Lysine	923.0 [b]	870.3 [c]	1019.3 [a]	1012.8 [a]	19.21	0.001
Methionine	394.3 [b]	434.3 [a]	301.8 [c]	285.0 [c]	16.87	0.001
Methionine + cysteine	554.5 [b]	627.5 [a]	513.5 [c]	490.0 [c]	30.01	0.002
Phenylalanine	679.8 [b]	744.5 [a]	537.8 [d]	618.3 [c]	32.51	0.001
Phenylalanine + tyrosine	1199.0 [a]	1237.8 [a]	1099.0 [b]	1137.8 [b]	30.01	0.002
Threonine	613.0 [b]	578.3 [b]	674.8 [a]	699.5 [a]	23.48	0.001
Tryptophan	147.0	145.5	132.0	148.0	9.04	0.094
Valine	781.8 [b]	815.3 [a]	707.8 [c]	673.8 [d]	15.91	0.001
EAAs	5634.0 [b]	5668.3 [a]	5458.0 [c]	5439.5 [c]	106.4	0.021
Alanine	713.0 [a]	737.3 [a]	651.0 [b]	558.0 [c]	20.89	0.001
Aspartic acid	1295.8 [a]	1310.5 [a]	1260.0 [ab]	1218.5 [b]	36.66	0.019
Glutamic acid	1687.0 [a]	1713.5 [a]	1642.8[a]	1555.3 [b]	37.32	0.001
Glycine	426.0 [ab]	441.3 [ab]	381.8 [b]	473.8 [a]	32.42	0.014
Proline	507.0 [b]	493.3 [b]	523.8 [a]	535.8 [a]	10.75	0.001
Serine	956.3 [c]	881.5 [d]	1099.8 [a]	1055.3 [b]	23.54	0.001
NEAAs	7033.5	6957.5	7247.0	7001.8	167.9	0.129
EAAs/NEAAs	0.801 [a]	0.815 [a]	0.753 [c]	0.777 [b]	0.0214	0.011

[a–d] means with varying superscripts differ significantly ($p < 0.05$); RMSE, Root mean square error; *p* value, probability level; EAAs, total essential amino acids; NEAAs, total non-essential amino acids; EAAs/NEAAs, total essential amino acids/total non-essential amino acids ratio.

Glycine, am NEAA, was higher in the D source than in the C source. Isoleucine, methionine, and methionine + cystine were significantly higher in the B source than in other sources. In addition, source A showed greater isoleucine, methionine, and methionine + cysteine than did sources C and D, although source C exhibited higher isoleucine than did source D.

Phenylalanine, an EAAs for the biosynthesis of norepinephrine and epinephrine, and valine, which is important amino acids for maintaining muscles, as well as for the regulation of the immune system were significantly higher in source B than in other sources. The other sources exhibited significant differences, showing a trend of B > A > D > C for phenylalanine and B > A > C > D for valine. Valine, along with leucine and isoleucine, are branched-chain amino acids and represented about two-thirds of amino acids in the body protein [27]. In whole edible parts of eggs, branched-chain amino acids amounted to ~45% of EAAs. Alanine, the 4th principal NEAAs, was similar in sources A and B, which was significantly greater than in sources C and D. The latter sources showed significant differences between them, with source C having higher. A similar trend was found for aspartic acid, but the differences between source C and other sources were not significant.

Glutamic acid, the chief NEAA, was significantly greater in sources A, B, and C than in source D. Serine, which is important in biosynthesis of purine and pyrimidines, was significantly higher in source C than in the other sources, which showed significant differences among them in the following order: D > A > B. These findings reflected to some extent difference in amino acids in egg albumen and egg yolk and indicated that eggs from various sources have different amino acid patterns, which could positively affect the nutritional values and consumer preference [4,27].

3.2. Crude Protein and Amino Acid Patterns of Albumen

Table 2 reports the crude protein and amino acid composition of the egg albumen. It was observed that the differences among the egg sources were significant for crude protein with sources A and B had usually higher values than sources C and D. Egg source shows a significant effect on some amino acids except for arginine, histidine (precursor for histamine and carnosine synthesis), lysine, threonine, tryptophan (important EAAs along with other EAAs in biosynthesis of protein), valine, alanine, aspartic, serine, and EAAs/NEAAs.

Table 2. Crude protein and amino acid patterns of albumen of eggs of different sources in the retail market.

Crude Protein (g/100 g)/Amino Acid (mg/100 g)	Source of Eggs				Statistical Analysis	
	A	B	C	D	RMSE	*p*-Value
Crude protein	12.33 [a]	12.45 [a]	11.73 [b]	11.83 [b]	0.536	0.036
Arginine	543.5	544.5	516.5	497.8	38.16	0.293
Histidine	221.8	222.8	194.8	198.8	19.16	0.130
Isoleucine	566.5	567.5	567.0	503.3	31.44	0.035
Leucine	845.0 [a]	845.8 [a]	818.3 [ab]	767.8 [b]	35.79	0.033
Lysine	745.8	746.8	743.5	738.0	25.48	0.962
Methionine	342.5 [a]	343.5 [a]	315.3 [a]	262.5 [b]	17.28	0.001
Methionine + Cystine	497.3 [a]	499.3 [a]	443.5 [b]	381.5 [c]	25.30	0.001
Phenylalanine	583.8 [a]	584.8 [a]	556.8 [ab]	519.5 [b]	25.24	0.012
Phenyalanine + Tyrosine	972.6 [a]	974.3 [a]	918.0 [b]	868.0 [c]	30.22	0.001
Threonine	454.5	455.5	427.3	424.0	27.11	0.251
Tryptophan	118.0	118.8	117.8	132.0	15.53	0.354
Valine	613.0	588.5	585.8	564.8	62.73	0.758
EAAs	4489.8 [a]	4473.5 [a]	4320.8 [b]	4110.8 [c]	76.45	0.001
Alanine	577.8	578.5	552.0	581.3	47.12	0.799
Aspartic acid	1023.5	1024.5	996.3	1035.3	29.22	0.325
Glutamic acid	1336.5 [a]	1337.8 [a]	1309.8 [a]	1066.5 [b]	34.65	0.001
Glycine	347.8 [b]	348.8 [b]	321.0 [b]	598.0 [a]	22.65	0.001
Proline	388.5 [a]	389.5 [a]	361.3 [b]	328.8 [c]	14.57	0.003
Serine	689.8	690.8	663.0	662.3	18.41	0.077
NEAAs	5450.8 [a]	5459.7 [a]	5209.8 [b]	5137.5 [b]	130.8	0.009
EAAs/NEAAs	0.824	0.819	0.829	0.800	0.0261	0.423

[a–c] means with varying superscripts differ significantly (*p* < 0.05); RMSE, Root mean square error; *p* value, probability level; EAAs, total essential amino acids; NEAAs, total non-essential amino acids; EAAs/NEAAs, total essential amino acids/total non-essential amino acids ratio.

Leucine and phenylalanine were similar in sources A and B and markedly (*p* ≤ 0.05) higher than in source D, while the C source did not significantly differ from the other sources. Methionine and glutamic acid were significantly greater of sources A, B, and C than that of source D. Methionine + cystine, phenylalanine + tyrosine, EAAs, and proline were similar in sources A and B and significantly lower in sources C and D. Source D exhibited significantly lower values than source C. Glycine was significantly higher in source D than in the other sources. Non-essential amino acids were similar in sources A and B and significantly lower in sources C and D, with no significant differences between the latter sources.

3.3. Crude Protein and Amino Acid Patterns of Egg Yolk

Table 3 shows the findings related to crude protein and amino acid profile of egg yolk. The difference in crude protein was not significant among various sources. The observed differences between the sources of eggs were significant ($p \leq 0.05$) for some amino acids except for arginine, lysine, phenylalanine + tyrosine, threonine, tryptophan, valine, serine, and EAAs/NEAAs ratio.

Table 3. Crude protein and amino acid patterns of the yolk of eggs of different sources in the retail market.

Crude Protein (g/100 g)/Amino Acid (mg/100 g)	Source of Eggs				Statistical Analysis	
	A	B	C	D	RMSE	*p*-Value
Crude protein	15.97	16.12	15.83	16.04	0.796	0.873
Arginine	1156.3	1150.0	1133.8	1131.3	24.78	0.440
Histidine	418.3 [a]	412.0 [a]	393.3 [ab]	375.5 [b]	16.68	0.016
Isoleucine	819.8 [a]	813.0 [a]	795.3 [a]	712.5 [b]	19.11	0.001
Leucine	1418.0 [a]	1411.8 [a]	1393.3 [a]	1317.5 [b]	15.49	0.001
Lysine	1284.3	1278.0	1236.8	1243.5	33.96	0.174
Methionine	401.0 [a]	394.8 [a]	353.5 [b]	348.8 [b]	19.27	0.004
Methionine + cysteine	531.8 [c]	678.8 [a]	598.8 [b]	624.5 [ab]	36.01	0.001
Phenylalanine	691.8 [a]	685.0 [a]	655.8 [b]	644.3 [b]	17.53	0.008
Phenylalanine + tyrosine	1412.5	1399.0	1363.0	1342.0	36.99	0.073
Threonine	860.3	854.0	822.3	888.0	31.62	0.082
Tryptophan	185.5	218.5	263.8	208.8	35.89	0.059
Valine	860.8	854.3	854.8	878.0	58.07	0.947
EAAs	6939.8 [a]	6921.5 [a]	6768.0 [ab]	6628.5 [b]	112.7	0.008
Alanine	831.0 [a]	824.8 [a]	788.5 [ab]	761.3 [b]	22.40	0.003
Aspartic acid	1580.8 [a]	1574.3 [a]	1547.8 [a]	1492.3 [b]	33.99	0.014
Glutamic acid	2051.8 [a]	2045.0 [a]	2024.3 [a]	1958.0 [b]	27.41	0.002
Glycine	499.8 [b]	493.3 [b]	464.8 [c]	537.8 [a]	14.36	0.002
Proline	674.3 [a]	668.0 [a]	632.5 [b]	626.8 [b]	20.74	0.015
Serine	1343.5	1377.0	1335.8	1315.3	38.71	0.212
NEAAs	8988.5 [ab]	9131.0 [a]	8864.5 [b]	8811.5 [b]	115.1	0.011
EAAs/NEAAs	0.772	0.758	0.763	0.752	0.0169	0.481

[a–c] Means with varying superscripts differ significantly ($p < 0.05$); RMSE, Root mean square error; p value, probability level; EAAs, total essential amino acids; NEAAs, total non-essential amino acids; EAAs/NEAAs, total essential amino acids/total non-essential amino acids ratio.

Isoleucine, leucine, aspartic acid, and glutamic acid were similar in sources A, B, and C, but significantly lower in source D. Methionine, phenylalanine, and proline were similar in eggs from sources A and B and in sources C and D, although differences between the former and the latter groups were remarkable ($p \leq 0.05$) in favor of the former group.

Methionine + cystine was significantly higher in source B than in sources A and C, while the C source had higher values than source A. Source D also had higher values than that of source A but did not significantly differ from sources B and C. Essential amino acids and alanine were significantly lower in source D than in sources A and B, while source C had intermediate values. In contrast, glycine was significantly higher in source D than in other sources. In addition, sources A and B showed similar values for glycine, which were greater than in source C. Glycine is a precursor to proteins and for the biosynthesis of collagen and considered the simplest amino acid that can be formed from serine and this reaction is reversible [28].

Essential amino acids were higher ($p \leq 0.05$) in egg yolk from sources A and B than in source D, while source C exhibited intermediate values. Source B had greater ($p \leq 0.05$) NEAAs than did sources C and D, while source A had intermediate values.

In general, the results indicated that EAAs were the highest in source A and B of both egg albumen and egg yolk. However, the EAAs/NEAAs ratio was not different among various sources of eggs. Leucine, followed by lysine and arginine, were the most abundant EAAs. Glutamic acid and aspartic acids were the predominant NEAAs in the whole edible parts of eggs and the egg yolk. The principal amino acids in egg albumen were leucine, lysine, and valine, with the glutamic acid and aspartic acid were the chief NEAAs. These results are similar to the results reported by [9], the sum of phenylalanine and tyrosine was found to have the highest values, whereas tryptophan displayed the lowest values in different parts of eggs. According to [3,5,29], animal protein is better digested and contains more EAAs and greater available AAs than vegetable protein. Besides, the protein quality of animal protein was higher than that of leguminous seeds [30,31].

3.4. Crude Protein and Amino Acid Patterns of Different Eggs Parts

Table 4. displays the crude protein and amino acid profile of different egg parts compared to the recommended daily allowances. Results showed variations ($p \leq 0.05$) among different parts, e.g., the whole edible parts of eggs (albumen + yolk), albumen, and yolk. The yolk had a greater ($p \leq 0.05$) concentration of crude protein and amino acids than the whole edible parts of eggs, which in turn had greater values ($p \leq 0.05$) than the albumen. However, for the EAAs/NEAAs, the trend was opposite, being highest in the albumen and lowest in the yolk. The values for the whole edible parts of the eggs were intermediate.

Table 4. Amino acid patterns of different egg parts in the retail market compared to recommended daily allowance standard Food and Agriculture Organization (FAO)/World Health Organization (WHO) [22] values of adults (70 kg).

Crude Protein (g/100 g)/Amino Acid (mg/100 g)	Egg Part			Statistical Analysis		RDA, mg/kg Day
	Whole Eggs	**Albumen**	**Yolk**	**RMSE**	***p*-Value**	
Crude protein	13.95 [b] (26.6%)	12.42 [c] (23.7%)	16.00 [a] (30.5%)	3.32	0.054	0.75 g/kg per day
Arginine	814.9 [b]	525.6 [c]	1142.8 [a]	43.93	0.001	–
Histidine	312.3[b] (44.6%)	209.5[c] (29.9%)	399.8 [a] (57.1%)	16.33	0.001	10 (700 mg/day)
Isoleucine	668.3 [b] (47.7%)	551.1 [c] (39.4%)	785.1 [a] (56.1%)	22.63	0.001	20 (1400 mg/day)
Leucine	1091.8 [b] (40.0%)	819.2 [c] (30.0%)	1385.1 [a] (70.7%)	31.42	0.001	39 (2730 mg/day)
Lysine	956.3 [b] (45.5%)	743.5 [c] (35.4%)	1260.6 [a] (60.0%)	27.13	0.001	30 (2100 mg/day)
Methionine	353.8 [b] (48.6%)	315.9 [c] (43.4)	374.5 [a] (51.4)	18.94	0.001	10.4 (728 mg/day)
Methionine + cysteine	546.4 [b] (52.0%)	455.4 [c] (43.4%)	608.4 [a] (57.9%)	29.99	0.001	15 (1050 mg/day)
Phenylalanine	645.1 [b]	516.2 [c]	669.2 [a]+	27.64	0.001	–
Phenyalanine + tyrosine	1168.4 [b] (66.8%)	933.1 [c] (53.3%)	1379.1 [a] (78.8%)	34.73	0.001	25 (1750 mg/day)
Threonine	614.4 [b] (58.5%)	440.3 [c] (41.9%)	856.1 [a] (81.5%)	27.87	0.001	15 (1050 mg/day)
Tryptophan	143.1 [b] (51.1%)	120.1 [c] (42.9%)	219.1 [a] (78.3%)	23.13	0.001	4 (280 mg/day)
Valine	744.6 [b] (40.9%)	588.0 [c] (32.3%)	864.7 [a] (47.5%)	49.06	0.001	26 (1820 mg/day)
EAAs	5549.9 [b]	4348.7 [c]	6814.4 [a]	104.2	0.001	–
Alanine	664.8 [b]	572.4 [c]	801.4 [a]	31.78	0.001	–
Aspartic acid	1271.2 [b]	1019.9 [c]	1548.8 [a]	33.79	0.001	–
Glutamic acid	1649.6 [b]	1262.6 [c]	2019.8 [a]	33.52	0.001	–
Glycine	430.7 [b]	378.8 [c]	498.9 [a]	23.81	0.001	–
Proline	514.9 [b]	367.1 [c]	650.4 [a]	15.51	0.001	–
Serine	998.2 [b]	676.4 [c]	1342.9 [a]	28.39	0.001	–
NEAAS	7059.9 [b]	5314.4 [c]	8948.9 [a]	136.2	0.001	–
EAAs/NEAAs	0.786 [b]	0.818 [a]	0.761 [c]	0.022	0.001	–

[a–c] Means with varying superscripts differ significantly ($p < 0.05$); RMSE, Root mean square error; p value, probability level; FAO/WHO, Food and Agriculture Organization/World Health Organization; RDA, recommended daily allowance; EAAs, total essential amino acids; NEAAs, total non-essential amino acids; EAAs/NEAAs, total essential amino acids/total non-essential amino acids ratio.

Eggs are well-known animal proteins with high protein/amino acid quality that fulfil the human RDA better than vegetable protein sources, and EAAs deficiency can depress growth and lead to many health problems [2,26]. In this regard, different egg sources showed different patterns in phenylalanine, methionine, lysine, isoleucine, valine, and threonine in the whole edible parts of eggs. Differences in these amino acids were 38.5, 52.5, 16.4, 17.5, 20.9, and 21.1%, respectively. Similar trends were found in the albumen and yolk of eggs. These results agree with those reported by [6,11], who reported that raw regular Kampung eggs (eggs that produced by village hens that allowed to roam in a building, room or open area that includes nest space and perches, but they did have access to the outdoors) and nutrient-enriched eggs have different amino acid patterns, with lysine, leucine, phenylalanine, and valine having the highest concentrations in eggs. The difference in protein and AA patterns of various eggs could be attributed to the impact of the diet composition on the level of CP and amino acids. Similar results were reported by [1,5–7].

Results indicate that the yolk fulfils the highest percentage of the RDA for adults (70 kg) between 19–30 years of age, followed by the whole egg and albumen. For the egg yolk, the highest percentages were for threonine (81.5%), phenylalanine + tyrosine (78.8%), and tryptophan (78.3%) while the lowest was for methionine (~51.4%). The corresponding values for the whole edible parts of eggs and albumen were 58.5% and 41.9% for threonine, respectively, 66.8 and 53.3% phenylalanine + tyrosine, respectively, and 51.1% and 42.9% for tryptophan, respectively.

From the results obtained for amino acid scores, phenylalanine + tyrosine was the most limiting, the amino acids that were deficient in protein and when supplemented resulted in the highest response, the amino acid in all sources of eggs, followed by histidine and lysine [32]. Limiting EAAs for humans that can't be synthesized at an adequate amount by human cells are phenylalanine, methionine, lysine, leucine, isoleucine, valine, threonine, and tryptophan [11,33]. In addition, histidine, arginine, cystine and tyrosine are regarded as EAAs for infants and growing children [34].

From a biochemical point of view, differences in amino acid patterns in egg proteins/amino acids could be attributed to the type and percentage of protein in the albumen and yolk, e.g., ovalbumin, ovotransferrin, lysozyme, ovomucoid, ovomucin, and immunoglobulin Y [2,35]. The ovalbumin is the major protein that amounts to greater than 50% of the protein in the albumen of eggs [36]. In this regard, lysine (509 mg/g) was the greatest amino acid in eggs, whereas cysteine (128 mg/g) was the lowest [37]. Arginine, serine, cysteine and iso-leucine amino acids in eggs were higher than those in soy protein, beef, casein, wheat flour, and egg white [38].

3.5. Amino Acids Ratios

Table 5 shows the data for amino acid scores, aromatic amino acids, and essential amino acids to TAAs, antioxidant amino acid protein of eggs from different sources. The results indicate that the amino acid scores for histidine and threonine were significantly different among egg sources, with the highest values from source D and the lowest from source B. Differences in amino acid scores were also significant for the amino acids valine, isoleucine, leucine lysine, methionine + cystine, and phenylalanine + tyrosine.

The present results show that source B had greater ($p \leq 0.05$) amino acid scores for valine, isoleucine, and methionine + cystine than did sources C and D. In addition, source C had greater scores ($p \leq 0.05$) for leucine, lysine, phenylalanine + tyrosine than did sources B and D, A and B and B, respectively. The results indicate that source C had the lowest quality of protein in terms of total aromatic amino acids (TAAAs) indices, but when tyrosine + tryptophan was considered as the main contributor to the antioxidant properties of eggs [4], source C had the highest value. Differences between the highest and lowest values for TAAAs and tyrosine + tryptophan were 7.4 and 8.5%, respectively.

The EAAs/TAAs and cystine/sulfur amino acids were also higher in source C and D compared to the resources A and B; the differences in these criteria reached 4.7 and 45%, respectively, showing higher variability in cystine/sulfur amino acids as indices. The results also show greater amino acid scores in sources A and B, particularly for valine, isoleucine, and methionine + cystine. In contrast, the P-PER based on the content of leucine and tyrosine showed insignificant differences among different egg sources, indicating that this measurement is not a suitable parameter for protein quality evaluation of eggs, as the amino acids involved in the equation of calculation did not reflect the abundant amino acids= in eggs, suggesting a need for precise equation.

The present results are in line with those cited by Adeyeye [6], who observed a TAAs of 10.0 and 10.1 mg/g of dry yolk in corn- and wheat-based diets, respectively. Moreover, differences in glutamic acid, glutamine, glycine, arginine, isoleucine, and ornithine were significant among dietary compositions. Similar to the present findings, EAAs accounted for about 46–47%, while TAAAs accounted for about 11% of the TAAs; arginine was the most copious amino acid, followed by glutamic acid and lysine [1,4]. Differences in the ratios of the amino acids of eggs from various sources could be due to the strains/breeds of layers and poultry species [5,6,39], dietary composition, cooking method, storage time, environment, [11,40,41] and egg parts [12,42,43].

The total aromatic amino acids, (TAAAs)/TAAs, as the index of antioxidant property of eggs [6], and EAAs/TAAs as an index of protein quality [42,43] were significantly different among various sources of eggs, with sources A and B having greater ($p \leq 0.05$) values than source C. In addition, TAAAs/TAAs were significantly greater in source D than in source C. Tyrosine + tryptophan, as an absolute value or relative to TAAAs or TAAs, was significantly different among different egg sources. The C sources had greater values ($p \leq 0.05$) than B sources. In addition, sources A and B showed similar values as source C except for tyrosine + tryptophan/TAAAs, which was higher ($p \leq 0.05$) in source C. No significant ($p \geq 0.05$) variations in the predicated protein efficiency ratio (P-PER) ratio among various egg sources. The cystine/sulfur amino acids ratio was significantly higher for eggs of source C.

Table 5. Amino acids score (%), aromatic amino acids, essential amino acids to the total amino acid ratio. Antioxidant amino acids of the whole edible egg parts (albumen + yolk) of different sources compared to standard Food and Agriculture Organization (FAO)/World Health Organization (WHO) [22] values.

Amino Acid (mg/100 g)	Source of Eggs				FAO/WHO, 2007 (g/100 g Protein)	Statistical Analysis	
	A	B	C	D		RMSE	*p*-Value
Amino Acid Score							
Histidine	95.3 [ab]	89.5 [b]	101.4 [ab]	106.5 [a]	1.5	3.26	0.001
Valine	136.2 [ab]	141.9 [a]	123.3 [b]	117.4 [b]	3.9	4.16	0.025
Isoleucine	152.0 [a]	156.8 [a]	139.8 [b]	133.5 [b]	3.0	5.14	0.004
Leucine	101.6 [a]	99.9 [b]	104.2 [a]	97.7 [b]	5.9	2.93	0.005
Lysine	97.0 [h]	91.5 [h]	107.2 [a]	106.5 [a]	4.5	3.07	0.002
Methionine + cysteine	292.4 [ab]	301.9 [a]	268.3 [b]	277.6 [b]	2.2	5.29	0.001
Phenylalanine + tyrosine	50.2 [ab]	47.7 [b]	54.3 [a]	50.3 [ab]	3.8	1.56	0.001
Threonine	109.9 [b]	105.5 [b]	121.1 [a]	125.5 [a]	2.3	4.56	0.003
Protein quality indices							
TAAAs	1643 [a]	1662 [a]	1547 [b]	1618 [ab]	—	46.5	0.024
TAAAs/TAAs ratio	0.1297 [a]	0.1317 [a]	0.1217 [b]	0.1300 [a]	—	0.003	0.001
Tyrosine + tryptophan	666.3 [ab]	638.8 [b]	693.3 [a]	667.5 [ab]	—	22.66	0.042
Tyrosine + tryptophan/TAAs ratio	0.0526 [a]	0.0506 [b]	0.0546 [a]	0.0536 [a]	—	0.001	0.001
Tyrosine + tryptophan/TAAAs ratio	0.406 [b]	0.384c	0.448 [a]	0.413 [b]	—	0.067	0.003
EAAs/TAAs ratio	0.445 [a]	0.449 [a]	0.429 [b]	0.437 [ab]	—	0.007	0.009
P-PER	3.15	3.10	3.21	2.99	—	1.53	0.196
Cystine/sulfur amino acids ratio	0.289 [b]	0.308 [b]	0.412 [a]	0.419 [a]	—	0.015	0.001

[a–c] means with varying superscripts differ significantly ($p < 0.05$); RMSE, Root mean square error; *p*-value, probability level; FAO/WHO, Food and Agriculture Organization/World Health Organization; EAAs, total essential amino acids; TAAAs, Total aromatic amino acids, TAAAs/TAAs, Total aromatic amino acids/total amino acids, EAAs/TAAs, essential amino acids/total amino acids, P-PER, predicated protein efficiency ratio.

4. Conclusions

In conclusion, the investigated eggs showed different EAAs/NEAAs ratio and antioxidant amino acids, indicating a potential for enhancing the nutritional values and extending the shelf life of eggs by different strategic nutritional approaches such as increasing antioxidants, and essential amino acid content of eggs.

Author Contributions: Y.A.A. conducted the experimental design, data collection, and first draft of the manuscript; M.A.A.-H. carried out the data collection and contribution to the experimental set-up, M.A.K. and M.H.S., lab work, and reading the manuscript; and Y.A.A. undertook the statistical analyses and proofreading of the manuscript. All authors have read and agreed to the published version of the manuscript.

Funding: This work was supported by the Deanship of Scientific Research (DSR), King Abdulaziz University, Jeddah, under grant No (D-156-155-1438 H).

Acknowledgments: The authors express their sincere gratitude and appreciation to DSR, King Abdulaziz University for technical and finical support.

Conflicts of Interest: The authors declare no conflicts of interest.

References

1. Li, F.; Yang, Y.; Yang, X.; Shan, M.; Gao, X.; Zhang, Y.; Hu, J.; Shan, A. Dietary intake of broiler breeder hens during the laying period affects amino acid and fatty acid profiles in eggs. *Rev. Bras. Zootec.* **2019**, *48*, e20180292. [CrossRef]
2. Bradley, F.A.; King, A. Egg Basics for the Consumer: Packaging, Storage and Nutritional Information. 2000. Available online: http://anrcatalog.ucdavis.edu/pdf/8154.pdf (accessed on 12 January 2018).
3. Al-Harthi, M.A.; El-Deek, A.A.; Attia, Y.A. Impacts of dried whole eggs on productive performance, quality of fresh and stored eggs, reproductive organs and lipid metabolism of laying hens. *Br. Poult. Sci.* **2011**, *52*, 333–344. [CrossRef] [PubMed]
4. Nimalaratne, C.D.; Lopes-Lutz, A.; Schieber, A.; Wu, J. Free aromatic amino acids in egg yolk show antioxidant properties. *Food Chem.* **2011**, *129*, 155–161. [CrossRef]
5. Mori, H.; Takaya, M.; Nishimura, K.; Goto, T. Breed and feed affect amino acid contents of egg yolk and eggshell color in chickens. *Poult. Sci.* **2020**, *99*, 172–178. [CrossRef] [PubMed]
6. Adeyeye, E.I. The comparison of the amino acids patterns of whole eggs of duck, francolin and turkey consumed in Nigeria. *Glob. J. Sci. Front. Res. Chem.* **2013**, *13*. [CrossRef]
7. Attia, Y.A.; Al-Harthi, M.A.; Shiboob, M.M. Evaluation of quality and nutrient contents of table eggs from different sources in the retail market. *Ital. J. Anim. Sci.* **2014**, *13*, 369–376. [CrossRef]
8. Carvalho, T.S.M.; Sousa, L.S.; Nogueira, F.A.; Vaz, D.P.; Saldanha, M.M.; Triginelli, M.V.; Pinto, M.F.V.S.; Baiao, N.C.; Lara, L.J.C. Digestible methionine + cysteine in the diet of commercial layers and its influence on the performance, quality, and amino acid profile of eggs and economic evaluation. *Poult. Sci.* **2018**, *97*, 2044–2052. [CrossRef]
9. Meza-Espinoza, L.; Sáyago-Ayerdi, S.G.; de Lourdes García-Magaña, M.; Tovar-Pérez, E.G.; Yahia, E.M.; Vallejo-Cordoba, B.; González-Córdova, A.F.; Hernández-Mendoza, A.; Montalvo-González, E. Antioxidant capacity of egg, milk and soy protein hydrolysates and biopeptides produced by *Bromelia penguin* and *Bromelia karatas*-derived proteases. *Emir. J. Food Agric.* **2018**, *30*, 122–130.
10. Liu, J.; Zhang, D.; Zhu, Y.; Wang, Y.; He, S.; Zhang, T. Enhancing the in vitro antioxidants capacities via the interaction of amino acids. *Emir. J. Food Agric.* **2018**, *30*, 224–231.
11. Ismail, M.; Mariod, A.; Pin, S.S. Effects of preparation methods on protein and amino acid contents of various eggs available in Malaysian local markets. *Acta Sci. Pol. Technol. Aliment.* **2013**, *12*, 21.
12. Msarah, M.; Ahmed, A.A. Protein digestibility and amino acid content of Malaysian local egg protein prepared by different methods. *Environ. Ecosyst. Sci.* **2018**, *2*, 7–9. [CrossRef]
13. Roe, M.; Pinchen, H.; Church, S.; Finglas, P. *Nutrient Analyses of Eggs*; Analytical Report Revised Version; Institute of Food Research: Norwich, UK, 2013.
14. Association of Official Analytical Chemists (AOAC). *Official Methods of Analyses*, 16th ed.; Association of Official Analytical Chemists: Washington, DC, USA, 1999.
15. Moore, S.; Spackman, D.H.; Stein, W.H. Chromatography of amino acids on sulfonated polystyrene resins. *Anal. Chem.* **1958**, *30*, 1185–1190. [CrossRef]
16. Moore, S. On the determination of cysteine as cystine acid. *J. Biol. Chem.* **1963**, *238*, 235–237.
17. Csomos, E.; Simon-Sarkadi, L. Characterisation of tokaj wines based on free amino acid and biogenic amine using ion-exchange chromatography. *Chromatographia* **2002**, *56*, 185–188. [CrossRef]
18. Concon, J.M. Rapid and simple method for the determination of tryptophan in cereal grains. *Anal. Biochem.* **1975**, *67*, 206–219. [CrossRef]
19. Ogunsua, A.O. Amino acid determination in conophor nut by gas-liquid chromatography. *Food Chem.* **1988**, *28*, 287–298. [CrossRef]
20. Alsmeyer, R.H.; Cunigham, A.E.; Hapich, M.I. Equations predicted PER from amino acid analysis. *Food Technol.* **1974**, *28*, 34–38.
21. Oshodi, A.A.; Esuoso, K.O.; Akintayo, E.T. Proximate and amino acid composition of some under-utilized Nigerian legume flour and protein concentrate. *Riv. Ital. Sustanze Grasse* **1998**, *75*, 409–412.
22. Food and Agriculture Organization (FAO); World Health Organization (WHO); United Nations University (UNU). *Protein and Amino Acid Requirements in Human Nutrition*; Report of a Joint WHO/FAO/UNU Expert Consultation; WHO Technical Report Series No. 935; World Health Organization (WHO): Geneva, Switzerland, 2007.

23. SAS Institute. *User's Guide*; SAS institute Inc.: Cary, NC, USA, 2009.

24. Secci, G.; Bovera, F.; Parisi, G.; Moniello, G. Quality of eggs and albumen technological properties as affected by *Hermetia illucens* larvae meal in hens' diet and hen age. *Animals* **2020**, *10*, 81. [CrossRef]

25. Secci, G.; Bovera, F.; Nizza, S.; Baronti, N.; Gasco, L.; Conte, G.; Serra, A.; Bonelli, A.; Parisi, G. Quality of eggs from Lohmann Brown Classic laying hens fed black soldier fly meal as substitute for soya bean. *Animal* **2018**, *12*, 2191–2197. [CrossRef]

26. Whitney, E.; Rolfes, S.R. *Understanding Nutrition*, 14th ed.; Cengage Learning: Boston, MA, USA, 2015.

27. King, M.W. Amino Acids and Proteins: Introduction to Amino Acid Metabolism. The Medical Biochemistry Page. 2015. Available online: http://themedicalbiochemistrypage.org/amino-acid-metabolism.php (accessed on 20 March 2017).

28. Nelson, D.L.; Cox, M.M. *Principles of Biochemistry*, 4th ed.; Nelson, D.L., Cox, M.C., Freeman, W.H., Eds.; Freeman and Company: New York, NY, USA, 2004; p. 127.

29. Sakanaka, S.; Kitahata, K.; Mitsuya, T.; Gutierrez, M.A. Protein quality determination of delipidated egg-yolk. *J. Food Comp. Anal.* **2000**, *13*, 773–781. [CrossRef]

30. Mahan, L.K.; Escott-Stump, S. *Food Nutrition and Diet Therapy*; Saunders Publication Co.: Philadelphia, PA, USA, 2004.

31. Baudoin, J.P.; Maquet, A. Improvement of protein and amino acid contents in seeds of food legumes: A case study in Phaseolus. *Biotechnol. Agron. Soc. Environ.* **1999**, *3*, 220–224.

32. Harper, A.E.; Yoshimura, N.N. Protein quality, amino balance, utilization and evaluation of diets containing amino acids astherapeutic agents. *Nutrition* **1993**, *9*, 460–469. [PubMed]

33. Young, V.R. Adult amino acid requirements, the case for a major revision in current recommendations. *J. Nutr.* **1994**, *124*, 1517–1523. [CrossRef] [PubMed]

34. Imura, K.; Okada, A. Amino acid metabolism in pediatric patients. *Nutrition* **1998**, *4*, 143–148. [CrossRef]

35. Belitz, H.D.; Grosch, W.; Schieberle, P. Eggs. In *Food Chemistry*; Springer Publishing Co.: New York, NY, USA, 2009; Chapter 11; pp. 546–561.

36. McWilliams, M. Eggs. In *Foods, Experimental Perspectives*; Prentice Hall: Upper Saddle River, NJ, USA, 2001; p. 357.

37. Sosulski, F.W.; Imafidon, G.I. Amino acid composition and nitrogen-to-protein conversion factors of animal and plant foods. *J. Agric. Food Chem.* **1990**, *38*, 1351–1356. [CrossRef]

38. Sarwar, G.; Christiansen, D.A.; Finlayson, A.T.; Friedman, M.; Hackler, L.R.; Mckenzie, S.L.; Pellet, P.L.; Tkachuk, R. Intra-and inter-laboratory variation in amino acids analyses. *J. Food Sci.* **1983**, *48*, 526–531. [CrossRef]

39. Cook, F.; Briggs, G.M. Nutritive value of eggs. In *Eggs Science and Technology*; Avi Publishing Co: Westport, CT, USA, 1997; p. 92.

40. Tee, E.S.; Noor, M.I.; Azudin, M.N.; Idris, K. *Nutrient Composition of Malaysian Foods*; Malaysian Food Composition Database Programme C/O; Institute for Medical Research: Kuala Lumpur, Malaysia, 1997.

41. Taş, N.G.; Gokmen, V. Profiling of the contents of amino acids, water-soluble vitamins, minerals, sugars and organic acids in Turkish hazelunt varieties. *Pol. J. Food Nutr. Sci.* **2018**, *68*, 223–234. [CrossRef]

42. National Institute of Health-Office of Dietary Supplements. *Dietary Reference Intakes for Energy, Carbohydrate: Fiber, Fat, Fatty Acids, Cholesterol, Protein, and Amino Acids RDA*; National Institute of Health-Office of Dietary Supplements: Bethesda, MD, USA, 2005.

43. Hester, K.L. *Eggs for Young America*; University Press of New England: Lebanon, NH, USA, 1997; Chapter 1.

Article

Evaluation of the Combined Use of Saccharomyces Cerevisiae and Aspergillus Oryzae with Phytase Fermentation Products on Growth, Inflammatory, and Intestinal Morphology in Broilers

Wen Yang. Chuang [1], Wei Chih. Lin [1], Yun Chen. Hsieh [1], Chung Ming. Huang [1], Shen Chang. Chang [2] and Tzu-Tai Lee [1,3,*]

[1] Department of Animal Science, National Chung Hsing University, Taichung 402, Taiwan; xssaazxssaaz@yahoo.com.tw (W.Y.C.); waynezi2227738@gmail.com (W.C.L.); richard840909@gmail.com (Y.C.H.); wxads508@gmail.com (C.M.H.)
[2] Kaohsiung Animal Propagation Station, Livestock Research Institute, Council of Agriculture, Pingtung 912, Taiwan; macawh@mail.tlri.gov.tw
[3] The iEGG and Animal Biotechnology Center, National Chung Hsing University, Taichung 402, Taiwan
* Correspondence: ttlee@dragon.nchu.edu.tw; Tel.: +886-4-22840366; Fax: +886-4-22860265

Received: 15 October 2019; Accepted: 21 November 2019; Published: 1 December 2019

Simple Summary: The stress and anti-nutrient effect caused by environmental problems and animal feed is an urgent problem in poultry production. As ancient probiotics, *Aspergillus oryzae* and *Saccharomyces cerevisiae* can effectively improve the immunity of animals. Furthermore, the anti-nutrient object, phytate, reduces nutrition absorption. Therefore, *S. cerevisiae* or *A. oryzae* with phytase co-fermentation may help solve these problems. Results show that the addition of a fermentation product can effectively reduce the inflammatory response and drop the number of harmful bacteria in the ileum of broilers. Among them, *A. oryzae* fermentation product has a better effect than *S. cerevisiae* fermentation product.

Abstract: *Saccharomyces cerevisiae* and *Aspergillus oryzae* are both ancient probiotic species traditionally used as microbes for brewing beer and soy sauce, respectively. This study investigated the effect of adding these two probiotics with phytase fermentation products to the broilers diet. Fermented products possess protease and cellulase, and the activities were 777.1 and 189.5 U/g dry matter (DM) on *S. cerevisiae* fermented products (SCFP) and 190 and 213.4 U/g DM on *A. oryzae* fermented products (AOFP), respectively. Liposaccharides stimulated PBMCs to produce nitric oxide to 120 μmol. Both SCFP and AOFP reduced lipopolysaccharides stimulated peripheral blood mononuclear cells (PBMCs) nitric oxide release to 40 and 60 μmol, respectively. Nevertheless, in an MTT (3-(4,5-dimethylthiazol-2-yl)-2,5-diphenyltetrazolium bromide) assay, SCFP and AOFP also increased the survival rate of lipopolysaccharides stimulated PBMCs by almost two-fold compared to the negative control. A total of 240 broilers were divided into four groups as Control, SCFP 0.1% (SCFP), SCFP 0.05% + AOFP 0.05% (SAFP), and AOFP 0.1% (AOFP) groups, respectively. Each group had 20 broilers, and three replicate pens. The results showed that the addition of SCFP, SAFP, and AOFP groups did not affect the growth performances, but increased the jejunum value of villus height and villus: crypt ratio on SAFP and AOFP groups compared to the control and SCFP groups. Furthermore, adding SCFP, SAFP, and AOFP significantly reduced the number of *Clostridium perfringens* in ileum chyme. SCFP, SAFP, and AOFP significantly reduced the amount of interleukin-1β, inducible nitric oxide synthases, interferon-γ, and nuclear factor kappa B mRNA expression in PBMCs, especially in the AOFP group. In summary, all the SCFP, SAFP, and AOFP groups can be suggested as a functional feed additive since they enhanced villus: crypt ratio and decreased inflammation-related mRNA expression, especially for AOFP group in broilers.

Animals **2019**, *9*, 1051

Keywords: *Saccharomyces cerevisiae*; *Aspergillus oryzae*; phytase; fermentation; broilers

---

1. Introduction

As the global climate changes, the price of commodity feed ingredients, such as corn and soybean, rise continuously. Therefore, it is imperative to find new agricultural by-products as substitutes. Every year, millions of tons of wheat are produced all over the world, and wheat bran (WB) is the main by-product [1]. WB is inexpensive and may become one of the replacement sources for animal feed ingredients. WB consists of about 15% crude protein, 4% crude fat, 28% carbohydrate, 42% insoluble fiber, and is low in energy content. WB contains many compounds, such as ferulic acid, tocopherol, and lutein [2], which are good antioxidants. However, it contains high fiber content, 1% phytate, about 84% phytate-p and the total-p ratio [3] and susceptibility to mycotoxin contaminations because of improper storage. These have negative impacts on the utilization of WB as a viable ingredient in animal feed.

It is well-known that yeast cell walls are rich in ß-glucan and mannan, which can regulate immunity and protect the host from mycotoxins damage [4]. Although past studies have indicated that the addition of *S. cerevisiae* powder does not improve the growth traits of animals, it can increase the villi height and reduce the number of *E. coli* in the gut [5]. *S. cerevisiae* also has considerable enzyme secretion capabilities, such as proteases, xylanases, and cellulases [6]. In addition to *S. cerevisiae*, a previous study also pointed out that *A. oryzae*, generally recognized as safe (GRAS) probiotic, can be used as a probiotic for fermenting; it also has good enzyme secretion ability and can increase the crude protein content of the fermented material [7]. Moreover, many studies indicated that WB could increase its utilization value after fermentation by probiotics such as *Bacillus amyloliquefaciens*, *S. cerevisiae*, or white rot fungi in broilers diet [5,8]. Furthermore, a previous study pointed out that probiotics can decrease the inflammation response in animals [9]. Inflammation can cause many chronic diseases that affect animal health and longevity [10]. The reduction of the inflammatory response can decrease energy loss and improve the cell survival rate [11].

It is well-known that phytate possess high content of phosphorus and can chelate sodium, calcium or amino acids, thus affecting the absorption and utilization of nutrients by animals; however, phytase addition could solve the above-mentioned problem, and increase the digestibility of amino acids such as Lys, Met, Cys, and Thr [12], and the weight of the tibia, as well as increase the retention of phosphorus [13].

Few studies have been carried out on co-fermenting WB with probiotics combined with phytase to examine inflammation in broilers. Therefore, this study aims to investigate the *S. cerevisiae* fermented products (SCFP) and *A. oryzae* fermented products (AOFP) additions in broiler diets on the growth characteristics, intestinal morphology, microbial morphology, and inflammation-related mRNA expression in broilers.

2. Materials and Methods

2.1. Probiotic Characteristics

S. cerevisiae was cultured in Yeast-Mald (YM) broth at 30 °C for 0, 6, 12, 18, and 24 h, and then the number of *S. cerevisiae* colony-forming unit (CFU) on YM agar (Sparks, MD 21152 USA) was counted after sequence dilution to measure the growth curve of *S. cerevisiae*. *A. oryzae* was cultured in potato dextrose broth (PDB, Neogen, Lansing, MI, USA) at 30 °C for 0, 6, 12, 18, and 24 h, and the content of dry mycelia weight of *A. oryzae* in PDB was measured for the growth curve of *A. oryzae*. Moreover, 10^8 CFU/mL *S. cerevisiae* or 10^6 spores/mL *A. oryzae* fluid were incubated in 85 °C hot water for 3 min for heat tolerance measurement. After heating, the surviving *S. cerevisiae* and *A. oryzae* were counted on the YM agar or potato dextrose agar (PDA, Neogen, Lansing, MI, USA) respectively.

Acid, base, and gastrointestinal fluid tolerance were measured by pH-adjusted phosphate buffered saline (PBS, pH 3.0, 4.0, or 13.0, adjusted by citric acid or NaOH) (Merck, Darmstadt, Germany), simulation intestinal fluid (pH 13.0, 0.3% bile acid (Merck, Darmstadt, Germany), and 0.1% trypsin (Merck, Darmstadt, Germany)), or gastric fluid (pH 3.0, 0.3% pepsin). Here, 10^8 CFU/mL *S. cerevisiae* or 10^6 spores/mL *A. oryzae* fluid were incubated in the pH adjusted fluid, simulation intestinal fluid, or gastric fluid for 0 or 3 h at 30 °C. After incubating, the surviving *S. cerevisiae* and *A. oryzae* were counted on the YM agar or PDA, respectively.

The adherence ability was measured by the method indicated by Annika et al. [14], briefly, sacrificing the broiler and removing its crop. After dissection of the crop, sterile pH 7.2 PBS was used for rinsing. After washing, the crop epithelial cells were scraped with a slide and *S. cerevisiae* or *A. oryzae* spore solution was added. The state of attachment was confirmed under an optical microscope.

2.2. S. cerevisiae Fermented Products (SCFP) and A. oryzae Fermented Products (AOFP) Preparation and Characteristics

Then 10^8 CFU/ mL of *S. cerevisiae* was added to dilute the fluid mentioned above into sterilized wheat bran at a ratio of 1 mL dilute fluid per 3 g of wheat bran. The moisture of *S. cerevisiae* pre-fermentation wheat bran was adjusted to 60%. After that, 500 phytase units (FTU) of phytase were added to 30 g *S. cerevisiae* pre-fermentation wheat bran. *S. cerevisiae* pre-fermentation wheat bran (contain phytase) was incubated at 30 °C for 5 days, and placed in a stove at 50 °C for 1 day. The *S. cerevisiae* fermented product (SCFP) was stored in a −20 °C refrigerator for the other test.

The spore counts of *A. oryzae* were diluted to 9×10^6 spores/mL with deionized water. The diluted fluid was added into sterilized WB at a ratio of 1 mL dilute fluid per 3 g wheat bran. The moisture of *A. oryzae* pre-fermentation wheat bran was adjusted to 50%. After that, 500 FTU was added to 30 g *A. oryzae* pre-fermentation wheat bran. *A. oryzae* pre-fermentation wheat bran (contain phytase) was incubated at 30 °C for 7 days, and placed in a stove at 50 °C for 1 day. The *A. oryzae* fermented product (AOFP) was stored in a −20 °C refrigerator for the other test. The xylanase [15], protease [16], cellulase, and ß-glucanase [17] activities in SCFP or AOFP were analyzed before the animal experiment.

2.3. Peripheral Blood Mononuclear Cell Isolation

The assay was conducted and modified according to Kaiser et al. [18]. Briefly, 5 mL whole blood was collected from a total of thirty-six 35-d-old broilers (3 for each pen, 9 for each treatment) in the Summax Single-use Containers for Human Venous Blood Specimen Collection (Lithium Heparin Tube), and centrifuged at 200× *g* for 10 min to separate blood cells and plasma. The supernatant was replaced with sterilization PBS, and the sample was mixed uniformly. The mixed sample was added to the same amount of ficol, and centrifuged at 200× *g* for 30 min. The separated PBMC were moved to the new RNase free Eppendorf tube (Gunster Biotech, Co., Ltd., Taipei, Taiwan), and washed by PBS for three times. After this, the sample was centrifuged at 200× *g* for 10 min to remove the suspension PBS. RPMI-1640 (Merck, Darmstadt, Germany) with 10% fetal bovine serum (Merck, Darmstadt, Germany) (for cell test) or PBS (for qPCR) were used to re-suspend PBMCs and dilute it to 10^7 cells/mL.

2.4. Nitric Oxide Assay

Isolation diluted PBMC (10^7 cells/mL) were cultured at 37 °C and 5% CO_2 for 2 h. Then 10 μL 1 ng/mL lipopolysaccharides (LPS) and 10 μL sample solution, which had been filtrated by the 0.22 μm filter, were added. Sterilized PBS was used as the control group. After co-incubation for 24 h, 100 μL Griess reagent was added, and the absorbance was detected at 540 nm.

2.5. 3-(4,5-Dimethylthiazol-2-yl)-2,5-Diphenyltetrazolium Bromide (MTT) Assay

Isolation diluted PBMC (10^7 cells/mL) were cultured at 37 °C and 5% CO_2 for 2 h. Then 10 μL 1 ng/mL lipopolysaccharides (LPS, Merck, Darmstadt, Germany) and 10 μL sample solution, which has been filtrated by 0.22 μm filter, were added. Sterilized PBS was used as the control group. After co-incubating for 48 h, 20 µL 0.5% MTT (Merck, Darmstadt, Germany) was added, and cultured for 4 h. After culturing, 100 µL dimethyl sulfoxide (DMSO, Merck, Darmstadt, Germany) was added, and the absorbance was detected at 570 nm.

2.6. Animal Experimental Design

To clarify the effect of SAFP, AOFP, and SAFP on broilers, the intestinal morphology, blood, and serum characteristics and mRNA expression in broilers' PBMC were measured. This experiment was conducted at National Chung Hsing University, Taiwan. All of the protocols for animal use were followed by the Animal Care and Use Committee (IACUC: 107-013). The methods of animal design, intestinal morphology, blood and serum characteristics, and microbial parameter in intestinal content were slightly modified according to Teng et al. [5]. A total of 240 male one-day-old broiler chickens (Ross 308) were used and divided into four groups: Control, 0.1% SCFP (SCFP), 0.05% SCFP + 0.05% AOFP (SAFP), and 0.1% AOFP (AOFP), respectively. Each pen had 20 broilers, and 3 replicate pens (a total of 60 birds per group). The average body weight (42.0 ± 0.5 g/bird) was similar among every pen initially. All birds were placed in the temperature control house. At 0–3 day-old, the temperature was 33 ± 0.5 °C, and then the temperature decreased as the broiler grew up. Temperature was controlled at 22 ± 1 °C after 30 days. Diets were divided into starter and finisher (Table 1). Both of them met or exceeded the nutrient requirements of broilers (NRC, 1994) with addition of SCFP and AOFP. The proximate composition was analyzed according to the methods of Association of Official Analytical Communities (AOAC) [19]. The starter and finisher diets were offered to the birds from 1–21 day-old and 22–35 day-old respectively. Body weight and feed consumption were recorded at 21 and 35 day-old. Body weight gain and feed conversion ratios (FCR) were calculated. Nine birds of each group were randomly selected to isolate PBMCs for qPCR and cell test.

2.7. Intestinal Morphology

Twenty-four 35-day-old broilers (2 of each pen, 6 for each treatment, fasting for 1 day) were used for the measurement of intestinal morphology. The birds were euthanized, and the middle of the jejunum and ileum were excised, and the content was flushed by PBS and fixed in 10% formalin. Each sample was embedded by paraffin and stained with hematoxylin and eosin. The slice was observed with a light microscope. The villus height, crypt depth, and tunica muscular were measured by Motic Image Plus 2.0 analysis system (Motic Instruments, Richmond, BC, Canada).

2.8. Blood and Serum Characteristics

Thirty six 35-day-old broilers (3 of each pen, 9 for each treatment) blood was collected randomly for the blood and serum characteristic measurement. Blood samples stood for about 4 to 5 h at 4 °C, and then centrifuged at 3000 rpm for 10 min at 4 °C. Blood and serum biochemical parameters analyses were measured by automatic biochemical analyzer (Hitachi, 7150 auto-analyzer, Tokyo, Japan).

2.9. Microbial Parameter in Intestinal Content

Twenty-four 35-day-old broilers (2 of each pen, 6 for each treatment) were randomly selected for microbial analysis. Birds were euthanized, and the contents of ileum and cecum were squeezed out. Next, 1 g chyme was put into 9 mL PBS to sequence a dilution for the microbial parameter. *Lactobacillus* spp. were cultured with DeMan, Rogosa, and Sharpe agar (Difco™ Lactobacilli MRS Agar, BD, Franklin Lakes, NJ); *Clostridium perfringens* were cultured with tryptose sulfite cycloserine agar (GranuCult™ TSCagar, Merck KGaA, 64271, Darmstadt, Germany). Each microbe mentioned above was cultured at 37 °C under anaerobic conditions for 48 h. After culturing, the number of CFU on the agar was counted.

Table 1. Composition and calculated analysis (g/kg as fed) of the basal diet for broilers (1–35 days) [1].

Ingredients	Starter Diet	Finisher Diet
	(1–21 Days)	(22–35 Days)
	g/kg	
Corn, yellow	485.5	535.6
Soybean meal, (CP 44 %)	348	281.4
Full-fat soybean meal	82.9	99.8
Soybean oil	36.5	42.4
Calcium carbonate	16.6	13.3
Monocalcium phosphate	18.2	16.3
DL-Methionine	1.97	1.33
L-Lysine-HCl	3.64	3.21
NaCl	3.88	3.81
Choline-Cl	0.83	0.79
Vitamin premix [2]	0.99	1.03
Mineral premix [3]	0.99	1.03
Total	1000	1000
Calculated nutrient value		
Dry matter, %	88.3	88.2
Crude protein, %DM	23	21
Crude fat, %	6.64	8.95
Calcium, %DM	1.05	0.9
Total Phosphorus, %DM	0.77	0.68
Available Phosphorus, %DM	0.5	0.45
Lysine, %DM	1.43	1.25
Methionine + Cysteine, %DM	1.07	0.96
ME, kcal/kg DM	3050	3175
Chemical analysis value		
Dry matter, %	88.0	88.7
Crude protein, %DM	23.1	20.9
Crude fat, %	6.58	8.78

[1] Maize–soybean base diet for all treatment and add SCFP, SAFP, and AOFP to basal diet for SCFP, SAFP, and AOFP groups, respectively. [2] Vitamin (premix content per kg diet): vit. A, 15,000 IU; vit. D3, 3000 IU; vit. E, 30 mg; vit. K3, 4 mg; thiamine, 3 mg; riboflavin, 8 mg; pyridoxine, 5 mg; vitamin B12, 25 µg; Ca-pantothenate, 19 mg; niacin, 50 mg; folic acid, 1.5 mg; and biotin, 60 µg. [3] Mineral (premix content per kg diet): Co ($CoCO_3$), 0.255 mg; Cu ($CuSO_4 \cdot 5H_2O$), 10.8 mg; Fe ($FeSO_4 \cdot H_2O$), 90 mg; Mn ($MnSO_4 \cdot H_2O$), 90 mg; Zn (ZnO), 68.4 mg; Se (Na_2SeO_3), 0.18 mg.

2.10. Total RNA Isolation and qPCR

Total RNA was isolated from PBMCs for determination of the mRNA expression level using SuperScript™ FirstStrand Synthesis System reagent (Thermo Fisher, Waltham, MA, USA) according to the manufacturer's protocol. The method of determination of total RNA purity, cDNA synthesis, and qPCR analysis was slightly modified from Hu et al. [20]. Briefly, 2× SYBR GREEN PCR Master Mix-ROX (Gunster Biotech, Co., Ltd., Taipei, Taiwan), cDNA, deionized water, and each primer were mixed at a ratio of 5:1.2:1.8:1. StepOnePlus™ Real-Time PCR System (Thermo Fisher, Waltham, MA, USA) was used to detect qRT-PCR performance, while the $2^{-\Delta\Delta Ct}$ method was used to calculate the relative mRNA expression level, and ß-actin was used as the housekeeping gene for normalization. Gene-specific primers are according to the genes of Gallus gallus (chicken) given as follows: *beta-actin (ß-actin,* 5′-CTGGCACCTAGCACAATGAA-3′ and 5′-ACATCTGCTGGAAGGTGGAC-3′, X00182.1); *interlukin-1 beta (IL-1ß,* 5′-GCTCTACATGTCGTGTGTGATGAG-3′ and 5′-TGTCGATGTCCC GCATGA-3′, NM_204524); *inducible nitric oxide synthase (iNOS,* 5′-TACTGCGTGTCCTTTCAACG -3′ and 5′-CCCATTCTTCTTCCAACCTC-3′, U46504); *interferon-gamma (IFN-γ* (5′-CTCCCGATGAACG

ACTTGAG-3′ and 5′-CTGAGACTGGCTCCTTTTCC-3′, Y07922); *Nuclear factor kappa light chain enhancement of activated B cell p65 (NFκB p65,* 5′-CCAGGTTGCCATCGTGTTCC-3′ and 5′-GCGTGCGTTTGCGCTTCT-3′, D13719.1).

2.11. Statistical Analysis

Data were analyzed for significance by analysis of variance (ANOVA) using SAS software (SAS® 9.4, 2018, SAS Institue Inc., Cary, NC, USA). Differences between treatment means were separated using Duncan's multiple range test with *p* value less than 0.05.

3. Results

3.1. The General Characteristics of the Probiotics and Fermented Products

After culturing for 24 h, the number of *S. cerevisiae* climbed to over 10^8 CFU/g DM, while *A. oryzae* dry mycelia weight increased to 176.8 mg/100 mL PDB. Both *S. cerevisiae* and *A. oryzae* survived in acid, base solution, and even gastrointestinal fluid, but *A. oryzae* did not survive at 85 °C. Both *S. cerevisiae* and *A. oryzae* adhered to broiler crop epithelial cells. (Figure 1).

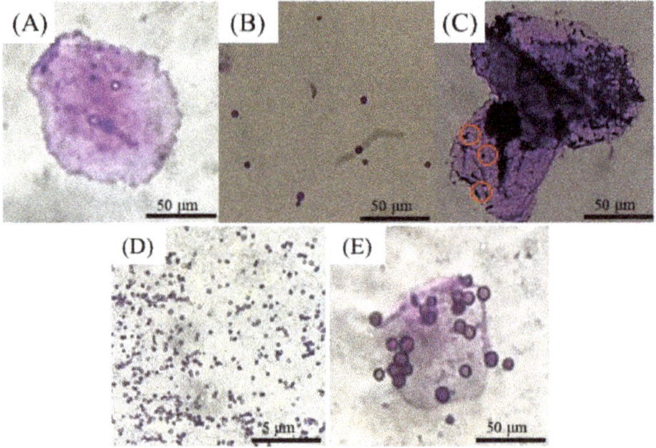

Figure 1. Photo of broiler crop epithelial cell (**A**), *Saccharomyces cerevisiae* (**B**), *Saccharomyces cerevisiae* adhering on the epithelial cell (red cycle) (**C**), spores of *Aspergillus oryzae* (**D**), and spores adhering on the epithelial cell (**E**), screened by an optical microscope.

The results of the enzyme activity of *S. cerevisiae* fermented products (SCFP) and *A. oryzae* fermented products (AOFP) extraction are shown in Table 2. Wheat bran had no detectable enzyme activities; however, after fermentation by *S. cerevisiae* or *A. oryzae*, xylanase, protease, cellulase, and ß-glucanase all exhibited a dramatic increase. Fermentation by *S. cerevisiae* or *A. oryzae*, xylanase, protease, cellulase, and ß-glucanase increased significantly. The enzyme activities of SCFP were 142.26, 777.13, 189.49, and 117.07 U/g DM, respectively, and AOFP are 120, 190, 213.38, and 120.21 U/g DM, respectively. Compared to AOFP, SCFP had a higher protease (777.1 vs. 190.0 U/g DM) and xylanase (142.3 vs. 120.0 U/g DM) production ability. Furthermore, both SCFP and AOFP could produce cellulase and hemicellulase to degrade the fiber content in wheat bran.

The results of nitric oxide (NO) release amounts are shown in Figure 2. After stimulating by LPS, the NO release amounts of PBMCs showed a significant increase. However, adding SCFP or AOFP extraction could reduce the NO release amount stimulated by LPS (*p* < 0.05). Among them, there were not significant differences in the 10 mg/mL LAOFP group, as well as the 50, 100 mg/mL LSCFP group, which means that AOFP has the same effect on reducing NO production of the LPS stimulated PBMC.

Table 2. The xylanase, protease, cellulase, and ß-glucanase activities of SCFP and AOFP extraction (U/g DM).

Enzyme	Products		
	WB [1]	SCFP [2]	AOFP [3]
Xylanase [4]	ND [5]	142.3	120.0
Protease [6]	ND	777.1	190.0
Cellulase [7]	ND	189.5	213.4
ß-glucanase [8]	ND	117.1	120.2

[1] WB: wheat bran. [2] SCFP: *Saccharomyces cerevisiae* fermented products. [3] AOFP: *Aspergillus oryzae* fermented products. [4] One unit of xylanase activity is defined as 1 μmol D-xylose generated from 10 mg/mL xylan in the condition of 37 °C and pH 5.5 in a minute. [5] Not detectable. [6] One unit of protease activity is defined as 1 μg L-tyrosine generated from 10 mg/mL casein in the condition of 40 °C and pH 7.5 in a minute. [7] One unit of cellulase activity is defined as 1 μmol, where reducing sugar is generated from 10 mg/mL CMC in the condition of 37 °C and pH 5.5 in a minute. [8] One unit of ß-glucanase activity is defined as 1 μmol, where reducing sugar is generated from 5 mg/mL ß-glucan in the conditions of 37 °C and pH 5.5 in one minute.

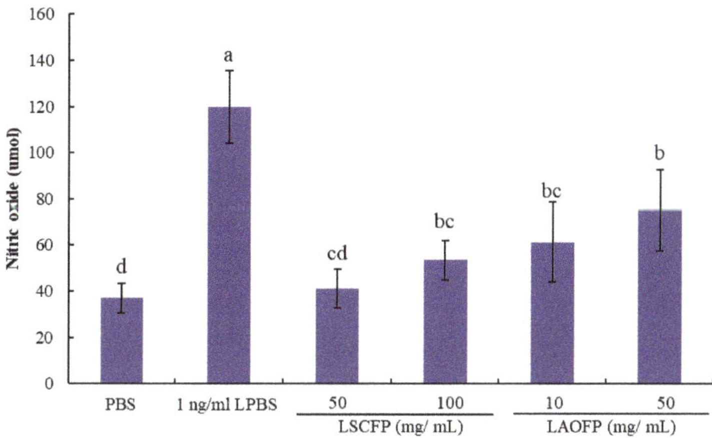

Figure 2. Nitric oxide release amount of phosphate buffer solution (PBS, control), 1 ng/mL lipopolysaccharide (LPBS, negative control), 50 mg/mL *Saccharomyces cerevisiae* fermented products extraction, and 1 ng/mL LPS include (50 mg/mL LSCFP), 100 mg/mL *Saccharomyces cerevisiae* fermented products extraction and 1 ng/mL LPS include (100 mg/mL LSCFP), 10 mg/mL *Aspergillus oryzae* fermented products extraction and 1 ng/mL LPS include (10 mg/mL LAOFP), and 25 mg/mL *Aspergillus oryzae* fermented products extraction and 1 ng/mL LPS include (25 mg/mL LAOFP). [a,b,c,d] Means within the same rows without the same superscript letter are significantly different ($p < 0.05$).

As the results shown in Figure 3, the MTT assay has an opposite result of NO release amount on the data. LPS stimulates PBMCs and causes the survival rate of PBMCs to decrease significantly. However, adding SCFP or AOFP extraction could increase the PBMCs survival rate after stimulation by LPS ($p < 0.05$). Among them, there were not significant differences in the LAOFP group and 100 mg/mL LSCFP group, meaning that AOFP had the same effect on increasing the survival rate in the MTT test with a lower concentration.

3.2. Animals Performance

Growth performances of SCFP, SAFP, and AOFP supplemented are shown in Table 3. All treatments exhibited no significant differences on body weight, feed consumption, weight gain, and feed conversion ratio (FCR) at 1–21 days, 22–35 days, and 1–35 days.

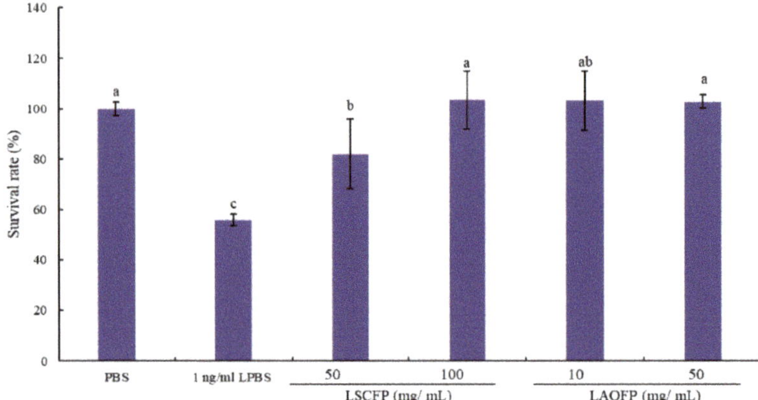

Figure 3. MTT (3-(4,5-dimethylthiazol-2-yl)-2,5-diphenyltetrazolium bromide) assay of phosphate buffer solution (PBS, control), PBS and 1 ng/mL lipopolysaccharide (LPBS, positive control), 50 mg/mL *Saccharomyces cerevisiae* fermented products extraction and 1 ng/mL LPS included (50 mg/mL LSCFP), 100 mg/mL *Saccharomyces cerevisiae* fermented products extraction and 1 ng/mL LPS included (100 mg/mL LSCFP), 10 mg/mL *Aspergillus oryzae* fermented products extraction and 1 ng/mL LPS include (10 mg/mL LAOFP), and 25 mg/mL *Aspergillus oryzae* fermented products extraction and 1 ng/mL LPS include (25 mg/mL LAOFP). [a,b,c] Means within the same rows without the same superscript letter are significantly different ($p < 0.05$).

Table 3. Effect of SCFP, SAFP, and AOFP supplemented in diet on growth performance of 1–35 day-old broilers (n = 3).

Item	Treatments					
	Control [1]	SCFP [1]	SAFP [2]	AOFP [3]	SEM [4]	*p* Value
1–21 days						
Body weight, g/bird	837.25	868.12	867.35	844.29	10.31	0.15
Feed consumption, g/bird	964.85	945.83	969.97	907.50	63.68	0.89
Weight gain, g/bird	794.25	825.12	824.35	801.29	10.32	0.15
FCR [5]	1.15	1.09	1.12	1.07	0.08	0.88
22–35 days						
Body weight, g/bird	2163.24	2164.70	2228.59	2168.29	28.90	0.37
Feed consumption, g/bird	2267.14	1915.24	2078.81	1984.48	124.15	0.28
Weight gain, g/bird	1325.99	1296.58	1361.24	1324.00	28.10	0.49
FCR	1.71	1.48	1.53	1.50	0.11	0.48
1–35 days						
Feed consumption, g/bird	3231.99	2861.08	3048.78	2891.98	157.51	0.38
Weight gain, g/bird	2120.24	2121.70	2185.59	2125.29	28.90	0.37
FCR	1.49	1.32	1.37	1.33	0.08	0.45

[1] SCFP: 0.5% *Saccharomyces cerevisiae* fermented products group. [2] SAFP: 0.5% *Saccharomyces cerevisiae* fermented products + 0.5% *Aspergillus oryzae* fermented products group. [3] AOFP: 0.5% *Aspergillus oryzae* fermented products group. [4] SEM: Standard error of the mean. [5] FCR: Feed conversion rate.

Although the results of growth performance showed no significant difference, the results of jejunum villus height and villus:crypt ratio on SAFP and AOFP both had a significant increase (1163.95 vs. 1508.93 and 1505.14 μm, 5.07 vs. 7.67 and 8.13, respectively, $p < 0.0001$). The crypt depth of the SCFP, SAFP, and AOFP groups showed a significant decrease (237.37 vs. 182.37, 203.43 and 188.42 μm, respectively, $p = 0.0002$). At the ileum, only SCFP and AOFP had a significant decrease on crypt depth ($p < 0.0001$), and AOFP had an increase in villus:crypt ratio ($p = 0.0005$) (Table 4). Photomicrography of jejunum and ileum of 35-day-old broilers are shown in Figure 4.

Table 4. Effect of SCFP, SAFP, and AOFP supplemented in diet on intestinal morphology of 35 d-old broilers (n = 6).

Item	Treatments					
	Control	SCFP [1]	SAFP [2]	AOFP [3]	SEM [4]	*p* Value
Jejunum						
Villus height (μm)	1163.95 [b]	1124.78 [b]	1508.93 [a]	1505.14 [a]	30.69	<0.0001
Crypt depth (μm)	237.37 [a]	182.37 [c]	203.43 [bc]	188.42 [c]	8.59	0.0002
Tunica muscularis (μm)	298.32 [a]	285.47 [b]	272.12 [b]	255.92 [b]	14.17	0.004
Villus:crypt	5.07 [c]	6.36 [b]	7.67 [a]	8.13 [a]	0.3	<0.0001
Ileum						
Villus height (μm)	1109.77	1055.33	1074	1050.13	19.04	0.1185
Crypt depth (μm)	209.25 [a]	173.25 [b]	201.29 [a]	157.32 [b]	6.82	<0.0001
Tunica muscularis (μm)	650.8	439.38	366.38	318.68	108.05	0.1453
Villus:crypt	5.52 [b]	6.25 [ab]	5.44 [b]	7.08 [a]	0.28	0.0005

[1] SCFP: 0.5% *Saccharomyces cerevisiae* fermented products group. [2] SAFP: 0.5% *Saccharomyces cerevisiae* fermented products + 0.5% *Aspergillus oryzae* fermented products group. [3] AOFP: 0.5% *Aspergillus oryzae* fermented products group. [4] SEM: Standard error of the mean. [a,b] Means within the same rows without the same superscript letter are significantly different (*p* < 0.05).

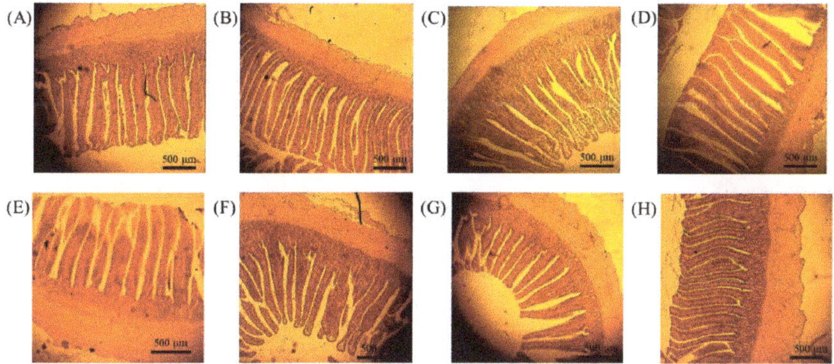

Figure 4. Photomicrography of jejunum and ileum of 35-day-old broiler feed with control, *Saccharomyces cerevisiae* fermented products (SCFP) or *Aspergillus oryzae* fermented products (AOFP). (**A–D**) jejunum, (**E–H**) ileum. Photos are respectively Control (**A,E**), 0.1% SCFP (**B,F**), 0.05% SCFP + 0.05% AOFP (**C,G**), and 1% AOFP (**D,H**). Hematoxylin and eosin stain 40×.

The white blood cells (WBC) increased significantly on SAFP (*p* < 0.0001) compared to other group. The concentration of uric acid (UA) significantly decreased following the treatment of SCFP and SAFP (*p* = 0.001). Furthermore, the SCFP group had a better effect on decreasing the UA content in serum compared to the AOFP group (3.36 vs. 4.69 mg/dL). However, there were no significant effects on triglyceride (TG), high-density lipoprotein (HDL), high-density lipoprotein (LDL), Ca, and P (Table 5).

The results of mRNA expression level of immunomodulatory genes in chicken PBMCs are shown in Figure 5. As seen, the *IL-1ß*, *iNOS*, *IFN-γ*, and *NFκB* mRNA expression significantly decreased in SCFP, SAFP, and AOFP (*p* < 0.05). Furthermore, it is worth noting that AOFP can reduce the inflammation-related mRNA content most effectively.

Table 5. Effect of SCFP, SAFP and AOFP supplemented in diet on serum characteristics of broilers (35 d) (n = 9).

Treatments	Items [1]							
	WBC	**UA**	**CHOL**	**TG**	**HDL-C**	**LDL-C**	**Ca**	**P**
Units	10^3/uL	mg/dL	mg/dL	mg/dL	mg/dL	mg/dL	mg/dL	mg/dL
Control	189.69 [b]	5.57 [a]	119.22	77	73.44	38.33	9.86	9.13
SCFP [2]	193.69 [b]	3.36 [c]	117.44	61.33	74.56	38.11	9.51	9
SAFP [3]	234.46 [a]	4.16 [bc]	133.22	89.11	81.33	41.78	9.99	8.94
AOFP [4]	185.40 [b]	4.69 [ab]	125.33	66.56	77.11	42.89	9.77	8.72
SEM [5]	4.35	0.35	5.84	8.19	3.67	2.95	0.12	0.34
p Value	<0.0001	0.001	0.2359	0.1017	0.4479	0.576	0.0641	0.8596

[1] WBC: white blood cell; UA: uric acid; CHOL: cholesterol; TG: triglycerides; HDL: cholesterol-high-density lipoprotein; LDL-C: cholesterol-low-density lipoprotein; Ca: calcium; P: phosphorus. [2] SCFP: 0.5% *Saccharomyces cerevisiae* fermented products group. [3] SAFP: 0.5% *Saccharomyces cerevisiae* fermented products + 0.5% *Aspergillus oryzae* fermented products group. [4] AOFP: 0.5% *Aspergillus oryzae* fermented products group. [5] SEM: standard error of the mean. [a,b,c] means within the same rows without the same superscript letter are significantly different (*p* < 0.05).

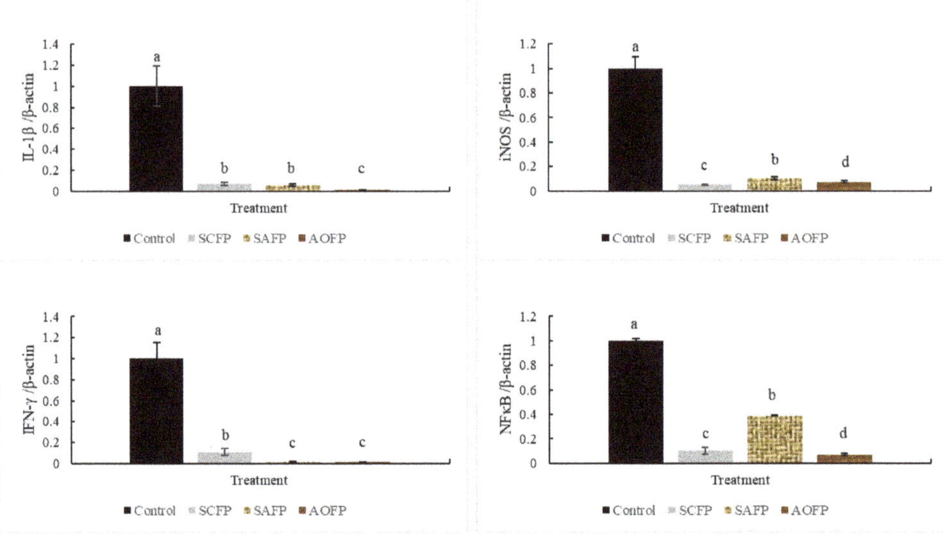

Figure 5. The mRNA expression level of immunomodulatory genes in chicken peripheral blood mononuclear cells at 35d. *IL-1ß* (top left), *iNOS* (top right), *IFN-γ* (bottom left), and *NFκB* (bottom right). The treatments are control, 0.5% *Saccharomyces cerevisiae* fermented products group (SCFP), 0.5% *Saccharomyces cerevisiae* fermented products + 0.5% *Aspergillus oryzae* fermented products group (SAFP), 0.5% *Aspergillus oryzae* fermented products group (AOFP), respectively. Data is presented in mean ± SE (n = 9). [a,b,c,d] Means within the same rows without the same superscript letter are significantly different (*p* < 0.05).

3.3. Microbial Parameter in Intestinal Content

The treatment of SCFP, SAFP, and AOFP significantly decrease the number of *Clostridium perfringens* (*p* = 0.0014), but did not increase the number of *Lactobacillus* spp. (Table 6). In the caecum, although the number of *C. perfringens* did not significantly decrease (*p* = 0.6770), the number of *C. perfringens* still decreased in the data (7.39, 7.00, 7.01, and 7.22 log CFU/g in the Control, SCFP, SAFP, and AOFP groups, respectively), especially for the SCFP group.

Table 6. Effect of SCFP, SAFP, and AOFP supplemented in diet on microbial parameter in the intestinal content of 35 d-old broilers (n = 3).

Microbial Parameter	Treatments					
	Control	SCFP [1]	SAFP [2]	AOFP [3]	SEM [4]	*p* Value
	log CFU/g					
Ileum						
Clostridium perfringens	7.84 [a]	7.02 [b]	7.05 [b]	7.22 [b]	0.12	0.0014
Lactobacillus spp.	8.35	8.84	8.25	8.91	0.39	0.1765
Caecum						
Clostridium perfringens	7.39	7.00	7.01	7.22	0.24	0.6770
Lactobacillus spp.	8.95	9.13	8.92	8.91	0.23	0.7039

[1] SCFP: 0.5% *Saccharomyces cerevisiae* fermented products group. [2] SAFP: 0.5% *Saccharomyces cerevisiae* fermented products + 0.5% *Aspergillus oryzae* fermented products group. [3] AOFP: 0.5% *Aspergillus oryzae* fermented products group. [4] SEM: Standard error of the mean. [a,b] Means within the same rows without the same superscript letter are significantly different ($p < 0.05$).

4. Discussion

The high content of insoluble fiber and phytate could reduce the utilization of nutrition in animals. SCFP and AOFP contain quite a large amount of enzymes, such as cellulase, ß-glucanase, xylanase, and protease, thereby improving the nutrient utilization of broilers [21]. Nevertheless, adding phytase to the fermentation process of SCFP and AOFP may also help to reduce the phytate content [12]. From the results, the addition of SCFP and AOFP did not improve body weight, feed consumption, weight gain, and FCR. However, Santos et al. [22] reported that phytase addition (500 FTU/kg) improved the body weight gain of 1–21 day-old broilers, but did not improve the body weight gain and FCR in 22–35 day-old broilers. Mountzouris et al. [23] pointed out that *S. cerevisiae* powder (50 mg (10^9 CFU)/kg diet) addition did not enhance the growth performance. Otherwise, there are few studies investigating the effects of *A. oryzae* addition on broilers growth performance. As a previous study shows, replacing soybean meal by *A. oryzae* fermented soybean meal (29.5% replacement in whole diet) can improve average daily gain and feed intake of broilers [24]. As the replacement is far higher than in the current study, it is speculated that the addition of *A. oryzae* should be sufficient to prove effective in growth performance.

The improvement of white blood cells (WBC) may be attributed to the joint effects of SCFP and AOFP because only SAFP increase WBC count in broilers blood. The SCFP and SAFP groups can significantly reduce the uric acid (UA) content in the blood, and the data number of SAFP is between SCFP and AOFP. This is because the protease yielded by SCFP was higher than AOFP and protease improved the protein utilization of the host, thereby decreasing the UA content in the blood. UA is a metabolite of protein that has an antioxidant function but is converted to a pro-oxidant in the cell or cytoplasm, and may be associated with cardiovascular disease when the concentration increases [25]. In addition, UA also promotes inflammation and is associated with insulin resistance and metabolic disorders [25]. Therefore, the addition of SCFP or AOFP to feed may prevent the development of cardiovascular disease, and reduce the oxidative stress of animal cells by the pathway of reduced UA.

In this study, villus height and the villus:crypt ratio in the jejunum were significantly increased and crypt depth significantly decreased on SAFP and AOFP, indicating that SAFP and AOFP may have an effect on improving nutrition absorption. One possible reason is that the enzyme yielded by SCFP and AOFP may degrade the anti-nutritional factor and make the energy utilization focus on villi growth and nutrient absorption [26]. Kalantar et al. [21] also reported that adding 0.1% commercial enzyme COMBO® (including 1000 FTU/g phytase, 200 U/g xylanase, 200 U/g β-glucanase, and 80 U/g hemicellulase) in broilers diet can decrease the crypt depth and increase the villus length/width ratio and villus length/crypt depth ratio. Gao et al. [27] showed that yeast supplement (2.5, 5.0, and 7.5 g/kg diet) can increase the villus height of the small intestine. Feng et al. [24] found that replacing soybean meal by *A. oryzae* fermented soybean meal can increase villus height on both duodenum and jejunum and decrease the crypt depth in jejunum. In addition, a decrease in the crypt depth indicates that the animal can spend less energy on the formation of the crypt, thereby reducing the energy loss [26]. Furthermore, the results showed that Tunica muscularis decreased significantly on jejunum after adding SCFP, SAFP, and AOFP. Tunica muscularis consists of smooth muscle and is related to digestion and absorption of chyme. Cil et al. [28] reported that inflammation causes the intestine to swell and increases Tunica muscularis weight as well as thickness. Therefore, by the anti-inflammatory effect of SCFP, SAFP, and AOFP confirmed by cell test and broilers PBMCs mRNA expression, the decreased thickness of Tunica muscularis is expected.

Previous studies pointed out that probiotic can reduce the number of harmful microbes in the gut [5]. Mountzouris et al. [23] indicated that *S. cerevisiae* powder (50 mg (1×10^9 CFU)/kg diet) can protect broilers from *Salmonella Enteritidis* (2×10^6 CFU/birds) challenge at day 15 and increase the growth performance of broilers. In this study, adding SCFP, SAFP, and AOFP can reduce *C. perfringens* in the ileum. Among them, the SCFP is better than the AOFP based on the data, but with no significant difference, while the SAFP effect was in the middle. The antibacterial effect may be due to the growth of probiotics that crowd out the growth resources of *C. perfringens*, or the glucan and mannan produced by the probiotics encapsulate the toxins yield from *C. perfringens* and reduce the competitiveness of *C. perfringens*. Furthermore, the number of *C. perfringens* is positively correlated with the histochemistry score and necrotizing enterocolitis [29], thereby decreasing the number of *C. perfringens* may decrease the amount of pro-inflammatory cytokine.

LPS is an extracellular structure of gram-negative bacteria that can cause inflammation and stimulate the release amount of nitric oxide (NO) from PBMCs [11]. However, SCFP and AOFP can protect PBMCs from LPS stimulus and decrease the NO production. The reason may be that the glucan and mannan in *S. cerevisiae* and *A. oryzae* cell wall acted as antitoxin [3]. In addition, it is well-known that low-grade inflammatory response can protect the host from pathogens; however, high-grade inflammatory causes cell damage and apoptosis, so it is important to decrease the level of inflammation to prevent unexpected cell death. It is worth noting that AOFP has a similar effect as SCFP on lowering the concentration on NO production. Therefore, AOFP has similar effects as SCFP on lowering the concentration, which may represent that AOFP has a better anti-inflammatory effect than SCFP does.

For broilers, inflammation is a double-edged sword depending of its level [11]. Lee et al. [30] indicated that LPS challenge on broilers (1 mg/kg body weight) causes significant weight loss because of the immune response and decrease in feed intake. In broilers, there are two ways to decrease the inflammation level: one is to decrease the stimulants, such as LPS or bacterial toxin, and the other way is to block the pro-inflammatory cytokine pathway. From the results, SCFP and AOFP reduced the inflammation-related stimulant, including UA, *C. perfringens*, and NO production from LPS stimulated PBMCs.

As a pro-inflammatory cytokine, *IL-1ß* mediates many pathways involved in apoptosis or inflammation [31]. It is well-known that *S. cerevisiae* and *A. oryzae* cell wall are rich in β-glucans [3], which can inhibit *IL-1ß* production [32] and thereby reduce *NF-κB*-mediated inflammatory responses [33]. Therefore, *IL-1ß* is positively correlated with *NF-κB*. *NF-κB* is a main inflammatory factor, which promotes reactive nitrogen species (RNS) content [11]. In normal conditions, *NF-κB* binds with the inhibitory protein named inhibitor of kappa B (IκB) and has no effect [33]. With stimulants such as heat stress or infection, *IκB* phosphorylates and releases the *NF-κB* as a nuclear transcription factor to induce an inflammatory response [33]. SCFP, SAFP, and AOFP addition can decrease the *IL-1ß* and *NF-κB* mRNA expression. Hegazy and Bedewy [9] had similar results, indicating that 1×10^{10} CFU of *Lactobacillus delbruekii* and *L. fermentum* can decrease the interleukin-6 (IL-6), tumor necrosis factor-α (TNF-α), and *NF-κB* expression. The results show that probiotics reduced inflammation by suppressing the *NF-κB* and *IL-1ß* expression. Furthermore, the expression of *iNOS* can increase NO production, and is associated with infection [34], while SCFP and AOFP reduced the *iNOS* mRNA expression. *NF-κB* can induce *iNOS* expression, and increase the RNS production [35]; therefore, decreasing the *NF-κB* content can also suppress *iNOS* expression. *IFN-γ* is associated with infection and high concentration of *IFN-γ* may contribute to autoimmune disease [36]. When animals suffer from infection, *IFN-γ* expression increases and induces *NF-κB* expression to produce reactive oxygen species (ROS) and RNS to resist pathogens [37]. The data shown above indicate that SCFP and AOFP can decrease the damage caused by LPS and the number of *C. perfringens* in ileum, which are both related to the infection. Therefore, *IFN-γ* suppressed by SCFP, SAFP, and AOFP in broilers also decreases the *NF-κB* expression and reduces the inflammatory response [37]. Decreasing the stimulant and blocking the inflammation-related mRNA expression mentioned above confirm that both SCFP and AOFP addition can reduce the inflammation, especially AOFP.

5. Conclusions

Based on the above results, this study confirmed that SCFP, SAFP, and AOFP have positive effects on a decrease in *C. perfringens* number in ileum and suppress the mRNA relation of *IL-1β*, *NF-κB*, *iNOS*, and *IFN-γ* on broilers' PBMCs. Among them, SCFP, SAFP, and AOFP have different effects. Results showed that the addition of AOFP is much more effective than the addition of SCFP, while the effect of SAFP is somewhere between them, and there is no multiplication effect. Therefore, compared to the SCFP and SAFP, the AOFP is suggested to be a functional feed additive.

Author Contributions: Conceptualization, W.Y.C. and T.-T.L.; materials analysis, W.Y.C. and C.M.H.; methodology, W.Y.C. and Y.C.H.; supervision, T.-T.L.; investigation, W.Y.C. and T.-T.L.; writing—original draft preparation, W.Y.C.; W.C.L., and T.-T.L.; writing—review and editing, W.Y.C.; W.C.L.; S.C.C., and T.-T.L.

Funding: This research received no external funding.

Acknowledgments: The authors thank the Ministry of Science and Technology (MOST 107-2313-B-005-037-MY2) and the iEGG and Animal Biotechnology Center from The Feature Areas Research Center Program within the framework of the Higher Education Sprout Project by the Ministry of Education (MOE) in Taiwan for supporting this study.

Conflicts of Interest: The authors declare no conflict of interest.

Abbreviations

SCFP	*Saccharomyces cerevisiae* fermented products
AOFP	*Aspergillus oryzae* fermented products
PBMC	Peripheral blood mononuclear cells
LPS	Lipopolysaccharides
SAFP	The mix of SCFP and AOFP at a ratio of 1:1;
WB	Wheat bran
GRAS	Generally recognized as safe
Lys	Lysine
Met	Methionine
Cys	Cysteine
Thr	Threonine
YM	Yeast-Mald
CFU	Colony-forming unit
PDB	Potato dextrose broth
PBS	Phosphate buffered saline
FTU	Phytase units
RPMI	Roswell park memorial institute
FBS	Fetal bovine serum
MTT	3-(4,5-Dimethylthiazol-2-yl)-2,5-Diphenyltetrazolium Bromide
DMSO	Dimethyl sulfoxide
MRS	Man, Rogosa, and Sharpe agar
TSC	Tryptose sulfite cycloserine
DM	Dry matter
NO	Nitric oxide

FCR	Feed conversion ratio
WBC	White blood cells
UA	Uric acid
TG	Triglyceride
HDL	High-density lipoprotein
LDL	High-density lipoprotein
IL-1β	Interleukin-1 beta
iNOS	Inducible nitric oxide synthase
IFN-γ	Interferon-gamma
NFκB	Nuclear factor kappa light chain enhancement of activated B cell
RNS	Reactive nitrogen species
IκB	Inhibitor of kappa B
IL-6	Interleukin-6
TNF-α	Tumor necrosis factor-α
ROS	Reactive oxygen species

References

1. Hemery, Y.; Rouau, X.; Lullien, V.; Barron, C.; Abécassis, J. Dry processes to develop wheat fractions and products with enhanced nutritional quality. *J. Cereal Sci.* **2007**, *46*, 327–347. [CrossRef]

2. Shenoy, A.; Prakash, J. Wheat bran (*Triticum aestivum*): Composition, functionality and incorporation in unleavened bread. *J. Food Qual.* **2007**, *25*, 197–211.

3. Eeckhout, W.; Paepe, M.D. Total phosphorus, phytate phosphorus and phytase activity in plant feedstuffs. *Anim. Feed Sci. Technol.* **1994**, *47*, 19–29. [CrossRef]

4. Awaad, M.; Atta, A.M.; El-Ghany, W.A.; Elmenawey, M.; Ahmed, A.; Hassan, A.A.; Nada, A.; Abdelaleem, G.A.; Kawkab, A.A. Effect of a specific combination of mannan-oligosaccharides and β-glucans extracted from yeast cell wall on the health status and growth performance of ochratoxicated broiler chickens. *J. Am. Sci.* **2011**, *7*, 82–96.

5. Teng, P.Y.; Chung, C.H.; Chao, Y.P.; Chiang, C.J.; Chang, S.C.; Yu, B.; Lee, T.T. Administration of *Bacillus amyloliquefaciens* and *Saccharomyces cerevisiae* as direct-fed microbials improves intestinal microflora and morphology in broiler chickens. *J. Poult. Sci.* **2017**, *54*, 134–141. [CrossRef]

6. Shirazi, S.H.; Rahman, S.R.; Rahman, M.M. Production of extracellular lipases by saccharomyces cerevisiae. *World J. Microbiol. Biotechnol.* **1998**, *14*, 595–597. [CrossRef]

7. Khempaka, S.; Thongkratok, R.; Okrathok, S.; Molee, W. An evaluation of cassava pulp feedstuff fermented with *A. oryzae*, on growth performance, nutrient digestibility and carcass quality of broilers. *J. Poult. Sci.* **2013**, *51*, 71–79. [CrossRef]

8. Wang, C.C.; Lin, L.J.; Chao, Y.P.; Chiang, C.J.; Lee, M.T.; Chang, S.C.; Yu, B.; Lee, T.T. Antioxidant molecular targets of wheat bran fermented by white rot fungi and its potential modulation of antioxidative status in broiler chickens. *Br. Poult. Sci.* **2017**, *58*, 262–271. [CrossRef]

9. Hegazy, S.K.; El-Bedewy, M.M. Effect of probiotics on pro-inflammatory cytokines and *NF-κB* activation in ulcerative colitis. *World J. Gastroenterol.* **2010**, *16*, 4145–4151. [CrossRef]

10. Sears, B. Anti-inflammatory Diets. *J. Am. Coll. Nutr.* **2015**, *34*, 14–21. [CrossRef]

11. Lee, Y.; Lee, S.; Gadde, U.D.; Oh, S.; Lillehoj, H.S. Relievable effect of dietary Allium hookeri on LPS induced intestinal inflammation response in young broiler chickens. *J. Immunol.* **2017**, *198* (Suppl. S1), 226.3.

12. Cowieson, A.J.; Ruckebusch, J.P.; Sorbara, J.O.B.; Wilson, J.W.; Guggenbuhl, P.; Roos, F.F. A systematic view on the effect of phytase on ileal amino acid digestibility in broilers. *Anim. Feed Sci. Technol.* **2017**, *225*, 182–194. [CrossRef]

13. Hamdi, M.; Perez, J.F.; L'etourneau-Montminy, M.P.; Franco-Rossell'o, R.; Aligue, R.; Sol'a-Oriol, D. The effects of microbisal phytase and dietary calcium and phosphorus levels on the productive performance and bone mineralization of broilers. *Anim. Feed Sci. Technol.* **2018**, *243*, 41–51. [CrossRef]

14. Annika, M.M.; Manninen, M.; Gylienberg, H. The adherence of lactic acid bacteria to the columnar epithelial cells of pigs and calves. *J. Appl. Bacteriol.* **1983**, *55*, 241–245.

15. Miller, G.L. Use of dinitrosalicylic acid reagent for determination of reducing sugar. *Anal. Chem.* **1959**, *31*, 426–428. [CrossRef]

16. Sandhya, C.; Alagarsamy, S.; Szakacs, G.; Pandey, A. Comparative evaluation of neutral protease production by *Aspergillus oryzae* in submerged and solid-state fermentation. *Process. Biochem.* **2005**, *40*, 2689–2694. [CrossRef]

17. Tian, S.; Wang, Z.; Fan, Z.; Zuo, L.; Wang, J. Determination methods of cellulase activity. *Afr. J. Biotechnol.* **2011**, *10*, 7122–7125.

18. Kaiser, M.G.; Cheeseman, J.H.; Kaiser, P.; Lamont, S.J. Cytokine expression in chicken peripheral blood mononuclear cells after in vitro exposure to *Salmonella enterica* serovar enteritidis. *Poult. Sci.* **2006**, *85*, 1907–1911. [CrossRef]

19. AOAC. *Official Methods of Analysis of AOAC International*, 19th ed.; AOAC International: Gaithersburg, MD, USA, 2012.

20. Hu, X.; Chi, Q.; Wang, D.; Chi, X.; Teng, X.; Li, S. Hydrogen sulfide inhalation-induced immune damage is involved in oxidative stress, inflammation, apoptosis and the Th1/Th2 imbalance in broiler bursa of Fabricius. *Ecotox Environ. Safe.* **2018**, *164*, 201–209. [CrossRef]

21. Kalantar, M.; Schreurs, N.M.; Raza, S.H.A.; Khan, R.; Ahmed, J.Z.; Yaghobfar, A.; Shah, M.A.; Kalantar, M.H.; Hosseini, S.M.; Rahman, S.U. Effect of different cereal-based diets supplemented with multi-enzyme blend on growth performance villus structure and gene expression (SGLT1, GLUT2, PepT1 and MUC2) in the small intestine of broiler chickens. *Gene Rep.* **2019**, *15*, 100367. [CrossRef]

22. Santos, T.T.; Srinongkote, S.; Bedford, M.R.; Walk, C.L. Effect of high phytase inclusion rates on performance of broilers fed diets not severely limited in available phosphorus. *Asian-Australas J. Anim. Sci.* **2013**, *26*, 227–232. [CrossRef] [PubMed]

23. Mountzouris, K.C.; Dalaka, E.; Palamidi, I.; Paraskeuas, V.; Demey, V.; Theodoropoulos, G.; Fegeros, K. Evaluation of yeast dietary supplementation in broilers challenged or not with Salmonella on growth performance, cecal microbiota composition and Salmonella in ceca, cloacae and carcass skin. *Poult. Sci.* **2015**, *94*, 2445–2455. [CrossRef] [PubMed]

24. Feng, J.; Liu, X.; Xu, Z.R.; Wang, Y.Z.; Liu, J.X. Effects of fermented soybean meal on digestive enzyme activities and intestinal morphology in broilers. *Poult. Sci.* **2007**, *86*, 1149–1154. [CrossRef] [PubMed]

25. Ndrepepa, G. Uric acid and cardiovascular disease. *Eur. Caediol.* **2018**, *484*, 150–163. [CrossRef]

26. Hong, K.J.; Lee, C.H.; Kim, S.W. *Aspergillus oryzae* GB-107 fermentation improves nutritional quality of food soybeans and feed soybean meals. *J. Med. Food.* **2004**, *7*, 430–434. [CrossRef]

27. Gao, J.H.; Zhang, J.; Yu, S.H.; Wu, S.G.; Yoon, I.; Quigley, J.; Gao, Y.P.; Qi, G.H. Effects of yeast culture in broiler diets on performance and immunomodulatory functions. *Poult. Sci.* **2008**, *87*, 1377–1384. [CrossRef]

28. Cil, O.; Phuan, P.W.; Gillespie, A.M.; Lee, S.; Tradtrantip, L.; Yin, J.; Tse, M.; Zachos, N.C.; Lin, R.; Donowitz, M.; et al. Benzopyrimido-pyrrolo-oxazine-dione CFTR inhibitor (R)-BPO-27 for anti-secretory therapy of diarrheas caused by bacterial enterotoxins. *FASEB J.* **2016**, *31*, 751–760. [CrossRef]

29. Lacey, J.A.; Stanley, D.; Keyburn, A.L.; Ford, M.; Chen, H.; Johanesen, P.; Lyras, D.; Moore, R.J. *Clostridium perfringens*-mediated necrotic enteritis is not influenced by the pre-existing microbiota but is promoted by large changes in the post-challenge microbiota. *Vet. Microbiol.* **2018**, *227*, 119–126. [CrossRef]

30. Lee, M.T.; Lin, W.C.; Lee, T.T. Potential crosstalk of oxidative stress and immune response in poultry through phytochemicals—A review. *Asian-Australas J. Anim. Sci.* **2019**, *32*, 309–319. [CrossRef]

31. Gabay, C.; Lamacchia, C.; Palmer, G. IL-1 pathways in inflammation and human diseases. *Nat. Rev. Rheumatol.* **2010**, *6*, 232–241. [CrossRef]

32. Municio, C.; Alvarez, Y.; Montero, O.; Hugo, E.; Rodri'guez, M.; Domingo, E.; Alonso, S.; Ferna'ndez, N.; Crespo, M.S. The response of human macrophages to ß-glucans depends on the inflammatory milieu. *PLoS ONE* **2013**, *8*, e62016. [CrossRef] [PubMed]

33. Huang, C.M.; Lee, T.T. Immunomodulatory effects of phytogenics in chickens and pigs—A review. *Asian-Australas. J. Anim. Sci.* **2017**, *31*, 617–627. [CrossRef] [PubMed]

34. Suschek, C.V.; Schnorr, O.; Bachofen, V.K. The role of *iNOS* in chronic inflammatory processes in vivo: Is it damage-promoting, protective, or active at all? *Curr. Mol. Med.* **2004**, *4*, 763–775. [CrossRef] [PubMed]

35. Arias-Salvatierra, D.; Silbergeld, E.K.; Acosta-Saavedra, L.C.; Calderon-Aranda, E.S. Role of nitric oxide produced by *iNOS* through *NF-κB* pathway in migration of cerebellar granule neurons induced by lipopolysaccharide. *Cell Signal.* **2011**, *23*, 425–435. [CrossRef] [PubMed]

36. Ivashkiv, B.; Donlin, L.T. Regulation of type I interferon responses. *Nat. Rev. Immunol.* **2014**, *14*, 36–49. [CrossRef] [PubMed]
37. Harada, K.; Isse, K.; Nakanuma, Y. Interferon gamma accelerates *NF-κB* activation of biliary epithelial cells induced by Toll-like receptor and ligand interaction. *J. Clin. Pathol.* **2006**, *59*, 184–190. [CrossRef]

Article

Evaluation of Waste Mushroom Compost as a Feed Supplement and Its Effects on the Fat Metabolism and Antioxidant Capacity of Broilers

Wen Yang Chuang [1], Chu Ling Liu [1], Chia Fen Tsai [1], Wei Chih Lin [1], Shen Chang Chang [2], Hsin Der Shih [3], Yi Ming Shy [4] and Tzu-Tai Lee [1,5,*]

[1] Department of Animal Science, National Chung Hsing University, Taichung 402, Taiwan; xssaazxssaaz@yahoo.com.tw (W.Y.C.); s04610215@gmail.com (C.L.L.); bbella1123tw@gmail.com (C.F.T.); waynezi2227738@gmail.com (W.C.L.)

[2] Kaohsiung Animal Propagation Station, Livestock Research Institute, Council of Agriculture, Tainan 71246, Taiwan; macawh@mail.tlri.gov.tw

[3] Taiwan Agricultural Research Institute Council of Agriculture, Executive Yuan, Taichung City 41362, Taiwan; tedshih@tari.gov.tw

[4] Hsinchu Branch, Taiwan Livestock Research Institute, Council of Agriculture, Executive Yuan, Tainan 71246, Taiwan; emshy@tlri.gov.tw

[5] The iEGG and Animal Biotechnology Center, National Chung Hsing University, Taichung 402, Taiwan

* Correspondence: ttlee@dragon.nchu.edu.tw; Tel.: +886-4-22840366; Fax: +886-4-22860265

Received: 13 February 2020; Accepted: 5 March 2020; Published: 6 March 2020

Simple Summary: Mushroom waste compost is the main byproduct when cultivating mushrooms. Due to its high mycelium content, mushroom waste compost may improve animal health by increasing antioxidant capacity. Furthermore, increasing evidence suggests that supplementing animal diets with fiber could improve body composition and health. The results showed that supplementation with mushroom waste compost accelerates adipolysis and enhances the antioxidant capacity of broilers. Among all treatment groups, broilers given dietary supplementation with 0.5% mushroom waste compost showed improved feed conversion rate and the highest adipose metabolism.

Abstract: *Pennisetum purpureum* Schum No. 2 waste mushroom compost (PWMC) is the main byproduct when cultivating *Pleurotus eryngii*. Due to the high mycelium levels in PWMC, it may have potential as a feed supplement for broilers. This study investigated the effects of PWMC supplementation on antioxidant capacity and adipose metabolism in broilers. In the study, 240 broilers were randomly allocated to one of four treatment groups: basal diet (control), 0.5%, 1%, or 2% PWMC supplementation. Each treatment group had 60 broilers, divided into three replicates. The results showed that supplementation with 0.5% PWMC decreased the feed conversion rate (FCR) from 1.36 to 1.28, compared to the control. Supplementation with 0.5% or 2% PWMC decreased glucose and triglyceride levels, compared to the control ($p < 0.0001$), the concentrations of adiponectin and oxytocin increased from 5948 to 5709, 11820, and 7938 ng/ mL; and 259 to 447, 873, and 963 pg/ mL, respectively. Toll-like receptor 4 was slightly increased in the 0.5% and 1% PWMC groups. Both interferon-γ (IFN-γ) and interleukin-1ß (IL-1ß) were significantly decreased, by about three to five times for IFN-γ ($p < 0.0001$) and 1.1 to 1.6 times for IL-1ß ($p = 0.0002$). All antioxidant-related mRNA, including nuclear factor erythroid 2–related factor 2 (Nrf-2) and superoxidase dismutase-1 (SOD-1), increased significantly following PWMC supplementation. Both claudin-1 and zonula occludens 1 increased, especially in the 2% PWMC group. Excitatory amino acid transporter 3 (EAAT3) significantly increased by about 5, 12, and 11 times in the 0.5%, 1%, and 2% PWMC groups. All adipolysis-related mRNA were induced in the PWMC treatment groups, further enhancing adipolysis. Overall, 0.5% PWMC supplementation was recommended due to its improving FCR, similar antioxidant capacity, and upregulated adipolysis.

Keywords: mushroom compost; fat metabolism; antioxidant; broilers

---

1. Introduction

Due to global climate change and rising crude oil prices, high-starch animal feed ingredients such as corn and cassava must be reconsidered in view of their value as raw materials for biomass energy [1]. Considering the cost of raising broilers, it is increasingly important to find cheap alternatives that can maintain animal growth and health. Crops are the main sources of raw feed materials, resulting in a large amount of high-fiber agricultural byproducts [2]. Traditionally, due to the limitations of animal digestion, high-fiber plant materials have not been considered for animal feed [3]. Unlike animals, mushrooms produce a large amount of cellulase and hemicellulase, and fiber is the main source of their growth [4]. As such, the culture medium used in mushroom cultivation is high in fiber and contains few minerals or crude proteins [4]. However, following cultivation, the medium is still high in crude fiber and may cause new environmental issues [5,6].

Due to its high-fiber characteristics, *Pennisetum* can be used as one of the raw materials or an adjustment material in the cultivation of mushrooms, reducing the environmental problems caused by the traditional use of wood chips [6]. In addition, mushroom waste compost produced by mushroom planting contains a lot of mycelium, which could improve animal antioxidant capacity and health [5]. Mahfuz et al. also indicated that although 2% mushroom waste compost addition would not increase the feed conversion rate or body weight of broilers, mushroom waste compost addition could decrease broilers' serum cholesterol content and improve immunomodulation [7].

In addition, previous researches pointed out that plant-based phenol-like compounds, such as epigallocatechin gallate (EGCG), could suppress fatty acid synthase and acetyl CoA carboxylase activities thereby decreasing the adipogenesis in broilers [8]. Lee et al. [9] also pointed out that the addition of mushroom waste compost could decrease crude fat in broiler meat and the results might be caused by the decrease of lipid peroxidation.

However, there is limited research on the reuse of mushroom waste compost or its effects when used as a supplement in broiler feed. This study therefore investigated the effect of mushroom waste compost on antioxidant capacity and how could it alter adipose metabolism in broilers.

2. Materials and Methods

2.1. The Collection and Characteristics of Pennisetum purpureum Schum No. 2 Waste Mushroom Compost (PWMC)

PWMC was the medium used to grow *Pleurotus eryngii*. It contained at least 70% *Pennisetum purpureum* Schum No. 2 (PP) and all mature *P. eryngii* were removed. The PWMC came from the Taiwan Agricultural Research Institute Council of Agriculture, Executive Yuan. Briefly, after removing *P. eryngii*, the medium with fresh residues had about 50% moisture and was immediately dried at 50 °C for 1 day before being crushed into powder. PWMC was stored at 4 °C until use.

The Folin–Ciocalteu method described by Tabart et al. [10] was used for the total phenol content analysis. Briefly, the sample was extracted by deionized water at 95 °C for 30 min and cooled down to room temperature before being used. Sample solution of 1 mL was mixed with 5 mL Folin–Ciocalteu reagent and 4 mL 7.5% sodium carbonate. The mixture was incubated in the room temperature for 30 min and measured the absorbance at 730 nm. The 0.01–0.1 mg/mL gallic acid was also measured by the methods described above for the standard curve. The presentation of total phenol content was described as gallic acid equivalent (GAE) mg/g sample.

PWMC was extracted by deionized water at 95 °C for 30 min and cooled down to room temperature, and filtrated by a 0.22 μm filter before used for the phenol-like compounds detection. HPLC (Hitachi, Kyoto, Japan) equipped with a pump (CM 5110), a column (C18-AR, 250 × 4.8 mm, maintained at

40 °C by the column oven (CM 5310)), an autosampler (L-2200) and a computer system with HPLC D-2000 Elite. The mobile phase conditions were (A) 0.05% *v/v* H_3PO_4 and (B) 3:2 *v/v* CH_3OH/CH_3CN solution, 1.0 mL/min, and UV detection (CM 5420) at 280nm. Also, 0.01~1.5 mg/ mL of gallocatechin (GC), epigallocatechin (EGC), catechin (CC), epicatechin (EC), gallic acid (GA), epigallocatechin gallate (EGCG), epicatechin gallate (ECG), catechin gallate (CG), and caffeic acid were measured by the methods described above for the standard curve.

The methods of total flavonoid content were described by Pourmorad et al. [11]. Briefly, a sample was extracted by methanol on ice for 30 min. The 1 mL sample solution, 3 mL methanol, 0.2 mL 10% $AlCl_3$, 0.2 mL 1M potassium acetate, and 5 mL distilled water were mixed and incubated at room temperature for 30 min. The absorbance of mixture was measured at 700 nm. Quercetin of 0.01–0.1 mg/mL was also measured by the methods described above for the standard curve.

The methods of crude triterpenoid content were described by Wei et al. [12]. Briefly, a 3 g sample was added to 100 mL of 50% ethanol and shaken (100 rpm) for 8 h at room temperature. The solution was centrifuged at 3000 rpm and the supernatant was removed to a new tube. Next, 100 mL 95% ethanol was added in the precipitate and shaken (100 rpm) for 12 h at room temperature. The new solution was centrifuged at 3000 rpm and the two supernatants were mixed. The new supernatant was filtrated by Advantec No.1 filter papers and stoved at 50 °C for three days. The 100 mL saturated $NaHCO_3$ (pH 3.5) and ethyl acetate were used to wash the remains to remove the impurity. The ethyl acetate solution was collected and dried to constant weight. The dry remains weight was measured and calculated as the formula described as follows.

$$\text{Crude triterpenoid content (\%)} = \text{Dry remains weight/ Sample weight} \times 100\% \tag{1}$$

2.2. Animal Experiment Design

PWMC was used to test its effect on broiler fat metabolism, nutrient absorption, antioxidant function, and inflammatory regulation. The experiment was carried out at National Chung Hsing University, Taiwan and all protocol followed that of the Animal Care and Use Committee in NCHU (IACUC: 108-049). Every broiler could drink and eat freely and had enough active space (about 20 broilers/9 m^2). If the chickens had become infected during rearing, they would have been quarantined and given antibiotics or medication at the discretion of the veterinarian, however every broiler remained healthy throughout the whole rearing period. Euthanasia was used to reduce broiler suffering. The methods for animal design, testing intestinal morphology, blood and serum characteristics, and mRNA expression were slightly modified from those of Chuang et al. [13]. Briefly, 240 male 1-day-old Ross 308 broilers were used for 35-day experiments and each was placed in one of four groups: basal diet (control), 0.5%, 1%, or 2% PWMC supplementation. Every group had 60 broilers with three replicates (20 per pen). The initial body weight of each chick was similar (48.0 ± 0.7 g/bird) among all groups. The temperature was kept at 33 °C for 1-day-old chicks then slowly decreased to 22 °C after 30 days. To achieve or exceed the nutritional requirements of the broilers, according to NRC 1994, all diets were recalculated and the proximate composition was analyzed according to the AOAC (2012) [14] to determine the gross energy and crude protein levels (Table 1). The starter diet was provided for the first 21 days, and the finisher diet provided for days 22–35. Body weight gain, feed intake, and feed conversion rate (FCR) were measured on 21- and 35-day-old broilers. On day 35, the ileum, abdominal fat, spleen, and liver were collected for qPCR analysis.

2.3. Serum Characteristics

At the end of the experiment, 5 mL whole blood was collected from the broilers (three from each pen, nine per treatment) and left to stand for 4 to 5 h at 4 °C before measurement. The samples were centrifuged at 3000 rpm for 10 min to separate the blood cells and serum, before the serum was frozen at −20 °C until use. The concentrations of oxytocin (OXT), corticosterone), tumor necrosis factor alpha (TNF-α), glutathione peroxidase (Gpx), superoxidase dismutase (SOD) A), and malondialdehyde

(MDA) in the broiler serum were measured. All analysis methods followed the manufacturer's protocol. Briefly, the protein analytic was bound by the antibody and the enzyme-linked immunosorbent assay reader was used to measure absorbance at wavelengths provided by the manufacturer. Other blood cells and serum characteristics were measured using the automatic biochemical analyzer (Hitachi, 7150 auto-analyzer, Tokyo, Japan).

Table 1. Composition and calculated analysis (g/kg as fed) of the basal diet for broilers (1–35 days) [1].

Ingredients	Starter Diet	Finisher Diet
	(1–21 Days)	(22–35 Days)
	g/kg	
Corn, yellow	520	572
Soybean meal (CP 44%)	369	247
Full-fat soybean meal	0.00	50.0
Soybean oil	40.6	41.4
Fish meal (CP 65%)	30.0	50.0
Calcium carbonate	16.5	16.5
Monocalcium phosphate	11.2	11.2
DL-Methionine	2.00	2.00
L-Lysine-HCl	3.70	3.70
NaCl	3.90	3.90
Choline-Cl	0.80	0.80
Vitamin premix [2]	1.00	1.00
Mineral premix [3]	1.00	1.00
Total	1000	1000
Calculated nutrient value		
Dry matter, %	88.1	88.2
Crude protein, %DM	23.0	21.0
Crude fiber, %DM [4]	4.71	3.55
Calcium, %DM	1.11	1.21
Total Phosphorus, %DM	0.68	0.72
Available Phosphorus, %DM	0.46	0.52
Methionine + Cysteine, %DM	0.92	0.89
ME, kcal/kg DM	3050	3175
Chemical analysis value		
Dry matter, %	88.7	89.3
Crude protein, %DM	23.1	21.0
Crude fat, DM%	6.85	7.96
Chemical analysis value	PWMC	
Neutral detergent fiber, %DM	60.1 ± 0.70	
Crude protein, %DM	6.50 ± 0.40	
Ash, %DM	10.2 ± 0.20	

[1] Con: basal diet for control group; PWMC: *Pennisetum purpureum* Schum No. 2 waste mushroom compost. PWMC was added directly to the basal diet at different percentages. [2] Vitamins (premix content per kg diet): Vit. A, 15,000 IU; Vit. D3, 3000 IU; Vit. E, 30 mg; Vit. K3, 4 mg; thiamine, 3 mg; riboflavin, 8 mg; pyridoxine, 5 mg; Vit. B12, 25 μg; Ca-pantothenate, 19 mg; niacin, 50 mg; folic acid, 1.5 mg; and biotin, 60 μg. [3] Minerals (premix content per kg diet): Co ($CoCO_3$), 0.255 mg; Cu ($CuSO_4 \cdot 5H_2O$), 10.8 mg; Fe ($FeSO_4 \cdot H_2O$), 90 mg; Mn ($MnSO_4 \cdot H_2O$), 90 mg; Zn (ZnO), 68.4 mg; Se (Na_2SeO_3), 0.18 mg. [4] The analysis of crude fiber content in the basal diet, 0.05, 1 and 2% PWMC addition treatments were 4.8, 5.1, 5.3, and 5.7% dry matter (DM) in the starter stage and 3.6, 3.8, 4.1, and 4.5% DM in the finisher stage, respectively.

2.4. Total RNA Isolation and qPCR

Total RNA isolation and qPCR were used to analyze mRNA expression in the liver, spleen, ileum, and abdominal fat. All mRNA was collected from 35-day-old broilers in each treatment group. The methods of total mRNA isolation followed the manufacturer's protocol (SuperScript™ FirstStrand Synthesis System reagent, Thermo Fisher, Waltham, MA, USA). The mRNA purity was determined by the absorbance ratio of 260/280 nm, and the methods for cDNA synthesis and qPCR analysis were as per Chuang et al. [13] Briefly, cDNA was mixed with 2× SYBR GREEN PCR Master Mix-ROX (Gunster Biotech, Co., Ltd., Taipei, Taiwan), deionized water, and each primer at a ratio of 5:1.2:1.8:1. A StepOnePlus™ Real-Time PCR System (Thermo Fisher, Waltham, MA, USA) was used to detect qRT-PCR performance. The $2^{-\Delta\Delta Ct}$ method was used to calculate the relative mRNA expression level, and ß-actin was used as the housekeeping gene for normalization. All primer sequences matched the genes of *Gallus gallus* (chicken) from Genbank (Table 2).

Table 2. The primer sequence of each gene according to Genbank or other research.

Gene Name [1]	Primer Sequence	Genbank No.
ß-actin	F: 5′-CTGGCACCTAGCACAATGAA-3′ R: 5′-ACATCTGCTGGAAGGTGGAC-3′	X00182.1
TNF-α	F: 5′-TGTGTATGTGCAGCAACCCGTAGT-3′ R: 5′-GGCATTGCAATTTGGACAGAAGT-3′	NM 204267
TLR4	F: 5′-TGCACAGGACAGAACATCTCTGGA-3′ R: 5′-AGCTCCTGCAGGGTATTCAAGTGT-3′	NM_001030693
NQO-1	F: 5′-AAGAAGATTGAAGCGGCTGA-3′ R: 5′-GCATGGCTTTCTTCTTCTGG-3′	NM_001277619.1
NFκB	F: 5′- CCAGGTTGCCATCGTGTTCC- 3′ R: 5′- GCGTGCGTTTGCGCTTCT -3′	D13719.1
iNOS	F: 5′-TACTGCGTGTCCTTTCAACG -3′ R: 5′-CCCATTCTTCTTCCAACCTC-3′	U46504
IFN-γ	F: 5′-CTCCCGATGAACGACTTGAG-3′ R: 5′-CTGAGACTGGCTCCTTTTCC-3′	Y07922
IL-1ß	F: 5′-GCTCTACATGTCGTGTGTGATGAG-3′ R: 5′-TGTCGATGTCCCGCATGA-3′	NM_204524
HO-1	F: 5′-GGTCCCGAATGAATGCCCTTG-3′ R: 5′-ACCGTTCTCCTGGCTCTTGG-3′	NM_205344.1
Nrf-2	F: 5′-GGAAGAAGGTGCGTTTCGGAGC-3′ R: 5′-GGGCAAGGCAGATCTCTTCCAA-3′	NM_205117.1
GCLC	F: 5′-CAGCACCCAGACTACAAGCA-3′ R: 5′-CTACCCCCAACAGTTCTGGA-3′	XM_419910.3
GPx	F: 5′- CAGCAAGAACCAGACACCAA-3′ R: 5′- CCAGGTTGGTTCTTCTCCAG-3′	NM_001163245.1
SOD-1	F: 5′- ATTACCGGCTTGTCTGATGG-3′ R: 5′- CCTCCCTTTGCAGTCACATT-3′	NM_205064.155
Claudin-1	F: 5′-GGAGGATGACCAGGTGAAGA-3′ R: 5′-TCTGGTGTTAACGGGTGTGA-3′	NM_001013611.2
MUC-2	F: 5′-GCTACAGGATCTGCCTTTGC-3′ R: 5′-AATGGGCCCTCTGAGTTTTT-3′	JX284122.1
Occludin	F: 5′-GTCTGTGGGTTCCTCATCGT-3′ R: 5′-GTTCTTCACCCACTCCTCCA-3′	NM_205128.1
ZO-1	F: 5′-AGGTGAAGTGTTTCGGGTTG-3′ R: 5′-CCTCCTGCTGTCTTTGGAAG-3′	XM_015278975.1
EAAT3	F: 5′- ACCCCTTCTGATCACCTCT-3′ R: 5′- TGAGCATGCTGATTCCAAAG-3′	XM_424930.6
FFAR2	F: 5′-GCCCCATAGCAAACTTCT-3′ R: 5′-GGGCAGCCATAAAGAGAG-3′	[15]
GLUT2	F: 5′- CCGCAGAAGGTGATAGAAGC-3′ R: 5′- ACACAGTGGGGTCCTCAAAG-3′	NM_207178.1
SGLT	F: 5′- CATCTTCCGAGATGCTGTCA-3′ R: 5′- CAGGTATCCGCACATCACAC-3′	NM_001293240.1
PEPT-1	F: 5′- CAGGGATCGAGATGGACACT-3′ R: 5′- CACTTGCAAAAGAGCAGCAG-3′	NM_204365.1
KCTD-15	F: 5′-TTAAAAACACCCCGTTCTGC-3′ R: 5′-AAAACAAACCAAGCGACCAC-3′	XM_004944237
Adiponectin	F: 5′-ACTTTCATGGGCTTCCTCCT-3′ R: 5′-GTCCCACGGAAGTCACTTGT-3′	NM_206991
ATGL	F: 5′-CAGCAGGACGTTTGGGTATT-3′ R: 5′-CCACGCAAAGTTGGAGGTAT-3′	EU240627.2
AMPK-α2	F: 5′-GGCATTGAGGAAATCAGGAA-3′ R: 5′-CCTGAACCAATGTGTGTTGC-3′	DQ340396
FAS	F: 5′-GCTGAGAGCTCCCTAGCAGA-3′ R: 5′-TCCTCTGCTGTCCCAGTCTT-3′	NM_205155
FABP4	F: 5′-CAGCATCAATGGTGATGTGA-3′ R: 5′-TCTCTTTGCCATCCCACTTC-3′	NM_204290
CEBPα	F: 5′-GGAGCAAGCCAACTTCTACG-3′ R: 5′-GTCGATGGAGTGCTCGTTCT-3′	NM_001031459
CPT-1	F: 5′-ATCCCAGCTGCAGTGAGTCT-3′ R: 5′-ATTCGCAAGTCAATCCCATC-3′	NM_001012898
IL-6	F: 5′-AGGACGAGATGTGCAAGAAGTTC-3′ R: 5′-TTGGGCAGGTTGAGGTTGTT-3′	NM_204628
PPAR-γ	F: 5′-GATCGCCCAGGTTTGTTAAA-3′ R: 5′-TGCACGTGTTCCGTTACAAT-3′	NM_001001460
PPAR-α	F: 5′-AGGCCAAGTTGAAAGCAGAA-3′ R: 5′-GTCTTCTCTGCCATGCACAA-3′	NM_001001464.1

[1] TNF-α: Tumor necrosis factor alpha; TLR4: Toll-like receptor 4; NQO-1: NADPH dehydrogenase 1; NFκB: Nuclear factor kappa B p 65; iNOS: Inducible nitric oxide synthases; IFN-γ: Interferon-γ; IL-1ß: Interleukin-1ß; HO-1: Heme oxygenase-1; Nrf-2: Nuclear factor erythroid 2–related factor 2; GCLC: Glutamate-cysteine ligase catalytic; Gpx: glutathione peroxidase; SOD-1: Superoxide dismutase-1; MUC-2: Mucin2; ZO-1: Zonula occludens 1; EAAT3: Excitatory amino acid transporter 3; FFAR2: Free fatty acid receptor 2; GLUT2: glucose transporter 2; SGLT: sodium-dependent glucose cotransporters 1; PEPT-1: Peptide transporter 1; KCTD-15: Potassium channel tetramerization domain-containing 15; ATGL: Adipose triglyceride lipase; AMPK-α2: 5′-AMP-activated protein kinase catalytic subunit alpha-2; FAS: Fatty acid synthase; FABP4: Fatty acid binding protein 4; CEBPα: CCAAT-enhancer-binding proteins-alpha; CPT-1: Carnitine palmitoyltransferase I; IL-6: Interleukin-6; PPAR-γ: Peroxisome proliferator-activated receptor gamma; PPAR-α: Peroxisome proliferator-activated receptor alpha.

2.5. Statistical Analysis

All data analysis was calculated using SAS software (SAS® 9.4, 2018, SAS Institute Inc., Cary, NC, USA). The difference between each experimental group was determined by the Tukey range test with a P value less than 0.05. The chemical analysis value present in the unit of "%DM" means analysis value/the percentage of dry matter (DM).

3. Results

3.1. The Characteristics of PWMC

Table 3 reports the levels of the different functional compounds in PWMC, including crude triterpenes, polyphenols, and flavonoids. The total phenol and flavonoids content were 1.84 GAE mg/g DM and 1.20 QE mg/g DM in PWMC. Among the phenol-like compounds, the concentrations of GA, GC, EGC, CC, caffeic acid, EC, EGCG, ECG, and CG in PWMC were 114, 1035, 1493, 7.9, 113, 230, 210, 78, and 401 µg/g DM.

Table 3. The functional chemical composition analysis in PWMC.

Items	PWMC
Functional component analysis	
Crude triterpenes (mg/ g DM [1])	6.25 ± 0.37
Total phenol contents (GAE [2] mg/ g DM)	1.84 ± 0.05
Total flavonoids (QE [3] mg/ g DM)	1.20 ± 0.26
Phenol-like chemical analysis (µg/ g DM)	
Gallic acid	114.0 ± 2.7
Gallocatechin	1035.0 ± 8.0
Epigallocatechin	1493.0 ± 14.0
Catechin	7.9 ± 0.6
Caffeic acid	113.0 ± 0.5
Epicatechin	229.8 ± 2.0
Epigallocatechin gallate	210.2 ± 0.1
Epicatechin gallate	78.1 ± 2.8
Catechin gallate	401.2 ± 5.1

[1] DM: Dry matter; [2] GAE: Gallic acid equivalent; [3] QE: Quercetin. (n = 5).

3.2. Broiler Growth Performance with PWMC Supplementation

Broiler growth performance for days 1–21 and 22–35 are shown in Table 4. Within these two evaluation periods, there was no significant difference in body weight, weight gain, and feed consumption. The best feed conversion rate (FCR) was seen in the 0.5% PWMC group during the starter stage, compared to the control (1.36 vs. 1.28, $p = 0.0374$). However, there were no significant changes in weight gain and FCR for days 22–35 and 1–35.

Table 4. The growth performance of 35-day-old broilers following dietary PWMC [1] supplementation.

Items	Treatments				SEM [2]	*p* Value
	Con	0.5% PWMC	1% PWMC	2% PWMC		
1–21 d						
Body weight, g/bird	925	956	939	919	30.2	0.848
Weight gain, g/bird	880	907	890	871	30.1	0.853
Feed consumption, g/bird	1197	1161	1184	1219	38.1	0.691
FCR	1.36 [a]	1.28 [b]	1.33 [ab]	1.40 [a]	0.03	0.037
22–35d						
Body weight, g/bird	2019	2103	2021	2009	30.7	0.184
Weight gain, g/bird	1091	1148	1081	1089	56.7	0.335
Feed consumption, g/bird	2007	2204	2142	2114	107	0.641
FCR	1.84	1.92	1.98	1.94	0.09	0.742
1–35 d						
Feed consumption, g/bird	3213	3123	3135	3176	30.9	0.190
Weight gain, g/bird	1971	2055	1972	1960	122	0.965
FCR	1.63	1.52	1.59	1.62	0.058	0.588

[1] Con: basal diet for control group; PWMC: *Pennisetum purpureum* Schum No. 2 waste mushroom compost. PWMC was added directly to the basal diet at different percentages. [2] SEM: Standard error of mean. [a,b] Means within the same rows without the same superscript letter are significantly different ($p < 0.05$).

3.3. Intestinal Barrier and Nutrient Absorption-Related mRNA Expression in 35-Day-Old Broilers

There was no significant difference in occludin (*OCCN*) content within each group ($p > 0.05$). However, mucin 2 (*MUC2*), the major protein that maintains the function and structure of mucus, increased about 2.5, 2.5, and 3.5 times in the 0.5%, 1%, and 2% PWMC groups, respectively (Figure 1A, $p < 0.0001$). Both claudin-1 (*CLDN-1*) and zonula occludens 1 (*ZO-1*), which are tight junction-related proteins, increased in the PWMC-supplemented groups, especially 2% PWMC (Figure 1A, $p < 0.0001$ and $=0.0245$, respectively). Among all the groups, only 2% PWMC supplementation enhanced *ZO-1* mRNA expression. The amount of PWMC supplementation was positively correlated to *CLDN1* mRNA expression but there was no significant difference between the 0.5% and 1% PWMC groups.

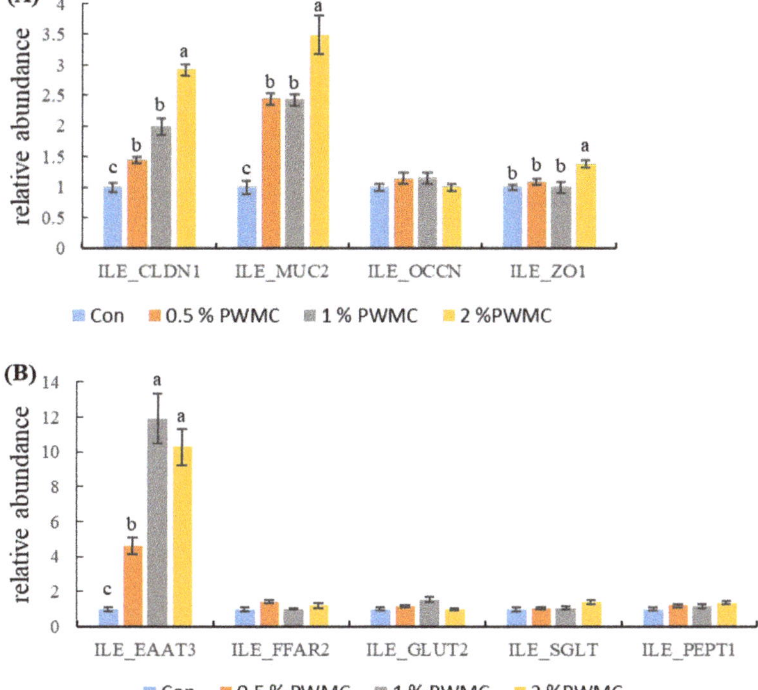

Figure 1. Intestinal barrier (**A**) and nutrient absorption-related (**B**) mRNA expression in the ileum (ILE) for 35-day-old broilers. CLDN: Claudin-1; MUC-2: Mucin2; OCCN: Occludin; ZO-1: Zonula occludens 1; EAAT3: Excitatory amino acid transporter 3; FFAR2: Free fatty acid receptor 2; GLUT2: Glucose transporter 2; SGLT: Sodium-dependent glucose cotransporter 1; PEPT-1: Peptide transporter 1. [a-c] Means within the same rows without the same superscript letter are significantly different ($p < 0.05$).

Most nutrient absorption is through transporters on the intestinal epithelial cells. By adding PWMC to broiler feed, excitatory amino acid transporter 3 (*EAAT3*) levels significantly increased by about 5, 12, and 11 times for the 0.5%, 1%, and 2% PWMC groups, respectively (Figure 1B, $p < 0.0001$). However, the expression of other nutrient absorption-related mRNA, like free fatty acid receptor 2 (*FFAR2*), glucose transporter 2 (*GLUT2*), sodium-dependent glucose cotransporter 1 (*SGLT*), and peptide transporter 1 (*PEPT1*), was not significantly different from the control (Figure 1B).

3.4. Intestinal Morphology in 35-Day-Old Broilers

The addition of PWMC did not significantly change the villus height or crypt depth in the jejunum or ileum of 35-day-old broilers, however the *Tunica muscularis* in the jejunum increased significantly by 1.47, 1.34, and 1.06 times in the 0.5%, 1%, and 2% PWMC groups ($p = 0.0104$) (Table 5). Only the 0.5% PWMC group was significantly different from the control in jejunum *Tunica muscularis* ($p = 0.010$), while the 2% PWMC group had a villus:crypt ratio about 1.22 times higher than the control.

Table 5. Jejunum and ileum morphology of 35-day-old broilers following PWMC [1] supplementation.

Items	Treatments				SEM [2]	*p* Value
	Con	0.5% PWMC	1% PWMC	2% PWMC		
Jejunum						
Villus height (μm)	1212	1301	1367	1467	69.9	0.109
Crypt depth (μm)	192	208	203	190	9.37	0.496
Tunica muscularis (μm)	152 [b]	224 [a]	204 [ab]	162[ab]	22.1	0.010
Villus:crypt	6.32 [b]	6.29 [b]	6.73 [b]	7.75[a]	0.24	0.002
Ileum						
Villus height (μm)	1166	1110	1102	1038	38.3	0.175
Crypt depth (μm)	171	159	168	158	10.8	0.798
Tunica muscularis (μm)	166	137	144	173	25.6	0.712
Villus:crypt	6.84	6.99	6.58	6.99	0.52	0.932

[1] Con: basal diet for control group; PWMC: *Pennisetum purpureum* Schum No. 2 waste mushroom compost. PWMC was added directly to the basal diet at different percentages. [2] SEM: Standard error of means. [a,b] Means within the same rows without the same superscript letter are significantly different ($p < 0.05$).

3.5. Adipogenesis- and Adipolysis-Related mRNA Expression in the Liver and Adipocytes of 35-Day-Old Broilers

The mRNA expression of potassium channel tetramerization domain-containing 15 (*KCTD-15*), adiponectin, adipose triglyceride lipase (*ATGL*), and 5′-AMP-activated protein kinase catalytic subunit alpha-2 (AMPK-α2) were significantly increased in the livers (L) of the treatment groups ($p < 0.0001$, < 0.0001, $=0.0024$ and $= 0.0015$, respectively), thereby increasing the rate of adipolysis (Figure 2A). There was no significant difference among the PWMC-supplemented groups. Similar results are shown in Figure 2C, where the adipolysis-related mRNA, including adiponectin, KCTD15, ATGL, and carnitine palmitoyltransferase 1 (*CPT-1*) (all of the $p < 0.0001$), increased in adipocytes (F) by at least four times. However, the adipogenesis-related mRNA shown in Figure 2B, including fatty acid synthase (*FAS*), fatty acid binding protein 4 (*FABP4*), and CCAAT-enhancer-binding proteins-alpha (*CEBPα*), had a compensatory one to three times increase in the adipocytes ($p < 0.0001$, <0.0001 and $= 0.0052$) (Figure 2D). When comparing adipogenesis- and adipolysis-related mRNA, the relative abundance of adipolysis-related mRNA expression was much higher (Figure 2). Nevertheless, interleukin-6 (IL-6) mRNA expression in adipocytes decreased ($p < 0.0001$), thereby decreasing inflammation.

3.6. Serum Characteristics in 35-Day-Old Broilers

As the data shows in Table 6, serum glucose content decreased in the 0.5% and 2% PWMC-supplemented groups ($p < 0.0001$), but there was no significant difference in MDA, SGOT, SGPT, or Alk-P levels. Gpx activity was not significantly different among the different groups ($p = 0.6412$), however SOD, the antioxidant-related enzyme, increased significantly from 523.9 mU/mL to 865.2 mU/mL, 842.6 mU/mL, and 879.3 mU/mL ($p < 0.0001$), respectively, and showed no dose effect. Serum TNF-α levels decreased 1.2 to two times ($p = 0.001$), and adiponectin increased the most in the 1% PWMC group (from 5948 ng/mL to 11820 ng/mL, compared to the control, $p < 0.0001$). OXT content also increased with a dose effect ($p = 0.0016$). The data showed that cholestenone (CHOL) levels, including high density lipoprotein-cholestenone (HDL-C) and low density lipoprotein- cholestenone (LDL-C), were not affected by the addition of PWMC, but triglyceride content decreased significantly ($p < 0.0001$).

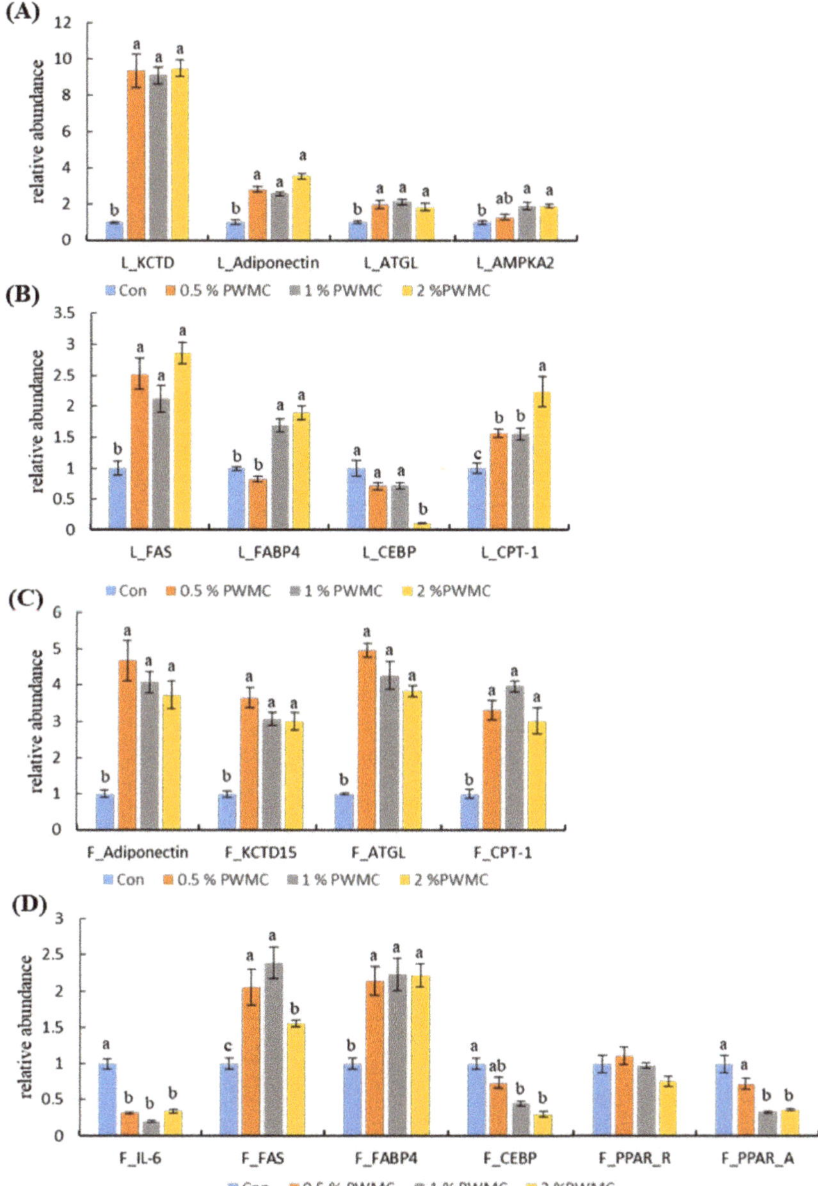

Figure 2. Adipolysis- and adipogenesis-related mRNA expression in the liver (L) and adipocytes (F) of 35-day-old broilers. (**A**) Potassium channel tetramerization domain-containing 15 (KCTD-15), adiponectin, adipose triglyceride lipase (ATGL) and 5′-AMP-activated protein kinase catalytic subunit alpha-2 (AMPK-α2) mRNA expression in L; (**B**) fatty acid synthase (FAS), fatty acid binding protein 4 (FABP4), CCAAT-enhancer-binding proteins-alpha (CEBPα), and carnitine palmitoyltransferase 1 (CPT-1) mRNA expression in L; (**C**) adiponectin, KCTD-15, ATGL, and CPT-1 mRNA expression in F; (**D**) interleukin-6 (IL-6), FAS, FABP4, CEBPα, peroxisome proliferator-activated receptor gamma (PPAR-γ), and peroxisome proliferator-activated receptor alpha (PPAR-α). [a,b] Means within the same rows without the same superscript letter are significantly different ($p < 0.05$).

Table 6. Serum characteristics of 35-day-old broilers following PWMC [1] supplementation.

Items [3]	Treatments				SEM [2]	*p* Value
	Con	0.5% PWMC	1% PWMC	2% PWMC		
GLU (mg/dL)	276 [a]	243 [b]	260 [ab]	224 [c]	6.10	<0.001
MDA (uM)	10.1	15.2	13.3	11.4	1.70	0.234
Protein						
SGOT (U/ L)	332	286	273	342	41.0	0.568
SGPT (U/ L)	4.20	4.00	4.50	6.00	0.690	0.184
Alk-P (IU/ L)	1086	937	1366	1282	277	0.694
Gpx (nmol/min/mL)	115	126	129	115	9.59	0.641
SOD (mU/ mL)	524 [b]	865 [a]	843 [a]	879 [a]	20.7	<0.001
TNF-α (pg/mL)	285 [a]	162 [c]	176 [c]	232 [b]	16.6	0.001
Oxytocin (pg/mL)	259 [b]	447 [b]	874 [a]	963 [a]	93.9	0.002
Adiponectin (ng/mL)	5948 [c]	5709 [c]	11820 [a]	7938 [b]	303	<0.001
Corticosterone (pg/mL)	133	121	126	118.2	3.45	0.064
Lipid						
CHOL (mg/dL)	125	124	137	122	7.00	0.379
TG (mg/dL)	54.8 [a]	38.3 [b]	43.7 [ab]	40.2 [b]	2.30	<0.001
HDL-C (mg/dL)	78.0	76.2	83.7	73.2	3.70	0.263
LDL-C (mg/dL)	42.2	42.5	48.0	42.5	3.40	0.577

[1] Con: basal diet for control group; PWMC: *Pennisetum purpureum* Schum No. 2 waste mushroom compost. PWMC was added directly to the basal diet at different percentages. [2] SEM: Standard error of means. [3] GLU: Glucose; MDA: malondialdehyde; SGOT: Serum glutamic-oxaloacetic transaminase; SGPT: Serum glutamic-pyruvic transaminase; Alk-P: Alkaline phosphatase; Gpx: Glutathione peroxidase; SOD: Superoxide dismutase; TNF-α: Tumor necrosis factor alpha; CHOL: Cholestenone; TG: Triglyceride; HDL-C: High density lipoprotein-cholestenone; LDL-C: Low density lipoprotein-cholestenone. [a,b,c] Means within the same rows without the same superscript letter are significantly different ($p < 0.05$).

3.7. Antioxidation and Inflammation-Related mRNA Expression in 35-Day-Old Broilers

When testing for inflammation-related mRNA expression in the spleens of 35-day-old broilers, there were no significant differences in TNF-α, NADPH dehydrogenase 1 (*NQO-1*), nuclear factor kappa B p 65 (*NFκB*), and inducible nitric oxide synthase (*iNOS*) levels ($p > 0.05$). However, the levels of toll-like receptor 4 (*TLR4*), a major lipopolysaccharide receptor, were higher in the 0.5% and 1% PWMC groups ($p = 0.015$). Both interferon-γ (*IFN-γ*) and interleukin-1ß (*IL-1ß*) levels decreased significantly ($p < 0.0001$ and $= 0.0002$, respectively), by about three to five times for *IFN-γ* and 1.1 to 1.6 times for *IL-1ß* (Figure 3A).

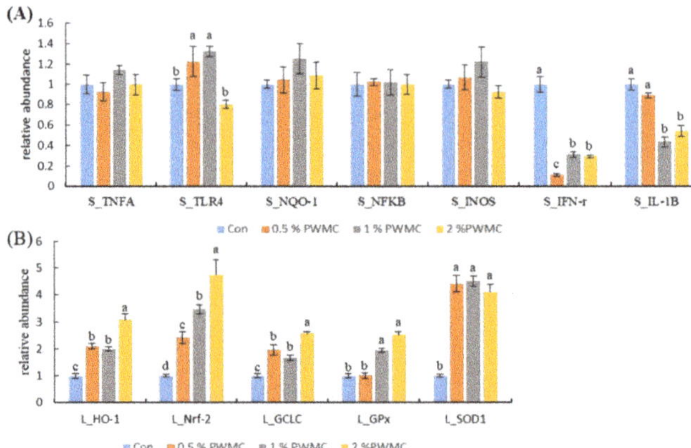

Figure 3. Inflammation- (**A**) and antioxidant-related (**B**) mRNA expression in the spleen (S) and liver (L) of 35-day-old broilers. TNF-α: Tumor necrosis factor alpha; TLR4: Toll-like receptor 4; NQO-1: NADPH dehydrogenase 1; NFκB: Nuclear factor kappa B p 65; iNOS: Inducible nitric oxide synthases; IFN-γ: Interferon-γ; IL-1ß: Interleukin-1ß; HO-1: Heme oxygenase-1; Nrf-2: Nuclear factor erythroid 2–related factor 2; GCLC: Glutamate-cysteine ligase catalytic; Gpx: glutathione peroxidase; SOD-1: Superoxide dismutase-1. [a-d] Means within the same rows without the same superscript letter are significantly different ($p < 0.05$).

The expression of all antioxidant-related mRNA, including heme oxygenase-1 (HO-1), nuclear factor erythroid 2–related factor 2 (Nrf-2), glutamate-cysteine ligase catalytic (GCLC), Gpx, and *SOD-1*, increased significantly following PWMC supplementation (Figure 3B), but while *HO-1, Nrf-2, GCLC,* and *Gpx* increased with PWMC dosage, *SOD-1* increased equally in each group (Figure 3B).

4. Discussion

Food intake is the only way for animals to gain energy, maintain growth, and survive. Traditionally, due to its low energy content, fiber has not been used in feed [3]. However, research proves that adding fiber to animal feed enhances gut barrier expression, increases antioxidant capacity, and decreases the inflammatory response [16]. There are many polyprotein structures on intestinal epithelial tissue. These structures can block exogenous toxins and bacteria in the intestinal lumen [17]. The immune response is triggered and macrophages are activated when these toxins invade [18]. Due to the complexity of gut microbiota, animals have developed a set of modes that can coexist with microorganisms. Within a single day, animals can adjust the composition of gut microbiota through their diet, further affecting their health and the state of the intestinal barrier [16,17]. Fiber supplementation in animal diets could significantly increase gut barrier function and alter metabolism [16,19]. After PWMC supplementation, the gut barrier in a broiler's ileum increases, by increments that are positively related to the amount of supplement provided. It was also determined that PWMC supplementation increases the expression of *EAAT3,* one of the major transporters of glutamate and cysteine. Furthermore, under the oxidant stress the *EAAT3* expression would decrease in broilers [20]. In this study, high-fiber PWMC supplementation had no negative impact on broiler growth, and even slightly increased FCR performance in the 0.5% PWMC group during the starter stage; there was no significant difference in the other groups.

In animals, the main tissues for adipose synthesis are the liver and adipocytes. CCAAT-enhancer-binding proteins (*CEBP*) and peroxisome proliferator-activated receptor (*PPAR*) are involved in the early differentiation of adipocytes [21]; however, both genes were downregulated in the PWMC-supplemented groups. *FAS* and *FABP4* are mainly involved in the late differentiation of fat cells and the accumulation of fatty oil droplets [22,23]. The former is an upstream gene for fat synthesis and the latter is one of the major proteins that transport fatty acids [22,23]. Therefore, PWMC could suppress adipocyte differentiation by blocking *CEBP* and *PPAR* but would not affect adipose synthesis. Both KCTD-15 and adiponectin promote adipose metabolism, but adiponectin also promotes leptin production [24,25]. It is well-known that leptin stimulates fat metabolism in animals and enhances muscle-cell development [26]. In this study, both mRNA and protein types adiponectin significantly improved with 1% and 2% PWMC supplementation. While both *ATGL* and *CPT-1* enhance adipose metabolism, the former separates glycerol from fatty acids, while the latter enhances fatty acid metabolism [27,28]. The increase of *ATGL* mRNA expression leads to a decrease in the triglyceride (TG) content in the serum of broilers.

PWMC has a high amount of phenolic compounds. Phenolic compounds, which exist in plant-based ingredients have been regarded as a functional antioxidant component for decades [9,29]. Among them, catechins were regarded as one the most functional compounds on both antioxidant and anti-adipogenesis grounds [8,29]. Past research has pointed out that catechins such as EGCG (210 µg/ g PWMC) can assist in adipose degradation and improve the antioxidant capacity of animals [29]. Traditionally, OXT has been considered a contractile stimulator of smooth muscle that also helps regulate animal mood [30]. However, recent studies have shown that OXT can also help muscle cells take up glucose, enhance fat metabolism, and promote muscle hyperplasia [31]. In addition, OXT can help solve the problem of fat infiltration caused by obesity [31].

PWMC can enhance adipolysis in both the liver and adipocytes, thus compensating for the gene for adipogenesis. As the increase in adipogenesis-related genes is much lower than adipolysis-related gene expression, the overall adipose content in broilers is reduced. Although adipose is rapidly metabolized, broiler weight was still not significantly different from that of the control. This may be due to the transfer of energy from the breakdown of fat to the production of muscle. Previous research also indicates that dietary fiber supplementation could increase broiler muscle content and the slaughter rate of poultry [32]. However, to our best knowledge, there are only a few studies into the cause of increased muscle weight. Although we only measured adipose metabolism-related mRNA and protein expression in this study, we still confirmed that adding extra fiber to the diet can improve the pattern of adipose metabolism, which may be the reason why high fiber diets can promote muscle mass gains in poultry [19].

Animals may suffer from oxidative stress caused by exogenous reactive oxygen species (ROS) which causes incomplete replication of DNA, and induce mutations or cell lesions [33]. After the mature *Pleurotus eryngii* is removed, the remaining culture medium, also known as PWMC, contains many mycelia. Mushrooms have superb antioxidant capacities and can increase antioxidant capacity in animals [5]. Wang et al. [5,34] indicated that mushroom compost can improve the antioxidant capacity of poultry whether in vivo or *in vitro*. Animal antioxidant systems can be classified as enzymatic or non-enzymatic systems. The former contains enzymes that reduce hydrogen peroxide, such as catalase, and the latter contains small molecules that can directly neutralize ROS, such as glutathione (GSH) [35]. Nrf2 is an upstream antioxidant gene that normally binds to Kelch-like ECH-associated protein 1 in the cytoplasm. When *Nrf2* is activated, Nrf2 enters the nucleus and forms a dimer with the Maf protein. It then reacts with the antioxidant responsive element (ARE) and promotes the expression of downstream antioxidant genes, including *GCLC* and *HO-1* [36,37]. Both *GCLC* and *HO-1* are involved in the antioxidant enzymatic system. *GCLC* is a rate-limiting enzyme in the GSH synthesis pathway [38]. GSH can receive electrons from ROS and become reduced glutathione disulfide (GSSG), which can then be reoxidized by Gpx into GSH. As a catalyst, *HO-1* can indirectly reduce the harm caused by H_2O_2 by cleaving hemoglobin into biliverdin [39]. Previous research determined that phenols have antioxidant capacities and will activate *Nrf2* expression [40], therefore mushroom additives can improve the enzymatic antioxidant system and enhance animals' adaptability to oxidative stress [34]. In addition, as a phenolic compound, epigallocatechin (1.493 mg/g PWMC) can enhance the antioxidant capacity of animals by chelating metal ions or increasing antioxidant enzymes [29]. In addition to ROS, Lee et al. [18] pointed out that inflammation is a double-edged sword. Although inflammation can help animals fight pathogens, it may also increase oxidative stress and cause cell damage [18]. Therefore, it is important to appropriately suppress the inflammatory response under non-disease conditions. *IL-1ß*, one of the most upstream inflammatory factors, is involved in upregulating the expression of inflammatory genes like *iNOS* and *NF-κB*, and may lead to the production of reactive nitrogen species [13]. However, suppressed *IL-1ß* performance would not alter the inflammatory response under normal conditions. *IFN-γ* levels are an indicator of infection. When an animal is infected with a virus or bacteria, *IFN-γ* levels increase and activate *NF-κB*, which further increases inflammation. The glucan content in PWMC might decrease *IFN-γ* expression [13]. Overall, PWMC could increase gut barrier function, antioxidant capacity, and alter the adipose metabolism pattern.

5. Conclusions

Dietary supplementation with PWMC can significantly change the adipose metabolism pattern in broilers and accelerate adipose degradation. Although higher amounts of PWMC could increase intestinal barrier function and antioxidant capacity, supplementation with 0.5% and 2% PWMC resulted in similar levels of adipose metabolism. Furthermore, the 0.5% PWMC group had better FCR. Therefore, to improve health and growth efficiency in broilers, our results suggest that supplementation with 0.5% PWMC is the most effective way to improve fat metabolism and antioxidant capacity. However, further research is recommended to detect the actual mechanism of fat metabolism and antioxidant capacity of PWMC in broilers.

Author Contributions: Conceptualization, W.Y.C. and T.-T.L.; materials analysis, W.Y.C., C.L.L. and H.D.S.; methodology, W.Y.C. and Y.M.S.; supervision, T.-T.L. and C.F.T.; investigation, W.Y.C. and T.-T.L.; writing—original draft preparation, W.Y.C., W.C.L. and T.-T.L.; writing—review and editing, W.Y.C., W.C.L., S.C.C. and T.-T.L. All authors have read and agreed to the published version of the manuscript.

Animals **2020**, *10*, 445

Funding: This research received no external funding.

Acknowledgments: The authors thank the Ministry of Science and Technology (MOST 107-2313-B-005-037-MY2) and the iEGG and Animal Biotechnology Center from The Feature Areas Research Center Program within the framework of the Higher Education Sprout Project by the Ministry of Education (MOE) in Taiwan for supporting this study.

Conflicts of Interest: The authors declare no conflicts of interest.

Abbreviations

PWMC	*Pennisetum purpureum* schum No.2 waste mushroom compost
PP	*Pennisetum purpureum* schum No.2
OXT	Oxytocin
TNF-α	Tumor necrosis factor alpha
Gpx	Glutathione peroxidase
SOD	Superoxidase dismutase
MDA	Malondialdehyde
TLR4	Toll-like receptor 4
NQO-1	NADPH dehydrogenase 1
NFκB	Nuclear factor kappa B p 65
iNOS	Inducible nitric oxide synthases
IFN-γ	Interferon-γ
IL-1ß	Interleukin-1ß
HO-1	Heme oxygenase-1
Nrf-2	Nuclear factor erythroid 2–related factor 2
GCLC	Glutamate-cysteine ligase catalytic
MUC2	Mucin2
ZO-1	Zonula occludens 1
EAAT3	Excitatory amino acid transporter 3
FFAR2	Free fatty acid receptor 2
GLUT2	Glucose transporter 2
SGLT	Sodium-dependent glucose cotransporters 1
PEPT-1	Peptide transporter 1
KCTD-15	Potassium channel tetramerization domain-containing 15
ATGL	Adipose triglyceride lipase
AMPK-α2	5′-AMP-activated protein kinase catalytic subunit alpha-2
FAS	Fatty acid synthase
FABP4	Fatty acid binding protein 4
CEBPα	CCAAT-enhancer-binding proteins-alpha
CPT-1	Carnitine palmitoyltransferase I
IL-6	Interleukin-6
PPAR-γ	Peroxisome proliferator-activated receptor gamma
PPAR-α	Peroxisome proliferator-activated receptor alpha
FCR	Feed conversion rate
SEM	Standard error of mean
GLU	Glucose
SGOT	Serum glutamic-oxaloacetic transaminase
SGPT	Serum glutamic-pyruvic transaminase
Alk-P	Alkaline phosphatase
CHOL	Cholestenone
TG	Triglyceride
HDL-C	High density lipoprotein-cholestenone
LDL-C	Low density lipoprotein-cholestenone
ROS	Reactive oxygen species
GSH	Glutathione
ARE	Anti-oxidant responsive element
GSSG	Glutathione disulfide

References

1. Al-Harthi, M.A.; Attia, Y.A.; Al-sagan, A.; Elgandy, M.F. Nutrients profile, protein quality and energy value of whole prosopis pods meal as a feedstuff for poultry feeding. *Ital. J. Anim. Sci.* **2018**, *18*, 30–38. [CrossRef]
2. Hemery, Y.; Rouau, X.; Lullien, V.; Barron, C.; Abécassis, J. Dry processes to develop wheat fractions and products with enhanced nutritional quality. *J. Cereal Sci.* **2007**, *46*, 327–347. [CrossRef]
3. Buxton, D.R.; Redfearn, D.D. Plant limitations to fiber digestion and utilization. *J. Nutr.* **1997**, *127*, 8145–8185. [CrossRef] [PubMed]
4. O'Brien, B.J.; Milligan, E.; Carver, J.; Roy, E.D. Integrating anaerobic co-digestion of dairy manure and food waste with cultivation of edible mushrooms for nutrient recovery. *Bioresour. Technol.* **2019**, *285*, 121312. [CrossRef] [PubMed]
5. Wang, C.C.; Lin, L.J.; Chao, Y.P.; Chiang, C.J.; Lee, M.T.; Chang, S.C.; Yu, B.; Lee, T.T. Antioxidant molecular targets of wheat bran fermented by white rot fungi and its potential modulation of antioxidative status in broiler chickens. *Br. Poult. Sci.* **2017**, *58*, 262–271. [CrossRef] [PubMed]
6. Maleko, D.; Mwilawa, A.; Msalya, G.; Pasape, L.; Mtei, K. Forage growth, yield and nutritional characteristics of four varieties of napier grass (Pennisetum purpureum Schumach) in the west Usambara highlands. *Afr. Crop Sci. J.* **2019**, *6*, e00214. [CrossRef]
7. Mahfuz, S.; He, T.; Liu, S.; Wu, D.; Long, S.; Piao, X. Dietary inclusion of mushroom (flammulina velutipes) stem waste on growth performance, antibody response, immune status, and serum cholesterol in broiler chickens. *Animals* **2019**, *9*, 692. [CrossRef]
8. Huang, J.B.; Zhang, Y.; Zhou, Y.B.; Wan, X.C.; Zhang, J.S. Effects of epigallocatechin gallate on lipid metabolism and its underlying molecular mechanism in broiler chickens. *J. Anim. Physiol. Anim. Nutr.* **2015**, *99*, 719–727. [CrossRef]
9. Lee, T.T.; Ciou, J.Y.; Chiang, C.J.; Chao, Y.P.; Yu, B. Effect of pleurotus eryngii stalk residue on the oxidative status and meat quality of broiler chickens. *J. Agric. Food Chem.* **2012**, *60*, 11157–11163. [CrossRef]
10. Tabart, J.; Kevers, C.; Pincemail, J.; Defraigne, J.O.; Dommes, J. Comparative antioxidant capacities of phenolic compounds measured by various tests. *Food Chem.* **2009**, *113*, 1226–1233. [CrossRef]
11. Pourmorad, F.; Hosseinimehr, S.J.; Shahabimajd, N. Antioxidant activity, phenol and flavonoid contents of some selected Iranian medicinal plants. *Afr. J. Biotechnol.* **2006**, *5*, 1142–1145.
12. Wei, L.; Zhang, W.; Yin, L.; Yan, F.; Xu, Y.; Chen, F. Extraction optimization of total triterpenoids from Jatropha curcas leaves using response surface methodology and evaluations of their antimicrobial and antioxidant capacities. *Electron. J. Biotechnol.* **2015**, *18*, 88–95. [CrossRef]
13. Chuang, W.Y.; Lin, W.C.; Hsieh, Y.C.; Huang, C.M.; Chang, S.C.; Lee, T.T. Evaluation of the combined use of Saccharomyces Cerevisiae and Aspergillus Oryzae with phytase fermentation products on growth, inflammatory, and intestinal morphology in broilers. *Animals* **2019**, *9*, E1051. [CrossRef] [PubMed]
14. AOAC. *Official Methods of Analysis of AOAC International*, 19th ed.; AOAC International: Gaithersburg, MD, USA, 2012.
15. Meslin, C.; Desert, C.; Callebaut, I.; Djari, A.; Klopp, C.; Pitel, F.; Leroux, S.; Martin, P.; Froment, P.; Guilbert, E. Expanding duplication of free fatty acid receptor-2 (GPR43) genes in the chicken genome. *Genome Biol. Evol.* **2015**, *7*, 1332–1348. [CrossRef] [PubMed]
16. Dos Santos, T.T.; Baal, S.C.S.; Lee, S.A.; E Silva, F.R.O.; Scheraiber, M.; da Silva, A.V.F. Influence of dietary fibre and betaine on mucus production and digesta and plasma osmolality of broiler chicks from hatch to 14 days of age. *Livest. Sci.* **2019**, *220*, 67–73. [CrossRef]
17. Barekatain, R.; Chrystal, P.V.; Howarth, G.S.; McLaughlan, C.J.; Gilani, S.; Nattrass, G.S. Performance, intestinal permeability, and gene expression of selected tight junction proteins in broiler chickens fed reduced protein diets supplemented with arginine, glutamine, and glycine subjected to a leaky gut model. *Poult. Sci.* **2019**, *98*, 6761–6771. [CrossRef] [PubMed]
18. Lee, M.T.; Lin, W.C.; Lee, T.T. Potential crosstalk of oxidative stress and immune response in poultry through phytochemicals—A review. *Asian-Australas. J. Anim. Sci.* **2019**, *32*, 309–319. [CrossRef]
19. Al-Harthi, M.A.; Attia, Y.A.; Al-sagan, A.; Elgandy, M.F. The effects of autoclaving or/and multi-Enzymes complex supplementation on performance, egg quality and profitability of laying hens fed whole Prosopis juliflora pods meal diet. *Europ. Poult. Sci.* **2018**, *82*. [CrossRef]

20. Ebrahimi, R.; Jahromi, M.F.; Liang, J.B.; Farjam, A.S.; Shokryazdan, P.; Idrus, Z. Effect of Dietary Lead on Intestinal Nutrient Transporters mRNA Expression in Broiler Chickens. *Biomed Res. Int.* **2015**, *1*, 149745. [CrossRef]

21. Barbosa-da-Silva, S.; Souza-Mello, V.; Magliano, D.C.; Marinho, T.S.; Aguila, M.B.; Mandarim-de-Lacerda, C.A. Singular effects of PPAR agonists on nonalcoholic fatty liver disease of diet-Induced obese mice. *Life Sci.* **2015**, *127*, 73–81. [CrossRef]

22. Wang, Q.; Guan, T.; Li, H.; Bernlohr, D.A. A novel polymorphism in the chicken adipocyte fatty acid-Binding protein gene (FABP4) that alters ligand-Binding and correlates with fatness. *Comp. Biochem. Physiol. B* **2009**, *154*, 298–302. [CrossRef] [PubMed]

23. De Silva, G.S.; Desai, K.; Darwech, M.; Naim, U.; Jin, X.; Adak, S.; Harroun, N.; Sanchez, L.A.; Semenkovich, C.F.; Zayed, M.A. Circulating serum fatty acid synthase is elevated in patients with diabetes and carotid artery stenosis and is LDL-Associated. *Atherosclerosis* **2019**, *287*, 38–45. [CrossRef] [PubMed]

24. Liang, S.S.; Ouyang, H.J.; Liu, J.; Chen, B.; Nie, Q.H.; Zhang, X.Q. Expression of variant transcripts of the potassium channel tetramerization domain-Containing 15 (KCTD15) gene and their association with fatness traits in chickens. *Domest. Anim.* **2015**, *50*, 65–71. [CrossRef] [PubMed]

25. Tahmoorespur, M.; Ghazanfari, S.; Nobari, K. Evaluation of adiponectin gene expression in the abdominal adipose tissue of broiler chickens: Feed restriction, dietary energy, and protein influences adiponectin messenger ribonucleic acid expression. *Poult. Sci.* **2010**, *89*, 2092–2100. [CrossRef] [PubMed]

26. Murase, D.; Namekawa, S.; Ohkubo, T. Leptin activates chicken growth hormone promoter without chicken STAT3 in vitro. *Comp. Biochem. Physiol. B* **2016**, *191*, 46–52. [CrossRef] [PubMed]

27. Habets, D.D.J.; Coumans, W.A.; Hasnaoui, M.E.; Zarrinpashneh, E.; Bertrand, L.; Viollet, B.; Kiens, B.; Jensen, T.E.; Richter, E.A.; Bonen, A.J.; et al. Crucial role for LKB1 to AMPKα2 axis in the regulation of CD36-Mediated long-Chain fatty acid uptake into cardiomyocytes. *BBA-Mol. Cell Biol. L.* **2009**, *1791*, 212–219. [CrossRef]

28. Andler, S.M.; Wang, L.S.; Goddard, J.M.; Rotello, V.M. Chapter One-Preparation of Biocatalytic Microparticles by Interfacial Self-Assembly of Enzyme–Nanoparticle Conjugates Around a Cross-Linkable Core. *Methods Enzymol.* **2016**, *516*, 1–17.

29. Yan, J.; Zhao, Y.; Suo, S.; Liu, Y.; Zhao, B. Green tea catechins ameliorate adipose insulin resistance by improving oxidative stress. *Free Radic. Biol. Med.* **2012**, *52*, 1648–1657. [CrossRef]

30. Arrowsmith, S. Oxytocin and vasopressin signaling and myometrial contraction. *Curr. Opin. Physiol.* **2020**, *13*, 62–70. [CrossRef]

31. Iwasa, T.; Matsuzaki, T.; Mayila, Y.; Yanagihara, R.; Yamamoto, Y.; Kawakita, T.; Kuwahara, A.; Irahara, M. Oxytocin treatment reduced food intake and body fat and ameliorated obesity in ovariectomized female rats. *Neuropeptides* **2019**, *75*, 49–57. [CrossRef]

32. Nassara, M.K.; Lyua, S.; Zentekc, J.; Brockmann, G.A. Dietary fiber content affects growth, body composition, and feed intake and their associations with a major growth locus in growing male chickens of an advanced intercross population. *Livest. Sci.* **2019**, *227*, 135–142. [CrossRef]

33. Poulsen, H.E. Oxidative DNA modifications. *Exp. Toxicol. Pathol.* **2005**, *57*, 161–169. [CrossRef] [PubMed]

34. Wang, C.C.; Chang, C.H.; Chang, S.C.; Fan, G.J.; Lin, M.J.; Yu, B.; Lee, T.T. In vitro free radicals scavenging activity and antioxidant capacity of solid-state fermented wheat bran and its potential modulation of antioxidative molecular targets in chicken PBMC. *R. Bras. Zootec.* **2016**, *45*, 451–457. [CrossRef]

35. Jung, K.A.; Kwak, M.K. The Nrf2 System as a potential target for the development of indirect antioxidants. *Molecules* **2010**, *15*, 7266–7291. [CrossRef] [PubMed]

36. Dickinson, D.; Levonen, A.L.; Moellering, D.R.; Arnold, E.K.; Zhang, H.; Darley-Usmer, V.M.; Forman, H.J. Human glutamate cysteine ligase gene regulation through the electrophile response element. *Arch. Biochem. Biophys.* **2004**, *423*, 116–125. [CrossRef]

37. Sahin, K.; Orhan, C.; Smith, M.O.; Sahin, N. Molecular targets of dietary phytochemicals for the alleviation of heat stress in poultry. *Poult. Sci.* **2013**, *69*, 113–124.

38. Teskey, G.; Abrahem, R.; Cao, R.; Gyurjian, K.; Islamoglu, H.; Lucero, M.; Martinez, A.; Paredes, E.; Salaiz, O.; Robinson, B.; et al. Glutathione as a marker for human disease. *Adv. Clin. Chem.* **2018**, *87*, 141–159.

39. Araujo, J.A.; Zhang, M.; Yin, F. Heme oxygenase-1, oxidation, inflammation, and atherosclerosis. *Front. Pharmacol.* **2012**, *3*, 1–17. [CrossRef]
40. Arredondo, F.; Echeverry, C.; Abin-Carriquiry, J.A.; Blasina, F.; Antunez, K.; Jones, D.P.; Go, Y.M.; Liang, Y.L. After cellular internalization, quercetin causes Nrf2 nuclear translocation, increases glutathione levels, and prevents neuronal death against an oxidative insult. *Free Radic. Biol. Med.* **2010**, *49*, 738–747. [CrossRef]

Article

Evaluation of Heavy Metal Content in Feed, Litter, Meat, Meat Products, Liver, and Table Eggs of Chickens

Mohamed A. Korish [1,*] and Youssef A. Attia [2,*]

[1] The Strategic Center to Kingdom Vision Realization, King Abdulaziz University, P.O. Box 80200, Jeddah 21589, Saudi Arabia

[2] Arid Land Agriculture Department, Faculty of Meteorology, Environment and Arid Land Agriculture, King Abdulaziz University, P.O. Box 80208, Jeddah 21589, Saudi Arabia

* Correspondence: mkorish@kau.edu.sa (M.A.K.); yaattia@kau.edu.sa (Y.A.A.)

Received: 26 February 2020; Accepted: 15 April 2020; Published: 22 April 2020

Simple Summary: Foods contain a wide range of trace elements, some of these are of important nutritional value, such as iron (Fe), copper (Cu), zinc (Zn), manganese (Mn), selenium (Se), cobalt (Co), and chromium (Cr), while others have toxic effects, such as lead (Pb), cadmium (Cd), and nickel (Ni). The food chain creates potential health hazards with regards to the transmission of toxic elements to animal tissues, and subsequently to humans. Therefore, the feeding strategy of animals is a useful tool for limiting health hazards through the food chain. We assessed the concentrations of Fe, Cu, Zn, Mn, Se, Co, Cr, Pb, Cd, and Ni in eggs and chicken meat products, as well as those in the feed and litter of chickens, to determine the possible risk posed to consumers from the consumption of chicken products. The results indicated that Cd, Pb, and Se were under detectable levels in chicken meat products and eggs, suggesting that there is no threat from toxic heavy metals. On the other hand, the greatest concentration of heavy metals was recorded in broiler liver except for Cr, Co, and Ni. Meat products exhibited higher Cd, Cu, Mn, Ni, Pb, and Co levels than raw meat and table eggs. Overall, findings indicated that levels of Pb and Ni were four times and seven times high than the tolerable upper limit showing a health threat to humans from the consumption of chicken meat products.

Abstract: We assessed the concentrations of Fe, Cu, Zn, Mn, Se, Co, Cr, Pb, Cd, and Ni in chicken meat and meat products, feed, and litter, as well as laying hens' eggs, feed and litter to monitor the quality of products on the market and their safety for human consumption as judged by recommended daily allowance (RDA) and tolerable upper levels. Samples were chosen as the most popular poultry products in Saudi Arabia. A total of 45 broiler samples of frozen or fresh meat, liver, burger, or frankfurter were chosen from the same brand. Additionally, 60 table eggs from four commercial brands were collected, and the edible parts of these were used to determine levels of minerals and toxic elements. Furthermore, 30 feed and litter samples were collected from the starter, grower, and layer diets of broilers and laying hens. The results indicated that there were significant levels of most of the trace elements and heavy metals in the different meat sources. Furthermore, the liver contained the highest levels of elements, except for Cr, Co, and Ni. The highest Cr level was detected in the fresh meat, followed by frozen meat. Trace elements (Mn and Co) and heavy metals (Ni and Pb) were not detected in either the frozen or the fresh meat. The chicken burger and the frankfurter exhibited similar trace-element and heavy-metal contents, except for Zn and Mn, as the frankfurter showed higher concentrations than the burger. Differences in most of the trace and toxic elements among the different sources of eggs were not found to be significant, except for Zn. Differences between the broiler meat and table eggs were only substantial for Fe and Zn. Fe was significantly higher in meat than in eggs, and the opposite trend was found for Zn. The liver contained higher heavy metals than the eggs, except for Cr. In addition, the burger had higher concentrations of essential (Cu and Co) and heavy metals (Pb and Ni) than the eggs but had lower levels of Zn and Cr. The frankfurter exhibited significantly higher levels of Fe, Cu, Mn, Co, Pb, and Ni than the eggs but lower levels of Zn and Cr. To summarize, Cd, Pb, As, and Se were not detected in the broiler meat or eggs, indicating no risks from these toxic elements. Conversely, the liver exhibited the highest content of heavy metals, except for Cr, indicating that the intake of Pb and Cd was above the recommended daily allowance (RDA) for adults. The meat products exhibited higher Pb, Cd, and Ni levels than the broiler meat and the table eggs, suggesting that they posed a health threat to humans, and the intake of Pb in the meat products was higher than the RDA. Thus, chicken meat and table eggs, which are primary protein sources, are safe sources of human nutrition, while liver and meat products may present potential health hazards through the food chain.

Keywords: heavy metals; chicken eggs; meat; meat products

---

1. Introduction

Poultry products, such as eggs, meat, liver, and meat products (i.e., burgers, luncheon meat, and frankfurters) are worldwide primary sources of protein, energy, vitamins, and minerals because they are nutritious, delicious, and affordable and provide much of the recommended daily allowance (RDA) of trace minerals, proteins, and energy [1–5]. However, poultry products may pose risks due to contamination with heavy metals from the environment or through the food chain [6–8]. The diet of broilers is commonly supplemented with microelements such as Fe, Cu, Zn, Mn, I, and As to sustain the growth and health of birds [3,4]. Minerals are also essential for maintaining correct metabolic activity, the balance of bodily functions, and immunity in living organisms [8,9]. However, trace minerals can exceed animal requirements and are consequently excreted in manure, which has a negative environmental impact [9].

In addition, other heavy metals, such as Pb, Cd, and Hg, are not necessary to the integrity and function of the body [10,11]. Generally, increasing the intake of toxic elements poses hazards to both animals and humans [12]. The mineral content of animal products is important for healthy human embryonic development [3,4,13,14] and has been suggested for use as a bioindicator of pollution by environmental elements [15]. However, animal products might have high concentrations of toxic elements that originate mainly from the feed, water, litter, and the environment [9,10]. Cd, As, and Pb are hazardous and can be transmitted through the food chain, having a toxic impact on the health of both animals and humans [16]. Lead can cause metabolic harm as a neurotoxin [16] and can adversely affect renal function and hemopoiesis, as well as the nervous and gastrointestinal systems [17,18]. Diet is a source of contamination by Cd, which originates from various food sources and the environment [17] and is transmitted through the food chain to animals and, consequently, humans, inducing kidney dysfunction, hypertension, and pulmonary and hepatocellular damage [18].

In the available literature, mineral pollution in animal feeds is unavoidable, as they may be contaminated by heavy metals due to environmental pollution and presence in supplements and concentrates or the machinery and equipment used during manufacture. Both in human and animals, liver and kidney are very important for detoxification and execration of toxic elements, and they are thus also the most damaged organs due to an excess of toxic elements in feed or food [15–18].

The As is stored in animal tissues depending on the type of feed intake [19]; As can induce nausea, headache, and severe gut irritation [20]. Similarly, Cu adversely affects kidney, brain, and liver function and can result in haemolytic crisis when consumed at high levels [21]. The assessment of essential trace minerals and heavy metal levels in poultry products are useful tools with regards to nutritional safety and environmental sustainability. Thus, the aim of the present study is to assess the concentrations of Fe, Cu, Zn, Mn, Se, Co, Cr, Pb, Cd, Ni, B, and Al in poultry diets, litter, eggs, liver, meat, and meat products, which is essential to determine the possible risk posed to consumers by the consumption of poultry products, as well as the environmental impact of the heavy metal content of litter.

2. Materials and Methods

The experiment was designed to mentor the trace and toxic elements in poultry products in relation to human wellbeing as judged by RDA and upper tolerable levels. The experimental protocol of this work was approved by the Deanship of Scientific Research (DSR), King Abdulaziz University, Jeddah, Saudi Arabia, under project no. 'G: 475-375-1440'. The work was carried out according to Royal Decree number M59 in 14/9/1431H and did not involve any living animals and/or human subjects.

2.1. Sample Collection and Analyses

The samples were collected from the retail market in Jeddah City, Saudi Arabia during June–July 2017.

The samples included meat and meat products; thus, a whole broiler carcass, equivalent to 45 samples of fresh (cold) carcasses, was collected to represent three brands locally produced in Saudi Arabia.

In addition, 45 samples of meat products such as burgers, frankfurters, and liver were obtained from the carcasses of the same brands produced locally in Saudi Arabia

The carcasses were from grade A with 1 kg weight class having similar production and expiry dates.

Further, 45 frozen carcasses were obtained from three imported, well-known international brands. The frozen carcasses were imported from Brazil, France and USA.

The carcasses were deboned and skinned, and the meat was cut into two pieces and minced using a meat mincer (Moulinex-HV8, Paris, France).

Additionally, 60 table eggs from four commercial brands (A, B, C, D) were collected. The egg samples were from four different commercial brands produced in Saudi Arabia. The eggs were cracked open, and the edible parts (albumen and yolk) were mixed and homogenized using a Moulinex-HV8 (France).

Fifteen samples of broilers were collected from broiler farms to represent the broiler starter, grower, and finisher commercial standard corn-soybean diets as five samples of each diet. The diets are commercial and consist mainly of corn and soybean meal. Broiler diets meet the nutrient requirements of broiler during different growth periods [22]. The composition of the starter broiler diet was metabolizable energy (ME) 3010 kcal/kg diet, crude protein (CP) 22.3%, methionine + cysteine 0.93%, lysine 1.25%, Ca 1%, and available phosphorus 0.51%. The grower broiler diet contained ME 3150 kcal/kg diet, CP 20.3%, methionine + cysteine 0.88%, lysine 1.20%, Ca 0.90%, and available phosphorus 0.45%. The finisher broiler diet consisted of ME 3260 kcal/kg diet, CP 18.1%, methionine + cysteine 0.73%, lysine 1.10%, Ca 0.83%, and available phosphorus 0.41%.

The same number of samples were collected from layer farms to represent starter, grower, and layer commercial standard diets and consisted mainly of corn and soybean meal as five samples of each diet. The laying diets were a soybean meal-based diet that meets nutrient requirements of white eggshell layers during the starter, grower, and laying periods [22]. The starter layer diet consisted of ME 2918 kcal/kg diet, CP 21.0%, methionine + cysteine 0.81%, lysine 1.11%, Ca 0.96%, and available phosphorus 0.45%. The grower layer diet consisted of ME 2710 kcal/kg diet, CP 15.8%, methionine + cysteine 0.68%, lysine 0.77%, Ca 0.93%, and available phosphorus 0.43%. The laying hens diet contained ME 2800 kcal/kg diet, CP 17.5%, methionine + cysteine 0.74%, lysine 0.85%, Ca 3.7%, and available phosphorus 0.36%.

In addition, 30 samples of litter from broilers and laying hens were collected from the same houses where feed samples were collected in the Jeddah area, Saudi Arabia. Thus, there were five samples of each type of litter of broilers and laying hens.

The rearing, farming, and feeding practices of the broilers and laying hens were in accordance with the standard breeder's management guide, but these details are not available from the producers.

Broilers and laying hens are usually fed according to the breeder guide that meets their nutrient requirements. They were offered ad libitum a pelleted diet and freshwater. The broilers were slaughtered in automatic slaughterhouses according to the Islamic method.

The meat, eggs, diet, and litter samples were dried in a force ventilated oven at 105 °C until a constant weight was achieved. Then, 2 g of each of the dried meat, egg, diet, and litter samples were placed in a flask and digested with a mixture of concentrated HNO_3 and H_2SO_4 (3:1 V/V) [23]. The digestion process was continued until the solution became clear. The samples were then reconstituted in an aqueous matrix containing 0.5% hydrochloric acid and 2% nitric acid, added to stabilize elements as an ionic solution. The samples were transferred to another flask and diluted to 25 mL with distilled water [24]. The trace elements and heavy metals contents were determined using a Varian ICP-Optical Emission Spectrometer (Varian 720-ES, ICP-OEM, Leuven, Belgium).

The working protocols of ICP-MS were adapted from Olajire and Ayodele [25]. The operation condition was Radio Frequency (RF) generator power 1150 W, observation height 12 mm, cooling gas flow 0.7 L/min, 105 auxiliary gas flow 0.5 L/min, analysis pump speed 50 rpm, high purity argon, and detection mass (m/z). The ICP-MS offers auto-dilution techniques to counteract the high total dissolved solids samples. The method has been validated for the determination of Al, As, Cd, Cr, Cu, Fe, Hg, Pb, Mn, Ni, Se, Tl, and Zn in various types of food, including composite diets, cereals, rice, fish and offal by CSL (http://randd.defra.gov.uk/Document.aspx?Document=11372_Appendix5TRACEprocedureformultielementanalysis.pdf).

The daily intake of each essential element and heavy metal in the edible parts of the eggs (yolk and albumen), meat, meat products, and liver was estimated using the concentration of each element in each product and an average of 100 g of the poultry products according to the following equation: daily intake = average consumption of any poultry products (100 g/person/day) × concentration of an element in each product.

The values were compared with the recommended and upper tolerable levels of the Food and Agriculture Organisation (FAO)/ World Health organization (WHO).

2.2. Statistical Analysis

The collected data were subjected to a one-way ANOVA using SAS software [26]. The statistical model included the effect of the types of meat or eggs according to the following model:

$$Yij = \mu + Di + eij \tag{1}$$

where Y = the dependent variables; μ = general mean; D = meat or egg source; and e = random error. Before analysis, all the percentages were subjected to logarithmic transformation (log10 × + 1) to normalize the data distribution. The differences among the means were determined using the Student Newman Keuls test [25]. Significance was considered when the *p* value was 0.05 or less, whereas when the *p* value was between 0.05 and 0.10 it was considered as a trend.

3. Results

Table 1 shows the concentrations of essential trace minerals and heavy metals in samples from different diets for broilers. Different broiler diet samples had similar traces of mineral and heavy metals content.

In general, B was detectable in the diets of broiler, but Co and Al were not detected in the diets of the broilers. The B concentrations were higher in the grower and finisher broiler diets than in the starter broiler diets.

Table 2 indicates there were no significant differences among the various diet samples for layers with regards to traces and heavy metals content, except for Zn, which was higher in starter than grower and laying diets.

Table 1. Essential trace element and heavy metal content of broiler diets on a dry matter basis compared with the maximum permitted concentration element.

Elements	Broiler Diets			Toxic Level [22]	RMSE	*p* Value
	Starter	Grower	Finisher			
Essential trace elements, ppm						
Iron	142.9	120.3	132.8	4500	58.2	0.764
Copper	11.93	8.94	11.47	50–806	3.16	0.562
Zinc	33.2	30.2	39.2	500–4000	4.89	0.211
Manganese	31.8	27.3	32.6	4000–4800	7.59	0.649
Selenium	2.30	2.21	0.861	5–20	1.85	0.577
Chromium	2.07	4.33	3.19	10–300	2.63	0.113
Cobalt	UDL	UDL	UDL	100–200	ND	ND
Heavy metals, ppm						
Lead	2.09	2.63	5.27	10–1000	3.09	0.136
Cadmium	0.097	0.110	0.111	12–40	0.077	0.077
Arsenic	2.76	2.98	1.48	100	3.29	0.829
Nickel	1.76	2.15	1.46	300–500	1.39	0.088
Boron	4.88	5.56	5.73	200–5000	3.62	0.394
Aluminium	UDL	UDL	UDL	500–3000	ND	ND

RMSE = root mean square error. MPC = maximum permitted concentration, UDL = undetectable level. ND = not done.

Table 2. Essential trace element and heavy metal content of layer chickens' diets on a dry matter basis compared with the maximum permitted concentration element.

Elements	Layer Diets			Toxic Level [22]	RMSE	*p* Value
	Starter	Grower	Layers			
Essential trace elements, ppm						
Iron	99.4	153.7	183.3	4500	67.5	0.774
Copper	9.05	11.14	16.61	50–806	7.32	0.983
Zinc	60.3 [a]	49.9 [b]	45.4 [b]	500–4000	10.8	0.001
Manganese	28.6	32.1	42.5	4000–4800	7.68	0.554
Selenium	2.11	2.86	4.24	5–20	1.48	0.515
Chromium	2.63	1.51	5.20	10–300	1.47	0.166
Cobalt	UDL	UDL	UDL	100–200	ND	ND
Heavy metals, ppm						
Lead	3.02	3.50	4.14	10–1000	5.05	0.181
Cadmium	0.055	0.061	0.096	12–40	0.136	0.373
Arsenic	2.03	3.13	2.78	100	2.36	0.986
Nickel	4.23	9.85	2.18	300–500	2.74	0.187
Boron	6.05	6.41	5.73	200–5000	3.47	0.802
Aluminium	UDL	UDL	UDL	500–3000	ND	ND

[a,b] means with different superscripts within a row are significantly different. RMSE = root mean square error. MPC = maximum permitted concentration, UDL = undetectable level. ND = not done.

The highest levels of Pb were from the laying hens' diet, but the level was 41.4% of the toxic level. The highest levels of As, Ni, and B were found in the growing diet of the layers. Co and Al were under detectable levels in the different types of layers' diets.

Table 3 displays the levels of essential trace elements and heavy metals of the different litters of the broiler at various stages of production. The litter of the broiler diets showed similar levels of trace elements and heavy metals with the exception of Cu ($p = 0.027$), where the highest level of Cu was during the starter phase.

Table 3. Essential trace elements and heavy metals contents of the litter of broiler chickens on a dry matter basis compared with values cited in the literature.

Element	Broiler Litter			Literature Values	RMSE	*p* Value
	Starter	Grower	Finisher			
Essential trace elements, ppm						
Iron	581.9	536.4	572.3	852 [27]	66.9	0.180
Copper	28.5 [a]	23.3 [b]	24.5 [a,b]	31.8–335 [28–30]	4.25	0.027
Zinc	134.8	99.4	112.3	196–845.1 [28,29]	43.5	0.641
Manganese	241.9	297.7	263.2	375 [27]	235.4	0.996
Selenium	0.00	0.230	0.315	1.1 [31]	0.317	0.955
Chromium	17.1	1.74	3.54	1.5–143 [28,30,32]	7.61	0.159
Cobalt	UDL	UDL	UDL	0.39 [33]	ND	ND
Toxic metals, ppm						
Lead	2.56	4.42	3.75	0.42–107.1 [28–30]	5.77	0.729
Cadmium	0.137	0.155	0.172	0.031–19 [28,30,33]	0.251	0.736
Arsenic	UDL	0.0003	0.0005	1.8–38 [30,31]	0.0003	0.123
Nickel	46.7	23.0	54.3	1.71–276 [29,32,33]	29.9	0.502
Boron	UDL	0.238	0.172	NA	0.274	0.851
Aluminium	UDL	3.134	1.153	NA	0.0049	0.812

[a,b] means with different superscripts within a row are significantly different. RMSE = root mean square error, UDL = undetectable level, ND = not done, NA = not available.

The litter of the broilers during the grower phase had the highest Pb level, followed by that of the finisher phase and then that of the starter phase. The highest levels of Cd, As, and Ni were found in the litter of the broilers during the finisher phase.

Co was not delectable in the broiler litter during different stages of production, and Al was detected in the litter only during the starter period, with the grower litter having 2.7 times more than the finisher litter.

Table 4 shows the levels of essential trace elements and heavy metals of different litters of laying hens at various stages of production. The litters showed similar levels of trace elements and heavy metals with the exception of Se ($p = 0.046$), with the highest level of Se noted in the litter of the laying hens and the lowest in the grower litter. In addition, the litter exhibited the highest levels of Cu and Zn.

The highest levels of heavy metals, such as Cd, As, Ni, B, and Al were found in the litter of the layers during the starter phase. The litter of the laying hens during the grower phase had the highest Pb level, followed by the starter diets.

Table 5 shows the essential trace mineral and heavy metal contents of various chicken meat products and liver. There were significant differences among the various meat sources in most of the trace elements and heavy metals except for Se and As, which were undetectable in all samples. Essential trace elements (Mn and Co) and heavy metals (Pb, Cd, and Ni) were undetectable in the frozen or fresh meat.

The heavy metal (Co, Pb, Cd, and Ni) contents in the liver, burger, and frankfurter were similar. Chicken burger and frankfurter showed similar levels of trace minerals and heavy metals except for Zn and Mn, whereas the frankfurter showed higher concentrations than the burger.

The liver had the highest concentrations of Fe, Cu, Zn, Mn, Pb, and Cd content but not Cr, Co, and Ni. The highest levels of Cr, Co, and Ni were found in fresh meat, burger, and frankfurter, respectively.

Table 6 shows the essential trace element and heavy metal contents of different sources of table eggs. No significant difference in the content of most of the trace elements and heavy metals between various sources of eggs was detected except for Zn, whereas sources B and D exhibited higher concentrations than sources A and C. Essential elements such as Mn, Se, and Co, and heavy metals such as Pb, Cd, As, and Ni, were undetectable in the different sources of table eggs. The Fe content in eggs ranged from 11.83 to 17.70 ppm, while the Cu content ranged from 1.17 to 1.64 ppm.

Table 4. Essential trace elements and heavy metals contents of the litter of layer chickens on a dry matter basis compared with values cited in the literature.

Element	Layer Litter			Literature Values	RMSE	*p* Value
	Starter	**Grower**	**Laying Hens**			
			Essential trace elements, ppm			
Iron	617.4	719.1	676.5	852 [27]	60.6	0.291
Copper	27.8	30.8	36.6	31.8–335 [28–30]	10.5	0.961
Zinc	90.6	84.0	275.3	196–845.1 [28,29]	77.3	0.679
Manganese	122.3	141.0	141.2	375 [27]	8.41	0.129
Selenium	0.615 [a,b]	0.451 [b]	0.691 [a]	1.1 [31]	0.087	0.046
Chromium	15.2	13.6	10.9	1.5–143 [28,30,32]	2.11	0.288
Cobalt	UDL	UDL	UDL	0.39 [33]	ND	ND
			Toxic metals, ppm			
Lead	9.21	11.77	4.53	0.42–107.1 [28–30]	4.49	0.409
Cadmium	0.629	0.388	0.278	0.031–19 [28,30,33]	0.27	0.605
Arsenic	0.00071	0.00048	0.00067	1.8–38 [30,31]	0.0003	0.978
Nickel	11.4	10.8	8.74	1.71–276 [29,32,33]	4.71	0.873
Boron	0.461	0.348	0.377	NA	0.057	0.175
Aluminium	7.29	4.27	4.06	NA	0.004	0.608

[ab] means with different superscripts within a row are significantly different. RMSE = root mean square error, UDL = undetectable level, ND = not done, NA = not available.

Table 5. Heavy metal contents of broiler meat, liver, burger, and frankfurter on a dry matter basis in the retail market.

Element	Broilers' Meat		Chicken Meat Products and Liver			RMSE	*p* Value
	Frozen	**Fresh**	**Burger**	**Frankfurter**	**Liver**		
			Essential trace elements, ppm				
Iron	87.8 [b]	63.1 [b,c]	38.1 [c]	72.8 [b,c]	288.2 [a]	45.2	0.001
Copper	0.036 [c]	0.056 [c]	6.66 [b]	7.80 [b]	19.24 [a]	2.61	0.001
Zinc	35.8 [b]	41.4 [b]	14.9 [d]	26.4 [c]	79.8 [a]	10.57	0.001
Manganese	UDL [d]	UDL [d]	6.41 [c]	9.77 [b]	18.2 [a]	3.40	0.001
Selenium	UDL	UDL	UDL	UDL	UDL	ND	ND
Chromium	7.01 [b]	9.75 [a]	2.54 [c]	2.46 [c]	2.79 [c]	2.38	0.001
Cobalt	UDL [b]	UDL [b]	2.72 [a]	2.22 [a]	2.39 [a]	0.761	0.001
			Heavy metals, ppm				
Lead	UDL [b]	UDL [b]	16.69 [a]	14.84 [a]	16.51 [a]	4.84	0.001
Cadmium	UDL [b]	UDL [b]	0.433 [a,b]	0.379 [a,b]	1.12 [a]	0.763	0.003
Arsenic	UDL	UDL	UDL	UDL	UDL	ND	ND
Nickel	UDL [b]	UDL [b]	7.10 [a]	7.28 [a]	6.68 [a]	3.04	0.001
Boron	UDL	UDL	UDL	UDL	UDL	ND	ND
Aluminium	UDL	UDL	UDL	UDL	UDL	ND	ND

[a, b, c, d] means with different superscripts within a row are significantly different, RMSE = root means square error, UDL = undetectable level, ND = not done.

Table 6. Essential trace elements and heavy metals contents of whole eggs on a dry matter basis of different commercial brands in the retail market.

Element	Egg Source				RMSE	*p* Value
	A	B	C	D		
Essential trace elements, ppm						
Iron	11.83	12.70	12.22	12.51	1.10	0.638
Copper	1.17	1.64	1.56	1.61	1.50	0.748
Zinc	58.6 [b]	64.7 [a]	60.2 [b]	68.3 [a]	3.18	0.001
Manganese	UDL	UDL	UDL	UDL	ND	ND
Selenium	UDL	UDL	UDL	UDL	ND	ND
Chromium	8.04	8.62	7.96	8.25	0.564	0.293
Cobalt	UDL	UDL	UDL	UDL	ND	ND
Heavy metals, ppm						
Lead	UDL	UDL	UDL	UDL	ND	ND
Cadmium	UDL	UDL	UDL	UDL	ND	ND
Arsenic	UDL	UDL	UDL	UDL	ND	ND
Nickel	UDL	UDL	UDL	UDL	ND	ND
Boron	UDL	UDL	UDL	UDL	ND	ND
Aluminium	UDL	UDL	UDL	UDL	ND	ND

[a, b] means with different superscripts with a row in similar treatment groups are significantly different, RMSE = root means square error, UDL = undetectable level, ND = not done.

Table 7 shows essential trace mineral and heavy metal contents of eggs, meat, meat products, and liver. Differences between the element content of broiler meat and table eggs were significant only for Fe and Zn, where the Fe level was significantly higher in meat than in eggs, while the Zn level was lower. In general, the liver had higher levels of toxic elements than the eggs, except for Cr. The burger and frankfurter also had higher concentrations of most of the trace elements than the eggs, except for Cr and Zn. On the other hand, Pb and Ni levels were higher in meat products (burger and frankfurter) than eggs.

Table 7. Heavy metals contents of commercial table eggs and broiler meat, burger, frankfurter, and liver on a dry matter basis in the retail market.

Element	Types of Poultry Products					RMSE	*p* Value
	Eggs	Meat	Burger	Frankfurter	Liver		
Essential trace elements, ppm							
Iron	12.3 [c]	75.9 [b]	38.1 [c]	72.8 [b]	288.2 [a]	39.4	0.001
Copper	1.49 [c]	UDL [c]	6.66 [b]	7.80 [b]	19.24 [a]	2.27	0.001
Zinc	62.9 [b]	38.5 [c]	14.9 [e]	26.4 [d]	79.8 [a]	9.51	0.001
Manganese	UDL [c]	UDL [c]	6.41 [c]	9.77 [b]	18.2 [a]	2.91	0.001
Selenium	UDL	UDL	UDL	UDL	UDL	ND	ND
Chromium	8.22 [a]	8.32 [a]	2.54 [b]	2.46 [b]	2.79 [b]	2.22	0.001
Cobalt	UDL [b]	UDL [b]	2.72 [a]	2.22 [a]	2.39 [a]	0.651	0.001
Heavy metals, ppm							
Lead	UDL [b]	UDL [b]	16.69 [a]	14.84 [a]	16.52 [a]	4.16	0.001
Cadmium	UDL [b]	UDL [b]	0.433 [b]	0.379 [b]	1.12 [a]	0.655	0.001
Arsenic	UDL	UDL	UDL	UDL	UDL	ND	ND
Nickel	UDL [b]	UDL [b]	7.10 [a]	7.28 [a]	6.68 [a]	2.61	0.001
Boron	UDL	UDL	UDL	UDL	UDL	ND	ND
Aluminium	UDL	UDL	UDL	UDL	UDL	ND	ND

[a, b, c, d,e] means with different superscripts within a row are significantly different, RMSE = root means square error, UDL = undetectable level, ND= not done.

The maximum permissible levels of essential elements and heavy metals from the literature in different poultry products are presented in Table 8. The results indicated that the permissible level for each element is different based on the nature of the element and its physiological body function. The maximum was for Zn, followed by B, and the lowest was for Cd.

Table 8. The maximum permissible level of metals (ppm) in poultry products set by international standards.

Elements	Types of Poultry Products				
	Eggs	Meat	Burger	Frankfurter	Liver
Essential minerals, ppm					
Iron	NA	NA	NA	NA	NA
Copper	10 [34,35]	1.0 [34]	NA	NA	1.0 [34]
Zinc	NA	20 [34–36]	NA	NA	20 [34–36]
Manganese	NA	0.5 [37]	NA	NA	0.5 [37]
Selenium	0.5 [38]	0.5 [38]	NA	NA	NA
Chromium	1.0 [34]	1.0 [34]	NA	NA	0.05 [37]
Cobalt	NA	NA	NA	NA	NA
Heavy metals, ppm					
Lead	0.50 [33]	0.10 [34]	NA	NA	0.10 [34]
Cadmium	0.05 [37]	0.05 [34]	NA	NA	0.50 [34]
Arsenic	0.1 [39]	0.1 [39]	NA	NA	NA
Nickel	NA	0.5 [39]	NA	NA	NA
Boron	10 [38]	10 [38]	NA	NA	NA
Aluminium	1 [40]	1 [40]	NA	NA	NA

NA, not available. [40], JECFA established a PTWI for Al of 1 mg/kg BW for all aluminum compounds in food.

Estimated Daily Intake

The estimated daily intake of the essential elements and heavy metals compared with the RDA are presented in Table 9. The consumption of liver can meet the RDA for all trace elements, while contributing a higher intake of Pb and Cd. From a safety point of view, eggs and poultry meat meet the RDA for trace elements and do not present a threat of toxicity from any heavy metals, except Cr. Consumption of meat products, such as burger and frankfurter can supply a considerable part of the RDA of trace minerals and are abundant sources compared with meat and eggs, but they can present a health threat from Pb.

Table 9. Estimated daily intake of essential elements and heavy metals per 100 g of poultry products compared to the recommended daily allowance for adults Food and Agriculture Organization/World Health Organization [FAO/WHO].

Elements	Types of Poultry Products					RDA (FAO/WHO), mg/AI	Tolerable Upper Intake Levels (U/L), mg
	Eggs	Meat	Burger	Frankfurter	Liver		
Essential elements, ppm [41]							
Iron	1.23	7.59	3.81	7.28	28.82	8	45
Copper	0.149	UDL	0.666	0.780	1.924	0.9	10
Zinc	6.29	3.85	1.49	2.64	7.98	11	40
Manganese	UDL	UDL	0.641	0.977	1.82	2.3	11
Selenium	UDL	UDL	UDL	UDL	UDL	0.55	0.4
Chromium	0.822	0.832	0.254	0.246	0.279	0.35	ND
Cobalt	UDL	UDL	0.272	0.222	0.239	ND	ND
Heavy metals, ppm [42]							
Lead	UDL	UDL	1.669	1.484	1.652	0.21	0.43 [43]
Cadmium	UDL	UDL	0.0433	0.0379	0.112	0.06	0.5-20 [44]
Arsenic	UDL	UDL	UDL	UDL	UDL	0.13	0.1 [39]
Nickel	UDL	UDL	0.710	0.728	0.668	ND [45]	0.1 [39]
Boron	UDL	UDL	UDL	UDL	UDL	NR [45]	NR
Aluminium	UDL	UDL	UDL	UDL	UDL	60 [46,47]	1 [40]

UDL = undetectable level. ND = not determined yet, RDA = recommended dietary allowances, AI = adequate intake, where no RDA has been established, but the amount is somewhat less firmly believed to be sufficient for everyone in the demographic group, UL = tolerable upper intake levels (UL). NR= no risk effect.

4. Discussion

In general, the harmful effects of heavy metals include deleterious functional and physiological effects influencing cell metabolism, having also oxidative impact on biological macromolecules that adversely influence nuclear proteins and DNA [29,48–60]. These metals are essential to the maintenance of different physiological and biochemical functions in living organisms in small amounts; however, they become harmful when they exceed specific standards, and it is acknowledged that heavy metals can cause cell malfunction and, ultimately, toxicity [29,59]. Sources of contamination by heavy metals are different from one element to another [60–64], and mainly depends on the type of soil, environment risks broadens, animal species and product feeds, and geographic area [65–69].

4.1. Essential Elements and Heavy Metals in Frozen and Fresh Meat and Meat Products

It should be mentioned that frozen and fresh meat had similar essential elements and heavy metals contents, except for Cr, which were higher (1.2%) in the fresh meat than frozen meat. It was evident that the type of poultry product had a pronounced effect on the Fe content, and the liver is the most abundant Fe source, according to other authors [43,70]. The values of Fe in different types of meat and liver found herein were higher than those obtained (6.77–7.49) by Elsharawy [70], (41.4–54.9 ppm) by Khan et al. [43] and (12.37–14.39 ppm) by Muhammad et al. [71] for liver and meat from different districts of Pakistan. As with the present results, Alturiqi and Albedair [72] observed that the Fe levels in meat products such as beef loin, pastrami, sausage, and luncheon meat were 175.7, 188.5, 242.4, and 203.1 ppm, respectively. The Fe content of chicken meat from different districts in Saudi Arabia ranged from 135.3 to 290.0 ppm.

The Fe content of frozen and fresh meat, liver, burgers, and frankfurters exceed the RDA for humans while burgers can supply a considerable amount (27.3–30.8% and 84.7–95.3%, respectively) of the RDA of Fe for children and adults, with the liver being the most abundant source. Meanwhile, the values obtained herein for Fe content of varying meat products are in line with those reported by Chowdhury et al. [73] for chicken meat (16.7–60.3 ppm) and meat products (11.4–290.1 ppm). In this context, the liver is the site of metabolism and storage of Fe, which is vital for animal and human nutrition as an essential part of haemoglobin [43,51,71]. Metabolism of proteins, lipids, and carbohydrates is facilitated by Fe, which plays a vital part in the survival and growth of living organisms, cytochrome oxidase, catalase, oxygen-transporting haemoglobin, and myoglobin, as well as the redox process [51]. Deficiency of Fe causes a high susceptibility to gut infections, myocardial infarctions, and nose bleeds [51,70]. The influence of toxic concentrations of Fe in animals includes coma, depression, cardiac arrest, respiratory failure, and convulsions. The maximum tolerable level of Fe for adult females/males (14–70 years old) and children (0–8 years old) is 45 and 40 ppm per day, respectively [41].

The maximum permissible levels for Cu in meat and meat products were reported to be 10 ppm and 1 ppm [35,37]. In addition, the allowable level for Cu in meat and offal in Egypt should not exceed 15 ppm [44]. Thus, fresh and frozen meat, burger, and frankfurter were found to be safe for human consumption and within the permissible level, but the liver samples showed the highest values of Cu and posed a hazard for humans. The Cu content of chicken meat from different districts in Saudi Arabia ranged from 2.30 to 7.88 ppm [72]. In accordance with the current findings, Cu levels in meat products, beef loin, pastrami, sausage, and luncheon meat have been found to be 14.84, 11.11, 18.51, and 13.78 ppm, respectively [73]. Cu is an essential element for various enzymes and is mostly stored in the liver and muscle and is involved in different body functions [22], but an increased Cu dose provokes stomach, nausea, jaundice, diarrhea, and severe colic, and liver and renal problems, as well as anemia, while excessive deposition of Cu in the gizzard, liver, eyes, and brain are characteristic of Wilson's disease [53,70]. Therefore, the consumption of animal products with increased Cu levels may pose a threat to public health [51,59,70,72].

It was shown that Zn residuals in frozen and fresh meat and meat products are higher than the permissible levels for human consumption (20 ppm) [34], while liver samples showed the highest vales and burgers showed the lowest values. Similarly to the present results, it was found [59,73] that the concentration of Zn in the liver was 22.2–74.84 ppm and was 4.46–168.7 ppm in meat. Furthermore, it was observed [73] that Zn in meat products ranged from 17.2 ppm for burgers to 138.4 for chicken wings. However, [71] found that Zn levels in meat and liver are similar (12.23 vs 13.93, ppm). In addition, it was revealed [43] that Zn levels in the liver, thigh, and breast meat from different districts in Pakistan were similar and ranged from 106.6 to 110.3 ppm. Zinc content in poultry products may vary due to geographical area and the type of product [74,75] for example, the values of residual Zn varied in meat products (30.3–73.9) and in meat (27.9–36.9) among different studies from Saudi districts [72], in Zambia [76], and in Pakistan [77], being similar to the present findings.

The amounts of Mn in frozen and fresh meat and different poultry products recorded herein agree with those noted by others [72,77] for chicken meat (0.0–9.98 ppm), and liver (0.24–4.32 ppm) [34,35]. The liver is the seat of metabolism, and thus, a high concentration of Mn in the liver would be expected. Likewise, previous studies [8,43,78] found that the Mn content in the liver (1.12–340 ppm) was higher than that in the muscle (0.696–102 ppm). In addition, Mn contents in meat products, beef loin, pastrami, sausage, and luncheon meat were 15.73, 11.97, 18.33, and 32.62 ppm, respectively [72]. The Mn value of chicken meat from different districts in Saudi Arabia ranged from 21.48 to 34.42 ppm. The toxic impact of Mn includes a decrease in fetal weight and retardation of the skeleton and internal organs [79], as well as a decrease in birth weight of term-born infants [80]. The toxicity of Mn may cause DNA damage, chromosomal aberrations, and result in a harmful influence on the embryo and fetus [81], due to accumulation in various brain regions [82], neurotoxicity [83], and Parkinson-like syndrome [84,85] and the generation of reactive oxygen species causing oxidative stress [86,87].

In this study, Se was not recorded in frozen and fresh meat, different meat products and liver, showing no residual impact on human health. It is worth mentioning that Se is an essential mineral for selenoproteins, living organisms, and is involved in metabolism, reproduction, immunological responses, and antioxidant balance [55,88,89]. Se was found to be 0.087–0.115 ppm in liver samples, 0.133–0.164 ppm in breast meat, and 0.169–0.200 ppm in thigh meat in [59]. It has been cited that Se levels are significantly higher in the breast muscle and liver of broiler and much higher in the liver than the other tissues [90,91]. However, excess Se may cause harmful effects, for example, nail changes and alopecia [92]. As selenosis progresses, decreased cognitive function, weakness, paralysis, and death can occur [93]. Post mortem, a blood selenium level > 1400 ppm is consistent with acute toxicity as the cause of death during the first day of exposure.

The toxicity of heavy metals negatively affects animal performance and human health through food chain and depends on the type of metals, metal intakes, age and health status of human [94,95]. In this study, differences in Cr contents of fresh and frozen meat were obvious, being higher in fresh than frozen meat. In addition, different meat products and liver had lower Cr than frozen meat. Cr contents of different meat and meat products and liver found herein were higher than tolerable levels, which are 1.0 ppm in meat and 0.5 ppm in liver [34,37]. The application of agriculture technology has resulted in the release of Cr into the environment, causing Cr hazards as a result of sewage, fertilizers, Cr dust, and using wastewater in irrigation, which influences the food chain. As with the present results, Cr levels have been found to be 0–0.69 ppm in chicken meat, while they were 0–4.33 ppm in meat products [73]. In addition, the Cr level was 0.06 ppm in meat and ranged between 0.08 and 0.11 in the liver in three districts in Pakistan [43]. Cr concentration was 0.061–0.111 ppm in meat samples and 0.086–0.092 ppm in liver samples [59]. Furthermore, the Cr concentration range was 0.15 ppm [71] and 0.064–0.073 ppm in the liver and 0.075 ppm in muscle [95]. It is worth mentioning that improvements in insulin sensitivity, blood glucose, insulin, lipids, hemoglobin, lean body mass, and related variables are seen in response to improved Cr nutrition [96,97], and further evidence was exhibited in broilers where Cr supplementation decreased the blood glucose level [98]. Cr toxicity increases reactive oxygen species [50,99–103] damaging proteins and DNA [104]. This is dependent on the type of Cr; Cr (VI) is considered to be carcinogenic and can cause problems in the liver, kidneys, neural tissues, and the circulatory system. Skin irritations and ulcers can also occur, as well as metabolic defects such as diabetes and heart problems [51,100–102].

The concentration of Co in frozen and fresh meat was undetectable and was lower than that in the liver, burgers, and frankfurters that had similar amounts (2.22–2.39 ppm), which is markedly higher than in meat. The residual Co in the liver samples indicated that the liver is the site of Co metabolism, whereas the content of Co in meat products, burgers, and frankfurters showed contamination with Co during the manufacturing process. Cobalt is an essential constituent of vitamin B_{12}; however, data relating to Co toxicity, to the best of our knowledge, is rare in the literature [105] and was established by [22] to be 100–200 ppm for poultry. The negative health effects of Co include endocrine deficits, neurological syndromes (e.g., visual impairment and hearing), and cardiovascular problems. The adverse health impact of Co does not occur at a Co blood level below 300 µg/L in healthy subjects, which is not connected with changes in concentrations of hemoglobin, red blood cell count, and hematocrit, nor with changes in neurological, cardiac, or thyroid function [105,106]

Frozen and fresh meat was free of Pb residual in the present study, but Pb in meat products and liver was similar and exceeded the tolerable upper intake [43]. The permissible limit for lead residues in meat and offal must not exceed 0.1 ppm for meat and liver [34,44]. Several researchers have reported similar values of Pb residuals in chicken meat, (0–3.94, ppm) [73]. In addition, the Pb concentrations in the flesh and liver samples were 0.25–0.26 and 0.31 ppm, respectively, being higher in the liver than the meat [70]. In this respect, Pb concentration ranged from 0.09–0.51 ppm in the broiler meat [94]. In addition, Pb residuals in chicken meat in different Saudi Arabia districts were in the range of 7.61–10.49 ppm [72], 0.055–0.116 ppm in meat and 0.068–0.093 ppm in the liver [59], 0.07 in meat and 0.056–4.15 ppm in the chickens' liver [71,76], and from 0–2.23 ppm in meat and 0–7.56 ppm in the liver [43,77].

Lead bio-accumulates in animal and human tissues, mainly in the liver and the bones leading to several diseases such as irritability, cardiovascular problems, auditory, neuropathy, wrist, and food drop, haemolytic anemia, atherosclerosis, and liver apoptosis [50,51]. The nervous, hematopoietic, and adrenal systems are the main systems that are sensitive to Pb toxicity [9,43]. Lead is one of the riskiest heavy metals when consumed through the food chain, and Pb has marked side effects on human health since it is transmitted through the food chain; nonetheless, it is not indispensable for biological function [50,59]. Heavy metals contaminations such as lead contamination can result from the use of foods like vegetables, meat, fruits, seafood, wine herbicides, chemical fertilizers, as well as the use of sewage resulting in soil and environmental pollution and is represented as biological biomarkers [53,72,94,107–109].

The results of Cd levels in frozen and fresh meat, meat products, and liver recorded herein agree with those reported in chicken meat (1.36–1.68 ppm) and meat products (3.06–4.08 ppm) [72]. The results showed that frozen and fresh meat is safer than meat products (0.379–0.438 ppm) and liver (1.12 ppm). The permissible limit for Cd in poultry meat and offal was determined to be 5 ppm for meat and 20 ppm for poultry offal [44]. According to these limits, most tested samples of meat products, except for the liver, were within the allowable levels and considered safe for human consumption. Additionally, the permissible limit for Cd in meat and liver was reported to be 0.05 ppm [34,37]. The value of Cd in the liver found herein represents a hazard level according to [107], which is estimated to be 1 ppm as the tolerable upper level, and the estimation of 0.06–0.07 ppm [43].

Similarly to the present findings, Cd has been found to be higher in the liver than in thigh meat, and the latter was higher than in breast meat [70,108]. The values of Cd published in the literature ranged from 0.001–0.002 ppm in meat [59], 0.006–0.23 ppm in different meat products [73], and 0.002–1.6 ppm in the liver [59,71,77,109]. Cadmium is chiefly found in the earth's crust and is easily absorbed by the organic substances that form the soil, thereby presenting a high risk due to transportation through the food chain from the earth, to food and animals and/or humans [94]. In addition, contamination of Cd can result from chemical fertilizers, particularly phosphate, in the soil, lakes, and groundwater supplies and can negatively impact animals and fish through the food chain [59,73]. In addition, Cd has greater adverse effects on children in whom Cd accumulates to a greater extent in the tissues than the adults. Cadmium is a dispensable metal, but increasing Cd intake above the tolerable level causes respiratory symptoms and lung damage, renal dysfunction, hepatic injury, hypertension [43,109], mental retardation, cardiovascular and auditory systems dysfunction [50,77], carcinogenesis, and mutagenesis diseases [51,76]. The most important negative impact of Cd toxicity is Itai-Itai disease in humans, which directly interferes with calcium and bone mineralization resulting in osteoporosis and osteomalacia [52,110]. Cd causes significant alterations in the detoxification of enzymes in the gizzard and liver [52].

Arsenic is a metalloid, which indicates that it has both non-metallic and metallic characteristics [94]. The residual of As was not observed herein in different meat samples, meat products, and liver, showing the safety of different chicken protein products. In other studies, As was recorded at 0–0.01 ppm in meat and meat products [73]. The values of As in meat and liver were 0.36–0.49 ppm and 0.77 ppm, respectively [70,111]. Other researchers obtained As values of 0.012–0.029 ppm in meat and 0.023–0.049 ppm in the liver [52,59]. Additionally, As levels were 0.003–0.09 ppm in various meats of chicken [94]. The permissible limit for As residues in poultry meat and offal has not yet been set, according to [44,70]; however, the allowable limit for As in poultry meat and offal was estimated to be 0.1 ppm for meat [39]. The contamination of the environment results from the chemical and glass industries, and the pollutants reach water resources, where they come into contact with marine life. As could enter the environment and water resources as a result of As application in medicine and livestock production [50,51]. The accumulation of As in meat is low, and the principal tissues involved in accumulation are the gizzards and the liver [70]. As is the main cause of acute heavy metal poisoning in adults. Exposure to As can induce liver disease, cardiovascular problems, diabetes, cancer, and skin disease.

There was no risk assessed for frozen and fresh meat and meat products due to Ni contamination in this study. The permissible limit for Ni in poultry meat was estimated to be 0.1 ppm in [39]. However, Ni was recorded at 0.057–0.106 ppm in meat and 0.003–0.277 ppm in the liver in [95], 0.036–0.069 ppm in meat and 0.051–0.059 ppm in the liver in [59], and 0.13 ppm in muscle and not noticeable in the liver in [71]. Nickel is an essential element for red blood cell formation, although when excess Ni enters the body via ingestion, inhalation, or absorption, Ni toxicity can be observed and affects foetal organs such as the larynx, nose, and lungs, and can also alter the heart and the prostate [71,95]. It seems that Ni toxicity or contamination only results from very high consumption of Ni, mainly when feeds and/or foods are cultivated in Ni-rich soils, thus contributing greater quantities of Ni to the food chain [59,95].

Frozen and fresh meat and different meat products and liver were free from B residual in the current study. The literature values of B in different poultry products are absent, and this is the first time that B levels in poultry meat, meat products, and liver have been reported. The permissible limit for B in poultry meat and offal was estimated to be 10 ppm for meat and eggs in [112]. The lethal dose of boric acid in one-day-old chicks was found to be 2.95 +/− 0.35 g/kg of body weight, which classifies this product as only slightly toxic to chickens [113]. Boron residue levels in the brain, kidney, liver, and white muscle were not significantly increased following a 15-day exposure period to 500 ppm or 1250 ppm boric acid in feed ad libitum chickens for three weeks; however, B markedly increased due to feeding with 2500 ppm or 5000 ppm boric acid. Boron was not accumulated in the soft tissues of the animals but did accumulate in the bone. Normal levels of B in soft tissues, urine, and blood generally ranged from less than 0.05 ppm to no more than 10 ppm [114]. In poisoning incidents, the amount of boric acid in the brain and the liver tissue has been found to be as high as 2000 ppm. Boron may contribute to decreased male fertility in rodents fed 9000 ppm of boric acid in feed [115]. Within a few days, B levels in the blood and most soft tissues quickly reached a plateau of about 15 ppm. Boron in bone did not appear to plateau, reaching 47 ppm after seven days on a diet. Cessation of exposure to dietary B resulted in a rapid drop in bone B [114]. B does not seem to be metabolized in humans and animals, owing to the massive energy needed for the breakdown of the B–O bond [115].

In the present study, the residual of Al was not recorded in the frozen and fresh meat and different meat products and liver. The permissible limit for Al residues in poultry meat is 1 ppm, according to [40]. Conclusively, there was no risk assessed for meat and meat products due to Al continuation in this study. Al interferes with most physical and cellular processes [116]. Al toxicity presents a threat to humans, animals, and plants and results in many diseases [117]. The toxicity of Al might be produced from the interaction between Al and the plasma membrane, affecting most physical and cellular processes in organisms [118]. In humans, Al^{3+} has been shown to replace Mg^{2+} and Fe^{3+} and resulted in disturbances of cellular growth and intercellular communication, as well as in neurotoxicity effects, and secretory functions [119]. The modifications that are induced in neurons by Al are similar to the degenerative lesions observed in Alzheimer patients [51,120].

4.2. Essential Elements and Heavy Metals in Eggs

From a risk assessment point of view, eggs were found to be the safest product for trace elements and heavy metals, and these agree with previous findings [43,69]. The average values of iron in different egg sources was in the range of 11.83–12.70 ppm, indicating that eggs are a rich source of Fe. The values observed herein are higher than those found in egg albumen (1.05–1.27 ppm) and yolk (3.19–3.36 ppm) and in eggs (1.47–2.03 ppm) [46,70].

The values of Cu contents of eggs of different sources in the present study were in the range of 1.17–1.64 ppm and were found to be safe for human consumption and within the permissible level. The maximum permissible level for Cu in eggs is reported to be 10 ppm [34,35]. The Cu content in eggs was in the range of 0.009–0.014 ppm in commercial tables eggs in Saudi Arabia [69].

The source of eggs markedly affects eggs' Zn content; the values ranged from 58.6 to 68.3 ppm with 14.2% difference. In literature, Zn levels found in egg albumen (1.97–2.05 ppm) and yolk (39.9–40.4 ppm) were similar in different districts of Pakistan; however, yolk had a higher Zn content than the albumen, according to Guyot and Nys [121], which indicates that egg yolk is the major contributor to iron and zinc supply. The Zn residual in eggs was in the range of 1–1.13 ppm in commercial tables eggs in Saudi Arabia [69]. Thus, it might be shown that Zn residuals in eggs found herein are higher than the permissible levels for human consumption [43]. Zn as an essential element is crucial for health at an appropriate level for appetite, taste, and smell, immunity, wound healing, and skin health [51]. Zinc deficiency delays the development of sex organs and causes retarded growth in young men [75].

Sources of Zn pollution include mining, purifying of Zn, Pb, and Cd ores, coal burning, steel production, and waste burning [75].

The Mn residuals in different sources of table eggs were absent, suggesting that eggs are safe for human consumption with regards to residual Mn. The tolerable upper level of Mn for eggs is also absent in the literature. The average Mn in the albumen and yolk of eggs was significantly different and was in the range of 0.19–0.31 ppm for albumen and 1.33–1.40 ppm for yolk [43].

The present results indicated a lack of Se residuals in eggs and thus no risk associated with their consumption. The maximum permissible concentration of Se in eggs is 0.5 ppm [112]. Similar to the present results, Se was not recorded in three sources of commercial eggs in Jeddah City in Saudi Arabia [69]. The importance of Se for human consumption and its deterioration effects were previously discussed in the abovementioned section.

Cr in eggs was in the range of 7.96–8.62 ppm with no differences among the four sources of table eggs, but Cr residuals exceeded the permissible concentrations (1 ppm) [34], being higher than those of the meat products and liver, which presented lower levels of risk than the eggs.

The levels of Co in table eggs are below dangerous levels for human consumption. The residual cobalt in different sources of table eggs was absent, confirming the safety of eggs for human consumption.

The concentration of Pb in different sources of table eggs was below the risk level for human consumption, which is 0.5 ppm [37]. In addition, Khan et al. [43] reported that the upper tolerance level of Pb is 0.43 ppm. In this respect, several researchers have reported similar values of Pb residuals in eggs (0.34–12.1, ppm) [73]. The egg albumen had higher Pb content (0.12–0.13, ppm) than yolk (0.06–0.09, ppm) and was lower than that in meat.

There were no Cd residues in eggs, which were below the risk level for human consumption, which is 0.05 ppm [34,37]. The absence of Cd in eggs agrees with the results reported by [77]. The values of Cd published in the literature ranged from 0 to 0.99 ppm in egg albumen and yolk [73]. Furthermore, the Cd level in eggs ranged from 0.51 to 0.68 ppm, and from 0.03 to 0.06 ppm in egg albumen and egg yolk from three districts in Pakistan [43]. The results showed the eggs are safer than meat products (0.379–0.438 ppm) and liver (1.12 ppm).

The level of arsenic in different egg sources was under detectable levels, suggesting there was no risk assessed for eggs due to As contamination. The permissible limit for As residues in eggs has not yet been set, according to [44,70]; however, the allowable limit for As in poultry eggs was estimated to be 0.1 ppm [39]. In addition, As was found to be at 0–0.01 ppm in eggs [73]; furthermore, As concentration was 0.01, 0.01, and 0.004 ppm in albumen, yolk, and whole eggs, respectively [94].

It was found that different sources of eggs showed no Ni residuals and thus are free of Ni contamination. Thus, there was no risk assessed due to B continuation in this study. The literature values of B in eggs are absent, and this is the first time that B levels in commercial eggs have been reported. The permissible limit for B in eggs was estimated to be 10 ppm [112]. The level of Al in eggs was under detectable levels. The permissible limit for Al residues in poultry eggs is 1 ppm, according to [40]. Conclusively, there was no risk assessed for eggs due to Al continuation in this study.

4.3. Heavy Metals in Poultry Diets and Litter

The results indicate that trace minerals in broilers' diets are in general agreement with recommended levels for broilers and layers during different ages [22]. The heavy metals were also below the toxic levels [22]. Essential minerals are important for the normal physiological functions of animals, and several minerals (Fe, Zn, Se) have been used recently in the production of functional foods for human health benefits [13,14]. It should be mentioned that differences in essential trace minerals found herein among different types of diets, and/or between the diets of broilers and layers, regardless of the feeding stage, are acceptable based on the differences in mineral content of different foodstuffs, diet composition, and/or type of mineral premix used, as well as the source of feeds.

Most of the heavy metals were recorded in diet samples used in this study except for Al, but the levels were found to be less than the toxic limit [22], showing the high quality of feeds and that there are few hazards of heavy metals toxicity. Compared with the results reported by Okoye et al. [62], the concentrations of heavy metals recorded herein were less than those for Pb, As, Cd, Cr, Ni, and Cu and were 0.12–0.293, 0.068–0.167, 0.281–0.379, 0.082–0.212, 0.039–0.172, and 0.069–0.205 ppm, respectively, suggesting low pollution by heavy metals in the present study. The toxic concentrations of heavy metals depend on the chemical form of these elements, the poultry species, and the age of the birds; hence, they are the subject of investigation [22,61].

The concentrations of heavy metals, such as Cr and Pb, in broiler and laying hen diets ranged from 1.51 to 5.20 ppm and 2.09 to 5.27 ppm, respectively. The upper tolerable levels of Cr and Pb were suggested as 10–300 and 10–1000 ppm, respectively. This indicates that Cr and Pb are present at alarming levels in the diets tested in the present research. The Pb contamination in feeds found here is mostly related to environmental pollution resulted from fuel brun herbicides, chemical fertilizers, and sewage causing soil and feed contamination [50,51,53,56]. The higher Cr contents in the feed mixture could be due to the use of fertilizers, sewage, Cr dust, and wastewater in irrigation of crops in Saudi Arabia due to limited water resources. In addition, Jeddah city is an industrial and urban area. This could influence the food chain and, thus, the production of reactive oxygen species [50,63,64]. The high Pb and Cr in the feed mixture concurred with increased contents of Pb and Cr in poultry products, which implies health concerns for humans. This is suggesting regular checking of heavy metals to monitor the food safety should be mandatory in this area of the research. Additionally, a correlation of a durable nature has been recorded between heavy metals in diet and excreted waste, which indicates that the poultry diet has the potential to contaminate several components of the environment [32]. Heavy metals may enter the production system of livestock by different routes, including land application of inorganic fertilizers, atmospheric deposition, agrochemicals, animal waste, and biosolids [65,66]. Furthermore, fish and poultry diets have shown various heavy metal content (e.g., Cd, Cr, and Pb), suggesting that the supply of feedstuffs requires greater governmental quality control [67].

In general, trace minerals and toxic elements in the litter of broilers and layers may reflect to some extent the types of feeds, stages of production, and/or types of chickens. These levels are reflected in the diets with some exceptions, particularly for Al, which was not found in feeds but was recorded in the litter, except for in the broiler starter phase, showing that Al is a litter contaminant at higher concentration in layers than in broilers.

It should be mentioned that except for Se, levels of essential minerals and toxic elements in poultry litters were lower than the values reported in the research of several authors [27–34]. Selenium contents in animals' feeds are influenced by soil Se content and Se supplementation to meet animal requirements [56]. Furthermore, increasing the Se content of animal diets was recently used for the production of functional food such as eggs, milk, and meat [15]. Heavy metals in animal manure lead to the accumulation of toxic elements in the soil and water due to the use of litter as a fertilizer and for soil amendments [67,68] and hence can be transmitted through feed chain to humans.

5. Conclusions

It can be concluded that Cd, Pb, As, and Se were not detected in chicken meat and eggs locally produced or imported, indicating a lack of hazards from these toxic elements for humans. However, the liver showed the highest concentration of heavy metals, except for Cr, and the intake of Pb and Cd from broilers' liver was above the RDA for adults. Meat products, such as burgers and frankfurters, showed higher Pb, Cd, and Ni concentrations than chicken meat and table eggs suggesting a possible health threat to humans. Thus, to improve the quality of poultry products for human consumption, appropriate legislation is required for monitoring the quality of poultry products, as well as the feeds/food and litter of chickens. Additionally, essential measurements should be used for detoxification of heavy metals from the waste. The relationship between the minerals within poultry production and between poultry diets and poultry litter remains fertile for further research.

Author Contributions: M.A.K. and Y.A.A. contributed equally to all stages of the work and in the preparation of the revision and approval of the manuscript for final submission. All authors have read and agreed to the published version of the manuscript.

Funding: This project was funded by the Deanship of Scientific Research (DSR), King Abdulaziz University, Jeddah, Saudi Arabia, under grant No."G: 475-375-1440". The authors, therefore, acknowledge with thanks DSR for technical and financial support.

Conflicts of Interest: The authors declare no conflict of interest.

References

1. Doyle, J.J.; Spaulding, J.E. Toxic and essential trace elements in meat-A review. *J. Anim. Sci.* **1978**, *47*, 398–419. [CrossRef] [PubMed]
2. Martz, W. The essential trace elements. *Science* **1981**, *213*, 1332–1337. [CrossRef] [PubMed]
3. Attia, Y.A.; Al-Harthi, M.A.; Korish, M.A.; Shiboob, M.M. Evaluation of the broiler's meat quality in the retail market, Effects of type and source of carcasses. *Rev. Mex. Cienc. Pec.* **2016**, *7*, 321–339. [CrossRef]
4. Attia, Y.A.; Al-Harthi, M.A.; Shiboob, M.M. Evaluation of quality and nutrient contents of table eggs from different sources in the retail market. *Ital. J. Anim. Sci.* **2014**, *13*, 369. [CrossRef]
5. Bamuwamye, M.; Ogwok, P.; Tumuhairwe, V. Cancer and Non-cancer Risks Associated With Heavy Metal Exposures from Street Foods, Evaluation of Roasted Meats in an Urban Setting. *J. Environ. Pollut. Hum. Health* **2015**, *3*, 24–30.
6. Attia, Y.A.; Abdalah, A.A.; Zeweil, H.S.; Bovera, F.; Tag El-Din, A.A.; Araft, M.A. Effect of inorganic or organic copper additions on reproductive performance, lipid metabolism and morphology of organs of dual-purpose breeding hens. *Arch. Geflügelkd.* **2011**, *75*, 169–178.
7. Abduljaleel, S.A.; Shuhaimi-Othman, M.; Babji, A. Assessment of Trace Metals Contents in Chicken (Gallus gallusdomesticus) and Quail (Coturnix coturnix japonica) Tissues from Selangor. *J. Environ. Sci. Technol.* **2012**, *5*, 441–451.
8. Rehman, K.; Andalib, S.; Ansar, M.; Bukhari, S.; Naeem, N.M.; Yousaf, K. Assessment of heavy metal in different tissues of broiler and domestic layers. *J. Glob. Vet.* **2012**, *9*, 32–37.
9. Eton, E.C.; Rufus, L.C.; Charles, L.M. Effects of broiler litter management practices on phosphorus, copper, zinc, manganese, and arsenic concentrations in Maryland coastal plain soils. *Commun. Soil. Sci. Plan.* **2008**, *39*, 1193–1205.
10. Ayar, A.; Sert, D.; Akin, N. The trace metal levels in milk and dairy products consumed in middle Anatolia-Turkey. *Environ. Monit. Assess.* **2009**, *152*, 1–12. [CrossRef]
11. Qin, L.Q.; Wang, X.P.; Li, W.; Tong, X.; Tong, W.J. The minerals and heavy metals in cow's milk from China and Japan. *J. Health Sci.* **2009**, *55*, 300–305. [CrossRef]
12. McCrory, L.Y.; Powel, D.F.; Saam, J.M.; Jackson, J.D. A survey of selected heavy metal concentrations in Wisconsin Dairy Feeds. *J. Dairy Sci.* **2005**, *88*, 2911–2922.
13. Surai, P.F.; Sparks, N.H.C. Designer egg, from improvement of egg composition to functional food. *Trends Food Sci. Tech.* **2013**, *12*, 7–16. [CrossRef]
14. Sparks, N.H.C. Hen's egg–Is its role in human nutrition changing? World's. *Poult. Sci. J.* **2006**, *62*, 308–315.
15. Pappas, A.C.; Karadas, F.; Surai, P.F.; Wood, N.A.R.; Cassey, P.; Bortolotti, G.R.; Speake, B.K. Interspecies variation in yolk selenium concentrations among eggs of free-living birds, The effect of phylogeny. *J. Trace Elem. Med. Biol.* **2006**, *20*, 155–160. [CrossRef]
16. Cunningham, W.P.; Saigo, B.W. *Environmental Science a Global Concern*, 4th ed.; WMC Brown Publisher: New York, NY, USA, 1997; p. 389.
17. Baykov, B.D.; Stoyanov, M.P.; Gugova, M.L. Cadmium and lead bioaccumulation in male chickens for high food concentrations. *Toxicol. Environ. Chem.* **1996**, *54*, 155–159. [CrossRef]
18. Daniel, B.; Edward, A.K. *Environmental Science, Earth as a Living Planet*; John Wiley and Sons, Inc.: New York, NY, USA, 1995; pp. 278–279.
19. Shils, M.E.; Olson, J.A.; Shike, M. *Modern Nutrition in Health and Disease*, 8th ed.; Part II; Lea & Febiger: Philadelphia, PA, USA, 1994; pp. 1597–1598.
20. Allan, G.; Robert, A.C.; Reilly, D.C.J.; Stewart, M.J.; James, S. *Clinical Biochemistry*, 2nd ed.; Harcourt Brace and Company Ltd.: Orlando, FL, USA, 1995; pp. 114–115.

21. Judith, R.T. Copper. In *Modern Nutrition in Health and Disease*, 8th ed.; Maurice, E.S., James, A.O., Moshe, S.L., Febiger, Eds.; Part I; Lea & Febiger: Philadelphia, PA, USA, 1994; pp. 237–240.
22. National Research Council, NRC. *Nutrient Requirements of Poultry*, 9th ed.; 8. Toxicity of Certain Inorganic Elements; National Academy Press: Washington, DC, USA, 1994; pp. 58–60.
23. Association of Official Analytical Chemists, AOAC. *Official Methods of Analysis*; Association of Official Analytical Chemists: Washington, DC, USA, 2004.
24. Pawan, R.S.; Pratima, S.; Chirika, S.T.; Pradeep, K.B. Studies and determination of heavy metals in waste tires and their impacts on the environment. *Pak. J. Anal. Envir. Chem.* **2006**, *7*, 70–76.
25. Olajire, A.; Ayodele, E.T. Contamination of roadside soil and grass with heavy metals. *Environ. Int.* **1997**, *23*, 91–101. [CrossRef]
26. SAS Institute. *SAS Statistical Guide for Personal Computer*; SAS Institute Inc.: Cary, NC, USA, 2002.
27. Vukobratović, M.; Vukobratović, Z.; Lončarić, Z.; Kerovac, D. Heavy metals in animal manure and effects of composting on it. *Acta Hortic.* **2014**, *1034*, 591–597. [CrossRef]
28. Arroyo, M.D.D.; Hornedo, R.M.D.; Peralta, F.A.; Almestre, C.R.; Sánchez, J.V.M. Heavy metals concentration in soil, plant, earthworm and leachate from poultry manure applied to agricultural land. *Rev. Int. Contam. Ambie.* **2014**, *30*, 43–50.
29. Ravindran, B.; Mupambwa, H.A.; Silwana, S.; Mnkeni, P.N.S. Assessment of nutrient quality, heavy metals and phytotoxic properties of chicken manure on selected commercial vegetable crops. *Heliyon* **2017**, *3*, e00493. [CrossRef] [PubMed]
30. Okeke, O.R.; Ujah, I.I.; Okoye, P.A.C.; Ajiwe, V.I.E.; Eze, C.P. Assessment of the heavy metal levels in feeds and litters of chickens rose with in Awka Metropolis and its environs. *IOSR J. Appl. Chem. (IOSR-JAC)* **2015**, *8*, 60–63.
31. Wang, W.; Zhang, W.; Wang, X.; Lei, C.; Tang, R.; Zhang, F.; Yang, Q.; Zhu, F. Tracing heavy metals in 'swine manure - maggot - chicken' production chain. *Sci. Rep.* **2017**, *7*, 1–9. [CrossRef] [PubMed]
32. Ukpe, R.A.; Chokor, A.A. Correlation between concentrations of some heavy metal in poultry feed and waste. *Open Access J. Toxicol. (OAJT)* **2018**, *3*, 4. [CrossRef]
33. Jaja, N.; Mbila, M.; Codling, E.E.; Reddy, S.S.; Reddy, C.K. Trace metal enrichment and distribution in a poultry litter-amended soil under different tillage practices. *Open Agric. J.* **2013**, *7*, 88–95. [CrossRef]
34. FAO/WHO. *Codex Alimentarius*; Schedule 1 of the Proposed Draft Codex General Standards for Contaminants and Toxins in Food. Joint FAO/WHO, Food Standards Programme; Reference CX/FAC 02/16; Codex Committee: Rotterdam, The Netherlands, 2002.
35. FAO/WHO; Roychowdhury, T.; Tokunaga, H.; Ando, M. Survey of arsenic and other heavy metals in food composites and drinking water and estimation of dietary intake by the villagers from an arsenic affected area of West Bengal. *India Sci. Total Environ.* **2003**, *308*, 15–35. [CrossRef]
36. Attia, Y.A.; Addeo, N.F.; Al-Hamid, A.; Bovera, F. Effects of phytase supplementation to diets with or without zinc addition on growth performance and zinc utilization of white pekin ducks. *Animals* **2019**, *9*, 280. [CrossRef]
37. FAO/WHO. *Report of the 32nd Session of the Codex Committee of the Food Additives. Contaminants*; Geneva (Switzerland) FAO/WHO Codex Alimentarius Commission: Rome, Italy, 2000.
38. *National Standard of the People's Republic of China GB 2762—2005 (Replaces GB 2762—1994, GB 4809—1984, etc)*; Maximum Levels of Contaminants in Foods Issued on January 25, 2005 Implemented on October 1, 2005; Ministry of Hygienic and the Standardization Administration of China: Beijing, China, 2005.
39. JECFA. Codex general standard for contaminants and toxins in food and feeds. In Proceedings of the 64th Meeting of the Joint FAO/WHO Expert Committee on Food Additives (JECFA), Rome, Italy, 8–17 February 2005.
40. JECFA. *Joint FAO/WHO Food Standards Programme Codex Committee on Contaminants in Foods Fifth Session*; The Joint FAO/WHO Expert Committee on Food Additives (JECFA): Rome, Italy, 2011.
41. *Dietary Reference Intakes (DRIs), Recommended Intakes for Individuals, Food and Nutrition Board*; Institute of Medicine, The National Academies Press: Washington, DC, USA, 2004.
42. JECFA. *Evaluations of the Joint FAO/WHO Expert Committee on Food Additives*; FAO: Rome, Italy, 2009.
43. Khan, Z.; Sultan, A.; Khan, R.; Khan, S.; Imran, U.D.L.; Kamran, F. Concentrations of heavy metals and minerals in poultry eggs and meat produced in Khyber Pakhtunkhwa, Pakistan. *Meat Sci. Vet. Public Health* **2016**, *1*, 4–10.

44. Egyptian Official Standard, EOS. 2010. Available online: https//www.slideshare.net/flash_hero/es-7136 (accessed on 26 March 2019).

45. Dietary Reference Intakes for Vitamin A, Vitamin K, Arsenic, Boron, Chromium, Copper, Iodine, Iron, Manganese, Molybdenum, Nickel, Silicon, Vanadium, and Zinc, Copyright 2001 by The National Academies. 2001. Available online: www.nap.edu (accessed on 20 March 2018).

46. World Health Organization [WHO]. *Aluminum in Drinking-Water*; Background Document for Preparation of WHO Guidelines for Drinking-Water Quality; (WHO/SDE/WSH/03.04/53); WHO: Geneva, Switzerland, 2003.

47. FSA. *Measurement of the Concentrations of Metals and Other Elements from the 2006 UK Total Diet Study*; Report No. 01/09; Food Standards Agency: London, UK, 2009.

48. Zmudzki, J.; Szkoda, J. Concentrations of trace elements in hen eggs in Polar. *Bromatol. I-Chem. Toksykol.* **1996**, *29*, 55–57.

49. Kan, C.A.; Meijer, G.A.L. The risk of contamination of food with toxic substances present in animal feed. *Anim. Feed Sci. Technol.* **2007**, *133*, 84–108. [CrossRef]

50. Mansour, S. Monitoring and health risk assessment of heavy metal contamination in Food. In *Practical Food Safety, Contemporary Issues and Future Directions*, 1st ed.; Bhat, R., Gomez-Lopez, V.M., Eds.; John Wiley & Sons, Ltd.: Hoboken, NJ, USA, 2014; pp. 235–255.

51. Jaishankar, M.; Tseten, T.; Anbalagan, N.; Mathew, B.B.; Beeregowda, K.N. Toxicity, mechanism and health effects of some heavy metals. *Interdiscip. Toxicol.* **2014**, *7*, 60–72. [CrossRef] [PubMed]

52. Akan, J.C.; Abdulrahman, F.I.; Sodipo, O.A.; Chiroma, Y.A. Distribution of Heavy Metals in the liver, Kidney and Meat of Beef, Mutton, Caprine and Chicken from KasuwanShanu Market in Maiduguri Metropolis, Borno State, Nigeria. *Res. J. Appl Sci. Eng. Tech.* **2010**, *2*, 743–748.

53. Ogwok, P.; Bamuwamye, M.; Apili, G.; Musalima, J.H. Health Risk Posed by Lead, Copper and Iron via Consumption of Organ Meats in Kampala City (Uganda). *J. Environ. Pollut. Hum. Health* **2014**, *2*, 69–73.

54. Järup, L. Hazards of heavy metal contamination. *Br. Med. Bull.* **2003**, *68*, 167–182. [CrossRef]

55. Attia, Y.A.; Qota, E.M.; Bovera, F.; Tag El-Din, A.E.; Mansour, S.A. Effect of amount and source of manganese and/or phytase supplementation on productive and reproductive performance and some physiological traits of dual purpose-cross-bred hens in the tropics. *Br. Poult. Sci.* **2010**, *51*, 235–245. [CrossRef]

56. Attia, Y.A.; Abdalah, A.A.; Zeweil, H.S.; Bovera, F.; Tag El-Din, A.A.; Araft, M.A. Effect of inorganic or organic selenium supplementation on productive performance, egg quality and some physiological traits of dual purpose breeding hens. *Cezh. J. Anim. Sci.* **2010**, *55*, 505–519. [CrossRef]

57. Attia, Y.A.; Qota, E.M.; Zeweil, H.S.; Bovera, F.; Abd Al-Hamid, A.E.; Sahledom, M.D. Effect of different dietary concentrations of inorganic and organic copper on growth performance and lipid metabolism of White Pekin male ducks. *Br. Poultry Sci.* **2012**, *53*, 77–88. [CrossRef]

58. Attia, Y.A.; Abd El-Hamid, A.E.; Zeweil, H.S.; Qota, E.M.; Bovera, F.; Monastra, M.; Sahledom, M.D. Effect of dietary amounts of organic and inorganic Zinc on productive and physiological traits of white peckin ducks. *Animal* **2013**, *7*, 695–700. [CrossRef]

59. Hu, Y.; Zhang, W.; Chen, G.; Cheng, H.; Taod, S. Public health risk of trace metals in fresh chicken meat products on the food markets of a major production region in southern China. *Environ. Pollut.* **2018**, *234*, 667–676. [CrossRef]

60. Saleh, A.A.; Ragab, M.M.; Ahmed, E.A.M.; Abudabos, A.M.; Ebeid, T.A. Effect of dietary zinc-methionine supplementationon growth performance, nutrient utilization, antioxidative properties and immune response in broiler chickens under high ambient temperature. *J. Appl. Anim Res.* **2018**, *46*, 820–827. [CrossRef]

61. Scott, M.L.; Nesheim, M.C.; Young, R.J. *Nutrition of the Chicken*, 3nd ed.; M.L. Scott & Associates: Ithaca, NY, USA, 1982.

62. Okoye, C.O.B.; Aneke, A.U.; Ibeto, C.N.; Ihedioha, J.N. Heavy Metals Analysis of Local and Exotic Poultry Meat. *IJAES* **2011**, *6*, 49–55.

63. Jothi, J.S.; Yeasmin, N.; Anka, I.Z.; Hashem, S. Chromium and lead contamination in commercial poultry feeds of Bangladesh. *Int. J. Agril. Res. Innov. Tech.* **2016**, *6*, 57–60. [CrossRef]

64. Nwude, D.O.; Okoye, P.A.C.; Babayemi, J.O. heavy metal levels in animal muscle tissue, a case Study of Nigerian raised cattle. *Res. J. Appl. Sci.* **2010**, *5*, 146–150. [CrossRef]

65. Nicholson, F.A.; Smith, S.R.; Alloway, B.J.; Carlton-Smith, C.; Chambers, B.J. An inventory of heavy metals inputs to agricultural soils in England and Wales. *Sci. Total Environ.* **2003**, *311*, 205–219. [CrossRef]

66. Alexieva, D.; Chobanova, S.; Ilchev, A. Study on the level of heavy metal contamination in feed materials and compound feed for pigs and poultry in Bulgaria. *Trakia J. Sci.* **2007**, *5*, 61–66.

67. Sarker, M.S.; Quadir, Q.F.; Hossen, M.Z.; Nazneen, T.; Rahman, A. Evaluation of commonly used fertilizers, fish and poultry feeds as potential sources of heavy metals contamination in food. *Asian Australas J. Food Saf. Secur.* **2017**, *1*, 74–81.

68. Oyewale, A.T.; Adesakin, T.A.; Adu, A.I. Environmental Impact of Heavy Metals from Poultry Waste Discharged into the Olosuru Stream, Ikire, Southwestern Nigeria. *J. Health Pollut.* **2019**, *9*, 1906–1907. [CrossRef]

69. Khan, S.A.; Khan, A.; Khan, S.A.; Beg, M.A.; Ali, A.; Damanhouri, G. Comparative study of fatty-acid composition of table eggs from the Jeddah food market and effect of value addition in omega-3 bio-fortified eggs. *Saudi J. Biol. Sci.* **2017**, *24*, 929–935. [CrossRef]

70. Elsharawy, N.T.M. Some Heavy Metals Residues in Chicken Meat and their Edible Offal in New Valley. In Proceedings of the 2nd Conference of Food Safety, Ismailia, Egypt, August 2015; Volume I, pp. 53–60.

71. Muhammad, N.U.; Ahmad, M.G.; Nuhu, T.; Nafi'u, A. Investigation of heavy metals in different tissues of domestic Chicken. *Inter. J. Basic Appl. Sci. IJBAS-IJENS* **2017**, *17*, 49–53.

72. Alturiqi, A.S.; Albedair, L.A. Evaluation of some heavy metals in certain fish, meat and meat products in Saudi Arabian markets. *Egypt. J. Aquat. Res.* **2012**, *38*, 45–49. [CrossRef]

73. Chowdhury, M.Z.A.; Siddique, Z.A.; Hossain, S.M.A.; Kazi, A.I.I.; Ahsan, A.A.; Ahmed, S.; Zaman, M.M. Determination of essential and toxic metals in meats, meat products and eggs by spectrophotometric method. *JBCS* **2011**, *24*, 165–172. [CrossRef]

74. Flores, É.M. de M.; Martins, A.F. Distribution of Trace Elements in Egg Samples Collected Near Coal Power Plants. *J. Environ. Qual.* **1996**, *26*, 744–748. [CrossRef]

75. Gerberding, J.L. *Toxicological Profile for Zinc*; Public Health Service; ATSDR: Atlanta GA, USA, 2005; pp. 22–23.

76. Jan, A.T.; Azam, M.; Siddiqui, K.; Ali, A.; Choi, I.; Haq, M.O.R. Heavy Metals and Human Health, Mechanistic Insight into Toxicity and Counter Defense System of Antioxidants. *Int. J. Mol. Sci.* **2015**, *16*, 29592–29630. [CrossRef] [PubMed]

77. Rehman, U.H.; Rehman, A.; Ullah, N.; Ullah, F.; Zeb, S.; Iqbal, T.; Ullah, R.; Azeem, T.; Rehman, N.; Farhan. Comparative study of heavy metals in different parts of domestic and broiler chickens. *Int. J. Pharm. Sci. Rev. Res.* **2013**, *23*, 151–154.

78. Oforka, N.C.; Osuji, L.C.; Onwuachu, U.I. Assessment of Heavy Metal Pollution in Muscles and Internal Organs of Chickens Raised in Rivers State, Nigeria. *JETEAS* **2012**, *3*, 406–411.

79. ATSDR. *Toxicological Profile for Manganese*; Agency for Toxic Substances and Disease Registry: Atlanta, GA, USA, 2000.

80. Grazuleviciene, R.; Nadisauskiene, R.; Buinauskiene, J.; Grazulevicius, T. Effects of elevated levels of manganese and iron in drinking water on birth outcomes. *Polish J. Environ. Stud.* **2009**, *18*, 819–825.

81. Gerber, G.B.; Leonard, A.; Hantson, P. Carcinogenicity, mutagenicity and teratogenicity of manganese compounds. *Crit. Rev. Oncol. Hematol.* **2002**, *42*, 25–34. [CrossRef]

82. Mergler, D.; Baldwin, M.; Bélanger, S.; Larribe, F.; Beuter, A.; Bowler, R.; Panisset, M.; Edwards, R.; de Geoffroy, A.; Sassine, M.P.; et al. Manganese neu- rotoxicity a continuum of dysfunction, results from a com- munity based study. *Neurotoxicology* **1999**, *20*, 327.

83. Montes, S.; Riojas-Rodriguez, H.; Sabidopedraza, E.; Rios, C. Biomarkers of manganese exposure in population living close to a mine and mineral processing plant in Mexico. *Environ. Res.* **2008**, *106*, 89–95. [CrossRef]

84. Hudnell, H.K. Effects from environmental Mn exposures, a review of the evidence from non-occupational exposure studies. *Neutrotoxicology* **1999**, *20*, 379–397.

85. Dobson, A.W.; Erikson, K.M.; Aschner, M. Manganese neurotoxicity. *Ann. N.Y. Acad. Sci.* **2004**, *1012*, 115–129. [CrossRef] [PubMed]

86. Erikson, K.M.; Dorman, D.C.; Fitsanakis, V.A.; Lash, L.H.; Aschner, M. Alterations of oxidative stress biomarkers due to in utero and neonatal exposures of airborne manganese. *Biol. Trace Res.* **2006**, *111*, 199–215. [CrossRef]

87. Taylor, M.D.; Erikson, K.M.; Dobson, A.W.; Fitsanakis, V.A.; Dorman, D.C.; Aschner, M. Effects of inhaled manganese biomarkers of oxidative stress in the rat brain. *Neurotoxicology* **2006**, *27*, 788–797. [CrossRef] [PubMed]

88. Hosnedlova, B.; Kepinska, M.; Skalickova, S.; Fernandez, C.; Ruttkay-Nedecky, B.; Peng, Q.; Baron, M.; Melcova, M.; Opatrilova, R.; Zidkova, J.; et al. Nano-selenium and its nanomedicine applications, a critical review. *Int. J. Nanomed.* **2018**, *13*, 2107–2128. [CrossRef]

89. Saleh, A.A.; Ebeid, T.A. Feeding sodium selenite and nano-selenium stimulates growth and oxidation resistance in broilers. *S. Afr. J. Anim. Sci.* **2019**, *49*, 176–184. [CrossRef]

90. El-Deep, M.; Ijiri, H.D.; Ebeid, T.A.; Otsuka, A. Effects of dietary nano-selenium supplementation on growth performance, antioxidative status, and immunity in broiler chickens under thermoneutral and high ambient temperature conditions. *J. Poult Sci.* **2016**, *43*, 255–265. [CrossRef]

91. Ebeid, T.A.; Zeweil, H.S.; Basyony, M.M.; Dosoky, W.M.; Badry, H. Fortification of rabbit diets with vitamin E or selenium affects growth performance, lipid peroxidation, oxidative status and immune response in growing rabbits. *Livest. Sci.* **2013**, *155*, 323–331. [CrossRef]

92. Yang, G.; Wang, S.; Zhou, R.; Sun, S. Endemic selenium intoxication of humans in China. *Am. J. Clin Nutr.* **1983**, *37*, 872–881. [CrossRef]

93. Nuttall, K.L. Review, Evaluating Selenium Poisoning. *Ann. Clin. Lab. Sci.* **2006**, *36*, 409–420.

94. Rashid, M.A.; Sarker, M.S.K.; Khatun, H.; Sarker, N.R.; Ali, M.Y.; Islam, M.N. Detection of heavy metals in poultry feed, meat and eggs. *Asian Australas J. Food Saf. Secur.* **2018**, *2*, 1–5.

95. Yabe, J.; Nakayama, S.M.; Ikenaka, Y.; Muzandu, K.; Choongo, K.; Mainda, G.; Kabeta, M.; Ishizuka, M.; Umemura, T. Metal distribution in tissues of free-range chickens near a lead–zinc mine in Kabwe, Zambia. *Environ. Toxicol. Chem.* **2013**, *32*, 189–192. [CrossRef] [PubMed]

96. Vincent, J.B. The bioinorganic chemistry of chromium (III). *Polyhedron* **2001**, *20*, 1–26. [CrossRef]

97. Anderson, R.A. Chromium and insulin resistance. *Nutr. Res. Rev.* **2003**, *16*, 267–275. [CrossRef] [PubMed]

98. Lien, T.F.; Horng, Y.M.; Yang, K.H. Performance, serum characteristics, carcase traits and lipid metabolism of broilers as a_ected by supplement of chromium picolinate. *Br. Poult. Sci.* **1999**, *40*, 357–363. [CrossRef]

99. Zheng, C.; Huang, Y.; Xiao, F.; Lin, X.; Lloyd, K. Effects of supplemental chromium source and concentration on growth, carcass characteristics, and serum lipid parameters of broilers reared under normal conditions. *Biol. Trace Elem. Res.* **2016**, *169*, 352–358. [CrossRef]

100. Sahin, K.; Sahin, N.; Onderci, M.; Gursu, F.; Cikim, G. Optimal dietary concentration of chromium for alleviating the e_ect of heat stress on growth, carcass qualities, and some serum metabolites of broiler chickens. *Biol. Trace Elem. Res.* **2002**, *89*, 53–64. [CrossRef]

101. Debski, B.; Zalewski, W.; Gralak, M.A.; Kosla, T. Chromium yeast supplementation of broilers in an industrial farming system. *J. Trace Elem. Med. Biol.* **2004**, *18*, 47–51. [CrossRef]

102. Toghyani, M.; Toghyani, M.; Shivazad, M.; Gheisari, A.; Bahadoran, R. Chromium Supplementation Can Alleviate the Negative Effects of Heat Stress on Growth Performance, Carcass Traits, and Meat Lipid Oxidation of Broiler Chicks without Any Adverse Impacts on Blood Constituents. *Biol. Trace. Elem. Res.* **2012**, *146*, 171–180. [CrossRef]

103. Duan, J.; Tan, J. Atmospheric heavy metals and Arsenic in China, Situation, sources and control policies. *Atmos. Environ.* **2013**, *74*, 93–101. [CrossRef]

104. Stohs, S.J.; Bagchi, D. Oxidative mechanisms in the toxicity of metal ions. *Free Radic. Biol. Med.* **1995**, *18*, 321–336. [CrossRef]

105. Leyssens, L.; Vinck, B.; Van Der Straeten, C.; Wuyts, F.; Maes, L. Cobalt toxicity in humans-A review of the potential sources and systemic health effects. *Toxicology* **2017**, *15*, 43–56. [CrossRef] [PubMed]

106. Tvermoes, B.E.; Paustenbach, D.J.; Kerger, B.D.; Finley, B.L.; Unice, K.M. Review of cobalt toxicokinetics following oral dosing, Implications for health risk assessments and metal-on-metal hip implant patients. *Crit Rev. Toxicol.* **2015**, *45*, 367–387. [CrossRef] [PubMed]

107. EC. *Commission Regulation No. 466/2001 of 8 March 2001*; OJEC 1.77/1; EC: Brussels, Belgium, 2001.

108. Hussain, R.T.; Ebraheem, M.K.; Hanady, M.M. Assessment of heavy metals (Cd, Pb and Zn) contents in the livers of chicken available in the local markets of Basrah city, Iraq. *Bas. J. Vet. Res.* **2012**, *11*, 43–51. [CrossRef]

109. Ismail, S.A.; Abolghait, S.K. Estimation of Lead and Cadmium residual levels in chicken giblets at retail markets in Ismailia city. Egypt. *Inter. J. Vet. Sci. Med.* **2013**, *1*, 109–112. [CrossRef]

110. Sadeghi, A.; Hashemi, M.; Jamali-Behnam, F.; Zohani, A.; Esmaily, H.; Dehghan, A. determination of chromium, lead and cadmium levels in edible organs of marketed chickens in mashhad, iran. *JFQHC* **2015**, *2*, 134–138.

111. ANZFA (Australia New Zealand Food Authority), Wellington NZ 6036 May. 2001. Available online: http//www.anzfa.gov.au (accessed on 20 August 2019).

112. *National Standard of the People's Republic of China GB 2762*; Ministry of Health of the People's Republic of China: Beijing, China, 2005.

113. Sander, J.E.; Dufour, L.; Wyatt, R.D.; Bush, P.B.; Page, R.K. Acute toxicity of boric acid and boron tissue residues after chronic exposure in broiler chickens. *Avian Dis.* **1991**, *35*, 745–749. [CrossRef]

114. Moseman, R.F. Chemical disposition of boron in animals and humans. *Environ. Health Perspect* **1994**, *102*, 113–117.

115. Murray, F.J. A comparative review of the pharmacokinetics of boric acid in rodents and humans. *Biol. Trace Elem. Res.* **1998**, *66*, 331–341. [CrossRef]

116. FAO/WHO. *Nutrition and Agriculture, Street Foods*; World Health Organization: Geneva, Switzerland, 1996.

117. Barabasz, W.; Albińska, D.; Jaśkowska, M.J. Lipiec Ecotoxicology of Aluminium. *Polish J. Environ. Stud.* **2002**, *11*, 199–203.

118. Kochian, L.V.; Piñeros, M.A.; Hoekenga, O.A. The physiology, genetics and molecular biology of plant aluminum resistance and toxicity. *Plant Soil* **2005**, *274*, 175–195. [CrossRef]

119. Vardar, F.; Ünal, M. Aluminum toxicity and resistance in higher plants. *Adv. Mol. Biol.* **2007**, *1*, 1–12.

120. Krewski, D.; Yokel, R.A.; Nieboer, E.; Borchelt, D.; Cohen, J.; Harry, J.; Kacew, S.; Lindsay, J.; Mahfouz, A.M.; Rondeau, V. Human health risk assessment for Aluminium, Aluminium oxide, and Aluminium hydroxide. *J. Toxicol. Environ. Health B Crit. Rev.* **2007**, *10*, 1–269. [CrossRef] [PubMed]

121. Réhault-Godbert, S.; Nicolas, G.; Nys, Y. The Golden Egg: Nutritional Value, Bioactivities, and Emerging Benefits for Human Health. *Nutrients* **2019**, *11*, 684. [CrossRef] [PubMed]

Article

The Anti-Oxidation and Mechanism of Essential Oil of *Paederia scandens* in the NAFLD Model of Chicken

Qiang Wu [1,2,†], **Huaqiao Tang** [3,†] **and Hongbin Wang** [1,*]

[1] College of Animal Medicine, Northeast Agricultural University, Harbin 150030, China; wuqiangfirst@163.com
[2] Agricultural College, Tongren Polytechnic College, Tongren 554300, China
[3] Department Pharmacy, Sichuan Agricultural University, Chengdu 611130, China; turtletang@163.com
[*] Correspondence: hbwang@neau.edu.cn; Tel.: +86-137-6562-6781
[†] Co-first authors.

Received: 10 September 2019; Accepted: 18 October 2019; Published: 22 October 2019

Simple Summary: The essential oil of *Paederia scandens* can remedy non-alcoholic fatty liver disease of chicken, but the mechanisms remain unclear. In this study, proteomics technology was used to declare the anti-non-alcoholic fatty liver disease mechanism of *Paederia scandens* essential oil. The results show that the essential oil of *Paederia scandens* significant decreased the oxidative stress of non-alcoholic fatty liver disease in chicken, which was mainly due to the center regulation protein of HSP7C being significantly inhibited.

Abstract: The aim of the study is to determine the underlying pathogenic mechanisms of oxidative stress and detect the anti-oxidative target of essential oil of *Paederia scandens* in non-alcoholic fatty liver disease (NAFLD). Chicken NAFLD was modeled by feeding with a high-capacity diet and *Paederia scandens* essential oil was used to treat the disease. The levels of hepatic reactive oxygen species (ROS), malondialdehyde (MDA), superoxide dismutase (SOD), and the differential proteins and network of protein–protein interactions were investigated in model and drug-treated groups. The results showed that essential oil of *Paederia scandens* down regulated the hepatic ROS and MDA level significantly ($p < 0.05$ and 0.01, respectively). The heat shock cognate 71 kDa protein (HSP7C) was down regulated significantly, which was in the center of the network and interacted with 22 other proteins. The results showed that oxidative stress played an important role in the pathogenesis of chicken NAFLD. The essential oil of *Paederia scandens* showed good anti-oxidation activity by down regulating the HSP7C protein, which can be used as a potential therapeutic target in chicken NAFLD.

Keywords: oxidative stress; essential oil of *Paederia scandens*; non-alcoholic fatty liver disease; HSP7C; chicken

---

1. Introduction

Non-alcoholic fatty liver disease (NAFLD) is the most common worldwide nutritional metabolic disease in the general population due to lifestyle changes, such as increased consumption of high-fat food and lack of exercise, which are also the main reasons for chronic diseases, including dyslipidemia, chronic kidney disease, obesity, and hyperglycemia [1,2]. Chickens are also prone to burst fatty liver disease due to factors such as nutrition, hormones, genetics, and environment. Chicken NAFLD affects the normal function of the liver, and even causes liver cell rupture, eventually leading to intrahepatic hemorrhage and death, which is an important reason for the lost in the poultry industry [3]. The estimated prevalence of chicken NAFLD is around 5%, but occasional outbreaks occur up to 20% and cause huge economic losses [4].

There is no sufficient explanation of the mechanisms which are associated with the development of NAFLD, including progression to other nosologic units. It is hypothesized that it depends on genetic and environmental factors, and eventual progression is the result of bilateral interactions [5]. The most widely accepted pathogenesis of NAFLD is the "two hits" hypotheses [6,7], in which lipid accumulation is considered as the first hit and increased oxidative stress is proposed to be the second hit. Lipid oxidation generates toxic metabolites including ROS and MDA, which can damage many cellular components such as DNA, protein, lipid, and mitochondria [8,9]. Meanwhile, ROS drives peroxidation of the accumulated hepatic lipids directly [10], which increases the formation of lipid peroxidation products, such as MDA, which is major aldehydic metabolites of lipid peroxidation and has been used to reflect lipid peroxidation [11] and stimulates further production of ROS [9].

Some studies have shown that oxidative stress may be the most critical pathogenesis of NAFLD leading to disease progression, and there can be an effective preventive and/or treatment strategy against chicken NAFLD by hepatoprotective effect or anti-oxidant activity. The recent study showed that essential oil of *Paederia scandens* has the activity of hepatoprotective and anti-oxidant effect [12]. However, the targets of hepatoprotective effects and anti-oxidant activity of essential oil of *Paederia scandens* are still unclear. This study was carried out on the intervention of a high-capacity diet and essential oil of *Paederia scandens*, which aimed to determine the underlying pathogenic mechanisms and to obtain the hepatoprotective target of this essential oil in chicken NAFLD.

2. Materials and Methods

2.1. Materials and Extraction

Plant materials were harvested from Guizhou (27.7183° N, 109.192° E, southwest China), in June 2015. All solvents and reagents were analytical grade. The harvested samples were washed with tap water and dried at 30 °C for 6 days by oven. A sample of 150 g crushed *Paederia scandens* was subjected to extraction by hydro distillation for 3 h in 500 mL distilled water using a clevenger type apparatus. The extracted oil was recovered and stored at 4 °C, and its extraction yield was calculated as the ratio of the weight of oil to the weight of fruits using the following equation:

$$\% \text{ yield of oil} = (\text{weight of oil/weight of dried materials}) \times 100\% \tag{1}$$

2.2. High Performance Liquid Chromatography (HPLC) Analysis

The main ingredients of linalool, L-α-terpineol, dextro-α-terpineol, methyl salicylate, camphor, borneol, eugenol, and isoeugenol were chosen as the standards to determine the constituent of the essential oil of *Paederia scandens*. The standards and extraction were analyzed by Agilent 1260 high performance liquid chromatography (HPLC) equipped with an Agilent Zorbax Eclipse XDB-C8 column (4.6 mm × 250 mm, 5 μm) (Agilent Technologies, Santa Clara, CA, USA). The test condition was acetonitrile–water (55:45); flow rate 1.0 mL/min; injection volume 10 μL; wavelength 230 nm; column temperature 30 °C.

2.3. Animals and Experimental Procedure

Seventy-five, one-day-old, healthy Ross 305 chicks (33–40 g) were selected from the breeding chicken farm of Sichuan Wenjiang China Tai Livestock Co., Ltd. (Chengdu, China). All chickens were housed in wood cages under the recommended environment. Chicks were brooded at 33 °C during the first week; the brooding temperature was reduced 3 °C/week to approximately 24 °C by week four of age. Light was provided continually using incandescent lamps. The environmental humidity was controlled at 60–65%.

The chickens were randomly divided into control group, model group, and drug treated group, 25 individuals per group. The control group was given a normal diet, and model and drug-treated groups were given a high-capacity diet. Cooked pigs' oil was the main calorie sources of the high-capacity diet.

The drug group chickens were treated with essential oil of *Paederia scandens* by nasal drops, 2 mg/kg per day continuously for 4 weeks. The dose of *Paederia scandens* essential oil was calculated by previous experiments, and 2 mg/kg was the safe and effective dose to treat NAFLD in chicken [12]. Drug treated chickens were given the same diet as the model group chickens. The normal diet and high-capacity diet formulations are shown in Table 1. All chickens were given free access to laboratory feed and water for 4 weeks, and the chickens had no access to food for 12 h before they were anesthetized and sacrificed. Five chickens were randomly selected and killed weekly in each group, and the livers were excised and removed promptly for further analysis. All the animal tests were in accordance with the Administration of Affairs Concerning Experimental Animals of the State Council of the People's Republic of China. The experiment protocol was approved by the Committee on Experimental Animal Management of Tongren Polytechnic college in Guizhou (with protocol No. 2016051407).

Table 1. Composition and nutrient levels of the diets.

Ingredients (%)	High-Capacity Diets	Normal Diets
Corn	53.00	62.48
Soybean meal	-	-
Fish meal	1.00	3.00
Fried meal	29.30	28.10
Rapeseeds	-	2.20
Wheat bran	-	-
Cooked pigs' oil	12.95	-
Additives	0.75	0.75
Lys	0.12	0.09
Met	0.12	0.18
$CaHPO_4$	1.11	1.22
Limestone meal	1.52	1.82
Salt	0.13	0.16
ME (MJ/kg)	16.07	13.39
CP (%)	16.00	18.40
Ca	0.90	1.00
AP	0.35	0.45
Lys	1.00	1.10
Met	0.38	0.50

The additives contains 0.50% of microelements, per kg dietary contains Fe 80 mg, Zn 40 mg, Cu 8 mg, Mn 60 mg, I 0.35 mg, Se 0.15 mg, 0.20% of choline, Va 1500 IU, VD_3 200 IU, VE 10 IU, VK_3 0.05 mg, VB_1 1.80 mg, VB_2 3.60 mg, VB_{12} 0.01 mg, VB_7 0.15 mg, VB_9 0.55 mg, VB_5 35 mg, VB_3 10 mg, VB_6 3.50 mg.

2.4. Oxidative Stress Assays

A small part of fresh liver was homogenized to prepare 10% tissue homogenate by adding 0.9% saline on ice in each group. The homogenate was centrifuged at 1500× g for 10 min at 0 °C. The supernatant was harvested and adjusted to 1% tissue homogenate by adding 0.9% saline. The levels of MDA, SOD and ROS were measured by assays kits from the NanJing Jiancheng bio-engineering research institute (NanJing, China).

2.5. Protein Separation and Identification

The liver (1.0 g) was suspended in 10 mL buffer consisting of 50 mM Tris and 1.0 mM phenylmethylsulphonyl fluoride for preparation of the total protein extract. The suspension was homogenized with the homogenizer for 1 min, sonicated for 30 s, and centrifuged at 14,000× g for 15 min in the buffer. The homogenate was poured into a glass beaker, which was placed on ice, and the homogenate was stirred gently for 30 min at 4 °C. Cell debris and other particulate matter were removed from the homogenate by centrifugation at 14,000× g for 20 min at 4 °C. After filtering the supernatant, the cells were washed by adding 10 mL phosphate buffer saline to the centrifuge tube. The cell pellet was resuspended by pipetting the mixture up and down. The cells were centrifuged again at 4800× g for 5 min at 4 °C. The supernatant was discarded without disturbing the pellet. A pipette was used to transfer the extraction buffer into the tube and vortex the tube briefly (10 s) to resuspend the pellet. Samples of 1.0 mg protein were applied on immobilized pH 3–10 nonlinear strips. Each sample was analyzed in triplicate. Focusing started at 200 V at 3 V/min and was kept constant for a further 24 h. The second-dimensional separation was performed in 12% SDS–polyacrylamide gels. The gels (18 × 18 × 0.15 cm) were run at 40 mA per gel. After protein fixation for 12 h in 40% methanol containing 5% phosphoric acid, the gels were stained with Coomassie blue R250 (0.5 g/L) for 24 h. The gels were scanned as tiff files and analyzed by PDQuest 8.01 (Bio-Rad, Berkeley, CA, USA). The significantly different pots were selected for MALDI-TOF/MS analysis. These proteins were identified by GPM-XE software (The Global Proteome Machine Organisation, USA, www.thegpm.org) and the network of protein–protein interactions was analyzed by Cytoscape v2.6.3 (U.S. National Institute of General Medical Sciences, USA, http://www.cytoscape.org).

2.6. Statistical Analysis

The statistical analyses were performed using the SPSS 19.0 software package (ICM, Armonk, NY, USA) and the data were presented as means and standard deviations (means ± SD), and group comparison was done by independent-sample *t*-tests.

3. Results

3.1. Paederia scandens Essential Oil Chemical composition

The extraction rate of essential oil was about 0.5% in *Paederia scandens*. Eight chemical ingredients were tested by HPLC, and the contents were shown in Table 2. The main ingredients were linalool and borneol, the contents were 261.142 and 118.784 mg/mL in the oil, the results showed in Table 2. These eight components account for 54.14% of essential oil. The HPLC chromatogram is shown in Figure 1.

Table 2. Contents of main components of essential oil in *Paederia scandens*.

No.	Ingredients	Raw Sample (mg/mL)
1	Linalool	261.142
2	L-α-terpineol	10.126
3	D-α-terpineol	11.233
4	Methyl salicylate	78.902
5	Camphor	15.234
6	Borneol	118.784
7	Eugenol	24.634
8	Isoeugenol	21.346

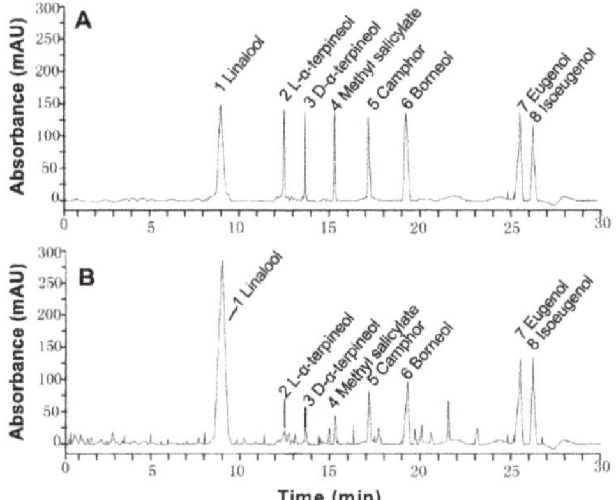

Figure 1. High performance liquid chromatogram of 100 µg/L mixed standard solution (**A**) and 500 µg/L in *Paederia scandens* essential oil (**B**).

3.2. Clinical Symptoms and Relative Weight of the Liver

The clinical manifestations were weight gain, unstable standing, mouth breathing, lethargy, unilateral lying, increased body temperature, and feces rot in most of the chickens treated with high-calorie diets. Even a small number of chickens died. There was no chicken death in the drug treatment group, no obvious clinical features, and the chickens grew well.

In the model group, most chickens showed a large amount of yellow–brown fat deposition in the abdominal cavity and mesentery. The liver was swollen (data shown in Table 3), deep yellow, soft, and brittle. Some livers lost their original shape. In the drug treatment group, the hepatic hemorrhage band was reduced or even disappeared, and some showed small bleeding spots. The overall liver appearance was similar to that of the normal group.

Table 3. The change of liver index (%) in broilers.

Time (Week)	Control Group	Model Group	Drug Group
2	2.20 ± 0.15	3.02 ± 0.18 **	3.02 ± 0.18 **
3	2.22 ± 0.21	3.15 ± 0.11 **	2.83 ± 0.21 **
4	2.22 ± 0.22	3.05 ± 0.30 **	2.84 ± 0.17

** means $p < 0.01$ vs. control group.

3.3. Hepatic ROS, MDA, and Superoxidase Dismutase (SOD) Levels

The results showed that hepatic ROS levels were significantly higher in the model and drug groups than in the control group during the whole experiment ($p < 0.01$ or $p < 0.05$), and hepatic ROS levels were lower in the drug group than in the model group in the early stage. Hepatic MDA levels were significantly higher in the model group than in the drug group and control group from 2 weeks to 4 weeks ($p < 0.01$ or $p < 0.05$). Hepatic SOD activity was significantly lower in the model group and drug group than in the control group from 3 weeks ($p < 0.01$ or $p < 0.05$). The essential oil of *Paederia scandens* significantly down regulated the hepatic ROS and MDA levels in NAFLD chicken. The results are listed in Table 4.

Table 4. The level of hepatic ROS, MDA, and SOD.

Item	Week	Control Group	Model Group	Drug Group
ROS (U/mg protein)	1	88.56 ± 1.44	113.34 ± 4.06 **	97.64 ± 1.62 *
	2	88.56 ± 2.20	106.66 ± 4.48 **	98.33 ± 1.39 *
	3	93.60 ± 3.55	107.49 ± 7.96 **	105.01 ± 6.20 *
	4	100.69 ± 6.33	125.00 ± 5.86 **	109.08 ± 4.46
MDA (nmol/mg protein)	1	6.57 ± 1.46	10.60 ± 1.34 **	9.04 ± 1.27 **
	2	8.32 ± 1.63	10.51 ± 1.29 **	9.47 ± 1.56 **
	3	8.82 ± 1.26	13.47 ± 1.66 **	10.42 ± 1.79 **
	4	9.57 ± 1.18	13.46 ± 1.42 **	9.94 ± 1.03
SOD (U/mg protein)	1	498.96 ± 35.83	461.32 ± 37.49	436.11 ± 43.69 *
	2	536.60 ± 31.67	501.19 ± 27.13	476.48 ± 26.99 **
	3	564.89 ± 26.22	492.36 ± 30.23 **	425.99 ± 26.58 **
	4	523.83 ± 29.05	443.22 ± 33.35 **	417.21 ± 30.79 **

$* p < 0.05$ and $** p < 0.01$ vs. control group.

3.4. Two-Dimensional (2D) Gel Electrophoresis

The total proteins were analyzed by 2D electrophoresis. The samples were analyzed on broad pH range immobilized pH gradient (IPG) strips and the spots were visualized following staining with Coomassie brilliant blue. The results showed a representative analysis of total liver proteins separated on a broad pH range 3–10 2D gel. On each gel, 1.0 mg of total protein amount was applied. The analysis of two-dimensional maps in PDQuest showed that there were about 400 spots detected by Commassie Brilliant Blue-stained in each group. There were 73 markedly differential spots in B gel compared with group A, and there were 22 markedly differential spots in C gel compared with B gel. These differential spots are shown on Figure 2.

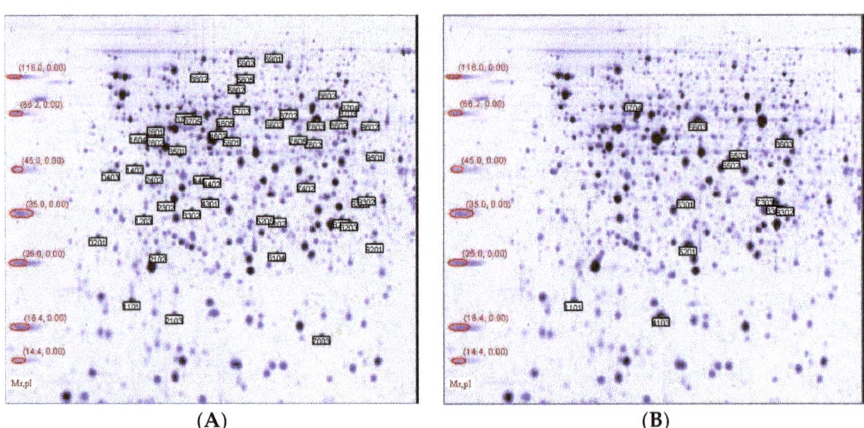

(**A**) (**B**)

Figure 2. Two-dimension gel electrophoresis of proteins from liver extract stained with Coomassie Brilliant Blue. pH 3–10 nonlinear first dimension, and 12% SDS–PAGE second dimension. (**A**) shows that there were 73 differential spots in the drug-treated group compared with model group; (**B**) shows that there were 22 differential spots in the drug-treated group compared with control group. Differentially expressed proteins are marked in each spot sustained an individual identification number (SSP).

3.5. Protein Separation and Identification

In total, 22 differential proteins were identified in C gel, eight of which were up-regulation, such as heat shock 70 kDa protein 5, heat shock 70 kDa protein 8, long-chain specific acyl-CoA dehydrogenase, etc. There were four proteins with the function of lipid metabolism and fatty acid beta-oxidation, such as 3-alpha-hydroxysteroid dehydrogenase, 3-hydroxy-3-methylglutaryl-Coenzyme A synthase 2, Carbonic anhydrase 3, and Catalase. Additionally, there were four proteins with the function of electron transport and cell communication, such as electron transfer flavoprotein subunit alpha, ATP synthase subunit d, Regucalcin, and Guanine nucleotide-binding protein subunit beta-2-like 1. The results are listed in Table 5.

Table 5. Different proteins expression in chicken liver.

SSP	Protein	Gene	MW/PI	Trend
1101	Uncharacterized protein	-	-	-
1203	Regucalcin	*Rgn*	33/5.3	↓
3704	Heat shock 70 kDa protein 5 (glucose-regulated protein, 78 kDa)	*Hspa5*	72/5.1	↑
4103	ATP synthase subunit d	*Atp5h*	19/6.2	↓
4301	Guanine nucleotide-binding protein G(I)/G(S)/G(T) subunit beta 1	*Gnb1*	35/7.6	↓
4401	S-adenosylmethionine synthase isoform type-1	*MAT1A*	44/5.6	↓
4402	Fumarylacetoacetase	*Fah*	46/6.7	↓
5201	Carbonic anhydrase 3	*Car3*	29/6.9	↓
5301	3-alpha-hydroxysteroid dehydrogenase	*Akr1c2*	37/6.7	↑
5603	3-hydroxy-3-methylglutaryl-Coenzyme A synthase 2	*Hmgcs2*	57/8.9	↑
5702	Catalase	*Cat*	60/7.1	↑

Table 5. *Cont.*

SSP	Protein	Gene	MW/PI	Trend
6104	Glutathione S-transferase Mu 1	*Gstm1*	26/8.3	↓
6502	Arginosuccinate synthase 1	*Ass1*	46/7.6	↓
6503	Long-chain specific acyl-CoA dehydrogenase	*Acadl*	48/7.6	↑
6702	Heat shock 70 kDa protein 8	*Hspa8*	71/5.4	↑
6901	Uncharacterized protein	-	-	-
7003	Beta-actin	*actb*	15/5.7	↑
7303	Uncharacterized protein	*Ote*	39/8.9	
8301	Electron transfer flavoprotein subunit alpha	*Etfa*	35/8.6	↓
8302	Glutathione S-transferase Mu 2	*Gstm2*	28/8.2	↓
9301	Glyceraldehyde-3-phosphate dehydrogenase-like	*Gapdh*	36/8.4	↓
9501	Betaine-homocysteine S-methyltransferase 1	*Bhmt*	45/8.0	↓
8603	Uncharacterized protein	-	-	-

The spots representing total proteins were analyzed by PDQuest 8.01 software and identified by mass spectrometry and GPM-XE software. "↓" means the proteins were down regulated; "↑" means the proteins were up-regulated. SSP means each spot sustained an individual identification number. MW/PI means the ratio of molecular weight and isoelectric point.

3.6. Analysis of the Network of Protein–Protein Interactions

The analysis of the network of protein–protein interactions was done by Cytoscape v2.6.3. The results showed that HSP7C, heat shock cognate 71 kDa protein, was in the center of the network and interaction among 22 other proteins, which showed that essential oil of *Paederia scandens* had the activity of hepatoprotective effect and anti-oxidant by up-regulation of HSP7C expression and affected other proteins by interaction. The results are listed in Table 6 and Figure 3.

Table 6. The analysis of the network among differential protein expression.

ID	Degree	Label	Gene	Name
146490	22	HSP7C	*HSPA8*	Heat shock cognate 71 kDa protein
145948	18	GRP78	*HSPA5*	78 kDa glucose-regulated protein
93344	16	PABP1	*PABPC1*	Emerin
71781	16	YBOX1	*YBX1*	Nuclease-sensitive element-binding protein 1
57072	16	PRKDC	*PRKDC*	Heterogeneous nuclear ribonucleoproteins A2/B1
51704	16	ACTB	*ACTB*	Polyadenylate-binding protein 1
50923	16	RO52	*TRIM21*	E3 ubiquitin-protein ligase TRIM21
50095	16	ROA2	*HNRNPA2B1*	DNA-dependent protein kinase catalytic subunit
15490	16	EMD	*EMD*	Actin, cytoplasmic 1
150257	16	G3P	*GAPDH*	Glyceraldehyde-3-phosphate dehydrogenase
89014	12	RS3	*RPS3*	40S ribosomal protein S9
88619	12	RS9	*RPS9*	40S ribosomal protein S3
5116	8	IF2B	*EIF2S2*	Eukaryotic translation initiation factor 2 subunit 2
50028	8	RLA0	*RPLP0*	60S acidic ribosomal protein P0
4817	8	IF2A	*EIF2S1*	Eukaryotic translation initiation factor 2 subunit 1
117387	8	NCBP1	*NCBP1*	Nuclear cap-binding protein subunit 1
95707	2	CITE1	*CITED1*	Cbp/p300-interacting trans-activator 1
15584	2	ERN1	*ERN1*	Serine/threonine-protein kinase/endoribonuclease IRE1

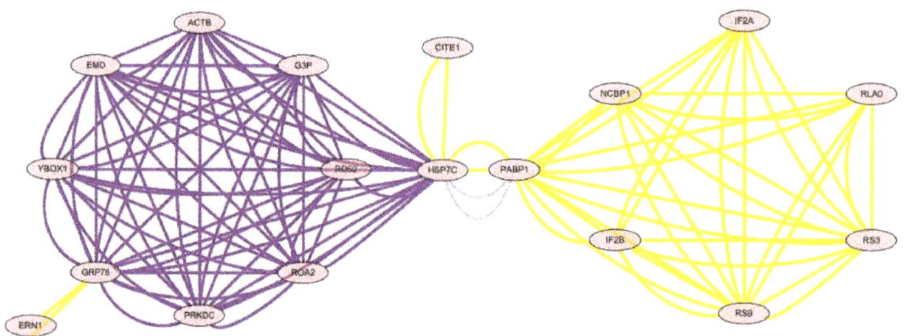

Figure 3. The network of differential protein expression.

4. Discussion

Oxidative stress can be described as a condition resulting from an uncontrolled increase in ROS and MDA or an insufficiency in the anti-oxidant system under certain pathological states [13]. The balance of oxidation and the anti-oxidation system plays an essential role in the progression and pathogenesis of NAFLD [14]. In the present study, we established the high-capacity induced NAFLD model in chicken and explored the degree and mechanisms of oxidative stress through detecting the hepatic ROS, MDA, and SOD. The key proteins were screened and the anti-oxidative target of essential oil of *Paederia scandens* was determined in chicken NAFLD.

The results showed that hepatic ROS and MDA levels were significantly higher in the model group than other groups, but SOD activity was significantly lower at the same time, which indicated that oxidative stress occurred and played a significant role in liver injury and disease progression in chicken NAFLD. Recent studies have shown that high levels of lipid accumulation exceeded the storage capacity of liver in NAFLD [15], which led to lipid peroxidation, thereby causing production of MDA and ROS with important toxic effects [16]. Excessive production of ROS and MDA may eventually overwhelm anti-oxidant defenses and generate highly toxic lipid peroxides [17]. The activity of SOD, sensitive to inactivation by ROS and MDA, was decreased in the NAFLD chicken [18,19]. Furthermore, to prevent oxidative stress, there was a continuous balance between intrahepatic antioxidants and oxidative stress, but an imbalance between the increased ROS and MDA level, and the decreased SOD activity occurred in the model group, which contributed to the pathogenesis of chicken NAFLD [20,21]. Meanwhile, hepatic ROS and MDA levels were significantly lower in the drug group than in the model group, which demonstrated the anti-oxidative activity of essential oil of *Paederia scandens*.

The differential proteins were analyzed by two-dimensional (2D) gel electrophoresis and mass spectrometry in chicken NAFLD and drug treated groups, which indicated that chicken NAFLD followed many differential protein expressions, and essential oil of *Paederia scandens* can up or down regulate some proteins, which is related to the hepatic ROS and MDA level regulation by the network of protein–protein interactions. There were 22 differential proteins in the drug-treated group compared with the model group. Of these, there were four proteins with the function of lipid metabolism and fatty acid beta-oxidation, such as 3-alpha-hydroxysteroid dehydrogenase, 3-hydroxy-3-methylglutaryl-Coenzyme A synthase 2, Carbonic anhydrase 3, and Catalase. Additionally, there were four proteins with the function of electron transport and cell communication, such as electron transfer flavoprotein subunit alpha, ATP synthase subunit d, Regucalcin, and Guanine nucleotide-binding protein subunit beta-2-like 1. Eight proteins were up-regulated, such as heat shock 70 kDa protein 5 (glucose-regulated protein, 78 kDa), heat shock 70 kDa protein 8, Long-chain specific acyl-CoA dehydrogenase, Beta-actin, and 3-hydroxy-3-methylglutaryl-Coenzyme A synthase 2.

To define the mechanism of anti-oxidation and further determine the targets of essential oil of *Paederia scandens* in treating chicken NAFLD, the protein network analysis results showed that HSP7C, heat shock cognate 71 kDa protein, was in the center of the network and interaction among the other 22 proteins, which indicated that HSP7C may be a core target for essential oil of *Paederia scandens* against chicken NAFLD. Some recent studies showed that HSP7C is a member of the heat shock protein 70 (HSP70) family and is involved in the folding and assembly of proteins in the endoplasmic reticulum (ER). As this protein interacts with many ER proteins, it may play a key role in monitoring protein transport through the cell [22,23]. Above all, essential oil of *Paederia scandens* has a hepatoprotective effect and anti-oxidant activity by up-regulation of HSP7C expression, which interacts with other proteins; similar results were reported by Attia et al. [24].

5. Conclusions

In conclusion, excessive production of ROS and MDA and decreased SOD activity overwhelmed antioxidant defenses and further generated highly toxic lipid peroxides, which induced the formation of oxidative stress in chicken NAFLD. The essential oil of *Paederia scandens* has a hepatoprotective effect and anti-oxidant activity by up-regulation of HSP7C protein expression, which may be a potential therapeutic target in treating NAFLD chicken.

Author Contributions: Conceptualization, Q.W.; methodology, H.T.; writing—original draft preparation, H.T.; writing—review and editing, Q.W.; supervision, H.W.; funding acquisition, Q.W.

Funding: This research was funded by the National Natural Science Foundation of China (Grant No. 31302134) and the National Natural Science Foundation of China's Guizhou (Grant No. 2017-1191).

Conflicts of Interest: The authors declare no conflict of interest.

References

1. Than, N.N.; Newsome, P.N. A concise review of non-alcoholic fatty liver disease. *Atherosclerosis* **2015**, *239*, 192–202. [CrossRef]
2. Zhou, Y.; Ding, Y.L.; Zhang, J.L.; Zhang, P.; Wang, J.Q.; Li, J.H. Alpinetin improved high fat diet-induced non-alcoholic fatty liver disease (NAFLD) through improving oxidative stress, inflammatory response and lipid metabolism. *Biomed. Pharmacother.* **2018**, *97*, 1397–1408. [CrossRef]
3. Polin, D.; Wolford, J.H. Role of estrogen as a cause of fatty liver hemorrhagic syndrome. *J. Nutr.* **1977**, *107*, 873–886. [CrossRef]
4. Miele, L.; Forgione, A.; Gasbarrini, G.; Grieco, A. Noninvasive assessment of fibrosis in non-alcoholic fatty liver disease (NAFLD) and non-alcoholic steatohepatitis (NASH). *Transl. Res.* **2007**, *149*, 114–125. [CrossRef] [PubMed]
5. Bannister, D.W. The biochemistry of fatty liver and kidney syndrome. Biotin-mediated restoration of hepatic gluconeogenesis in vitro and its relationship to pyruvate carboxylase activity. *Biochem. J.* **1976**, *156*, 167–173. [CrossRef] [PubMed]

6. Day, C.P.; James, O.F. Steatohepatitis: A tale of two "hits"? *Gastroenterology* **1998**, *114*, 842–845. [CrossRef]

7. Byrne, C.D. Fatty liver: Role of inflammation and fatty acid nutrition. *Prostaglandins Leukot. Essent. Fat. Acids* **2010**, *82*, 265–271. [CrossRef]

8. Gaté, L.; Paul, J.; Ba, G.N.; Tew, K.D.; Tapiero, H. Oxidative stress induced in pathologies: The role of antioxidants. *Biomed. Pharm.* **1999**, *53*, 169–180.

9. Fromenty, B.; Robin, M.A.; Igoudjil, A.; Mansouri, A.; Pessayre, D. The ins and outs of mitochondrial dysfunction in NASH. *Diabetes Metab.* **2004**, *30*, 121–138. [CrossRef]

10. Albano, E.; Mottaran, E.; Occhino, G.; Reale, E.; Vidali, M. Review article: Role of oxidative stress in the progression of non-alcoholic steatosis. *Aliment. Pharm. Aliment. Pharm.* **2005**, *22*, 71–73. [CrossRef]

11. Andreadou, I.; Iliodromitis, E.K.; Mikros, E.; Bofilis, E.; Zoga, A.; Constantinou, M.; Tsantili-Kakoulidou, A.; Kremastinos, D.T. Melatonin does not prevent the protection of ischemic preconditioning in vivo despite its antioxidant effect against oxidative stress. *Free Radic Biol. Med.* **2004**, *37*, 500–510. [CrossRef] [PubMed]

12. Wu, Q.; Yu, J.; Cui, H.; Li, Y. Role of essential oil oil of white *Paederia scandens* in endotoxin of Salmonella Enteritidis. *Chin. Vet. Sci.* **2012**, *42*, 758–764.

13. Dhibi, M.; Brahmi, F.; Mnari, A.; Houas, Z.; Chargui, I.; Bchir, L.; Gazzah, N.; Alsaif, M.A.; Hammami, M. The intake of high fat diet with different trans fatty acid levels differentially induces oxidative stress and non alcoholic fatty liver disease (NAFLD) in rats. *Nutr. Metab.* **2011**, *8*, 65. [CrossRef] [PubMed]

14. Das, S.K.; Vasudevan, D.M. Monitoring oxidative stress in patients with non-alcoholic and alcoholic liver diseases. *Indian J. Clin. Biochem.* **2005**, *20*, 24–28. [CrossRef] [PubMed]

15. Kusminski, C.M.; Shetty, S.; Orci, L.; Unger, R.H.; Scherer, P.E. Diabetes and apoptosis: Lipotoxicity. *Apoptosis* **2009**, *14*, 1484–1495. [CrossRef]

16. Henson, P.M.; Bratton, D.L.; Fadok, V.A. The phosphatidylserine receptor: A crucial molecular switch? *Nat. Rev. Mol. Cell Biol.* **2001**, *2*, 627–633. [CrossRef]

17. Landar, A.; Shiva, S.; Levonen, A.L.; Oh, J.Y.; Zaragoza, C.; Johnson, M.S.; Darley-Usmar, V.M. Induction of the permeability transition and cytochrome c release by 15-deoxy-Delta12, 14-prostaglandin J2 in mitochondria. *Biochem. J.* **2006**, *394*, 185–195. [CrossRef]

18. Bailey, S.M.; Patel, V.B.; Young, T.A.; Asayama, K.; Cunningham, C.C. Chronic ethanol consumption alters the glutathione/glutathione peroxidase-1 system and protein oxidation status in rat liver. *Alcohol. Clin. Exp. Res.* **2001**, *25*, 726–733. [CrossRef]

19. Polavarapu, R.; Spitz, D.R.; Sim, J.E.; Follansbee, M.H.; Oberley, L.W.; Rahemtulla, A.; Nanji, A.A. Increased lipid peroxidation and impaired antioxidant enzyme function is associated with pathological liver injury in experimental alcoholic liver disease in rats fed diets high in corn oil and fish oil. *Hepatology* **1998**, *27*, 1317–1323. [CrossRef]

20. Chaudière, J.; Ferrari-Iliou, R. Intracellular antioxidants: From chemical to biochemical mechanisms. *Food Chem. Toxicol.* **1999**, *37*, 949–962. [CrossRef]

21. García-Trevijano, E.R.; Iraburu, M.J.; Fontana, L.; Dominguez-Rosales, J.A.; Auster, A.S.; Covarrubias-Pinedo, A.; Rojkind, M. Transforming growth factor beta1 induces the expression of alpha1(I) procollagen mRNA by a hydrogen peroxide-C/EBPbeta-dependent mechanism in rat hepatic stellate cells. *Hepatology* **1999**, *29*, 960–970.

22. Booth, L.; Roberts, J.L.; Dent, P. HSPA5/Dna K may be a useful target for human disease therapies. *DNA Cell Biol.* **2015**, *34*, 153–158. [CrossRef] [PubMed]

23. Lee, W.S.; Sung, M.S.; Lee, E.G.; Yoo, H.G.; Cheon, Y.H.; Chae, H.J.; Yoo, W.H. A pathogenic role for ER stress-induced autophagy and ER chaperone GRP78/BiP in T lymphocyte systemic lupus erythematosus. *J. Leukoc. Biol.* **2015**, *97*, 425–433. [CrossRef] [PubMed]

24. Attia, Y.A.; Bakhashwain, A.A.; Bertu, N.K. Thyme oil (Thyme vulgaris L.) as a natural growth promoter for broiler chickens reared under hot climate. *Ital. J. Anim. Sci.* **2017**, *16*, 275–282. [CrossRef]

Article

Fatty Acid Signatures in Different Tissues of Mediterranean Yellowtail, *Seriola dumerili* (Risso, 1810), Fed Diets Containing Different Levels of Vegetable and Fish Oils

Francesco Bordignon [1], Ana Tomás-Vidal [2], Angela Trocino [1,*], Maria C. Milián Sorribes [2], Miguel Jover-Cerdá [2] and Silvia Martínez-Llorens [2]

[1] Department of Comparative Biomedicine and Food Science (BCA), University of Padova, Viale dell'Università 16, I-35020 Legnaro, Italy; francesco.bordignon.3@phd.unipd.it

[2] Institute of Animal Science and Technology, Group of Aquaculture and Biodiversity, Universitat Politècnica de València, Camino de Vera, 14, 46071 València, Spain; atomasv@dca.upv.es (A.T.-V.); mamisor@etsia.upv.es (M.C.M.S.); mjover@dca.upv.es (M.J.-C.); silmarll@dca.upv.es (S.M.-L.)

* Correspondence: angela.trocino@unipd.it; Tel.: +39-0498272583

Received: 28 December 2019; Accepted: 21 January 2020; Published: 24 January 2020

Simple Summary: Most of the studies performed to date mainly investigated on the effects of dietary substitution of fish oil with vegetable oils on growth and fatty acid composition of edible muscle tissues. On the other hand, a few assessed how dietary lipids are retained in other tissues, such as brain, liver, and adipose tissue, which would provide further insights into the fatty acid requirements of new farmed marine fish species such as *Seriola dumerili*. Thus, this study evaluated how the replacement of fish oil with different proportions of vegetable oils in diets affects the tissue-specific fatty acid composition (also known as signature) of brain, muscle, liver, and visceral fat of *S. dumerili*. The fatty acid composition of the diet had a strong effect on the fatty acid signature of muscle, liver, and visceral fat, whereas the brain signature was less sensitive to dietary changes. These new insights contribute to identify the essential fatty acids requirements of Mediterranean yellowtail and to define the conditions under which the physiological functions in these fish are preserved when they are fed diets with low fish oil levels to guarantee the sustainability of their production and welfare.

Abstract: The study aimed to evaluate how replacing different proportions of fish oil (FO) with vegetable oils (VO) in the diet of Mediterranean yellowtail, *Seriola dumerili* (Risso, 1810), affects the fatty acids (FA) signature, i.e., overall FA profile, in different tissues. A total of 225 Mediterranean yellowtail juveniles (initial live weight: 176 ± 3.62 g) were fed for 109 days with one of three diets: A control diet (FO 100), with FO as the only lipid source, or diets with 75% and 100% of FO replaced with a VO mixture. At the end of the feeding trial, the brains, muscles, livers, and visceral fat were sampled in four fish per tank (12 per treatment), and their fat were extracted and used for FA analysis. The FA signatures of red and white muscle, liver, and visceral fat tissues changed when the dietary FA source changed, whereas FA signatures in the brain were rather robust to such dietary changes. These new insights might help evaluate whether key physiological functions are preserved when fish are fed diets with low FO levels, as well as define the dietary FA requirements of Mediterranean yellowtail to improve the sustainability of the production and welfare of the fish.

Keywords: brain; muscle; liver; greater amberjack; EPA; DHA

---

1. Introduction

The Mediterranean yellowtail, *Seriola dumerili* (Risso, 1810), also called the greater amberjack, is a new, high-value candidate for production in marine aquaculture. This cosmopolitan fish is mainly produced in Japan [1], Spain [2], Italy [3], and recently in Vietnam, Korea, and China [4]. It has high growth rates (reaching 6 kg within 2.5 years of culture), and exceptionally high consumer acceptance [4,5].

Since the Mediterranean yellowtail is an emerging species in aquaculture, its nutritional requirements have been defined in terms of major nutrients (47%–50% crude protein (CP) and 12%–14% crude lipid (CL)) [6–10]; however, little information is available on its fatty acid requirements [11], which thus require further investigation. Only one previous study assessed the effects of dietary fatty acids on Mediterranean yellowtail body composition [11].

Fatty acids (FA) play key roles in fish health and nutrition. They maintain the structural and functional integrity of cell membranes, provide metabolic energy through their oxidative metabolism, contribute to visual and brain development, and are precursors of a group of paracrine hormones with relevant biological roles known as eicosanoids [12–14]. Fish health, growth, and reproduction are strongly dependent on n-3 and n-6 polyunsaturated fatty acids (PUFA), especially arachidonic acid (AA, C20:4 n-6), eicosapentaenoic acid (EPA, C20:5 n-3), and docosahexaenoic acid (DHA, C22:6 n-3) [13,15–17]. However, marine carnivorous fish such as *S. dumerili* have limited ability to bio-convert the essential precursors of PUFA, such as linoleic acid (LA, C18:2 n-6) and alpha-linolenic acid (ALA, C18:3 n-3), into these essential FA [18,19].

The overall fatty acid profile within a given tissue, also known as its FA signature [20], is strongly dependent on the physiological function(s) of the tissue itself [21]. The liver and muscles are the main sites of β-oxidation [22], whereas the brain stores n-3 LC-PUFA, mainly DHA, which perform neurological functions [23–25]. Nevertheless, in nature, the FA signatures of prey tissues can be transferred to their predators with little modification and in a predictable manner [20,26,27]. Thus, in recent decades, FA have been extensively used as biomarkers in riverine and marine ecosystems [20,28,29]. Additionally, under farming conditions [30–32], the available literature collectively supports the conclusion that there is a close association between the diet and fillet FA composition in farmed fish, as reviewed by Turchini et al. [19].

As in other species [33], understanding the FA distributions and signatures within Mediterranean yellowtail tissues might help us understand whether the physiological needs and essential fatty acid (EFA) requirements of fish are satisfied under farming conditions. This is particularly crucial when diets that contain low levels of fish oil (FO) are used in fish farming. In fact, to ensure the long-term sustainability of the aquaculture sector and to improve its competitiveness, FO has been increasingly replaced by vegetable oils (VO) in fish diets because FO has a high cost and is available in finite and limited amounts [19]. Compared to FO, plant seed oils are rich in C:18 PUFA [19], but lacking in n-3 highly unsaturated fatty acids (HUFA), which are essential for marine fish [13].

Most studies performed in farmed fish to date have focused mainly on the effects of dietary lipids on the FA composition of edible muscle tissues [19]. Only one study [33], which investigated FA signatures in gilthead seabream (*Sparus aurata* L.; 1758), assessed how dietary lipids were retained in different tissues (brain, liver, and mesenteric adipose tissue) to obtain further insights into the FA requirements of farmed marine fish.

Therefore, this study was performed to evaluate how the replacement of FO with different proportions of VO in the diet affects the tissue-specific FA signatures and their robustness in the brain, muscle, liver, and visceral fat tissues of Mediterranean yellowtail (*S. dumerili*).

2. Materials and Methods

A feeding trial was performed at the Fish Nutrition Laboratory (LAC) of the Institute of Animal Science and Technology (ICTA) of the Universitat Politècnica de València (Polytechnic University of Valencia, Spain). The experimental protocol was approved by the Committee of Ethics and Animal

Welfare of the Polytechnic University of Valencia, following the Spanish Royal Decree 53/2013 on the protection of animals used for scientific purposes.

The facility used a thermoregulated recirculating seawater system (65 m^3 capacity), with a rotary drum-type filter and a mechanical gravity biofilter with a volume of 2 m^3, equipped with 9 cylindrical fiberglass tanks with a capacity of 1750 L each with aeration.

2.1. Experimental Diets

Three isoproteic (59% CP, 50% digestible protein (DP)), and isolipidic (15% CL) extruded diets were formulated, with increasing levels of vegetable oils used to replace the fish oil in the diets as follows: 0% (control diet, FO 100), 75% (FO 25), and 100% (FO 0) replacement of FO with VO (Table 1). The mixture of vegetable oils used consisted of linseed oil, sunflower oil, and palm oil (4:3:3). Diets were prepared using a cooking-extrusion processor with a semi-industrial twin-screw extruder (CLEXTRAL BC-45; Firmity, St Etienne, France), at a screw speed of 100 rpm, temperature of 110 °C, and pressure of 40–50 atm, to obtain pellets 2–3 mm in diameter.

Table 1. Ingredients (g kg^{-1} as fed) and proximate composition (% dry matter; DM) of the experimental diets.

	Diet		
	FO 100	**FO 25**	**FO 0**
Fish meal	350	350	350
Wheat	100	100	100
Wheat gluten	140	140	140
Defatted soybean meal	185	185	185
Iberian meat meal	110	110	110
Fish oil	95	24	0
Linseed oil	-	28	38
Sunflower oil	-	21	28
Palm oil	-	22	29
Multivitamin and minerals mix	20	20	20
Proximate composition			
Dry matter, %	87.4	88.8	89.1
Crude protein, % DM	58.8	60.6	58.8
Crude lipid, % DM	15.9	15.1	16.6
Ash, %DM	8.4	10.3	8.3
Gross energy, MJ kg^{-1} DM	24.3	23.8	24.4

Fish oil (FO) 100 diet: Diet formulated with fish oil as lipid source; FO 25 diet: Diet in which fish oil was included at a content of 25%; FO 0 diet: Diet in which fish oil was totally substituted with vegetable oil. [1] Vitamins and mineral mixture (values are g kg^{-1}): Premix, 25; Hill, 10; DL-α-tocopherol, 5; ascorbic acid, 5; (PO$_4$)$_2$Ca$_3$, 5. Premix composition (values are IU kg^{-1}): Retinol acetate, 1,000,000; calciferol, 500; DL-α-tocopherol, 10; menadione sodium bisulphite, 0.8; hidroclorhidrate thiamine, 2.3; riboflavin, 2.3; pyridoxine hydrochloride, 15; cyanocobalamin, 25; nicotinamide, 15; pantothenic acid, 6; folic acid, 0.65; biotin, 0.07; ascorbic acid, 75; inositol, 15; betaine, 100; polypeptides, 12.

The FA profile of the experimental diets changed depending on the relative proportions of the two lipid sources included in them. Diets containing more FO contained higher proportions of highly unsaturated fatty acids from the n-3 series, whereas diets containing more vegetable oils had higher proportions of LA and ALA (Table 2). In all diets, C16:0 accounted for the bulk of the saturated fatty acids (SFA) present, C18:1n-9 for most of the monounsaturated fatty acids (MUFA), and EPA and ARA for most of the n-3 long-chain polyunsaturated fatty acids (PUFA). The inclusion of VO as a lipid source increased the dietary proportion of C18:1 n-9 (27.1%, 29.6%, and 32.7%), C18:1 n-6 (12.7%, 13.4%, and 14.9%), and ALA (14.1%, 14.3%, and 15.4%), whereas it decreased the proportions of total SFA, EPA, and docosahexaenoic acid (DHA) in the diets (SFA: 27.8%, 27.1%, and 25.5% in the FO 100, FO 25, and FO 0 diets, respectively; EPA: 5.81%, 4.34%, and 2.77%; DHA: 13.0%, 7.6%, and 4.3%) (Table 2).

Table 2. Fatty acid composition (% of total fatty acid content) of the experimental diets.

	Diet		
	FO 100	**FO 25**	**FO 0**
14:0	3.28	2.38	1.65
16:0	18.88	19.58	18.93
17:0	0.53	0.24	0.16
18:0	5.07	4.88	4.71
Σ SFA	27.79	27.12	25.46
16:1 n-9	4.22	2.77	1.82
18:1 n-9	27.14	29.64	32.67
18:1 n-7	3.94	2.94	2.44
22:1 n-9	0.32	0.04	0.06
Σ MUFA	35.62	35.39	36.99
18:2 n-6	12.66	13.36	14.86
18:3 n-6	0.10	0.09	0.09
20:3 n-6	0.10	0.04	0.04
20:4 n-6	1.02	0.58	0.35
22:4 n-6	0.24	0.19	0.09
Σ n-6 PUFA	14.12	14.26	15.43
18:3 n-3	2.24	10.48	14.60
20:3 n-3	0.15	0.08	0.06
20:5 n-3	5.81	4.34	2.77
22:5 n-3	1.29	0.73	0.42
22:6 n-3	12.98	7.61	4.28
Σ n-3 PUFA	22.47	23.24	22.13
Σ PUFA	36.59	37.50	37.56
Σ n-6/Σ n-3	0.63	0.61	0.70
DHA/EPA	2.23	1.76	1.54

FO 100 diet: Diet formulated with fish oil as lipid source; FO 25 diet: Diet in which fish oil was included at a content of 25%; FO 0 diet: Diet in which fish oil was totally substituted with vegetable oil. Abbreviations: SFA: Saturated fatty acids; MUFA: Monounsaturated fatty acids; PUFA: Polyunsaturated fatty acids; DHA/EPA: 22:6 n-3/20:5 n-3.

2.2. Fish and Experimental Design

A total of 225 Mediterranean yellowtail juveniles obtained from a private hatchery (Futuna Blue S.A.; Cádiz, Spain) were transported to the Fish Nutrition Laboratory of the Universitat Politècnica de València for use in the feeding trial. After four weeks of acclimation, fish were individually weighed (initial weight: 176 ± 3.62 g), randomly distributed among nine tanks (25 fish per tank), and fed the experimental diets (FO 100, FO 25, or FO 0, with three tanks per diet) for 109 days.

During the trial, the water temperature averaged 21.5 ± 2.4 °C, the salinity was 31.5 ± 4.1 g L^{-1}, and the dissolved oxygen content was 8.0 ± 0.4 mg L^{-1}. The water pH ranged from 7.5 to 8.0, and the levels of nitrogenous compounds in the water were kept within the limits recommended for marine species.

2.3. In Vivo Recordings

Feed was distributed by hand, twice a day (at 09:00 and 16:00) for six days per week, until apparent satiation. Feed intake was recorded at each administration. Mortality was checked every day. At the beginning and at the end of the trial, fish were individually weighed after one day of feed deprivation and under light anaesthesia (10 mg L^{-1} clove oil containing 87% eugenol; Guinama®, Valencia, Spain). Fish health during weighing was assessed by direct observation.

After 109 days of feeding, four fish per tank (12 per treatment, 36 in total) were randomly sacrificed by lethal immersion in clove oil (150 mg L^{-1}). Fish were dissected to sample their brain, white and red muscle, liver, and visceral fat tissues, which were then frozen in liquid nitrogen and stored at −80 °C until subsequent analyses.

2.4. Chemical Analyses of Diets and Fish Tissues

Diets were ground up and analysed according to AOAC procedures [34] to determine their dry matter (by heating at 105 °C until a constant weight was attained), ash (by incinerating them at 550

°C for 5 h), crude protein (AOAC official method 990.03, by the DUMAS direct combustion method, using a LECO CN628 apparatus, LECO, ST. Joseph, MI, USA), and crude lipid content (by extraction with methyl ether using an ANKOMXT10 extractor, ANKOM Technology, Macedon, NY, USA). Gross energy content (GE) was calculated according to Brouwer [35], from the C (g) and N (g) balance in the feed (GE = 51.8 × C − 19.4 × N). The carbon and nitrogen content were analysed based on the Dumas principle (TruSpec CN; Leco Corporation, St Joseph, MI, USA). All analyses were performed in triplicate.

After thawing, tissues were minced and homogenised. The crude fats in the brain, muscle, liver, and visceral fat tissue sampled were then extracted [36]. A direct method of fatty acid methyl ester (FAME) synthesis was used for this procedure [37]. The analysis of brain, muscle, liver, and visceral fat tissues was carried out using 10–30 mg of extracted crude fat from each tissue. First, 1 mL of tridecanoic acid (C13:0) was used as internal standard. Then, 0.7 mL of 10 N KOH and 5.3 mL of HPLC (high-performance liquid chromatography)-quality methanol were added. Tubes were incubated at 55 °C in a thermoblock for 1.5 h, and underwent vigorous shaking for 5 s every 20 min. After cooling at ambient temperature in a water bath, 1.5 mL of HPLC-quality hexane was added to the reaction tubes, which were then vortex-mixed and centrifuged at 1006× *g* for 5 min. After this, the hexane layer, containing the FAMEs, was placed into vials for analysis by gas chromatography. The vials were kept at −80 °C until gas chromatography was performed. The FAMEs were analysed in a Focus gas chromatograph (Thermo, Milan, Italy) equipped with a split/splitless injector and a flame ionisation detector. Separation of the methyl esters was performed in a SPTM 2560 fused silica capillary column (Supelco, PA, USA) (100 m × 0.25 mm × 0.2 µm film thickness). Helium was used as the carrier gas at a flow rate of 20 cm s^{-1}. Samples were injected with a split ratio of 1:100.

The initial oven temperature, set at 140 °C, was held constant for 5 min, and then increased by 4 °C min^{-1} to 240 °C, at which this temperature was then maintained for 30 min. The FA were identified by comparing their retention times with those of the standards supplied by Supelco. The content of each type of FA was expressed as the percentage of the total FA content.

2.5. Statistical Analyses

The fish live weight, tissue fat content, and FA composition data were compared by analysis of variance (ANOVA), with diet included as the main effect and tank included as a random effect. The PROC MIXED procedure of the Statistical Analysis System (SAS) software [38] was used for all analyses. Adjusted means were compared among treatment levels using Bonferroni's t-test. Differences between means with $p \leq 0.05$ were considered statistically significant.

3. Results

In the present study of *S. dumerili* juveniles (176 g of initial live weight), diets containing only VO as the lipid source had no effect on fish growth (Figure 1). In fact, fish fed FO 100, FO 25, and FO 0 diets reached a final weight of 423, 409, and 419 g, respectively ($p > 0.05$). The survival rate was lower in fish fed diets without fish oil compared with those in fish fed diets containing 100% (FO 100) or 25% (FO 25) fish oil (80.3% vs. 89.7% and 92.7%, respectively; $p < 0.05$) (Figure 1).

3.1. Brain

With regard to the brain, the dietary treatment did not affect the proportions of total SFA (32.9% of the total FA on average), MUFA (29.2%), PUFA (37.9%), n-3 PUFA (29.6%), or n-6 PUFA (8.0%) present in the tissue ($p > 0.05$) (Table 3). Nevertheless, the replacement of FO with VO decreased the relative content of EPA (3.01% vs. 2.81% and 2.85% in FO 100 vs. FO 25 and FO 0, respectively; $p < 0.001$) and DHA (23.8% vs. 21.4% in FO 100 and FO 25 vs. FO 0; $p < 0.01$), whereas it increased the content of ALA (0.81% vs. 1.63% vs. 2.66% in FO 100 vs. FO 25 vs. FO 0; $p < 0.001$).

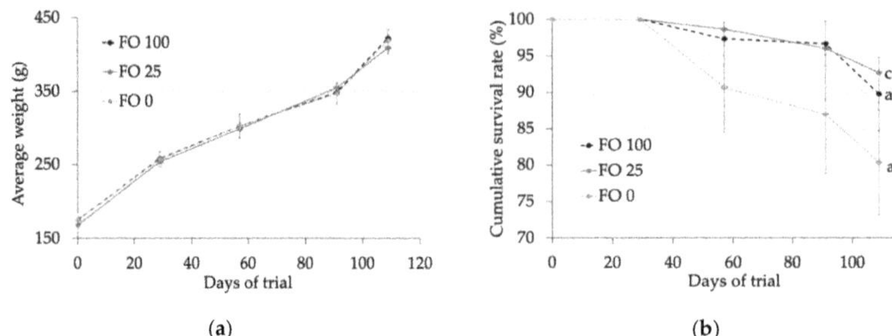

Figure 1. (a) Fish live weight (g) and (b) survival (%) of fish fed experimental diets during the 109 days of trial. Values are expressed as means ± standard error.

Table 3. Fat content and fatty acid composition (% of total fatty acid content) of the brain in Mediterranean yellowtail fed the experimental diets ($n = 12$ per diet). Values are expressed as least square (LS)means.

	Diet			*p*-Value	RSD
	FO 100	FO 25	FO 0		
Fat, % WW	3.22	4.23	3.52	0.999	0.951
Fatty acids, %					
10:0	0.42	0.46	0.57	0.083	0.138
14:0	0.76	0.64	0.68	0.134	0.141
15:0	0.12 [b]	0.10 [a]	0.09 [a]	<0.001	0.000
16:0	16.31	16.54	16.21	0.378	0.563
17:0	0.19 [b]	0.14 [a]	0.14 [a]	0.006	0.000
18:0	11.42	11.76	11.45	0.367	0.625
20:0	0.46	0.44	0.44	0.475	0.045
22:0	0.79	0.76	0.75	0.618	0.105
24:0	2.41	2.35	2.20	0.387	0.318
Σ SFA	32.90	33.17	32.51	0.278	0.932
16:1 n-9	2.67 [b]	2.40 [a]	2.34 [a]	<0.001	0.114
17:1 n-10	0.55	0.49	0.52	0.824	0.235
18:1 n-9	20.33 [ab]	19.93 [a]	21.49 [b]	0.009	1.050
18:1 n-7	2.80 [b]	2.58 [ab]	2.46 [a]	0.005	0.148
20:1 n-9	0.70 [b]	0.59 [a]	0.59 [a]	<0.001	0.063
22:1 n-9	0.32	0.32	0.28	0.271	0.045
24:1 n-9	2.16	2.07	1.98	0.378	0.261
Σ MUFA	29.47	28.44	29.66	0.139	1.336
18:2 n-6	4.25 [a]	4.39 [a]	5.55 [b]	0.016	1.080
20:2 n-6	0.26	0.23	0.25	0.298	0.055
20:4 n-6	2.65 [b]	2.51 [b]	2.27 [a]	<0.001	0.187
22:2 n-6	0.23 [b]	0.19 [ab]	0.18 [a]	0.028	0.045
22:4 n-6	0.22	0.22	0.18	0.150	0.055
Σ PUFA n-6	7.74	7.69	8.54	0.077	0.967
18:3 n-3	0.81 [a]	1.63 [b]	2.66 [c]	<0.001	0.644
20:3 n-3	0.10 [a]	0.16 [b]	0.19 [b]	<0.001	0.032
20:5 n-3	3.01 [b]	2.81 [a]	2.85 [a]	<0.001	0.118
22:5 n-3	2.13	2.26	2.14	0.483	0.219
22:6 n-3	23.83 [b]	23.83 [b]	21.43 [a]	0.006	1.668
Σ PUFA n-3	29.89	30.71	29.27	0.120	1.127
Σ PUFA	37.63	38.32	37.83	0.176	0.823
Σ n-6/Σ n-3	0.26	0.25	0.30	0.098	0.041
DHA/EPA	7.95 [ab]	8.50 [b]	7.54 [a]	0.023	0.780

FO 100 diet: Diet formulated with fish oil as lipid source; FO 25 diet: Diet in which fish oil was included at a content of 25%; FO 0 diet: Diet in which fish oil was totally substituted with vegetable oil. Abbreviations: WW: Wet weight; RSD: Residual standard deviation; SFA: Saturated fatty acids; MUFA: Monounsaturated fatty acids; PUFA: Polyunsaturated fatty acids; DHA/EPA: 22:6 n-3/20:5 n-3. The FA < 0.1% of the total FA (i.e., C12:0, C14:1, C18:3 n-6, and C30:3 n-6) are not given in the table. Different superscript letters indicate significant statistical differences among diets ($p \leq 0.05$).

3.2. Liver

In the liver, the replacement of FO with VO in the diet significantly decreased ($p < 0.001$) the proportion of the total SFA (24.3% and 23.3% vs. 20.2% of the total FA in FO 100 and FO 25 vs. FO 0), which was due to the lower proportion of C16:0 in VO than FO (13.9% and 13.4% vs. 11.6%) (Table 4).

Table 4. Fat content and fatty acid composition (% of total fatty acid content) of the liver in Mediterranean yellowtail fed the experimental diets ($n = 12$ per diet). Values are expressed as LS means.

	Diet			*p*-Value	RSD
	FO 100	**FO 25**	**FO 0**		
Fat, % WW	26.6	30.1	25.7	0.721	9.296
Fatty acids, %					
14:0	1.55 [c]	1.08 [b]	0.79 [a]	<0.001	0.129
15:0	0.26 [c]	0.15 [b]	0.10 [a]	<0.001	0.026
16:0	13.88 [b]	13.40 [b]	11.57 [a]	<0.001	0.885
17:0	0.49 [c]	0.32 [b]	0.24 [a]	<0.001	0.037
18:0	7.68 [b]	7.82 [b]	7.01 [a]	0.008	0.621
20:0	0.31 [c]	0.25 [b]	0.21 [a]	<0.001	0.014
24:0	0.06	0.07	0.05	0.369	0.020
Σ SFA	24.32 [b]	23.26 [b]	20.07 [a]	<0.001	1.317
16:1 n-9	3.26 [c]	2.23 [b]	1.64 [a]	<0.001	0.153
17:1 n-10	0.26	0.18	0.19	0.323	0.139
18:1 n-9	32.02 [a]	35.18 [b]	37.55 [c]	<0.001	1.371
18:1 n-7	7.30 [c]	5.77 [b]	4.95 [a]	<0.001	0.289
20:1 n-9	1.88 [b]	1.17 [a]	0.90 [a]	<0.001	0.314
22:1 n-9	0.37 [b]	0.24 [a]	0.16 [a]	<0.001	0.040
24:1 n-9	0.21 [b]	0.12 [a]	0.09 [a]	<0.001	0.063
Σ MUFA	45.29	44.90	45.47	0.613	1.431
18:2 n-6	15.23 [a]	15.21 [a]	16.96 [b]	0.002	1.10
18:3 n-6	0.12	0.11	0.11	0.482	0.026
20:2 n-6	1.23 [b]	0.95 [a]	0.88 [a]	<0.001	0.097
20:3 n-6	0.21 [c]	0.13 [b]	0.08 [a]	<0.001	0.021
20:4 n-6	0.89 [c]	0.53 [b]	0.39 [a]	<0.001	0.097
22:2 n-6	0.79 [c]	0.42 [b]	0.26 [a]	<0.001	0.052
22:4 n-6	0.26 [b]	0.08 [a]	0.05 [a]	<0.001	0.049
Σ PUFA n-6	18.73 [b]	17.43 [a]	18.74 [b]	0.036	1.207
18:3 n-3	2.48 [a]	8.60 [b]	11.27 [c]	<0.001	0.562
20:3 n-3	0.36 [a]	0.81 [b]	0.95 [c]	<0.001	0.117
20:5 n-3	2.21 [c]	1.45 [b]	1.10 [a]	<0.001	0.200
22:5 n-3	2.35 [c]	1.35 [b]	0.83 [a]	<0.001	0.199
22:6 n-3	4.26 [b]	2.29 [a]	1.57 [a]	<0.001	0.607
Σ PUFA n-3	11.66 [a]	14.45 [b]	15.72 [c]	<0.001	0.924
Σ PUFA	30.39 [a]	31.84 [a]	34.46 [b]	<0.001	1.703
Σ n-6/Σ n-3	1.62 [b]	1.20 [a]	1.20 [a]	<0.001	0.113
DHA/EPA	1.93 [b]	1.59 [ab]	1.36 [a]	<0.001	0.249

FO 100 diet: Diet formulated with fish oil as lipid source; FO 25 diet: Diet in which fish oil was included at a content of 25%; FO 0 diet: Diet in which fish oil was totally substituted with vegetable oil. Abbreviations: WW: Wet weight; RSD: Residual standard deviation; SFA: Saturated fatty acids; MUFA: Monounsaturated fatty acids; PUFA: Polyunsaturated fatty acids; DHA/EPA: 22:6 n-3/20:5 n-3. The FA < 0.1% of the total FA (i.e., C12:0, C14:1, C22:0, and C24:0) are not given in the table. Different superscript letters indicate significant statistical differences among diets ($p \leq 0.05$).

Moreover, the substitution of FO with VO in the diet decreased the relative content of C16:1 n-9 (3.26% vs. 2.23% vs. 1.64% in FO 100 vs. FO 25 vs. FO 0) and C17:1 n-7 (7.30% vs. 5.77% vs. 4.95%), whereas it increased that of C18:1 n-9 (32.0% vs. 35.2% vs. 37.6%) ($p < 0.001$). Thus, the proportion of the total MUFA (45.2% on average) was not affected by the dietary treatment ($p > 0.05$). With regard to n-3 PUFA, the inclusion of less FO in the diet decreased the liver tissue's relative content of EPA (2.21% vs. 1.45% vs. 1.10% in FO 100 vs. FO 25 vs. FO 0), C22:5 n-3 (2.35% vs. 1.35% vs. 0.83%), and DHA (4.26% vs. 2.29% and 1.57%), whereas it greatly increased ($p < 0.001$) its content of ALA (2.48% vs. 8.60% vs. 11.3%), and thus the total n-3 PUFA content. As for n-6 PUFA, the substitution of FO with VO decreased the AA proportion of liver tissue (0.89% vs. 0.53% vs. 0.39%; $p < 0.001$) and increased that of LA (15.2% and 15.2% vs. 17.0% in FO 100 and FO 25 vs. FO 0; $p < 0.01$).

3.3. Visceral Fat

In the visceral fat, the total substitution of FO with VO in the diet significantly decreased ($p <$ 0.001) the proportion of the total SFA (23.6% and 23.8% vs. 22.5% of the total FA in FO 100 and FO 25 vs. FO 0) (Table 5). Moreover, the replacement of FO with VO decreased the C16:1 n-9 (4.07% vs. 2.98% vs. 2.48% in FO 100 vs. FO 25 vs. FO 0) and 18:1 n-7 (4.89% vs. 4.11% vs. 3.64%) proportions in the visceral fat, whereas it increased that of C18:1 n-9 (27.8% vs. 30.0% vs. 32.6%) ($p <$ 0.001). With regard to n-3 PUFA, the inclusion of less FO in the diet decreased the EPA (3.68% vs. 2.76% vs. 2.07%), C22:5 n-3 (1.82% vs. 1.24% vs. 0.9%), and DHA (9.57% vs. 5.72% vs. 4.01%) proportions in the visceral fat, and increased that of ALA therein (3.76% vs. 9.44% vs. 12.0%) ($p <$ 0.001), whereas the proportion of total n-3 PUFA (19.2% on average) was not affected by diet ($p >$ 0.05). As for n-6 PUFA, the substitution of FO with VO increased the LA relative content (14.6% vs. 16.2% vs. 16.9%), which affected the total proportion of n-6 PUFA in this tissue (17.4% vs. 17.9% vs. 18.2%) ($p <$ 0.001).

Table 5. Fat content and fatty acid composition (% of total fatty acid content) of the visceral fat in Mediterranean yellowtails fed the experimental diets (n = 12 per diet). Values are expressed as LS means.

	Diet			*p*-Value	RSD
	FO 100	FO 25	FO 0		
Fat, % WW	37.90	36.90	34.91	0.920	9.529
Fatty acids, %					
14:0	2.55 [c]	1.99 [b]	1.64	<0.001	0.094
15:0	0.32 [c]	0.22 [b]	0.16 [a]	<0.001	0.018
16:0	14.57 [a]	15.00 [b]	14.36 [a]	<0.001	0.293
17:0	0.44 [c]	0.32 [b]	0.26 [a]	<0.001	0.023
18:0	5.11 [a]	5.64 [b]	5.46 [b]	<0.001	0.207
20:0	0.36 [c]	0.32 [b]	0.29 [a]	<0.001	0.009
22:0	0.14 [a]	0.18 [b]	0.19 [b]	0.001	0.011
24:0	0.10	0.11	0.12	0.307	0.026
Σ SFA	23.63 [b]	23.80 [b]	22.53 [a]	<0.001	0.415
16:1 n-9	4.07 [c]	2.98 [b]	2.48 [a]	<0.001	0.184
17:1 n-10	0.37 [b]	0.20 [a]	0.16 [a]	<0.001	0.049
18:1 n-9	27.8 [a]	30.0 [b]	32.6 [c]	<0.001	0.547
18:1 n-7	4.89 [c]	4.11 [b]	3.64 [a]	<0.001	0.154
20:1 n-9	1.96 [c]	1.11 [b]	0.79 [a]	<0.001	0.116
22:1 n-9	0.40 [c]	0.22 [b]	0.15 [a]	<0.001	0.046
24:1 n-9	0.49 [c]	0.30 [b]	0.23 [a]	<0.001	0.038
Σ MUFA	39.95 [b]	38.87 [a]	40.07 [b]	<0.001	0.495
18:2 n-6c	14.64 [a]	16.17 [b]	16.89 [c]	<0.001	0.534
18:3 n-6	0.14 [b]	0.12 [a]	0.12 [a]	<0.001	0.003
20:2 n-6	0.92 [c]	0.64 [b]	0.48 [a]	<0.001	0.043
20:3 n-6	0.15 [b]	0.07 [ab]	0.06 [a]	0.023	0.075
20:4 n-6	0.67 [c]	0.43 [b]	0.32 [a]	<0.001	0.044
22:2 n-6	0.52 [c]	0.31 [b]	0.26 [a]	<0.001	0.044
22:4 n-6	0.34 [c]	0.18 [b]	0.10 [a]	<0.001	0.046
Σ PUFA n-6	17.37 [a]	17.92 [ab]	18.19 [b]	0.003	0.534
18:3 n-3	3.76 [a]	9.44 [b]	12.0 [c]	<0.001	0.843
20:3 n-3	0.22 [a]	0.26 [b]	0.28 [b]	<0.001	0.022
20:5 n-3	3.68 [c]	2.76 [b]	2.07 [a]	<0.001	0.150
22:5 n-3	1.82 [c]	1.24 [b]	0.90 [a]	<0.001	0.010
22:6 n-3	9.57 [c]	5.72 [b]	4.01 [a]	<0.001	0.634
Σ PUFA n-3	19.05	19.40	19.21	0.266	0.372
Σ PUFA	36.42 [a]	37.33 [b]	37.40 [b]	0.002	0.613
Σ n-6/Σ n-3	0.91	0.92	0.95	0.104	0.035
DHA/EPA	2.60 [b]	2.07 [a]	1.94 [a]	<0.001	0.153

FO 100 diet: Diet formulated with fish oil as lipid source; FO 25 diet: Diet in which fish oil was included at a content of 25%; FO 0 diet: Diet in which fish oil was totally substituted with vegetable oil. Abbreviations: WW: Wet weight; RSD: Residual standard deviation; SFA: Saturated fatty acids; MUFA: Monounsaturated fatty acids; PUFA: Polyunsaturated fatty acids; DHA/EPA: 22:6 n-3/20:5 n-3. The FA < 0.1% of the total FA (i.e., C10:0, C12:0, C14:1) are not given in the table. Different superscript letters indicate significant statistical differences among diets ($p \leq$ 0.05).

3.4. Muscle

In the red muscle, the replacement of FO with VO did not affect the proportion of total SFA (24.0% of the total FA on average) (Table 6). Among MUFA, the replacement of FO with VO increased the relative content of C18:1 n-9 (27.2% vs. 29.0% vs. 31.8% in FO 100 vs. FO 25 vs. FO 0), whereas it decreased that of C18:1 n-7 (4.21% vs. 3.65% vs. 3.26% in FO 100 vs. FO 25 vs. FO 0) and C20:1 n-9 (1.91% vs. 1.05% vs. 0.74%) ($p < 0.001$). For n-3 PUFA, the inclusion of less FO in the diet decreased the proportions of EPA (3.22% vs. 2.60% vs. 2.14%), C22:5 n-3 (2.12% vs. 1.58% vs. 1.29%), and DHA (14.0% vs. 9.38% vs. 7.47%) in red muscle tissue, whereas it increased that of ALA (2.99% vs. 7.68% vs. 9.84%) ($p < 0.001$). In addition, the partial and total substitution of FO with VO in the diet increased the total n-6 PUFA content of red muscle (14.7% vs. 16.0% and 16.7%) due to changes in the proportions of LA ($p < 0.001$).

Table 6. Fat content and fatty acid composition (% of total fatty acid content) of the red muscle in Mediterranean yellowtails fed the experimental diets ($n = 12$ per diet). Values are expressed as LS means.

	Diet			*p*-Value	RSD
	FO 100	FO 25	FO 0		
Fat, % WW	4.37	4.89	4.49	0.500	0.816
Fatty acids, %					
14:0	1.81	1.81	1.26	0.105	0.602
16:0	14.41	15.01	14.40	0.125	0.455
17:0	0.44 [c]	0.30 [b]	0.24 [a]	<0.001	0.035
18:0	6.53 [a]	6.91 [b]	6.83 [b]	<0.001	0.184
20:0	0.38 [b]	0.33 [a]	0.32 [a]	<0.001	0.017
22:0	0.12 [a]	0.15 [b]	0.16 [b]	<0.001	0.015
Σ SFA	23.88	24.65	23.33	0.061	0.769
14:1 n-9	0.11	0.12	0.10	0.956	0.100
16:1 n-9	3.14 [b]	2.57 [ab]	2.05 [a]	0.005	0.125
17:1 n-10	0.31 [b]	0.21 [a]	0.16 [a]	<0.001	0.011
18:1 n-9	27.21 [a]	29.04 [b]	31.76 [c]	<0.001	0.509
18:1 n-7	4.21 [c]	3.65 [b]	3.26 [a]	<0.001	0.208
20:1 n-9	1.91 [c]	1.05 [b]	0.74 [a]	<0.001	0.134
22:1 n-9	0.36 [b]	0.18 [a]	0.13 [a]	<0.001	0.088
24:1 n-9	0.41 [c]	0.26 [b]	0.20 [a]	<0.001	0.038
Σ MUFA	37.62 [ab]	37.08 [a]	38.37 [b]	0.044	0.651
18:2 n-6c	13.05 [a]	14.83 [b]	15.76 [c]	<0.001	0.552
18:3 n-6	0.12	0.11	0.10	0.053	0.015
20:2 n-6	0.78 [b]	0.58 [a]	0.45 [a]	<0.001	0.027
20:4 n-6	0.91 [c]	0.65 [b]	0.54 [a]	<0.001	0.055
22:2 n-6	0.40 [c]	0.24 [b]	0.16 [a]	<0.001	0.031
22:4 n-6	0.51 [c]	0.29 [b]	0.21 [a]	<0.001	0.040
Σ PUFA n-6	14.70 [a]	15.95 [b]	16.68 [b]	<0.001	0.513
18:3 n-3	2.99 [a]	7.68 [b]	9.84 [c]	<0.001	0.705
20:3 n-3	0.20 [a]	0.25 [b]	0.27 [b]	<0.001	0.021
20:5 n-3	3.22 [c]	2.60 [b]	2.14 [a]	<0.001	0.116
22:5 n-3	2.12 [c]	1.58 [b]	1.29 [a]	<0.001	0.155
22:6 n-3	14.00 [c]	9.38 [b]	7.47 [a]	<0.001	0.946
Σ PUFA n-3	22.56 [b]	21.50 [ab]	21.01 [a]	0.036	0.884
Σ PUFA	38.50	38.27	38.30	0.925	0.881
Σ n-6/Σ n-3	1.54 [b]	1.35 [a]	1.26 [a]	<0.001	0.083
DHA/EPA	4.34 [b]	3.61 [a]	3.49 [a]	0.001	0.280

FO 100 diet: Diet formulated with fish oil as lipid source; FO 25 diet: Diet in which fish oil was included at a content of 25%; FO 0 diet: Diet in which fish oil was totally substituted with vegetable oil. Abbreviations: WW: Wet weight; RSD: Residual standard deviation; SFA: Saturated fatty acids; MUFA: Monounsaturated fatty acids; PUFA: Polyunsaturated fatty acids; DHA/EPA: 22:6 n-3/20:5 n-3. The FA < 0.1% of the total FA (i.e., C12:0, C15:0, C20:3 n-6, and C24:0) are not given in the table. Different superscript letters indicate significant statistical differences among diets ($p \leq 0.05$).

In the white muscle, the substitution of FO with VO in the diet decreased the proportion of total SFA (24.0% and 23.8% vs. 22.6% in FO 100 vs. FO 25 vs. FO 0) and increased that of C18:1 n-9 (27.6% vs. 29.0% vs. 31.7%); this also decreased the relative content of C16:1 n-9 (3.78% vs. 2.90% vs. 2.36%)

and C20:1 n-9 (1.76% vs. 1.09% vs. 0.76%) in white muscle tissue (Table 7). For n-3 PUFA, the inclusion of less FO in the diet decreased the EPA (3.41% vs. 2.87% vs. 2.26%), C22:5 n-3 (1.71% vs. 1.38% vs. 1.05%), and DHA (11.6% vs. 8.59% vs. 6.38%) content in white muscle tissue, whereas it increased the ALA content (3.76% vs. 9.44% vs. 12.0%) ($p < 0.001$). Nevertheless, the proportion of total n-3 PUFA (21.0% on average) was not affected ($p > 0.05$) by dietary treatment. Finally, the LA relative content in the white muscle was increased by this treatment (14.2% vs. 15.6% and 16.5%), which affected the total n-6 PUFA content (15.6% vs. 16.7% vs. 17.3%) ($p < 0.001$).

Table 7. Fat content and fatty acid composition (% of total fatty acid content) of the white muscle in Mediterranean yellowtails fed the experimental diets ($n = 12$ per diet). Values are expressed as LS means.

	Diet			*p*-Value	RSD
	FO 100	FO 25	FO 0		
Fat, % WW	5.84	5.94	6.03	0.941	0.770
Fatty acids, %					
14:0	2.28 [c]	1.80 [b]	1.47 [a]	<0.001	0.132
16:0	14.97 [b]	14.78 [b]	14.22 [a]	0.035	0.366
17:0	0.39 [c]	0.30 [b]	0.23 [a]	<0.001	0.037
18:0	5.89	6.21	6.13	0.050	0.171
20:0	0.36 [c]	0.33 [b]	0.32 [a]	<0.001	0.013
22:0	0.12 [a]	0.15 [b]	0.16 [b]	0.020	0.032
Σ SFA	23.94 [b]	23.80 [b]	22.63 [a]	0.009	0.494
14:1 n-9	0.28 [c]	0.22 [b]	0.15 [a]	<0.001	0.038
16:1 n-9	3.78 [c]	2.90 [b]	2.36 [a]	<0.001	0.200
17:1 n-10	0.33 [c]	0.22 [b]	0.17 [a]	<0.001	0.026
18:1 n-9	27.55 [a]	29.02 [b]	31.72 [c]	<0.001	0.759
18:1 n-7	4.15 [c]	3.56 [b]	3.20 [a]	<0.001	0.165
20:1 n-9	1.76 [c]	1.09 [b]	0.76 [a]	<0.001	0.175
22:1 n-9	0.29 [c]	0.19 [b]	0.12 [a]	<0.001	0.045
24:1 n-9	0.35 [c]	0.27 [b]	0.19 [a]	<0.001	0.048
Σ MUFA	38.63	37.50	38.63	0.050	0.615
18:2 n-6c	14.15 [a]	15.59 [b]	16.46 [b]	<0.001	0.552
18:3 n-6	0.14 [b]	0.13 [a]	0.12 [a]	0.001	0.009
20:2 n-6	0.86 [c]	0.63 [b]	0.49 [a]	<0.001	0.059
20:4 n-6	0.79 [c]	0.61 [b]	0.48 [a]	<0.001	0.076
22:2 n-6	0.42 [c]	0.28 [b]	0.19 [a]	<0.001	0.037
22:4 n-6	0.41 [c]	0.26 [b]	0.18 [a]	<0.001	0.052
Σ PUFA n-6	16.45 [a]	17.20 [ab]	17.80 [b]	0.001	0.493
18:3 n-3	3.93 [a]	8.25 [b]	10.84 [c]	<0.001	1.141
20:3 n-3	0.20 [a]	0.23 [b]	0.25 [c]	<0.001	0.019
20:5 n-3	3.41 [c]	2.87 [b]	2.26 [a]	<0.001	0.198
22:5 n-3	1.71 [c]	1.38 [b]	1.05 [a]	<0.001	0.136
22:6 n-3	11.63 [c]	8.59 [b]	6.38 [a]	<0.001	1.196
Σ PUFA n-3	20.60	21.21	20.77	0.830	0.773
Σ PUFA	37.48	38.70	38.74	0.500	0.825
Σ n-6/Σ n-3	0.80	0.81	0.86	0.086	0.069
DHA/EPA	3.43 [b]	2.95 [a]	2.82 [a]	0.019	0.234

FO 100 diet: Diet formulated with fish oil as lipid source; FO 25 diet: Diet in which fish oil was included at a content of 25%; FO 0 diet: Diet in which fish oil was totally substituted with vegetable oil. Abbreviations: WW: Wet weight; RSD: Residual standard deviation; SFA: Saturated fatty acids; MUFA: Monounsaturated fatty acids; PUFA: Polyunsaturated fatty acids; DHA/EPA: 22:6 n-3/20:5 n-3. The FA < 0.1% of the total FA (i.e., C12:0, C15:0, C20:3 n-6, and C24:0) are not given in the table. Different superscript letters indicate significant statistical differences among diets ($p \leq 0.05$).

4. Discussion

Successful growth performance may be achieved in fish fed with diets containing low levels of FO as long as their minimum EFA requirements are met [39]. In the larval stages, DHA plays a key role in promoting the development of neural membranes and eyes [40], improving growth, vitality, and survival [1,14,41], and preventing swimming disorders such as spinning and disorientation [41]. The accumulation of DHA in the brain during fish development has been measured in several marine species [14], which also show very low rates of DHA biosynthesis. Thus, any DHA deficiency during larval development will have serious consequences for the successful performance of fish larvae [14]. Moreover, DHA is essential for the development of schooling behaviour in the Mediterranean yellowtail [42–44]. Data in the published literature show that larvae (aged 3–7 days after hatching (dah)) of *Seriola dumerili* fed with rotifers [1] require a supply of DAH equal to 1.5% of the dry matter (DM) in their diet. Larvae (7 mm in length) of the related Japanese amberjack (*S. quinqueradiata* Temminck and Schlegel, 1845) fed with *Artemia* required a diet containing 3.9% of n-3 HUFA (2.5% and 1.3% DM of EPA and DHA, respectively) in a previous study [41]. Further, in *S. rivoliana* Valenciennes, 1833 larvae (aged 30 to 50 dah) had higher survival rates and better stress resistance when fed microdiets containing 3.2% DM of DHA [45].

In this experiment, the dietary treatments tested provided a minimum n-3 HUFA supply of 0.75% DM, which should be sufficient to meet the EFA requirements of most marine species (minimum value: 0.5% DM in juveniles and subadults of marine species [39]). Under different conditions, the total EFA requirements for juveniles (39 to 387 g of live weight in a 154-day feeding trial) of Mediterranean yellowtail have been estimated to be 1.2% DM [11]. In fact, juvenile fish, although they still need EFA, are likely to require lower levels of EPA and DHA in their diet than those needed in the larval stage. Nevertheless, dietary DHA deprivation in juveniles of pelagic marine species, such as carangids and tunids, can be particularly deleterious because of their fast growth rates [14], especially when they are fed diets containing low levels of FO.

During the grow-out phase, EFA deficiencies may reduce fish growth. In this regard, a substitution of 60% of the FO with VO in the diet is considered acceptable [46]. For carangid species, the growth performance of Japanese yellowtail (*S. quinqueradiata*) (average initial weight: 252 g; final weight 412 g) was not affected by the full replacement of FO with olive oil in a previous short-term feeding trial (40 days) [47]. On the contrary, yellowtail kingfish (*S. lalandi* Valenciennes, 1833) showed impaired growth rates when FO was totally substituted with canola or sunflower oil in their diet during a five-week-long feeding trial (with a weight change from 96 to 260 g) [47].

In the present study of *S. dumerili* (grown from 176 to 417 g of live weight, with 109 days of feeding), diets containing only VO as the lipid source had no detrimental impacts on the growth performance of these fish. Nevertheless, the survival rate was lower in fish fed diets without fish oil compared with those in fish fed diets containing 100% (FO 100) or 25% (FO 25) fish oil.

Indeed, Monge-Ortiz et al. [11] reported that diets including at least 525 g kg^{-1} of fish meal (FM) are likely to supply enough LC-PUFA to meet the needs of fish, even with the complete substitution of FO (with fish reaching a final weight of 397 g). In fact, FM usually contains up to 8%–15% DM of crude lipid, 30%–35% of which is composed of n-3 LC-PUFA [19]. In the present study, the experimental diets contained 350 g kg^{-1} of FM, which likely met the FA requirements of the studied fish during the grow-out phase.

The present study measured fat storage in different tissues, including the muscles, brain, liver, and visceral fat. Indeed, the locations of fat storage differ within and between fish species [15,48,49]. In a previous study [50], a high lipid content (53% DM) was found in the liver of farmed *S. dumerili*, which was consistent with that recorded in the present study (60% DM) and higher than the fat content of the muscle tissue (12% DM) [50].

The replacement of FO with VO in fish diets may lead to the accumulation of fat in the fish liver, generating a fatty liver syndrome, which may be the result of inefficient nutrient utilisation and increased rates of lipid peroxidation [51,52]. Nevertheless, we did not find differences in the liver fat content of the tested fish, even when they were fed with diets containing only VO as the lipid source, which was consistent with the findings of previous studies done on European seabass [53], gilthead sea bream [54], and turbot (*Psetta maxima* (L.; 1758)) [55].

Regardless of whether FO replacement affects fish growth and the lipid content of their tissues, its impact on fatty acid composition will vary depending on the dietary lipid content and source, as well as on the tissues considered [19]. The fatty acid signatures of fish tissues are closely related to their dietary FA composition [19]. Complete or partial FO replacement with VO blends is known to affect the FA compositions of muscles, fish organs, and fat storage tissues [19]. However, the magnitude of the changes in FA signatures varies among different fish species and tissues [56–58].

The FA signature of several wild and farmed fish was previously measured in muscle and liver tissues, but to our knowledge, few studies have investigated the FA signatures of other fatty tissues, such as those of the visceral fat and brain. Specifically, in wild and farmed Mediterranean yellowtail, the FA composition has been analysed in the muscles, liver, ovary [10,11,50,59,60], and eggs [61]. Other studies have been performed in other *Seriola* species, such as *S. lalandi* [47,62,63], *S. quinqueradiata* [64,65], *S. dorsalis* (Gill, 1863) [66,67], and *S. rivoliana* [45], which mainly focused on the FA composition of muscle tissues.

In the liver of Mediterranean yellowtail, we found a high proportion of MUFA and a low proportion of SFA, likely because their C16:0, C18:0, and C22:0 content was sufficient to maintain or even exceed the requirements of liver metabolic functions. In the brain and visceral fat, PUFA were highly represented (representing on average 37.9% and 37.1% of the total FA content in the brain and visceral fat, respectively). In fact, PUFA (especially C18 FA, such as LA and ALA) are generally stored in non-lipogenic tissues, especially the visceral fat, and are likely used as metabolic substrates for β-oxidation [58,68,69]. Moreover, the brain contained a high proportion of DHA (on average 23.0% of the total FA), which is known to regulate membrane fluidity, the blood-brain barrier, and the activities of certain enzymes, such as ionic channel proteins and nerve growth factors, both in mammals [70,71] and fish [1,14,45]. In salmonids, Atlantic cod, and flatfishes, DHA is also present at high levels in the brain [12,15,72], which agrees with our findings.

Consistent with other studies [56,73,74], the muscle, liver, and visceral fat FA signatures of Mediterranean yellowtail found in our study reflected the dietary FA profiles of these fish, with the FA signatures of these tissues thus showing a low robustness to changes in the dietary lipid sources from FO to VO. In contrast, the brain FA signature was less strongly affected by the dietary FA composition, whereas only ALA, EPA, and DHA changed. Nevertheless, in European seabass, brain lipids appeared to be sensitive to dietary lipid inputs [75]. On the contrary, a more recent study in juvenile European seabass fed n-3 LC-PUFA-deficient diets showed that polar lipids in their neural tissues had the highest capacity to regulate and preserve their DHA content [76]. In the present study, fish fed the diet with 100% VO showed a 10% lower DHA content in the brain than those fed diets with fish oil (FO 100 and FO 25) and a lower survival, which could have been related to some effects of EFA deficiency in them. In contrast, the administration of the FO 25 diet did not decrease the DHA proportion in the brain, which suggests that the substitution of 75% of the FO in the diet with VO could be suitable for use in diets for rearing Mediterranean yellowtail juveniles from 176 to 417 g of live weight. Further studies are necessary to confirm these results over longer feeding periods.

5. Conclusions

This study provided new findings on the FA signatures in different tissues of *Seriola dumerili* fed diets with decreasing levels of FO included in them. The FA composition of the diet had a strong effect on the FA compositions of muscle, liver, and visceral fat tissues. The brain had a FA signature that was more robust to dietary changes, whereas only some EFA (EPA and DHA) decreased when fish oil was totally replaced by vegetable oil.

Author Contributions: Conceptualization, S.M.-L.; A.T.-V.; and M.J.-C.; Formal analysis, F.B. and M.C.M.S.; Investigation, F.B. and M.C.M.S.; Data curation, A.T.-V.; F.B.; and A.T.; Writing—original draft preparation, F.B.; A.T.; A.T.-V.; and S.M.-L.; Writing—review and editing, F.B.; A.T.; A.T.-V.; S.M.-L.; and M.J.-C. All authors have read and agreed to the published version of the manuscript.

Funding: The Ph.D. grant held by Francesco Bordignon is funded by the ECCEAQUA project (MIUR; CUP: C26C18000030004).

Acknowledgments: The authors gratefully acknowledge V. J. Moya Salvador and L. Ródenas for their technical support during chemical analyses of fats and fatty acids.

Conflicts of Interest: The authors declare no conflict of interest.

References

1. Matsunari, H.; Hashimoto, H.; Oda, K.; Masuda, Y.; Imaizumi, H.; Teruya, K.; Furuita, H.; Yamamoto, T.; Hamada, K.; Mushiake, K. Effects of docosahexaenoic acid on growth, survival and swim bladder inflation of larval amberjack (*Seriola dumerili*, Risso). *Aquac. Res.* **2013**, *44*, 1696–1705. [CrossRef]
2. Grau, A.; Riera, F.; Carbonell, E. Some protozoan and metazoan parasites of the amberjack from the Balearic Sea (western Mediterranean). *Aquac. Int.* **1999**, *7*, 307–317. [CrossRef]
3. Andaloro, F.; Pipitone, C. Food and feeding habits of the amberjack, *Seriola dumerili* in the Central Mediterranean Sea during the spawning season. *Cah. Biol. Mar.* **1997**, *38*, 91–96.
4. Sicuro, B.; Luzzana, U. The State of Seriola spp. other than yellowtail (*S. quinqueradiata*) farming in the World. *Rev. Fish. Sci. Aquac.* **2016**, *24*, 314–325. [CrossRef]
5. Mazzola, A.; Favaloro, E.; Sarà, G. Cultivation of the Mediterranean amberjack, *Seriola dumerili* (Risso, 1810), in submerged cages in the Western Mediterranean Sea. *Aquaculture* **2000**, *181*, 257–268. [CrossRef]
6. Jover, M.; García-Gómez, A.; Tomás, A.; De La Gándara, F.; Pérez, L. Growth of Mediterranean yellowtail (*Seriola dumerili*) fed extruded diets containing different levels of protein and lipid. *Aquaculture* **1999**, *179*, 25–33. [CrossRef]
7. Takakuwa, F.; Fukada, H.; Hosokawa, H.; Masumoto, T. Optimum digestible protein and energy levels and ratio for greater amberjack *Seriola dumerili* (Risso) fingerling. *Aquac. Res.* **2006**, *37*, 1532–1539. [CrossRef]
8. Vidal, A.T.; De La Gándara, F.; Gómez, A.G.; Cerdá, M.J. Effect of the protein/energy ratio on the growth of Mediterranean yellowtail (*Seriola dumerili*). *Aquac. Res.* **2008**, *39*, 1141–1148. [CrossRef]
9. Papadakis, I.E.; Chatzifotis, S.; Divanach, P.; Kentouri, M. Weaning of greater amberjack (*Seriola dumerili* Risso 1810) juveniles from moist to dry pellet. *Aquac. Int.* **2008**, *16*, 13–25. [CrossRef]
10. Haouas, W.G.; Zayene, N.; Guerbej, H.; Hammami, M.; Achour, L. Fatty acids distribution in different tissues of wild and reared *Seriola dumerili*. *Int. J. Food Sci. Technol.* **2010**, *45*, 1478–1485. [CrossRef]
11. Monge-Ortiz, R.; Tomás-Vidal, A.; Rodriguez-Barreto, D.; Martínez-Llorens, S.; Pérez, J.A.; Jover-Cerdá, M.; Lorenzo, A. Replacement of fish oil with vegetable oil blends in feeds for greater amberjack (*Seriola dumerili*) juveniles: Effect on growth performance, feed efficiency, tissue fatty acid composition and flesh nutritional value. *Aquac. Nutr.* **2018**, *24*, 605–615. [CrossRef]
12. Mourente, G.; Tocher, D.R.; Sargent, J.R. Specific accumulation of docosahexaenoic acid (22:6n-3) in brain lipids during development of juvenile turbot *Scophthalmus maximus* L. *Lipids* **1991**, *26*, 871–877. [CrossRef]
13. Sargent, J.; Bell, G.; McEvoy, L.; Tocher, D.; Estevez, A. Recent developments in the essential fatty acid nutrition of fish. *Aquaculture* **1999**, *77*, 191–199. [CrossRef]
14. Mourente, G. Accumulation of DHA (docosahexaenoic acid; 22:6n-3) in larval and juvenile fish brain. In Proceedings of the 26th Annual Larval Fish Conference, Os, Norway, 22–26 July 2002; Brownman, H., Skiftesvik, A.B., Eds.; Institute of Marine Research: Bergen, Norway, 2003; pp. 239–248.
15. Tocher, D.R.; Harvie, D.G. Fatty acid compositions of the major phosphoglycerides from fish neural tissues; (n-3) and (n-6) polyunsaturated fatty acids in rainbow trout (*Salmo gairdneri*) and cod (*Gadus morhua*) brains and retinas. *Fish. Physiol. Biochem.* **1988**, *5*, 229–239. [CrossRef] [PubMed]
16. Bell, J.G.; Castell, J.D.; Tocher, D.R.; MacDonald, F.M.; Sargent, J.R. Effects of different dietary arachidonic acid: Docosahexaenoic acid ratios on phospholipid fatty acid compositions and prostaglandin production in juvenile turbot (*Scophthalmus maximus*). *Fish. Physiol. Biochem.* **1995**, *14*, 139–151. [CrossRef] [PubMed]
17. Sargent, J.R.; McEvoy, L.A.; Bell, J.G. Requirements, presentation and sources of polyunsaturated fatty acids in marine fish larval feeds. *Aquaculture* **1997**, *155*, 117–127. [CrossRef]
18. Tocher, D.R.; Ghioni, C. Fatty acid metabolism in marine fish: Low activity of fatty acyl delta5 desaturation in gilthead sea bream (*Sparus aurata*) cells. *Lipids* **1999**, *34*, 433–440. [CrossRef]
19. Turchini, G.M.; Torstensen, B.E.; Ng, W.K. Fish oil replacement in finfish nutrition. *Rev. Aquac.* **2009**, *1*, 10–57. [CrossRef]
20. Iverson, S.J.; Field, C.; Bowen, W.D.; Blanchard, W. Quantitative fatty acid signature analysis: A new method. *Ecol. Monogr.* **2004**, *74*, 211–235. [CrossRef]
21. Sargent, J.R.; Bell, J.G.; Bell, M.V.; Henderson, R.J.; Tocher, D.R. The metabolism of phospholipids and polyunsaturated fatty acids in fish. In *Aquaculture: Fundamental and Applied Research*; Lahlou, B., Vitiello, P., Eds.; Coastal and Estuarine Studies; American Geophysical Union: Washington, DC, USA, 1993; Volume 43, pp. 103–124.

22. Henderson, J.R.; Tocher, D.R. The lipid composition and biochemistry of freshwater fish. *Prog. Lipid Res.* **1987**, *26*, 281–347. [CrossRef]

23. Furuita, H.; Takeuchi, T.; Uematsu, K. Effects of eicosapentaenoic and docosahexaenoic acids on growth, survival and brain development of larval Japanese flounder (*Paralichthys olivaceus*). *Aquaculture* **1998**, *161*, 269–279. [CrossRef]

24. Anderson, G.J.; Connoro, W.E.; Corliss, J.D. Docosahexaenoic acid is the preferred dietary n-3 fatty acid for the development of the brain and retina. *Pediatr. Res.* **1990**, *27*, 89–97. [CrossRef] [PubMed]

25. Bianconi, S.; Santillán, M.E.; del Rosario Solís, M.; Martini, A.C.; Ponzio, M.F.; Vincenti, L.M.; Schiöth, H.B.; Carlini, V.P.; Stutz, G. Effects of dietary omega-3 PUFAs on growth and development: Somatic, neurobiological and reproductive functions in a murine model. *J. Nutr. Biochem.* **2018**, *61*, 82–90. [CrossRef] [PubMed]

26. Bergé, J.P.; Barnathan, G. Fatty acids from lipids of marine organisms: Molecular biodiversity, roles as biomarkers, biologically active compounds, and economical aspects. *Adv. Biochem. Eng. Biotechnol.* **2005**, *96*, 49–125.

27. Thiemann, G.W. Using fatty acid signatures to study bear foraging: Technical considerations and future applications. *Ursus* **2008**, *19*, 59–72. [CrossRef]

28. Kaushik, S.J.; Corraze, G.; Radunz-Neto, J.; Larroquet, L.; Dumas, J. Fatty acid profiles of wild brown trout and Atlantic salmon juveniles in the Nivelle basin. *J. Fish. Biol.* **2006**, *68*, 1376–1387. [CrossRef]

29. Stowasser, G.; McAllen, R.; Pierce, G.J.; Collins, M.A.; Moffat, C.F.; Priede, I.G.; Pond, D. Trophic position of deep-sea fish-Assessment through fatty acid and stable isotope analyses. *Deep Res. Part I Oceanogr. Res. Pap.* **2009**, *56*, 812–826. [CrossRef]

30. Budge, S.M.; Penney, S.N.; Lall, S.P. Estimating diets of Atlantic salmon (*Salmo salar*) using fatty acid signature analyses; validation with controlled feeding studies. *Can. J. Fish. Aquat. Sci.* **2012**, *69*, 1033–1046. [CrossRef]

31. Magnone, L.; Bessonart, M.; Rocamora, M.; Gadea, J.; Salhi, M. Diet estimation of *Paralichthys orbignyanus* in a coastal lagoon via quantitative fatty acid signature analysis. *J. Exp. Mar. Biol. Ecol.* **2015**, *462*, 36–49. [CrossRef]

32. Happel, A.; Stratton, L.; Pattridge, R.; Rinchard, J.; Czesny, S. Fatty-acid profiles of juvenile lake trout reflect experimental diets consisting of natural prey. *Freshw. Biol.* **2016**, *61*, 1466–1476. [CrossRef]

33. Benedito-Palos, L.; Navarro, J.C.; Kaushik, S.; Pérez-Sánchez, J. Tissue-specific robustness of fatty acid signatures in cultured gilthead sea bream (*Sparus aurata* L.) fed practical diets with a combined high replacement of fish meal and fish oil. *J. Anim. Sci.* **2010**, *88*, 1759–1770. [CrossRef]

34. AOAC. *International Official Methods of Analysis of AOAC International*, 17th ed.; Association of Official Analytical Chemists: Arlington, VA, USA, 1990.

35. Brouwer, E. Report of Sub-Committee on Constants and Factors. In *Proceedings of the 3rd Symposium on Energy Metabolism*; Blaxter, K.L., Ed.; London Academic Press: London, UK, 1965; pp. 441–443.

36. Folch, J.; Lees, M.; Sloane Stanley, G.H. A simple method for the isolation and purification of total lipids from animal tissues. *J. Biol. Chem.* **1957**, *226*, 497–509.

37. O'Fallon, J.V.; Busboom, J.R.; Nelson, M.L.; Gaskins, C.T. A direct method for fatty acid methyl ester synthesis: Application to wet meat tissues, oils, and feedstuffs. *J. Anim. Sci.* **2007**, *85*, 1511–1521. [CrossRef]

38. SAS (Statistical Analysis System Institute, Inc.). *SAS/STAT(R) 9.3 User's Guide*, 3rd ed.; SAS Institute: Cary, NC, USA, 2013; Available online: https://support.sas.com/documentation/cdl/en/statug/63962/HTML/default/viewer.htm#titlepage.htm (accessed on 1 May 2019).

39. Tocher, D.R. Fatty acid requirements in ontogeny of marine and freshwater fish. *Aquac. Res.* **2010**, *41*, 717–732. [CrossRef]

40. Ishizaki, Y.; Masuda, R.; Uematsu, K.; Shimizu, K.; Arimoto, M.; Takeuchi, T. The effect of dietary docosahexaenoic acid on schooling behaviour and brain development in larval yellowtail. *J. Fish. Biol.* **2001**, *58*, 1691–1703. [CrossRef]

41. Furuita, H.; Takeuchi, T.; Watanabe, T.; Fujimoto, H.; Sekiya, S.; Imaizumi, K. Requirements of larval yellowtail for eicosapentanoic acid, docosahaenoic acid, and n-3 highly unsaturated fatty acid. *Fish. Sci.* **1996**, *62*, 372–379. [CrossRef]

42. Masuda, R.; Takeuchi, T.; Tsukamoto, K.; Sato, H.; Shimizu, K.; Imaizumi, K. Incorporation of dietary docosahexaenoic acid into the central nervous system of the yellowtail *Seriola quinqueradiata*. *Brain Behav. Evol.* **1999**, *53*, 173–179. [CrossRef]

43. Masuda, R.; Takeuchi, T.; Takeuchi, T.; Tsukamoto, K.; Tsukamoto, K.; Ishizaki, Y.; Kanematsu, M.; Imaizum, K. Critical involvement of dietary docosahexaenoic acid in the ontogeny of schooling behaviour in the yellowtail. *J. Fish. Biol.* **1998**, *53*, 471–484. [CrossRef]

44. Masuda, R.; Tsukamoto, K. School formation and concurrent developmental changes in carangid fish with reference to dietary conditions. *Environ. Biol. Fish.* **1999**, *56*, 243–252. [CrossRef]

45. Mesa-Rodriguez, A.; Hernández-Cruz, C.M.; Betancor, M.B.; Fernández-Palacios, H.; Izquierdo, M.S.; Roo, J. Effect of increasing docosahexaenoic acid content in weaning diets on survival, growth and skeletal anomalies of longfin yellowtail (*Seriola rivoliana*, Valenciennes 1833). *Aquac. Res.* **2018**, *49*, 1200–1209. [CrossRef]

46. Nasopoulou, C.; Zabetakis, I. Benefits of fish oil replacement by plant originated oils in compounded fish feeds. A review. *LWT-Food Sci. Technol.* **2012**, *47*, 217–224. [CrossRef]

47. Bowyer, J.N.; Qin, J.G.; Smullen, R.P.; Stone, D.A.J. Replacement of fish oil by poultry oil and canola oil in yellowtail kingfish (*Seriola lalandi*) at optimal and suboptimal temperatures. *Aquaculture* **2012**, *356–357*, 211–222. [CrossRef]

48. Bell, J.G.; Strachan, F.; Good, J.E.; Tocher, D.R. Effect of dietary echium oil on growth, fatty acid composition and metabolism, gill prostaglandin production and macrophage activity in Atlantic cod (*Gadus morhua* L.). *Aquac. Res.* **2006**, *37*, 606–617. [CrossRef]

49. Stoknes, I.S.; Økland, H.M.W.; Falch, E.; Synnes, M. Fatty acid and lipid class composition in eyes and brain from teleosts and elasmobranchs. *Comp. Biochem. Physiol.-B Biochem. Mol. Biol.* **2004**, *138*, 183–191. [CrossRef]

50. Rodríguez-Barreto, D.; Jerez, S.; Cejas, J.R.; Martin, M.V.; Acosta, N.G.; Bolaños, A.; Lorenzo, A. Comparative study of lipid and fatty acid composition in different tissues of wild and cultured female broodstock of greater amberjack (*Seriola dumerili*). *Aquaculture* **2012**, *360–361*, 1–9. [CrossRef]

51. Benedito-Palos, L.; Navarro, J.C.; Sitjà-Bobadilla, A.; Bell, G.J.; Kaushik, S.; Pérez-Sánchez, J. High levels of vegetable oils in plant protein-rich diets fed to gilthead sea bream (*Sparus aurata* L.): Growth performance, muscle fatty acid profiles and histological alterations of target tissues. *Br. J. Nutr.* **2008**, *100*, 992–1003. [CrossRef]

52. Piedecausa, M.A.; Mazón, M.J.; García García, B.; Hernández, M.D. Effects of total replacement of fish oil by vegetable oils in the diets of sharpsnout seabream (*Diplodus puntazzo*). *Aquaculture* **2007**, *263*, 211–219. [CrossRef]

53. Richard, N.; Mourente, G.; Kaushik, S.; Corraze, G. Replacement of a large portion of fish oil by vegetable oils does not affect lipogenesis, lipid transport and tissue lipid uptake in European seabass (*Dicentrarchus labrax* L.). *Aquaculture* **2006**, *261*, 1077–1087. [CrossRef]

54. Bouraoui, L.; Sánchez-Gurmaches, J.; Cruz-Garcia, L.; Gutiérrez, J.; Benedito-Palos, L.; Pérez-Sánchez, J.; Navarro, I. Effect of dietary fish meal and fish oil replacement on lipogenic and lipoprotein lipase activities and plasma insulin in gilthead sea bream (*Sparus aurata*). *Aquac. Nutr.* **2011**, *17*, 54–63. [CrossRef]

55. Regost, C.; Arzel, J.; Robin, J.; Rosenlund, G.; Kaushik, S.J. Total replacement of fish oil by soybean or linseed oil with a return to fish oil in turbot (*Psetta maxima*) 1. Growth performance, flesh fatty acid profile, and lipid metabolism. *Aquaculture* **2003**, *217*, 465–482. [CrossRef]

56. Bell, J.G.; McEvoy, J.; Tocher, D.R.; McGhee, F.; Campbell, P.J.; Sargent, J.R. Replacement of fish oil with rapeseed oil in diets of Atlantic Salmon (*Salmo salar*) affects tissue lipid compositions and hepatocyte fatty acid metabolism. *J. Nutr.* **2001**, *131*, 1535–1543. [CrossRef]

57. Bell, J.G.; Sargent, J.R. Arachidonic acid in aquaculture feeds: Current status and future opportunities. *Aquaculture* **2003**, *218*, 491–499. [CrossRef]

58. Torstensen, B.; Frøyland, L.; Lie, Ø. Replacing dietary fish oil with increasing levels of rapeseed oil and olive oil–effects on Atlantic salmon (*Salmo salar* L.) tissue and lipoprotein lipid composition and lipogenic enzyme activities. *Aquac. Nutr.* **2004**, *10*, 175–192. [CrossRef]

59. Saito, H. Lipid characteristics of two subtropical Seriola fishes, *Seriola dumerili* and *Seriola rivoliana*, with differences between cultured and wild varieties. *Food Chem.* **2012**, *135*, 1718–1729. [CrossRef]

60. Rodríguez-Barreto, D.; Jerez, S.; Cejas, J.R.; Martin, M.V.; Acosta, N.G.; Bolaños, A. Effect of different rearing conditions on body lipid composition of greater amberjack broodstock (*Seriola dumerili*). *Aquac. Res.* **2017**, *48*, 505–520. [CrossRef]

61. Rodríguez-Barreto, D.; Jerez, S.; Cejas, J.R.; Martin, M.V.; Acosta, N.G.; Bolaños, A.; Lorenzo, A. Ovary and egg fatty acid composition of greater amberjack broodstock (*Seriola dumerili*) fed different dietary fatty acids profiles. *Eur. J. Lipid Sci. Technol.* **2014**, *116*, 584–595. [CrossRef]

62. O'Neill, B.; Le Roux, A.; Hoffman, L.C. Comparative study of the nutritional composition of wild versus farmed yellowtail (*Seriola lalandi*). *Aquaculture* **2015**, *448*, 169–175. [CrossRef]

63. Rombenso, A.N.; Trushenski, J.T.; Drawbridge, M. Saturated lipids are more effective than others in juvenile California yellowtail feeds—Understanding and harnessing LC-PUFA sparing for fish oil replacement. *Aquaculture* **2018**, *493*, 192–203. [CrossRef]

64. Seno-O, A.; Takakuwa, F.; Hashiguchi, T.; Morioka, K.; Masumoto, T.; Fukada, H. Replacement of dietary fish oil with olive oil in young yellowtail *Seriola quinqueradiata*: Effects on growth, muscular fatty acid composition and prevention of dark muscle discoloration during refrigerated storage. *Fish. Sci.* **2008**, *74*, 1297–1306. [CrossRef]

65. Fukada, H.; Taniguchi, E.; Morioka, K.; Masumoto, T. Effects of replacing fish oil with canola oil on the growth performance, fatty acid composition and metabolic enzyme activity of juvenile yellowtail *Seriola quinqueradiata* (Temminck & Schlegel, 1845). *Aquac. Res.* **2017**, *48*, 5928–5939.

66. Bergman, A.M.; Trushenski, J.T.; Drawbridge, M. Replacing fish oil with hydrogenated soybean oils in feeds for yellowtail. *N. Am. J. Aquac.* **2018**, *80*, 141–152. [CrossRef]

67. Stuart, K.; Johnson, R.; Armbruster, L.; Drawbridge, M. Arachidonic acid in the diet of captive yellowtail and its effects on egg quality. *N. Am. J. Aquac.* **2018**, *80*, 97–106. [CrossRef]

68. Bell, J.G.; McGhee, F.; Campbell, P.J.; Sargent, J.R. Rapeseed oil as an alternative to marine fish oil in diets of post-smolt Atlantic salmon (*Salmo salar*): Changes in flesh fatty acid composition and effectiveness of subsequent fish oil "wash out". *Aquaculture* **2003**, *218*, 515–528. [CrossRef]

69. Stubhaug, I.; Lie, Ø.; Torstensen, B.E. Fatty acid productive value and β-oxidation capacity in Atlantic salmon (*Salmo salar* L.) fed on different lipid sources along the whole growth period. *Aquac. Nutr.* **2007**, *13*, 145–155. [CrossRef]

70. Ikemoto, A.; Nitta, A.; Furukawa, S.; Ohishi, M.; Nakamura, A.; Fujii, Y.; Okuyama, H. Dietary n-3 fatty acid deficiency decreases nerve growth factor content in rat hippocampus. *Neurosci. Lett.* **2000**, *285*, 99–102. [CrossRef]

71. Kitajka, K.; Puskas, L.G.; Zvara, A.; Hackler, L.; Barcelo-Coblijn, G.; Yeo, Y.K.; Farkas, T. The role of n-3 polyunsaturated fatty acids in brain: Modulation of rat brain gene expression by dietary n-3 fatty acids. *Proc. Natl. Acad. Sci. USA* **2002**, *99*, 2619–2624. [CrossRef]

72. Kreps, E.M.; Chebotarëva, M.A.; Akulin, V.N. Fatty acid composition of brain and body phospholipids of the anadromous salmon, *Oncorhynchus nerka*, from fresh-water and marine habitat. *Comp. Biochem. Physiol.* **1969**, *31*, 419–430. [CrossRef]

73. Caballero, M.J.; Obach, A.; Rosenlund, G.; Montero, D.; Gisvold, M.; Izquierdo, M.S. Impact of different dietary lipid sources on growth, lipid digestibility, tissue fatty acid composition and histology of rainbow trout, *Oncorhynchus mykiss*. *Aquaculture* **2002**, *214*, 253–271. [CrossRef]

74. Campos, I.; Matos, E.; Maia, M.R.G.; Marques, A.; Valente, L.M.P. Partial and total replacement of fish oil by poultry fat in diets for European seabass (*Dicentrarchus labrax*) juveniles: Effects on nutrient utilization, growth performance, tissue composition and lipid metabolism. *Aquaculture* **2019**, *502*, 107–120. [CrossRef]

75. Pagliarani, A.; Pirini, M.; Trigari, G.; Ventrella, V. Effect of diets containing different oils on brain fatty acid composition in sea bass (*Dicentrarchus labrax* L.). *Comp. Biochem. Physiol. Part B Biochem.* **1986**, *83*, 277–282. [CrossRef]

76. Skalli, A.; Robin, J.H.; Le Bayon, N.; Le Delliou, H.; Person-Le Ruyet, J. Impact of essential fatty acid deficiency and temperature on tissues' fatty acid composition of European sea bass (*Dicentrarchus labrax*). *Aquaculture* **2006**, *255*, 223–232. [CrossRef]

Article

Improvement of the Water Quality in Rainbow Trout Farming by Means of the Feeding Type and Management over 10 Years (2009–2019)

Elisa Fiordelmondo [1,*], **Gian Enrico Magi** [1], **Francesca Mariotti** [1], **Rigers Bakiu** [2] and **Alessandra Roncarati** [1]

[1] School of Bioscience and Veterinary Medicine, University of Camerino, Circonvallazione 93-95, 62034 Matelica, Italy; gianenrico.magi@unicam.it (G.E.M.); francesca.mariotti@unicam.it (F.M.); alessandra.roncarati@unicam.it (A.R.)

[2] Department of Aquaculture, University of Tirana, Place Mother Tereza 183, 695581 Tirana, Albania; bakiurigers@gmail.com

* Correspondence: elisa.fiordelmondo@unicam.it; Tel.: +39-0737-403416

Received: 11 June 2020; Accepted: 26 August 2020; Published: 1 September 2020

Simple Summary: The European Union Water Framework Directive has set the objective to develop a good status for water bodies and showed the importance of promoting the sustainable use of water resources in inland anthropic activities, such as freshwater aquaculture, performed in strict accordance with natural waters. In aquafeed, two elements have mostly contributed to improving aspects of change in fish feeding among rainbow trout—the use of the feed extrusion technique and the adoption of restrictive environmental rules. In this context, the main physico-chemical parameters of water quality were investigated in 2019 and the values obtained were then compared to the parameters obtained 10 years before (2009) in order to show if there were differences in the water quality of an outlet farm. Considering this, the present study aimed to evaluate the suitability of changes adopted in rearing and feeding techniques to improve growth performance, sustainability and the water quality environment. This study was made possible by considering data sampled on a yearly basis in the outlet and compared with the inlet water.

Abstract: Background: In Europe, rainbow trout is one of the main fresh water fish farmed in a constantly developing environment that requires innovative studies to improve farm management, fish welfare and environmental sustainability. The aim of this paper is to investigate the trend of water quality parameters over 10 years, after a feeding strategy change from pellet to extruded feed. Methods: The study was conducted on a farm in central Italy, based on parallel raceways. The cycle started from young rainbow trout (90 ± 2 g) that were grown until they reached market size. A water sample of 500 cm^3 was collected monthly from 2009 to 2019 from the lagoon basin in order to investigate the trends of the total suspended solids (TSS), biochemical oxygen demand (BOD$_5$), chemical oxygen demand (COD), nitrites (NO$_2$-N), nitrates (NO$_3$-N), total ammonia nitrogen (TAN), total phosphorus (TP) and pH. Results: All of the studied parameters (TSS, BOD$_5$, COD, NO$_2$-N, NO$_3$-N, TAN and TP) showed a significant improvement from 2009 to 2019. The pH parameter did not display notable variation during the studied period. The feed conversion ratio (FCR) was also investigated and exhibited a significant improvement from 1.4 to 1.1. Conclusion: Based on the decrease of all the investigated parameters, it is possible to say that extrusion is currently an excellent processing feed technique in aquaculture with a good level of respect for the environment.

Keywords: rainbow trout farming; water quality; feed extrusion technology; total suspended solids

---

1. Introduction

In the last two decades, new strategies for improving feeding techniques for the main fish species have been developing in order to reach global sustainability, in particular in an environmentally responsible manner. At the start of the new millennium, the European Union adopted the Water Framework Directive (2000/60/EC), which introduced a transnational vision of the management of the freshwater environment able to protect, manage and improve the quality of water resources across the EU. In 2018, the European Environment Agency [1] showed the efforts carried out by countries to monitor and assess the general health status of water bodies and stressed the importance of also promoting the sustainable use of available water resources in inland anthropic activities of the primary sector, such as agriculture and aquaculture, performed in strict accordance with natural waters. Many papers [2–5] have focused on the environmental impact of trout farms when uneaten food, fish catabolites (metabolites, ammonia gill excretion and carbon dioxide) and chemical treatments are not controlled and are discharged into the natural receiving waters. In the case of aquaculture specialized in the growing phase of rainbow trout (*Oncorhynchus mykiss*), the rearing technique is based on the use of flow-through systems which consist of raceways or concrete tanks with water constantly flowing down basins and the removal of waste at the outlet by gravity and the water current [6].

In aquafeed, two elements have mostly contributed to improving aspects of change concerning fish feeding, in particular, in rainbow trout: The use of the feed extrusion technique and the adoption of restrictive environmental rules. First, regarding rainbow trout feeding, the best technique is based on the fact that the feed distribution must be carried out until reaching a level close to satiety [7]. In fact, the feed conversion index and the protein efficiency index improve when rainbow trout are fed with a rationing level equal to 70% of the "ad libitum technique" [7]. This technique drastically reduces the amount of food not eaten by trout.

Feed quality also directly affects water quality, because a proportion of the feed intake by fish is returned to the environment as metabolites or soluble by-products of metabolism [2]. In this context, two main levels of measures (legislative and productive) have been adopted. Restrictive environmental rules of EU countries have imposed settling areas to eliminate solid waste, whilst new feeding technologies have also been considered. The evaluation of the impact of rainbow trout farming on the receiving water quality needs to take into consideration the water quality monitoring over a long period, based on a decade [8] or two years, in different flow-through farms [4]. Concerning feeding technologies, nowadays, modern systems use extruded feed instead of pellet feed. In particular, the metabolite status of fish depends on the degree of gelatinization of the feeding starch. In the modern aquatic system, the removal of metabolites can be decreased using formulations with ingredients able to help bind metabolite matter, allowing these particles to be more easily and thoroughly removed from the water [9–11].

Furthermore, small feed portions not consumed are decreased by a smart diet formulation and processing; nowadays, extruded feed represents the best solution, instead of the pellet one, which was previously the common feeding strategy [9–11]. The digestibility of the extruded feed increases up to 96% relative to the raw wheat starch, while, in the case of pelleted feed, this raw material is digestible to approximately 54% [12]. Moreover, the catabolic residues emitted by rainbow trout, fed with extruded feed with gelatinized starch, are easy to remove from the water, thanks to their sedimentation [11]. Therefore, extruded feed also enables high environmental sustainability due to the increase of digestibility and stability in water, with a consequent reduction of suspended solids and nutrients. With regard to the rearing environment and its traits, it is appropriate to consider how it influences the conditions of well-being. Correct water exchange is essential: Water dilutes and removes the catabolites of fish, as well as feed residues, thus reducing the exposure of the farmed subjects to dangerous nitrogen compounds which are also able to negatively influence the state of well-being.

To check that the water quality is suitable for the farming of rainbow trout, a multitude of physical and chemical parameters must be considered [13]. In the case of suspended solids, their concentration is particularly important, because they directly influence the water turbidity, which can prevent the vision of the fish and finally compromise their life cycle. According to Boyd [3] and Becke [14], water turbidity values higher than 400 mg/L can cause thickening and deformation of the gill filaments, with trout consequently suffering. At the same time, dissolved organic substances, including the sedimentable (undigested portion of the ration and food residues) and non-sedimentable (product of endogenous metabolism) solids in suspension, must be taken into consideration.

For these reasons, checking the water quality is of primary importance for the welfare of trout. Inappropriate rearing conditions, such as inadequate space, excessive densities and poor feeding, can have strong negative repercussions for farmed fish species. Damaged, eroded or hemorrhagic fins are not only correlated with pathological events but also with inadequate environmental factors, connected to stress-related aspects such as a fish stocking density that is too high with a non-optimal water quality [15].

Considering the use of diets administered to salmonids, many papers have focused on the emission of catabolites in the external environment. According to European Environment Agency [16], 15–25% of the total food energy is lost in ammonia and urea through the gills and is released into the environment.

Most of the papers on this topic have focused their attention on relationships between feeding strategies, animal welfare and environmental sustainability and in the current literature, the correlation between water quality and productive performances of farmed rainbow trout is still a key point of discussion.

Based on these considerations, a study focusing on the trend of the most important water quality parameters in rainbow trout farming of central Apennine in terms of long-term activity (2009–2019) was carried out. Before the decade focused on in this study, the first historical monitoring of water quality took place in a trial carried out in 2004, when the owners started to re-think the farming technique adopted until then, based on high stocking densities (40 kg/m^3), in order to produce a higher fish welfare status [17]. Considering this, the present study aimed to evaluate the suitability of changes adopted in rearing and feeding techniques to improve growth performance, sustainability and the water quality environment. The raceway water quality was monitored in terms of the Total Suspended Solids (TSS), Biochemical Oxygen Demand (BOD$_5$), Chemical Oxygen Demand (COD), Total Ammonia Nitrogen (TAN), Nitrites (NO$_2$-N), Nitrates (NO$_3$-N), pH and Total Phosphorus (TP). These parameters were investigated over a decade, starting in 2009 and were compared to the respective annual values of the 10 years after, until the year 2019, in order to show differences in the water quality and feed conversion rate.

2. Materials and Methods

2.1. Trout Farm Description and Water Quality Sampling

A study on the water quality trend was performed on a rainbow trout farm, located in the Apennine area of central Italy, based on raceways in parallel (120 m^3 each).

During the time of the study, the genetic line of rainbow trout did not change and it was directly controlled by the farm's owners, who had a broodstock farm in a different area from which fishes for fattening came and were selected.

On the rainbow trout farm in which the study was conducted, the feeding technique was the same during the entire experimental period (2009–2019) and the feeding rate was the same for fish at the same life cycle stage. The feed was distributed with a semi-moving wagon up to the level close to satiety and at the same time during the day, twice a day.

The water supply system used on the farm allowed at least one complete daily water change. The inlet water came to the farm system with a constant velocity of 0.25 m/s and flowed through four parallel raceways.

A water quality sample was obtained monthly during the last decade (2009–2019) in correspondence with the lagoon basin (outlet water) below the raceways and receiving downstream.

All samples were collected in the early morning before feeding using a polypropylene bottle with a screw cap [18].

In order to monitor the water quality, different physicochemical parameters were investigated. The dissolved oxygen, temperature and pH were measured using portable electronic devices (YSI mod. 55 and 60). At the same time, five samples of 500 cm^3 water were collected for laboratory-based determination of the following parameters: total suspended solids (TSS), biochemical oxygen demand (BOD$_5$), chemical oxygen demand (COD), total ammonia nitrogen (TAN), nitrites (NO$_2$-N), nitrates (NO$_3$-N) and total phosphorus (TP). Nitrogen (N) compounds and TP were determined using a spectrophotometer (Hach mod-2005, Hach Company, Loveland, USA), following the American Water Works Association and Water Pollution Control Federation of American Public Health Association (APHA) standard methods [19]. TSS was recorded following official methods [20]. BOD$_5$ and COD were determined according to IRSA-CNR [20] and ISPRA [18] methods, respectively.

In order to quantify the results of the waste water of the rainbow trout farm, the quality of the inlet water was also assessed by analyzing the values of TSS, COD, BOD$_5$, NO$_2$-N, NO$_3$-N and TAN, using the same laboratory methodology as described for the outlet water. Each sample of inlet water was collected using a sampler that conducted an average sampling process over three hours. During each year of the study, inlet water was collected seven times, in the months of January, April, May, June, August, October and December. The inlet water showed constant values for every parameter from 2009 to 2019, with the following averages: TSS, 5 mg/L; COD, 5 mg/L; BOD$_5$, 5 mg/L; NO$_2$-N, 0.09 mg/L; NO$_3$-N, 1.4 mg/L; TAN, 1.5 mg/L. Since these values were always suitable for farming rainbow trout, additional investigations and more frequent water samples were not necessary, allowing us to focus our attention on the analysis of the outlet water.

2.2. Experimental Design and Rearing of the Rainbow Trout

In order to summarize the methodology that was followed, Figure 1 shows the experimental design step by step. With reference to the first year (2009) and the final year (2019) of the considered decade, fish growth and water quality assessments were also evaluated by considering the balance of nitrogen and phosphorus released in waters. In a fish plant, the productive cycle starts from young rainbow trout, with an average body weight of 90 ± 2 g, which are grown at a stocking density of 20 kg/m^3 until reaching the market size (350 g). In particular, at the beginning of the life cycle, the fish density is low, about 8 kg/m^3 of water, in order to limit the stress of young trout and prevent gill diseases or bacterial infections. This density then increases with the increasing size of the trout until the end of the cycle, when it reaches 20 kg/m^3 of water.

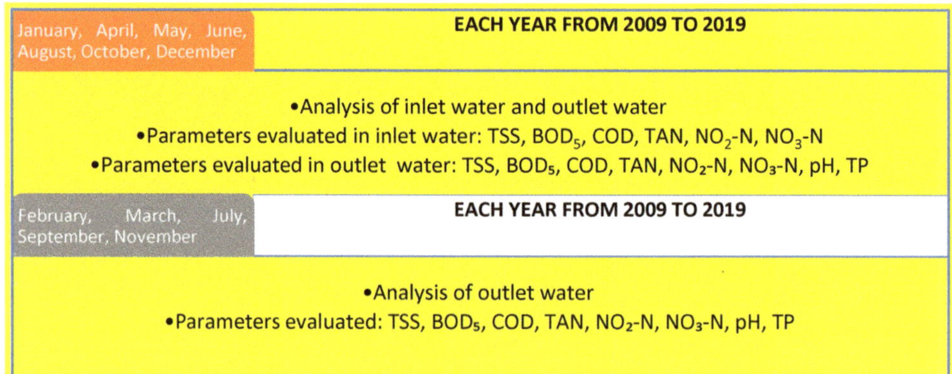

Figure 1. *Cont.*

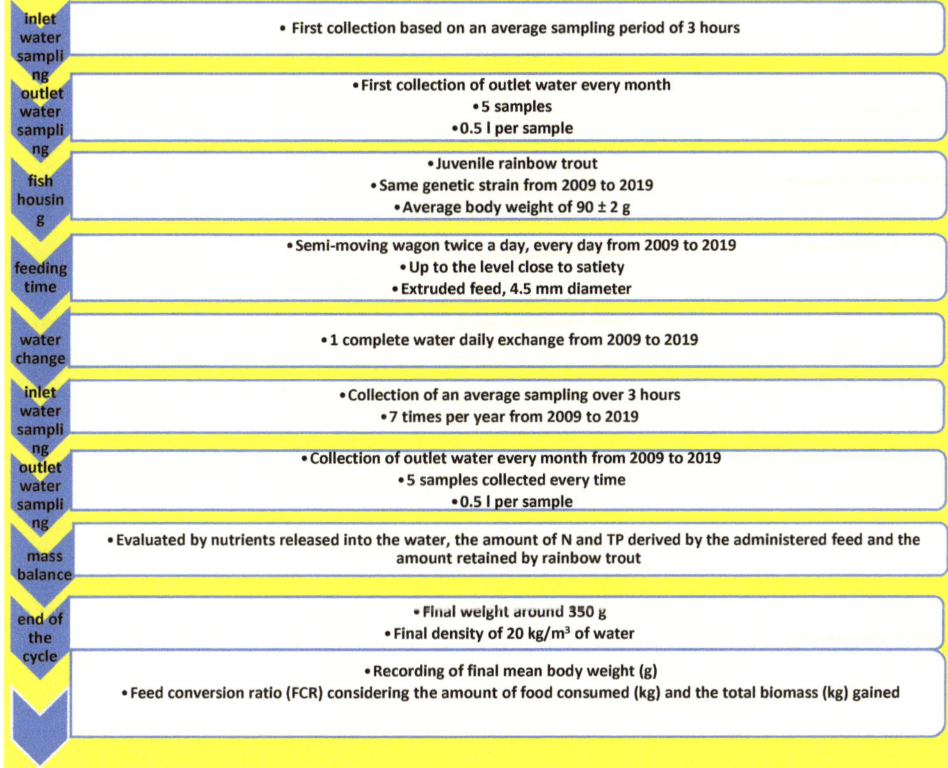

Figure 1. Diagram showing the experimental design adopted during the 10 years of the study.

Fish received extruded feed (closed formula) that was 4.5 mm in diameter and was manufactured by the same fish farmer who had vegetable feedstuffs available at his own land farm. The proximate composition of the two feeding types (pellet, extruded) employed to feed rainbow trout during the decade is reported in Table 1.

Table 1. Proximate composition of the feeds employed for the rainbow trout growing over the decade of water quality monitoring (2009–2019).

Feed	Pellet	Extruded
Chemical composition (%)		
Moisture	6.8	5.5
Crude protein	45.7	44.8
Crude lipid	16.0	21.0
Ash	6.7	8.4
Gross energy (MJ kg)	21.16	18.38

The final mean body weight (g) was recorded and the feed conversion ratio (FCR) was calculated from the amount of food consumed (kg) and the total biomass (kg) gained:

FCR = kg food consumed/kg final biomass − (kg initial biomass + kg sampled fish) + mortalities.

In order to evaluate the mass balance of nutrients released into the water, the amount of N and TP derived from the administered feed and the amount retained by rainbow trout were considered, as shown in Table 2, by comparing the budget of these compounds, expressed as the seasonal mean of the first year (2009) and the last year (2019) of study and applying the coefficients indicated by Bureau et al. [21] and applied by other authors [4].

Table 2. Total ammonia nitrogen (TAN) and phosphorus (TP) budget in the lagoon basin outlet water (mean ± standard deviation).

Year	TAN (mg/L)	TP (mg/L)
2009	0.55 ± 0.03	0.011 ± 0.001
2019	0.46 ± 0.02	0.009 ± 0.002

2.3. Statistical Analysis

All collected data on the outlet water quality were analyzed to determine whether there were significant differences over the 10 considered years. For this aim, the months of the year were divided to define the four seasons: winter included the months of December, January, February and March; spring was represented by April, May and June; summer included July, August and September; and autumn included October and November.

After all seasonal data were collected, the mean of each parameter was calculated season by season for every year and data were finally organized as graphics.

The seasonal mean of the investigated parameters (TSS, BOD_5, COD, NO_2-N, NO_3-N, TAN, TP and pH), recorded per year, was subjected to one-way analysis of variance (ANOVA) using the General Model Procedure of SPSS 25 (IBM Corp., New York, NY, USA) [22], in order to assess if data means were statistically different within the season of the different years. Significance was considered if $p < 0.05$ and the means were compared using the Student-Newman-Keuls (SNK) test.

3. Results

The results of the water quality parameters measured in the outlet lagoon basin in the 10 year-study are reported in Figures 2–9, comparing the means of the same season of the 10 years.

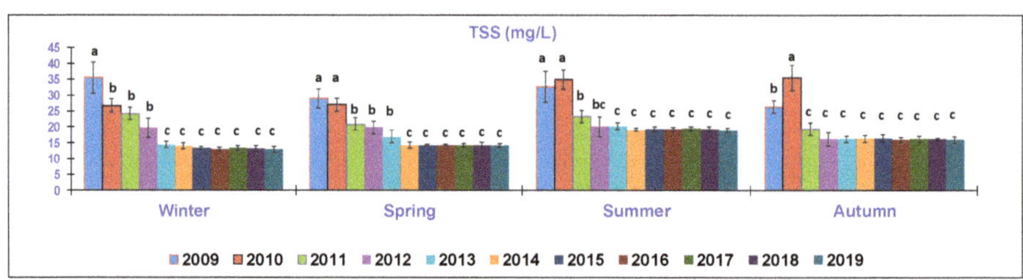

Figure 2. Trend of outlet water Total Suspended Solids (TSS) (mean ± standard deviation) seasonally determined during the 10 year-study. Different letters (a, b, c) per season show significant differences ($p < 0.05$) among the 10 years of sampling.

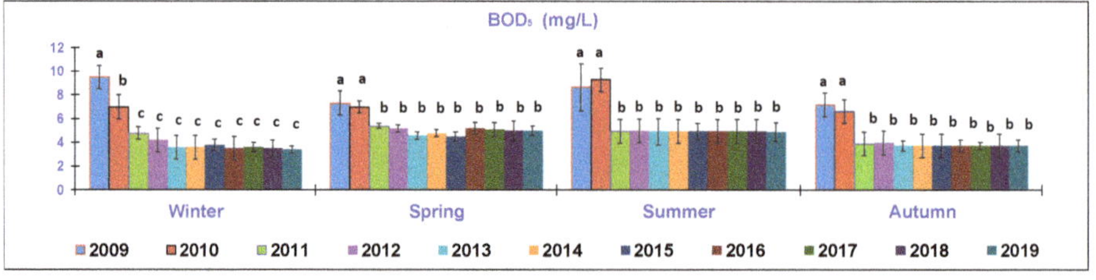

Figure 3. Trend of outlet water Biochemical Oxygen Demand (BOD_5) (mg/L) (mean ± standard deviation) seasonally determined during the 10 year-study. Different letters (a, b, c) per season show significant differences ($p < 0.05$) among the 10 years of sampling.

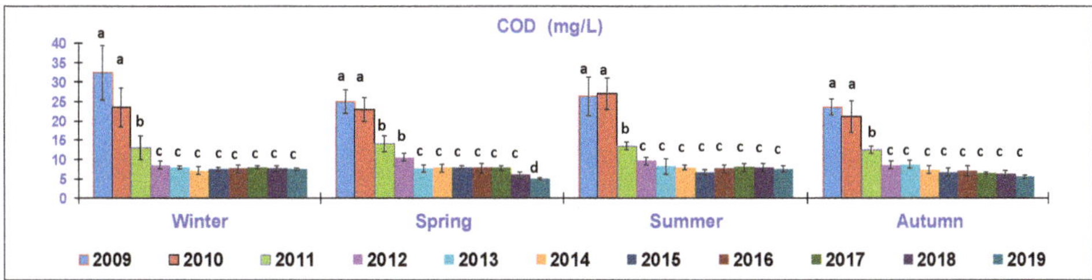

Figure 4. Trend of outlet water Chemical Oxygen Demand (COD) (mg/L) (mean ± standard deviation) seasonally determined during the 10 year-study. Different letters (a, b, c, d) per season show significant differences ($p < 0.05$) among the 10 years of sampling.

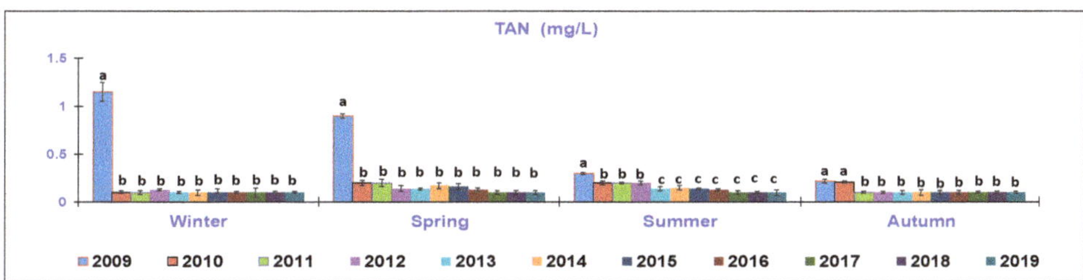

Figure 5. Trend of outlet water Total Ammonia Nitrogen (TAN) (mg/L) (mean ± standard deviation) seasonally determined during the 10 year-study. Different letters (a, b, c) per season show significant differences ($p < 0.05$) among the 10 years of sampling.

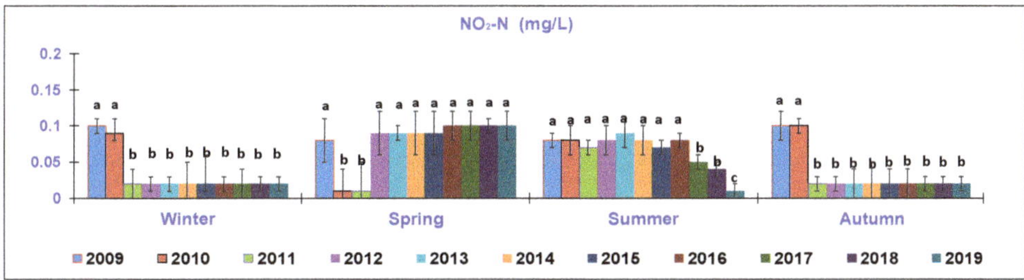

Figure 6. Trend of outlet water Nitrites (NO$_2$-N) (mg/L) (mean ± standard deviation) seasonally determined during the 10 year-study. Different letters (a, b, c) per season show significant differences ($p < 0.05$) among the 10 years of sampling.

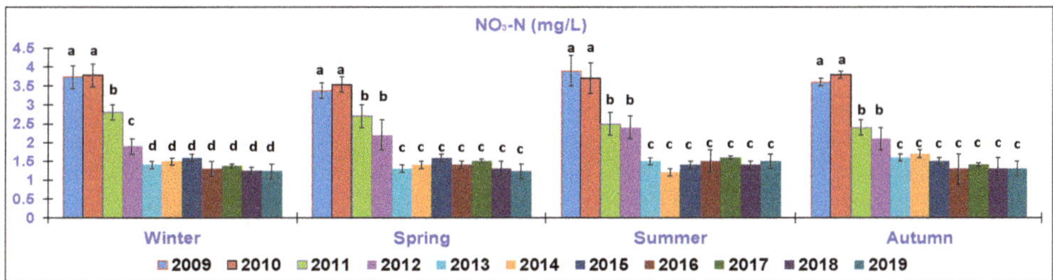

Figure 7. Trend of outlet water Nitrates (NO$_3$-N) (mg/L) (mean ± standard deviation) seasonally determined during the 10 year-study. Different letters (a, b, c, d) per season show significant differences ($p < 0.05$) among the 10 years of sampling.

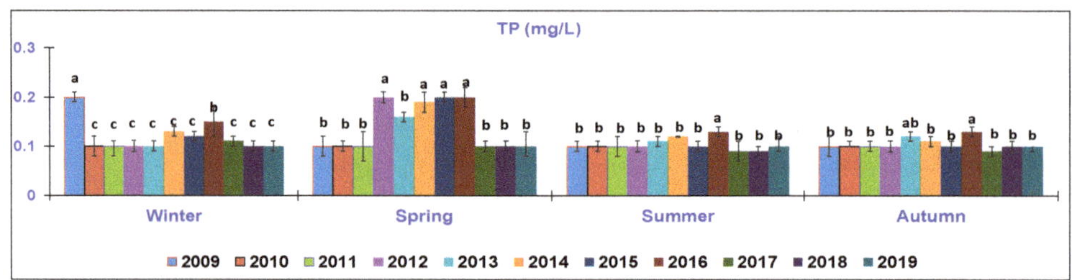

Figure 8. Trend of outlet water Total Phosphorus (TP) (mg/L) (mean ± standard deviation) seasonally determined during the 10 year-study. Different letters (a, b, c) per season show significant differences ($p < 0.05$) among the 10 years of sampling.

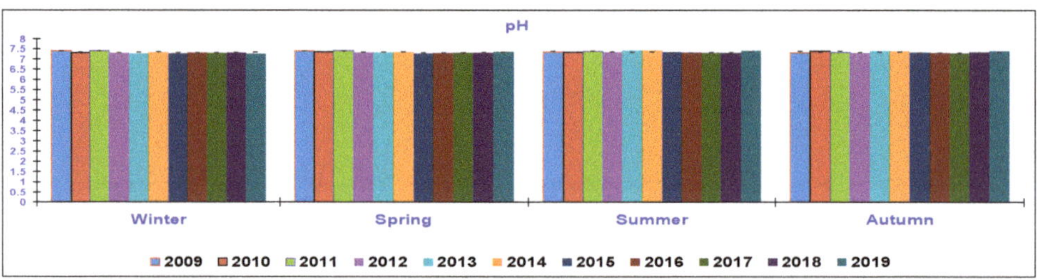

Figure 9. Trend of outlet water pH (mean ± standard deviation) seasonally determined during the 10 year-study.

The TSS (Figure 2) content showed a significant decrease when passing from the first seasons of sampling to the last years. In winter, TSS exhibited a significantly decreased mean value from 35.5 ± 5 mg/L in 2009 to 13 ± 3 mg/L in 2019, with an intermediate reduction notably shown in 2010–2012 (26.75–19.8 mg/L); in 2013 (14.5 ± 2 mg/L), TSS reached the lowest values until 2019. In spring, the same trend was observed, when the means decreased from 29 ± 3 and 27.3 ± 2.6 mg/L (2009–2010, respectively) to around 14.3 mg/L, which was maintained from 2014 to 2019. In the summer season, the recorded means were 37.20 ± 5 mg/L in 2009 and 35.4 ± 4 mg/L in 2010, without significant differences, whereas a marked reduction was observed from the year after (2011, 23.4 ± 3 mg/L), gradually diminishing from 2012 and remaining at around 16 mg/L until 2019. In Autumn, the highest TSS content was recorded in 2010 (35.5 ± 3.9 mg/L), being significantly different from the analyses of the first year of monitoring (2009, 26.5 ± 2 mg/L) but from 2011, the mean value was significantly lower and was maintained at around 16.00 ± mg/L until the end of the study. In particular, considering the seasonal average of TSS detected in Autumn, there was a reduction of almost 55% of this parameter.

Considering the trend of BOD$_5$ (Figure 3), there was a marked improvement in 2019 compared to the 10 years before. In winter, BOD$_5$ recorded a significant decrease, passing from the mean value of 9.5 ± 0.9 mg/L in 2009 to that of 7 ± 0.8 mg/L in 2010 and further decreased to around 3.7 ± 2 mg/L in the following years, until 2019. In spring, the highest levels were determined in 2009 (7.33 ± 1 mg/L) and 2010 (7.06 ± 0.7 mg/L); after this, the means significantly stayed at around 5.00 ± 0.7 mg/L until 2019. In summer, the highest levels of BOD$_5$ were observed in 2009–2010 (8.7–9.3 mg/L), which were significantly different from all of the following means, being around 4.9 ± 1 mg/L until the end of the monitoring period. In autumn, a similar trend was also observed, with a notable difference between the first two years of sampling (7.18–6.72 mg/L in 2009–2010) and the other following years, being around 3.7 ± 0.7 mg/L until the end of the study.

In terms of COD determination (Figure 4), all four seasons showed a similar trend, resulting in a sharp reduction through the years. In winter, the highest levels were recorded in the first two years (2009, 32.5 ± 5.1 mg/L; 2010, 23.5 ± 4.4 mg/L), significantly decreasing in 2011 (13 ± 3.1 mg/L) and reaching the lowest mean in 2012, when the average of around 7.5 mg/L was maintained until 2019.

In spring, the highest values occurred over the first two years (25 ± 1.3 mg/L) and then decreased in 2012–2013 (25–23.1 mg/L), until reaching the significantly lowest level in the last year (5 ± 0.2 mg/L). In summer, the first two years exhibited the highest levels (26.3–27 ± 1 mg/L). The mean value was decreased in 2011 (13.5 ± 0.3 mg/L) but was significantly higher than that in the following years, being around 7.5 ± 0.6 mg/L until the end of the study. In autumn, the mean of this parameter was 23.52 ± 0.5 mg/L in 2009 and 21.14 ± 0.7 in 2010; then, it decreased to 12.5 ± 0.3 mg/L in 2011 and reached 6.8–5.51 mg/L in the last years, without notable differences.

Regarding the TAN parameter (Figure 5), a clear reduction was observed in the range of the years taken into consideration. In the winter season, the data showed a significant decrease from the first year of study, from 1.15 ± 0.3 mg/L in 2009 to 0.1 ± 0.02 mg/L in 2019. In spring, TAN significantly decreased from 0.90 ± 0.2 mg/L in 2009 to 0.10 ± 0.02 mg/L in 2019. During summer time, a significant reduction was observed, that is, the value was 0.30 ± 0.01 mg/L in 2009 and significantly decreased from 2010–2012 (0.21–0.2 mg/L) and then further decreased to 0.1 ± 0.02 mg/L in 2019. In autumn, the TAN value decreased from 0.21–0.22 mg/L in 2009 and 2010 in the following years (2011–2019), when a mean value of 0.1 ± 0.02 mg/L was constantly recorded.

In terms of the NO_2-N parameter (Figure 6), in winter, the mean values (0.1–0.09 mg/L), detected in 2009–2010, significantly decreased to 0.02 mg/L throughout the following years. In spring, an opposite trend was noted, with the lowest concentrations (0.01 mg/L) occurring in 2010–2011 in comparison to all other years (0.1 mg/L). In summer, a similar trend was observed, although the last year of monitoring showed the lowest mean (0.01 ± 0.002 mg/L) recorded among the years. On the contrary, in autumn, the TAN level detected in the two first years (0.1 mg/L) significantly decreased to 0.02 ± 0.001 mg/L in 2019.

In terms of the NO_3-N parameter (Figure 7), a similar trend was observed in all the seasons, with a notable reduction emerging in 2013 and being maintained throughout the last years. In winter, the content dropped from 3.73–3.78 mg/L in 2009–2010 to only 1.22 ± 0.3 mg/L in 2019. In spring, the value significantly changed from 3.37 ± 0.5 mg/L in 2009 to 1.23 ± 0.3 mg/L in 2019. In summer, the value ranged from 3.90 ± 0.5 mg/L in 2009 to 1.50 ± 0.3 mg/L in 2019. In autumn, the concentration ranged from 3.60 ± 0.4 mg/L in 2009 to 1.30 ± 0.2 mg/L in 2019.

Considering the TP parameter (Figure 8), in winter, the values significantly decreased from 0.2 ± 0.01 mg/L in 2009 to 0.1 ± 0.002 mg/L in 2019. In spring, the highest values (0.19–0.2 mg/L) were recorded in the central part of the study (2014–2016); then, a reduction was observed, that is, 0.1 mg/L. In summer and winter, although there was a significantly higher level in 2016 (0.13 ± 00.2 mg/L), TP remained at a low value (0.1 mg/L) until 2019.

The pH parameter (Figure 9) did not show significant variation, remaining approximate neutral during the entire studied period and the obtained values were always in accordance with the range considered appropriate for rainbow trout [8,17].

At the end of the study, the FCR was 1.1 and significantly improved in comparison with the FCR obtained in 2009 (1.4). The survival rate was maintained at around 97% from 2012 onwards compared to the survival rate of the years before (94%).

In Table 2, the nutrient budget of outlet water is reported. In 2009, TAN was approximately 0.55 mg/L, whereas in 2019, the budget had an average value of 0.46 mg/L. The same trend was observed for the TP budget, which ranged from 0.011 mg/L (2009) to 0.009 mg/L over the 10 years.

4. Discussion

The aim of the current study was to investigate the impact of a new type of feed on water quality. Based on a comparison of the values of the analyzed parameters from 2009 to 2019, a notable decrease was observed in the TSS, BOD_5, COD, TAN, NO_2-N, NO_3-N, pH and TP.

As other studies have shown [8,23,24], BOD_5 is a very important parameter for assessing the water quality because it indicates the consumption of oxygen in the processes of indirect oxidation by the metabolic systems of aerobic microorganisms present in the water. Aside from the TSS level [24],

BOD_5 determination is essential for detecting the oxygen demand of the aerobic microbial flora required for the decomposition of organic substances present at a certain temperature in a defined time range. Therefore, it is an indicator that increases as the amount of organic substance to be mineralized increases.

The low values of BOD_5 detected during 2019, together with the decreased COD levels, indicate that aerobic microorganisms prevailed over anaerobic ones and therefore, the self-purifying capacity of the watercourse can be considered good. The reduction of the COD parameter that exceeded 76.5% is proof that the technical and management interventions implemented at the breeding rainbow trout farm effectively resulted in an improvement in the environmental protection. This result is in agreement with data presented by Galezan [24], while Tahar [8] reported lower BOD_5 values.

Based on analysis of the main water quality parameters of the trout farm over the last decade, a significant improvement in the rearing environment was shown. The significant reduction registered can be attributed to the different type of feed adopted, which changed from pelleted to extruded feed. Together with adequate quantities of oxygen dissolved in the water and through the nitrification process, a decreased amount of TAN can be converted into nitrites and nitrates. Nitrates are less toxic to fish over a long time period throughout the nitrification process [24]. A trend of the reduction of undesirable physicochemical traits was clearly observed when the new feeding type was introduced in 2011. Based on a comparison of the pellet feed with the extrusion technology, the digestible level of the diet was shown to have improved. This extruded manufacture drastically reduces the amount of food not eaten by trout and increases the feed efficiency [25].

The property of the extruded feed, which floats more than the pellet feed, increased its availability to the trout for a longer time in the water column before falling to the bottom of the raceways [9,26]. Fish can quickly intercept food without its dispersion and waste production, which occurs more frequently when using pellet feed because it sinks and degrades more easily and quickly in the water column.

In particular, concerning the two different feed techniques, that is, the pellet feed and the extruded feed, the processing style changed from the first to the second in 2011 but the raw materials remained the same. The fish meal and fish oil feedstuff were the same and both were of a high quality. The owner of the farm where the study was conducted monitored the feed quality through periodic laboratory analysis to ensure that the composition of the feed remained unchanged.

During the decade of the study, no differences in feed quality were shown and for the most important nutrients, no raw material changes were reported. For this reason, in the current study, the specific chemical analysis of the feed was not taken into consideration in order to stress the topic of water quality.

NO_2-N represents an important stage in the oxidation of organic substances containing nitrogen. It is therefore a transitory form of nitrogen which is transformed into nitrates due to the bacterial activity in the presence of optimal quantities of dissolved oxygen in the water tanks.

Nitrates represent the final indicator of the degradation protein processes; the main sources are correlated with anthropic activity and in surface waters, the trend is usually seasonal.

Taking into consideration the TSS parameters detected in 2009 and comparing them with those of 2019, a significant reduction was clear, especially the trend observed during winter and Autumn.

According to a number of studies that have focused their attention on the relationship between feeding management and water quality [2,3,11,23,24], the significant improvement in water quality at this farm was due to the adoption of the modern type of feed based on the extrusion technique. In fact, compared to the pellet food, the new extruded feeding technique showed a greater stability in water and it allowed the food to be available to trout for a longer period of time. There are other positive elements to consider: The extruded feed had a higher fat absorption capacity and it was possible to know its specific weight. Therefore, it was easy to control the fish feeding and, as a consequence, the composition of the meat obtained from this fish. Our study supported the idea that extrusion is nowadays the best processing feed technique in aquaculture, as previously shown by other studies.

Welker [26] compared the extrusion technique with the pellet feed by analyzing the water stability, fecal durability and digestibility and found the best results with the use of extruded feed. Similar results were reported by Tyapkova [9].

Moreover, in aquaculture, the extrusion technique positively affected the water quality: Food waste due to dust, breaks and "leaking/leaching" was decreased, improving the availability of nutrients for fish and minimizing the environmental impact. Another element of the extruded feed is that it is characterized by a low sedimentation rate. To ensure feed suitable for trout, a low sedimentation rate is one of the main physical properties to consider because it means the extruded feed is available to the fish for a long time; consequently, the fish can rapidly intercept the food without its dispersion and waste production. On the other hand, the use of pellet food means that feed is crumbled, which involves the fragmentation of food cylinders. As a consequence, this could lead to a loss of food (even if decreased). The feed was pulverized and this waste could remain at the bottom of the raceway water, becoming a possible source of water degradation, as well as microbial contamination.

Based on data concerning the monitoring of the rearing waters and considering the loads of N and TP of extruded feed, which took place in the last 10 years, it is possible to say that the improvement of the feed, administered to rainbow trout, is proof of the excellent quality of the ingredients used in fish diets [27,28].

Our data are in accordance with previous studies [3–5,8] that estimated ammonia and phosphorus emissions in water tanks from dietary analysis. In particular, we found lower TAN values than those reported by Moraes [5] and Aubin [4], which is a sign of the high feed quality used on the farm involved in our study. Concerning the p values, our results are in line with those reported by Moraes [5] and are lower than the values reported by Aubin [4] and Dalsgaard [23].

In this study, the amount of TAN and TP present in the tanks that in the past had hosted trout that received pelleted feed was compared to current tanks that host trout fed with extruded feed. The significantly decreased TAN and TP loads are proof of an improvement in diet quality, with a consequent benefit for the environment and the health of the trout.

This study provides insights into the connection between fish feed and water quality, that is, the fish habitat, which has to be suitable for producing healthy fish and as waste water, which returns to natural water bodies, taking into consideration specific parameters to investigate the water quality. In particular, in our study, an improvement trend was mostly demonstrated by the TSS, confirming that feed manufactured by the extrusion technique was of high quality. In particular, a notable positive change in the general trend was demonstrated after the new extruded food processing style was introduced, which occurred in 2011. In fact, all parameters showed an important decrease after the adoption of the extruded feed. More specifically, in the last years of the analyzed decade, a more favorable feed conversion rate was shown with respect to when pellet feed was administrated, resulting in better growing performances exhibited by trout whilst saving approximately 40% of the feed. In fact, the feed conversion index and the protein efficiency index improve when rainbow trout are fed with a rationing level equal to 70% of the "ad libitum technique" [7].

Another important change, which could have contributed to improving the efficiency feed conversion, could be justified by the fact that, on the farm, only trout larger than 90 g are reared, which do not require meal or crumbled feed. The fragmentation of crumbled diets leads to a more pulverized feed. In this case, the uneaten extruded feed, over 4 mm in size, is easier to remove than the pelleted diet.

Regarding the feedstuffs of the extruded feed, the inclusion of vegetable sources (soybean, wheat and pea) from the owners' farm, located close to the fish plant, can be reported as an example of sustainability with positive effects on the production cost. On the fish farm, the fish stocking density was decreased (20 kg/m^3) with respect to the first years of the decade, when rainbow trout were stocked at double the biomass. The choice to limit the culture density is the basis of the application of the "multisite" rearing technique, which aims to increase biosecurity, preventing diseases due to a vertical transmission of pathogens [29]. The decision of this plant to rear only rainbow trout starting from pre-fattened fish showed more advantages in terms of the survival rate, which was more satisfactory in the last years.

Concerning, more specifically, the fish farming density, before the decade focused on in this study and in 2004, the owners decreased the fish density in their raceways to improve the welfare of the rainbow trout. From 2009 to 2019, the fish density remained the same due to the effective improvement in animal welfare.

As previously shown by Welker [11], the correct management of tanks is essential to ensure good water conditions for trout and the appropriate use of water oxygenation systems contributes to maintaining optimal living conditions for these fish. This aspect could be connected to the improvement of the outlet basin receiving water from the raceways before being introduced into the stream.

Similar to what was observed by Becke [14], the pH was always close to neutrality and it was kept as stable as possible. This helped to minimize environmental stress and allowed nitrifying bacteria to effectively remove the nitrogen that accumulated in the sediments. Conversely, in other studies, the pH values were closer to 6 [5] or between 6.5 and 7.5 [8], which means that the effluent water was more acidic than the water considered here.

On the farm, the workers carried out normal system control every day, in order to ensure an oxygen level in the water that never dropped below 85% of the saturation level and its concentration was never less than 5 mg/L. The water supply system used on the farm allowed at least one complete daily water change, which directly affected the water chemical–physical parameters. Furthermore, the receiving water body has benefitted from these modified techniques, in particular the outlet basin, which was populated by floating macrophyte duckweed (*Lemna* spp.) that are known to be suitable for wastewater treatment and for food for pigs and poultry [30–32]. The duckweed filters nutrients, with a reduction of the eutrophic load of the water; it is an excellent purifying plant and absorbs the nitrogen in the water by eliminating nitrates [30,31].

Another aspect to consider is that, by comparing the parameters of the inlet water with those of the outlet water, which are the focus of the current study and due to the fact that the inlet water showed the same values over the 10 years of the study, it was possible to affirm that there was an improvement in the farming water quality. In fact, as shown before, the values of TSS, BOD_5, COD, TAN, NO_2-N and NO_3-N were lower in 2019 than 10 years before.

Many papers have focused on the environmental impact of trout farms on the natural receiving waters in terms of uneaten food, fish metabolites and chemical treatments. For example, in a study carried out by Tahar et al., in 2018 [8], inlet and outlet concentrations of water parameters for four consecutive flow-through rainbow trout farms over a ten-year period were analyzed in Ireland to characterize the impact of each fish farm on the water quality as a function of their production and to identify any seasonal variability.

5. Conclusions

In the present trial, water parameters were investigated to determine whether the new feeding strategy based on extruded feed changed the water composition. The analyzed parameters (TSS, BOD_5, COD, NO_2-N, NO_3-N, TAN and TP) showed an important improvement from 2009 to 2019; the pH parameter did not show important variation during the studied period.

Water is effectively the habitat where rainbow trout live so it must be monitored and preserved as best as possible in terms of temperature and physico-chemical parameters and quality. All these aspects are the basis of ensuring high-quality rainbow trout farming. Sustainable management, together with a genetic program of rainbow trout specimens employed, based on selected fish showing the best performance in feed efficiency, will be the next challenge to further improve the fish performance and environment.

Considering what has been discussed, the feedback of these results should be considered a significant index of improvement in the quality of the environment and the adoption of the modern formulation certainly contributed to obtaining this result.

Author Contributions: Conceptualization, A.R. and E.F.; methodology, E.F. and A.R.; formal analysis, all authors; data elaboration, E.F.; writing, review and editing, all authors; funding acquisition, A.R. All authors have read and agreed to the published version of the manuscript.

Funding: This research was partially supported by the Fishery Office—Province of Macerata and Eureka Project 2019.

Acknowledgments: We are grateful to the Erede Rossi Company for their valuable collaboration and technical assistance during the water sampling activities.

Conflicts of Interest: The authors declare no conflict of interest. The funders had no role in the design of the study; in the collection, analyses or interpretation of data; in the writing of the manuscript; or in the decision to publish the results.

References

1. European Environment Agency. *European Waters—Assessment of Status and Pressures*; European Environment Agency: Copenhagen, Denmark, 2018; pp. 21–29. ISBN 978-92-9213-947-6.
2. Thorpe, J.E.; Cho, C.Y. Minimising waste through bioenergetically and behaviourally based feeding strategies. *Water Sci. Technol.* **1995**, *31*, 29–40. [CrossRef]
3. Boyd, C.E.; Queiroz, J.F. Nitrogen and phosphorus loads by system, USEPA should consider system variables in setting new effluent rules. *Advocate* **2001**, *1*, 85–86.
4. Aubin, J.; Tocqueville, A.; Kaushik, S.J. Characterisation of waste output from flow-through trout farms in France: Comparison of nutrient mass-balance modelling and hydrological methods. *Aquat. Living Resour.* **2011**, *24*, 63–70. [CrossRef]
5. Moraes, M.D.A.B.; Carmo, C.F.D.; Ishikawa, C.M.; Tabata, Y.A.; Mercante, C.T.J. Daily mass balance of phosphorus and nitrogen in effluents of production sectors of trout farming system. *Acta Limnol. Bras.* **2015**, *27*, 330–340. [CrossRef]
6. Parisi, G.; Terova, G.; Gasco, L.; Piccolo, G.; Roncarati, A.; Moretti, V.M.; Pais, A. Current status and future perspectives of Italian finfish aquaculture. *Fish Biol. Fish.* **2013**, *24*, 15–73. [CrossRef]
7. Bureau, D.P.; Hua, K.; Cho, C.Y. Effect of feeding level on growth and nutrient deposition in rainbow trout (Oncorhynchus mykiss Walbaum) growing from 150 to 600 g. *Aquac. Res.* **2006**, *37*, 1090–1098. [CrossRef]
8. Tahar, A.; Kennedy, A.M.; Fitzgerald, R.D.; Clifford, E.; Rowan, N. Longitudinal evaluation of the impact of traditional rainbow trout farming on receiving water quality in Ireland. *Peer. J.* **2018**, *6*, e5281. [CrossRef]
9. Tyapkova, O.; Osen, R.; Wagenstaller, M.; Baier, B.; Specht, F.; Zacherl, C. Replacing fishmeal with oilseed cakes in fish feed—A study on the influence of processing parameters on the extrusion behavior and quality properties of the feed pellets. *J. Food Eng.* **2016**, *191*, 28–36. [CrossRef]
10. Funk, D.; Rokey, G.; Russell, K.; Wenger Manufacturing. Laboratory analysis and lab scale demonstrations highlight extrusion processing conditions critical for high performance RAS feed production. *Aquafeed Adv. Process. Formul.* **2019**, *11*, 15–17.
11. Welker, T.L.; Overturf, K.; Abernathy, J. Effect of aeration and oxygenation on growth and survival of rainbow trout in a commercial serial-pass, flow-through raceway system. *Aquac. Rep.* **2019**, *14*, 100194. [CrossRef]
12. Burel, C.; Boujard, T.; Tulli, F.; Kaushik, S.J. Digestibility of extruded peas, extruded lupin, and rapeseed meal in rainbow trout (Oncorhynchus mykiss) and turbot (Psetta maxima). *Aquaculture* **2000**, *188*, 285–298. [CrossRef]
13. Alzieu, C. Water—The medium for culture. In *Aquaculture*; Barnabé, G., Ed.; Ellis Horwood Ltd.: London, UK, 1990; Volume 1, pp. 37–60.
14. Becke, C.; Schumann, M.; Steinhagen, D.; Rojas-Tirado, P.; Geist, J.; Brinker, A. Effects of unionized ammonia and suspended solids on rainbow trout (Oncorhynchus mykiss) in recirculating aquaculture systems. *Aquaculture* **2019**, *499*, 348–357. [CrossRef]
15. Alabaster, J.S.; Lloyd, R. *Water Quality Criteria for Freshwater Fish*, 2nd ed.; Elsevier: Amsterdam, The Netherlands, 1982; ISBN 9781483163116.
16. European Environment Agency. *Annual Report*; European Environment Agency: Copenhagen, Denmark, 2004; ISBN 92-9167-761-2, ISSN 1561-2120.
17. Melotti, P.; Roncarati, A.; Angellotti, A.; Dees, A.; Magi, G.E.; Mazzini, C.; Bianchi, C.; Casciano, R. Effects of rearing density on rainbow trout welfare, determined by plasmatic and tissue parameters. *Ital. J. Anim. Sci.* **2004**, *3*, 393–400. [CrossRef]

18. ISPRA. *Procedura di Misurazione Per La Determinazione Della Richiesta Chimica Di Ossigeno (COD) Mediante Test in Cuvetta: Metodo 5135. Manuali e Linee Guida 117/2004*; ISPRA: Lombardy, Italy, 2014; ISBN 978-88-448-0678-1.

19. APHA (American Public Health Association, American Water Works Association and Water Pollution Control Federation). *Standard Methods for the Examination of Water and Wastewater*; A.P.H.A.: Washington, DC, USA, 1995.

20. IRSA-CNR; APAT. *Metodi Analitici Per Le Acque. Manuali e Linee Guida 29/2003*; IRSA-CNR/APAT: Rome, Italy, 2003; Volume 1, pp. 161–170.

21. Bureau, D.P.; Harris, A.M.; Cho, C.Y. Apparent digestibility of rendered animal protein ingredients for rainbow trout (*Oncorhynchus mykiss*). *Aquaculture* **1999**, *180*, 345–358. [CrossRef]

22. IBM Corp. *IBM SPSS Statistics for Windows, Version 25.0*; IBM Corp: Armonk, NY, USA, 2017.

23. Dalsgaard, J.; Pedersen, P.B. Solid and suspended/dissolved waste (N, P, O) from rainbow trout (*Oncorynchus mykiss*). *Aquaculture* **2011**, *313*, 92–99. [CrossRef]

24. Galezan, F.H.; Bayati, M.R.; Safari, O.; Rohani, A. Modeling oxygen and organic matter concentration in the intensive rainbow trout (Oncorhynchus mykiss) rearing system. *Environ. Monit. Assess.* **2020**, *192*. [CrossRef]

25. Kaushik, S.J. Feed management and on-farm feeding practices of temperate fish with special reference to salmonids. In *On-Farm Feeding and Feed Management in Aquaculture*; FAO Fisheries and Aquaculture Technical Paper No. 583; Hasan, M.R., New, M.B., Eds.; FAO: Rome, Italy, 2013; pp. 519–551.

26. Welker, T.L.; Overturf, K.; Snyder, S.; Liu, K.; Abernathy, J.; Frost, J.; Barrows, F.T. Effects of feed processing method (extrusion and expansion-compression pelleting) on water quality and growth of rainbow trout in a commercial setting. *J. Appl. Aquac.* **2018**, *30*, 97–124. [CrossRef]

27. Islam, M.S. Nitrogen and phosphorus budget in coastal and marine cage aquaculture and impacts of effuent loading on ecosystem: Review and analysis towards model development. *Mar. Pollut. Bull.* **2005**, *50*, 48–61. [CrossRef]

28. Jia, B.; Tang, Y.; Tian, L.; Franz, L.; Alewell, C.; Huang, J.-H. Impact of Fish Farming on Phosphorus in Reservoir Sediments. *Sci. Rep.* **2015**, *5*, 16617. [CrossRef]

29. Delabbio, J.L.; Johnson, G.R.; Murphy, B.R.; Hallerman, E.; Woart, A.; McMullin, S.L. Fish Disease and Biosecurity: Attitudes, Beliefs, and Perceptions of Managers and Owners of Commercial Finfish Recirculating Facilities in the United States and Canada. *J. Aquat. Anim. Health* **2005**, *17*, 153–159. [CrossRef]

30. FAO. *Duckweed: A Tiny Aquatic Plant with Enormous Potential for Agriculture and Environment*; Food and Agricultural Organization of the United Nations: Geneva, Switzerland, 2010.

31. Stadtlander, T.; Forster, S.; Rosskothen, D.; Leiber, F. Slurry-grown duckweed (Spirodela polyrhiza) as a means to recycle nitrogen into feed for rainbow trout fry. *J. Clean. Prod.* **2019**, *228*, 86–93. [CrossRef]

32. Chakrabarti, R.; Clark, W.D.; Sharma, J.G.; Goswami, R.K.; Shrivastav, A.K.; Tocher, D.R. Mass Production of Lemna minor and Its Amino Acid and Fatty Acid Profiles. *Front. Chem.* **2018**, *6*, 479. [CrossRef] [PubMed]

Article

Effect of Two Nutritional Strategies to Balance Energy and Protein Supply in Fattening Heifers on Performance, Ruminal Metabolism, and Carcass Characteristics

Rodrigo A. Arias [1,2,]*, Gonzalo Guajardo [3], Stefan Kunick [3], Christian Alvarado-Gilis [1] and Juan Pablo Keim [1]

[1] Instituto de Producción Animal, Universidad Austral de Chile, Valdivia-Chile, Valdivia 5090000, Chile; calvarado@uach.cl (C.A.-G.); juan.keim@uach.cl (J.P.K.)

[2] Centro de Investigación de Suelos Volcánicos, Universidad Austral de Chile, Valdivia 5090000, Chile

[3] Escuela de Graduados, Facultad de Ciencias Agrarias y Alimentarias, Universidad Austral de Chile, Valdivia 5090000, Chile; gonzalo.guajardo.c@gmail.com (G.G.); stefan93_@live.cl (S.K.)

* Correspondence: rodrigo.arias@uach.cl; Tel.: +56-63-229-3337

Received: 20 April 2020; Accepted: 11 May 2020; Published: 14 May 2020

Simple Summary: Beef production has been under strong scrutiny during the last half-century. First, because of its supposed negative impact on human health, and more recently, due to the negative impact on the environment, mainly from nitrogen and greenhouse gases. We conducted an experiment to assess the effects of a diet formulated based on the metabolizable protein system and synchronicity between energy and protein, on the nitrogen losses to the environment, performance, and carcass characteristics of the fattening heifers. Our results show that a diet combining a high synchrony index with the metabolizable protein system increases the nitrogen use efficiency without negatively affecting animal performance or carcass characteristics compared to heifers fed a diet without a balanced protein but with a high synchrony index.

Abstract: Latin America is an important contributor to the worldwide beef business, but in general, there are limited studies considering strategies to reduce nitrogen contamination in their production systems. The study's goal was to assess the effect of two nutritional strategies to balance energy and protein supply in fattening heifers on performance, ruminal metabolism, and carcass characteristics. A total of 24 crossbred heifers (initial body weight 'BW 'of 372 ± 36 kg) were used to create two blocks (based on live weight) of two pens each, that were equipped with individual feeders. Within each block, half of the animals were assigned to a diet based on tabular Crude Protein (CP) requirements denominated Crude Protein Diet 'CPD' but without a ruminal degradable protein balance. The other half received a diet denominated Metabolizable Protein Diet 'MPD', formulated with the metabolizable protein system, balanced for the ruminal degradable protein. Both diets had the same ingredients and as well as similar synchrony indexes (0.80 and 0.83, respectively). For nitrogen concentration in feces and urine as well as microbial crude protein synthesis, a total of 12 heifers (three per pen) were randomly selected to collect samples. The dataset was analyzed as a randomized complete block design with a 5% significance. No diet × time interaction was observed for Average Daily Gain 'ADG' ($p = 0.89$), but there was an effect of the time on ADG ($p \leq 0.001$). No differences were observed neither for final weight, dry matter intake 'DMI', and feed conversion rate ($p > 0.05$). Heifers fed with CPD showed greater cold carcass weight ($p = 0.041$), but without differences in ribeye area, backfat thickness, pH, dressing %, and marbling ($p > 0.05$). Differences between diets were observed for the in vitro parameters as well as for the Total Volatile Fatty Acids 'VFA' and NH_3 ($p < 0.05$). Total N concentrations (urine + feces) of heifers fed with MDP was lower than in those fed with the CPD ($p < 0.01$), but no differences were observed in microbial protein, purine derivatives, and creatinine ($p > 0.05$). We conclude that the combination of synchrony and the metabolizable protein system achieve greater efficiency in the use of nitrogen, without negatively affecting animals' performance or the quality of the carcass.

Keywords: nitrogen efficiency; synchrony index; metabolizable protein system

1. Introduction

In many countries, traditional beef production systems have a low level of intensification. Though, the implementation of new technologies in herd management, such as the type and form of feeding, would have a positive response in productive efficiency, which directly impacts the final utility of the producer [1]. Nevertheless, due to the growing world beef demand in conjunction with more environmental restrictions, it is expected that the increase in production is mainly due to more intensification [2]. This new scenario should aim not only to achieve more beef and of better quality but also improve animal welfare and, at the same time, cause minimum environmental impact. However, there is a risk that greater intensification is a source of environmental pollution since it has been associated with highly intensive animal production systems [3]. This is mainly due to nitrogen (N) and phosphorus (P), which pollute groundwater, surface water, and air. The same authors detail that the main reasons for contamination are excretions (urine and feces) that are discharged into surface waters, or by volatilization of N in the form of NH_3. It also has been suggested that new feeding strategies should be adopted to improve the use of N and reduce its losses. In this sense, Klemesrud et al. [4] state that satisfying the animal's amino acid requirements, without falling into excess protein in the rations, it allows to reduce N excretions. Thus, adequate management of nutrition should not only allow digesting and utilizing the fiber from forages and promote microbial protein synthesis but also reduce N losses at the ruminal level, which may be excreted to the environment. Farmers and nutritionists should try to maximize microbial protein production in the rumen because it has been established that microbial protein can provide from 50 to 100% of the metabolizable protein (MP) required by the animal, which depends on the amount of ruminal degradable protein and fermentable carbohydrates in the diet [5].

Owen and Sapienza [6] pointed out that to achieve maximum efficiency and production of microbial protein, factors that affect its synthesis should be considered, such as the availability of energy, ammonium or other N sources, minerals, vitamins, and factors related to animal growth. The synchrony between energy and degradable protein in the rumen is a very important factor in the optimization of microbial protein synthesis [7]. Therefore, not only the contribution or availability of energy and nitrogen sources at the ruminal level should be necessary [8,9], but also synchrony [10]. Other authors consider that the amount or availability of energy in the rumen, in a certain nitrogen range, is usually more important than synchrony as such for the synthesis of microbial protein [9,11]. According to Sinclair [12], ruminal synchrony would improve the rumen fermentation, digestibility of nutrients, and the microorganism's population [13], as well as bacterial protein synthesis, nitrogen retention, the animal's productive response, and consequently, the N use efficiency (NUE) [14].

Ruminal synchrony could be affected by different factors such as diet, which is the main determinant of the quantity and quality of nutrients supplied to microorganisms [10]. Mention has also been made to the chemical composition of the feedstuffs [15], the ingredients used in the ration [16], grains processing [17], and the degradation characteristics of organic matter and protein in food [13,18]. Other factors related to the animal, such as health, nutritional requirements, and the interaction with the type of management and the environment, also alter the use of nutrients, causing ruminal asynchrony [10]. As mentioned, the synchrony between energy and protein available in the rumen theoretically should allow more efficient use of nutrients, improving the production of microbial products to increase the supply of nutrients, and thus improve the animal's performance. However, after years of studies and analysis of this concept, it has not been fully demonstrated in practice.

Likewise, asynchronous nutrient diets have generated results as good or higher than synchronous diets in growing beef and dairy cattle in confinement [10]. Duarte et al. [19] evaluated the effects of different levels of energy and non-degradable protein in the rumen on consumption, growth, carcass characteristics, and meat quality in fattening heifers. The authors concluded that there was a higher average daily gain (ADG) with high levels of rumen undegradable protein (RUP). However, levels of RUP had no effect on carcass characteristics and composition, but there was an effect on the energy level with a greater area of the eye of the back (longissimus muscle).

At present, the MP system widely used by beef nutritionists, predicts protein requirements accurately, being effective in providing protein levels even at or near the predicted requirements, ensuring good animal performance [20]. Erickson et al. [21] evaluated the effect of the use of diets balanced with the conventional method (based on CP) vs. others based on the MP system in fattening of steers and calves under highly intensive systems in the USA. They determined that there were no differences in the animal's productive responses. However, the excretion of N was reduced with experimental diets. Nowadays, there is limited information about the implementation of the MP system in the South American beef production systems. Most of these production systems are much less intensive than North American feedlots, such as those studied by Erickson et al. [21]. In this context, we hypothesize that a diet formulated with the MP system will decrease the environmental impact by reducing nitrogen losses to the environment, without negatively affecting in vitro rumen fermentation, the productive performance and the carcass characteristics of the fattening heifers.

2. Materials and Methods

All the procedures, including animal care and handling procedures, followed national legislation (Law No. 20,380 on Protection of Animals; Decree No. 29 about regulation on the protection of animals during their industrial production, their commercialization and in other areas to hold animals), whose application is supervised by the National Service of Agriculture and Livestock (SAG), the competent authority in this matter.

2.1. Animals and Facilities

The study was conducted at the Austral Agricultural Research Station (EEAA) of the Universidad Austral de Chile, located 8 km North of Valdivia, Los Ríos region (39° 46′28″ S 73° 14′11″ W), from September 6th, 2016 until December 6th, 2016. A total of 24 crossbred heifers (Hereford × Angus) with an average initial BW of 372 ± 36 kg were used (68% of mature weight). All heifers were dewormed with 1% ivermectin for the control of gastrointestinal parasites. Upon arrival, animals were grouped based on their initial live weight conforming two blocks (light: 345 ± 23.32 kg and heavy: 399 ± 25.10 kg) and were located in four pens that were randomly assigned to each block. Each block consisted of two pens with a space allowance of 26.25 m² per head. In addition, each pen was equipped with six individual semi-automatic feeders (American Calan Inc., Northwood, NH, USA), which opened through an electronic key that hangs on the neck of each animal. This allowed the control of individual food intake, considering each animal as an experimental and observational unit. Feeders were under a shed to protect feeds from rain, with a cement surface (platform) of 3.5 m wide. Cattle had access to water through two shared drinking fountains of 600 L, each with a float system to ensure their filling, each located between two adjacent pens.

Heifers were weighed approximately every 20 d with an electronic scale (Iconix FX31, Iconix NZ Ltd., Palmerston North, New Zealand). The weighing was performed during the morning before the feed was supplied to avoid variations due to gastrointestinal content. Average daily weight gain (ADG) was estimated at intervals corresponding to weighing dates. Additionally, feed conversion of the whole trial was determined by dividing the total amount of food consumed by the total kg of body live weight (LW) gained. Finally, the LW of heifers was used to calculate the synthesis of microbial protein.

There was a training period of approximately 20 d that was carried out so that the animals got used to the facilities and the feeders, as was previously described by Arias et al. [22]. During this time, heifers were fed ad libitum with a 100% pasture haylage and water. Animals were fed once a day (09:00 to 11:00 am). The ingredients were weighed and mixed daily before supplying the diet in each feeder, and orts were collected and weighed individually (per feeder) once a week. Simultaneously, the preference of each animal for the feeders was observed in order to assign the respective key. Finally, once assigned and the keys placed on the neck of the heifers, the electrical system was connected to block them and, the experimental phase was started, which took place over a period of 64 d. Animals were fed once a day (09:00 to 11:00 am).

2.2. Dietary Treatments and Nutrient Concentration Analyses

The chemical composition of feed ingredients and their inclusion in the dietary treatments are reported in Tables 1 and 2. The animals within each block were randomly assigned to one of two dietary nutritional strategies. Diets were formulated to be isoenergetic. Each diet was designed according to the ruminal degradation parameters of their respective ingredients in order to obtain similar synchrony between energy and protein in the rumen based on the synchrony index (SI) proposed by Sinclair et al. [12]. The SI of both diets was calculated based on the ruminal degradation parameters of OM and CP of each ingredient of the diets adopting the methodology of Verbič et al. [23]. Both diets were calculated to obtain an ADG of 1.0 kg. The first treatment was formulated based on heifers CP requirements and was defined as the 'Crude Protein diet' (CPD). It was designed to provide the amount of daily metabolizable energy and crude protein needed for heifers with an SI = 0.80. This diet was analyzed using the Beef Cattle Nutrient Requirement Model (BCNRM) software [24], contrasting the values obtained in the balance for predicted ruminal degradable protein (RDP). According to the National Academies of Sciences, Engineering, and Medicine (NASEM) [24], there was an imbalance of the RDP supply (+240 g d^{-1}) since it was calculated based on the CP system. The second dietary treatment 'Metabolizable Protein diet' (MPD) was formulated by using the MP system, providing an adequate supply of ruminal undegradable protein (RUP; -10 g d^{-1}) and with a similar SI than the CPD (0.83).

Table 1. Nutritional characteristics of the ingredients utilized in the study.

Item	Pasture Haylage *	Canola Meal	Sugar Beet Pulp	Triticale Grain
Dry Matter, %	49.67	91.70	88.90	90.35
Crude Protein, %	14.31	33.48	8.69	8.13
Soluble Protein, %	8.10	8.56	0.17	1.79
Neutral Detergent Fiber, %	53.68	32.15	50.22	20.32
Acid Detergent Fiber, %	30.58	21.51	24.65	3.79
DOMD, %	76.20	74.40	90.12	92.52
TDN, %	76.50	74.83	88.97	91.17
Organic matter, %	92.98	94.19	94.75	98.40
NFC, %	21.34	16.97	35.11	68.43
Ether Extract, %	3.65	11.59	0.73	1.52
Ash, %	7.02	5.81	5.25	1.60
ME, Mcal kg^{-1} DM	2.76	2.70	3.21	3.29
NEm, Mcal kg^{-1} DM	1.83	1.78	2.20	2.27
NEg, Mcal kg^{-1} DM	1.20	1.16	1.52	1.57

* Purchased to local farmers, made mainly of *Lolium spp.* but also included a lower proportion of *Dactylis glomerata*, *Bromus spp.*, and *Trifolium spp.* DOMD = digestible organic matter on a dry matter basis; TDN = Total Digestible Nutrients estimated as ((ME/0.82)/4.4) × 100; NFC = Non-fiber carbohydrates estimated as 100—(%NDF + %CP + %Ash + %EE); ME = Metabolizable energy; NEm = Net energy of maintenance, and NEg = Net energy of gain.

Dry matter content was measured by weighing the samples before and after drying with a forced-air oven, initially at 60 °C for 48 h, and then at 105 °C for 12 h. The CP concentration was determined by combustion (Leco Model FP-428, Leco Corporation, St Joseph, MI, USA) based on the DUMAS method (N × 6.25), digestible organic matter on a dry matter basis (DOMD) was measured according to Tilley and Terry [25], neutral detergent fiber (aNDF) was measured by using a heat-stable amylase [26], and ash and ether extract (EE) were analyzed according to AOAC [27] (Methods ID 942.05 and ID 920.39 for ash and EE, respectively).

Table 2. Nutritional characteristics of the diets, ingredient inclusion, and synchronize index per treatment.

Item	Treatment	
	CPC	**MPD**
Synchrony index (SI) [†]	0.80	0.83
Ingredient inclusion in the diet		
Pasture haylage, %	50.0	58. 0
Canola meal, %	15.0	4.0
Sugar beet pulp, %	20.0	0.0
Triticale grain, %	15.0	38.0
Dry Matter, %	64.1	61.3
Crude Protein, %	15.1	12.7
Soluble Protein, %	5.6	5.7
Neutral Detergent Fiber, %	44.8	40.1
Acid Detergent Fiber, %	24.0	21.5
Digestibility value, %	81.2	74.4
TDN, %	66.2	68.8
Organic Matter, %	94.3	95.1
NFC, %	30.5	39.1
Ether Extract, %	3.9	3.2
ME, Mcal kg^{-1} DM	2.9	3.0
NEm, Mcal kg^{-1} DM	2.0	2.0
NEg, Mcal kg^{-1} DM	1.3	1.4
RDP, (% CP) [‡]	86.0	83.3
RUP, (% CP) [‡]	14.0	16.8
MPB, (g/d) [‡]	5.8	0.1
RDPB, (g/d) [‡]	240.0	−10.0

CPD: Crude protein diet; MPD: metabolizable protein diet; TDN = Total Digestible Nutrients estimated as ((ME/0.82)/4.4) × 100; NFC = Non-fiber carbohydrates estimated as 100—(% NDF + % CP + % Ash + % EE); ME = Metabolizable energy; NEm = Net energy of maintenance, and NEg = Net energy of gain; RDP = Ruminal degradable protein; RUP = Ruminal undegradable protein; MPB: Metabolizable protein balance; RDPB = Ruminal degradable protein balance. [†] Calculated from in situ degradation parameters and Verbič et al. [23] equations. [‡] Data obtained from the BCNRM software v1.0.37.14. [24]. Both diets were supplemented with 80 g d^{-1} per heifer of mineral salts.

2.3. Organic Matter and Crude Protein in Situ Degradation Parameters of Feed Ingredients

Food samples were lyophilized and grounded through a 5 mm sieve. Polyester bags of 10 × 20 cm (40–60 μm porosity) contained 4 g of sample achieving a ratio of 16 mg of sample per cm^2. Two replicates (bags) by incubation period were introduced in a fistulated cow (Holstein breed in a state of maintenance) in reverse order to remove all the bags from the rumen at the same time. Incubation times corresponded to 0, 2, 4, 8, 10, 14, 24, and 48, and 72 h were used. The cow was fed a ration formulated according to its nutritional requirements and managed in the EEAA housing yard. Prior to ruminal incubation, the bags were placed in 20 × 30 cm porous laundry bags and soaked in warm water (30 °C) for 20 min and then introduced into the donor's cow rumen. Once removed from the rumen, bags were washed with running water until the water was clear and then frozen at −20 °C for 24 h to stop any fermentation activity. Thereafter, samples were thawed and washed in a conventional washing machine for 10 min and dried in a forced-air oven at 60 °C for 48 h. The time 0 h was not incubated in the rumen and was used to determine the soluble fraction of the DM and CP. Residues after ruminal incubation were weighed to determine the amount of feed degraded at a certain incubation time.

Ruminal degradation parameters of OM and CP were determined using the model proposed by Ørskov and McDonald [28], through the nonlinear procedure of the GraphPad Prism v6.0 software, using the exponential model without a lag phase [28]:

$$PD = A + B\,(1 - e^{-kt}) \tag{1}$$

where *A* is the soluble fraction (g kg^{-1} fraction of bags washed at time 0 h), *B* is the insoluble but potentially degradable fraction (g kg^{-1}), *k* is the degradation rate constant (% h^{-1}), and *t* is the incubation time (h).

Effective degradability (ED) was calculated from the aforementioned parameters assuming fractional passage rates (kp) of 5% h^{-1} according to McDonald [29]:

$$ED\ (g\ kg^{-1}) = A + B[k\,/\,(k + kp)]e^{-kp} \tag{2}$$

where *A* (g kg^{-1}) is the soluble fraction; *B* (g kg^{-1}) is the insoluble but potentially degradable fraction; *k* and *kp* (h^{-1}) are the ruminal degradation rate constant and passage rate constant (5%), respectively, according to AFRC [30].

A correction for small particle losses was made according to Hvelplund and Weisbjerg [31]. Thus, samples of each food (1.5 g) were placed in containers to which 40 mL of water was added and stored at room temperature (20 °C) for an hour. Then they were filtered and washed eight times with 20 mL of water; the residue was dried in a forced-air oven at 60 °C for 48 h, and then weighed, calculating the water soluble (SOL) fraction. Assuming that, losses of small particles are degraded similarly to the particles left in the bag, corrections can be made for the loss:

$$PDcor\ (ti) = PD\ (ti) - P\ [1 - ((PD\ (ti) - (P + SOL))\,/\,(1 - (P + SOL)))] \tag{3}$$

$$EDcor = SOL + [((1 - SOL)\,/\,(1 - (P + SOL))) \times (ED - (P + SOL))] \tag{4}$$

$$acor = A - P \tag{5}$$

$$bcor = B + P\ [B\,/\,(1 - (P + SOL))] \tag{6}$$

$$ccor = c \tag{7}$$

where: *PDcor (ti)* is the degradability corrected at the time of incubation *ti*, *EDcor* is the corrected effective degradation, *P* are the losses of small particles, and *SOL* is the water solubility.

The values of the in-situ degradation characteristics of OM and CP are shown in Table S1, which was mainly used for the formulation of treatments and to determine the SI. The adjusted degradation curves of the OM and CP are shown in Figure S1.

2.4. In Vitro Fermentation

In vitro gas production was evaluated according to Theodorou et al. [32]. Three replicates (120 mL bottles) were used, including two "control" bottles (a standard concentrate with a known gas production pattern to compare and ensure the proper fermentation process) and two blanks (bottles without substrate) in three incubation runs. One gram of dried lyophilized substrate was used in each bottle. Then, they were filled with 85 mL of Goering–Van Soest medium, gasified with CO_2, and closed with rubber stoppers, and left at 4 °C overnight. The next day, 4 mL of reducing agent (Distilled water, sodium sulfate, cysteine HCl, and 1 mol/L NaOH) were added; each bottle was gasified again, completely closed with the rubber stopper, and sealed with aluminum. Then they were placed in a water bath at 39 °C. Rumen fluid was obtained directly from heifers in the experiment, randomly selected from the two treatments of each block. The extraction of ruminal liquor was conducted in the morning before heifers were fed, using an oro-ruminal probe (FLORA Ruminator; Profs-Products, Guelph, ON, Canada), stored immediately in thermos flasks and transported to the laboratory. The ruminal liquid was filtered through a cheesecloth, mixed and placed under constant CO_2 gasification. Later, ruminal liquor from heifers fed their corresponding diet was added (10 mL) to bottles containing the same diet. Once the rumen fluid was inoculated, the initial gas was extracted from the bottles. After inoculation, the bottles were placed in a water bath at 39 °C under continuous horizontal movement at 50 rpm. The gas pressure in the headspace of the bottles, above atmospheric pressure, was measured manually with a pressure transducer (PCE Instruments, Tobarra, Albacete, Spain) at 1, 2, 3, 4, 5, 6, 8, 10, 12, 15, 18, 24, 36, and 48 h, and the volume of gas produced was measured by extraction, using syringes connected through a three-way Luer valve from the bottles until the visual display of the transducer read zero. Once the volume of gas produced was recorded, it was eliminated. Fermentations were stopped after 48 h by placing the bottles on ice. For the in vitro gas production kinetics, after correcting for the white gas production, it was adjusted by the Michaelis–Menten model without lag phase [33] to obtain the corresponding fermentation parameters of the treatments and each of the feeds.

$$GP = A \times [T^n / (T^n + K^n)] \tag{8}$$

where *GP* is the production of gas at time *T*; *A* is the volume of asymptotic gas (mL g^{-1} MS); n is the coefficient that determines the shape of the curve in the function, and *K* is the time in which half of *A* (h^{-1}) occurs. The other parameters were calculated according to Groot et al. [34] and France et al. [33]:

$$C = n / (2 \times K) \tag{9}$$

$$MDR = (n - 1) ((n - 1) / n) / K \tag{10}$$

$$ta, tb \text{ and } tc = K \times (((X / (1 - X)) (1 / n)) \tag{11}$$

where *C* is the degradation rate in the middle of the asymptote; *MDR* is the maximum degradation rate; *ta*, *tb*, and *tc* (h^{-1}) correspond to the time at which 25, 75, and 90% of *A* occurs; and *X* varies between 0.25, 0.75, and 0.90.

Volatile fatty acids (VFA) and ammonium (NH_3) concentrations were measured only for diets. For this, samples were obtained from the bottles at 4 and 48 h post-incubation, plus the two samples of inoculum to calculate net VFA and NH_3 production. Then, 5 mL was extracted from the supernatant, and 0.1 mL of concentrated hydrochloric acid (37%) were added. These were stored frozen (−20 °C) and were subsequently analyzed to determine the concentration (mmol L^{-1}) and proportion (%) of VFA (acetic acid (C2), propionic acid (C3), butyric acid (C4), isobutyric, valeric and isovaleric [VFAscr]), and the concentration of NH_3 (mg L^{-1}).

2.5. The Concentration of N in Feces, Purine and N Derivatives in Urine

A total of 12 heifers (three per pen) were randomly selected to collect fecal samples during the experiment, but one of them was dismissed from the study because of pregnancy. Samples were obtained three times during the experimental period (one day after the 1st, 3rd, and 4th weighing) and always from the same animals. The fecal samples were collected manually from the rectum, then frozen at −20 °C and then lyophilized. The determination of the concentration of N was made with the LECO FP528 nitrogen analyzer described above.

Additionally, urine samples were also obtained from the same 11 heifers by manual stimulation of the vulva. Urine was collected in a 1.0 L container covered with a mesh of lingerie to avoid solid remains coming from around the vulva. A minimum volume of 60 mL of urine was collected for the analysis of the N concentration, plus another 20 mL to determine purine derivatives in the urine. Previously, each bottle was filled with H_2SO_4 (10% v v^{-1}) of the collected volume. Samples were stored in a cooled container, then frozen at −20 °C. Samples destined to obtain N were lyophilized and subsequently analyzed through the LECO FP528 nitrogen analyzer. Samples for purine derivatives analyses (n = 10) were thawed and analyzed by using the HPLC-UV technique described by Vlassa et al. [35]. Urine volume was estimated according to Al-Khalidi [36], whereas daily total creatinine excretion was calculated according to Chizzotti [37]. Daily excretion of purine derivatives was estimated with the equation of Faichney et al. [38]. The creatinine daily excretion coefficient (mg d^{-1} K) = 113.12 $\times$ LW$^{-0.25}$ used was based on Ørskov et al. [39]. Daily purine absorption and the contribution of microbial N were calculated according to Chen and Gomes [40]. Finally, the total microbial protein (TMP) was obtained by the following equation:

$$TMP = MN \times 6.25 \tag{12}$$

where TMP: Total microbial protein (g d^{-1}); MN: Microbial Nitrogen (g d^{-1}); and 6.25 is the N content in proteins. These results were utilized to estimate the daily synthesis of microbial protein.

2.6. Nitrogen Use Efficiency (NUE)

Calculation of NUE was made based on the results of the N balance according to the BCNRM software v1.0.37.14. [24] and the equations proposed by Cole et al. [41] to estimate the retained N based on ADG and LW. The software estimates the amount of N excreted in the urine and feces separately, in addition to the N retained by the animal. Likewise, it estimates the N intake from the diet CP, considering 16% of N. The absorbed N was estimated according to the amount of MP, and the N retained according to the net growth protein. In addition, the software considers both excretions and endogenous N from the animal. Estimations of NUE were made for two periods (Period 1 = first 43 d, and Period 2 = the last 21 d) since consumption was different during these periods. Therefore, the nitrogen consumed was also different at each stage.

$$NUE = (N \text{ retained} / N \text{ intake}) \times 100 \tag{13}$$

2.7. Carcass Characteristics

Heifers were transported into a slaughterhouse for processing, where the following data were collected: cold carcass weight, carcass pH (measured directly 24 h postmortem), backfat thickness, ribeye area (measured in the cross-section of the *Longissimus dorsi* muscle, between the 9th and 10th ribs), and marbling (on the surface of the 9th rib muscle, determined by using USDA standards.

2.8. Data Analysis

The experimental model corresponded to a randomized complete block design (live weight as a block factor) with a univariate treatment structure with two levels (diets). For performance and carcass quality, each animal was considered as an experimental and observational unit. For variables derived from purine in urine, there were 10 observational units (four PCD and six MPD) for the variables of N in urine and feces n = 11 (five CPD and six MPD). The ANOVA model was

$$Y_{ijk} = \mu + \beta_j + \alpha_j + \varepsilon_{ijk} \tag{14}$$

where μ corresponds to the general mean; βi represents the effect of the ith block; α_j represents the effect of the jth diet, and ε_{ijk} the experimental error associated to the kth animal of the jth diet in the ith block. For multiple comparisons, the Tukey test was conducted with a significance = 0.05 when corresponded. In the case of the microbial protein synthesis, N in urine and feces, and results of in vitro fermentation, a repeated measure in the time model was used ($Y_{ijk} = \mu + \alpha_i + S_{k(i)} + \beta_j + [\alpha\beta_{ij}] + [\beta \times S_{jk(i)}] + \varepsilon_{ijk}$, where Y_{ijk} is the observation of the ith level diet and the jth level of time for subject k; μ is the general mean, α_i is the fixed effect level i of diet, $S_{k(i)}$ is the effect of subject nested within a block, β_j is the fixed effect of level j of factor time, $\alpha\beta_{ij}$ is the interaction effect between diet and time, ε_{ijk} is the random error effect). Since there was no replication for each combination of subject and factor time, the [$\beta \times S$] interaction effect cannot be separated from the error term and must be assumed to be zero. Animal performance and carcass data were analyzed with the statistical package JMP v14.0 (SAS Institute Inc.,

Cary, NC, USA). Repeated measures variables were analyzed with the SAS statistical package version 9.4 PROC MIXED (SAS Institute Inc.).

3. Results

3.1. Animal Performance and Carcass Characteristics

No diet effects for final LW, DMI, ADG, and Feed Conversion (Table 3) were observed. Likewise, no diet × time interaction was observed for ADG ($p = 0.89$), but there was an effect of the time on ADG ($p \leq 0.001$). During the first 21 d, ADG was as expected (near 1.0 kg/d), while in the second period (from 21 to 43 d), there was a slight increase for both diets. Finally, in the last period (43 to 64 d), there was a steep decrease in ADG in both treatments. In addition, no differences ($p > 0.05$) were observed for the weights recorded in the second and third weightings. However, in the last period (43 to 64 d), a decrease in the growth response for both treatments was observed for both the ADG and LW variables.

Table 3. Least square means for live weight, dry matter consumption, feed conversion, and daily weight gain per diet.

Variable	CPD ± SEM	MPD ± SEM	*p*-Value
Initial LW, kg	390.7 ± 6.09	386.6 ± 5.51	0.302
Final LW, kg	445.9 ± 4.72	437.5 ± 4.33	0.207
ADG, kg d^{-1}			
d 0 21	1.01 ± 0.15	0.90 ± 0.14	0.604
d 21–43	1.24 ± 0.15	1.02 ± 0.14	0.308
d 43–64	0.36 ± 0.15	0.30 ± 0.14	0.774
d 0–43	1.11 ± 0.10	0.95 ± 0.09	0.234
d 0–64	0.87 ± 0.07	0.74 ± 0.06	0.200
DMI, kg d^{-1}	8.56 ± 0.04	8.49 ± 0.03	0.192
Feed conversion rate, kg DM kg LW gain $^{-1}$	10.7 ± 1.42	12.3 ± 1.59	0.381

CPD: Crude Protein Diet; MPD: Metabolizable Protein Diet; ADG = Average daily gain; DMI = Dry matter intake; LW = Live body weight.

Heifers fed with the CPD showed greater cold carcass weight (Table 4), approximately 12 kg more than those fed with the MPD diet ($p = 0.04$). However, no differences were observed in the other carcass characteristics ($p > 0.10$), even when the rib eye area was numerically higher in heifers fed with the MPD.

Table 4. Least square means for carcass characteristics variables measured in heifers fed with two different diets.

Variable	CPD	MPD	*p*-Value
Cold carcass weight, kg [†]	240.2 ± 4.12 [a]	228.3 ± 3.55 [b]	0.041
Ribeye area, cm^2	88.1 ± 3.10	92.6 ± 2.68	0.281
Back fat thickness, mm	7.0 ± 1.10	6.5 ± 0.91	0.704
Marbling [‡]	1.9 ± 0.20	1.8 ± 0.17	0.728
Muscle pH	5.0 ± 0.10	5.1 ± 0.05	0.135
Dressing, %	54.0	53.0	0.150

CPD: Crude Protein Diet; MPD: Metabolizable Protein Diet; [†] Different letters between columns indicate statistical difference, Tukey test; [‡] Marbling values scale from 1 to 3 (1 = Choice +; 2 = Choice; and 3 = Choice −).

3.2. In Vitro Fermentation Products

Parameters of in vitro gas production kinetics for both diets are shown in Table 5. No differences in asymptotic gas and 48 h GP were observed. The MPD exceeded the CPD in the MDR and C, whereas the time to produce 25, 50, 75, and 90% of asymptotic production was greater for CPD than MPD.

Table 5. Least square means of the parameters of the in vitro gas production kinetics of diets.

Variable	Diets		SEM	*p*-Value
	CPD	MPD		
GP48	248.6	257.4	9.33	0.125
A	266.7	269.1	10.84	0.715
n	1.8	1.8	0.06	0.128
K	12.2	11.1	0.42	0.008
C	0.1	0.1	0	0.003
MDR	0.1	0.1	0	0.004
ta	6.6	6.1	0.36	0.039
tb	22.6	20.2	0.44	0.004
tc	42.1	36.6	1.22	0.004

CPD = Crude Protein Diet; MPD = Metabolizable Protein Diet; GP48: Gas volume produced at 48 h of incubation $(mL\ g^{-1}\ DM)$; A = Asymptote gas volume $(mL\ g^{-1}\ DM)$; n = coefficient that determines the curve in the function; K = time to produce 50% of A (h), C: fractional rate of gas production in the middle of the asymptote (h^{-1}), MDR = maximum degradation rate (h^{-1}); ta, tb y tc = time to produce 0.25, 0.75 y 0.90 of A (h^{-1}).

The production and proportion of VFAs and NH_3 concentrations are presented in Table 6. There was a diet by time interaction for the concentration of NH_3, being similar at 4 h after incubation but at 48 h, the concentrations were 72% higher in the CPD. The total VFA production was greater for CPD when compared to MPD. However, the proportions of propionic acid and VFAscr were higher for MPD, whereas no differences were observed for the proportions of acetic and butyric acids, nor for the acetic: propionic and (acetic + butyric): propionic ratios.

Table 6. Least square means for the production and concentration of volatile fatty acids (VFAs) and the concentration of ammonium (NH_3) for both diets.

Variable	Diets			Time			*p*-Values		
	CPD	MPD	SEM	4 h	48 h	SEM	Diet	Time	D × T
NH_3, mg L^{-1}	89.9	66.5	4.6	67.8	88.6	4.48	<0.05	<0.01	<0.01
Total VFA, mmol L^{-1}	21.0	15.8	0.9	4.1	32.7	1.28	<0.01	<0.01	NS
Acetate, mol 100 mol^{-1}	58.0	52.4	2.5	63.6	46.8	2.83	NS	<0.05	NS
Propionate, mol 100 mol^{-1}	29.8	32.9	1.9	27.0	35.7	2.27	<0.05	<0.05	NS
Butyrate, mol 100 mol^{-1}	9.9	11.3	2.1	8.9	12.3	1.84	NS	NS	NS
VFAscr, mol 100 mol^{-1}	2.3	3.5	0.1	0.6	5.2	0.25	<0.01	<0.01	NS
C2:C3	2.1	1.7	0.2	2.5	1.3	0.29	NS	<0.05	NS
(C2 + C4):C3	2.4	2.1	0.3	2.8	1.7	0.33	NS	<0.05	NS

CPD = Crude Protein Diet; MPD = Metabolizable Protein Diet; SEM = Standard error of the mean; NH_3 = ammonia concentration; VFAscr = Branched chain volatile fatty acids production (isobutiric + valeric + isovaleric); C2:C3 = ratio acetic: propionic acid; (C2 + C4)/C3 = ratio ketogenic acids: glucogenic acid.

3.3. Nitrogen Balance

There was an increase in TMP synthesis over time for both diets (Table 7; $p < 0.01$). In addition, D1, D43, and D64 showed no differences between treatments in the daily amount of TMP produced. Likewise, there was no diet effect on the synthesis of TMP ($p > 0.05$), and the interaction between diet and time was also not significant ($p > 0.05$). There were no differences in time ($p > 0.05$) in the three dates when urinary N was evaluated. The N concentration in the urine was higher in the CPD ($p < 0.01$). Therefore, heifers that received the MPD had a lower concentration of nitrogen in the urine during the fattening period. In addition, there was no interaction between treatments and time ($p > 0.05$).

Table 7. Least square means for microbial protein, purine derivatives, creatinine, urine, and feces nitrogen concentration for both diets.

Variable	Diets				Time				p-Values		
	CPD	SEM	MPD	SEM	Day 1	Day 43	Day 64	SEM	Diet	Time	D × T
TMP, g d^{-1}	581.2	47.8	643.4	52.3	420.4	670.3	746.2	44.6	NS	<0.01	NS
N$_{Mic}$, g d^{-1}	93	7.6	102.9	8.4	67.3	197.2	119.4	7.1	NS	<0.01	NS
A, mmol L^{-1}	11.4	2.2	13.9	2.4	14.2	11.2	12.6	2.2	NS	NS	NS
UA, mmol L^{-1}	1.7	0.3	2.0	0.3	2.6	2.3	0.5	0.2	NS	<0.01	NS
C, mmol L^{-1}	11.5	2.1	14.7	2.3	20.1	9.9	9.2	2.1	NS	<0.01	NS
TPD, mmol L^{-1}	13.0	2.4	15.9	2.6	16.8	13.5	13.1	2.3	NS	NS	NS
N in urine, %	7.4	0.4	4.9	0.3	6.1	6.5	6.3	0.5	<0.01	NS	NS
N in feces, %	3.3	0.1	2.7	0.1	2.9	2.8	3.0	0.1	<0.01	NS	<0.05

CPD = Crude Protein Diet; MPD = Metabolizable Protein Diet; SEM = Standard error of the mean; TMP = Total microbial protein; N$_{Mic}$ = Microbial nitrogen; A = Allantoin; UA = Uric acid; C = Creatinine; TDP = Total purine derivatives (allantoin + uric acid); N urine = Concentration of nitrogen in urine; N feces = Nitrogen concentration in feces; NS = Not significant.

A significant interaction between diet and time ($p < 0.05$) was observed for the N concentration in the feces, with CPD showing a higher percentage of N over time. Therefore, total N concentrations (urine + feces) of heifers fed a balanced diet in terms of ruminal degradable protein was lower than in those who received an unbalanced ruminal degradable protein diet. Finally, estimation of NUE using the BCNRM software was higher in the MPD (17.44%) when compared to CPD (14.27%) for the whole period. However, when NUE was estimated according to Cole et al. [41], lower values were obtained (8.36% and 9.19%, respectively) decreasing even more in the second period (43 to 64 d), but a without difference between diets ($p > 0.05$).

4. Discussion

Nutritional management of ruminants is a relevant issue to a better understanding of the use of nutrients and for optimization purposes, especially in the framework of where environmental, economic, and social matters acquire more relevance [42–46]. The present experiment was conducted to assess the hypothesis that balancing the diet by the MP system would improve N metabolism, rumen fermentation, and microbial protein synthesis in heifers. To test this hypothesis, experimental diets were formulated to have similar SI but differences in RUP and MP balances. To our knowledge, this is the first study that jointly evaluate the SI with the MP system in beef production systems in South America.

In vitro, GP is mainly used to analyze the fermentation kinetics in feed for ruminants [32]. In addition, it has been shown that there is a positive relationship between in vitro GP and digestibility [47]. Although we found no differences in GP, there were differences in VFA production between the two diets ($p < 0.01$). When comparing diets, we observed that CPD produced more total VFA, but that did not reflect a greater production of microbial protein. A possible explanation is that the greater amount of protein was transformed into NH_3, leaving the carbonated chains that make it up; these would be fermented, producing a greater amount of VFA compared to MPD. However, MPD presented a lower value of the K parameter when compared to CPD, which implies faster fermentation, is in accordance with the MDR being higher for MPD.

On the other hand, increasing CP content in the diet (or N intake) increased the amount of NH_3 exponentially [3,48]. In addition, the saturation of microbial ammonia uptake has been reported to occur between 10 and 14% CP [49]. Our results agree with these authors, since, an interaction between diet and time was found, with CPD achieving a higher concentration of NH_3 at 48 h, which would be explained by the higher consumption of N (15.13% vs. 12.73% of CP in CPD and MPD, respectively). Consequently, more N was lost in CPD, which was not used to synthesize microbial protein. In this study, a similar amount of microbial protein was synthesized in both diets, but N in urine (%) was higher in CPD compared to MPD. In this context, Firkins et al. [50] pointed out that there is a higher microbial protein synthesis efficiency when the NH_3 concentration in the rumen is lower. However, in our study, no differences were observed in the synthesis of microbial protein between diets that showed different concentrations of NH_3 in the rumen.

Chumpawadee et al. [13] showed that as the SI increases, the greater the synthesis of microbial protein that reaches the small intestine since it seeks to synchronize the source of carbohydrates with N to maximize the efficiency of microbial growth [51]. In our trial, both diets had a similar rate of synchrony, both close to 1.0. This would explain that the synthesis of the microbial protein has not been different between both dietary strategies. In a similar way, the increase in the synthesis of the microbial protein over time could be explained by the same concept. In fact, in the first measurement (day 1), heifers were not yet consuming a synchronized or balanced diet. Thus, as time progressed, the microorganisms adapted to the simultaneous availability of carbohydrates and N sources, which was reflected on measurements done on days 43 and 64. Therefore, microbial growth for both diets was more efficient.

The MP system allows to better adjust the nutrient needs by separating the requirements of the animal from the microbes of the rumen. Consequently, an improvement of some productive parameters and a significant decrease in the excretions of N into the environment is achieved [52]. In this same sense, Henning et al. [9] concluded that the synchrony between energy and the availability of N might be of less importance for microbial growth. However, the energy supply can improve the efficiency of microbial growth. Other authors support this idea [10], since the animal does not have an endogenous system to ensure the supply of energy, but in the case of N, the animal would adapt to ensure the availability of the nutrient through recycling, particularly when it is low in the diet. It has also been reported that at a higher percentage of TDN, which is an indicator of energy in the diet, there is a greater amount of synthesized microbial protein [6]. In this experiment, TDN energy values of both diets were similar (difference of 2.64%), which would confirm the similarity in the synthesis of microbial protein in both dietary strategies.

Seo et al. [53] reported a greater daily microbial growth with higher SI (0.81 and 0.83) when compared with an SI of 0.77. These authors attributed to the fact that their diet with the lower SI had greater amounts of non-structural carbohydrates; therefore, greater production of lactic acid, and consequently, an effect on rumen pH that affects the synthesis of microbial protein [54]. However, in their results, Seo et al. [53] did not report differences in rumen pH. In our study, no differences in pH (data not shown) were observed either, so it follows that it was not a determining factor affecting microbial protein synthesis. Seo et al. [55] suggest that the lower ruminal NH_3-N concentration correlates with higher utilization of NH_3-N for microbial protein synthesis. This is in accordance with our study, as MPD had a lower CP concentration and N intake showed a lower concentration of NH_3 and resulted in similar microbial protein synthesis.

The amount of microbial N produced was within the values described by the literature that used similar breeds [56–58], but when compared with breeds destined for milk production, these values were much lower, mainly due to the passage rates compared to those seen in dairy cattle, which were higher and also had higher consumption than beef production cattle [59]. The optimal range of NH_3 for microbial production is 5 to 8 mg 100 mL^{-1}. In our study, NH_3 was measured in vitro, and for none of the diets, it was limiting microbial growth.

The lowest concentration of N on the three sampling dates was observed in MPD. This would be mainly explained by the balance of the protein through the MP system established by the BCNRM software. The foregoing given that both treatments had a similar SI (0.80 and 0.83), and the difference between them was that it was balanced to adjust the availability of the rumen degradable protein and MP. Meanwhile, CPD was calculated considering the total demand for CP. In our study, a decrease in the total CP concentration of the diet was achieved when it was balanced by the MP system (MPD), reaching 12.73% CP, while the conventional CPD diet was 15.13% CP. Therefore, MPD resulted in a lower excretion of N, as observed in the N concentrations in the urine and feces. Several authors have reported similar results, that is, the higher the content of N consumed, the greater the excretions of N, both for urine and feces [60–62]. This is mainly because, by increasing the consumption of CP or rather the RDP, an excess of NH_3 is produced in the rumen, which is converted to urea in the liver and subsequently eliminated in the urine [63].

On the other hand, Erickson and Klopfenstein [52] found differences in the excretions of N for both one-year-old calves and calves when using a nutritional strategy that incorporates the feeding in phases balanced by the MP system. The authors reported that lower excretions of N were observed in treatments balanced with the MP system. Likewise, as occurred in our study, they pointed out that the contents of CP in the diet balanced by MP were lower. Finally, they reported that adopting this method could reduce excretions in steers in the evaluated fattening period (137 d) by up to 6.1 kg of N. Cole et al. [64] reported similar results regarding the concentration of N in feces and urine, evaluating diets with different CP contents (11.5%, 13.0%, and 14.5%). They observed that the diet with the highest concentration of CP was higher in the N concentration in feces and urine, coinciding with our results.

Carcass characteristics were not affected by the treatments, with the exception of cold carcass weight. This was not expected due to the similarity in the final LW of both dietary treatments. Nevertheless, another study reported that the weight of the hot carcass responded to the increase in the concentration of PC in the diet [65]. On the other hand, Cole et al. [64] and Gleghorn et al. [66] reported a lack of differences in the degree of marbling and fat coverage regarding the change of protein offered in the diet. In contrast, there was a trend for a smaller ribeye area in diets with higher CP, that is, 14% vs. 12% (Cole et al. 2003). Nevertheless, our results must be analyzed with caution because of the low number of replicates in the study.

The values of the NUE herein reported are within the ranges reported in the literature [21,41,45]. The N consumed in MPD was lower compared to CPD, because CP content was lower, and the RUP balance was better. The latter resulted in a lower concentration of NH_3 in MPD, and when there is less ruminal NH_3, NUE use to be higher [54]. In addition, the concentration of N in the feces and urine was lower in heifers fed with MPD. Therefore, they consumed less N and utilized it more efficiently than heifers fed the CPD. Although both, MP system and the SI, seek to improve NUE [12,44], using them together can produce better results than using each one separately. Galyean [67] promotes the use of the MP system since the protein intake is calculated according to the requirements of the microorganisms, thus avoiding excesses of N in the diets and fewer excretions of N into the environment. Therefore, using the MP system can reduce N excretions and therefore improve NUE [44,45].

By using the equations of Cole et al. [41], a numerically lower NUE was obtained in both periods but showing the same behavior as the values obtained from the BCNRM software. The lower NUE observed in the second period was due to animals decreasing their ADG caused by the decrease in DMI. Therefore, they were less efficient in retaining the ingested N. This scenario was not reflected with the values obtained using the BCNRM software.

Although there is no official data in Chile, some reports [68], as well as the empirical experience, indicate values of CP ranging from 10.5 to 20.0%, whereas energy metabolizable energy (ME) range between 2.60 to 2.90 Mcal/kg DM. This variation is explained in part because many production systems, particularly in Southern Chile, are grass-fed and use strategic supplementation with concentrates (usually small cereal grains and byproducts). Meanwhile, feed yards are closer to 13.5% CP and 2.75 Mcal ME/kg DM, but still showing great variability among them. Ruminal protein balance (RPB) has been proposed as a new trait to be used in diet formulation by Sauvant and Nozière [69] in the INRA (Institute National de Recherche Agronomique) 2018 system because it induces several interesting responses of protein and energy efficiency. These authors also indicated a negative relation of RPB with microbial growth efficiency and highlighted the relevance of degradable CP and energy in the rumen. Also, it has been demonstrated a decrease of OM digestibility when RPB decrease under zero [69,70].

Most of the actuals systems (NASEM, INRA 2018, DVE/OEB, NorFor, AFRC, CSIRO) have a mechanistic approach that looks at improving animal performance, product quality, the efficiency of feed use, and at the same time considers the environmental impacts by reducing N excretion and CH_4 emissions. The last release of INRA system (2018), has considered as a relevant criterion not only to evaluate the balance between degradable N and the energy available in the rumen but also to integrate the quantitative effects of the energy × nitrogen interactions in digestive processes, as previously described by Sauvant and Nozière [69]. A major focus in the new model was fractional passage rates of liquids, forages, and concentrates (proportion of concentrate in the diet), as well as the level of feeding. In our case, we use a fractional passage rate constant of 0.05%/h, but based on equations of the INRA 2018, it should be 0.04%/h. This, as mentioned before, affected the digestibility of the OM and the results associated with the NUE.

5. Conclusions

A dietary strategy that combines a high synchrony index with the MP system increases the NUE without negatively affecting animal performance or carcass characteristics compared to heifers fed a diet formulated based on their CP requirements. The lower N excretions show that the MPD is more environmentally friendly. In addition, the lower CP content in the MP diet may increase the beef production system's profitability because protein is one of the highest costs in the ration. Finally, we accepted the hypothesis, since the combination of synchrony and the MP system achieve greater efficiency in the use of nitrogen, without negatively affecting animals' performance neither the quality of the carcass.

Supplementary Materials: The following are available online at http://www.mdpi.com/2076-2615/10/5/852/s1, Figure S1: Adjusted degradation curves of the OM and CP, Table S1: Values of the in situ degradation characteristics of OM and CP.

Author Contributions: R.A.A. conceived and supervised the study and wrote the first draft. S.K. and G.G. executed the experiment, collected and analyzed the data. J.P.K. supervised in vitro and in situ incubations, collaborated in the statistical analysis, and contributed to editing the manuscript. C.A.-G. was the Laboratory responsible and contributed the final review of the manuscript. All authors have read and agreed to the published version of the manuscript.

Funding: This research received no external funding.

Acknowledgments: The authors thank all the members of the Animal Nutrition Laboratory at Universidad Austral de Chile for their generous collaboration in this research.

Conflicts of Interest: The authors declare no conflict of interest.

References

1. Arias, R.A. Intensificación de los sistemas ganaderos y la contaminación ambiental, el caso de los gases efecto invernadero y el nitrógeno. In Proceedings of the XXII Jornadas de Extensión Agrícola "Producción de carne: Aspectos técnicos para enfrentar las demandas de calidad y sustentabilidad en un clima cambiante", Temuco, Chile, 31 May 2011; p. 11. (In Spanish).
2. Alfaro, M.; Salazar, F. Ganadería y contaminación difusa, implicancias para el sur de Chile. *Agricultura Técnica* **2005**, *65*, 330–340. (In Spanish) [CrossRef]
3. Vasconcelos, J.T.; Tedeschi, L.O.; Fox, D.G.; Galyean, M.L.; Greene, L.W. Feeding nitrogen and phosphorus in beef cattle feedlot production to mitigate environmental impacts. *Prof. Anim. Sci.* **2007**, *23*, 8–17. [CrossRef]
4. Klemesrud, M.J.; Klopfenstein, T.J.; Lewis, A.J. Metabolizable methionine and lysine requirements of growing cattle. *J. Anim. Sci.* **2000**, *78*, 199–206. [CrossRef] [PubMed]
5. NRC. *Nutrient Requirement of Beef Cattle, update 2000*; National Academy Press: Washington DC, USA, 2000; p. 232.
6. Owens, F.N.; Qi, S.; Sapienza, D.A. Invited Review: Applied protein nutrition of ruminants—Current status and future directions. *PAS* **2014**, *30*, 150–179. [CrossRef]
7. Herrera-Saldana, R.; Gomez-Alarcon, R.; Torabi, M.; Huber, J.T. Influence of Synchronizing Protein and Starch Degradation in the Rumen on Nutrient Utilization and Microbial Protein Synthesis. *J. Dairy Sci.* **1990**, *73*, 142–148. [CrossRef]
8. Clark, J.H.; Klusmeyer, T.H.; Cameron, M.R. Microbial Protein-Synthesis and Flows of Nitrogen Fractions to the Duodenum of Dairy-Cows. *J. Dairy Sci.* **1992**, *75*, 2304–2323. [CrossRef]
9. Henning, P.H.; Steyn, D.G.; Meissner, H.H. The effect of energy and nitrogen supply pattern on rumen bacterial growth in vitro. *Anim. Sci. J.* **1991**, *53*, 165–175. [CrossRef]

10. Hall, M.B.; Huntington, G.B. Nutrient synchrony: Sound in theory, elusive in practice. *J. Anim. Sci.* **2008**, *86*, E287–E292. [CrossRef]

11. Keim, J.P.; Anrique, R. Nutritional strategies to improve nitrogen use efficiency by grazing dairy cows. *Chil. J. Agric. Res.* **2011**, *71*, 623–633. [CrossRef]

12. Sinclair, L.A.; Garnsworth, P.C.; Newbold, J.R.; Buttery, P.J. Effect of synchronizing the rate of dietary energy and nitrogen release on rumen fermentation and microbial protein synthesis in sheep. *J. Agric. Sci.* **1993**, *120*, 251–263. [CrossRef]

13. Chumpawadee, S.; Sommart, K.; Vongpralub, T.; Pattarajinda, V. Effects of synchronizing the rate of dietary energy and nitrogen release on ruminal fermentation, microbial protein synthesis, blood urea nitrogen and nutrient digestibility in beef cattle. *Asian Australas. J. Anim. Sci.* **2006**, *19*, 181–188. [CrossRef]

14. Yang, J.Y.; Seo, J.; Kim, H.J.; Seo, S.; Ha, J.K. Nutrient Synchrony: Is it a Suitable Strategy to Improve Nitrogen Utilization and Animal Performance? *Asian Australas J. Anim. Sci.* **2010**, *23*, 972–979. [CrossRef]

15. Berthiaume, R.; Benchaar, C.; Chaves, A.V.; Tremblay, G.F.; Castonguay, Y.; Bertrand, A.; Bélanger, G.; Michaud, R.; Lafrenière, C.; McAllister, T.A.; et al. Effects of nonstructural carbohydrate concentration in alfalfa on fermentation and microbial protein synthesis in continuous culture1. *J. Dairy Sci.* **2010**, *93*, 693–700. [CrossRef] [PubMed]

16. Dewhurst, R.J.; Davies, D.R.; Merry, R.J. Microbial protein supply from the rumen. *Anim. Feed Sci. Technol.* **2000**, *85*, 1–21. [CrossRef]

17. Huntington, G.B. Starch utilization by ruminants: From basics to the bunk. *J. Anim. Sci.* **1997**, *75*, 852–867. [CrossRef]

18. Chumpawadee, S.; Sommart, K.; Vongpralub, T.; Pattarajinda, V. Effect of synchronizing the rate of degradation of dietary energy and nitrogen release on growth performance in Brahman beef cattle. *Suranaree J. Sci. Technol.* **2006**, *28*, 59–70.

19. Duarte, M.d.S.; Paulino, P.V.R.; Filho, S.d.C.V.; Paulino, M.F.; Detmann, E.; Zervoudakis, J.T.; Monnerat, J.P.I.d.S.; Viana, G.d.S.; Silva, L.H.P.; Serão, N.V.L. Performance and meat quality traits of beef heifers fed with two levels of concentrate and ruminally undegradable protein. *Trop. Anim. Health Prod.* **2011**, *43*, 877–886. [CrossRef]

20. Cooper, R.J.; Milton, T.; Klopfenstein, T.J. *Phase-Feeding Metabolizable Protein for Finishing Steers*; University of Nebraska: Lincoln, NE, USA, 2000; pp. 63–65.

21. Erickson, G.E.; Klopfenstein, T.J.; Milton, T.; Herold, D. *Effects of Matching Protein Requirements on Performance and Waste Management in the Feedlot*; Nebraska Beef Reports; University of Nebraska: Lincoln, NE, USA, 1 January 1999 ; pp. 60–63.

22. Arias, R.A.; Keim, J.P.; Gandarillas, M.; Velasquez, A.; Alvarado-Gilis, C.; Mader, T.L. Performance and carcass characteristics of steers fed with two levels of metabolizable energy intake during summer and winter season. *Animal* **2019**, *13*, 221–230. [CrossRef]

23. Verbič, J.; Ørskov, E.R.; Žgajnar, J.; Chen, X.B.; Žnidaršič-Pongrac, V. The effect of method of forage preservation on the protein degradability and microbial protein synthesis in the rumen. *Anim. Feed Sci. Technol.* **1999**, *82*, 195–212. [CrossRef]

24. National Academies of Sciences Engineering and Medicine (NASEM). *Nutrient Requirements of Beef Cattle: Eighth Revised Edition*; The National Academies Press: Washington, DC, USA, 2016; p. 494. [CrossRef]

25. Tilley, J.M.A.; Terry, R.A. A two-stage technique for the in vitro digestion of forage crops. *Grass Forage Sci.* **1963**, *18*, 104–111. [CrossRef]

26. Van Soest, P.J.; Robertson, J.B.; Lewis, B.A. Methods for Dietary Fiber, Neutral Detergent Fiber, and Nonstarch Polysaccharides in Relation to Animal Nutrition. *J. Dairy Sci.* **1991**, *74*, 3583–3597. [CrossRef]

27. Association of Official Analytical Chemists (AOAC). *Official Methods of Analysis*, 16th ed.; Association of Official Analytical Chemists: Washington, DC, USA, 1996.

28. Ørskov, E.R.; McDonald, I. The estimation of protein degradability in the rumen from incubation measurements weighted according to rate of passage. *J. Agr. Sci. Cambridge* **1979**, *92*, 499–503. [CrossRef]

29. Mcdonald, I. A Revised Model for the Estimation of Protein Degradability in the Rumen. *J. Agric. Sci.* **1981**, *96*, 251–252. [CrossRef]

30. AFRC. *Energy and Protein Requirments of Ruminants*; CAB International: Wallingford, UK, 1993.

31. Hvelplund, T.; Weisbjerg, M.R. In situ techniques for the estimation of protein degradability and postrumen availability. In *Forage Evaluation in Ruminant Nutrition*; Givens, D.I., O'wen, E., Axford, R.F.E., Omed, H.M., Eds.; CAB international: Wallingford, UK, 2000; pp. 233–258. [CrossRef]

32. Theodorou, M.K.; Williams, B.A.; Dhanoa, M.S.; McAllan, A.B.; France, J. A simple gas production method using a pressure transducer to determine the fermentation kinetics of ruminant feeds. *Anim. Feed Sci. Technol.* **1994**, *48*, 185–197. [CrossRef]

33. France, J.; Dijkstra, J.; Dhanoa, M.S.; López, S.; Bannink, A. Estimating the extent of degradation of ruminant feeds from a description of their gas production profiles observed in vitro: Derivation of models and other mathematical considerations. *Br. J. Nutr.* **2000**, *83*, 143–150. [CrossRef]

34. Groot, J.C.J.; Cone, J.W.; Williams, B.A.; Debersaques, F.M.A.; Lantinga, E.A. Multiphasic analysis of gas production kinetics for in vitro fermentation of ruminant feeds. *Anim. Feed Sci. Technol.* **1996**, *64*, 77–89. [CrossRef]

35. Vlassa, M.; Filip, M. Determination of purine derivatives in bovine urine using rapid chromatographic techniques. *Arch. Zootech.* **2009**, *12*, 59–70.

36. Al-Khalidi, U.A.S.; Chaglassian, T.H. The species distribution of xanthine oxidase. *Biochem. J.* **1965**, *97*, 318–320. [CrossRef]

37. Chizzotti, M.L.; Valadares Filho, S.C.; Valadares, R.F.D.; Chizzotti, F.H.M.; Tedeschi, L.O. Determination of creatinine excretion and evaluation of spot urine sampling in Holstein cattle. *Livest. Sci.* **2008**, *113*, 218–225. [CrossRef]

38. Faichney, G.J.; Welch, R.J.; Brown, G.H. Prediction of the excretion of allantoin and total purine derivatives by sheep from the 'creatinine coefficient'. *J. Agric. Sci.* **1995**, *125*, 425–428. [CrossRef]

39. Ørskov, E.R.; MacLeod, N.A.; Nakashima, Y. Effect of different volatile fatty acids mixtures on energy metabolism in cattle. *J. Anim. Sci.* **1991**, *69*, 3389–3397. [CrossRef]

40. Chen, X.B.; Gomes, M.J. *Estimation of Microbial Protein Supply to Sheep and Cattle based on Urinary Excretion of Purine Derivatives—An Overview of the Technical Details*; International Feed Resources Unit: Aberdeen, UK, September 1995; pp. 21–25.

41. Cole, N.A.; Defoor, P.J.; Galyean, M.L.; Duff, G.C.; Gleghorn, J.F. Effects of phase-feeding of crude protein on performance, carcass characteristics, serum urea nitrogen concentrations, and manure nitrogen of finishing beef steers. *J. Anim. Sci.* **2006**, *84*, 3421–3432. [CrossRef]

42. Erickson, G.E.; Klopfenstein, T.J.; Milton, T.C. Dietary protein effects on nitrogen excretion and volatilization in open-dirt feedlots. In Proceedings of the 8th International Symposium on Animals, Agriculture, and Food Processing Wastes, St. Joseph, MO, USA, 9–11 October 2000; pp. 297–304.

43. Hersom, M.J. Opportunities to enhance performance and efficiency through nutrient synchrony in forage-fed ruminants. *J. Anim. Sci.* **2008**, *86*, E306–E317. [CrossRef]

44. Erickson, G.; Klopfenstein, T. Nutritional and management methods to decrease nitrogen losses from beef feedlots. *J. Anim. Sci.* **2010**, *88*, E172–E180. [CrossRef]

45. Arias, R.A.; Velasquez, A.; Toneatti, M. Simulation of the nitrogen use efficiency in pasture-finished steers in southern Chile. *Arch. Med. Vet.* **2013**, *45*, 125–134. [CrossRef]

46. Waldrip, H.M.; Cole, N.A.; Todd, R.W. Nitrogen sustainability and beef-cattle feedyards: I. Introduction and influence of pen surface conditions and diet. *PAS* **2015**, *31*, 89–100. [CrossRef]

47. Xu, M.; Rinker, M.; McLeod, K.R.; Harmon, D.L. *Yucca schidigera* extract decreases in vitro methane production in a variety of forages and diets. *Anim. Feed Sci. Technol.* **2010**, *159*, 18–26. [CrossRef]

48. Detmann, E.; Valente, É.E.L.; Batista, E.D.; Huhtanen, P. An evaluation of the performance and efficiency of nitrogen utilization in cattle fed tropical grass pastures with supplementation. *Livest. Sci.* **2014**, *162*, 141–153. [CrossRef]

49. Detmann, E.; Paulino, M.F.; Mantovani, H.C.; Valadares, S.D.; Sampaio, C.B.; de Souza, M.A.; Lazzarini, I.; Detmann, K.S.C. Parameterization of ruminal fibre degradation in low-quality tropical forage using Michaelis-Menten kinetics. *Livest. Sci.* **2009**, *126*, 136–146. [CrossRef]

50. Firkins, J.L.; Berger, L.L.; Merchen, N.R.; Fahey, G.C., Jr. Effects of forage particle size, level of feed intake and supplemental protein degradability on microbial protein synthesis and site of nutrient digestion in steers. *J. Anim. Sci.* **1986**, *62*, 1081–1094. [CrossRef]

51. Reynolds, C.K.; Kristensen, N.B. Nitrogen recycling through the gut and the nitrogen economy of ruminants: An asynchronous symbiosis. *J. Anim. Sci.* **2008**, *86*, E293–E305. [CrossRef]

52. Erickson, G.E.; Klopfenstein, T.J. Managing N inputs and the effect on N losses following excretion in open-dirt feedlots in Nebraska. *Sci. World J.* **2001**, *2*, 830–835. [CrossRef]

53. Seo, J.K.; Yang, J.; Kim, H.J.; Upadhaya, S.D.; Cho, W.M.; Ha, J.K. Effects of Synchronization of Carbohydrate and Protein Supply on Ruminal Fermentation, Nitrogen Metabolism and Microbial Protein Synthesis in Holstein Steers. *Asian Australas. J. Anim. Sci.* **2010**, *23*, 1455–1461. [CrossRef]

54. Bach, A.; Calsamiglia, S.; Stern, M.D. Nitrogen Metabolism in the Rumen. *J. Dairy Sci.* **2005**, *88*, E9–E21. [CrossRef]

55. Seo, J.K.; Kim, M.H.; Yang, J.Y.; Kim, H.J.; Lee, C.H.; Kim, K.H.; Ha, J.K. Effects of Synchronicity of Carbohydrate and Protein Degradation on Rumen Fermentation Characteristics and Microbial Protein Synthesis. *Asian Australas. J. Anim. Sci.* **2013**, *26*, 358–365. [CrossRef]

56. Prendergast, S.L.; Gibbs, S.J. A comparison of microbial protein synthesis in beef steers fed ad libitum winter ryegrass or fodder beet. *Proc. New. Zeal. Soc. An.* **2015**, *75*, 251–256.

57. Koenig, K.M.; Beauchemin, K.A. Nitrogen metabolism and route of excretion in beef feedlot cattle fed barley-based finishing diets varying in protein concentration and rumen degradability. *J. Anim. Sci.* **2013**, *91*, 2310–2320. [CrossRef]

58. Davies, K.L.; McKinnon, J.J.; Mutsvangwa, T. Effects of dietary ruminally degradable starch and ruminally degradable protein levels on urea recycling, microbial protein production, nitrogen balance, and duodenal nutrient flow in beef heifers fed low crude protein diets. *Can. J. Anim. Sci.* **2013**, *93*, 123–136. [CrossRef]

59. Gehman, A.M.; Kononoff, P.J. Nitrogen utilization, nutrient digestibility, and excretion of purine derivatives in dairy cattle consuming rations containing corn milling co-products. *J. Dairy Sci.* **2010**, *93*, 3641–3651. [CrossRef]

60. Yan, T.; Frost, J.P.; Keady, T.W.J.; Agnew, R.E.; Mayne, C.S. Prediction of nitrogen excretion in feces and urine of beef cattle offered diets containing grass silage. *J. Anim. Sci.* **2007**, *85*, 1982–1989. [CrossRef]

61. Dijkstra, J.; Reynolds, C.K.; Kebreab, E.; Bannink, A.; Ellis, J.L.; France, J.; van Vuuren, A.M. Challenges in ruminant nutrition: Towards minimal nitrogen losses in cattle. In *Energy and Protein Metabolism and Nutrition in Sustainable Animal Production*; Oltjen, J., Kebreab, E., Lapierre, H., Eds.; Wageningen Academic Publishers: Wageningen, The Netherlands, 2013; pp. 47–58.

62. Cantalapiedra-Hijar, G.; Peyraud, J.L.; Lemosquet, S.; Molina-Alcaide, E.; Boudra, H.; Noziere, P.; Ortigues-Marty, I. Dietary carbohydrate composition modifies the milk N efficiency in late lactation cows fed low crude protein diets. *Animal* **2014**, *8*, 275–285. [CrossRef]

63. Erickson, G.; Watson, A.K.; MacDonald, J.C.; Klopfenstein, T.J. Protein nutrition evaluation and application to growing and finishing cattle. In Proceedings of the 27th annual Florida Ruminant Nutrition Symposium, Gainesville, FL, USA, 15–17 February 2016; pp. 91–102.

64. Cole, N.A.; Greene, L.W.; McCollum, F.T.; Montgomery, T.; McBride, K. Influence of oscillating dietary crude protein concentration on performance, acid-base balance, and nitrogen excretion of steers. *J. Anim. Sci.* **2003**, *81*, 2660–2668. [CrossRef]

65. Boonsaen, P.; Soe, N.W.; Maitreejet, W.; Majarune, S.; Reungprim, T.; Sawanon, S. Effects of protein levels and energy sources in total mixed ration on feedlot performance and carcass quality of Kamphaeng Saen steers. *Agric. Nat. Resour.* **2017**, *51*, 57–61. [CrossRef]

66. Gleghorn, J.F.; Elam, N.A.; Galyean, M.L.; Duff, G.C.; Cole, N.A.; Rivera, J.D. Effects of crude protein concentration and degradability on performance, carcass characteristics, and serum urea nitrogen concentrations in finishing beef steers. *J. Anim. Sci.* **2004**, *82*, 2705–2717. [CrossRef]

67. Galyean, M.L. Protein levels in beef cattle finishing diets: Industry application, university research, and systems results. *J. Anim. Sci.* **1996**, *74*, 2860–2870. [CrossRef]

68. Morales, R.; Folch, C.; Iraira, S.; Teuber, N.; Realini, C.E. Nutritional quality of beef produced in Chile from different production systems. *Chil. J. Agric. Res.* **2012**, *72*, 80–86. [CrossRef]

69. Sauvant, D.; Nozière, P. Quantification of the main digestive processes in ruminants: The equations involved in the renewed energy and protein feed evaluation systems. *Animal* **2016**, *10*, 755–770. [CrossRef]

70. Sauvant, D.; Nozière, P. The Rumen Protein Balance as a key trait to model ruminant responses to dietary proteins. In Proceedings of the 6th EAAP International Symposium on Energy and Protein Metabolism and Nutrition, Belo Horizonte, Brazil, 9–12 September 2019; p. 4.

Article

Effects of the Diet Inclusion of Common Vetch Hay Versus Alfalfa Hay on the Body Weight Gain, Nitrogen Utilization Efficiency, Energy Balance, and Enteric Methane Emissions of Crossbred Simmental Cattle

Wuchen Du [1], Fujiang Hou [2,*], Atsushi Tsunekawa [3,*], Nobuyuki Kobayashi [3], Toshiyoshi Ichinohe [4] and Fei Peng [5,6]

[1] The United Graduate School of Agricultural Sciences, Tottori University, Tottori 680-8550, Japan; duwuchen1990@126.com

[2] State Key Laboratory of Grassland Agro-Ecosystems, Key Laboratory of Grassland Livestock Industry Innovation, Ministry of Agriculture, College of Pastoral Agriculture Science and Technology, Lanzhou University, Lanzhou 730020, China

[3] Arid Land Research Center, Tottori University, Tottori 680-0001, Japan; kobayashi.nobuyuki@alrc.tottori-u.ac.jp

[4] Faculty of Life and Environmental Science, Shimane University, Matsue 690-8504, Japan; toshi@life.shimane-u.ac.jp

[5] International Platform for Dryland Research and Education, Tottori University, Tottori 680-0001, Japan; pengfei@tottori-u.ac.jp

[6] Key Laboratory of Desert and Desertification, Northwest Institute of Eco-Environment and Resources, Chinese Academy of Sciences, Lanzhou 730000, China

* Correspondence: cyhoufj@lzu.edu.cn (F.H.); tsunekawa@tottori-u.ac.jp (A.T.); Tel.: +86-0931-8913047 (F.H.); +81-0857-23-3411 (A.T.)

Received: 10 October 2019; Accepted: 14 November 2019; Published: 18 November 2019

Simple Summary: Nitrogen utilization efficiency and enteric methane emission from ruminants remain the primary concerns when developing ruminant feed globally. Nitrogen utilization efficiency is the ratio of retained nitrogen in body tissue to the total nitrogen intake, which is the main factor in the body weight gain of ruminants, and usually range from 15% to 40%. The methane emissions of ruminants are an inevitable by-product when feeds have been fermented in the rumen and represents a 2% to 12% loss of diet energy. The low nitrogen utilization of ruminants can damage air quality and lead to soil nitrification and acidification, whereas high methane emissions from ruminants can increase global warming. Our study investigated the effects of two kinds of legumes (alfalfa and common vetch) with different levels (20% vs. 40%) of total dry matter allowance on body weight gain, nutrient digestibility, nitrogen utilization efficiency, and enteric methane emissions for crossbred Simmental cattle. Our results suggested that nitrogen utilization efficiency and methane emissions are significantly affected by the legume species and proportions. These results could be beneficial for the development of regional or national ruminant feeding systems, thereby improving nitrogen utilization efficiency and reducing methane emissions.

Abstract: A low nitrogen utilization efficiency (NUE, the ratio of retained N to N intake) and high methane (CH_4) emissions of ruminants can lead to potentially high diet protein wastage and directly contribute to global warming. Diet manipulation is the most effective way to improve NUE or reduce CH_4 emissions. This study investigated how replacing oat hay with alfalfa hay (AH) or common vetch hay (CVH) with different proportions (20% (20) and 40% (40) of the total dry matter (DM) allowance) affects the body weight gain (BWG), NUE, and CH_4 emissions of crossbred Simmental cattle. The forage dry matter intake (DMI) and the total DMI of cattle fed on a CVH40 diet were significantly higher than the values for those fed on AH20 or AH40 diets ($p < 0.05$). There were no differences in the BWG for the four treatments observed, however, nutrient digestibility significantly decreased in the AH40 diet as compared with the AH20 diet ($p < 0.05$). The NUE was significantly lower in AH40 than in CVH20. The CH_4 emissions were significantly lower for the CVH40 diet than with the AH20 diet ($p < 0.05$). Our findings suggest that a 20% AH and 40% CVH substitution for oat hay are the optimal proportions to maintain the BWG, NUE, nutrient digestibility, and reduce the CH_4 emissions of crossbred Simmental cattle. Overall, CVH has a greater potential to reduce CH_4 emissions than AH.

Keywords: leguminous forage; digestibility; energy utilization efficiency; nitrogen metabolism; dryland

1. Introduction

The impacts of the low nitrogen utilization efficiency (NUE) and high enteric methane (CH_4) emissions of beef cattle remain the primary concerns in the development of ruminant feeding systems [1]. A low NUE could contribute more ammonia emissions to the air and more manure N outputs to the soil [2], which could damage air quality [3] and lead to soil nitrification and acidification [2]. The enteric CH_4 emissions from ruminants not only represent a loss of diet energy [4] but could also contribute to global warming [5]. The development of a diet that can improve the NUE and reduce enteric CH_4 emissions is in demand and would be beneficial to both animal husbandry and in facing global environmental challenges [6,7].

Grass occupies an important role in the ruminant feeding system as it represents a low-cost and abundant source of dry matter (DM). However, grass only is not capable of sustaining the required levels of animal production due to its low feeding value [8]. Hence, the interest in supplementing legumes into a grass-based diet because they are rich in protein and energy [9]. Previous studies in sheep have shown that the intake of organic matter (OM) and crude protein (CP), as well as ruminal ammonia nitrogen (N) concentrations increased with a 3:1 grass/legume mixture diet as compared to a diet of only grass [10]; the total tract digestibility of CP and digestible CP was significantly higher in a 1:1 alfalfa/oat mixture diet than oat only hay diet [11]. Moreover, alfalfa (78%) and grass (22%) pastures could reduce energy loss through CH_4 emissions of cows as compared to grass-only pastures [12]. However, these studies focused on diets where supplementation of legumes was the only factor considered. Few studies explained the effects of diets with different levels of legumes on feed intake, digestibility, and CH_4 emissions. In another study, the inclusion of 30% common vetch hay (CVH) was more optimal in reducing CH_4 emissions than 0%, 10%, and 20% CVH diets but significantly depressed digestibility as compared to a 20% CVH diet [13]. Recently, Kobayashi et al. [14,15] concluded that 8% to 14% alfalfa hay (AH) in the warm season and 8% to 21% in the cool season were optimal when considering body weight gain (BWG), metabolizable energy (ME) intake, and increased economic benefits of growing beef cattle on a corn- and straw-based diet.

Alfalfa (*Medicago sativa* L.) is the mostly widely planted perennial legume crop in the world and has been studied for many years [12,14]. Common vetch (*Vicia sativa* L.), a multipurpose annual cereal legume for livestock feed [16], not only plays an important role in dryland mixed farming systems [16] for grazing [17] or cutting for hay [18] but also meets the structural forage deficit in winter, which is linked to the seasonality of other feed sources [9]. Previous studies have shown that CP, digestible OM intake, and in vitro OM digestibility are significantly higher with the oat and common vetch mixture diet than with the oat-only diet for cattle [18], and the growth performance of animals is significantly higher with common vetch supplementation than without [19]. However, until now, there has been no available information on whether common vetch could substitute alfalfa in the ruminant feeding system or whether optimal proportions of common vetch could replace alfalfa. Therefore, the objective of this study was to investigate how CVH versus AH affects BWG, N metabolism (i.e., N digestibility, ruminal ammonia-N, and blood urea N (BUN) concentrations), and CH_4 emissions associated with ruminal fermentation parameters using two different proportions (20% (20) and 40% (40) of the total DM allowance) for growing crossbred Simmental cattle in dryland environments at similar CP and predicted ME levels.

2. Materials and Methods

The Animal Ethics Committee of Gansu Province, China, approved the experimental protocols (file No. 2010-1 and 2010-2). This study was conducted in Linze Grassland Agriculture Trial Station, Lanzhou University, Zhangye City, Gansu Province, China (latitude 39.24°N, longitude 00.06°E, 1390 m a.s.l.), which is characterized as a typical temperature continental climate because its average annual precipitation is 121.5 mm and annual average temperature is 7.7 °C. In this study, the AH was second cut, and common vetch (*Vicia sativa* L.cv. Lanjian No. 3) was harvested at the flowering stage and restored as common vetch hay (CVH). Oat hay (OH; *Avena sativa* L.) was purchased from a forage company (Sanbao Agricultural Company, Zhangye, Gansu, China). The ingredients for the concentrate (maize, soybean meal, and wheat bran) were acquired from a local source. The chemical composition of the forage and ingredients of the concentrate are shown in Table 1.

Table 1. Chemical composition of alfalfa hay, oat hay, common vetch hay, and ingredients of the concentrate used in the experimental diets.

Item [†]	Alfalfa Hay	Oat Hay	Common Vetch Hay	Soybean Meal	Wheat Bran	Maize
OM, g/kg DM	905	942	918	935	931	983
CP, g/kg DM	168	60	177	465	182	83
NDF, g/kg DM	458	559	413	166	454	100
ADF, g/kg DM	347	407	302	102	186	20
Ether extract, g/kg DM	22	18	23	26	55	44
GE, MJ/kg DM	17.9	16.8	17.7	19.6	19.4	18.5
MEC [§], MJ/kg DM	8.7	9.0	9.5	13.0	10.9	13.4
MPC [¶], g/kg DM	62	68	71	87	73	90

[†] OM, organic matter; CP, crude protein; NDF, neutral detergent fiber; ADF, acid detergent fiber; GE, gross energy; MEC, metabolizable energy concentration; MPC, metabolizable protein concentration; [§, ¶] They were calculated by the Agricultural and Food Research Council (1993) and the Chinese Feeding Standard for Beef Cattle (2004), see details in Methods and Materials.

2.1. Animals, Treatments, and Diets

The Animal Ethics Committee of Gansu Province, China, approved the experimental protocols. This experiment involved 16 crossbred male Simmental cattle (Simmental × local cattle) with initial body weights (BWs) of 216 ± 24.4 kg (mean ± standard deviation, 10 months of age) at the start of the experimental period. The experiment used a randomized block experimental design with a 2 × 2 factorial arrangement of diets. All 16 cattle were allocated to one of the 4 treatments. The forage to concentrate ratio was fixed (60:40, DM basis) for all diets. Diet treatments used two kinds of legumes (AH and CVH) and two different OH-to-AH/CVH ratios in the diet (40:20 or 20:40, DM basis), indicated as follows: 20% CVH and 40% OH (CVH20), 40% CVH and 20% OH (CVH40), 20% AH and 40% OH (AH20), and 40% AH and 20% OH (AH40). This experiment consisted of 2 feeding periods. Each period consisted of a 14 day diet adaptation in the cowshed and 32 day data collection period in the chambers.

The target BWG for each cattle was set at 1.5 kg/d. All experimental diets were formulated to provide sufficient ME and metabolizable protein (MP) to meet the target BWG for cattle according to the published estimation equations and values of the Agricultural and Food Research Council [20] and BW of cattle (measured every 8 days). The diet composition required to fulfill the ME and MP requirements was calculated based on the tabulated values of digestible energy and ruminal CP degradation parameters for OH, AH, and the concentrate ingredients established by the Chinese

Feeding Standard for Beef Cattle [21]. The digestibility of ruminal CP and energy and ruminal degradation parameters for CVH were taken from Larbi et al. [16]. The CP, ME, and MP levels of all diets are shown in Table 2. Throughout this experimental period of 8 weeks, all cattle were given free access to water and 10 g/day of mineral mixture containing (minimum values in mg) manganese, 720; copper, 30; biotin, 0.05; folic acid, 0.4; vitamin B_1, 50; vitamin B_2, 2.5; vitamin B_6, 0.5; and vitamin B_{12}, 0.1. The daily mixed forage was divided into two equal parts and offered as separate meals twice a day (08:00 and 19:00). The mixed concentrate was fed once a day (14:00).

Table 2. Composition of the feed ingredients and the target metabolizable energy concentration and metabolizable protein concentration of all diets.

Feed Formula	Experimental Diet [†]			
	CVH20	**CVH40**	**AH20**	**AH40**
Forage				
Leguminous forage (g/kg DM)	200	400	200	400
Oat hay (g/kg DM)	400	200	400	200
Concentrate				
Maize (g/kg DM)	30	80	48	120
Soybean meal (g/kg DM)	92	25	107	56
Wheat bran (g/kg DM)	278	295	245	224
Nutrient value [‡]				
CP (g/kg DM)	156.3	156.4	156.4	156.4
MEC [§] (MJ/kg DM)	10.05	10.05	10.05	10.05
MPC [¶] (g/kg DM)	102.9	94.6	106.1	101.4

[†] CVH20, 20% common vetch + 40% oat hay; CVH40, 40% common vetch + 20% oat hay; AH20, 20% alfalfa + 40% oat hay; AH40, 40% alfalfa + 20% oat hay. [‡] CP, crude protein, MEC, metabolizable energy concentration, MPC, metabolizable protein concentration. [§,¶] These values were calculated by the Agricultural and Food Research Council (1993) and the Chinese Feeding Standard for Beef Cattle (2004); see details in Methods and Materials.

2.2. Chamber Description

The four indirect open-respiration calorimeter chambers used in the present study were equipped with a computer-controlled air-handling system with air conditioning units set to a temperature of 18 ± 1 °C and relative humidity of $60\% \pm 10\%$. The calorimeter chambers were built with double Perspex walls fitted in aluminum frames [22], with a total volume of approximately 18 m³ (4.2 m long, 1.95 m wide, and 2.2 m high). Each chamber was equipped with a gas flow meter (GFM57, Aalborg, Orangeburg, New York, NY, USA) at the outflow site for recording the total airflow and an engine to ensure a slight negative pressure within the chamber. All chambers were ventilated by suction pumps with a flow rate of 45 to 50 m³/h. The exhaust air was removed for volume, temperature, and humidity measurement and analysis in the bottom, middle, and upper areas, inside each chamber. The concentrations of CO_2, CH_4, and O_2 in the air moving into and out of each chamber were measured every 16 min (the interval for each chamber) using a multigas analyzer (VA-3000, Horiba Ltd., Tokyo, Japan) in a general control room. The analyzer was calibrated using standard gases (O_2-free N_2 and a known quantity of CH_4, CO_2 and O_2, Dalian Special Gases Co., Ltd., Liaoning, China) at the beginning of the gas exchange collection period in each experiment. The determined concentrations were in an absolute range of 0–500 µL/L for CH_4, 0–2000 µL/L for CO_2, and 0%–25% (v/v) for O_2 (with linearity within this range). The recovery rate of CH_4 was determined by comparing the CH_4 release into the chamber with a given concentration as well as the CH_4 concentration at the outlet. The gas recovery rate was approximately $100\% \pm 2\%$ for all chambers, as highlighted recently by Gerrits et al. [23]. Each chamber was designed with a dedicated door, which was located next to the animal trough. The staff only opened the door to feed the animal immediately after the completion of each data collection in the chamber during the 3 day gas exchange data collection period. This minimized the effects of feeding activity (less than 1 min) on the gas concentrations inside. The CH_4 emissions were expressed as the average CH_4 emissions (g/day) from 3-day measurements divided by BWG, which was calculated from the BW change between moving in and moving out the chamber.

2.3. Energy Balance

ME intake (MEI) was calculated as the difference between GEI, excreted fecal energy (FE), and the sum of UE and CH_4 energy (CH_4-E) output. Retained energy (RE) was calculated using the equation MEI − heat production (HP). CH_4-E was calculated from CH_4 emissions (L/day) and the conversion coefficient (39.54 kJ/L; [24]). The CH_4 emissions was converted to grams from the CH_4 emissions (L/day) using the conversion coefficient (0.716 g/L, [24]). HP (kJ/day) was calculated with the following equation [24]:

$$HP \ (kJ/day) = 16.18 \times O_2 \ \text{consumption} \ (L/day) + 5.02 \times CO_2 \ \text{production} \ (L/day)$$
$$- 2.17 \times CH_4 \ \text{production} \ (L/day) - 5.99 \times N \ \text{excretion} \ (\text{urinary N, g/day})$$

2.4. Sample Collection and Procedures

The amount of offered forage and concentrate and all leftovers was weighed daily throughout the experimental period to calculate the daily DM intake (DMI) for individual cattle. On day 15 of the experimental period, after the 14 day acclimation period for target feeds, one cattle was randomly selected from each diet group and moved to one of the four chambers for 8 days. On day 22, these cattle were moved to the individual pens in the cowshed, and another 4 cattle, randomly selected from the remaining cattle of the four diet groups, entered the chambers and left on day 30. This process continued until day 46 for the first feeding period, when all 16 cattle had completed 8 days of measurement. These acclimation and chamber measurements for metabolism and gas exchange were repeated for another 46 days with the 16 cattle randomly allocated to the four diets. The BWs of all cattle were measured in the morning with an empty stomach to calculate CH_4 emissions, energy, and N balance based on the metabolic BW when exchanging cattle between the chambers and the cowshed. The BWG (kg/day) was calculated by the difference of the BW at the start and end time of each feeding period. During the 8 days of measurements in the chamber, the cattle were kept for acclimation for the first 2 days. We collected the digestibility data over the following 3 days and gas exchange data (O_2 consumption, CH_4, and CO_2 emission) over the remaining 3 days. During the digestibility data collection period, the total weight of the daily excreted feces and urine was recorded. Feces, which were excreted onto a plastic mat placed under the cattle, were collected right after excretion with a shovel and placed into a plastic container (around 15 times a day but varied according to individuals) and weighed, mixed, and sampled once per day. A total of 10% of each feces sample was stored at −20 °C for later chemical analysis. All urine was collected once a day through a handmade urine bag into a bucket containing 200 mL 10% (*v/v*) H_2SO_4 to reduce ammonia loss. Acidified urine was checked for pH with a portable pH instrument (PHBJ-260, Shanghai INESA Scientific Instrument Co., Ltd., Shanghai, China). A total of 20% of the daily urine was removed with a 500 mL cylinder (deviation ± 5 mL) and stored at −20 °C for chemical analysis.

Rumen fluid samples were taken from each cattle 4 h post forage supply every morning using stomach tubing on the last 2 days of each feeding period after these cattle were moved to the cowshed. The collected samples were immediately measured for pH using a portable pH meter (PHBJ-260, Shanghai INESA Scientific Instrument Co., Ltd., Shanghai, China), strained through two layers of muslin (mesh size 1 mm^2) and stored at −20 °C for volatile fatty acid (VFA) analysis. An additional 1 mL of strained rumen fluid was deproteinated by adding 0.2 mL metaphosphoric acid (215 g/L) and 0.1 mL internal standard (Crotonic acid), and the VFA concentrations were determined by a gas chromatograph (Trace1300, Thermo Ltd., Rodano Milan, Italy) fitted with a polar capillary column. The plasma urea N concentration was assumed to be equivalent to the BUN concentration in the serum, since urea diffuses freely into and out of blood cells [25].

2.5. Chemical Analysis

After the chamber measurement, the stored feces samples were thawed at room temperature for 12 h, and the feces samples from each cattle over the three days were then mixed. A portion of the thawed feces sample was used for the N measurement according to the Association of Official Analytical Chemists methods, method 976.05 [26]. The CP concentration was calculated by multiplying the N concentration by 6.25. The remaining samples were oven dried at 65 °C for 48 h and then ground to pass through a 1 mm screen. A portion of each dried sample, mixed forage, and concentrate samples were used to measure ash by combustion using a muffle furnace at 550 °C for 10 h (method 942.05 [26]). The organic matter (OM) content (g/kg DM) was calculated by 1000 ash content (g/kg DM). Another portion of each dried sample was finely ground to measure gross energy (GE), neutral detergent fibers (NDFs), and acid detergent fibers (ADFs). The GE of the samples was determined with an automatic isoperibol calorimeter (6400, PARR Instrument Company, Moline, IL, USA). The NDF and ADF concentrations were analyzed sequentially in an ANKOM 2000 fiber analyzer (ANKOM Technology, Fairport, NY, USA) following the protocol described by Van Soest [27]. The ash was included to facilitate the NDF and ADF analysis of all the forage, concentrate, and feces samples. The α-amylase for NDF analysis was used only for the concentrate samples. The urine samples from each cattle over the three days were also thawed at room temperature for 12 h and then mixed before determining their urinary energy (UE) by using an automatic isoperibol calorimeter (see above), and N was determined using Kjeldahl procedure described previously by the Association of Official Analytical Chemists [26]. For the UE measurement, 4 mL fully mixed urine was taken and absorbed by a filter paper of a known weight, and then the total energy of the filter paper with a urine sample was measured by an automatic isoperibol calorimeter after it became dry at room temperature. There were another 5 samples using the same filter paper (known weight) to be measured for energy content (MJ/kg), which was used to calculate the UE. The measurements of CP, NDF, and GE of the forage and concentrate of the diets also followed the above methods and instruments. The ether extract of the forage and concentrate was analyzed using an ANKOM XT15 Extractor (ANKOM Technology, Fairport, NY, USA).

2.6. Statistical Analysis

A one-way analysis of variance (ANOVA) and generalized linear model analysis were used to investigate the effects of legume species (LS), legume proportion (LP), and their interactions (LS × LP) on DMI, BWG, nutrient digestibility, energy balance, N metabolism, and energy/N utilization efficiency. Differences among the means were considered to be significant at a $p \leq 0.05$ on the basis of Tukey's test, unless otherwise stated. The statistical program used in the current study was IBM SPSS Statistics for Windows, version 19.0 (IBM Corp., Armonk, NY, USA).

3. Results

3.1. Feed Intake, Apparent Nutrient Digestibility, and BWG

LS significantly influenced the forage DMI and total DMI ($p < 0.05$, Table 3). In detail, the forage DMI and total DMI of cattle were significantly higher when fed on a CVH40 diet than on AH20 and AH40 diets ($p < 0.05$, Figure 1a). However, no significant differences were found in the concentrate DMI under LS ($p > 0.05$, Table 3). In addition, there were no significant differences in the forage DMI, concentrate DMI, and total DMI of cattle under LP ($p > 0.05$, Table 3, Figure 1a).

Table 3. A general linear model analysis of legume species (LS), legume proportion (LP), and their interaction effect on feed intake, digestibility, growth performance, and CH_4 emissions (n = 8).

Item [†]	LS [‡]	LP [‡]	LS × LP [‡]
Dry matter intake (DMI)			
Forage DMI (g/kg BW$^{0.75}$/day)	5.783 *	0.932	0.498
Concentrate DMI (g/kg BW$^{0.75}$/day)	1.108	1.189	0.001
Total DMI (g/kg BW$^{0.75}$/day)	5.207 *	0.109	0.598
Digestibility			
DM digestibility (%)	0.215	5.671 *	1.303
OM digestibility (%)	0.306	6.744 *	1.582
NDF digestibility (%)	1.177	18.476 ***	0.001
Apparent N digestibility (%)	5.515 *	5.949 *	0.265
Growth performance			
BWG (kg/day)	0.205	0.403	1.389
FCR (kg BWG/kg DMI)	0.077	2.515	5.796 *
CH_4 emissions			
CH_4 emissions (g/kg BW$^{0.75}$/24 h)	5.907 *	7.056 *	0.815
CH_4 emissions (g/kg DMI/24 h)	1.698	5.604 *	0.000

[†] DMI, dry matter intake; DM, dry matter; OM, organic matter; NDF, neutral detergent fiber; BWG, body weight gain; FCR, feed conversion ratio (ratio of BWG divided by the total DMI). [‡] values are the F value, * $p < 0.05$, and *** $p < 0.001$.

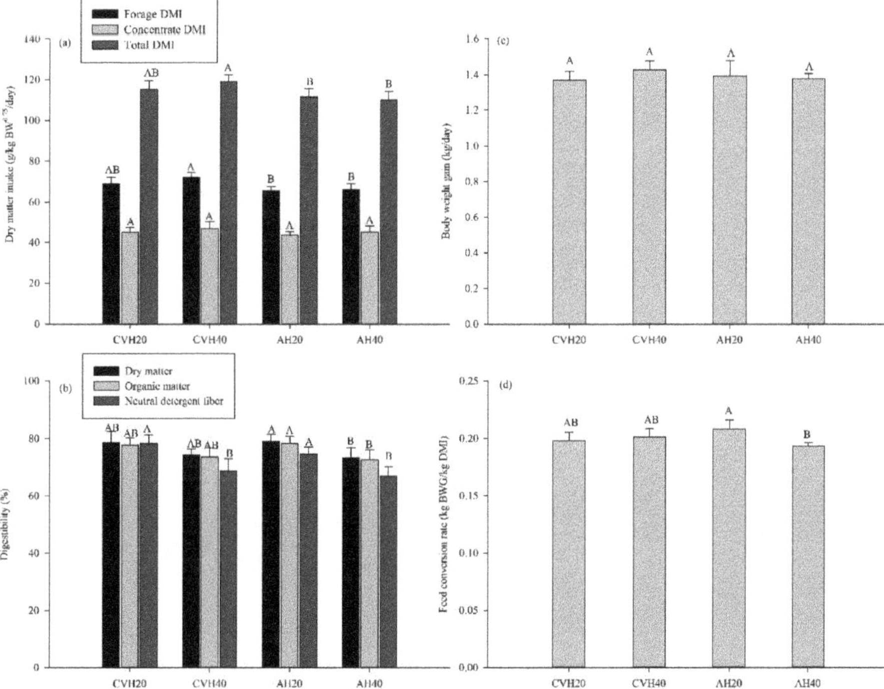

Figure 1. The dry matter intake (DMI) (**a**), digestibility (**b**), body weight gain (BWG) (**c**) and feed conversion rate (**d**) of cattle among the four diet groups. Values are presented as the mean ± standard deviation (SD). Uppercase letters only represent the difference among the four diet groups.

LP significantly affected the nutrient digestibility of cattle, including the digestibility of DM, OM, NDF, and apparent N ($p < 0.05$, Table 3). Specifically, the digestibility of the DM, OM, and NDF of cattle when fed on an AH40 diet were significantly lower than those on an AH20 diet ($p < 0.05$, Figure 1b).

In the CVH diet groups, only NDF digestibility was significantly lower in the CVH40 diet group than in the CVH20 diet group ($p < 0.05$, Figure 1b).

Both LS and LP did not significantly influence BWG and the feed conversion rate (FCR) of cattle ($p > 0.05$, Table 3), but the interaction between LS and LP had a significant effect on the FCR of cattle ($p < 0.05$, Table 3). In detail, the AH40 diet group had a significantly lower FCR than that in the AH20 diet group ($p < 0.05$, Figure 1d), whereas there was no difference between the CVH20 and CVH40 diet groups ($p > 0.05$, Figure 1d).

3.2. Enteric CH$_4$ Emission, Energy Balance, and Energy Utilization Efficiency

CH$_4$ emissions, expressed on a milligram scale every 15 min per kilogram of metabolic BW and on a gram scale per kilogram DMI over 24 h post feeding, are shown in Figure 2a,b, respectively. There were intermittent peaks throughout the day and it was apparent that the peaks occurred a short time after feed supply. Moreover, the peaks of CH$_4$ emissions (mg/kg BW$^{0.75}$ or g/kg DMI) were relatively higher after the concentrate supply than after the forage supply (Figure 2a,b).

Both LS and LP could significantly affect CH$_4$ emissions (g/kg BW$^{0.75}$) over 24 h ($p < 0.05$, Table 3). Individually, the CVH diet groups had lower accumulated CH$_4$ emissions (g/kg BW$^{0.75}$) than the AH diet groups (Figure 2c), and the CVH40 and AH40 diet groups had relatively lower accumulated CH$_4$ emissions (g/kg BW$^{0.75}$) than the CVH20 and AH20 diet groups (Figure 2c). In addition, the accumulated CH$_4$ emissions (g/kg BW$^{0.75}$) were significantly lower in the CVH40 diet group than in the AH20 diet group ($p < 0.05$, Figure 2c). For CH$_4$ emissions per kilogram DMI in a 24 h period, LP had a significant effect ($p < 0.05$, Table 3). In summary, the CVH40 diet group had significantly lower accumulated CH$_4$ emission (g/kg DMI) than the AH20 diet group ($p < 0.05$, Figure 2c).

Accumulated CH$_4$ emissions per metabolic BW and per kilogram DMI after forage and concentrate supply in a 24 h period are shown in Figure 2d–f. Accumulated CH$_4$ emissions (g/kg BW$^{0.75}$) after forage supply in the morning were significantly lower in the CVH40 and AH40 diet groups than in the CVH20 and AH20 diet groups, respectively ($p < 0.05$, Figure 2d), and the CVH40 diet group had a significantly lower value than the AH20 diet group, regardless of forage supply in the morning or night ($p < 0.05$, Figure 2d,f). There was a similar trend between the accumulated CH$_4$ emissions in grams per kilogram of metabolic BW and grams per kilogram DMI (Figure 2d). There were no significant differences in accumulated CH$_4$ emissions (g/kg DMI) among the four diet groups ($p > 0.05$, Figure 2e) after concentrate supply, whereas the accumulated CH$_4$ emissions (g/kg BW$^{0.75}$) were significantly higher in the CVH20 diet group than in the CVH40 diet group after concentrate supply ($p < 0.05$, Figure 2e).

LS only significantly affected MEI and HP ($p < 0.05$, Table 4). In particular, the CVH diet groups had higher MEI and HP than the AH diet groups (Figure 3a,e). Within the legume diet groups, LP only significantly influenced FE output ($p < 0.05$, Table 4). The CVH40 and AH40 diet groups had higher FE output than the CVH20 and AH20 diet groups (Figure 3d), whereas FE only significantly differed between the AH20 and AH40 diet groups ($p < 0.05$, Figure 3d). For energy utilization efficiency, LP only significantly influenced the ratio of FE to GEI ($p < 0.05$, Table 4). In detail, this ratio was significantly higher in the AH40 diet group than in the CVH20 and AH20 diet groups ($p < 0.05$, Figure 3d).

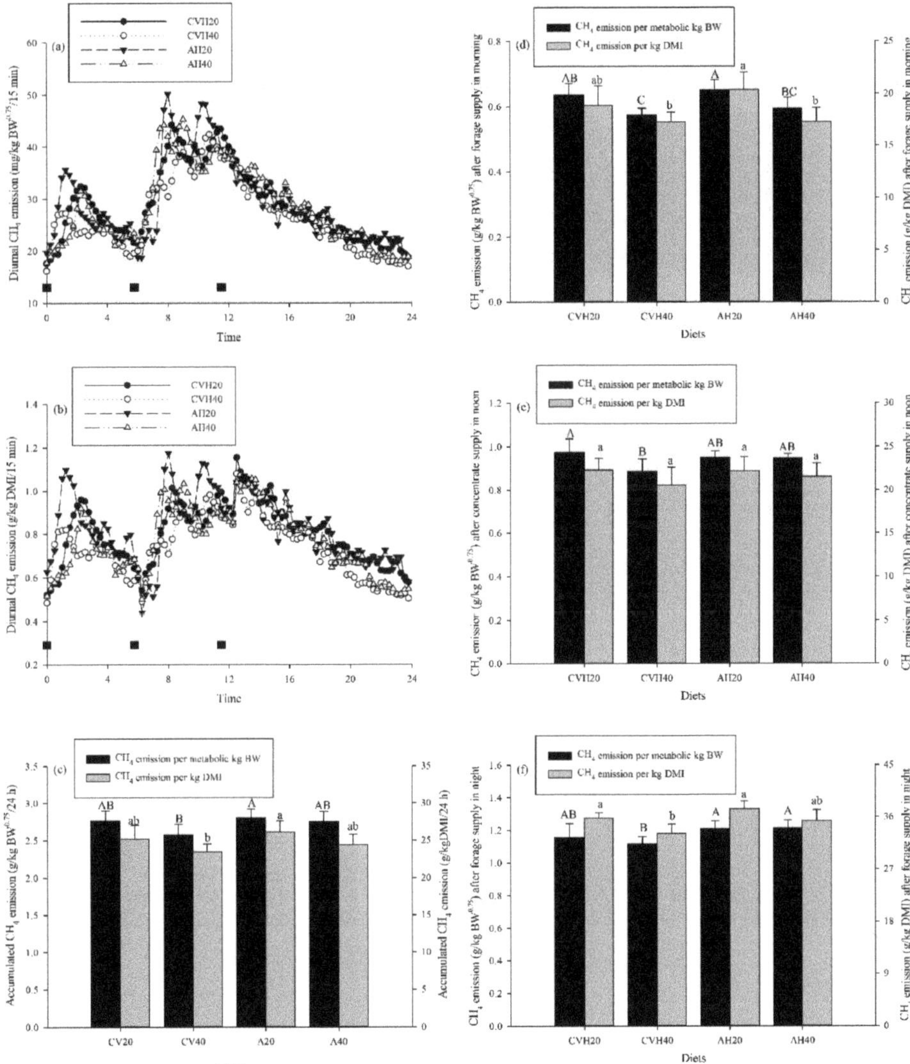

Figure 2. Diurnal CH_4 emissions (**a**) and (**b**), accumulated CH_4 emissions (**c**) after forage supply in the morning (**d**), after the concentrate supply in noon (**e**) and after forage supply in the night (**f**) for cattle among the four diet groups. Values are presented as the mean ± standard deviation (SD). Uppercase letters represent the differences among the four diet groups per kilogram of metabolic body weight, and lowercase letters represent the difference among the four diet groups per kilogram dry matter intake (DMI).

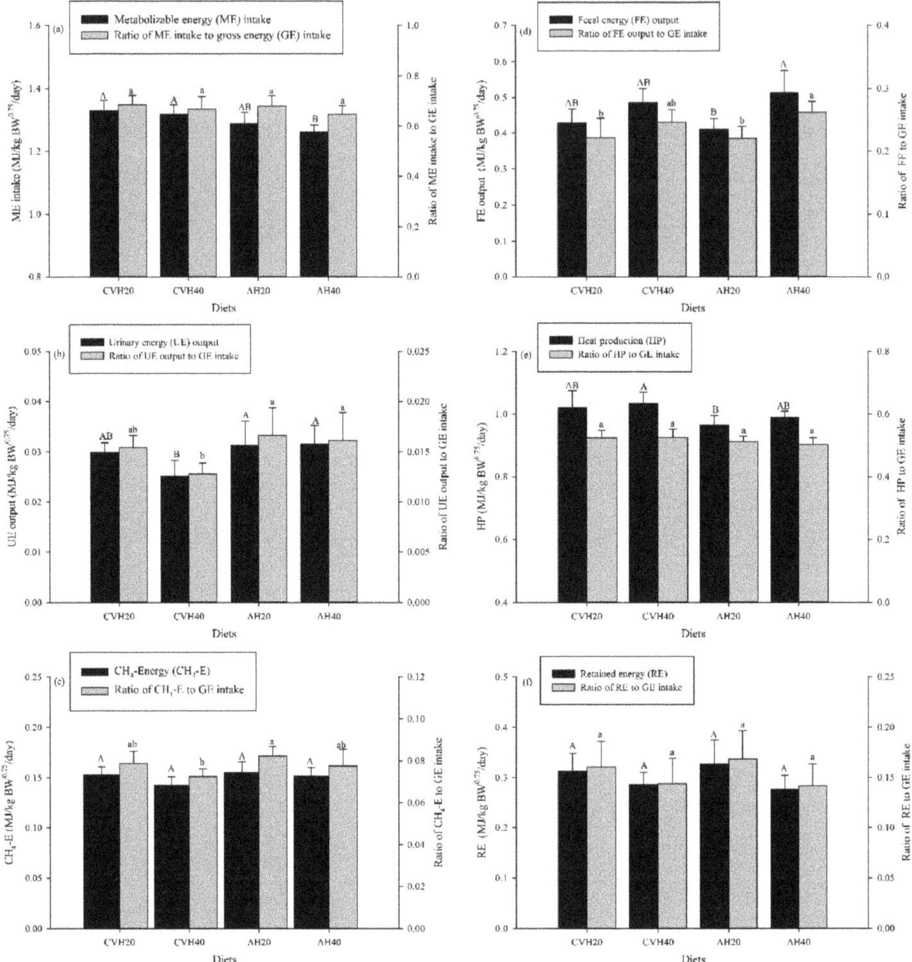

Figure 3. The energy balance and utilization efficiency of cattle among the four diet groups. Values are presented as the mean ± standard deviation (SD). Uppercase letters represent the differences among the four diet groups in energy balance, and lowercase letters represent the differences among the four diet groups in energy utilization. (**a–f**) represent ME intake, UE output, CH$_4$-E, FE output, HP, and RE, respectively, as well as their proportion of GE intake.

Table 4. A general linear model analysis of legume species (LS), legume proportion (LP) and their interaction effects on energy balance/nitrogen balance and energy/nitrogen utilization efficiency (n = 8).

Item [†]	LS [‡]	LP [‡]	LS × LP [‡]
Energy balance			
GE intake (MJ/kg BW$^{0.75}$/day)	1.302	2.783	0.126
ME intake (MJ/kg BW$^{0.75}$/day)	6.749 [*]	1.132	0.127
FE output (MJ/kg BW$^{0.75}$/day)	0.042	13.739 [**]	1.054
UE output (MJ/kg BW$^{0.75}$/day)	4.675	1.584	1.992
CH$_4$-E (MJ/kg BW$^{0.75}$/day)	1.604	2.225	0.684
HP (MJ/kg BW$^{0.75}$/day)	6.208 [**]	1.198	0.170
RE (MJ/kg BW$^{0.75}$/day)	0.012	4.758	0.469

Table 4. *Cont.*

Item [†]	LS [‡]	LP [‡]	LS $\times$ LP [‡]
Energy utilization efficiency			
Ratio of ME intake to GE intake (MJ/MJ)	0.436	1.589	0.224
Ratio of FE output to GE intake (MJ/MJ)	0.392	8.630 [*]	0.504
Ratio of UE output to GE intake (MJ/MJ)	4.647	2.254	1.025
Ratio of HP to GE intake (MJ/MJ)	2.189	0.148	0.171
Ratio of CH_4-E to GE intake (MJ/MJ)	2.332	3.644	0.066
Ratio of RE to GE intake (MJ/MJ)	0.051	2.993	0.178
Nitrogen balance			
N intake (g/kg $BW^{0.75}$/day)	2.956	1.317	0.168
FN output (g/kg $BW^{0.75}$/day)	8.792 [*]	21.653 [***]	0.207
UN output (g/kg $BW^{0.75}$/day)	9.602 [**]	0.046	0.176
RN (g/kg $BW^{0.75}$/day)	21.681 [***]	3.876	3.038
N metabolism			
Ruminal ammonia N (mmol/L)	2.044	12.989 [**]	1.685
Blood urea N (mmol/L)	14.243 [**]	6.884 [*]	0.970
Urinary ammonia N (mmol/L)	0.241	1.420	0.140
Nitrogen utilization efficiency			
Ratio of FN output to N intake (g/g)	3.464	12.862 [**]	0.459
Ratio of UN output to N intake (g/g)	16.116 [**]	0.311	0.398
Ratio of RN to N intake (g/g)	5.992 [*]	4.759	1.252

[†] GE, gross energy; ME, metabolizable energy; FE, fecal energy; UE, urinary energy; CH_4-E, methane energy; HP, heat production; RE, retained energy; N intake, nitrogen intake; FN, fecal N; UN, urinary N; RN, retained N. [‡] Values are the F value, [*] $p < 0.05$, [**] $p < 0.01$, and [***] $p < 0.001$.

3.3. N Balance, N Metabolism, and NUE

LS did not affect the NI of cattle, but it significantly affected the FN, UN and RN outputs in N balance ($p < 0.05$, Table 4). Although the UN output of the CVH20 and CVH40 diet groups was significantly lower than that in the AH40 diet group ($p < 0.05$, Figure 4b), the CVH diet groups had a relatively higher FN output, especially between the CVH40 and AH20 diet groups ($p < 0.05$, Figure 4c). As a consequence, the RN of cattle in the CVH20 and CVH40 diet groups was significantly higher than that in the AH40 diet group ($p < 0.05$, Figure 4d). For the effect of LP on N balance, the CVH40 and AH40 diet groups had relatively higher FE outputs than the CVH20 and AH20 diet groups, but FE output only significantly differed between the AH20 and AH40 diet groups ($p < 0.05$, Table 4, Figure 4c).

LP could significantly influence ruminal ammonia-N concentration ($p < 0.05$, Table 4), which was significantly lower in the AH40 diet group than in the AH20 and CVH20 diet groups ($p < 0.05$, Figure 4e). Both LS and LP significantly affected BUN concentration ($p < 0.05$, Table 4), and AH20 had a significantly lower BUN than the AH20, CVH20, and CVH40 diet groups ($p < 0.05$, Figure 4f). No differences were found for the urinary ammonia-N concentration among the four diet groups ($p > 0.05$, Table 4).

LP significantly affected the ratio of FN to NI ($p < 0.05$, Table 4). The AH20 diet group had a significantly lower FN to NI than that in the AH40 and CVH40 diet groups ($p < 0.05$, Figure 4c). Moreover, LS significantly influenced the ratio of UN to NI ($p < 0.05$, Table 4) and the ratio of RN to NI ($p < 0.05$, Table 4). The CVH40 diet group had a significantly lower UN:NI than the AH20 and AH40 diet groups ($p < 0.05$, Figure 4b), and the CVH20 diet group had a significantly higher RN:NI than the AH40 diet group ($p < 0.05$, Figure 4d).

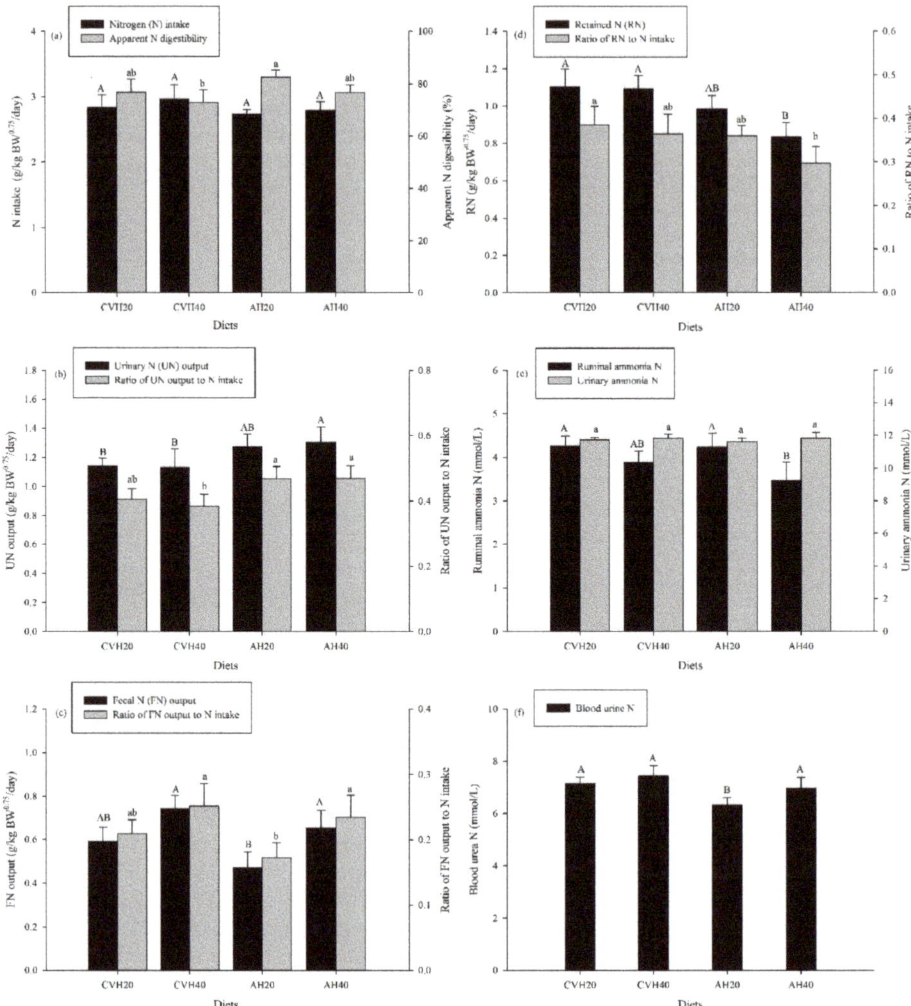

Figure 4. Nitrogen balance and utilization efficiency of cattle among the four diet groups. Values are presented as the mean ± standard deviation (SD). Uppercase letters represent the differences among the four diet groups in nitrogen balance and lowercase letters represent the differences among the four diet groups in nitrogen utilization. (**a**) represents N intake and apparent N digestibility; (**b–d**) represent UN output, FN output, and RN, respectively, as well as their proportion of N intake; (**e**) represents ruminal ammonia N and urinary ammonia N concentrations; (**f**) represents blood urea N.

3.4. Ruminal Fermentation Parameters

The total VFA and pH of the ruminal fluid did not significantly differ among the four diet treatments ($p > 0.05$, Table 5). However, the molar proportion of acetate was significantly lower in the CVH40 and AH40 diet groups than in the CVH20 and AH20 diet groups, respectively ($p < 0.05$, Table 5). Additionally, the molar proportion of acetate was also significantly lower in the CVH diet groups than in the AH diet groups ($p < 0.05$, Table 5). The molar proportions of propionate in the CVH40 and AH40 diet groups were significantly higher than those in the CVH20 and AH20 diet groups, respectively

($p < 0.05$, Table 5). As a consequence, the ratio of acetate to propionate was significantly lower in the CVH40 and AH40 diet groups than in the CVH20 and AH20 diet groups ($p < 0.05$, Table 5).

Table 5. Effects of different diets on the ruminal fermentation parameters in Simmental crossbred cattle.

Item	Experimental Diet [†]				Variance Analysis [‡]		
	CVH20	CVH40	AH20	AH40	LS	LP	LS × LS
Total VFA, mmol/L	75.4 ± 6.73	72.5 ± 7.22	77.8 ± 3.32	75.7 ± 9.98	0.536	0.423	0.011
pH	6.07 ± 0.16	6.12 ± 0.25	6.05 ± 0.08	6.01 ± 0.06	0.686	0.009	0.293
Molar proportions (mol/100 mol)							
Acetate	72.3 ± 1.24	70.8 ± 0.56	73.8 ± 0.64	72.7 ± 1.13	11.967 **	6.503 *	0.122
Propionate	14.4 ± 0.24	15.7 ± 1.08	13.9 ± 0.76	15.2 ± 0.75	1.382	10.576 **	0.007
Butyrate	10.2 ± 1.11	10.2 ± 1.44	9.2 ± 0.72	8.9 ± 0.46	4.747	0.072	0.042
Iso-butyrate	1.1 ± 0.06	1.2 ± 0.19	1.1 ± 0.20	1.1 ± 0.15	0.173	0.640	0.539
Valerate	0.7 ± 0.12	0.6 ± 0.08	0.6 ± 0.18	0.6 ± 0.05	1.444	1.950	0.544
Iso-valerate	1.23 ± 0.09	1.44 ± 0.24	1.32 ± 0.23	1.37 ± 0.18	0.003	1.780	0.679
Acetate/propionate ratio	5.01 ± 0.11	4.53 ± 0.30	5.32 ± 0.30	4.78 ± 0.31	3.987	11.522 **	0.036

[†] CVH20, 20% common vetch + 40% oat hay; CVH40, 40% common vetch + 20% oat hay; AH20, 20% alfalfa + 40% oat hay; AH40, 40% alfalfa + 20% oat hay. [‡] Values are presented as the mean ± standard deviation (SD); LS, legume species; LP, legume proportion; LS × LP, interaction between LS and LP. Values are the F value; * $p < 0.05$; ** $p < 0.01$.

4. Discussion

4.1. Feed Intake, Nutrient Digestibility, and BWG

In general, feed intake is restricted by the capacity of the rumen [22] and NDF content, which is a measure of cell wall content, and the digestibility of forage [28,29]. In this study, the higher forage DMI in the CVH diet groups than in the AH diet groups ($p < 0.05$, Table 3, Figure 1a) could be attributed to the lower NDF content in CVH (Table 1). This indicates that feeds equal in digestibility but different in NDF content would result in different intakes [22]. The similar DM digestibility (Figure 1b) in CVH20 and AH20, as well as CVH40 and AH40 (Figure 1b), along with the higher DMI in the CVH diet groups, confirm the above deduction.

The digestibility of mixed feed is affected by the feed's chemical composition [22]. For example, forage intake with increasing legume proportions could promote the passage rate of feedstuff in the rumen [12] because legumes have lower fiber content than grass, which reduces the retention time of forage in the rumen [22,30]. In this study, the lower NDF digestibility in diets with higher legume proportions than those in the lower legume proportion diets (Table 3, Figure 1b) confirms the above finding. Compared to grasses, highly lignified cell walls could decrease cell wall digestion in legumes and then decrease OM digestion in the rumen [31]. The lower OM digestibility in higher legume proportion diets as compared to those with lower proportions supports the previous finding (Figure 1b).

4.2. Enteric CH₄ Emission and Ruminal Fermentation

There is a clear relationship between forage type, concentrate feed or starch intake, OM digestibility, and patterns of ruminal fermentation [4]. In this study, the lower CH_4 emissions in the diets with higher proportions of legumes than in those with lower ones, regardless of the per kilogram metabolic BW or per kilogram DMI ($p < 0.05$, Table 3, Figure 2c), indicate that a diet with a higher proportion of legumes could decrease CH_4 emissions. This is consistent with the findings of Lee et al. [32], who reported that increasing the percentage of white clover feed with perennial ryegrass could decrease CH_4 emissions. This could be attributed to the polyphenolic compounds in legumes, such as condensed tannins, which have been previously shown to be negatively correlated with CH_4 emissions [33]. For rumen fermentation, there is a negative relationship between CH_4 emissions and propionate formation in the rumen, which could depress the activity of methanogens [4,5]. In this study, the lower ratios of acetate to propionate (Table 5) correspond to lower CH_4 emissions after forage supply (Figure 2d) in diet groups with a higher proportion of legumes and this is consistent with the above finding. In addition, it has been reported that lipid supplementation could reduce CH_4 emissions [34,35]. In this study, legumes have higher concentrations of crude fat (ether extract) than grasses (Table 2), which led to a higher crude fat concentration per unit of DM in the diets with higher proportions of legume than in those with lower proportions. The lower CH_4 emissions in the CVH40 and AH40 diet groups than in the CVH20 and AH20 groups could also be explained in the same way. More importantly, feed intake is the single most important determinant of CH_4 emissions [36]. In this study, there was no difference in the DMI between the diets with higher or lower proportions of legume (Figure 1a), but the lower OM digestibility (1b) of the diet with a higher proportion of legumes and a higher passage rate [12] leaves less time for microorganisms to ferment the feedstuff in the rumen [31]. Therefore, lower CH_4 emissions were observed in CVH40 and AH40 than in CVH20 and AH20 (Figure 2c).

In addition to the effects of LP on CH_4 emissions, LS also affected CH_4 emissions, especially on the basis of per kilogram metabolic BW (Table 3). The relatively lower CH_4 emissions (g/kg $BW^{0.75}$) of the CVH diet groups as compared to the AH diet groups at the same LP (Figure 2c) indicate that CVH has better potential to inhibit CH_4 emissions than AH, especially after forage supply (Figure 2d,f). This might be due to the lower content of NDF and ADF in CVH than in AH, which is in agreement with Beauchemin [35], who reported that lower CH_4 emissions for animals fed legumes could often be explained by the lower fiber content of their diets. Moreover, the production of propionate over acetate in the rumen could also reduce CH_4 emissions in the rumen [35]. In this study, these changes in propionate and acetate also confirmed that the acetate molar proportion was lower in the CVH diet groups than in the AH diet groups ($p < 0.05$, Table 5), although there were no differences in the propionate molar proportion (Table 5). The ratio of acetate to propionate was around 4.77 in the CVH diet groups (Table 5), which was higher than the result of the study by Calabrò et al. [36], who reported a value of 2.28 for an OH and CVH mixture diet using an in vitro gas production technique. This could be attributed to differences that may exist in vivo and in vitro. For example, increases of in vivo rumen propionate concentrations were lower than those observed in vitro [37].

Increasing the inclusion of concentrates in the diet, especially starch content, was regarded as another way to reduce CH_4 emissions [34,38]. In this study, the CVH40 and AH40 diet groups have a relatively higher proportion of maize than the CVH20 and AH20 diet groups (Table 2). As a consequence, lower CH_4 emissions were observed in the CVH40 and AH40 diet groups, even though this value only significantly differed between the CVH20 and CVH40 diet groups per kilogram metabolic BW (Figure 2e). Moreover, CH_4 emissions still tended to be lower in the higher proportion of maize diet groups per kilogram DMI, although this was not considered significant (Figure 2e). These results suggest that starch intake could suppress CH_4 emissions, in accordance with previous studies [4,31].

4.3. Energy Balance

In ruminants, energy is lost in the form of feces, urine, and methane emissions [39]. In this study, FE output and the ratio of FE output to GE intake were greater in the CVH40 and AH40 diet groups than in the CVH20 and AH20 diet groups (Figure 3d). This could be explained by the higher passage rate of the diets with a higher proportion of legumes in the rumen [12], as well as decreased DM digestibility (Figure 1b) because the greater the DM excretion, the greater the FE loss. The ratio of UE output to GE intake, which in previous studies was found to range from 0.9% to 4.8% [39,40], is an indispensable element of energy loss, and high UE loss is more common when animals are fed a silage diet [41]. In this study, the mean 1.4% for the ratio of UE output to GE intake fell within the lower range of the quoted studies, but LP did not significantly influence UE output or the ratio of UE output to GE intake (Figure 3b). The relatively lower values of the ratio of CH_4-E to GE intake in the CVH40 and AH40 diet groups could be explained by the lower OM digestibility (Figure 1b), which reduced the retention time of the feedstuff in the rumen.

ME intake, expressed as per kilogram metabolic BW, was higher in the CVH diet groups than in the AH diet groups (Table 3, Figure 3a), which could be attributed to the higher forage DMI in the

CVH diet groups (Figure 1a) because CVH had a higher ME concentration (MEC) than AH (Table 1). However, no differences were found for the ratio of ME intake to GE intake among the four diet groups (Figure 3a). Nevertheless, the higher ratio of FE output to GE intake in the diet with a higher proportion of legumes (Table 4 and Figure 3d), which accounted for the largest part of the feed energy that could not be utilized by the animals [19], still tended to be lower in the CVH40 and AH40 than in the CVH20 and AH20 diet groups (Figure 3a). Additionally, the ratio of ME intake to GE intake for the crossbred Simmental cattle, in this study, was around 0.67, which was higher than the previously reported 0.47 for mature Simmental cows [42]. This could be attributed to a higher OM digestibility (averaged 75.4%) in this study ascompared to that (62.4%) in the previous study [42]. The higher ME intake (Table 3 and Figure 3a) alongside a lack of differences in RE (Figure 3f) in the CVH diet groups as compared to the AH diet groups could be attributed to an increased HP for the CVH diet groups than the AH diet groups (Table 4 and Figure 3e). This is consistent with the finding of Ferrell and Jenkins [43] that HP increased alongside increasing ME intake for crossbred beef cattle.

4.4. N Balance, N Metabolism, and N Utilization Efficiency (NUE)

N excretion in feces and urine represents a considerable N loss from ruminant husbandry [7,22]. In this study, N losses were affected by LS and LP, although LS and LP did not influence total N intake (Table 3). For example, the significantly higher FN output and the ratio of FN output to N intake corresponded with a higher proportion of legume (CVH40 vs. CVH20 and AH40 vs. AH20, Figure 4c). These were likely caused by the decreased nutrient digestibility (Figure 1b), as well as decreased apparent N digestibility (Figure 4a), which usually lead to more N being excreted in feces. As a result, the higher FN output (Figure 4c) (but no different UN output, Figure 4b) in the diet with a higher proportion of legumes (Figure 4b) led to a reduced RN in the lower proportion of the legume diets ($p = 0.073$, Table 3, and Figure 4d). The UN, FN, and RN outputs were influenced by LS (Table 3). The UN output in the CVH diet groups was lower than that in the AH diet groups (Figure 4b), whereas the FN output presented an opposite result (Figure 4c). The greater shift of N excretion from urine to feces in the CVH diet groups than in the AH diet groups was regarded as a way to reduce the impact of volatile N excretion on the environment [6] because urinary urea is rapidly hydrolyzed to ammonium and then converted to ammonia which is readily volatilized and lost from the farm system to the environment [44]. By contrast, fecal ammonia production is generally low due to the slow mineralization rates of organic nitrogenous compounds [3,7]. As a consequence, the RN in the CVH diet groups was higher than that in the AH diet groups (Figure 4d). Therefore, the CVH diet has a greater potential to reduce the effects of volatile N excretions on the environment than the AH diet.

Generally, high ruminal ammonia-N concentrations for optimal OM degradation will result in an increase in the loss of N through urine [45]. In this study, ammonia-N concentrations in the rumen tended to be lower with a higher proportion of legumes, especially in AH diets (Figure 4e). This difference could possibly be due to the relatively higher passage rate of feedstuff in the rumen with increasing legume proportions [12], thereby, yielding a lower OM digestibility (Figure 1b) and ammonia-N concentrations in the diets with higher proportions of legumes than in the diets with lower proportions of legumes.

In addition, BUN levels reflected the protein status of cattle and positively corresponded with changes in the ammonia-N concentration in rumen fluid [46]. In this study, BUN tended to be higher in the diets with a higher proportion of legumes (Figure 4f), which was inconsistent with ruminal ammonia-N concentrations (Figure 4e). This might be attributable to the lowest pH in AH40 (Table 5), which depressed the transport of ammonia across the rumen wall. Studies have shown that the permeability of the rumen wall for ammonia is pH dependent, and has a positive correlation with pH [47]. Additionally, although the ruminal ammonia-N concentration tended to be lower in the diets with higher proportions of legumes than in the diets with lower proportions of legumes, there was no reduction in BWG (Figure 1c). This suggests that adequate ruminal available N was provided from the diet to maximize microbial fermentation in the rumen under a ruminal ammonia-N concentration of around 4.0 mmol/L.

5. Conclusions

The results of this study suggest the following: (1) a higher proportion of legumes in the diet could reduce CH_4 emissions and minimize the impact of volatile N excretion to the environment; (2) increasing legume proportions in the diet could reduce nutrient digestibility, whereas the degree of reduction differs between common vetch hay and alfalfa hay; and (3) common vetch hay has great potential to minimize the negative effects of CH_4 emissions and N excretion into the environment. Therefore, an opportunity for strategic feeding exists by using alfalfa hay (20%) and common vetch hay (40%) to reduce the direct impact of volatile N excretion and CH_4 emissions on the environment while maintaining BWG, as well as nutrient digestibility for crossbred Simmental cattle, in dryland environments.

Author Contributions: Conceptualization, A.T., F.H., and T.I.; methodology, A.T., F.H., N.K., T.I., and W.D.; data curation, W.D.; writing—original draft preparation, W.D.; writing—review and editing, F.P., and T.I.; visualization, W.D.; supervision, A.T. and F.P.; project administration, A.T. and F.H.; funding acquisition, A.T. and F.H.

Funding: This research was funded by the Marginal Region Agriculture Project of Tottori University, the Strategic Priority Research Program of Chinese Academy of Sciences of China (grant no. XDA20100102), the National Natural Science Foundation of China (no. 31672472), and the Program for Changjiang Scholars and Innovative Research Team at the University of China (IRT_17R50).

Acknowledgments: The authors thank Chang Shenghua, Zhang Cheng, and the students of the College of Pastoral Agriculture Science and Technology (Lanzhou University, China) for supporting the operation of the respiration chambers and for analyzing the feed, fecal, and urinary samples. The authors also wish to thank the Academician Nan Zhibiao for supporting common vetch seeds (*Vicia sativa* L.cv. Lanjian No. 3) for this study.

Conflicts of Interest: The authors declare no conflict of interest.

References

1. National Research Council (NRC). *Air Emissions from Animal Feeding Operations: Current Knowledge, Future Needs*; National Academies Press: Washington, DC, USA, 2003.

2. Montes, F.; Meinen, R.; Dell, C.; Rotz, A.; Hristov, A.N.; Oh, J.; Waghorn, G.; Gerber, P.J.; Henderson, B.; Makkar, H.P.S.; et al. Special topics—Mitigation of methane and nitrous oxide emissions from animal operations: II. A review of manure management mitigation options. *J. Anim. Sci.* **2013**, *91*, 5070–5094. [CrossRef]

3. Kebreab, E.; Dijkstra, J.; Bannink, A.; France, J. Recent advances in modeling nutrient utilization in ruminants. *J. Anim. Sci.* **2009**, *87*, E111–E122. [CrossRef] [PubMed]

4. Hristov, A.N.; Oh, J.; Firkins, J.L.; Dijkstra, J.; Kebreab, E.; Waghorn, G.; Makkar, H.P.S.; Adesogan, A.T.; Yang, W.; Lee, C.; et al. Special topics—Mitigation of methane and nitrous oxide emissions from animal operations: I. A review of enteric methane mitigation options. *J. Anim. Sci.* **2013**, *91*, 5045–5069. [CrossRef] [PubMed]

5. Dong, L.; Li, B.; Diao, Q. Effects of Dietary Forage Proportion on Feed Intake, Growth Performance, Nutrient Digestibility, and Enteric Methane Emissions of Holstein Heifers at Various Growth Stages. *Animals* **2019**, *9*, 725. [CrossRef] [PubMed]

6. Yan, T.; Frost, J.P.; Keady, T.W.J.; Agnew, R.E.; Mayne, C.S. Prediction of nitrogen excretion in feces and urine of beef cattle offered diets containing grass silage. *J. Anim. Sci.* **2007**, *85*, 1982–1989. [CrossRef]

7. Waldrip, H.M.; Todd, R.W.; Cole, N.A.H. Prediction of nitrogen excretion by beef cattle: A meta-analysis. *J. Anim. Sci.* **2013**, *91*, 4290–4302. [CrossRef]

8. Givens, D.I.; Owen, E.; Omed, H.M.; Axford, R.F.E. *Forage Evaluation in Ruminant Nutrition*; CABI Publishing: Wallingford, UK, 2000; pp. 9–10.

9. Graham, P.H.; Vance, C.P. Legumes: Importance and constraints to greater use. *Plant Physiol.* **2003**, *131*, 872–877. [CrossRef]

10. Abreu, A.; Carulla, J.E.; Lascano, C.E.; Diaz, T.E.; Kreuzer, M.; Hess, H.D. Effects of Sapindus saponaria fruits on ruminal fermentation and duodenal nitrogen flow of sheep fed a tropical grass diet with and without legume. *J. Anim. Sci.* **2004**, *82*, 1392–1400. [CrossRef]

11. Doran, M.P.; Laca, E.A.; Sainz, R.D. Total tract and rumen digestibility of mulberry foliage (*Morus alba*), alfalfa hay and oat hay in sheep. *Anim. Feed. Sci. Technol.* **2007**, *138*, 239–253. [CrossRef]

12. McCaughey, W.P.; Wittenberg, K.; Corrigan, D. Impact of pasture type on methane production by lactating beef cows. *Can. J. Anim. Sci.* **1999**, *79*, 221–226. [CrossRef]

13. Wuchen, D.; Fujiang, H.; Atsushi, T.; Kobayashi, N.; Fei, P.; Ichinohe, T. Effects of oat hay and leguminous forage mixture feeding on body weight gain, enteric methane emission and energy utilization in crossbred Simmental male beef cattle. *Anim. Sci. J.* **2019**. under review.

14. Kobayashi, N.; Hou, F.J.; Tsunekawa, A.; Chen, X.J.; Yan, T.; Ichinohe, T. Effects of substituting alfalfa hay for concentrate on energy utilization and feeding cost of crossbred Simmental male calves in Gansu Province, China. *Grassl. Sci.* **2017**, *63*, 245–254. [CrossRef]

15. Kobayashi, N.; Hou, F.J.; Tsunekawa, A.; Chen, X.J.; Yan, T.; Ichinohe, T. Appropriate level of alfalfa hay in diets for rearing Simmental crossbred cattle in dryland China. *Asian Australas. J. Anim. Sci.* **2018**, *31*, 1881–1889. [CrossRef] [PubMed]

16. Larbi, A.; El-Moneim, A.A.; Nakkoul, H.; Jammal, B.; Hassan, S. Intra-species variations in yield and quality determinants in Vicia species: 3. Common vetch (*Vicia sativa ssp. sativa* L.). *Anim. Feed Sci. Technol.* **2011**, *164*, 241–251. [CrossRef]

17. Rihawi, S.; Iniguez, L.; Knaus, W.F.; Zaklouta, M.; Wurzinger, M.; Soelkner, J.; Larbi, A.; Bomfim, M.A.D. Fattening performance of lambs of different Awassi genotypes, fed under cost-reducing diets and contrasting housing conditions. *Small Rumin. Res.* **2010**, *94*, 38–44. [CrossRef]

18. Assefa, G.; Ledin, I. Effect of variety, soil type and fertiliser on the establishment, growth, forage yield, quality and voluntary intake by cattle of oats and vetches cultivated in pure stands and mixtures. *Anim. Feed Sci.Technol.* **2001**, *92*, 95–111. [CrossRef]

19. Hernández-Ortega, M.; Heredia-Nava, D.; Espinoza-Ortega, A.; Sánchez-Vera, E.; Arriaga-Jordán, C.M. Effect of silage from ryegrass intercropped with winter or common vetch for grazing dairy cows in small-scale dairy systems in Mexico. *Trop. Anim. Health Prod.* **2011**, *43*, 947–954. [CrossRef]

20. Agricultural and Food Research Council. *Energy and Protein Requirements of Ruminants. An Advisory Manual Prepared by the AFRC Technical Committee on Responses to Nutrients*; Centre for Agriculture and Bioscience International: Wallingford, UK, 1993; pp. 1–38.

21. Chinese Feeding Standard for Beef Cattle (CFSBC). Ministry of Agriculture of the People's Republic of China. 2004. Available online: https://wenku.baidu.com/view/154d59bdf121dd36a32d8269.html (accessed on 4 October 2019).

22. Zhao, Y.G.; Aubry, A.; O'Connell, N.E.; Annett, R.; Yan, T. Effects of breed, sex, and concentrate supplementation on digestibility, enteric methane emissions, and nitrogen utilization efficiency in growing lambs offered fresh grass. *J. Anim. Sci.* **2015**, *93*, 5764–5773. [CrossRef]

23. Gerrits, W.; Labussiere, E.; Dijkstra, J.; Reynolds, C.; Metges, C.; Kuhla, B.; Lund, P.; Weisbjerg, M.R. Letter to the Editors: Recovery test results as a prerequisite for publication of gaseous exchange measurements. *Animal* **2018**, *12*, 4. [CrossRef]

24. Brouwer, E. Report of sub-committee on constants and factors. In Proceedings of the 3rd Symposium on Energy Metabolism of Farm Animals of European Association for Animal Production; Blaxter, K.L., Ed.; Academic Press: Scotland, UK, 1965.

25. Kohn, R.A.; Dinneen, M.M.; Russek-Cohen, E. Using blood urea nitrogen to predict nitrogen excretion and efficiency of nitrogen utilization in cattle, sheep, goats, horses, pigs, and rats. *J. Anim. Sci.* **2005**, *83*, 879–889. [CrossRef]

26. Association of Official Analytical Chemists (AOAC). *Official Methods of Analysis of the Association of Official Analytical Chemists*, 15th ed; Association of Official Analytical Chemists: Arlington, VA, USA, 1990.

27. Van Soest, P.J.; Robertson, J.B.; Lewis, B.A. Methods for dietary fiber, neutral detergent fiber, and nonstarch polysaccharides in relation to animal nutrition. *J. Dairy Sci.* **1991**, *74*, 3583–3597. [CrossRef]

28. Buxton, D.R. Quality-related characteristics of forages as influenced by plant environment and agronomic factors. *Anim. Feed Sci. Technol.* **1996**, *59*, 37–49. [CrossRef]

29. Karabulut, A.; Canbolat, O.; Kalkan, H.; Gurbuzol, F.; Sucu, E.; Filya, I. Comparison of in vitro gas production, metabolizable energy, organic matter digestibility and microbial protein production of some legume hays. *Asian Australas. J. Anim. Sci.* **2007**, *20*, 517–522. [CrossRef]

30. Patra, A.K. Effects of supplementing low–quality roughages with tree foliages on digestibility, nitrogen utilization and rumen characteristics in sheep: A meta–analysis. *J. Anim. Physiol. Anim. Nutr.* **2010**, *94*, 338–353. [CrossRef] [PubMed]

31. Archimède, H.; Eugène, M.; Marie Magdeleine, C.; Boval, M.; Martin, C.; Morgavi, D.P.; Lecomte, P.; Doreau, M. Comparison of methane production between C3 and C4 grasses and legumes. *Anim. Feed Sci. Technol.* **2011**, *166*, 59–64. [CrossRef]

32. Lee, J.M.; Woodward, S.L.; Waghorn, G.C.; Clark, D.A. Methane emissions by dairy cows fed increasing proportions of white clover (*Trifolium repens*) in pasture. *Proc. N. Z. Grassl. Assoc.* **2004**, *66*, 151–155.

33. Guglielmelli, A.; Calabrò, S.; Primi, R.; Carone, F.; Cutrignelli, M.I.; Tudisco, R.; Piccolo, G.; Ronchi, B.; Danieli, P.P. In vitro fermentation patterns and methane production of sainfoin (*Onobrychis viciifolia* Scop.) hay with different condensed tannin contents. *Grass Forage Sci.* **2011**, *66*, 488–500. [CrossRef]

34. Grainger, C.; Beauchemin, K.A. Can enteric methane emissions from ruminants be lowered without lowering their production? *Anim. Feed Sci. Technol.* **2011**, *166*, 308–320. [CrossRef]

35. Beauchemin, K.A.; Kreuzer, M.; O'Mara, F.; McAllister, T.A. Nutritional management for enteric methane abatement: A review. *Aust. J. Exp. Agric.* **2008**, *48*, 21–27. [CrossRef]

36. Calabrò, S.; Carone, F.; Cutrignelli, M.; D'Urso, S.; Piccolo, G.; Tudisco, R.; Angelino, G.; Infascelli, F. The effect of haymaking on the neutral detergent soluble fraction of two intercropped forages cut at different growth stages. *Ital. J. Anim. Sci.* **2006**, *5*, 327–339. [CrossRef]

37. Fievez, V.; Dohme, F.; Danneels, M.; Raes, K.; Demeyer, D. Fish oils as potent rumen methane inhibitors and associated effects on rumen fermentation in vitro and in vivo. *Anim. Feed Sci. Technol.* **2003**, *104*, 41–58. [CrossRef]

38. Yan, T.; Agnew, R.E.; Gordon, F.J.; Porter, M.G. Prediction of methane energy output in dairy and beef cattle offered grass silage-based diets. *Livest. Prod. Sci.* **2000**, *64*, 253–263. [CrossRef]

39. Chaokaur, A.; Nishida, T.; Phaowphaisal, I.; Sommart, K. Effects of feeding level on methane emissions and energy utilization of Brahman cattle in the tropics. *Agric. Ecosyst. Environ.* **2015**, *199*, 225–230. [CrossRef]

40. Zou, C.X.; Lively, F.O.; Wylie, A.R.G.; Yan, T. Estimation of the maintenance energy requirements, methane emissions and nitrogen utilization efficiency of two suckler cow genotypes. *Animal* **2016**, *10*, 616–622. [CrossRef]

41. Kirkpatrick, D.E.; Steen, R.W.J.; Unsworth, E.F. The effect of differing forage: Concentrate ratio and restricting feed intake on the energy and nitrogen utilization by beef cattle. *Livest. Prod. Sci.* **1997**, *51*, 151–164. [CrossRef]

42. Estermann, B.L.; Sutter, F.; Schlegel, P.O.; Erdin, D.; Wettstein, H.R.; Kreuzer, M. Effect of calf age and dam breed on intake, energy expenditure, and excretion of nitrogen, phosphorus, and methane of beef cows with cattle. *J. Anim. Sci.* **2002**, *80*, 1124–1134. [CrossRef]

43. Ferrell, C.L.; Jenkins, T.G. Body composition and energy utilization by steers of diverse genotypes fed a high-concentrate diet during the finishing period: I.; Angus, Belgian Blue, Hereford, and Piedmontese sires. *J. Anim. Sci.* **1998**, *76*, 637. [CrossRef]

44. Koenig, K.M.; Beauchemin, K.A. Effect of feeding condensed tannins in high protein finishing diets containing corn distillers grains on ruminal fermentation, nutrient digestibility, and route of nitrogen excretion in beef cattle. *J. Anim. Sci.* **2018**, *96*, 4398–4413. [CrossRef]

45. Ipharraguerre, I.R.; Clark, J.H. Varying protein and starch in the diet of dairy cows. II. Effects on performance and nitrogen utilization for milk production. *J. Dairy Sci.* **2005**, *88*, 2556–2570. [CrossRef]

46. Dong, R.L.; Zhao, G.Y.; Chai, L.L.; Beauchemin, K.A. Prediction of urinary and fecal nitrogen excretion by beef cattle. *J. Anim. Sci.* **2014**, *92*, 4669–4681. [CrossRef]

47. Abdoun, K.; Stumpff, F.; Martens, H. Ammonia and urea transport across the rumen epithelium: A review. *Anim. Health Res. Rev.* **2006**, *7*, 43–59. [CrossRef] [PubMed]

Article

Iodine Supplemented Diet Positively Affect Immune Response and Dairy Product Quality in Fresian Cow

Marco Iannaccone [1,†], Andrea Ianni [2,†], Ramy Elgendy [3], Camillo Martino [4], Mery Giantin [5], Lorenzo Cerretani [6], Mauro Dacasto [5] and Giuseppe Martino [1,*]

[1] Faculty of Bioscience and Technology for Food, Agriculture, and Environment, University of Teramo, Via R. Balzarini 1, 64100 Teramo, Italy; m.iannaccone@unina.it

[2] Department of Medical, Oral and Biotechnological Sciences, "G. d'Annunzio" University Chieti-Pescara, Via dei Vestini 31, 66100 Chieti, Italy; andreaianni@hotmail.it

[3] Department of Immunology, Genetics and Pathology, Uppsala University, Uppsala 75185, Sweden; ramy.elgendy@igp.uu.se

[4] Istituto Zooprofilattico Sperimentale dell'Abruzzo e del Molise "G. Caporale" Via Campo Boario, 64100 Teramo, Italy; c.martino@izs.it

[5] Department of Comparative Biomedicine and Food Science, University of Padua, Viale dell'Università 16, 35020 Legnaro (PD), Italy; mery.giantin@unipd.it (M.G.); mauro.dacasto@unipd.it (M.D.)

[6] Pizzoli SPA, Via Zenzalino Nord, 40054 1 Budrio (BO), Italy; l.cerretani@pizzoli.it

* Correspondence: gmartino@unite.it; Tel.: +39-0861-266950

† These authors contributed equally to this work.

Received: 25 September 2019; Accepted: 23 October 2019; Published: 25 October 2019

Simple Summary: Iodine represents an important micronutrient and plays a fundamental role in animal biology. This trace element is currently supplied to animal diet to investigate its potential effects on productive and reproductive performances. However, little is known about its role in the regulation of gene expression in ruminants. In this study, the dietary iodine supplementation in dairy cows showed effective modification of the expression of several molecular targets, with an improvement of the pathways involved in immune response and oxidative stress and undoubted positive repercussions on animal health.

Abstract: The effects of iodine supplementation on the whole-transcriptome of dairy cow using RNA sequencing has been investigated in this study. Iodine did not influence the milk composition, while an improvement was observed in the immune response as well as in the quality of dairy product. Indeed, the iodine intake specifically influenced the expression of 525 genes and the pathway analysis demonstrated that the most affected among them were related to immune response and oxidative stress. As a consequence, we indirectly showed a better response to bacterial infection because of the reduction of somatic cell counts; furthermore, an improvement of dairy product quality was observed since lipid oxidation reduced in fresh cheese. Such findings, together with the higher milk iodine content, clearly demonstrated that iodine supplementation in dairy cow could represent a beneficial practice to preserve animal health and to improve the nutraceutical properties of milk and its derived products.

Keywords: Iodine; immune response; oxidative stress; somatic cell count; fresh cheese; transcriptomics

---

1. Introduction

Micronutrients are essential to orchestrate all physiological functions. Among them, iodine (I) plays a unique role because it is the main component of the thyroid hormones, i.e., thyroxine (T4) and triiodothyronine (T3) [1]. Since the thyroid gland regulates many metabolic processes, the extent of I requirement strongly depends on the age and stage of development of the individual [2]. Thus, when the I demand is not satisfied, reduced functionality of the thyroid gland could occur (hypothyroidism) with negative consequences for proper mental development, body growth, and fertility. For these reason, diet is often integrated with I supplements, generally provided through iodized salt [3]. For infants, milk constitutes the only I source; therefore, the dietary calibration of this micronutrients in dairy animals assumes relevant importance in order to obtain I-rich milk without inducing variations in animal performances [4].

In animal husbandry, I supplementation as calcium iodate, sodium iodide, and other iodine compounds is needed since the native iodine content of plant straight feed-stuffs is low; moreover, the increasing use of rapeseed meal (RSM) in livestock diets is associated with the intake of glucosinolates, which are known to be iodine antagonists inhibiting the activity of sodium iodide symporter [5]. For this reason, the European community has recently brought to 5 mg/kg the maximum level of I supplementation for milk-producing ruminants (milk intended for human consumption), while it remained at 10 mg/kg of complete feed for other ruminant categories [6]. Consequently, several studies have been carried out to evaluate the productive performances in animals fed the I supplementation. Weiss et al. [7] showed that I concentration increased in serum but not in milk after supplementation of this element in diets of dairy cows. In contrast, studies performed in small ruminants showed that I supplementation doubled the milk iodine content when compared with the control group, even though no evident effect was observed in the milk gross composition [8]. I supplementation showed also beneficial effects on healthy status. Indeed, early study on feedlot cattle demonstrated an increased resistance to foot rot [9], which is due to improved phagocytic cell function [10]. Moreover, it was shown in lambs that high-dose potassium iodide supplementation can be effective to decrease the severity of airway viral infections, supposedly through the augmentation of mucosal oxidative defenses [11,12].

Recently, the RNA-sequencing approach was shown to be useful to elucidate which molecular pathways are affected by I supplementation in sheeps [13] and showed the positive effects of olive pomace-supplemented diet on inflammation and cholesterol in laying hens [14]. However, to date, little is known about the effects of I supplementation on transcriptomic profiles in dairy cows. Moreover, we tried to correlate the information concerning the signaling pathways influenced by I supplementation, with the qualitative parameters of milk and derivatives, taking into account those studies in which it was shown that dietary iodine supplementation in ruminants contributes to an improvement in the quality of dairy products [15].

2. Materials and Methods

2.1. Animal and Study Design

The study design was approved by the Teramo University Institutional Animal Care and Use Committee. Animals were managed according to Directive 2010/63/EU of the European Parliament regarding the protection of animals used for experimentation or other scientific purposes [16].

Twenty-two Friesian cattle, homogenous for age (range between 39 and 42 months), number of births (2 calves), and lactation length (70 ± 5 days), have been enrolled in this study. Animals, belonging to the same farm and bred in the same way were randomly divided in 2 groups of 11 cows each. During the 3-week acclimatization period (21 days), both the control (CTR) and the experimental iodine groups (IG) received a basal diet that mainly consisted of alfalfa hay plus a custom-formulated concentrate supplemented with 20 mg/day/animal I in order to guarantee the daily micronutrient requirement for each animal; then, the IG animals were fed for 56 days (during April and May) with a custom-formulated concentrate supplemented with additional 65 mg/day/animal of I in order to obtain a total intake of about 85 mg; this amount has been set not to exceed the maximum level allowed by law [17]. Ingredients and composition of total mixed ration (TMR) administered to the animals during the experimental period are reported in Table 1.

Table 1. Ingredients and composition of total mixed ration (TMR) administered to each animal of both group of study.

Ingredients of TMR	
Corn silage, %	23.7
First cut, alfalfa hay, %	5.3
Corn meal, %	3.4
Soybean, meal, %	3.2
Fine bran, %	3.0
Barley, meal, %	1.9
$CaCO_3$, %	0.2
Vitamins and minerals, %	0.4
Kg of dry matter/head per day	22.41
Chemical Composition of TMR	
Dry Matter, %	56.76
Crude protein [1], %	15.34
Ether extract [1], %	2.97
Ash [1], %	5.31
Neutral detergent fiber [1], %	32.51
Acid detergent fiber [1], %	20.03
Starch [1], %	27.02
Iodine (mg/head/day) [1]	20 (+65) *

[1] On a dry matter (DM) basis; * In brackets is the amount of iodine added to the diet for the experimental group.

2.2. Blood and Milk Sampling

Individual whole blood (WB) samples were collected at the beginning (T0) and at the end of the dietary supplementation for the evaluation of the hematochemical parameters. In the case of the RNA-Seq analysis, 2.5 mL of jugular venous blood was collected in duplicate from each animal only at the end of the experimental period. In this case, samples were collected in PAXgene™ tubes (Qiagen SpA, Milan, MI, Italy), stored overnight at room temperature, and then placed at −20 °C until RNA isolation, following the manufacturer's instructions.

Regarding milk, at the beginning (T0) and after 8 weeks (T8) of dietary supplementation, 50 mL of individual samples was collected from each group in triplicate in 80 mL polypropylene tubes containing a bronopol/sodium azide preservative solution, useful to extend the storage time before the analysis up to 48–72 h. Milking was performed through mechanical support (DeLaval, Milan, Italy), and teat disinfection occurred both before and after milking by using iodine-free solutions (DeLaval, Milan, Italy).

2.3. Blood Analysis

Complete blood cell count with leukocyte formula (total white blood cells, monocyte, lymphocyte, basophils, neutrophils, and eosinophils) for both the CTR and IG groups at T0 and T8 were performed at the Veterinary and Public Health Institute (Teramo, Italy) using a laser-based hematology analyzer with software applications for animal species (ADVIA 120 hematology system, Siemens, Munich, Germany). Plasma samples were analyzed for thyroid hormones (Thyroid-stimulating hormone (TSH), T3, and T4) with an automatic biochemistry analyzer (ILAB 650, Instrumentation Laboratory-Werfen, Milan, Italy) and following the routine procedure of the institute (Veterinary and Public Health Institute "G. Caporale", Teramo, Italy).

2.4. Chemical Analysis of Milk

Chemical composition of milk (fat, protein, casein, lactose, and urea) was determined by MilkoScan FT 6000 (Foss Integrator IMT; Foss, Hillerød, Denmark), whereas the somatic cells count (SCC) was performed using the Fossomatic FC (Foss).

2.5. Iodine Determination in Milk

The amount of I in the milk at T0 and T8 was determined according to Fecher et al. (1998) with some modifications [18]. Briefly, for each sample, 0.5 g of milk were homogenized with tetramethylammonium hydroxide (0.25 M) and 2 mL of deionized water (30%). Then, samples were heated in a microwave (800 W) at 170 °C for 30 min. After cooling, samples were transferred into a sterile tube and diluted with distilled water to a final volume of 15 mL. After centrifugation at 12,000 rpm $\times$ min^{-1} at room temperature for 10 min, samples were filtered through polytetra-fluoroethylene (PTFE) syringe filters (0.45 μm) and, finally, stored at 4 °C until analysis. A standard calibration curve was created using six calibration points equals to concentrations of 0, 5, 10, 25, 50, and 100 mg/L of I in tetramethylammonium hydroxide. For the carrier and gas formation, argon gas was used at flow rates of 1.05 and 0.2 L/min, respectively. The iodine content was determined by an inductively coupled plasma mass spectrometer Agilent 7500ce ICP-MS (Agilent Technologies, Palo Alto, CA, USA) at m/z = 127 and a total acquisition time of 21 s. Before the sequence analysis, the ICP-MS was auto-tuned by a solution containing 1 ppb of different metals (Li, Y, Ce, Tl, and Co).

2.6. Library Preparation and RNA-Seq Analysis

Total RNA was isolated from blood using the PAXgene blood RNA kit (Qiagen, Milan, Italy) as per the manufacturer's instructions. Total RNA concentration was determined by the NanoDrop ND-1000 spectrophotometer (NanoDrop Technologies Inc., Wilmington, DE, USA), and its quality was measured by the 2100 Bioanalyzer and RNA 6000 Nano kit (Agilent Technologies, Santa Clara, CA, USA). Strand-specific RNA-Seq libraries were prepared using the SureSelect strand-specific mRNA library preparation kit (Agilent Technologies, Santa Clara, CA, USA) as per the manufacturer's protocol. In brief, poly(A) RNA was purified from 1 μg of total RNA using two serial rounds of binding to oligo(dT) magnetic particles and, then, fragmented and reverse transcribed to generate cDNA. Illumina-specific adaptor was sequentially ligated to the 3' end of cDNA fragments and purified using the AMPure XP beads (Beckman Coulter, Brea, CA, USA), and finally PCR-amplified (13 cycles) using an appropriate indexing primer to allow further samples multiplexing. The PCR-amplified libraries were purified with the AMPure XP beads (Beckman Coulter, Brea, CA, USA) and then assessed for their quality and fragments distribution using the 2100 Bioanalyzer DNA 1000 assay (Agilent Technologies, Santa Clara, CA, USA). In the presence of adaptor-dimers (Electropherogram's peak at 100 to 150-bp), another round of magnetic beads purification was performed. Libraries were quantified by both the Qubit® Fluorometer (Life Technologies, Carlsbad, CA, USA) and the qPCR-based NEBNext library quantification kit (New England BioLabs, Hitchin, UK). Finally, libraries were pooled and then sequenced by an Illumina HiSeq 2500 for 50 sequencing cycles.

The raw 50-bp single-end sequences (Sanger/Illumina 1.9 encoding) were quality controlled using FastQC (v.0.11.4; Babraham Institute, Cambridge, UK), and the low-quality bases (quality scores < 30) and adaptor contamination (if present) were removed by Trimmomatic v.0.36. The high-quality reads were mapped by HISAT v.2.0.5 against the Bos taurus reference genome (Ensembl Bos_taurus_UMD_3.1.1). The uniquely mapped reads aligned to exons were counted with HTSeq v.0.6.1 [19] and then tested by the DESeq2 R package v.1.14.1 [20] for the presence of differentially expressed genes (DEGs) in the IG group compared with the CTR one (e.g., T8 I vs T8 CTR). Genes with a false discovery rate (FDR) less than 0.05 were considered as DEGs. All analyses were performed using the software Artificial Intelligence RNA-seq (A.I.R; developed by Sequentia Biotech, Barcelona, Spain) and the sequencing data (FASTQ files) associated with this project are deposited in the GenBank's Sequence Read Archive (SRA) under the accession number PRJNA516565.

2.7. Enriched Pathway Analysis

The STRING software (Version 11.0, http://string-db.org/) was used to identify canonical pathways using the dataset of 525 DEGs identified between the IG group and the CTR one with a FDR < 0.05.

We set the interaction score as 0.9, the highest value permitted by the software to avoid false positives. The significance of the canonical pathway was measured with the *p*-value and the ratio of DEG/number of genes in the pathway.

2.8. Ricotta Cheese-Making Procedure and Lipid Oxidation by Thiobarbituric Acid Reactive Substance Test

At the end of the supplementation period, an aliquot of about 150 L of bulk milk from each experimental group was separately collected and manipulated according to common cheese-making procedures. The whey separately collected from each experimental group during the cheese-making was used to produce "ricotta" cheese. This was obtained by acid-thermal coagulation of whey, through heating up to 80 °C and adding 50 g of lactic acid. The whey flocs were collected in holed baskets with a total capacity equal to 0.5 L and left to drain for 30 min. Then, baskets still containing the ricotta were transferred in a refrigerator and stored under chilling conditions (4 °C). From each experimental group, 9 ricotta forms (single weight = about 400 g) were obtained. In order to evaluate lipid peroxidation, thiobarbituric acid reactive substances (TBARS) were measured at day 0 and day 7. The analysis was performed according to the procedure reported by Ianni et al. with slight modifications [21]: 4.5 g of frozen ricotta cheese were mixed within 2 min of sample withdrawal from the freezer, with 450 μL of 0.1% of butylated hydroxytoluene in methanol with the aim to stop the oxidation process. The mixture was homogenized with UltraTurrax T-25 high speed homogenizer (IKA, Staufen, Germany) in 40 mL of an aqueous solution of 7% trichloroacetic acid and, then, distilled; 1.5 mL of each distillate was mixed with an equal volume of a 0.02 M thiobarbituric acid (TBA) in 90% acetic acid, and the solution was kept for one hour in a thermostatic bath at 80 °C. Only after cooling, the absorbance at 534 nm was evaluated with a spectrophotometer (JENWAY 6305 UV/vis, Jenway, Essex, UK). The amount of malondialdehyde (MDA) of each sample was calculated by using a calibration curve ranging from 0 to 100 ppm ($R^2 = 0.986$), and results were expressed in μg of MDA per g of cheese.

2.9. Statistics

GraphPad Prism version 6.0 (GraphPad Software, La Jolla, CA, USA) was used for statistical analysis. Differences in milk parameters, as well as the amount of thyroid hormone in blood sera and malondialdehyde (MDA) concentrations in ricotta fresh cheese were statistically evaluated by using ordinary two-way ANOVA.

3. Results

3.1. Serum Thyroid Hormone and Iodine Concentrations

All animals maintained a good state of health for the entire duration of the trial, and no significant variations in milk yield have been observed between the two groups at the end of the experimental period.

Because I affects production and secretion of thyroid hormones, we quantified the levels of thyrotropin (TSH), triiodothyronine (T3), and thyroxine (T4) at the beginning (T0) and at the end (T8) of the supplementation in the serum blood samples. As shown in Figure 1, no differences in hormone levels were ever noticed between groups, suggesting that the thyroid functionality was not affected by the iodine supplementation and that the CTR group was not in iodine-deficiency condition.

As expected (Figure 2), we appreciate a higher amount of iodine in the I group, indicating that consumption of milk from dairy cows fed with high iodine intake is helpful to integrate diets where physiological stages like infancy and/or pregnancy require higher iodine intake [22].

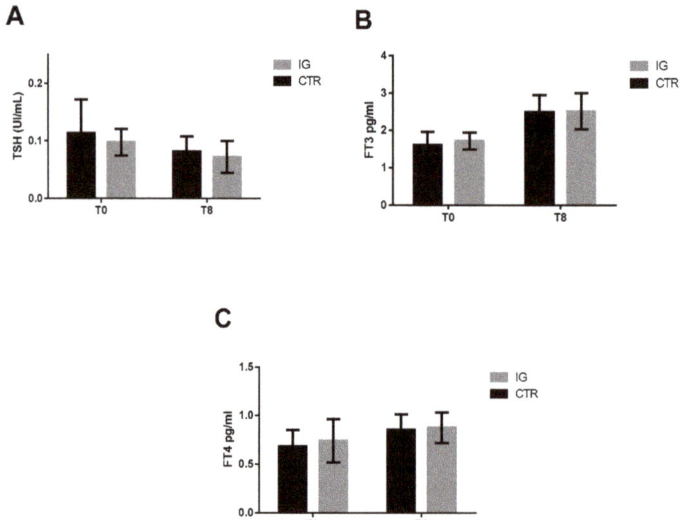

Figure 1. Thyroid hormone concentrations: Thyroid-stimulating hormone (**A**), free thyroxine (**B**), and free triiodothyronine (**C**) were measured in serum samples from control (CTR) ($n = 11$) and iodine group (IG) ($n = 11$) at the beginning and at the end of iodine supplementation. Data are expressed as mean ± SD, and differences were assessed using 2-way ANOVA.

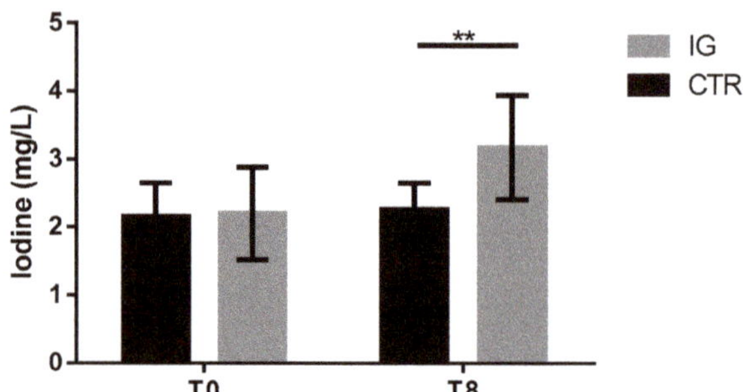

Figure 2. Iodine quantification in milk samples: Iodine was quantified in milk samples of both CTR ($n = 11$) and IG ($n = 11$) at the beginning and at the end of iodine supplementation. Data are shown as mean ± SD, and differences were assessed using 2-way ANOVA. ** p-value < 0.01.

However, it is worth mentioning that milk parameters were not influenced by iodine supplementation (Table S1).

3.2. Influence of I-supplemented Diet on Blood Transcriptome

To identify the molecular networks associated to I supplementation, we collected peripheral blood from both groups at the end of I supplementation period and performed a transcriptomic analysis by RNA-Seq. After trimming and quality control, on average, 96% of reads resulted of high-quality reads (Table S2) and were mapped against Bos taurus reference genome (Bos_taurus_UMD_3.1.1).

Filtering our data by using a FDR < 0.05, we identified 525 DEGs (Table S3); in particular, 274 and 248 genes were respectively down- and upregulated in the I group compared to CTR; moreover, they enabled us to discriminate the two groups not only on a heat map scale but also on a hierarchical clustering analysis, thereby indicating the robustness of analysis (Figures 3 and 4).

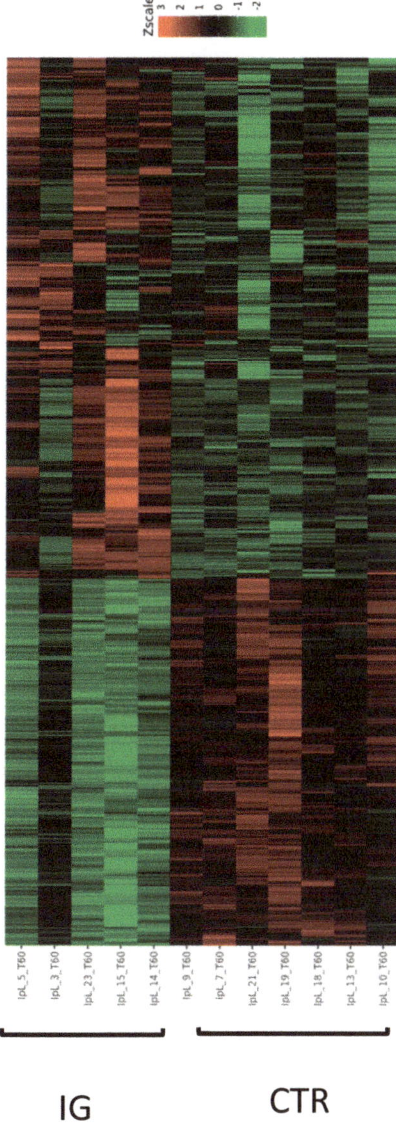

Figure 3. Heat map analysis of the 525 differentially expressed genes identified at the end of iodine supplementation period.

In order to identify the enriched pathways and putative interactions among DEGs, we interrogated the STRING software by using the up- and downregulated genes. To increase stringency and confidence in our data set, we applied the highest available interaction score (0.9), and in Table 2 (downregulated) and Table 3 (upregulated) are listed the most significant pathways (FDR < 0.05). The highest enriched pathway was "Fc gamma R-mediated phagocytosis" (FDR: 2.63×10^{-6}), which was associated with clearance of opsonized particles, suggesting that iodine supplementation could help the immune system against invading bacteria [23]. Interestingly, also, the "oxidative phosphorylation" pathway was dysregulated by iodine supplementation, indicating a positive effect of iodine on mitochondrial activity with consequent reduction of oxidative stress, which could potentially act on quality dairy products.

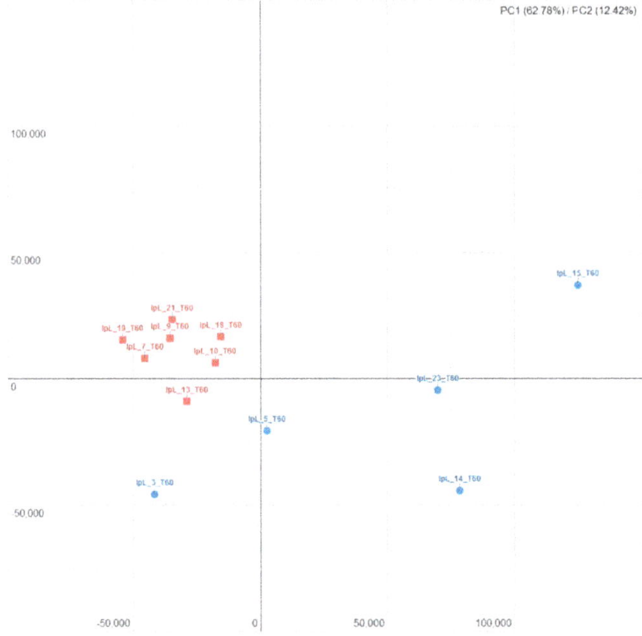

Figure 4. Principal components analysis using the 525 differentially expressed gene identified at the end of the iodine supplementation period: The blue circles identify the iodine samples while the red squares identify CTR samples.

Table 2. List of significantly enriched pathways obtained using the downregulated genes following iodine supplementation.

Pathway	FDR	Genes
regulation of transcription, DNA-templated	1.30×10^{-5}	ZNF93, SBNO1, ZBTB6, SNAPC3, ZFX, ZNF12, LOC530973, ZNF655, LOC104968476, ZNF175, MAP3K7, ZNF184, JADE1, ZNF182, ZNF148, RBAK, ZNF248, ZNF286A, DNTTIP2, MYNN, ZNF605, ZNF572, ZNF436
cellular response to DNA damage stimulus	7.45×10^{-4}	SHPRH, HELB, ZMAT3, FMR1, RNF168, SPRTN, USP16, NEK4
transcription, DNA-templated	0.0057	CCNT2, ZKSCAN8, ZBTB6, SNAPC3, ZNF131, ZFX, SCAI, ZNF518A, MAP3K7, ZNF184, JADE1, ZNF148, PSIP1, DNTTIP2, USP16, MYNN, ZNF572
RNA processing	0.0153	DHX29, U2SURP, YTHDC2, DHX36
negative regulation of transforming growth factor beta receptor signaling pathway	0.0186	ZNF451, LEMD3, SMURF2, SIRT1
cell division	0.0473	CDC40, USP37, USP16, CENPJ, SPICE1, SMC3

FDR: false discovery rate.

Table 3. List of significantly enriched pathways obtained using the upregulated genes following iodine supplementation.

Pathway	FDR	Genes
Fc gamma R-mediated phagocytosis	2.63×10^{-6}	AKT1, AKT2, ARF6, ARPC1B, CFL1, CRKL, GSN, LIMK1, RAC2, SCIN, VASP
Non-alcoholic fatty liver disease (NAFLD)	1.33×10^{-5}	AKT1, AKT2, COX8A, GSK3A, MAP3K11, NDUFA7, NDUFB7, NDUFS6, NDUFS7, NDUFS8, SDHA, TGFB1, UQCR11
Oxidative phosphorylation	0.0005	ATP5D, ATP5G2, COX8A, NDUFA7, NDUFB7, NDUFS6, NDUFS7, NDUFS8, SDHA, UQCR11
Rap1 signaling pathway	0.0008	ACTB, ACTG1, AKT1, AKT2, CRKL, MAP2K3, RAC2, RAPGEF1, RASSF5, SIPA1, TLN1, VASP
Bacterial invasion of epithelial cells	0.0016	ACTB, ACTG1, ARPC1B, CRKL, PXN, RHOG, SEPT9
Proteasome	0.0078	PSMB10, PSMB4, PSMB8, PSMB9, PSMD4
Carbon metabolism	0.0079	ENO1, GPI, HK1, IDH2, PFKL, SDHA, TPI1
Focal adhesion	0.0079	ACTB, ACTG1, AKT1, AKT2, CRKL, PXN, RAC2, RAPGEF1, TLN1, VASP
Regulation of actin cytoskeleton	0.0079	ACTB, ACTG1, ARPC1B, CFL1, CRKL, GSN, LIMK1, PXN, RAC2, SCIN
B cell receptor signaling pathway	0.0024	AKT1, AKT2, CD81, NFKBIB, RAC2

FDR: false discovery rate.

3.3. Effects of I-Supplementation on Quality of Dairy Products

To test the hypothesis if I supplementation could positively influence quality of dairy products, we evaluated the milk somatic cell count (SCC) at the beginning T0 and the end T8 of the supplementation. Clearly, in Figure 5, we demonstrated a significant reduction of SCC, indicating that the I treatment ameliorated immune response against bacteria responsible for infectious diseases like mastitis [24].

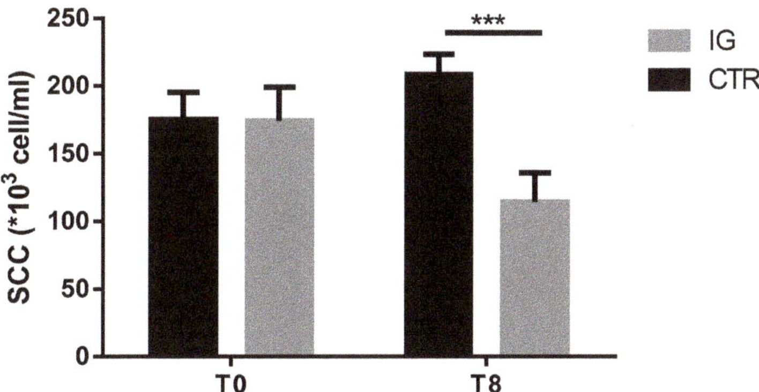

Figure 5. Somatic cell count in individual milk samples from CTR ($n = 11$) and IG ($n = 11$): Data are shown as mean ± SD, and differences were assessed using 2-way ANOVA. *** p-value < 0.001.

Then, we evaluated also the quality of fresh dairy cheese (ricotta) obtained using milk derived by the CTR and I groups. Lipid oxidation is strongly influenced by the redox homeostasis, and in our data set, iodine influences mitochondrial activity, which is highly involved in controlling oxidative stress. Thus, we measured the time-dependent variation (at day 0 and day 7) in malondialdehyde

(MDA) levels in fresh ricotta cheese kept at 4 °C. As showed in Figure 6, ricotta from the I group showed statistically lower amounts of MDA, which confirms the favorable effect of I supplementation on cheese quality.

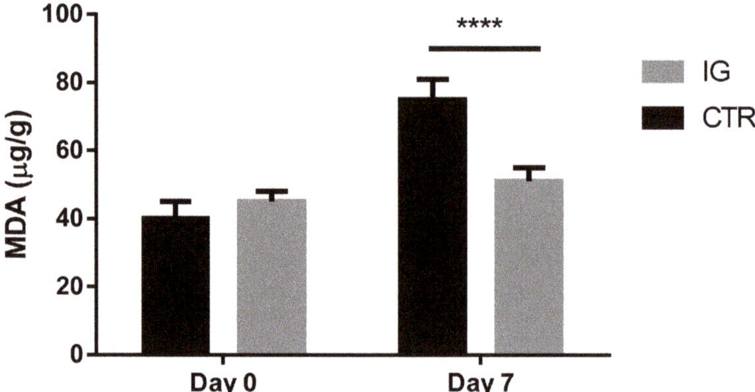

Figure 6. Malondialdehyde (MDA) levels at 0 (40 ± 5 vs. 45 ± 3) and 7 (75 ± 6 vs. 51 ± 4) days from fresh cheese ricotta stored at 4 °C: Data represent mean ± SD, **** $p < 0.0001$, two-way ANOVA (5 samples/group).

4. Discussion

In this study, we provide evidence that dairy cows fed an I-supplemented diet for a limited time showed transcriptional changes related to immune response and oxidative stress. In our experience, whole blood is a good starting point to understand in ruminants the effects of different diet supplements such as agro-industrial by-products and microelements [25–27]. To avoid that differences identified in gene expression in this study that could be influenced by different composition in white blood cell, we measured the complete blood cell count both in CTR and IG at the beginning T0 and at the end of supplementation, and we did not find any difference (Table S4). Then, because I is the major component of thyroid hormones, we measured the free hormone levels in the sera of both groups. In agreement with previous studies, the thyroid hormone levels did not differ between the two groups, clearly indicating that I supplement used in this study does not affect thyroid functionality [28].

In our study, RNA-sequencing analysis confirmed that I supplementation deregulated the expression of numerous genes, showing a significant biological connection confirmed by a very small p-value for protein–protein interaction, indicating no random nodes within our data set (PPI enrichment p-value $< 1.0 \times 10^{-16}$). More in detail, we identified many pathways related to immune response (Table 2), and the most significant one was that of "Fc gamma R-mediated phagocytosis" (FDR: 2.36 $\times 10^{-6}$) and, as a consequence, also of "B cell receptor signaling pathway" (FDR: 0.0024), which is in agreement with previous study in which I exposure produced an increase in immunoglobulin synthesis by lymphocytes [29]. Moreover, phagocytes are the lymphocyte subsets which express the higher level of sodium iodide symporter [28]. Thus, iodide supplementation reinforces immune response via strengthened antibody production and phagocytosis. Moreover, iodide also interacts with myeloperoxidase and H2O2, reinforcing the killing of bacterial activity [30]. Altogether, these positive effects are potentially associated with pathogen clearance and we have demonstrated that milk collected from animals belonging to the experimental group showed a reduced somatic cell count, which has directly associated with reduced bacterial load [24].

Also, oxidative phosphorylation is one of the activated pathways in our study (4.75×10^{-4}). The effects of I and thyroid hormones have been previously fully reviewed [31]; thus, it is not surprising to find the activation of genes belonging to the multimeric protein ATP synthase (ATP5G2 and ATP5D), which is responsible for ATP production from ADP. Indeed, it is been demonstrated that increased production of ATP was inversely associated with production of cellular damage enzyme associated biomarkers (Serum lactate dehydrogenase, LHD; creatine phosphokinase, CPK; Alkaline phosphatase, ALP) following I supplementation. Also, several genes encoding for different subunits of the NADH: ubiquinone oxidoreductase were upregulated in our data set. These results further confirm the effect of I-supplementation on oxidative phosphorylation with a potential favourable energy balance.

In addition, I has been considered, in an evolutionary sense, one of the most ancient antioxidant because it can easily interact with reactive oxygen species (ROS) [32]. In our study, milk samples collected from the experimental groups have shown increased I content in agreement with previous studies on I supplementation in ruminants [13,33]. Thus, when we analyzed ricotta cheese produced by using milk from the CTR and I groups, we appreciated a reduced production of MDA in IG & CTR, indirectly confirming a reduction in lipid oxidation.

In conclusion, in our study, we have demonstrated that iodine supplementation has multiple positive effects on animal health and dairy products quality. Indeed, on one side, I supplementation ameliorates immune response and resistance to infectious disease, improving phagocytosis, and on the other side, it improves the quality of fresh cheese, reducing lipid oxidation with beneficial effects also on consumer's health. Moreover, because of the higher I content in milk, make potentially this product suitable also for subjects that, for particular physiological conditions (i.e., pregnancy) or life stage (i.e., infant), need more I intake.

Supplementary Materials: The following are available online at http://www.mdpi.com/2076-2615/9/11/866/s1, Table S1: Milk parameter comparison between control and I-supplemented group, Table S2: Quality control of reads after trimming, Table S3: Complete list of all significant DEGs following the iodine supplementation period, Table S4: Blood cell count comparison between control and I supplementation group.

Author Contributions: Conceptualization, G.M., M.I., and A.I.; methodology, R.E., C.M., and M.G.; validation, M.I., A.I., and R.E.; formal analysis, M.I. and A.I.; investigation, M.I., A.I., and R.E.; resources, G.M., L.C., and M.D.; data curation, M.I. and A.I.; writing—original draft preparation, M.I. and A.I.; writing—review and editing, R.E., M.D., and G.M.; supervision, G.M. and M.D.; project administration, G.M.; funding acquisition, G.M.

Funding: This work is part of the project "Innovazione della filiera bovina da latte in Abruzzo per produzioni lattiero-casearie ad elevato contenuto salutistico ed ecosostenibile," supported by a grant from Rural Development Plan 2007–2013–MISURA 1.2.4-Regione Abruzzo (Italy), project manager: Giuseppe Martino.

Conflicts of Interest: The authors declare no conflict of interest. The funder had no role in the design of the study; in the collection, analyses, or interpretation of data; in the writing of the manuscript; or in the decision to publish the results.

References

1. Zimmermann, M.B. Iodine deficiency. *Endocr. Rev.* **2009**, *30*, 376–408. [CrossRef] [PubMed]

2. Gunnarsdottir, I.; Dahl, L. Iodine intake in human nutrition: A systematic literature review. *Food Nutr. Res.* **2012**, *56*. [CrossRef] [PubMed]

3. Tetens, I. EFSA Panel on Dietetic Products Nutrition and Allergies. Scientific Opinion on Dietary Reference Values for iodine. *EFSA J.* **2014**, *12*, 3660.

4. Ghirri, P.; Lunardi, S.; Boldrini, A. Iodine Supplementation in the Newborn. *Nutrients* **2014**, *6*, 382–390. [CrossRef] [PubMed]

5. Flachowsky, G.; Franke, K.; Meyer, U.; Leiterer, M.; Schöne, F. Influencing factors on iodine content of cow milk. *Eur. J. Nutr.* **2014**, *53*, 351–365. [CrossRef] [PubMed]

6. EU. Commission Implementing Regulation 2015/861 of 3 June 2015 concerning the authorisation of potassium iodide, calcium iodate anhydrous and coated granulated calcium iodate anhydrous as feed additives for all animal species. *Off. J. Eur. Union.* **2015**, *L137*, 1–7.

7. Weiss, W.P.; Wyatt, D.J.; Kleinschmit, D.H.; Socha, M.T. Effect of including canola meal and supplemental iodine in diets of dairy cows on short-term changes in iodine concentrations in milk. *J. Dairy Sci.* **2015**, *98*, 4841–4849. [CrossRef] [PubMed]

8. Nudda, A.; Battacone, G.; Decandia, M.; Acciaro, M.; Aghini-Lombardi, F.; Frigeri, M.; Pulina, G. The effect of dietary iodine supplementation in dairy goats on milk production traits and milk iodine content. *J. Dairy Sci.* **2009**, *92*, 5133–5138. [CrossRef]

9. Maas, J.; Davis, L.E.; Hempstead, C.; Berg, J.N.; Hoffman, K.A. Efficacy of ethylenediamine dihydriodide in the prevention of naturally occurring foot rot in cattle. *Am. J. Vet. Res.* **1984**, *45*, 2347–2350.

10. Kulow, M.; Merkatoris, P.; Anklam, K.S.; Rieman, J.; Larson, C.; Branine, M.; Döpfer, D. Evaluation of the prevalence of digital dermatitis and the effects on performance in beef feedlot cattle under organic trace mineral supplementation. *J. Anim. Sci.* **2017**, *95*, 3435–3444. [CrossRef]

11. Antaya, N.T.; Soder, K.J.; Kraft, J.; Whitehouse, N.L.; Guindon, N.E.; Erickson, P.S.; Conroy, A.B.; Brito, A.F. Incremental amounts of *Ascophyllum nodosum* meal do not improve animal performance but do increase milk iodine output in early lactation dairy cows fed high-forage diets. *J. Dairy Sci.* **2015**, *98*, 1991–2004. [CrossRef] [PubMed]

12. Derscheid, R.J.; van Geelen, A.; Berkebile, A.R.; Gallup, J.M.; Hostetter, S.J.; Banfi, B.; McCray, P.B., Jr.; Ackermann, M.R. Increased Concentration of Iodide in Airway Secretions Is Associated with Reduced Respiratory Syncytial Virus Disease Severity. *Am. J. Respir. Cell Mol. Biol.* **2014**, *50*, 389–397. [CrossRef] [PubMed]

13. Iannaccone, M.; Elgendy, R.; Ianni, A.; Martino, C.; Palazzo, F.; Giantin, M.; Grotta, L.; Dacasto, M.; Martino, G. Whole-transcriptome profiling of sheep fed with a high iodine-supplemented diet. *Animal* **2019**. [CrossRef] [PubMed]

14. Iannaccone, M.; Ianni, A.; Ramazzotti, S.; Grotta, L.; Marone, E.; Cichelli, A.; Martino, G. Whole Blood Transcriptome Analysis Reveals Positive Effects of Dried Olive Pomace-Supplemented Diet on Inflammation and Cholesterol in Laying Hens. *Animals* **2019**, *9*, 427. [CrossRef] [PubMed]

15. Lopez, C.C.; Serio, A.; Rossi, C.; Mazzarrino, G.; Marchetti, S.; Castellani, F.; Grotta, L.; Fiorentino, F.P.; Paparella, A.; Martino, G. Effect of diet supplementation with *Ascophyllum nodosum* on cow milk composition and microbiota. *J. Dairy Sci.* **2016**, *99*, 6285–6297. [CrossRef] [PubMed]

16. EU. Directive 2010/63/EU of the European Parliament and of the Council of 22 September 2010 on the protection of animals used for scientific purposes. *Off. J. Eur. Union.* **2010**, *L276*, 33–79.

17. EC. No 1459/2005 of 8 September 2005 amending the conditions for authorisation of a number of feed additives belonging to the group of trace elements. *Off. J. Eur. Union.* **2005**, *L233*, 8–10.

18. Fecher, P.A.; Goldmann, I.; Nagengast, A. Determination of iodine in food samples by inductively coupled plasma mass spectrometry after alkaline extraction. *J. Anal. At. Spectrom.* **1998**, *13*, 977–982. [CrossRef]

19. Anders, S.; Pyl, P.T.; Huber, W. HTSeq-A Python framework to work with high-throughput sequencing data. *Bioinformatics* **2015**, *31*, 166–169. [CrossRef]

20. Love, M.I.; Huber, W.; Anders, S. Moderated estimation of fold change and dispersion for RNA-seq data with DESeq2. *Genome Biol.* **2014**, *15*, 550. [CrossRef]

21. Ianni, A.; Innosa, D.; Martino, C.; Grotta, L.; Bennato, F.; Martino, G. Zinc supplementation of Friesian cows: Effect on chemical-nutritional composition and aromatic profile of dairy products. *J. Dairy Sci.* **2019**, *102*, 2918–2927. [CrossRef] [PubMed]

22. Zimmermann, M.B. The Effects of Iodine Deficiency in Pregnancy and Infancy. *Paediatr. Perinat. Epidemiol.* **2012**, *26*, 108–117. [CrossRef] [PubMed]

23. De Vito, P.; Incerpi, S.; Pedersen, J.Z.; Luly, P.; Davis, F.B.; Davis, P.J. Thyroid Hormones as Modulators of Immune Activities at the Cellular Level. *Thyroid* **2011**, *21*, 879–890. [CrossRef] [PubMed]

24. Sharma, N.; Singh, N.K.; Bhadwal, M.S. Relationship of Somatic Cell Count and Mastitis: An Overview. *Asian-Australas. J. Anim. Sci.* **2011**, *24*, 429–438. [CrossRef]

25. Iannaccone, M.; Elgendy, R.; Giantin, M.; Martino, C.; Giansante, D.; Ianni, A.; Dacasto, M.; Martino, G. RNA Sequencing-Based Whole-Transcriptome Analysis of Friesian Cattle Fed with Grape Pomace-Supplemented Diet. *Animals* **2018**, *8*, 188. [CrossRef]

26. Elgendy, R.; Giantin, M.; Castellani, F.; Grotta, L.; Palazzo, F.; Dacasto, M.; Martino, G. Transcriptomic signature of high dietary organic selenium supplementation in sheep: A nutrigenomic insight using a custom microarray platform and gene set enrichment analysis. *J. Anim. Sci.* **2016**, *94*, 3169–3184. [CrossRef]

27. Elgendy, R.; Palazzo, F.; Castellani, F.; Giantin, M.; Grotta, L.; Cerretani, L.; Dacasto, M.; Martino, G. Transcriptome profiling and functional analysis of sheep fed with high zinc-supplemented diet: A nutrigenomic approach. *Anim. Feed Sci. Tech.* **2017**, *234*, 195–2042. [CrossRef]

28. Bilal, M.Y.; Dambaeva, S.; Kwak-Kim, J.; Gilman-Sachs, A.; Beaman, K. A Role for Iodide and Thyroglobulin in Modulating the Function of Human Immune Cells. *Front. Immunol.* **2017**, *8*, 1–13. [CrossRef]
29. Weetman, A.P.; Mcgregor, A.M.; Campbell, H.; Lazarus, J.H.; Ibbertson, H.K.; Hall, R. Iodide enhances IgG synthesis by human peripheral blood lymphocytes in vitro. *Eur. J. Endocrinol.* **1983**, *103*, 210–215. [CrossRef]
30. Klebanoff, S.J.; Kettle, A.J.; Rosen, H.; Winterbourn, C.C.; Nauseef, W.M. Myeloperoxidase: A front-line defender against phagocytosed microorganisms. *J. Leukoc. Biol.* **2013**, *93*, 185–198. [CrossRef]
31. Soboll, S. Thyroid hormone action on mitochondrial energy transfer. *Biochim. Biophis. Acta Bioenerg.* **1993**, *1144*, 1–16. [CrossRef]
32. Venturi, S. Evolutionary Significance of Iodine. *Curr. Chem. Biol.* **2011**, *5*, 155–162. [CrossRef]
33. Grace, N.D.; Waghorn, G.C. Impact of iodine supplementation of dairy cows on milk production and iodine concentrations in milk. *N. Z. Vet. J.* **2005**, *53*, 10–13. [CrossRef] [PubMed]

Article

Comparison of Reconstituted, Acidified Reconstituted Milk or Acidified Fresh Milk on Growth Performance, Diarrhea Rate, and Hematological Parameters in Preweaning Dairy Calves

Lingyan Li [1,†], Jiachen Qu [2,†], Xiaoyue Xin [1], Shuxin Yin [1] and Yongli Qu [1,*]

[1] Key Laboratory of Efficient Utilization of Feed Resources and Nutrition Regulation in Cold Region of Heilongjiang Province, College of Animal Science and Veterinary Medicine, Heilongjiang Bayi Agricultural University, Daqing 163319, China; llytiger@163.com (L.L.); 15147012655@163.com (X.X.); SY20183040619@cau.edu.cn (S.Y)

[2] College of Animal Science and Technology, Inner Mongolia Agricultural University, Hohhot 010018, China; qjc990313@163.com

* Correspondence: Ylqu007@126.com; Tel.: +86-138-3696-1030

† These authors have contributed equally to this work.

Received: 23 August 2019; Accepted: 9 October 2019; Published: 10 October 2019

Simple Summary: The preweaning phase is the period for the rapid growth and development of dairy calves. During this period, dairy calves receive their nutrients through milk. Feeding hygienic milk is of great benefit to optimum growth rate and health status of dairy calves. Previous studies focused on the effects of hygienic milk by acidification on dairy calves' health and growth. Reconstituted milk, as the common source of milk, is being used in dairy calves feeding. However, no previous studies reported the effects of feeding acidified reconstituted milk on dairy calves' health and growth. Our study will provide the evidence that the acidification of reconstituted milk had positive effects on growth performance and health status of preweaning dairy calves.

Abstract: The present experiment was carried out to assess the effects of reconstituted milk (RM), acidified reconstituted milk (ARM), and acidified fresh milk (AFM) on growth performance, diarrhea rate, and hematological parameters of preweaning dairy calves. For this purpose, a total of 27 Holstein female calves (one month of age) with initial body weight of (67.46 ± 4.08) kg were divided into three groups in such a way that each group contained nine calves. Calves were housed individually, and starter was offered ad libitum to each calf. The dietary treatments were RM, ARM, and AFM. The highest milk intake was observed in calves receiving AFM as compared to other treatments ($p < 0.01$). Calves fed AFM had more feed intake than those fed ARM and RM ($p < 0.01$). Feed efficiency was significantly lower for calves offered ARM than those offered RM and AFM ($p < 0.01$). A lower withers height growth was found for calves fed RM than those fed ARM and AFM ($p < 0.05$). Diarrhea rate and white blood cell (WBC) and lymphocytes (LYM) counts were greater for calves fed RM than those fed ARM and AFM ($p < 0.05$). These findings suggested that ARM and AFM had positive effects on growth performance and health status of the preweaning dairy calves.

Keywords: reconstituted milk; acidified reconstituted milk; acidified fresh milk; growth performance; diarrhea rate; hematological parameters; calves

---

1. Introduction

The preweaning phase is the period for the rapid growth and development of female dairy calves [1]. Providing calves with sufficient nutrients to achieve the optimum growth rate in preweaning phase is beneficial for promoting growth performance after adulthood, which positively associated with increased productivity in later life [2].

On large scale dairy farms, pre-weaning dairy calves are usually reared by milk or milk replacer [3–5]. Delay of feeding fresh milk to calves results in the multiplication of bacteria in the milk due to favorable conditions. Therefore, preservation of milk is recommended at large-scale dairy farms to overcome the consequences of bacterial load in fresh milk. It has been reported that acidification of milk to a pH value between 4.0 and 4.5 can inhibit pathogenic bacterial growth and enable milk to be preserved without refrigeration for a short time prior to feeding to calves [6,7]. It has also been reported that acidified milk protects calves from diseases caused by bacteria growth and reduces the incidence of diarrhea [8].

Reconstituted milk is resulted from the addition of water to dried or condensed milk powder in the amount necessary to reestablish the specified water:solids ratio. Reconstituted milk is also being used on the dairy farms for preweaning dairy calves [9]. However, to our knowledge, the acidification of reconstituted milk in preweaning growing calves has not been investigated. Previous studies only focused on acidified milk or acidified milk replacer in dairy calves [10–12]. Therefore, a comparative study was undertaken to compare the growth performance, diarrhea rate, and hematological parameters of preweaning dairy calves fed reconstituted milk, acidified reconstituted milk, or acidified fresh milk.

2. Materials and Methods

2.1. Calves, Feeds and Management

The animal care and experimental procedures were conducted by the Animal Welfare and Ethics Committee of Heilongjiang Bayi Agriculture University. Animal care and handling were followed the guidelines by the regulations for the Administration of Affairs Concerning Experimental Animals (The State Science and Technology Commission of China, 1988). Twenty-seven Holstein female calves procured from large dairy herd of same age with initial body weight of (67.46 ± 4.08) kg were selected and randomly divided into three groups in such a way that each group contained nine calves. Three types of milk were applied to different treatment groups—(1) reconstituted milk (RM), which was made by mixing milk powder with warm water (35–40 °C) in a weight ratio of 1:7; (2) acidified reconstituted milk (ARM), which was obtained by adding 30 mL formic acid (food grade) of 8.5% concentration in 1L reconstituted milk at temperature 5–10 °C; and (3) acidified milk made by using fresh milk (AFM), which was obtained by adding 30 mL formic acid (food grade) of 8.5% concentration in 1L fresh milk at temperature 5–10 °C. The fresh milk was procured from general and healthy herd and not pasteurized before feeding. Milk samples of 50 mL were collected every 10 days to detect the compositions. A milk analyzer (Foss Milkoscan 4000, Foss, Hillerod, Danmark) was used to determine the composition of milk, and milk composition are shown in Table 1. The pH value of acidified milk was determined by a pH meter (FE20, Mettler Toledo, Zurich, Switzerland). Milk samples of 50 mL were successively diluted (10^{-1} to 10^{-4}) using distilled water, and a volume of 1 mL of each dilution was inoculated onto the surface of plate count agar (PCA), violet red bile agar (VRBA), and MRS agar for the cultivation of total bacteria, lactobacillus, and *Escherichia coli*, respectively. Total bacteria and *Escherichia coli* were incubated at 37 °C for 48 and 24 h. Lactobacillus was incubated anaerobically at 37 °C for 72 h. After incubation, the numbers of total bacteria, lactobacillus and Escherichia coli colony-forming units (cfu) were calculated. Bacterial numbers of different types of milk are shown in Table 2.

Prior to onset of trial, all the calves were offered the same basal ration from day 1 to 21. Experimental treatments were applied from day 22 onward. Calves were given a nine-day adaptation period before the start of data collection. No abnormal behaviors were observed during the adaptation period, and the initial data were collected on day 30. Calves were housed individually in calf hutches, and 6 L milk were offered twice daily at 5:00 and 16:30. All calves had free-choice access to clean, fresh water and starter throughout the experiment. The chemical compositions of the starter diets are shown in Table 3.

The experiment period lasted for 40 days until weaning. Standard management and environmental conditions were ensured to avoid any stress as described in recent researches [13,14].

Table 1. Composition of different types of milk.

Items	ARM	RM	AFM
Milk lactose, %	4.08	4.48	4.63
Milk fat, %	4.30	4.55	4.74
Milk protein, %	2.86	2.94	3.13
Total solids, %	11.53	12.29	12.84
pH	4.26	6.87	4.32

ARM: acidified reconstituted milk. RM: reconstituted milk. AFM: acidified milk made by using fresh milk.

Table 2. Bacterial numbers of different types of milk.

Item	Diet			SEM	*p*-Value
	ARM	RM	AFM		
Total Bacterial Count (10^4 cfu/mL)	43.67 [a]	322.33 [b]	38.00 [a]	44.6	0.001
Lactobacillus (10^3 cfu/mL)	114.33	52.17	115.00	30.4	0.137
Escherichia coli (10^1 cfu/mL)	154.83 [a]	458.67 [b]	155.67 [a]	39.1	0.000

ARM: acidified reconstituted milk. RM: reconstituted milk. AFM: acidified milk made by using fresh milk. [a,b] Mean values within a row with different superscript letter differ significantly ($p < 0.05$). SEM: Standard Error of Mean.

Table 3. Composition and nutrient levels of the starter (DM basis, %).

Ingredient	Content	Nutrient Levels	Content
Corn	33.00	NEL (Mcal/kg) [3]	1.69
Soybean meal	23.50	DM [4]	88.07
Expanded corn	10.00	CP [5]	20.95
DDGS [1]	8.00	EE [6]	3.31
Corn husk	5.07	NDF [7]	16.34
Extruded soybean	4.00	ADF [8]	5.48
Corn germ meal	9.00	Ash	7.40
Limestone	1.50	Ca [9]	0.95
MDCP [2]	0.15	P [10]	0.51
NaCl	0.77		
Beer yeast	2.00		
Elancoban	0.01		
Glucose	2.00		
Premix [11]	1.00		

[1] DDGS: Distillers Dried Grains with Solubles; [2] MDCP:mono-calcium and di-calcium phosphate; [3] NEL: net energy for lactation; [4] DM: dry matter; [5] CP: crude protein; [6] EE: ether extract; [7] NDF: neutral detergent fiber; [8] ADF: acid detergent fibre; [9] Ca: Calcium; [10] P: phosphorus; [11] The premix provided the following per kilogram of the diet: V_A 12,000 IU, V_D 4000 IU, V_E 2900IU, Cu 5000 mg, Fe 9000 mg, Mn 6000 mg, Se 67 mg, I 227 mg, Co 20 mg, Mg 9800 mg.

2.2. Growth Performance

Calves were weighed before the morning feeding for three consecutive days at the start and end of the experiment to calculate the initial body weight, final body weight, and average daily gain (ADG). Milk and starter consumption were recorded daily. Feed efficiency was determined from total feed consumption and ADG.

$$\text{Feed efficiency} = \frac{\text{Total feed intake}\,(\text{kg}/\text{day DM})}{\text{ADG}\,(\text{kg}/\text{day})} \tag{1}$$

In addition, body measurements of each calf including body length, withers height, heart girth, and shin circumference were taken at the beginning and end of the trial according to the study made by Li et al. (2014) [15].

2.3. Fecal Score and Diarrhea Rate

Fecal consistency was scored everyday on a scale of 1 to 4, where 1 = normal-firm but not hard, 2 = soft-does not hold form, 3 = runny-spreads easily, and 4 = devoid of solid matter. Fecal score of 3 or 4 were considered as diarrhea. The diarrhea rate was calculated according to the procedure described in the recent study [16] and formula is given below.

$$\text{Diarrhea rate} = \frac{\text{number of calves with diarrhea} \times \text{days of diarrhea}}{\text{total number of calves} \times \text{examined days}} \times 100 \tag{2}$$

2.4. Hematological Parameters

Blood samples (about 10 mL) were taken from the jugular vein in the morning, approximately 3h post feeding on the first and final day of the experiment. Blood was anticoagulated with EDTA for hematological analysis within three hours after collection. Hematological parameters values including white blood cell (WBC), lymphocytes (LYM), monocytes (MO), red blood cell (RBC), hemoglobin (Hb), and hematocrit (HCT) were detected by using automatic hematology analyzer (XN9000, Sysmex, Kobe, Japan).

2.5. Statistical Analysis

All the data were analyzed by PROC MIXED of SAS version 9.4 (SAS Institute Inc., Cary, NC, USA), the statistical model was

$$Y_{ij} = \mu + D_i + C_j + \varepsilon_{ij} \tag{3}$$

where Y_{ij} is the observed variable, μ is the overall mean, D_i is the fixed effect of diet treatment, C_j is the random effect of calves, and ε_{ij} denotes the residual error. Significance was declared for $p < 0.05$, and trends were reported at $0.05 < p < 0.10$. When a significant effect of treatment was detected ($p < 0.05$), differences between the means were tested using Bonferroni multiple comparison test.

3. Results

3.1. Growth Performance

The results of growth performance of dairy calves are shown in Table 4. The initial BW, final BW, ADG, and starter intake were not different among treatments. Milk fed amount in liters were not different among different treatments. However dry matter intake from milk differed among treatments. Lowest dry matter intake from milk was observed in calves fed ARM, while the highest dry matter intake from milk was observed in calves fed AFM ($p < 0.01$). Calves fed AFM had increased total feed intake compared to those fed ARM and RM ($p < 0.01$). The value of feed efficiency was significantly lower for calves offered ARM than those offered RM and AFM ($p < 0.01$), which suggested that the feed efficiency was better for calves fed ARM than RM and AFM.

Table 4. Growth performance of dairy calves fed different types of milk.

Item	Diet			SEM	*p*-Value
	ARM	**RM**	**AFM**		
Initial BW (kg)	67.60	67.60	67.20	2.33	0.513
Final BW (kg)	104.40	102.20	101.80	3.79	0.766
ADG (kg/day)	0.92	0.88	0.96	0.25	0.952
Milk intake (L/day)	11.28	11.17	11.41	0.25	0.637
Milk intake (kg/day DM)	1.34 [a]	1.41 [b]	1.51 [c]	0.03	0.002
Starter intake (kg/day DM)	0.43	0.40	0.42	0.08	0.994
Total feed intake (kg/day DM)	1.76 [a]	1.81 [a]	1.92 [b]	0.03	0.002
Feed efficiency (kg intake/kg gain)	1.85 [a]	2.05 [b]	2.00 [b]	0.03	0.000

ARM: acidified reconstituted milk. RM: reconstituted milk. AFM: acidified milk made by using fresh milk. [a,b,c] Mean values within a row with different superscript letter differ significantly ($p < 0.05$). SEM: Standard Error of Mean.

Body measurements data of different treatments are presented in Table 5. Body measurements were not different at the beginning of the experiment. Smaller heart girth was observed in the end of the experiment for calves fed AFM than those fed ARM and RM ($p < 0.05$). Compared with calves fed ARM and AFM, a lower withers height growth could be found for calves fed RM ($p < 0.05$). Similarly, final withers height was lower for calves fed RM as compare to ARM and AFM ($p < 0.05$).

Table 5. Body measurements of dairy calves fed different types of milk.

Item	Diet			SEM	*p*-Value
	ARM	**RM**	**AFM**		
Initial					
Body length (cm)	83.80	83.60	82.60	0.76	0.275
Withers height (cm)	83.60	83.80	83.00	0.87	0.634
Heart girth (cm)	94.00	93.60	92.20	1.37	0.411
Shin circumference (cm)	12.00	11.80	11.60	0.26	0.335
Final					
Body length (cm)	94.60	94.80	93.80	1.17	0.671
Withers height (cm)	95.20 [a]	91.00 [b]	94.60 [a]	1.44	0.026
Heart girth (cm)	107.20 [a]	108.00 [a]	103.00 [b]	1.78	0.034
Shin circumference (cm)	12.70	12.80	12.40	0.44	0.649
Growth					
Body length (cm)	10.8	11.2	11.2	1.22	0.931
Withers height (cm)	11.6 [a]	7.20 [b]	11.6 [a]	1.45	0.015
Heart girth (cm)	13.2	14.4	10.80	1.76	0.157
Shin circumference (cm)	0.70	1.00	0.80	0.29	0.597

ARM: acidified reconstituted milk. RM: reconstituted milk. AFM: acidified milk made by using fresh milk. [a,b] Mean values within a row with different superscript letter differ significantly ($p < 0.05$). SEM: Standard Error of Mean.

3.2. Fecal Score and Diarrhea Rate

Fecal score and diarrhea rate of dairy calves were presented in Table 6. Average fecal consistency score and diarrhea rate during 30–50 day, 51–60 day, and 61–70 day were greater for calves fed RM than those fed ARM and AFM ($p < 0.01$).

Table 6. Fecal score and diarrhea rate of dairy calves fed different types of milk.

Item	Diet			SEM	*p*-Value
	ARM	**RM**	**AFM**		
Fecal consistency score					
30–50 day	1.12 [a]	1.48 [b]	1.21 [a]	0.10	0.003
51–60 day	1.20	1.45	1.30	0.13	0.207
61–70 day	1.29 [a]	1.82 [b]	1.63 [b]	0.15	0.006
Diarrhea rate (%)					
30–50 day	0.72 [a]	11.07 [b]	0.74 [a]	0.03	0.000
51–60 day	0.17 [a]	10.87 [b]	0.23 [a]	0.20	0.000
61–70 day	2.90 [a]	9.40 [b]	2.48 [a]	0.41	0.000

ARM: acidified reconstituted milk. RM: reconstituted milk. AFM: acidified milk made by using fresh milk. [a,b] Mean values within a row with different superscript letter differ significantly ($p < 0.05$). SEM: Standard Error of Mean.

3.3. Hematological Parameters

Hematological parameters of different treatments are presented in Table 7. Hematological parameters values of WBC, LYM, MO, RBC, HB, and HCT were not different among the treatments at the start of the experiment. At the end of the trial, higher white blood cell and lymphocytes counts were found in calves offered RM ($p < 0.05$).

Table 7. Hematological parameters of dairy calves fed different types of milk.

Item	Diet			SEM	*p*-Value
	ARM	**RM**	**AFM**		
Initial					
WBC [1] (10^9/L)	10.57	11.06	10.81	0.98	0.887
LYM [2] (10^9/L)	3.29	5.04	3.38	1.62	0.496
MO [3] (10^9/L)	1.69	1.70	2.57	0.51	0.185
RBC [4] (10^{12}/L)	4.49	4.69	4.57	0.48	0.921
Hb [5] (g/L)	130.60	135.60	131.40	5.21	0.601
HCT [6] (L/L)	0.22	0.23	0.22	0.01	0.383
Final					
WBC (10^9/L)	12.17 [a]	28.55 [b]	12.52 [a]	4.88	0.008
LYM (10^9/L)	3.21 [a]	4.83 [b]	3.00 [a]	0.71	0.049
MO (10^9/L)	1.28	1.36	1.26	0.31	0.947
RBC (10^{12}/L)	5.52	6.11	5.78	0.40	0.365
Hb (g/L)	116.20	121.40	118.00	4.61	0.537
HCT (L/L)	0.23	0.26	0.25	0.02	0.279

ARM: acidified reconstituted milk. RM: reconstituted milk. AFM: acidified milk made by using fresh milk. [a,b] Mean values within a row with different superscript letter differ significantly ($p < 0.05$). [1] WBC: white blood cell; [2] LYM: lymphocytes; [3] MO: monocytes; [4] RBC: red blood cell; [5] Hb: hemoglobin; [6] HCT: hematocrit. SEM: Standard Error of Mean.

4. Discussion

4.1. Growth Performance

Preweaning growth of female dairy calves is considered an important factor of milk production in their later stage of life. Feed intake, ADG, and feed efficiency are important indicators for growth performance of female dairy calves. Feed intake affects the level of nutrient intake, thus affecting ADG and final BW. ADG and final BW play an important role in assessing the feed efficiency and overall health status of dairy female calves [17]. Akins reported that female diary calves should gain 0.75 kg per day to reach optimum weight of weaning at eight weeks of age [18]. In the present study, ADG of calves in different treatment groups exceeded 0.75 kg/day. Although ADG were in favor of calves fed ARM and AFM compared to those fed RM (0.92, 0.96 vs. 0.88 kg/day), however the differences were not significant. Eren found that the calves offered acidified milk had increased ADG than those offered whole milk [19]. Zhang et al. reported that acidified milk replacers contribute to the improvement of growth rate of calves [12]. Dry matter intakes (DMI) from milk was the highest for calves fed AFM. This was expected as the highest total solid concentration of AFM which resulted in the higher consumption of DM from milk by calves. DMI from starter were not different for calves fed ARM, RM, and AFM, which were 0.43, 0.40, and 0.42 kg/day DM, respectively. The results are consistent with other research [10–12,20]. However, Sun et al. found a reduction in the intake of starter for the calves fed acidified milk with butyric acid addition [16]. Interestingly, in the current study, starter feed DMI was similar among all the group, which could be explained by different acidifier and experimental treatments as reported in the previous studies. Overall, ARM, RM, and AFM behave similarly on the intake of starter feed in the current study.

Body measurements usually reflect body growth and development, which can be used to estimate liveweight [21]. Calves assigned to the ARM and AFM treatment groups exhibited greater withers height gain in the body measurements than the RM calves. The results suggested that calves reared using ARM and AFM had better skeletal growth. Todd et al. reported the calves fed acidified milk replacer were associated with greater preweaning structural growth [5]. However, some studies found that body measurements of calves were not affected by feeding acidified milk or acidified milk replacer [3,10,16].

4.2. Fecal Score and Diarrhea Rate

Diarrhea is the most important cause of disease of calves, with high morbidity and mortality, which threatens the health and growth performance of calves and is a major cause of economic loss for cattle raisers [22]. Results in the present study showed that calves fed ARM and AFM exhibited lower fecal consistency scores and diarrhea rate. It has been reported that lower pH values inhibit the growth of total bacteria and *Escherichia coli* [23], and the current study provided the evidence that ARM and AFM contained lower number of total bacteria and *Escherichia coli* (Table 2). The lower number of total bacteria and *Escherichia coli* reduces the growth and reproduction of saprophytic bacteria that harm the health of gastro-intestinal tract and cause diarrhea. On the other hand, feeding acidified milk can increase the acidity in the intestinal tract of calves, as reported by Woodford et al. [24]. Relatively higher acidity of intestinal tract is likely to exhibit a bacterial influence, therefore reducing the incidence of diarrhea among calves. Therefore, the lower incidence of diarrhea in ARM and AFM could be explained by the lower pH of intestine caused by ARM and AFM and by the lower number of total bacteria and *Escherichia coli*.

4.3. Hematological Parameters

Hematological parameters such as WBC, LYM, and RBC counts and Hb concentration are considered as important clinical indicators that are widely used to reflect health and disease status [25]. Diseases of the preweaning calves is one of the major cause of economic loss in dairy production [26]. Thus, the measurement of hematological parameters could be used to evaluate the health and physiological status of the calves [27]. WBC are the major part of the immune system that helps fight infection and defend the body against other foreign materials. Different types of WBC are involved in recognizing intruders, killing harmful bacteria, and creating antibodies to protect the body against future exposure to some bacteria and viruses [28,29]. LYM are one of several different types of white blood cells that are also important in the immune system, with T cells being responsible for directly killing many foreign invaders and B cells producing antibodies to guard the body against infection [30]. Some research reported that mean value of WBC and LYM were 8.08–12.75 (10^9/L) and 2.67–4.77 (10^9/L) of weaning Hanwoo and Holstein calves [31–33]. In the present study, mean value of WBC was 28.55 (10^9/L) and LYM was 4.83 (10^9/L) for calves fed RM, which were much higher than that of the research mentioned above. However, mean value of WBC and LYM for calves offered ARM and AFM were very close to the values of the research we have reviewed. Therefore, it can be concluded that the calves offered RM were susceptibility to infection, which may result in detrimental effects on calf health.

5. Conclusions

Results from this research indicated that calves fed acidified reconstituted milk or acidified fresh milk had greater average daily gain and withers height growth, lower diarrhea rate, and white blood cell and lymphocytes counts. We concluded that calves offered reconstituted milk were susceptible to infection, which may have a negative effect on calf health. Acidification of reconstituted milk as well as fresh milk had positive effects on growth performance and health status of the preweaning calves.

Author Contributions: Conceptualization, L.L., J.Q., and Y.Q.; methodology, X.X. and Y.Q.; formal analysis, L.L. and J.Q.; investigation, X.X. and S.Y.; data curation, L.L. and S.Y.; writing—original draft preparation, L.L., J.Q., X.X., and S.Y.; writing—review and editing, L.L. and Y.Q.; visualization, L.L.; project administration, L.L. and Y.Q.; funding acquisition, Y.Q.

Funding: This research was funded by the National Natural Science Foundation of China (Grant No. 31902186), the Talents Introduction Foundation of Heilongjiang Bayi Agricultural University (Grant No. XYB2015-06), the Natural Science Foundation of Heilongjiang Province of China (Grant No. C2017044), the Research Program of Heilongjiang Agricultural Reclamation Administration (Grant No. HKKYZD190302), and the Heilongjiang Bayi Agricultural University Support Program for San Heng San Zong (Grant No. TDJH201805).

Acknowledgments: The authors would like to acknowledge Guang Shao, the dairy farm manager, and the workers of Yuda Animal Husbandry Co., Ltd, for their help on collecting experimental data.

Conflicts of Interest: The authors declare that there are no conflicts of interest.

References

1. Diao, Q.Y.; Zhang, R.; Yan, T. Current research progresses on calf rearing and nutrition in China. *J. Integr. Agr.* **2017**, *16*, 2805–2814. [CrossRef]
2. Drackley, J.K. Calf nutrition from birth to breeding. *Vet. Clin. North Am. Food Anim. Pract.* **2008**, *24*, 55–86. [CrossRef] [PubMed]
3. Bayram, B.; Yanar, M.; Güler, O.; Metin, J. Growth performance, health and behavioural characteristics of Brown Swiss calves fed a limited amount of acidified whole milk. *Ital. J. Anim. Sci.* **2007**, *6*, 273–279. [CrossRef]
4. Khan, M.; Weary, D.; Von Keyserlingk, M. Invited review: Effects of milk ration on solid feed intake, weaning, and performance in dairy heifers. *J. Dairy Sci.* **2011**, *94*, 1071–1081. [CrossRef] [PubMed]
5. Todd, C.; Leslie, K.; Millman, S.; Bielmann, V.; Anderson, N.; Sargeant, J.; DeVries, T. Clinical trial on the effects of a free-access acidified milk replacer feeding program on the health and growth of dairy replacement heifers and veal calves. *J. Dairy Sci.* **2017**, *100*, 713–725. [CrossRef] [PubMed]
6. Collings, L.; Proudfoot, K.; Veira, D. The effects of feeding untreated and formic acid-treated colostrum ad libitum on intake and immunoglobulin levels in dairy calves. *Can. J. Anim. Sci.* **2011**, *91*, 55–59. [CrossRef]
7. Todd, C.; Millman, S.; Leslie, K.; Anderson, N.; Sargeant, J.; DeVries, T. Effects of milk replacer acidification and free-access feeding on early life feeding, oral, and lying behavior of dairy calves. *J. Dairy Sci.* **2018**, *101*, 8236–8247. [CrossRef]
8. Güler, O.; Yanar, M.; Bayrum, B.; Metin, J. Performance and health of dairy calves fed limited amounts of acidified milk replacer. *S. Afr. J. Anim. Sci.* **2006**, *36*, 149–154. [CrossRef]
9. Jenkins, K.; Bona, A. Performance of calves fed combinations of whole milk and reconstituted skim milk powder. *J. Dairy Sci.* **1987**, *70*, 2091–2094. [CrossRef]
10. Yanar, M.; Güler, O.; Bayram, B.; METİN, J. Effects of feeding acidified milk replacer on the growth, health and behavioural characteristics of Holstein Friesian calves. *Turk. J. Vet. Anim. Sci.* **2006**, *30*, 235–241.
11. Hill, T.; Bateman II, H.; Aldrich, J.; Quigley, J.; Schlotterbeck, R. Evaluation of ad libitum acidified milk replacer programs for dairy calves. *J. Dairy Sci.* **2013**, *96*, 3153–3162. [CrossRef] [PubMed]
12. Zhang, R.; Diao, Q.-Y.; Zhou, Y.; Yun, Q.; Deng, K.-D.; Qi, D.; Tu, Y. Decreasing the pH of milk replacer containing soy flour affects nutrient digestibility, digesta pH, and gastrointestinal development of preweaned calves. *J. Dairy Sci.* **2017**, *100*, 236–243. [CrossRef] [PubMed]

13. ur Rahman, M.A.; Chuanqi, X.; Huawei, S.; Binghai, C. Effects of hay grass level and its physical form (full length vs. chopped) on standing time, drinking time, and social behavior of calves. *J. Vet Behav.* **2017**, *21*, 7–12. [CrossRef]

14. ur Rahman, M.A.; Qi, X.C.; Binghai, C. Nutrient intake, feeding patterns and abnormal behavior of growing bulls fed different concentrate levels and a single fiber source (corn stover silage). *J. Vet Behav.* **2019**, *33*, 46–53. [CrossRef]

15. Li, L.; Zhu, Y.; Wang, X.; He, Y.; Cao, B. Effects of different dietary energy and protein levels and sex on growth performance, carcass characteristics and meat quality of F1 Angus× Chinese Xiangxi yellow cattle. *J. Anim. Sci. Biotechnol.* **2014**, *5*, 21. [CrossRef] [PubMed]

16. Sun, Y.; Li, J.; Meng, Q.; Wu, D.; Xu, M. Effects of butyric acid supplementation of acidified milk on digestive function and weaning stress of cattle calves. *Livest Sci.* **2019**, *225*, 78–84. [CrossRef]

17. Dingwell, R.; Wallace, M.; McLaren, C.; Leslie, C.; Leslie, K. An evaluation of two indirect methods of estimating body weight in Holstein calves and heifers. *J. Dairy Sci.* **2006**, *89*, 3992–3998. [CrossRef]

18. Akins, M.S. Dairy heifer development and nutrition management. *Vet. Clin. North Am. Food Anim. Pract.* **2016**, *32*, 303–317. [CrossRef] [PubMed]

19. Eren, V. The effect of acidified milk on body weight gain, some blood parameters and health in calves. *Yüzüncü Yıl Üniversitesi Veteriner Fakültesi Dergisi* **2009**, *20*, 17–21.

20. Jaster, E.; McCoy, G.; Tomkins, T.; Davis, C. Feeding acidified or sweet milk replacer to dairy calves. *J. Dairy Sci.* **1990**, *73*, 3563–3566. [CrossRef]

21. Lukuyu, M.N.; Gibson, J.P.; Savage, D.; Duncan, A.J.; Mujibi, F.; Okeyo, A. Use of body linear measurements to estimate liveweight of crossbred dairy cattle in smallholder farms in Kenya. *Springer Plus.* **2016**, *5*, 63. [CrossRef] [PubMed]

22. Gomez, D.E.; Weese, J.S. Viral enteritis in calves. *Can. Vet. J.* **2017**, *58*, 1267–1274. [PubMed]

23. Stewart, S.; Godden, S.; Bey, R.; Rapnicki, P.; Fetrow, J.; Farnsworth, R.; Scanlon, M.; Arnold, Y.; Clow, L.; Mueller, K. Preventing bacterial contamination and proliferation during the harvest, storage, and feeding of fresh bovine colostrum. *J. Dairy Sci.* **2005**, *88*, 2571–2578. [CrossRef]

24. Woodford, S.; Whetstone, H.; Murphy, M.; Davis, C. Abomasal pH, nutrient digestibility, and growth of Holstein bull calves fed acidified milk replacer. *J. Dairy Sci.* **1987**, *70*, 888–891. [CrossRef]

25. Kelada, S.N.; Aylor, D.L.; Peck, B.C.; Ryan, J.F.; Tavarez, U.; Buus, R.J.; Miller, D.R.; Chesler, E.J.; Threadgill, D.W.; Churchill, G.A. Genetic analysis of hematological parameters in incipient lines of the collaborative cross. *G3 Genes Genomes Genet.* **2012**, *2*, 157–165. [CrossRef] [PubMed]

26. Mohri, M.; Sharifi, K.; Eidi, S. Hematology and serum biochemistry of Holstein dairy calves: Age related changes and comparison with blood composition in adults. *Res. Vet. Sci.* **2007**, *83*, 30–39. [CrossRef] [PubMed]

27. Roland, L.; Drillich, M.; Iwersen, M. Hematology as a diagnostic tool in bovine medicine. *J. Vet. Diagn. Invest.* **2014**, *26*, 592–598. [CrossRef] [PubMed]

28. Hoffman, W.; Lakkis, F.G.; Chalasani, G. B cells, antibodies, and more. *Clin. J. Am. Soc. Nephrol.* **2016**, *11*, 137–154. [CrossRef] [PubMed]

29. Karlmark, K.; Tacke, F.; Dunay, I. Monocytes in health and disease—Minireview. *Eur. J. Microbiol. Immunol.* **2012**, *2*, 97–102. [CrossRef] [PubMed]

30. Görgens, A.; Radtke, S.; Horn, P.; Giebel, B. New relationships of human hematopoietic lineages facilitate detection of multipotent hematopoietic stem and progenitor cells. *Cell Cycle* **2013**, *12*, 3478–3482. [CrossRef]

31. Roodposhti, P.M.; Dabiri, N. Effects of probiotic and prebiotic on average daily gain, fecal shedding of Escherichia coli, and immune system status in newborn female calves. *Asian-Australas. J. Anim. Sci.* **2012**, *25*, 1255. [CrossRef] [PubMed]

32. Kim, M.-H.; Yang, J.-Y.; Upadhaya, S.D.; Lee, H.-J.; Yun, C.-H.; Ha, J.K. The stress of weaning influences serum levels of acute-phase proteins, iron-binding proteins, inflammatory cytokines, cortisol, and leukocyte subsets in Holstein calves. *J. Vet. Sci.* **2011**, *12*, 151–157. [CrossRef] [PubMed]

33. Sarker, M.; Ko, S.; Lee, S.; Kim, G.; Choi, J.; Yang, C. Effect of different feed additives on growth performance and blood profiles of Korean Hanwoo calves. *Asian-Australas. J. Anim. Sci.* **2009**, *23*, 52–60. [CrossRef]

Article

Nutrigenomic Effects of Long-Term Grape Pomace Supplementation in Dairy Cows

Marianna Pauletto [1,†], Ramy Elgendy [1,2,†], Andrea Ianni [3], Elettra Marone [3], Mery Giantin [1], Lisa Grotta [3], Solange Ramazzotti [3], Francesca Bennato [3], Mauro Dacasto [1,*] and Giuseppe Martino [3]

[1] Department of Comparative Biomedicine and Food Science, University of Padua, Viale dell'Università 16, 35020 Legnaro (PD), Italy; marianna.pauletto@unipd.it (M.P.); ramy.elgendy@igp.uu.se (R.E.); mery.giantin@unipd.it (M.G.)

[2] Department of Pharmacology, Faculty of Veterinary Medicine, Suez Canal University, Ismailia 41522, Egypt

[3] Faculty of Bioscience and Technology for Food, Agriculture and Environment, University of Teramo, 64100 Teramo (TE), Italy; aianni@unite.it (A.I.); emarone@unite.it (E.M.); lgrotta@unite.it (L.G.); sramazzotti@unite.it (S.R.); fbennato@unite.it (F.B.); gmartino@unite.it (G.M.)

* Correspondence: mauro.dacasto@unipd.it; Tel.: +39-049-827-2935

† These Authors contributed equally to this work.

Received: 26 March 2020; Accepted: 16 April 2020; Published: 19 April 2020

Simple Summary: The aim of this study was to evaluate the effect of grape pomace (GP), the polyphenol-rich agricultural by-product, on dairy cows' whole-blood transcriptome, milk production and composition. Twelve lactating Holstein-Friesian cows were randomly assigned to two groups; the first received a GP-supplemented diet for 60 days (group GP), whereas the second was given only a basal diet (CTR). The results reveal 40 protein-coding genes differentially expressed in the GP group when compared with the CTR group, but no effects were noticed on milk production, concentrations of crude protein, fat, casein, lactose and urea, or somatic cell count. Compared to CTR, GP had a transcriptomic signature mainly reflecting a reinforced immunogenic response.

Abstract: The increasing demand for more animal products put pressure on improving livestock production efficiency and sustainability. In this context, advanced animal nutrition studies appear indispensable. Here, the effect of grape pomace (GP), the polyphenol-rich agricultural by-product, was evaluated on Holstein-Friesian cows' whole-blood transcriptome, milk production and composition. Two experimental groups were set up. The first one received a basal diet and served as a control, while the second one received a 7.5% GP-supplemented diet for a total of 60 days. Milk production and composition were not different between the group; however, the transcriptome analysis revealed a total of 40 genes significantly affected by GP supplementation. Among the most interesting down-regulated genes, we found the DnaJ heat-shock protein family member A1 (DNAJA1), the mitochondrial fission factor (MFF), and the impact RWD domain protein (IMPACT) genes. The gene set enrichment analysis evidenced the positive enrichment of 'interferon alpha (IFN-α) and IFN-γ response', 'IL6-JAK-STAT3 signaling' and 'complement' genes. Moreover, the functional analysis denoted positive enrichment of the 'response to protozoan' and 'negative regulation of viral genome replication' biological processes. Our data provide an overall view of the blood transcriptomic signature after a 60-day GP supplementation in dairy cows which mainly reflects a GP-induced immunomodulatory effect.

Keywords: nutrigenomics; grape pomace; polyphenols; dairy cows; RNA-sequencing

---

1. Introduction

Historically, human populations largely depend on domestic farm animals for the production of animal food-products (e.g., meat, fat, milk and other dairy products). The recent and increasing demand for more animal products, especially in developing countries with rising living standards, puts pressure on global livestock productivity. With a rapidly growing world population, this increasing demand will put a strain on our natural resources and the whole environment as livestock directly compete with humans for the available arable land [1]. Nonetheless, the high demand for these animal proteins has to deal with other challenges such as climate change. To improve further animal production efficiency and limit the global land use for livestock production, advanced animal nutrition studies are today indispensable.

In this context, a thorough understanding of animal gene expression driven by dietary nutrients can be regarded as the bottom line of advanced animal nutrition research. Nutrigenomics (including transcriptomics) is a new branch of animal nutrition at the basic molecular biological level. In other words, it studies the effects of dietary nutrients on cellular gene expression and ultimately, phenotypic changes in living animals. Among the transcriptomic approaches, RNA-sequencing (RNA-seq) simultaneously measures the differential expression of thousands of genes in any biological matrix. The resulting information can be used to determine the effects of a certain condition (e.g., a supplemented diet) from the perspective of the biological processes or molecular functions involved rather than from the expression levels of individual genes [2].

In the European Union (EU), agro-industrial by-products are being considered a key element in the EU's 2020 Environment Action Programme concerning waste management, recycling, and reuse [3]. The reuse of agricultural by-products to produce bio-liquids, oils [4] and bioactive compounds such as carotenoids and polyphenols [5,6] or as animal feed and land fertilizers [5–7], has valorized them as sustainable resources. Indeed, the use of several agricultural by-products in animal nutrition has been largely explored [7]. In addition to the 'traditional' agricultural residues such as oil meals, bran, brewers' grains, beet pulp and molasses, other 'alternative' by-product residues, resulting from fruit and vegetable processing, have become attractive as animal feedstuffs [8].

Grape pomace (GP) is the residue (by-product) that remains after crushing the grapes to collect its juice. It makes about one third of the total volume of the grapes used in wine production [9]. Although biodegradable, GP requires time to mineralize; thus, if left unexploited, it might represent a potential source of pollution to the environment [10]. However, GP is rich in phenolic compounds [9] that hold substantial antioxidant properties [11,12]. Moreover, due to its low cost and high fiber content, GP was proposed earlier as an alternative feed ingredient to 'partially' replace the forage portion in the diet of ruminants [9].

Recently, there has been growing interest in studying the effects of the dietary inclusion of GP on the overall productive performances and production aspects of food-producing animals. For instance, GP increased the biodiversity degree of intestinal microflora in broiler chicks [13], and improved the gain-to-feed ratio and overall performance in pigs [14,15]. Further, GP altered the nitrogen metabolism and decreased the ruminal ammonia production in male sheep [16], was shown to modify the rumen microbial population involved in methane metabolism [17], enhance the growth of facultative probiotic bacteria and inhibit the growth of pathogenic ones in lambs [18], but had no effect on milk yield and composition of dairy ewes [19]. In dairy cows, few studies, with varying results, have investigated the effects of GP dietary inclusion on milk production, composition and total polyphenols content as well as on the overall performance of animals [10,20–23]. We have previously described the transcriptomic signature of veal calves fed with a GP-supplemented diet [24]. Nevertheless, to our knowledge, few papers concerning the nutrigenomic effect of long-term dietary GP supplementation in dairy cows are available. It is conceivable that the dietary inclusion of GP (5% of total diet) would have a noticeable transcriptomic signature that could reflect its claimed antioxidant effects in dairy cows. Therefore, in the present study, a whole-transcriptome profiling, using the RNA-seq technology, was performed in dairy cows fed with a GP-supplemented diet and the obtained results are described herein.

2. Materials and Methods

2.1. Animals and Study Design

Animals were managed as per the Directive 2010/63 [25] of the European Parliament regarding the protection of animals used for experimentation or other scientific purposes. Blood sampling was performed concurrently with planned blood withdrawal for prophylaxis. Dairy cows had not been subjected to breeding practices different than those normally envisaged; thus, ethical declaration was not necessary.

Twelve lactating Holstein-Friesian cows (~163 days in milk), with an average calving number of 2.55 and an average body condition score of 2.97 were used as experimental animals for this study. The cows had an acclimation period of 2 weeks; then, they were randomly allocated into 2 experimental groups of 6 animals each. The first group received a basal diet and served as a control (CTR), while the other group received a diet supplemented with 7.5% GP-supplemented diet on a dry matter basis (GP group). The experimentation was conducted for a period of 60 days, in which all animals were housed in 2 separate areas of free housing with access to an identical feeding area in which each animal had an individual feeding bin, with water freely available all throughout the study. The GP, obtained from different wineries and distilleries in the region of Abruzzo (Italy), was prepared according to the procedure described in [26], and analyzed for its chemical composition before being added to the animals' diet. Weekly from the start of the trial, the daily feed intake (dry matter) has been calculated for each animal by subtracting the refused meal to the administered diet.

2.2. Extraction and Determination of GP Total Polyphenols Content (TPC)

In order to extract the GP polyphenols, 1 g of pomace was added to 5 mL of methanol and the mixture was stirred for 4 h at room temperature. The solution was then transferred in a 15 mL test tube and placed in an ultrasonic bath for 10 min. At the end of this procedure, centrifugation (4000 rpm) was carried out for 20 min. The supernatant was recovered and filtered with paper filters (Whatman 41) in order to remove the interfering substances; this operation was repeated three times. Total Phenolic Content (TPC) was estimated by using the Folin-Ciocalteu colorimetric method, according to the procedure reported by [27]. The chemical composition and total polyphenols content of the GP extract are reported in Table 1.

Table 1. Chemical composition and main total polyphenols content of the GP extract.

Chemical Composition of GP	
Dry Matter, %	92.20
Crude protein [1], %	12.40
Ether extract [1], %	4.20
Crude fibre [1], %	31.80
Neutral-detergent fibre [1], %	42.20
Acid detergent fibre [1], %	41.30
Acid-detergent lignin [1], %	28.80
Starch [1], %	-
Ash [1], %	8.50
Total Polyphenols Content (TCP) in GP	
TPC (GAE [2] mg/g)	16.07 ± 1.41

[1] On a dry matter (DM) basis; [2] Gallic Acid Equivalents.

2.3. Sampling

Milk, feed, and whole-blood (WB) samples were individually collected from each group. For the RNA-seq analysis, 2.5 mL of jugular venous blood were collected at T60 (2 groups × 6 animals each). Duplicate WB samples were collected in PAXgene[TM] tubes (Qiagen SpA, Milan, Italy), stored at room temperature overnight and, then, at −20 °C until RNA isolation. Milk yield (morning milking) was recorded for each cow and individual milk samples were collected from all animals and stored at −20 °C for further analysis.

2.4. Milk and Feed Composition Analysis

Milk samples were analyzed for their protein, casein, lactose, fat, urea, pH, and total solids content (%) using Fourier Transformed InfraRed (FTIR) spectrophotometry (MilkoScanTM FT 6000; Foss, Hillerød, Denmark) as per the manufacturer's instructions. The milk somatic cell count (SCC) was analyzed by a fluoro-opto-electronic counter (FossomaticTM FC; Foss, Hillerød, Denmark). The method is based on flow cytometry technology that counts somatic cells, following the mixing of milk samples (1 to 10 μL, at 30–42 °C) with a DNA molecule-binding staining solution. The administered diets were analyzed according to the methods described by AOAC International [28] in order to obtain information about dry matter (method 930.15), crude protein (method 954.01), fat content (method 920.39), crude fibre (method 962.09) and ash (method 942.05). The procedure reported by Goering and Van Soest [29] was instead used for the evaluation of Neutral Detergent Fibre (NDF) and Acid Detergent Fibre (ADF). The composition of the basal and supplemental diets is reported in Table 2.

Table 2. Ingredients and chemical composition of the two diets.

	Diets	
Ingredients of the diets	CTR/basal	GP
Corn silage, %	45.70	45.50
Second-cut alfalfa hay, %	20.30	20.20
Wheat straw, %	3.80	1.30
Corn grain meal, %	13.50	13.10
Soybean meal, %	8.40	7.60
Barley meal, %	2.50	2.50
Cotton seed, %	3.80	2.80
Dried grape pomace, %	-	5.00
Vitamins and minerals, %	2.00	2.00
Nutrient composition		
Dry Matter, %	61.50	61.70
Crude protein [1], %	15.30	15.90
Ether extract [1], %	2.30	2.20
Crude fibre [1], %	17.80	18.90
Neutral-detergent fibre [1], %	34.40	36.30
Acid detergent fibre [1], %	20.30	22.50
Acid-detergent lignin [1], %	4.60	5.90
Starch [1], %	25.20	24.80
Ash [1], %	4.50	5.10
Net energy (Milk UF/kg DM)	0.89	0.91

[1] On a dry matter (DM) basis.

2.5. Library Preparation and RNA-Seq Analysis

Total RNA was isolated from blood samples by using the PAXgene blood RNA kit (Qiagen, Milan, Italy) as per the manufacturer's instructions. To reduce the possible presence of genomic DNA contamination, a 15-minutes on-column DNase digestion step was included in the RNA isolation protocol. Total RNA concentration was determined by the NanoDrop ND-1000 spectrophotometer (NanoDrop Technologies Inc., Wilmington, DE, USA), and its quality was measured by the 2100 Bioanalyzer and RNA 6000 Nano kit (Agilent Technologies, Santa Clara, CA, USA). Strand-specific RNA-seq libraries were prepared using the SureSelect strand-specific mRNA library preparation kit (Agilent Technologies, Santa Clara, CA, USA) as per the manufacturer's protocol. In brief, poly(A) RNA was purified from 1 μg of total RNA using two serial rounds of binding to oligo(dT) magnetic particles;

then, the nucleic acid was fragmented and reverse transcribed to generate cDNA. Illumina-specific adaptor was sequentially ligated to the 3' end of cDNA fragments, purified using the AMPure XP beads (Beckman Coulter, Brea, CA, USA) and, finally, PCR-amplified (13 cycles) using an appropriate indexing primer to allow further samples multiplexing. The PCR-amplified libraries were purified with the AMPure XP beads (Beckman Coulter, Brea, CA, USA) and then assessed for their quality and fragments distribution using the 2100 Bioanalyzer DNA 1000 assay (Agilent Technologies, Santa Clara, CA, USA). In the presence of adaptor-dimers (electropherogram's peak at 100 to 150-bp), another round of magnetic beads purification was performed. Libraries were quantified by both the Qubit®Fluorometer (Life Technologies, Monza, Italy) and the qPCR-based NEBNext library quantification kit (New England BioLabs, Hitchin, UK). Finally, equimolar amounts of each 6 index-tagged libraries were multiplexed together in one pool (total of 4 pools, 24 single libraries) and then sequenced by an Illumina HiSeq 2500 for 50 sequencing cycles.

The raw 50 bp single-end sequences (Sanger/Illumina 1.9 encoding) were quality-controlled using FastQC v.0.11.4 [30], and the low-quality bases (Phred quality score <30) and adaptor contamination (if present) were removed by Trimmomatic v.0.36 [31], using the parameters 'LEADING:3 SLIDINGWINDOW:4:20 MINLEN:25'. The high-quality reads were mapped by HISAT v. 2.0.4 [32] against the *Bos taurus* reference genome (Ensembl UMD 3.1). The uniquely-mapped reads aligned to exons were counted with HTSeq v.0.6.1 [33], then tested—by the DESeq2 R package v.1.14.1 [34]—for the presence of differentially expressed genes (DEGs) between CTR and GP groups after the 60-day supplementation period. All genes with a false discovery rate (FDR) less than 0.1 were considered DEGs regardless of their fold-change (FC) value. The sequencing data (FASTQ files) associated with this project are deposited in the GenBank's Sequence Read Archive (SRA) under the accession number SRP105401.

2.6. Functional Analysis and Statistics

The Gene Set Enrichment Analysis (GSEA) was used to examine the significantly enriched pathways in the GP group—in comparison with the CTR group. GSEA is a computational method that identifies shared differential gene expression of predefined, functionally related gene sets representing biological pathways. This is quantified by using a different type of Enrichment Score (ES), a weighted Kolmogorov–Smirnov-like statistic that evaluates if the members of the pathway are randomly distributed or found at the extremes (top or bottom) of the list [35]. In the present study, the GSEA pre-ranked option was used to analyze the deregulated pathways in the GP-supplemented group (GP T60 vs. CTR T60). All the Ensembl gene IDs were collapsed to their corresponding HUGO gene symbols; then, the entire normalized transcriptome dataset was ranked by the logarithm transformed (base 2) FC, where the up- and down-regulated genes were assigned positive and negative values, respectively. The pre-ranked dataset was analyzed (1000 permutations) against the Hallmarks (h.all.v7.0), i.e., the curated canonical KEGG pathways (c2.cp.kegg.v7.0.) and the gene ontology biological process (c5.bp.v7.0) catalogs from the Molecular Signatures Database (MsigDB) [35]. The GSEA output reported in the present study represents the most negatively- or positively-enriched hallmarks, pathways or biological processes (highest normalized enrichment score; NES) in the GP group compared with CTR. To visualize the gene expression data of GP and CTR samples, principal component analysis (PCA) and hierarchical clustering plots were generated using the ClustVis R package [36].

3. Results

The health status of all the experimental animals was satisfactory all throughout the experiment, with no visible signs of illness or supplementation-related stress. The dairy cows produced an average daily amount of 17 ± 2.4 liters of milk. The effect of the GP supplementation on the daily amount of milk produced as well as on the milk biochemical profile is shown in Supplementary Figure S1. Long-term GP supplementation did not result in differences between the GP and CTR groups either in the daily amount of milk produced than in milk's biochemical profile, i.e., the percentage of fat, protein, casein, lactose, urea and solids. In addition to this, and in strict accordance with what was observed for the milk yield, no significant variations were highlighted in feed intake for the entire duration of the trial, with a daily dry matter intake (DMI) equal to 19.3 ± 1.2 and 19.5 ± 0.9 kg in CTR and GP group respectively.

Regarding RNA-seq, the sequencing of 12 samples (2 groups × 6 animals/group) resulted in an average of ~21 million reads per sample. A summary of the sequencing output is reported in Supplementary Table S1. After the assumption of a GP-supplemented diet for 60 days, 40 protein-coding genes were found to be differentially expressed in the GP group, of which 10 and 30 genes were down- and up-regulated, respectively (Supplementary Table S2). The most down-regulated gene in the GP group (FDR < 0.05) was the DnaJ heat shock protein family (HSP40) member A1. Additional interesting genes down-regulated by GP supplementation were the mitochondrial fission factor (MFF), the impact RWD domain protein (IMPACT), the ATP binding cassette subfamily A member 13 (ABCA13), the regulator of G-protein signaling 2 (RGS2), and the integrin subunit alpha 3 (ITGA3). The top 100 deregulated genes (using their normalized RNA-seq counts) were able to discriminate the two experimental groups either through a principal component analysis (PCA) plot (with the first 2 components accounting for > 60% of the total variability; Figure 1a) or by a hierarchical clustering heatmap (Figure 1b).

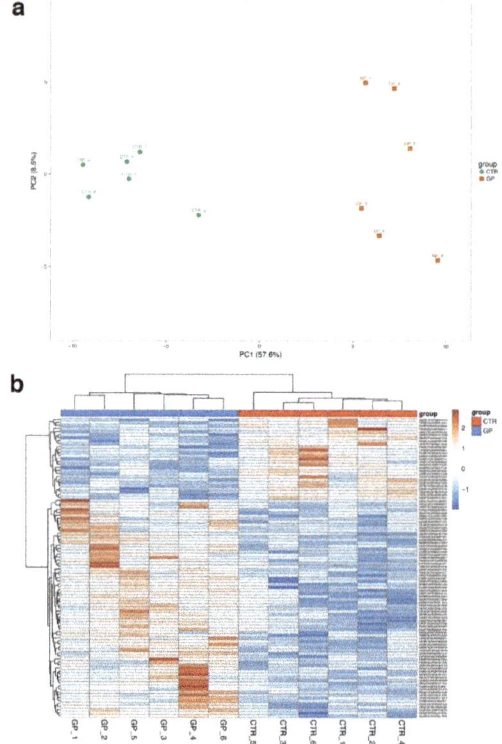

Figure 1. Principal Component Analysis (PCA) and hierarchical clustering heatmap. PCA and heatmap of the top 100 deregulated genes in dairy cows fed for 60 days with a GP-supplemented diet (GP T60 vs. CTR T60). (**a**) PCA plot of the GP (orange squares, right side) and CTR individuals (green circles, left side) after 60 days of GP dietary supplementation. (**b**) Heatmap generated from the DEGs using the hierarchical clustering (Euclidean distance clustering algorithm) of the top 100 deregulated genes of the GP group.

To further pinpoint the collective effect of the transcriptomic changes, we used GSEA to identify long-term GP-supplementation-associated pathways and biological processes. The GSEA analysis revealed that most of the core-enriched genes in the GP group, such as 'interferon alpha (IFN-α) response' (NES = 2.02), 'IFN-γ response' (NES = 1.94), 'IL6-JAK-STAT3 signaling' (NES = 1.74) or 'complement' hallmark genes (NES = 1.57), fell into the immune system-related hallmark gene sets (Figure 2). Moreover, some biological processes related with immune defense mechanisms were positively enriched in the GP group. Among these ones, the most enriched processes were the 'response to protozoan' (NES = 1.99) and the 'negative regulation of viral replication' (NES = 1.96; Figure 3). A summary of the GSEA analysis output is reported in Supplementary Table S3.

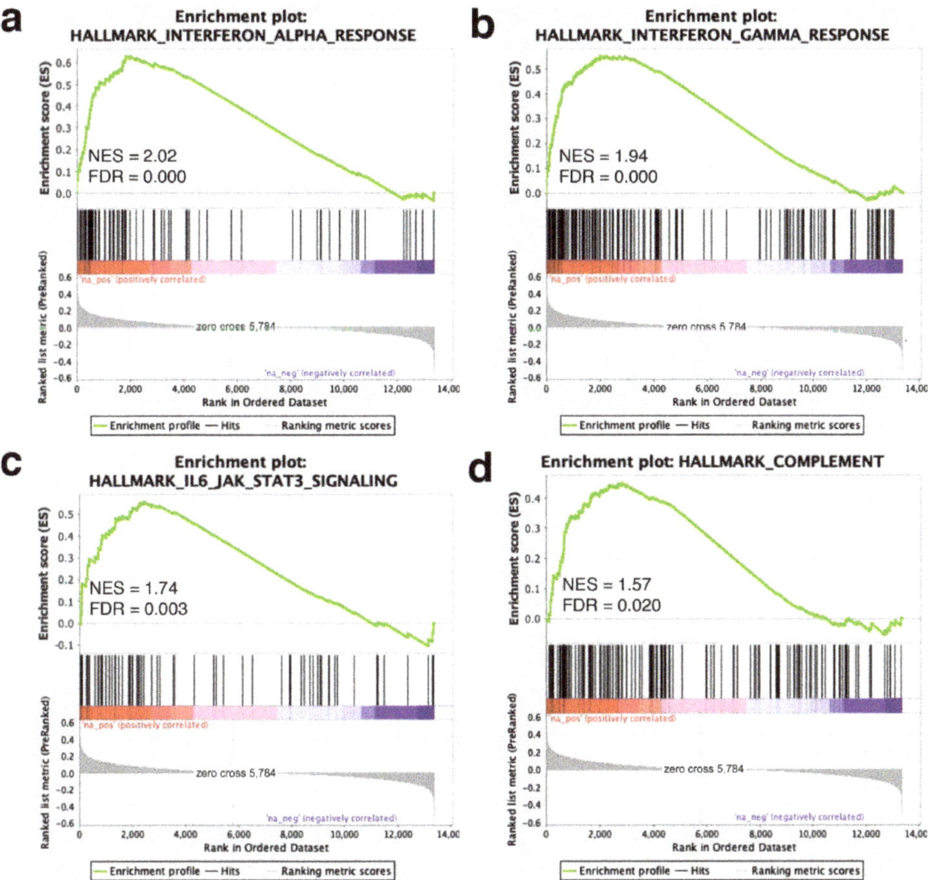

Figure 2. Gene set enrichment analysis (GSEA) plot (score curves) for enriched hallmarks. The GSEA analysis was performed with the hallmarks gene sets of GSEA Molecular Signatures Database. The "Signal-to-Noise" ratio (SNR) statistic was used to rank the genes per their correlation with the dietary GP supplementation (red) or the un-supplemented CTR phenotype (blue). The heatmap on the right side of each panel visualize the genes mostly contributing to the enriched gene set. For the fully detailed list, see Supplementary Table S3. The green curve corresponds to the Enrichment Score (ES) curve, which is the running sum of the weighed ES obtained from the GSEA software, while the Normalized Enrichment Score (NES) and the corresponding *p*-value are reported within each graph. Panels (**a**–**d**) denote the most enriched (significant) pathways (i.e., gene sets).

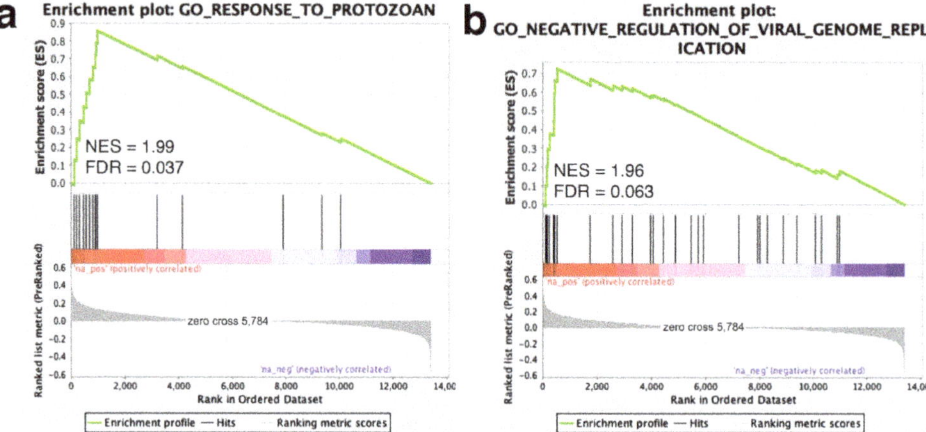

Figure 3. Gene set enrichment analysis (GSEA) plot (score curves) for enriched biological processes. The GSEA analysis was performed with the biological process gene sets of GSEA Molecular Signatures Database. The "Signal-to-Noise" ratio (SNR) statistic was used to rank the genes per their correlation with the dietary GP supplementation (red) or the un-supplemented CTR phenotype (blue). The heatmap on the right side of each panel visualize the genes mostly contributing to the enriched gene set. For the fully detailed list, see Supplementary Table S3. The green curve corresponds to the Enrichment Score (ES) curve, which is the running sum of the weighed ES obtained from the GSEA software, while the Normalized Enrichment Score (NES) and the corresponding *p*-value are reported within each graph. Panels (**a**,**b**) denote the most enriched (significant) biological processes (i.e., gene sets).

4. Discussion

The objective of this study was to examine the transcriptomic signature of a long-term dietary GP supplementation in dairy cows and to evaluate whether this signature reflects known GP-associated potential health benefits [37]. The main finding was that a 60-day dietary GP supplementation can modulate the expression of a considerable, albeit less than expected, number of genes in dairy cows and induce a noticeable transcriptomic signature that mainly reflects a GP-induced immunomodulatory effect.

Several studies assessed the effects of GP on milk production and composition in dairy cows. In two recent and independent studies, a dietary addition of ~10% GP (~2 kg GP/head/day for~60 days) did not show differences in milk composition [38,39]. Moreover, different concentrations (50, 75 and 100 g/kg of DM) of grape residue silage did not affect milk production nor the concentrations of CP, fat or lactose in dairy cows throughout a 21 days experimental period [10]. Further, the milk of dairy cows fed for three months with a diet containing 15% GP preserved the normal levels of fat, protein and caseins [22]. Finally, in a 4-week trial, dietary grape marc did not affect milk yield, milk protein or milk fat content of mid-lactation Holstein cows [40]. Overall, the present data further support the evidence that dietary GP does not alter milk yield or composition even after long-term supplementation.

Grape pomace accounts for ~25%–35% of the total weight of processed grapes; it contains significant amounts of dietary fiber, for which it may partially replace the forage portion in ruminants' diet [22,37]. Far more interesting, GP is enriched with potent bioactive compounds, especially polyphenols (~2%–6.5%). Grape polyphenolics, possessing different chemical structures and activities, are essentially categorized into two major classes of compounds: flavonoids (the most abundant polyphenols) and non-flavonoids [37,39,41]. Even though the biological activity of GP extracts mostly relate to their antioxidant properties, the dietary intake of these bioactive compounds results in pleiotropic health-promoting responses, e.g., cardio- and neuroprotective, anti-diabetic, anti-cancer, anti-aging, anti-adipogenetic as well as prebiotic and anti-microbial beneficial effects [39,41–43]. Overall, these evidences would suggest a link between polyphenol-rich food consumption and reduction in the incidence of numerous chronic disorders effect [42]. A separate discussion deserves the effects of polyphenols on the immune system, which is the second most important mechanism of action (after the antioxidant activity). Several studies have shown how polyphenols target multiple inflammatory components and lead to anti-inflammatory mechanisms and/or inflammation antagonism. Further, polyphenols today represent also promising candidates for the therapy of autoimmune diseases [43–46].

In this study, the change in expression of more than 40 protein-coding genes indicates that dietary GP has a traceable transcriptomic signature. Grape pomace contains lipid, proteins, fiber, minerals and large amounts of polyphenols (e.g., 10.4–64.8 g gallic acid equivalents/kg); moreover, polyphenolics show a linear correlation with in vitro antioxidant activity [47]. Therefore, one can safely presume that present transcriptional changes are due to GP polyphenols content. However, in the present experimental condition, only a few DEGs were significantly modulated by the dietary GP supplementation, and no genes were markedly over-expressed (in terms of FC). This was not surprising as in transcriptomics (e.g., microarray, RNA-seq), a large part of biologically meaningful DEGs possess FC less than 2. As a whole, these DEGs are not a minor population; rather, a significant fraction of transcriptional changes occurring in that sample (tissue, blood, etc.) [48]. This begs the question of whether such small changes reflect biologically meaningful events. In nutrigenomic studies, nutritionally relevant concentrations of bioactive compounds may elicit subtle changes in gene expression, resulting in critically important biological insights even though difficult to be detected reliably [24,49–51]. Moreover, small transcriptional variations are somehow expected when investigating long-term effects of a supplemented diet. In our opinion, such an event should not be ignored. Noteworthy, some interesting genes such as DNAJA1, MFF, IMPACT, ABCA13, RGS2, and ITGA were significantly downregulated in animals fed with a GP-supplemented diet (FDR < 0.05).

The top downregulated gene, DNAJA1, also known as heat shock protein (HSP) 40, plays a fundamental role during both physiological and stress conditions as it functions in mediating inappropriate proteins aggregation and folding [52]. Several studies have suggested that some heat shock proteins (both HSP70 and HSP40) are induced by wide ranges of stressful and pathological conditions [53]. Intriguingly, the natural flavonoid quercetin, containing a polyphenolic chemical structure and possessing healthy antioxidant properties, was shown to inhibit the expression of some HSPs at basal levels in human tissues [54–56]. Therefore, we might suggest that a lower expression of DNAJA1 in the GP group of animals compared to controls could be associated to healthier conditions of cows fed a GP-supplemented diet.

In living organisms, mitochondria continually change their morphology to maintain cell homeostasis. This process (mitochondrial dynamics) consists of two interconnected events, i.e., fission and fusion [57,58]. Mitochondrial fission is mediated by dynamin-related protein 1 (Drp1), a cytosolic highly conserved guanosine triphosphatase of the dynamin superfamily of proteins [57,59,60]. In mammals, the recruitment of Drp1 to mitochondria is mediated by MFF, a member of an array of Drp1-binding receptor factors localized at the mitochondrial outer membrane [57,58,60,61]. Fission is a key mechanistic step in apoptosis induction. Furthermore, Drp1 and MFF may contribute to apoptotic phenomena resulting from oxidative stress and reactive oxygen species (ROS) overproduction [57,58,60,62,63]. Some xenobiotics trigger oxidative stress, mitochondrial disfunction and ultimately apoptosis by increasing the expression of Drp1 (mostly) and MFF; among these ones ammonia [63], T2-toxin [64], cadmium [65], and microcystin [66]. However, curcumin, resveratrol, anthocyanins, quercetin and polyphenols inhibit mitochondrial fission through transcriptional and post-translational modifications of Drp-1, MFF as well as of other factors contributing to mitochondrial dynamics, thus preventing cells from oxidative stress [67–69]. Therefore, present MFF downregulation might be viewed as a protective effect provoked by GP supplementation against cellular oxidative stress, resulting in minor recruitment and transfer of Drp-1 inside mitochondria and, consequently, no mitochondrial fission/fusion imbalance and apoptosis. Concomitant inhibition of Drp-1 (i.e., DNM1L) was expected but no effects were observed for this gene.

IMPACT is a protein-coding gene highly expressed in neuronal tissue of mammals, where it promotes neuronal development [70–72]. Noteworthily, it substantially contributes to the adaptive response to nutritional stress and dietary restriction (i.e., a reduction in food intake without malnutrition) [70]. Under these stress conditions (e.g., amino acids depletion, proteasome inhibition, lack of glucose), IMPACT can inhibit a serine/threonine kinase coded by the General Control Non-derepressible 2 (GCN2) gene [70–74]. This latter gene, which is not listed among the DEGs, confers an adaptive response through distinct biochemical reactions (e.g., phosphorylation) and stress response genes activation. This cascade of events constitutes the Integrated Stress Response (ISR) pathway [70]. In addition to GCN2 inhibition, IMPACT overexpression also abrogates the phosphorylation of another GCN2 molecular target, i.e., the eukaryotic initiation factor 2, α subunit (eIF2α) [70,73]. Interestingly, IMPACT, GCN2 and ISR "molecular sensors" contribute substantially to immune system regulation [71,72,74,75]. This crosstalk between nutrient metabolism, gut immune system and host microbiota potentially shapes the overall gut responses to nutrients, and today represents a "hot" research topic [76,77]. As to IMPACT, KO mice are leaner than wild-type ones, especially when fed a high-fat diet. Furthermore, they show a defect in the control body temperature in response to starvation. This evidence, albeit not completely independent from GCN2, would suggest that IMPACT protein is involved in the maintenance of energy homeostasis [72]. The IMPACT gene downregulation we observed in the present study might be hypothetically interpreted as the result of a positive effect of GP on cattle nutritional status and immune system.

RGS2 is a conserved regulator of G protein-coupled receptors (GPCRs) signaling pathway. It belongs to the RGS superfamily of proteins, whose expression is regulated by epigenetic, transcriptional and post-translational mechanisms. These proteins determine the intensity and duration of GPCR-mediated effects through binding to the active Gα subunit of G proteins and activating GTPase. This protein-coding gene is expressed in several tissues such as CNS, heart, vasculature, kidney, immune system, lungs, bone and ovaries [78,79]. Interestingly, oxidative stress conditions increase RGS2 mRNA levels [79,80]; moreover, RGS2 contributes to the regulation of some enzymes involved in antioxidant defense, namely glyoxalase-1 and glutathione reductase-1 [81]. Finally, RGS2 has been shown to be involved in the pathophysiology of gastrointestinal inflammation and visceral pain, mostly regulating T-cell immunity [79,82,83]. In our opinion, the observed RGS2 gene downregulation can be viewed as the consequence of an indirect positive effect of GP on the overall cattle gastrointestinal tract homeostasis.

Limited information is currently available for the remaining two DEGs, i.e., ABCA13 and ITGA3. The former is a member of the ATP-binding cassette (ABC) superfamily of efflux drug transporters, encoding for the largest protein of the same family (5,058 residues) [84]. This gene is expressed in trachea, testis, bone marrow and blood-derived cells [85,86]. Interestingly, broilers fed diets supplemented with trace minerals (Zn, Cu, and Mn) showed reduced footpad lesions and improved wound healing process. At the molecular level, upregulation of genes involved in collagen synthesis, deposition and organization, cell migration, matrix remodelling, and angiogenesis, including ITGA2 and ITGA3, was noticed [87]. Despite this, no information about a possible modulatory effect of polyphenols on ABCA13 and ITGA3 are actually available.

In the present study, most of the core-enriched genes in the GP group (e.g., IFN-α, IFN-γ, and IL6-JAK-STAT3) belong to hallmarks and biological processes related to the immune response. All this was expected. Widespread evidence shows that polyphenols possess predominantly anti-oxidant, anti-inflammatory and immunomodulatory properties in both humans and farm animal species including cattle; hence, they are beneficial all-purpose nutraceuticals or supplements [45,46,88]. Polyphenols inhibit NF-κB, PI3K/Akt, MAPKs, and arachidonic acid-dependent signaling pathways. Regarding their immunosuppressant properties, polyphenols modulate T-cell function (e.g., suppressing T helper 2 activation and promoting the development of regulatory T cells), inhibit mast cell degranulation, and downregulate inflammatory cytokine responses [45,46,76]. Owing to these immunomodulatory properties, polyphenols have emerged as potential tools for the treatment of autoimmune disorders [46,89].

Additional biological processes positively enriched in GP group were the 'response to protozoan' and the 'negative regulation of viral replication'. Again, our results corroborate recent evidence about the use of polyphenolic compounds as potential therapeutic tools against protozoa and viruses, albeit scarce information is available on farm animals. Some dietary polyphenols (e.g., quercetin, resveratrol, rutin) show a certain efficacy against leishmaniosis [90], and giardiasis [91]; furthermore, resveratrol itself has been suggested as a potential restorative agent against toxoplasmosis brain disorders [92] and *Schistosoma mansoni* infections [93]. As far as the effects on viruses, human and animal evidence suggest polyphenolic compounds (e.g., resveratrol, tea polyphenols) possess potential preventive and therapeutic effects, and might be used as adjuvant therapy for the management of viral infection, e.g., influenza, herpes simplex, pseudorabies, hepatitis C, porcine reproductive and respiratory syndrome, and even Ebola [88,94–97].

5. Conclusions

This study provides evidence on the transcriptomic signature of long-term dietary GP supplementation in dairy cows. The present results strengthen the concept of using a nutrigenomic approach to assess the beneficial nor detrimental effects of farm animals' diets supplemented with grape polyphenolics and, in general, with undervalued food-related by-products and sources such as aromatic plants, culinary herbs, spices, vegetables and fruit trimmings. In our experimental conditions, the GP dietary supplementation resulted in a differential transcriptomic signature mostly reflecting an immunomodulatory effect. Overall, these data contribute to the growing body of nutrigenomics research in dairy cows and, to a wider extent, in ruminants.

Although additional studies should be performed to ascertain benefits and drawbacks of GP supplementation in dairy cows, and to define the optimal feeding schemes, the results here obtained let us suggest that this agricultural by-product has an overall positive impact on immune status of cows, without affecting milk quality.

The present study comes with some shortcomings. First, the number of DEGs was lower than expected; moreover, most of them were downregulated and showed lower FC. Nevertheless, we were expecting to find a higher number of DEGs, possibly confirming (e.g., gene upregulation) the antioxidant and anti-inflammatory properties of grape polyphenolics. Actually, the transcriptomic signature of GP-supplemented animals (namely, the list of DEGs and GSEA outputs) is suggestive of improved dairy cows' general health conditions, thereby confirming the GP wide range of health-promoting activities, e.g., anti-inflammatory, anti-aging, anti-oxidant, anti-microbial and antiviral properties. Moreover, it is important to highlight that feed might exert its biological effect at different levels. Thus, it would be possible that GP targets mechanisms other than transcription, such as mRNA silencing, mRNA translation, and enzyme activity/functionality. To this respect, the previous literature demonstrated that GP supplementation affects catalase activity in lambs [18], superoxide dismutase and glutathione peroxidase activity in piglets [98]. To our knowledge, proteomic studies investigating the effect of GP have not been so far performed in farmed animals. However, it is likely that protein expression may be a target of GP. Indeed, dietary inclusion of other bioactive components have been demonstrated to significantly affect the intestinal mucosa proteome of growing pigs [99] and the skeletal muscle proteins of pre-weaning calves [100].

Another possible limitation of the study is having measured the GP-dependent transcriptomic changes in dairy cows WB. Many experiments aiming at the characterization of the beneficial effect of nutraceuticals on the individual's transcriptome involve the use of gastrointestinal biopsies (e.g., in rodents, chicken, fish). Apart from the ethical considerations, there is evidence that the WB is a suitable surrogate tissue for transcriptomics studies. The final potential bias is intrinsic to grape polyphenolics composition and the corresponding biological activity. In general, the varying composition of GP originating from different grapes and locations are likely to show remarkable differences in their biological activity. As a consequence, this "biodiversity" impacts GP bioavailability, metabolism, bioactivity and, ultimately the animal's transcriptomics signature. This is still a major criticism of the use of grape polyphenols in food-producing animals as well as in other species including humans. In perspective, crude or highly purified forms of GP extract might be used to characterize the most of beneficial effect of a GP-supplemented diet. Despite these potential biases, the present results corroborate the health benefits, sustainability, and environmental impact of GP utilization in dairy animals and in general in food-producing species. Clearly, further investigations are recommended to improve our knowledge on the best use of polyphenolic compounds as nutraceuticals, e.g., evaluating different feeding schemes and/or percentage of GP supplementation, exploring the now in fashion "host-drug-nutraceutical-microbiota interactions".

Supplementary Materials: The following are available online at http://www.mdpi.com/2076-2615/10/4/714/s1. Figure S1. Effects of dietary GP supplementation on milk production. Table S1. Sequencing and alignment summary. Table S2. Differentially Expressed Genes (DEGs). Table S3. Summary of the GSEA analysis output.

Author Contributions: Conceptualization, M.G., M.D. and G.M.; Data curation, R.E., A.I. and E.M.; Formal analysis, R.E. and E.M.; Funding acquisition, G.M.; Investigation, M.P., A.I. and S.R.; Methodology, R.E., A.I., L.G., S.R. and F.B.; Project administration, G.M.; Resources, G.M.; Supervision, M.D. and G.M.; Writing—original draft, M.P., R.E. and M.D.; Writing—review and editing, A.I. and M.G. All authors have read and agreed to the published version of the manuscript.

Funding: This work is part of the project 'PROmozione della Salute del consumatore: valorizzazione nutrizionale dei prodotti agroalimentari della tradizione ITaliana (PROS.IT)' supported by grant, project ID: CTN01_00230_413096, Ministero dell'Istruzione dell'Università e della Ricerca (Italy). This work was also funded with grants to M.D. from "Regione Veneto" in the context of "POR FESR VENETO 2014–2020", Action 1.1.4 DGR n. 1139 (19/07/2017), Project 3S-4H, Call ID: 10065201.

Conflicts of Interest: The authors declare no conflict of interest. The funders had no role in the design of the study; in the collection, analyses, or interpretation of data; in the writing of the manuscript, or in the decision to publish the results.

References

1. Van Kernebeek, H.R.J.; Oosting, S.J.; Van Ittersum, M.K.; Bikker, P.; De Boer, I.J.M. Saving land to feed a growing population: Consequences for consumption of crop and livestock products. *Int. J. Life Cycle Assess.* **2016**, *21*, 677–687. [CrossRef]

2. Hasan, M.S.; Feugang, J.M.; Liao, S.F. A Nutrigenomics Approach Using RNA Sequencing Technology to Study Nutrient-Gene Interactions in Agricultural Animals. *Curr. Dev. Nutr.* **2019**, *3*, nzz082. [CrossRef]

3. EU. A General Union Environment Action Programme to 2020 "Living well, within the limits of our planet". *Off. J. Eur. Union* **2013**, *L354*, 171–200.

4. Leiva-Candia, D.E.; Pinzi, S.; Redel-Macías, M.D.; Koutinas, A.; Webb, C.; Dorado, M.P. The potential for agro-industrial waste utilization using oleaginous yeast for the production of biodiesel. *Fuel* **2014**, *123*, 33–42. [CrossRef]

5. Schieber, A.; Stintzing, F.C.; Carle, R. By-products of plant food processing as a source of functional compounds—Recent developments. *Trends Food Sci. Technol.* **2001**, *12*, 401–413. [CrossRef]

6. Vauzour, D.; Rodriguez-Mateos, A.; Corona, G.; Oruna-Concha, M.J.; Spencer, J.P.E. Polyphenols and human health: Prevention of disease and mechanisms of action. *Nutrients* **2010**, *2*, 1106–1131. [CrossRef]

7. Federici, F.; Fava, F.; Kalogerakis, N.; Mantzavinos, D. Valorisation of agro-industrial by-products, effluents and waste: Concept, opportunities and the case of olive mill waste waters. *J. Chem. Technol. Biotechnol.* **2009**, *84*, 895–900. [CrossRef]

8. Mirzaei-Aghsaghali, A.; Maheri-Sis, N. Nutritive value of some agro-industrial by-products for ruminants—A review. *World J. Zool.* **2008**, *3*, 40–46.

9. Makris, D.P.; Boskou, G.; Andrikopoulos, N.K. Polyphenolic content and in vitro antioxidant characteristics of wine industry and other agri-food solid waste extracts. *J. Food Compos. Anal.* **2007**, *20*, 125–132. [CrossRef]

10. Santos, N.W.; Santos, G.T.D.; Silva-Kazama, D.C.; Grande, P.A.; Pintro, P.M.; de Marchi, F.E.; Jobim, C.C.; Petit, H.V. Production, composition and antioxidants in milk of dairy cows fed diets containing soybean oil and grape residue silage. *Livest. Sci.* **2014**, *159*, 37–45. [CrossRef]

11. Llobera, A.; Cañellas, J. Antioxidant activity and dietary fibre of Prensal Blanc white grape (Vitis vinifera) by-products. *Int. J. Food Sci. Technol.* **2008**, *43*, 1953–1959. [CrossRef]

12. Negro, C.; Tommasi, L.; Miceli, A. Phenolic compounds and antioxidant activity from red grape marc extracts. *Bioresour. Technol.* **2003**, *87*, 41–44. [CrossRef]

13. Viveros, A.; Chamorro, S.; Pizarro, M.; Arija, I.; Centeno, C.; Brenes, A. Effects of dietary polyphenol-rich grape products on intestinal microflora and gut morphology in broiler chicks. *Poult. Sci.* **2011**, *90*, 566–578. [CrossRef] [PubMed]

14. Gessner, D.K.; Fiesel, A.; Most, E.; Dinges, J.; Wen, G.; Ringseis, R.; Eder, K. Supplementation of a grape seed and grape marc meal extract decreases activities of the oxidative stress-responsive transcription factors NF-κB and Nrf2 in the duodenal mucosa of pigs. *Acta Vet. Scand.* **2013**, *55*, 18. [CrossRef]

15. Fiesel, A.; Gessner, D.K.; Most, E.; Eder, K. Effects of dietary polyphenol-rich plant products from grape or hop on pro-inflammatory gene expression in the intestine, nutrient digestibility and faecal microbiota of weaned pigs. *BMC Vet. Res.* **2014**, *10*, 196. [CrossRef]

16. Ishida, K.; Kishi, Y.; Oishi, K.; Hirooka, H.; Kumagai, H. Effects of feeding polyphenol-rich winery wastes on digestibility, nitrogen utilization, ruminal fermentation, antioxidant status and oxidative stress in wethers. *Anim. Sci. J.* **2015**, *86*, 260–269. [CrossRef] [PubMed]

17. Biscarini, F.; Palazzo, F.; Castellani, F.; Masetti, G.; Grotta, L.; Cichelli, A.; Martino, G. Rumen microbiome in dairy calves fed copper and grape-pomace dietary supplementations: Composition and predicted functional profile. *PLoS ONE* **2018**, *13*, e0205670. [CrossRef]

18. Kafantaris, I.; Kotsampasi, B.; Christodoulou, V.; Kokka, E.; Kouka, P.; Terzopoulou, Z.; Gerasopoulos, K.; Stagos, D.; Mitsagga, C.; Giavasis, I.; et al. Grape pomace improves antioxidant capacity and faecal microflora of lambs. *J. Anim. Physiol. Anim. Nutr.* **2017**, *101*, e108–e121. [CrossRef] [PubMed]

19. Nudda, A.; Correddu, F.; Marzano, A.; Battacone, G.; Nicolussi, P.; Bonelli, P.; Pulina, G. Effects of diets containing grape seed, linseed, or both on milk production traits, liver and kidney activities, and immunity of lactating dairy ewes. *J. Dairy Sci.* **2015**, *98*, 1157–1166. [CrossRef]

20. Gessner, D.K.; Koch, C.; Romberg, F.J.; Winkler, A.; Dusel, G.; Herzog, E.; Most, E.; Eder, K. The effect of grape seed and grape marc meal extract on milk performance and the expression of genes of endoplasmic reticulum stress and inflammation in the liver of dairy cows in early lactation. *J. Dairy Sci.* **2015**, *98*, 8856–8868. [CrossRef]

21. Manso, T.; Gallardo, B.; Salvá, A.; Guerra-Rivas, C.; Mantecón, A.R.; Lavín, P.; de la Fuente, M.A. Influence of dietary grape pomace combined with linseed oil on fatty acid profile and milk composition. *J. Dairy Sci.* **2016**, *99*, 1111–1120. [CrossRef]

22. Chedea, V.S.; Pelmus, R.S.; Lazar, C.; Pistol, G.C.; Calin, L.G.; Toma, S.M.; Dragomir, C.; Taranu, I. Effects of a diet containing dried grape pomace on blood metabolites and milk composition of dairy cows. *J. Sci. Food Agric.* **2017**, *97*, 2516–2523. [CrossRef] [PubMed]

23. Ianni, A.; Martino, G. Dietary Grape Pomace Supplementation in Dairy Cows: Effect on Nutritional Quality of Milk and Its Derived Dairy Products. *Foods* **2020**, *9*, 168. [CrossRef]

24. Iannaccone, M.; Elgendy, R.; Giantin, M.; Martino, C.; Giansante, D.; Ianni, A.; Dacasto, M.; Martino, G. RNA sequencing-based whole-transcriptome analysis of friesian cattle fed with grape pomace-supplemented diet. *Animals* **2018**, *8*, 188. [CrossRef]

25. European Union. Directive 2010/63/EU of the European Parliament and of the Council of 22 September 2010 on the protection of animals used for scientific purposes. *Off. J. Eur. Union* **2010**, *53*, 33–79.

26. Ianni, A.; Di Luca, A.; Martino, C.; Bennato, F.; Marone, E.; Grotta, L.; Cichelli, A.; Martino, G. Dietary supplementation of dried grape pomace increases the amount of linoleic acid in beef, reduces the lipid oxidation and modifies the volatile profile. *Animals* **2019**, *9*, 578. [CrossRef] [PubMed]

27. Singleton, V.L.; Rossi, J. Colorimetry of total phenolics with phosphomolybdic. *Am. J. Enol. Vitic.* **1965**, *16*, 144–158.

28. Helrick, K. *Official Methods of Analysis (AOAC)*; Association of Official Analytical Chemist: Washington, DC, USA, 1990; ISBN 8589439488.

29. Goering, H.K.; Van Soest, P.J. *Forage Fiber Analyses (Apparatus, Reagent, Procedures and Some Applications): Agriculture Handbook No. 379*; Agricultural Research Service, US Department of Agriculture: Washington, DC, USA, 1970.

30. Andrews, S.; Krueger, F.; Seconds-Pichon, A.; Biggins, F.; Wingett, S. FastQC. A Quality Control Tool for High Throughput Sequence Data. Babraham Bioinformatics. Available online: http://www.bioinformatics.babraham.ac.uk/projects/fastqc/ (accessed on 14 April 2020).

31. Bolger, A.M.; Lohse, M.; Usadel, B. Trimmomatic: A flexible trimmer for Illumina sequence data. *Bioinformatics* **2014**, *30*, 2114–2120. [CrossRef]

32. Kim, D.; Langmead, B.; Salzberg, S.L. HISAT: A fast spliced aligner with low memory requirements. *Nat. Methods* **2015**, *12*, 357–360. [CrossRef]

33. Anders, S.; Pyl, P.T.; Huber, W. HTSeq-A Python framework to work with high-throughput sequencing data. *Bioinformatics* **2015**, *31*, 166–169. [CrossRef] [PubMed]

34. Love, M.I.; Huber, W.; Anders, S. Moderated estimation of fold change and dispersion for RNA-seq data with DESeq2. *Genome Biol.* **2014**, *15*, 550. [CrossRef] [PubMed]

35. Subramanian, A.; Tamayo, P.; Mootha, V.K.; Mukherjee, S.; Ebert, B.L.; Gillette, M.A.; Paulovich, A.; Pomeroy, S.L.; Golub, T.R.; Lander, E.S.; et al. Gene set enrichment analysis: A knowledge-based approach for interpreting genome-wide expression profiles. *Proc. Natl. Acad. Sci. USA* **2005**, *102*, 15545–15550. [CrossRef]

36. Metsalu, T.; Vilo, J. ClustVis: A web tool for visualizing clustering of multivariate data using Principal Component Analysis and heatmap. *Nucleic Acids Res.* **2015**, *43*, W566–W570. [CrossRef] [PubMed]

37. Averilla, J.N.; Oh, J.; Kim, H.J.; Kim, J.S.; Kim, J.S. Potential health benefits of phenolic compounds in grape processing by-products. *Food Sci. Biotechnol.* **2019**, *28*, 1607–1615. [CrossRef]

38. Ianni, A.; Di Maio, G.; Pittia, P.; Grotta, L.; Perpetuini, G.; Tofalo, R.; Cichelli, A.; Martino, G. Chemical–nutritional quality and oxidative stability of milk and dairy products obtained from Friesian cows fed with a dietary supplementation of dried grape pomace. *J. Sci. Food Agric.* **2019**, *99*, 3635–3643. [CrossRef] [PubMed]

39. Ianni, A.; Innosa, D.; Martino, C.; Bennato, F.; Martino, G. Short communication: Compositional characteristics and aromatic profile of caciotta cheese obtained from Friesian cows fed with a dietary supplementation of dried grape pomace. *J. Dairy Sci.* **2019**, *102*, 1025–1032. [CrossRef] [PubMed]

40. Scuderi, R.A.; Ebenstein, D.B.; Lam, Y.W.; Kraft, J.; Greenwood, S.L. Inclusion of grape marc in dairy cattle rations alters the bovine milk proteome. *J. Dairy Res.* **2019**, *86*, 154–161. [CrossRef]

41. Cotoras, M.; Vivanco, H.; Melo, R.; Aguirre, M.; Silva, E.; Mendoza, L. In vitro and in vivo evaluation of the antioxidant and prooxidant activity of phenolic compounds obtained from grape (*Vitis vinifera*) pomace. *Molecules* **2014**, *19*, 21154–21167. [CrossRef]

42. Singh, A.K.; Cabral, C.; Kumar, R.; Ganguly, R.; Rana, H.K.; Gupta, A.; Lauro, M.R.; Carbone, C.; Reis, F.; Pandey, A.K. Beneficial effects of dietary polyphenols on gut microbiota and strategies to improve delivery efficiency. *Nutrients* **2019**, *11*, 2216. [CrossRef]

43. Zhang, H.; Tsao, R. Dietary polyphenols, oxidative stress and antioxidant and anti-inflammatory effects. *Curr. Opin. Food Sci.* **2016**, *8*, 33–42. [CrossRef]

44. Yahfoufi, N.; Alsadi, N.; Jambi, M.; Matar, C. The immunomodulatory and anti-inflammatory role of polyphenols. *Nutrients* **2018**, *10*, 1618. [CrossRef] [PubMed]

45. Focaccetti, C.; Izzi, V.; Benvenuto, M.; Fazi, S.; Ciuffa, S.; Giganti, M.G.; Potenza, V.; Manzari, V.; Modesti, A.; Bei, R. Polyphenols as immunomodulatory compounds in the tumor microenvironment: Friends or foes? *Int. J. Mol. Sci.* **2019**, *20*, 1714. [CrossRef] [PubMed]

46. Khan, H.; Sureda, A.; Belwal, T.; Çetinkaya, S.; Süntar, İ.; Tejada, S.; Devkota, H.P.; Ullah, H.; Aschner, M. Polyphenols in the treatment of autoimmune diseases. *Autoimmun. Rev.* **2019**, *18*, 647–657. [CrossRef]

47. Jin, Q.; O'Hair, J.; Stewart, A.C.; O'Keefe, S.F.; Neilson, A.P.; Kim, Y.T.; McGuire, M.; Lee, A.; Wilder, G.; Huang, H. Compositional characterization of different industrial white and red grape pomaces in Virginia and the potential valorization of the major components. *Foods* **2019**, *8*, 667. [CrossRef] [PubMed]

48. St. Laurent, G.; Shtokalo, D.; Tackett, M.R.; Yang, Z.; Vyatkin, Y.; Milos, P.M.; Seilheimer, B.; McCaffrey, T.A.; Kapranov, P. On the importance of small changes in RNA expression. *Methods* **2013**, *63*, 18–24.

49. Garosi, P.; De Filippo, C.; van Erk, M.; Rocca-Serra, P.; Sansone, S.-A.; Elliott, R. Defining best practice for microarray analyses in nutrigenomic studies. *Br. J. Nutr.* **2005**, *93*, 425–432. [CrossRef]

50. Pagmantidis, V.; Méplan, C.; Van Schothorst, E.M.; Keijer, J.; Hesketh, J.E. Supplementation of healthy volunteers with nutritionally relevant amounts of selenium increases the expression of lymphocyte protein biosynthesis genes. *Am. J. Clin. Nutr.* **2008**, *87*, 181–189. [CrossRef]

51. Elgendy, R.; Giantin, M.; Castellani, F.; Grotta, L.; Palazzo, F.; Dacasto, M.; Martino, G. Transcriptomic signature of high dietary organic selenium supplementation in sheep: A nutrigenomic insight using a custom microarray platform and gene set enrichment analysis. *J. Anim. Sci.* **2016**, *94*, 3169–3184. [CrossRef]

52. Kim, Y.E.; Hipp, M.S.; Bracher, A.; Hayer-Hartl, M.; Ulrich Hartl, F. Molecular Chaperone Functions in Protein Folding and Proteostasis. *Annu. Rev. Biochem.* **2013**, *82*, 323–355. [CrossRef]

53. Liu, Q.; Liang, C.; Zhou, L. Structural and functional analysis of the Hsp70/Hsp40 chaperone system. *Protein Sci.* **2020**, *29*, 378–390. [CrossRef]

54. Elia, G.; Santoro, M.G. Regulation of heat shock protein synthesis by quercetin in human erythroleukaemia cells. *Biochem. J.* **1994**, *300*, 201–209. [CrossRef] [PubMed]

55. Nagai, N.; Nakai, A.; Nagata, K. Quercetin suppresses heat shock response by down regulation of HSF1. *Biochem. Biophys. Res. Commun.* **1995**, *208*, 1099–1105. [CrossRef] [PubMed]

56. Aalinkeel, R.; Bindukumar, B.; Reynolds, J.L.; Sykes, D.E.; Mahajan, S.D.; Chadha, K.C.; Schwartz, S.A. The dietary bioflavonoid, quercetin, selectively induces apoptosis of prostate cancer cells by down-regulating the expression of heat shock protein 90. *Prostate* **2008**, *68*, 1773–1789. [CrossRef] [PubMed]

57. Youle, R.J.; Van Der Bliek, A.M. Mitochondrial fission, fusion, and stress. *Science* **2012**, *337*, 1062–1065. [CrossRef] [PubMed]

58. Gilkerson, R. A disturbance in the force: Cellular stress sensing by the mitochondrial network. *Antioxidants* **2018**, *7*, 126. [CrossRef] [PubMed]

59. Inoue-Yamauchi, A.; Oda, H. Depletion of mitochondrial fission factor DRP1 causes increased apoptosis in human colon cancer cells. *Biochem. Biophys. Res. Commun.* **2012**, *421*, 81–85. [CrossRef]

60. Tak, H.; Eun, J.W.; Kim, J.; Park, S.J.; Kim, C.; Ji, E.; Lee, H.; Kang, H.; Cho, D.H.; Lee, K.; et al. T-cell-restricted intracellular antigen 1 facilitates mitochondrial fragmentation by enhancing the expression of mitochondrial fission factor. *Cell Death Differ.* **2017**, *24*, 49–58. [CrossRef]

61. Prieto, J.; León, M.; Ponsoda, X.; García-García, F.; Bort, R.; Serna, E.; Barneo-Muñoz, M.; Palau, F.; Dopazo, J.; López-García, C.; et al. Dysfunctional mitochondrial fission impairs cell reprogramming. *Cell Cycle* **2016**, *15*, 3240–3250. [CrossRef]

62. Toyama, E.Q.; Herzig, S.; Courchet, J.; Lewis, T.L., Jr.; Losón, O.C.; Hellberg, K.; Young, N.P.; Chen, H.; Polleux, F.; Chan, D.C.; et al. AMP-activated protein kinase mediates mitochondrial fission in response to energy stress. *Science* **2016**, *351*, 275–281. [CrossRef]

63. Shah, S.W.A.; Chen, J.; Han, Q.; Xu, Y.; Ishfaq, M.; Teng, X. Ammonia inhalation impaired immune function and mitochondrial integrity in the broilers bursa of fabricius: Implication of oxidative stress and apoptosis. *Ecotoxicol. Environ. Saf.* **2020**, *190*, 110078. [CrossRef]

64. Yang, J.; Guo, W.; Wang, J.; Yang, X.; Zhang, Z.; Zhao, Z. T-2 toxin-induced oxidative stress leads to imbalance of mitochondrial fission and fusion to activate cellular apoptosis in the human liver 7702 cell line. *Toxins* **2020**, *12*, 43. [CrossRef] [PubMed]

65. Ge, J.; Zhang, C.; Sun, Y.C.; Zhang, Q.; Lv, M.W.; Guo, K.; Li, J.L. Cadmium exposure triggers mitochondrial dysfunction and oxidative stress in chicken (*Gallus gallus*) kidney via mitochondrial UPR inhibition and Nrf2-mediated antioxidant defense activation. *Sci. Total Environ.* **2019**, *689*, 1160–1171. [CrossRef] [PubMed]

66. Zhang, C.; Wang, J.; Zhu, J.; Chen, Y.; Han, X. Microcystin-leucine-arginine induced neurotoxicity by initiating mitochondrial fission in hippocampal neurons. *Sci. Total Environ.* **2020**, *703*, 134702. [CrossRef] [PubMed]

67. Chen, C.; Huang, J.; Shen, J.; Bai, Q. Quercetin improves endothelial insulin sensitivity in obese mice by inhibiting Drp1 phosphorylation at serine 616 and mitochondrial fragmentation. *Acta Biochim. Biophys. Sin.* **2019**, *51*, 1250–1257. [CrossRef] [PubMed]

68. Li, A.; Zhang, S.; Li, J.; Liu, K.; Huang, F.; Liu, B. Metformin and resveratrol inhibit Drp1-mediated mitochondrial fission and prevent ER stress-associated NLRP3 inflammasome activation in the adipose tissue of diabetic mice. *Mol. Cell. Endocrinol.* **2016**, *434*, 36–47. [CrossRef] [PubMed]

69. Naoi, M.; Wu, Y.; Shamoto-Nagai, M.; Maruyama, W. Mitochondria in Neuroprotection by Phytochemicals: Bioactive Polyphenols Modulate Mitochondrial Apoptosis System, Function and Structure. *Int. J. Mol. Sci.* **2019**, *20*, 2451. [CrossRef]

70. Ferraz, R.C.; Camara, H.; De-Souza, E.A.; Pinto, S.; Pinca, A.P.F.; Silva, R.C.; Sato, V.N.; Castilho, B.A.; Mori, M.A. IMPACT is a GCN2 inhibitor that limits lifespan in Caenorhabditis elegans. *BMC Biol.* **2016**, *14*, 87. [CrossRef]

71. Cambiaghi, T.D.; Pereira, C.M.; Shanmugam, R.; Bolech, M.; Wek, R.C.; Sattlegger, E.; Castilho, B.A. Evolutionarily conserved IMPACT impairs various stress responses that require GCN1 for activating the eIF2 kinase GCN2. *Biochem. Biophys. Res. Commun.* **2014**, *443*, 592–597. [CrossRef]

72. Pereira, C.M.; Filev, R.; Dubiela, F.P.; Brandão, B.B.; Queiroz, C.M.; Ludwig, R.G.; Hipolide, D.; Longo, B.M.; Mello, L.E.; Mori, M.A.; et al. The GCN2 inhibitor IMPACT contributes to diet-induced obesity and body temperature control. *PLoS ONE* **2019**, *14*, e0217287. [CrossRef]

73. Castilho, B.A.; Shanmugam, R.; Silva, R.C.; Ramesh, R.; Himme, B.M.; Sattlegger, E. Keeping the eIF2 alpha kinase Gcn2 in check. *Biochim. Biophys. Acta Mol. Cell Res.* **2014**, *1843*, 1948–1968. [CrossRef]

74. Ravishankara, B.; Liua, H.; Shindea, R.; Chaudharya, K.; Xiaoa, W.; Bradleya, J.; Koritzinsky, M.; Madaio, M.P.; McGaha, T.L. The amino acid sensor GCN2 inhibits inflammatory responses to apoptotic cells promoting tolerance and suppressing systemic autoimmunity. *Proc. Natl. Acad. Sci. USA* **2015**, *112*, 10774–10779. [CrossRef] [PubMed]

75. Murguía, J.R.; Serrano, R. New functions of protein kinase Gcn2 in yeast and mammals. *IUBMB Life* **2012**, *64*, 971–974. [CrossRef] [PubMed]

76. Hachimura, S.; Totsuka, M.; Hosono, A. Immunomodulation by food: Impact on gut immunity and immune cell function. *Biosci. Biotechnol. Biochem.* **2018**, *82*, 584–599. [CrossRef] [PubMed]

77. Revelo, X.S.; Winer, S.; Winer, D.A. Starving Intestinal Inflammation with the Amino Acid Sensor GCN2. *Cell Metab.* **2016**, *23*, 763–765. [CrossRef]

78. Alqinyah, M.; Hooks, S.B. Regulating the regulators: Epigenetic, transcriptional, and post-translational regulation of RGS proteins. *Cell. Signal.* **2018**, *42*, 77–87. [CrossRef]

79. Ota, A.; Sawai, M.; Sakurai, H. Stress-induced transcription of regulator of G protein signaling 2 (RGS2) by heat shock transcription factor HSF1. *Biochimie* **2013**, *95*, 1432–1436. [CrossRef]

80. Zmijewski, J.W.; Song, L.; Jope, R.S.; Harkins, L.; Cobbs, C.S. Oxidative stress and heat shock stimulate RGS2 expression in 1321N1 astrocytoma cells. *Arch. Biochem. Biophys.* **2001**, *392*, 192–196. [CrossRef]

81. Salim, S.; Asghar, M.; Taneja, M.; Hovatta, I.; Wu, Y.L.; Saha, K.; Sarraj, N.; Hite, B. Novel role of RGS2 in regulation of antioxidant homeostasis in neuronal cells. *FEBS Lett.* **2011**, *585*, 1375–1381. [CrossRef] [PubMed]

82. Salaga, M.; Storr, M.; Martemyanov, K.A.; Fichna, J. RGS proteins as targets in the treatment of intestinal inflammation and visceral pain: New insights and future perspectives. *BioEssays* **2016**, *38*, 344–354. [CrossRef]

83. Wang, D. The essential role of G protein-coupled receptor (GPCR) signaling in regulating T cell immunity. *Immunopharmacol. Immunotoxicol.* **2018**, *40*, 187–192. [CrossRef]

84. Vasiliou, V.; Vasiliou, K.; Nebert, D.W. Human ATP-binding cassette (ABC) transporter family. *Hum. Genomics* **2009**, *3*, 281. [CrossRef]

85. Chaiworapongsa, T.; Romero, R.; Whitten, A.; Tarca, A.L.; Bhatti, G.; Draghici, S.; Chaemsaithong, P.; Miranda, J.; Hassan, S.S. Differences and similarities in the transcriptional profile of peripheral whole blood in early and lateonset preeclampsia: Insights into the molecular basis of the phenotype of preeclampsia. *J. Perinat. Med.* **2013**, *41*, 485–504. [CrossRef] [PubMed]

86. Dréan, A.; Rosenberg, S.; Lejeune, F.X.; Goli, L.; Nadaradjane, A.A.; Guehennec, J.; Schmitt, C.; Verreault, M.; Bielle, F.; Mokhtari, K.; et al. ATP binding cassette (ABC) transporters: Expression and clinical value in glioblastoma. *J. Neurooncol.* **2018**, *138*, 479–486. [CrossRef] [PubMed]

87. Chen, J.; Tellez, G.; Escobar, J.; Vazquez-Anon, M. Impact of Trace Minerals on Wound Healing of Footpad Dermatitis in Broilers. *Sci. Rep.* **2017**, *7*, 1–9. [CrossRef] [PubMed]

88. Gessner, D.K.; Ringseis, R.; Eder, K. Potential of plant polyphenols to combat oxidative stress and inflammatory processes in farm animals. *J. Anim. Physiol. Anim. Nutr.* **2017**, *101*, 605–628. [CrossRef]

89. Ding, S.; Jiang, H.; Fang, J. Regulation of immune function by polyphenols. *J. Immunol. Res.* **2018**. [CrossRef]

90. Da Silva, E.R.; Brogi, S.; Lucon-Júnior, J.F.; Campiani, G.; Gemma, S.; Maquiaveli, C.D.C. Dietary polyphenols rutin, taxifolin and quercetin related compounds target: Leishmania amazonensis arginase. *Food Funct.* **2019**, *10*, 3172–3180. [CrossRef]

91. Anthony, J.P.; Fyfe, L.; Stewart, D.; McDougall, G.J. Differential effectiveness of berry polyphenols as anti-giardial agents. *Parasitology* **2011**, *138*, 1110–1116. [CrossRef]

92. Bottari, N.B.; Schetinger, M.R.C.; Pillat, M.M.; Palma, T.V.; Ulrich, H.; Alves, M.S.; Morsch, V.M.; Melazzo, C.; de Barros, L.D.; Garcia, J.L.; et al. Resveratrol as a Therapy to Restore Neurogliogenesis of Neural Progenitor Cells Infected by Toxoplasma gondii. *Mol. Neurobiol.* **2019**, *56*, 2328–2338. [CrossRef]

93. Soliman, R.H.; Ismail, O.A.; Badr, M.S.; Nasr, S.M. Resveratrol ameliorates oxidative stress and organ dysfunction in Schistosoma mansoni infected mice. *Exp. Parasitol.* **2017**, *174*, 52–58. [CrossRef]

94. Bahramsoltani, R.; Sodagari, H.R.; Farzaei, M.H.; Abdolghaffari, A.H.; Gooshe, M.; Rezaei, N. The preventive and therapeutic potential of natural polyphenols on influenza. *Expert Rev. Anti. Infect. Ther.* **2016**, *14*, 57–80. [CrossRef] [PubMed]

95. Chowdhury, P.; Sahuc, M.E.; Rouillé, Y.; Rivière, C.; Bonneau, N.; Vandeputte, A.; Brodin, P.; Goswami, M.; Bandyopadhyay, T.; Dubuisson, J.; et al. Theaflavins, polyphenols of black tea, inhibit entry of hepatitis C virus in cell culture. *PLoS ONE* **2018**, *13*, e0198226. [CrossRef] [PubMed]

96. Zhang, M.; Wu, Q.; Chen, Y.; Duan, M.; Tian, G.; Deng, X.; Sun, Y.; Zhou, T.; Zhang, G.; Chen, W.; et al. Inhibition of proanthocyanidin A2 on porcine reproductive and respiratory syndrome virus replication in vitro. *PLoS ONE* **2018**, *13*, e0193309. [CrossRef] [PubMed]

97. Zhao, X.; Cui, Q.; Fu, Q.; Song, X.; Jia, R.; Yang, Y.; Zou, Y.; Li, L.; He, C.; Liang, X.; et al. Antiviral properties of resveratrol against pseudorabies virus are associated with the inhibition of IκB kinase activation. *Sci. Rep.* **2017**, *7*, 1–11. [CrossRef]

98. Chedea, V.S.; Palade, L.M.; Pelmus, R.S.; Dragomir, C.; Taranu, I. Red grape pomace rich in polyphenols diet increases the antioxidant status in key organs—Kidneys, liver, and spleen of piglets. *Animals* **2019**, *9*, 149. [CrossRef] [PubMed]

99. Herosimczyk, A.; Lepczyński, A.; Ożgo, M.; Tuśnio, A.; Taciak, M.; Barszcz, M. Effect of dietary inclusion of 1% or 3% of native chicory inulin on the large intestinal mucosa proteome of growing pigs. *Animal* **2020**. [CrossRef] [PubMed]

100. Yu, K.; Matzapetakis, M.; Horvatić, A.; Terré, M.; Bach, A.; Kuleš, J.; Yeste, N.; Gómez, N.; Arroyo, L.; Rodríguez-Tomàs, E.; et al. Metabolome and proteome changes in skeletal muscle and blood of pre-weaning calves fed leucine and threonine supplemented diets. *J. Proteomics* **2020**, *216*, 103677. [CrossRef] [PubMed]

Article

Chemical Characterisation and in Vitro Gas Production Kinetics of Eight Faba Bean Varieties

Alessandra Pelagalli [1], Nadia Musco [2], Nikita Trotta [3], Monica I. Cutrignelli [2,*],
Antonio Di Francia [4], Federico Infascelli [2], Raffaella Tudisco [2], Pietro Lombardi [2],
Alessandro Vastolo [2] and Serena Calabrò [2]

[1] Department of Advanced Biomedical Sciences, University of Napoli Federico II, 80100 Napoli, Italy;
alessandra.pelagalli@unina.it

[2] Department of Veterinary Medicine and Animal Production, University of Napoli Federico II, 80100 Napoli,
Italy; nadia.musco@unina.it (N.M.); federico.infascelli@unina.it (F.I.); raffaella.tudisco@unina.it (R.T.);
pietro.lombardi@unina.it (P.L.); alessandro.vastolo@unina.it (A.V.); serena.calabro@unina.it (S.C.)

[3] Centro di ricerca Politiche e Bio-economia R, CREA, Consiglio per la ricerca in agricoltura e l'analisi
economica agraria (CREA), 80143 Napoli, Italy; nikita.trotta@crea.gov.it

[4] Department of Agricultural Sciences, University of Napoli Federico II, 80100 Napoli, Italy;
antonio.difrancia@unina.it

* Correspondence: monica.cutrignelli@unina.it

Received: 15 January 2020; Accepted: 26 February 2020; Published: 28 February 2020

Simple Summary: The demand of vegetable protein is currently very high, both for human and animal nutrition. Soybean meal is the most used protein source in ruminant nutrition. Many Leguminosae seeds (i.e., faba bean, lupin, proteic pea), rich in protein and energy, are considered a valid alternative, especially in the organic production system. In this paper the physical and nutritional characteristics of eight varieties of *Vicia Faba* bean (four local and four commercial) were evaluated. To evaluate the digestive utilization an in vitro trial was carried out, incubating each substrate with an inoculum made up of bovine buffered rumen liquor for 48 h at 39 °C under anaerobiosis. The gas produced within the incubation period was registered, the dry matter digestibility and volatile fatty acid at the end of fermentation were determined. The results of this investigation confirm the possibility of using local faba bean varieties in ruminant nutrition with the advantages that, being local natural resources, they are better adapted to the climate and agronomic conditions and limit the environmental impact.

Abstract: Faba bean is an important vegetable protein source for ruminant diets. This research aimed to compare the nutritional characteristics of four commercial and four local cultivars in order to better characterise the local ones and promote their use in animal nutrition. The seeds' weight and the chemical composition, including starch and the energy, was evaluated. The in vitro fermentation characteristics were studied for 48 h using bull's rumen fluid as inoculum. All the varieties showed the values' weight corresponding to the specific botanical typology. The varieties significantly differed for protein, starch and lignin ($p < 0.01$) and structural carbohydrates ($p < 0.05$) concentration. No significant differences were observed for energy content. All the in vitro fermentation parameters resulted significantly different among the varieties. Organic matter degradability ranged between 89.9% and 85.1% and the potential gas production from 367 to 325 mL/g. The Pearson's analysis showed significant correlation between morphological characteristics, chemical data and in vitro fermentation parameters. In conclusion, this investigation confirms the possibility of using local faba bean varieties (i.e., Aquino, Castrocielo, 13#5, 4#4) in ruminant nutrition with the advantage that, being local natural resources, they are better adapted to the climate and agronomic conditions and limit environmental impact.

Animals **2020**, *10*, 398

Keywords: protein source; ruminant; in vitro fermentation; degradability; volatile fatty acids

1. Introduction

Grain legumes, both rich in protein and energy, have numerous and valuable uses in feed materials, as well as in food production [1]. In animal nutrition, many Leguminosae seeds (i.e., faba bean, lupin, proteic pea) are mainly considered a valid protein source, but the high starch content should also be taken into consideration [2]. These seeds are used as an alternative to soybean meal in the organic production system [3] and their cultivation is promoted by the European Union [4].

Faba bean (*Vicia faba* spp.) is grown worldwide as a source of protein and starch for humans and animals [5]. Its production is widespread in China, followed by Australia, France and Egypt [6,7]. The European contribution to global production is only 14.4 %, but the average yield registered in the Mediterranean countries is nearly double that of anywhere else. *Vicia faba* L. is an auto-diploid plant and it is classified into three main botanical types according to seed dimensions (g/n of seeds): major Herz, higher than 1000/1000, minor Beck, up to 700/1000 and equina Pers. 700–1000/1000 [8]. The first two types are gene pools of the Central and Northwest European varieties, while equina is more present in Mediterranean European varieties. Many other systematic classifications (e.g., based on the taxonomic levels) are known, but none of them has been completely defined. Moreover, it is important to consider how the environmental conditions as well as the origin can modify seeds by altering their dimensions and reducing the differences among varieties [6]. Like others legume seeds, faba bean improves soil fertility and reduces the nitrogen fertilization requirements by fixing atmospheric nitrogen in symbiosis with the soil rhizobia bacteria [9], thus playing an important role in crop rotation. The faba bean has been successfully used in Mediterranean areas as a high-protein concentrate for ruminants. Brunschwig and Lamy [10] showed that the use of 30% of ground faba beans in the concentrate for dairy cows did not alter feed consumption and milk production, in terms of yield, and milk quality (i.e., protein, fat, etc.). It is suitable also for growing animals: Leitgeb and Lettner [11] reported that the use of faba beans did not decrease feed consumption and did not alter animal growth and the carcass composition. In previous investigations, the faba bean is shown contributing to improving the local animal products (i.e., lower cholesterol and saturated fatty acid contents) such as meat from buffalo [12] and from Marchigiana young bulls [13,14].

In the last few years, according to EU strategy [15], many projects have been promoted in Italy in order to reduce the continuous loss of agronomical biodiversity. Moreover, the local production of some Leguminosae seeds should be valorised. The aim of the present research was to compare the morphological and nutritional characteristics of eight faba bean varieties, including four local varieties grown in the Mediterranean area, and four commercial varieties recorded in the National Agricultural Varieties Register. All the varieties were evaluated for morphological aspects, chemical composition and in vitro fermentation patterns. The hypothesis was that the local faba bean varieties could present nutritional characteristics comparable to the commercial ones and, consequently, be useful for ruminant nutrition.

2. Materials and Methods

2.1. Experimental Design

The eight varieties of faba bean investigated in the present study were selected in different Italian Regions among the three main types (equina, minor and major) by the researchers of the Centro di Ricerca Orticoltura e Florovivaismo (CREA-ORT, Pontecagnano, SA, Italy): four local varieties (Aquino, Castrocielo, 13#5, 4#4) and four commercial ones (Bolivia, Chiaro di Torre Lama, Sikelia and Aguadulce Supersimonia), as reported in Table 1. The commercial varieties were chosen as control for their similarity in morphological aspects with the local varieties. At CREA-ORT (40°38′36″60 N,

14°52′27″48 E, 34 m a.s.l.) in a 192 m^2 (32.0 m × 6.0 m), a field experiment was conducted to produce grains to be evaluated. For each variety, four randomized plots (5 m^2) were sown with 30 seeds positioned at 6–8 cm of depth in rows of 45 cm. During plant cultivation no fertilizing or pesticide treatments were carried out, except during the first growth stage, when an insecticide treatment, based on pirimicarb, was used. The seeds produced by the field trial were harvested from each plot and weighed in order to evaluate the seed yield and size. Subsequently, in the laboratory mixing two-by-two the seeds for each variety (two plots/variety), two aliquots were obtained. Then, the obtained samples were dried and characterized for chemical composition and the in vitro fermentation characteristics.

Table 1. Harvesting area of the *Vicia faba* L. varieties used for the field experiment.

Variety	Type	Province and Region	Latitude North	Longitude East	Altitude [m a.s.l.]
Local					
Aquino	*Equina*	Frosinone, Lazio	41°29′	13°42′	110
Castrocielo	*Minor*	Frosinone, Lazio	41°31′	13°41′	226
4#4	*Major*	Salerno, Campania	40°40′	14°46′	54.3
13#5	*Major*	Salerno, Campania	40°40′	14°46	54.3
Commercial					
Bolivia	*Major*	Roma, Lazio	41°94′	12°67′	20
Chiaro di Torre Lama	*Minor*	Massa Carrara, Toscana	43°25′	13°25′	350
Sikelia	*Minor*	Catania, Sicilia	37°51′	15°07′	7
Aguadulce Supersimonia	*Major*	Arezzo, Toscana	43°49′	11.63′	296

2.2. Seed Size

The morphological analysis of the faba seeds was performed according to the International Union for the protection of New Plant Varieties [16], evaluating the weight of 1000 seeds (g).

2.3. Nutritional Evaluation of Seeds

At the Department of Veterinary Medicine and Animal Production (Napoli, Italy) the seeds were milled at 1.1 mm screen (SM 100, Retsch, Haan, Germany) to determine dry matter (DM), crude protein (CP) and ash contents according to official procedures [17]: ID members 934.01, 954.01 and 942.05, respectively. Moreover, the neutral detergent fibre (aNDFom, adding sodium sulphite and heat resistant alpha-amylase, Ankom Macedon NY), the acid detergent fibre (ADF) and the lignin (ADL) were determined [18]. The starch content was determined through polarimetric detection (Polax L, Atago, Tokyo, Japan) according to the official procedure [19]. The energy content as metabolizable energy (ME, MJ/kg) was estimated according to INRA system as following [20]:

$$(43.3 \times CP \times degCP) + (77.2 \times EE \times degEE) + (35.9 \times CF \times degCF) + (36.3 \times NFE \times degNFE) \quad (1)$$

where: EE, CF and NFE are crude protein, ether extract, crude fiber and nitrogen free extract content (%), respectively; deg represent the degradability coefficients of each parameter proposed by the INRA system.

2.4. In Vitro Fermentation Characteristics

The fermentation characteristics and kinetics were studied using the in vitro gas production technique as proposed by Theodorou [21]. Two gas runs were conducted under similar experimental conditions within two weeks. All the samples (varieties) were incubated as substrates at 39 °C under anaerobic conditions with buffered rumen fluid [22]. Faba bean samples were weighed (1.000 g ± 0.002) in triplicate in 120 mL serum flasks with anaerobic medium (74 mL). Three flasks without substrate were incubated to correct organic matter degradability and gas volume. The rumen fluid was collected in a pre-warmed thermos from six/run bovine fasted young bulls (*Bos taurus*), regularly slaughtered in an authorized facility in accordance with current animal welfare legislation [23]. For 15 days before slaughter young bulls were fed 13 kg of dry matter/die of standard diet (NDF 45.5 and CP 12.0% DM) containing on dry matter basis 20% of corn silage, 20% of mixed hay and 40% of commercial concentrate (i.e., beet pulp, wheat middling, wheat flour middling, maize gluten, soya bean meal) [24]. All procedures involving animals were approved by the Ethical Animal Care and Use Committee of the University of Napoli Federico II (Prot. 2019/0013729 of 08/02/2019). The collected material was rapidly transported to a pre-warmed thermos in the laboratory, where it was pooled, flushed with CO_2, filtered through cheesecloth and added to each flask (5 mL) within 1 h of collection. All the steps were carried out at 39 °C and under insufflations of CO_2 to maintain anaerobic conditions. Gas production was recorded 17 times, during the 48 h of incubation, using a manual pressure transducer (Cole and Parmer Instrument Co., Vernon Hills, IL, USA). At this time the fermentation was stopped by cooling (4 °C) and the degraded organic matter (dOM, %) was determined by the differences between the incubated OM and the residual OM, obtained filtering the fermentation liquor throughout pre-weighed sintered glass crucibles (Schott Duran, Mainz, Germany, porosity #2) and successively burnt at 550 °C. An aliquot of fermentation liquor was taken to measure the pH using a pH-meter (ThermoOrion 720 A+, Fort Collins, CO, USA). Volatile fatty acids (VFA, mM/g) were determined through gas chromatography (GC Focus AI 3000, Thermo Scientific, Waltham, MA, USA) equipped with a fused silica capillary column SACTM-5, Supelco, 30 m x 0.25 mm x 0.25 µm film thickness using external standard solution (acetate, propionate, butyrate, iso-butyrate, valerate and iso-valerate) [24].

2.5. Statistical Analysis

The cumulative volume of gas produced at 48 h was related to incubate OM (OMCV, mL/g). For each flask the gas production profiles were processed with the following model [25]:

$$G = \frac{A}{\left[1 + \left(\frac{C^B}{t^B}\right)\right]} \tag{2}$$

where, G is the total gas produced (mL/g of incubated OM) at time t (h), A the asymptotic gas production (mL/g of incubated OM), B (h) the time at which one-half of the asymptote is reached, and C the curve switch. Maximum fermentation rate (R_{max}, mL/h) and time at which it occurs (T_{max}, h) were also calculated:

$$Rmax = \frac{A * B^C * B * Tmax^{(B-1)}}{\left[1 + (C^B * Tmax^{-B})^2\right]} \tag{3}$$

$$Tmax = C * \left[\frac{(B-1)}{(B+1)}\right]^{1/B}, \tag{4}$$

Seed size, chemical data, fermentation characteristics, and model parameters were statistically compared between faba bean varieties [JMP®1989–2007, Version 9.0, Cary (NC): SAS Institute Inc.] using the model:

$$y_{ij} = \mu + Var_i + \varepsilon_{ij}, \tag{5}$$

where y is the single datum, µ the mean, Var the variety (substrate) effect (i = 8) and ε the error term. The *t* test was used to verify the differences between means, considering for each variety four subsamples (e.g., plots) for seed size; two subsamples (mixing two-per-two plots) for chemical composition and four subsamples (the same samples of chemical composition used for two gas runs). The correlations among seed characteristics, chemical composition and in vitro data ($p < 0.05$) were also studied using the same software.

3. Results

On average, the seeds' yield in field for all the cultivated faba bean varieties was 395 ± 58 kg/ha (at 10% moisture). No significant differences ($p > 0.05$) were found between varieties.

3.1. Seeds Size

Comparing the weights, which referred to 1000 seeds (Figure 1), it was possible to highlight that all the analysed varieties showed characteristics corresponding to the specific botanical typology: Chiaro di Torre Lama, Castrocielo and Sikelia weighed less than 700 g, corresponding to the minor type; the Aquino varieties showed a weight of 856 g, corresponding to the equina type while all the remaining varieties showed weights higher than 1000 g. The weight of the Bolivia variety (2633 g) was particularly high.

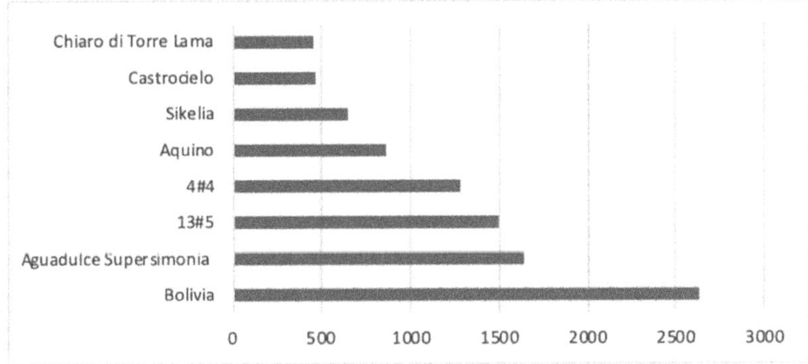

Figure 1. Weight of 1000 seeds of faba bean varieties (g).

3.2. Chemical Composition

The varieties significantly ($p < 0.01$, $p < 0.05$) differed in their chemical composition (Table 2). In particular, on DM basis the protein content was similar among the varieties, even if Aguadulce Supersimonia and 4#4 showed the highest ($p < 0.01$) values (293 and 290 g/kg, respectively). With regard to the structural carbohydrates, NDF on DM basis ranged from 145 and 228 g/kg in Bolivia and Sikelia, respectively and showed few statistical differences, whereas the lignin was significantly ($p < 0.01$) higher in 4#4 (46.7 g/kg). Instead, the starch content varied at a wider range ($p < 0.01$): Sikelia showed the highest value (438 g/kg) and 4#4 the lowest one (221 g/kg). The ash content slightly varied among varieties ranging from 35.6 g/kg in 13#5 and 45.1 g/kg in Bolivia. No significant differences were observed for energy content which varied from 10.9 to 11.9 MJ/kg.

Table 2. Chemical composition on dry matter basis of faba bean varieties.

Variety	CP	NDF	ADF	ADL	Starch	Ash	ME
	g/kg						MJ/kg
Local							
Aquino	273 [B]	207 [ab]	92.9 [EF]	8.0 [B]	390 [B]	42.6 [AB]	11.3
Castrocielo	264 [BC]	222 [a]	142 [A]	6.7 [B]	352 [C]	37.5 [EF]	11.5
13#5	250 [D]	201 [ab]	146 [A]	10.1 [B]	378 [B]	35.6 [F]	11.9
4#4	290 [A]	201 [ab]	120 [B]	46.7 [A]	221 [E]	42.7 [B]	11.2
Commercial							
Bolivia	254 [CD]	145 [c]	98.6 [DE]	7.6 [B]	298 [D]	45.2 [A]	11.8
Chiaro di Torre Lama	260 [C]	199 [ab]	88.1 [E]	9.7 [B]	375 [BC]	38.1 [DE]	11.2
Sikelia	245 [D]	229 [a]	104 [CD]	8.7 [B]	438 [A]	41.6 [BC]	10.9
Aguadulce Supersimonia	293 [A]	168 [bc]	111 [BC]	9.1 [B]	383 [B]	40.0 [CD]	11.9
RMSE	4.4	19.7	3.86	3.62	9.95	0.95	1.57
p value	<0.001	0.038	<0.001	<0.001	<0.001	<0.001	0.195

CP = crude protein; NDF = neutral detergent fiber; ADF = acid detergent fiber; ADL = acid detergent lignin; ME = metabolizable energy. Along the column: a–c = $p < 0.05$ and A–F = $p < 0.01$; RMSE = root mean square error.

3.3. In Vitro Fermentation Characteristics and Volatile Fatty Acids Production

All the fermentation parameters resulted as being significantly ($p < 0.01$) different among the varieties (Figure 2 and Table 3). All the varieties showed a percentage of degradability higher than 85% albeit by some significant differences ($p < 0.01$) among varieties were observed. The Bolivia variety showed the highest dOM value (89.9%) whereas the Chiaro di Torre Lama was the least degradable (dOM: 85.1%). The other varieties showed intermediate values. Regarding the gas production, 4#4 showed the lowest value (OMCV: 310 mL/g) whereas Bolivia and Sikelia the highest ones (351 and 332 mL/g, respectively). In most of the varieties, with the exception of the 4#4 and Sikelia a linear relation between dOM and OMCV was observed. The potential gas production varied little among the studied faba bean varieties, even if 4#4 presented the lowest value (325 mL/g; $p < 0.01$). The time at which the A/2 was reached (B parameter) varied from 13.7 h of the Aquino to 15.5 h of the 4#4 variety showing a mean value of 14.1 h ± 0.5. The maximum fermentation rate of Bolivia and Aquino reached the same maximum fermentation rate (R_{max}: 18.9 ml/h) in a similar time (T_{max}: 9.7 and 9.8 h, respectively). Sikelia showed a similar T_{max} value (9.4 h) with the lowest R_{max} (16.2 mL/h).

The pH values ranged from 6.89 to 6.94 in Castrocielo and Chiaro di Torre Lama, respectively, indicating that after 48 h the incubation was carried out adequately.

The volatile fatty acid (VFA) production (Table 4) was also significantly ($p < 0.01$) affected by the variety. With regard to the total VFA, Chiaro di Torre Lama showed the lowest value (75.8 mM/g; $p < 0.01$) mainly due to the lower production of acetate, isovalerate and butyrate, whereas Bolivia showed the highest VFA (98.0 mM/g) mainly due to the higher acetate, propionate and butyrate production in comparison to the other varieties. Overall, for the other varieties these parameters were similar.

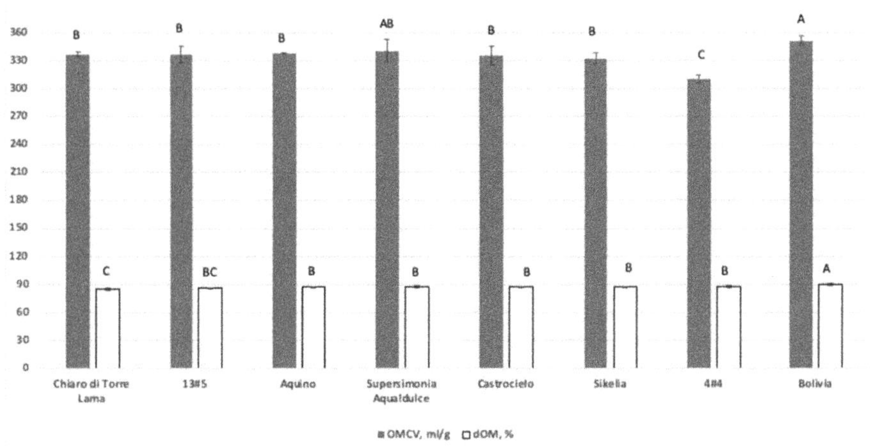

Figure 2. In vitro fermentation characteristics of faba bean varieties; dOM = organic matter degradability (% of incubated OM; p = 0.0007); OMCV = cumulative volume of gas related to incubated OM (mL/g; p = 0.0004). For each parameter: A–C = p < 0.01. Error bar = standard deviation.

Table 3. In vitro fermentation characteristics of faba bean varieties.

Variety	A	B	T_{max}	R_{max}
	mL/g	**h**	**h**	**mL/h**
Local				
Aquino	350 ^{B}	13.7 ^{DE}	9.8 ^{ab}	18.9 ^{A}
Castrocielo	355 ^{AB}	14.4 ^{B}	9.3 ^{bc}	17.0 ^{CDE}
13#5	353 ^{B}	14.0 ^{CD}	9.1 ^{c}	17.3 ^{BCD}
4#4	325 ^{C}	13.5 ^{E}	9.0 ^{c}	16.8 ^{DE}
Commercial				
Bolivia	367 ^{A}	14.0 ^{CD}	9.7 ^{abc}	18.9 ^{A}
Chiaro di Torre Lama	353 ^{B}	14.2 ^{BC}	10.1 ^{a}	18.3 ^{AB}
Sikelia	360 ^{AB}	15.0 ^{A}	9.4 ^{bc}	16.2 ^{E}
Aguadulce Supersimonia	360 ^{AB}	14.0 ^{CD}	9.2 ^{bc}	17.9 ^{ABC}
RMSE	8.4	0.3	0.4	0.8
p value	<0.001	<0.001	0.033	<0.001

A = potential gas production; B = time at which A/2 is formed; T_{max} = time at which maximum fermentation rate was reached; R_{max} = maximum fermentation rate. Along the column: a–c = p < 0.05 and A–E = p < 0.01. RMSE = root mean square error.

The Pearson's analysis (r) showed a significant correlation between the morphological characteristics, chemical data and in vitro fermentation parameters. Seed size affected positively the energy content (0.68; p < 0.05) and negatively NDF level (−0.89; p < 0.01). The structural carbohydrates negatively (p < 0.01) influenced the fermentation kinetics parameters (NDF vs. R_{max}: −0.67 and ADF vs. T_{max}: −0.80). Despite the low percentage of lignin in all the samples, this parameter negatively affected the OMCV (−0.88, p < 0.01). The starch fraction was positively correlated (p < 0.05) with A (0.63) and B (0.67).

Table 4. pH values and concentration of volatile fatty acids after 48 h of in vitro incubation.

Variety	pH	Acetate	Propionate	Isobutyrate	Butyrate	Isovalerate	Valerate	VFA
				mM/g				
Local								
Aquino	6.92 [B]	47.8 [A]	15.8 [A]	2.26 [AB]	19.3 [AB]	4.64 [B]	2.51 [AB]	92.3 [AB]
Castrocielo	6.89 [D]	46.3 [AB]	13.7 [BC]	2.00 [BC]	20.1 [AB]	4.30 [BC]	2.00 [AB]	88.4 [AB]
13#5	6.92 [BC]	47.5 [AB]	15.0 [AB]	2.05 [BC]	17.9 [B]	4.16 [BCD]	2.40 [AB]	89.1 [AB]
4#4	6.90 [D]	48.6 [A]	14.8 [AB]	1.95 [C]	18.5 [AB]	3.89 [CD]	1.94 [B]	89.7 [AB]
Commercial								
Bolivia	6.91 [CD]	51.0 [A]	15.5 [AB]	2.39 [A]	22.0 [A]	5.23 [A]	1.95 [AB]	98.0 [A]
Chiaro di Torre Lama	6.94 [A]	42.3 [B]	12.0 [C]	1.79 [C]	13.5 [C]	3.67 [D]	2.55 [A]	75.8 [C]
Sikelia	6.92 [BC]	47.8 [A]	14.8 [AB]	2.06 [BC]	19.9 [AB]	4.13 [BCD]	2.38 [AB]	91.0 [AB]
Aguadulce Supersimonia	6.93 [AB]	45.9 [AB]	14.0 [B]	1.98 [C]	19.1 [AB]	4.03 [CD]	1.93 [B]	87.0 [B]
RMSE	0.90	4.49	1.59	0.23	3.14	0.47	0.52	9.35
p value	<0.001	0.105	0.005	0.003	0.036	<0.01	0.132	0.022

VFA = volatile fatty acids. Along the column: A–D = $p < 0.01$. RMSE = root mean square error.

4. Discussion

Like other legume seeds, *Vicia faba* spp. can be considered a valuable protein source for ruminants albeit the different varieties need to be better evaluated for their more appropriate use. For the varieties tested in our study, the seed size, chemical composition, as well as the in vitro fermentation characteristics, fall into the range reported in the literature [26,27]. Our data concerning protein content are in line with those reported by other authors [28,29], ranging from 24.2 to 37.2%. With the in sacco method, Faurie et al. [30] showed that more than 85% of the nitrogen is degraded in the rumen in 2 h resulting in a mean theoretical nitrogen degradability of 92% for three different faba bean cultivars. For this reason, in a previous study [13] we suggested associating a protein source richer in rumen un-degradable protein to faba beans immediately after weaning. The structural carbohydrates of faba beans had a lower NDF content (mean value: 19.6% DM) compared to a previous study [27] on six faba bean varieties (22.7% DM). In this study, the NDF content was quite variable (from 14.5% to 22.9% DM) in function of the varieties, similarly to the data obtained in 74 faba bean varieties (NDF from 13.4% to 26.4% DM) [29]. Except for the very high value recorded in variety 4#4, the ADL contents were quite low and similar to those registered in a previous study [27]. According to other authors [31,32], the starch content was quite high (from 22.1% to 43.8% DM), even if Duc et al. [29] found higher values (37.0% to 50.5% DM). The high energy content and the low structural carbohydrate level in the tested faba beans grains favored in vitro fermentation patterns in terms of degradability, kinetics, gas and VFA production. In a previous study [27], incubating in vitro six different varieties of faba beans (i.e., Irene, Lady, Scuro di Torre Lama, Chiaro di Torre Lama, ProtHABAT69 and Siconia) with rumen fluid from buffalo, higher degradability values (dOM mean values: 91.8% vs. 87.3%), but slower fermentation kinetics (T_{max} mean value: 13.2 vs. 9.5 h; R_{max} mean value: 8.7 vs. 17.7 mL/h) were registered. Azarfar et al. [33], incubating for 72 h processed grains of a variety of Vicia faba minor using in vitro gas production technique with rumen liquor from dairy cows, found a lower gas production value (OMCV mean value: 194 vs. 335 mL/g) and a lower volatile fatty acid production (acetate, propionate and butyrate: 68.0 vs. 79.7 mM/g). These data indicate that many factors could influence the in vitro fermentation pathway (i.e., varieties, donor animal species, incubation time). As reported in a previous study [2], the carbohydrates' fractions differently affect the fermentation kinetics: starch promotes a more intense and rapid process, conversely structural carbohydrates cause a slower and less consistent fermentation. The particularly slow in vitro fermentation kinetics of the 4#4 variety (potential gas production lowest value, $p < 0.001$) was probably due to the high ADL content that has limited the access to the cell content by micro-organisms, reducing nutrients' degradability and slowing down the fermentation rate; as evidenced, also by the negative correlation between lignin content and OMCV. No significant correlation was observed between crude protein and OMCV and dOM in contrast with a previous study [2] on lupine seeds. The different results could be probably due to more balanced ratio between protein and carbohydrates of faba bean compared to lupine.

The significant correlation between the seed size and some chemical parameters (NDF and energy) testify the lower incidence of structural carbohydrates in the mass units (1 kg) and the higher energy value of the varieties bigger in size. However, no correlation was observed between seed size and in vitro parameters. All these findings confirm the nutritional characteristics of *Vicia faba*, endorsing the results of several authors obtained in vivo on growing and dairy ruminants [10–14].

Comparing the local varieties with their morphologically homologous commercial ones (Aquino vs. Sikelia; Castrocielo vs. Chiaro di Torre Lama; 13#5 and 4#4 vs. Bolivia and Aguadulce Supersimonia), some interesting results emerged from the nutritional point of view. The enhancement of local varieties already evidenced by historical documentation [34] assumes greater importance. For example, the Aquino variety showed specific characteristics (i.e., high protein, starch and energy contents, low fiber amounts, high in vitro fermentability and VFA production).

5. Conclusions

The results of this investigation confirm the possibility of using local faba bean varieties in ruminant nutrition with the advantage that, being local natural resources, they are better adapted to the climate and agronomic conditions and limit the environmental impact. These varieties, such as the commercial ones, present different nutritional characteristics, that affect the in vitro fermentation kinetics, and could also influence their in vivo utilization.

In any event, their use as alternative protein sources to the most common extruded soybean meal needs further studies aimed at evaluating the possible presence of anti-nutritional factors (i.e., tannins, vicine, etc.) that could negatively affect their digestive utilization and/or positively influence the environmental sustainability reducing green-house gases (e.g., methane) production.

Author Contributions: A.P., N.T. and S.C., conceived and designed the study, analyzed the data, and wrote the manuscript. N.M., R.T., and A.V. carried out experiments, collected data, and carried out data analyses. F.I., P.L., A.D.F., and M.I.C. contributed to critically revise the manuscript and to editing the manuscript. All authors have read and agreed to the published version of the manuscript.

Funding: This research received no external funding.

Acknowledgments: The authors would like to thank Maria Ferrara (Department of Veterinary Medicine and Animal Production) and Rosa Pepe (CREA-ORT) for their technical support. The research was carried out thanks to Departmental funds for individual research (Years 2016).

Conflicts of Interest: The authors declare no conflict of interest.

References

1. Crèpon, K.; Marget, P.; Peyronnet, C.; Carrouèe, B.; Arese, P.; Duc, G. Nutritional value of faba bean (*Vicia faba* L.) seeds for feed and food. *Field Crops Res.* **2010**, *115*, 329–339. [CrossRef]

2. Musco, N.; Cutrignelli, M.I.; Calabrò, S.; Tudisco, R.; Infascelli, F.; Grazioli, R.; Lo Presti, V.; Gresta, F.; Chiofalo, B. Comparison of nutritional and antinutritional traits among different species (*Lupinus albus* L., *Lupinus luteus* L., *Lupinus angustifolius* L.) and varieties of lupin seeds. *J. Anim. Physiol. Anim. Nutr.* **2017**, *101*, 1227–1241. [CrossRef] [PubMed]

3. EC Council Regulation 834/2007 on Organic production and labelling of organic products and repealing Regulation (EEC) No. 2092/91. *Off. J. Eur. Union* **2007**, *L 189/1*, 1–23.

4. EU Council Regulation 1535/2007 on Applicazione degli articoli 87 e 88 del trattato CE agli aiuti de minimis nel settore della produzione dei prodotti agricoli. *Gazz. Uff. Unione Eur.* **2007**, *L 337/35*.

5. Abreu, J.M.F.; Bruno-Soares, A.M. Chemical composition, organic matter digestibility and gas production of nine legume grains. *Anim. Feed Sci. Technol.* **1998**, *70*, 49–57. [CrossRef]

6. Maalouf, F.; Nawar, M.; Hamwieh, A.; Amri, A.; Zong, X.; Bao, S.; Tao, Y. 5. Faba Bean. In *Genetic and Genomic Resources of Grain Legume Improvement*; Singh, M., Upadhyaya, H.D., Singh Bisht, I., Eds.; Elsevier: Oxford, UK, 1998; pp. 113–136.

7. Hou, W.W.; Zhang, X.J.; Shi, J.B.; Liu, Y.J. Genetic diversity analysis of faba bean (*Vicia faba L.*) germplasms using sodium dodecyl sulfate-polyacrylamide gel electrophoresis. *Genet. Mol. Res.* **2015**, *14*, 3945–3953. [CrossRef] [PubMed]

8. Terzopoulos, P.J.; Bebeli, P.J. Genetic diversity analysis of Mediterranean faba bean (*Vicia faba* L.) with ISSR markers. *Field Crops Res.* **2008**, *108*, 39–44. [CrossRef]

9. Jensen, E.S.; Peoples, M.B.; Hauggaard-Nielsen, H. Faba bean in cropping systems. *Field Crops Res.* **2010**, *115*, 203–216. [CrossRef]

10. Brunschwig, P.; Lamy, J.M. Utilisation de féverole ou de tourteau de tournesol comme sources protéiques dans l'alimentation des vaches laitières. *Renc. Rech. Rum.* **2002**, *9*, 316.

11. Leitgeb, R.; Lettner, F. Use of faba beans in growing bulls. In *Association européenne de recherche sur les protéagineux editors. Recueil de communications 1re Conférence Européenne sur les Protéagineux*; Institut technique des céréales et des fourrages: Angers, France, 1992; pp. 493–494.

12. Calabrò, S.; Cutrignelli, M.I.; Gonzalez, O.J.; Chiofalo, B.; Grossi, M.; Tudisco, R.; Panetta, C.; Infascelli, F. Meat quality of buffalo young bulls fed faba bean as protein source. *Meat Sci.* **2014**, *96*, 591–596. [CrossRef]

13. Cutrignelli, M.I.; Piccolo, G.; Bovera, F.; Calabrò, S.; D'Urso, S.; Tudisco, R.; Infascelli, F. Effects of two protein sources and energy level of diet on the performance of young Marchigiana bulls. 1. Infra vitam performance and carcass quality. *Ital. J. Anim. Sci.* **2008**, *7*, 259–270. [CrossRef]

14. Cutrignelli, M.I.; Calabrò, S.; Bovera, F.; Tudisco, R.; D'Urso, S.; Marchiello, M.; Piccolo, V.; Infascelli, F. Effects of two protein sources and energy level of diet on the performance of young Marchigiana bulls. 2. Meat quality. *Ital. J. Anim. Sci.* **2008**, *7*, 271–285. [CrossRef]

15. European Commission. *The Common Agricultural Policy CAP towards 2020: Meeting the Food, Natural Resources and Territorial Challenges of the Future*; European Parliament: Bruxelles, Belgium, 2010; p. 672.

16. UPOV. *Description for Broad Bean*; TG/206/1 and TG/8/6; The International Union for the Protection of New Plant Varieties: Geneva, Switzerland, 2003.

17. AOAC. *Official Methods of Analysis*, 18th ed.; Association of Official Analytical Chemist: Arlington, VA, USA, 2005.

18. Van Soest, P.J.; Robertson, J.B.; Lewis, B.A. Methods for dietary fiber, neutral detergent fiber, and nonstarch polysaccharides in relation to animal nutrition. *J. Dairy Sci.* **1991**, *74*, 3583–3597. [CrossRef]

19. ISO 6493:2000. Animal Feeding Stuffs. Determination of Starch Content. Available online: https://www.iso. org/obp/ui/#iso:std:iso:6493:ed-1:v1:en (accessed on 18 May 2018).

20. Institut National De La Recherche Agronomique (France). *Feeding System for Ruminants*; Wageningen Academic Publisher: Wageningen, The Netherlands, 2018. [CrossRef]

21. Theodorou, M.K.; Williams, B.A.; Dhanoa, M.S.; McAllan, A.B.; France, J. A simple gas production method using a pressure transducer to determine the fermentation kinetics of ruminant feeds. *Anim. Feed Sci. Technol.* **1994**, *48*, 185–197. [CrossRef]

22. Boussaada, A.; Arhab, R.; Calabrò, S.; Grazioli, R.; Ferrara, M.; Musco, N.; Thlidjane, M.; Cutrignelli, M.I. Effect of Eucalyptus globulus leaves extracts on in vitro rumen fermentation, methanogenesis, degradability and protozoa population. *Ann. Anim. Sci.* **2018**, *18*, 753–767. [CrossRef]

23. EC Council Regulation 882/2004 on Official controls performed to ensure verification of compliance with feed and food law, animal health and animal welfare rules. *Off J Eur. Union* **2004**, *L191/1*, 1–52.

24. Calabrò, S.; Cutrignelli, M.I.; Bovera, F.; Piccolo, G.; Infascelli, F. In vitro fermentation kinetics of carbohydrate fractions of fresh forage, silage and hay of Avena sativa. *J. Sci. Food Agric.* **2005**, *85*, 1838–1844. [CrossRef]

25. Groot, J.C.J.; Cone, J.W.; William, B.A.; Debersaque, F.M.A. Multiphasic analysis of gas production kinetics for in vitro fermentation of ruminant feedstuff. *Anim. Feed Sci. Technol.* **1996**, *64*, 77–89. [CrossRef]

26. Trotta, N. A statistical analysis of the distinctness between the vegetable varieties. *Minerva Biotec.* **2011**, *23*, 19–27.

27. Calabrò, S.; Tudisco, R.; Balestrieri, A.; Piccolo, G.; Infascelli, F.; Cutrignelli, M.I. Fermentation characteristics of different grain legumes cultivars with the in vitro gas production technique. *Ital. J. Anim. Sci.* **2009**, *8*, 280–282. [CrossRef]

28. Abdulla, J.M.; Rose, S.P.; Mackenzie, A.M.; Pirgozliev, V.R. Feeding value of field beans (*Vicia faba* L. var. *minor*) with and without enzyme containing tannase, pectinase and xylanase activities for broilers. *Arch. Anim. Nutr.* **2017**, *71*, 150–164. [PubMed]

29. Duc, G.; Marget, P.; Esnault, R.; Le Guen, J.; Bastianelli, D. Genetic variability for feeding value of faba bean seeds (*Vicia faba*): Comparative chemical composition of isogenics involving zero-tannin and zero-vicine genes. *J. Agric. Sci.* **1999**, *133*, 185–196. [CrossRef]

30. Faurie, F.; Schametz, C.; Tisserand, J.L. Nitrogen content degradability of any proteaginous seeds. In *Association Européenne de Recherche sur les Protéagineux Editors. Recueil de communications 1re Conférence Européenne sur les Protéagineux. 1–3 June 1992*; Institut technique des céréales et des fourrages: Angers, France, 1992; pp. 515–516.

31. Sauvant, D.; Perez, J.M.; Tran, G. *Table de Composition et de Valeur Nutritionnelle des Matiéres Premiéres Destinées aux Animaux d'élevage*; 2ème INRA Editions Versailles: Paris, France, 2004; p. 306.

32. Guglielmelli, A.; Calabrò, S.; Cutrignelli, M.I.; Gonzalez, O.J.; Infascelli, F.; Tudisco, R.; Piccolo, V.; Infascelli, F.; Tudisco, R.; Piccolo, V. In vitro fermentation and methane production of fava and soy beans. Energy and protein metabolism and nutrition. In *Energy and Protein Metabolism and Nutrition*; Crovetto, G.M., Ed.; Wageningen Academic Publishers: Parma, Italy, 1992; pp. 457–460.

33. Azarfar, A.; Tamminga, S.; Pellikaan, W.F.; van der Poel, A.F.B. In vitro gas production profiles and fermentation end-products in processed peas, lupins and faba beans. *J. Sci. Food Agric.* **2008**, *88*, 1997–2010. [CrossRef]

34. Pelagalli, A. *Le Fave dei Morti tra Storia e Antichi riti. Una Tradizione del Passato ad Aquino*; Enzo Albano: Napoli, Italy, 2013.

 *animals* 

Article

Methane Emissions and Milk Fatty Acid Profiles in Dairy Cows Fed Linseed, Measured at the Group Level in a Naturally Ventilated Housing and Individually in Respiration Chambers

Jernej Poteko [1,2,†], Sabine Schrade [2], Kerstin Zeyer [3], Joachim Mohn [3], Michael Zaehner [2], Johanna O. Zeitz [1,4], Michael Kreuzer [1] and Angela Schwarm [1,5,*]

[1] ETH Zürich, Institute of Agricultural Sciences, 8092 Zurich, Switzerland; jernej.poteko@lfl.bayern.de (J.P.); jzeitz@gmx.de (J.O.Z.); michael.kreuzer@usys.ethz.ch (M.K.)

[2] Agroscope, Ruminants Research Unit, 8356 Ettenhausen, Switzerland; sabine.schrade@agroscope.admin.ch (S.S.); michael.zaehner@agroscope.admin.ch (M.Z.)

[3] Empa, Laboratory for Air Pollution/Environmental Technology, 8600 Duebendorf, Switzerland; kerstin.zeyer@empa.ch (K.Z.); joachim.mohn@empa.ch (J.M.)

[4] Laboklin, Klinische Labordiagnostik, 97688 Bad Kissingen, Germany

[5] Department of Animal and Aquacultural Sciences, Norwegian University of Life Sciences, P.O. Box 5003, 1432 Ås, Norway

[*] Correspondence: angela.schwarm@nmbu.no; Tel.: +47-67-23-26-17

[†] Current address: Bavarian State Research Center for Agriculture, Institute for Agricultural Engineering and Animal Husbandry, 85586 Grub-Poing, Germany.

Received: 8 May 2020; Accepted: 22 June 2020; Published: 24 June 2020

Simple Summary: Cows emit the greenhouse gas methane (CH_4) as a result of microbial feed digestion. Methane emissions can be reduced by adopting nutritional strategies, such as dietary supplementation of linseed. Additionally, the oil in linseed increases the proportion of favorable fatty acids in milk fat. This study evaluated the effect of linseed on CH_4 emission and milk fatty acid composition measured in a group of cows in a naturally ventilated barn and in individual cows in respiration chambers. The substantially higher proportions of favorable fatty acids in the milk of linseed-fed cows were detected in individual milk samples and in the milk of the herd. Therefore, the analysis of bulk milk could be a suitable control instrument for retailers. Visualizing the course of CH_4 emissions over a whole day showed slightly lower CH_4 values in linseed-supplemented individuals and groups. However, we found no significant reduction of CH_4 as a result of linseed supplementation. Feed supplements in concentrations that are effective in reducing CH_4 must show whether the reduction potential is comparable when determined at the group and individual levels.

Abstract: The present study evaluated the effects of linseed supplementation on CH_4 emission and milk fatty acid composition in dairy cows measured at the group level in an experimental dairy loose housing using a tracer gas technique and individually in tied stalls and respiration chambers. Cows (2×20) were maintained in two separate sections under loose-housing conditions and received a diet supplemented with extruded linseed (L) lipids ($29~g \cdot kg^{-1}$ dry matter) or a control (C) diet containing corn flour. Subsequently, 2×6 cows per dietary group were investigated in a tied-housing system and respiration chambers. Substantially higher proportions of favorable milk fatty acids were recovered in L cows when compared with C cows at the group level, making the analysis of bulk milk a suitable control instrument for retailers. Linseed supplementation resulted in a slightly lower diurnal course of CH_4 emission intensity than the control at the group and individual levels. However, we found no more than a trend for a CH_4 mitigating effect, unlike in other studies supplementing similar linseed lipid levels. Feed supplements in concentrations that lead to a significant reduction in CH_4 emissions must show whether the reduction potential determined at the group and individual levels is comparable.

Keywords: methanogenesis; methane mitigation; lipid; ruminant; cattle; emission measurement

---

1. Introduction

Ruminants emit methane (CH_4) as a result of enteric fermentation. Ruminal CH_4 emissions can be mitigated by adopting several nutritional strategies [1,2]. These are advantageous over animal breeding measures as they are immediately effective [3]. In particular, dietary supplementation of crushed or extruded oilseeds has been demonstrated to mitigate enteric CH_4 emission [4,5]. In this context, supplementing the diet with >50 g of linseed lipids per kg feed dry matter (DM) particularly decreased CH_4 emission, but diet digestibility was concomitantly impaired [5,6]. At supplementation levels of ≤20 g linseed lipids per kg feed DM, there was no reduction in the CH_4 emissions [7–9]. Thus, supplementing feed with moderate levels ranging from 26–36 g linseed lipids kg^{-1} feed DM appears to be the most promising approach to achieve a clear CH_4 reduction with marginal effects on digestion and intake [6,10,11]. In addition, extruded linseed increases the proportion of n–3 polyunsaturated fatty acids (FAs) in milk fat, which is considered beneficial to human health (e.g., [12]). Therefore, some retailers in France and Switzerland are currently paying higher prices to farmers for milk with a higher proportion of n–3 FAs in milk fat. Label programs may require evidence of CH_4 reduction at the herd level or changes in the FA profile of bulk milk.

Traditionally, CH_4 emissions are measured continuously on individual animals to quantify the reduction potential of dietary interventions. Continuous measurements in respiration chambers determine total emissions, but locomotion is restrained. The sulfur hexafluoride (SF_6) bolus technique allows free locomotion of the cows during continuous measurements [13] but detects only exhaled CH_4. These limitations do not apply to group measurements, which concomitantly allow more natural behavior of the cows [14,15] and total emission assessment. Furthermore, climate and housing-based effects, such as slurry storage and area soiling, on CH_4 emissions are considered. These differences could promote or reduce the effects of diet on CH_4 emissions. However, studies at the group level in relation to dietary interventions are rare, and none of the reports studied the effect of linseed supplementation on total CH_4 emission at the group level. Schmithausen et al. [16] quantified CH_4 emissions from cows that were fed diets with different levels of condensed tannins in a naturally ventilated experimental dairy housing using the carbon dioxide (CO_2) balance method. Several studies compared CH_4 emissions obtained on individual animals in respiration chambers with those determined at the practical scale using either SF_6 permeation tubes (e.g., [17]), the GreenFeed system (e.g., [18]) or other ventilated hoods (e.g., [13]). To our knowledge, there are currently no reported data on CH_4 emissions for individual animals in respiration chambers vs. groups of animals in naturally ventilated dairy housing.

Therefore, this study aimed to evaluate the effect of extruded linseed supplementation on the CH_4 emission and milk FA composition both at the group level on a practical scale in naturally ventilated housing and in respiration chambers individually. The following hypothesis was tested: moderate levels of lipids exert effects on the CH_4 emission and milk FA composition, which are detectable and similar in extent at the individual animal and group levels. In addition, the effect on feeding behavior was investigated.

2. Materials and Methods

The experiment was carried out from December 2015 to March 2016. Group level measurements were carried out in a naturally ventilated experimental loose-housing system for dairy cows at Agroscope (Waldegg, Switzerland; 47°29′22″ N, 8°55′10″ E, 550 m above sea level). Then, cows were measured individually at the AgroVet-Strickhof in a tied-housing barn (Wülflingen, Switzerland; 47°30′58″ N, 8°41′49″ E) and in respiration chambers (Eschikon, Switzerland; 47°26′53″ N, 8°40′43″ E). The cows were transported over a distance of 30 km from the loose- to the tied-housing and over 13 km from the tied-housing to the respiration chambers. The experimental protocol complied with the Swiss legislation on Animal Welfare and was approved (ZH091/15) by the Cantonal Veterinary Office of Zürich.

2.1. Animals and Experimental Design

2.1.1. Measurements at the Group Level in the Loose-Housing System

For group measurements, 40 primiparous and multiparous lactating cows of Brown Swiss and Swiss Fleckvieh breeds were selected and stratified into two groups in a completely randomized manner, balanced as best as possible for lactation number, breed, body weight (BW), days in milk, and milk yield. The control (C) group consisted of 14 Brown Swiss and six Swiss Fleckvieh cows, and the linseed (L) group consisted of 11 Brown Swiss and nine Swiss Fleckvieh cows. Group level measurements were conducted over 25 days, of which 21 days were used to allow the cows to adapt to the housing and diet, and emissions were measured for four consecutive days. The adaptation period was initiated with a proportionate increase of either corn flour (in the C cows) or linseed (in the L cows) supplementation at the group level during the first 5 days at 0.2, 0.4, 0.6, 0.8, and 1.0 of the final amounts. The feed intake of the two groups was assessed at the group level, whereas feeding behavior (eating and rumination time) was determined individually for ten cows per group.

2.1.2. Measurements at the Individual Level in the Tied-Housing System and in the Respiration Chambers

We randomly selected 2×6 exclusively multiparous cows from the 2×20 cows in the experimental groups for the experiments where the cows were housed in a tied-housing system. All but one cow were of the Brown Swiss breed. Individual measurements were obtained directly after the measurements at the group level. Cows were given the C and L diets (details on diet ingredients and nutrient composition are given in Section 2.2 and Table 1). The mixed ration was prepared daily at Agroscope (loose housing) and transported to the tied stall barn and respiration chambers. A period of 7 days was set to allow the cows to adapt to the tied-housing system. This was followed by a 7–day period of data and sample collection. Sampling was carried out in two blocks of 2×3 cows each. Within the sampling period, cows were moved in pairs, with one cow per experimental diet, into respiration chambers for an adaptation period of 6 h and gas exchange measurement period of 2 days. The tied-housing systems in the tied stall barn and in the respiration chambers were equipped with rubber mats and feces collection trays. Bedding material was omitted during sampling to avoid contamination by feces.

2.2. Feeding

The mixed ration consisted of grass silage, corn silage, hay, beet pulp, and concentrate (Table 1). The linseed product used (TradiLin 135; Trinova AG, Wangen, Switzerland) consisted of an extruded mixture of linseed and mill byproducts (0.6:0.4). The latter absorbs the oil released from the linseed during processing. This linseed product was added on top of the mixed ration, providing 112 g TradiLin 135 per kg of DM, equivalent to 67 g linseed and 29 g (as analyzed) of extra lipids. The cows in the C group received corn flour instead of linseed in an iso-energetic manner. The amount of linseed and corn flour was adjusted to the amount of mixed ration, ensuring that the relative intake of linseed and corn flour remained the same. In the loose-housing system, cows were fixed in the feeding barrier during the first 30 min of new feed provision to enable complete consumption of the linseed or corn flour without disturbance by dominant animals. In addition, concentrates that were rich in energy and protein were allocated according to milk yield following recommendations by Agroscope [19] by an automatic feeder (loose housing) or given manually after each feeding event (tied housing) (Table 1). Each animal received 50 g·day^{-1} of sodium chloride and 100 g·day^{-1} of a commercial mineral-vitamin mixture (Minex 976, UFA AG, Sursee, Switzerland) consisting (per kg) of 100 g calcium, 80 g phosphorus, 75 g magnesium, 20 g sodium, 4 g zinc, 3.8 g manganese, 1 g cooper, 1 g sulfur, 100 mg iodine, 40 mg cobalt, 40 mg selenium, 1,000,000 IU vitamin A, 200,000 IU.

Table 1. Ingredients and their nutrient compositions and realized average intakes with the control and the extruded linseed diets [7] in the loose and tied-housing systems (means ± standard deviation).

Component	Mixed Part of the Diet				Supplementary Concentrates				Linseed
	Grass Silage	Corn Silage	Hay	Beet Pulp	Concentrate 1	Rich in Protein [2]	Rich in Energy [3]	Corn Flour	Product
Dry matter (g·kg^{-1})	459	417	899	260	882	863	865	859	933
Nutrients (g·kg^{-1} of dry matter)									
Organic matter	917	970	914	900	917	919	923	989	957
Crude protein	94	62	109	83	563	413	182	79	180
Neutral detergent fiber	463	328	506	435	221	127	161	161	398
Acid detergent fiber	303	193	295	230	135	95	85	52	215
Acid detergent lignin	37.7	33.8	37.1	29.2	50.4	32.5	32.5	22.1	103
Ether extract [4]	20.8	29.6	17.5	3.9	42.0	37.4	62.1	35.7	255
Realized DM intake (kg·day^{-1} per cow)									
Loose housing (group observations) [5]									
Control	6.38 ± 0.13	3.30 ± 0.07	2.32 ± 0.05	0.94 ± 0.02	1.50 ± 0.03	1.55 ± 0.53	1.05 ± 1.72	2.16 ± 0.05	-
Linseed	6.40 ± 0.14	3.31 ± 0.07	2.32 ± 0.05	0.94 ± 0.02	1.50 ± 0.03	1.73	1.14 ± 1.39	-	2.19 ± 0.05
Tied housing (individual observations) [6]									
Control	6.60 ± 0.80	3.41 ± 0.42	2.40 ± 0.29	0.97 ± 0.12	1.55 ± 0.19	1.73	1.14 ± 1.62	2.23 ± 0.27	-
Linseed	6.40 ± 0.82	3.31 ± 0.43	2.33 ± 0.30	0.94 ± 0.12	1.51 ± 0.19	1.73	1.15 ± 1.67	-	2.19 ± 0.28

[1] Composed of soybean meal, dried distillers grains with solubles, rapeseed cake, rapeseed meal, milling byproducts, malt culms, corn gluten meal, and minerals (UFA AG, Herzogenbuchsee, Switzerland). [2] Composed of soybean meal, corn gluten meal, rapeseed cake, dextrose, sugar beet molasses, vitamin–mineral mixture, fruit syrup, vegetable fat, and minerals (Thurtalfutter AG, Frauenfeld, Switzerland). [3] Composed of barley, triticale, maize, rapeseed cake, soybean meal, peas, wheat, oat, sugar beet molasses, vegetable fat, vitamin-mineral mixture, corn gluten meal, fruit syrup, and minerals (Thurtalfutter AG, Frauenfeld, Switzerland). [4] Proportions of the fatty acids as analyzed (g·100 g^{-1} of total analyzed fatty acids) C16:0, C18:0, C18:1 n-9, C18:2 n-6, and C18:3 n-3 were 16.4, 1.98, 15.5, 33.0, and 24.8 in the total mixed part of the diet; 10.9, 2.18, 26.0, 54.0, and 4.51 in corn flour; and 7.04, 4.10, 16.8, 17.7, and 52.1 in linseed, respectively. [5] Average of group observations in the loose-housing system (two treatments × 4 days). Dry matter intake (DMI) per cow was calculated by dividing the group DMI by the number of cows. [6] Average of individual observations in the tied-housing system (two treatments × six animals). [7] Contents (g/kg dry matter) of organic matter, crude protein, neutral detergent fiber, acid detergent fiber, acid detergent lignin, and ether extract were 934, 153, 351, 212, 35, and 28 for the complete control diet and 930, 165, 377, 230, 44, and 53 for the complete linseed diet, respectively.

Vitamin D_3, 3 g vitamin E, and 100 mg biotin. New portions of diets provided with ad libitum access were offered at 16:45 after milking at the group level. Specialized equipment automatically moved feed towards the cows 18 times a day. At the individual level, feeding took place at 11:00 (10:00 in the respiration chambers), 13:00, 16:30, 23:00, and 05:30. The feed amounts offered to the two groups and to the individual cows were recorded daily. Leftovers were removed and weighed once daily. All the cows were given permanent access to water. The cows were milked at 05:30 and 16:30 at all experimental sites.

2.3. Measurement of Methane and Carbon Dioxide Emissions

2.3.1. Emission Measurement in the Experimental Dairy Loose-Housing System (Group Level)

At the group level, emissions were measured as described in detail by Mohn et al. [20]. Briefly, the housing system provided two spatially separated experimental sections with three rows of cubicles with straw. A milking parlor and analytical devices were located between the two sections. The covered underground slurry storages were separated for each section. Management routines such as milking, feeding, and 12 dung removals per day remained the same for both groups. Each group was fed separately to avoid cross-contamination. Curtains in the facades remained closed, ensuring that wind and temperature conditions in the two sections were equivalent. Adequate ventilation was ensured by spaced boards in the upper part of the longitudinal facades. There were no other natural (e.g., ruminants) or technical emission sources near the housing. The position of the housing axis was orthogonal to prevailing wind directions, minimizing cross-contamination between the experimental sections. The dual tracer ratio method [20,21] was applied to quantify emissions for each experimental section independently and detect potential cross-contamination. It involved constant dosing of the tracer gases SF_6 and trifluoromethyl sulfur pentafluoride (SF_5CF_3), one per section, at floor level using mass flow controllers (Contrec AG, Switzerland) to regulate the total flow and critical steel orifices to achieve homogenous spatial distribution. Representative air sampling in each section was accomplished with critical glass orifices (250 μm in diameter 2.5 m above the ground; Thermo-Instruments, Germany, and Louwers, The Netherlands). Concentrations of the tracer gases and target gases (CH_4 and CO_2) were analyzed in real time by gas chromatography with electron capture detection (GC-ECD, model 7890A, Agilent Technologies AG, Switzerland) and by cavity ring-down spectrometry (CRDS, CH_4, CO_2, model G2301, Picarro Inc., USA). More details on the implemented analytical technique and its performance with respect to suitability for point/areal sources, sensitivity, and uncertainty have been described in Mohn et al. [20]. The applied measurement sequence provided emission data with a temporal resolution of 10 min per section. Milking times were excluded from the analysis, as only part of the group was present in the sections during these periods.

2.3.2. Emission Measurement in the Respiration Chambers (Individual Level)

The two open-circuit respiration chambers used in the present study had a volume of 19.3 m^3 each [22]. The chambers were air-conditioned to 15 °C and 55% relative humidity at an air pressure of −60 Pa (relative to ambient). The airflow was set to 700 $L \cdot min^{-1}$ (Promethion FG 1000 flow generator, Sable Systems Europe GmbH, Berlin, Germany). A light:dark cycle of 16:8 was used. The CH_4, oxygen (O_2), and CO_2 concentrations in the spent air were measured with a Promethion GA 4 gas analyzer (Sable Systems, Las Vegas, USA), alternating between in- and outgoing air at 1-min intervals. Calibrations of the gas analyzer were initiated automatically before each 2-day measurement period using pure (99.999%) nitrogen (N_2) and a reference gas (19.8% O_2, 1.0% CO_2, 0.1% CH_4, in N_2). Recovery, averaging at 105%, was assessed before and after the two experimental blocks by burning propane gas. For statistical evaluation, emissions were averaged across the 2 days of measurement. The chambers were entered through the airlock for feeding, milking, and urine collection during the measurements. For feces collection, the back door was opened after 24 and 48 h of measurement.

2.4. Performance and Feeding Behavior (Individual Measurements in the Loose and Tied-Housing Systems)

We individually assessed the cows' BWs after evening milking on the days before and after the 4-day measuring period in the loose housing and on the first and the last day of the 7-day sampling period in the tied-housing system. The BW was directly measured with a cattle balance (Modell FX21, Iconix, Rotorua, New Zealand) in the tied housing and estimated from girth size determinations with a measuring tape in the loose- (and tied-) housing system. The close correlation of 0.88 (n = 24, $p < 0.001$) determined between actual weights and estimates with the measuring tape allowed the application of the latter in the loose housing. During all measurement periods, milk yield was recorded individually at each milking using a milk meter (EasyFlow, Fullwood, Ellesmere, United Kingdom) in the loose housing and by weighing the buckets in the tied housing. Chewing and ruminating activity was monitored with nose band pressure sensors of the RumiWatch System equipped with the Converter V.0.7.3.36 (Itin+Hoch GmbH, Liestal, Switzerland) as described by Werner et al. [23]. This was accomplished for ten multiparous cows per group, including those subjected later to individual measurements during the 4-day measuring period at the group level and at the individual level for 5 days during the sampling period (excluding days were the cows were moved). The times of eating and ruminating and the number of chews during eating and ruminating were evaluated. In the loose-housing system, data from one cow in the C treatment group were excluded due to technical problems with the sensor.

2.5. Sampling

Grass silage, corn silage, and hay samples were collected at six times, i.e., at the beginning and end of the measurements in the loose- and tied-housing systems (separately in the two animal blocks). Samples of the sugar beet pulp and the three types of concentrates were taken twice during the entire study. The linseed and corn flour were sampled daily and pooled for the measurement period in the loose housing and analyzed separately for the two blocks of animals in the tied-housing system. In the loose housing, samples of leftovers were collected daily per group. In the tied-housing system, the daily feed leftovers were sampled individually and in proportion to the total amount of leftover. These samples were immediately frozen. In the tied-housing system, urine was separated from the feces by attaching urinals fixed by hook-and-loop fastener straps around the vulva of the cows. These straps were glued (ergo 5011, Kisling, Wetzikon, Switzerland) onto shaved skin. The urine was collected in a large and a small container, the latter containing 5 M sulfuric acid to avoid N losses. The total amount of feces and urine was recorded during the 7–day sampling period. Urine and feces subsamples were collected once and twice per day, respectively; samples were a constant fraction of the amounts excreted. The subsamples were frozen and later pooled for each cow. Before analysis, the feed, leftover feed, and fecal samples were dried at 60 °C until a constant mass was achieved and then ground to pass a 1-mm screen of either a cutting mill or a centrifugal mill (extruded linseed, corn flour, and concentrates). In the loose-housing system, milk samples were obtained individually at each milking on sampling days 1 and 4. A constant fraction of the yield at each milking was collected and the samples were pooled per group. In the tied-housing system, samples were collected on each sampling day from each milking, and a constant fraction of the yield at each milking was composited for each cow. Aliquots of individual and pooled milk samples were either frozen without additive or preserved with Bronopol® (D&F Inc., Dublin, CA, USA).

2.6. Laboratory Analyses

Feeds, leftovers, and feces were analyzed according to standard procedures [24]. The DM and total ash contents (AOAC method 942.05) were determined with a thermo-gravimetric device (TGA 701, Leco Corporation, St Joseph, MI). For fiber analyses (AOAC method 973.18), a Fibertec System M (1020 Hot Extractor and 1021 Cold Extractor; Tecator, Foss Hillerød, Denmark) was used. Neutral detergent fiber (NDF) analysis was performed with the addition of heat stable α-amylase, but without sodium sulfite [25]. Acid detergent lignin was determined sequentially after acid detergent fiber (ADF) analysis by treatment with sulfuric acid (72%) for 3 h. Fiber data were corrected for ash content. Combustion energy in the feed and feces items to calculate energy intake and energy excretion with feces was quantified by a bomb calorimeter (C7000, IKA-Werke GmbH & Co. KG, Staufen, Germany). Feeds, leftovers, previously frozen milk, non-dried feces, and acidified urine were analyzed for N content (AOAC method 968.06) with a C/N analyzer (TruMac CN, Leco Corporation, St. Joseph, MI). Crude protein in the feed was defined as 6.25 × nitrogen (N). Carbon (C) content in the non-acidified urine was determined on the same C/N analyzer in order to be able to calculate urine energy (as described in Grandl et al. [26]). Ether extract (AOAC method 963.15) of feed items, leftovers, and feces was determined with a Soxhlet extraction system (model Extraktionsapparatur B-811, Büchi, Flawil, Switzerland). Bronopol-preserved milk samples were analyzed for fat, protein, lactose, and milk urea nitrogen (MUN) contents using a Fourier-transform infrared spectrophotometer (FTIR; MilkoScan FT6000, Foss, Hillerød, Denmark) at the Swiss routine milk analysis laboratory (Suisselab AG, Zollikofen, Switzerland). The MUN content was also determined in previously frozen milk by an enzymatic method [27].

For quantification of the proportions of FAs in the lipids of extruded linseed and corn flour, FAs were extracted by a solvent extractor (ASE 200, Dionex Corporation, Sunnyvale, CA) using a hexane:propane-2-ol mixture (3:2 v/v). The FAs were transformed to FA methyl esters (FAME) according to the International Union of Pure and Applied Chemistry (IUPAC) [28] method 2.301. Cleaning was performed as described in Wettstein et al. [29]. For FAME analysis, a gas chromatograph (model HP 6890 equipped with a flame ionization detector, Hewlett Packard, Palo Alto, CA, USA) and a CP7421 column (200 m × 0.25 mm, 0.25 μm; Varian Inc., Darmstadt, Germany) were used. Split injection (1:5) was applied. The internal FA standard was C11:0 (Fluka, Steinheim, Germany), and the external standard for the response factor was sunflower oil. A volume of 1 μL was injected with a constant hydrogen flow of 1.7 mL·min^{-1}. The temperature program was set to 170 °C for 60 min, increased to 230 °C at a rate of 5 °C·min^{-1}, ramped for 32 min, increased to 250 °C at a rate of 5 °C·min^{-1}, and ramped for 15 min. For analysis of the FA in milk, samples were thawed and gently mixed to disperse milk fat. Internal standards (5 mL of *n*-heptane containing triundecanoin, tetradecenoic methylate, and trivaleranoin) were mixed with 0.5 mL of milk. Cold transesterification to FAME was done with sodium methylate [30]. The response factors obtained from C6:0, C13:0, and C19:0 triglyceride standards were used to adjust individual FA data. The same gas chromatograph and column and operational conditions were applied as for FA from feeds, except for the temperature regime which was as follows: initial temperature of 60 °C, ramped for 12 min, increased to 170 °C at the rate of 5 °C·min^{-1}, ramped for 60 min, increased to 250 °C at the rate of 5 °C·min^{-1}, and ramped for 20 min. The FAME was identified using a Supelco 37 component standard (Supelco Inc., Bellefonte, PA, USA). Peaks were identified using chromatograms from Collomb and Bühler [31].

2.7. Calculations and Statistical Analysis

In the loose- and tied-housing systems, the realized dry matter intake (DMI) of each component of the mixed ration was calculated based on the assumptions that the proportions of each component were the same in the offered feed and in the leftovers, and that the extruded linseed and corn flour were consumed completely (as confirmed by observations). One livestock unit (LU), defined as 500 kg BW [32], was used to adjust for differences in BW resulting from using only the multiparous cows in the tied stall system. The energy-corrected milk (ECM) (kg·d^{-1}) was calculated as milk (kg·day^{-1}) × [0.038 × fat (g·kg^{-1}) + 0.024 × protein (g·kg^{-1}) + 0.017 × lactose (g·kg^{-1})]/3.14. [19]. The efficiency traits calculated were milk production efficiency (ECM/BW) and feed conversion efficiency (ECM/DMI). In the loose-housing system, the individual DMI was calculated by dividing the measured total DMI of each group by the number of animals in the group (n = 20). The energy-related variables were calculated with the standard equations used in energy balance studies as listed in detail in Grandl et al. [26], where the two different methods of calculating the proportionate utilization of metabolizable energy (ME) for milk energy formation as used in the present study are described.

In the loose-housing system, the mass flow of the target gases ($\dot{m}_{target}$) was calculated from the ratio between the background-corrected target gas concentration (c_{target}) and tracer gas concentration (c_{tracer}) multiplied with the known mass flow of the tracer gas ($\dot{m}_{tracer}$), as $\dot{m}_{target} = \dot{m}_{tracer} \times c_{target}/c_{tracer}$. For the loose- and tied-housing systems, the CH_4 yield was described as the absolute CH_4 amount in relation to the DMI, digestible organic matter (OM) intake, digestible NDF intake (dNDFI), and gross energy (GE) intake (GEI; CH_4 conversion factor, Y_m). The methane emission intensity was assessed by relating CH_4 to ECM and BW. Methane emissions measured during the 2-day period in the respiration chambers under the tied-housing conditions were related to intake and ECM yield obtained across the entire 7-day sampling periods in the tied stall barn.

Data were subjected to analysis of variance to compare dietary treatments with TIBCO Spotfire+®(version 8.2 for Windows), considering diet as the fixed effect and either day (group level; [33,34]) or animal (individual level) as the experimental units. Spearman correlation coefficients were calculated for some variables. The distributional data properties were visually checked using a normal quantile-quantile plot of residuals. Significance was set at $p < 0.05$, and tendency at $0.05 \leq p < 0.10$. Descriptive statistics were used to evaluate the data and effects of linseed supplementation in the loose- and tied-housing systems.

3. Results

3.1. Individual Performance, Digestibility, and Feeding Behavior (Loose- and Tied-Housing Systems)

The average BW did not differ between the C and L cows in any of the housing systems, but the average BW was approximately 80 kg higher in the subgroup of the groups used for the individual assessment (Table 2). The daily ECM and milk fat yields did not differ between C and L cows but were about 20% lower on average in the cows selected for the tied stall system compared with cows kept in the loose-housing system. Furthermore, there were no significant differences in the DM and OM intakes between the C and L cows. For C cows, the DMI was about 1 kg·day^{-1} higher at the individual level (tied housing) than at the group level (loose housing). The NDF and ADF intakes were higher ($p < 0.05$) in L cows at the group level and the ADF intake was higher at the individual level in L cows than in C cows. The NDF intake was slightly higher at the group level than at the individual level. Milk yield (per BW) and feed conversion efficiencies were not different between the L cows and C cows.

Table 2. Performance of cows from control and linseed treatments measured in the loose- and tied-housing phases (means ± standard deviation).

Housing	Loose (Group)			Tied (Individual)		
Item	Control	Linseed	*p*–Value	Control	Linseed	*p*–Value
Body weight (BW, kg) [1]	681 ± 93	682 ± 84	0.971	756 ± 71	761 ± 38	0.883
Energy-corrected milk (ECM, kg·day^{-1}) [1]	33.2 ± 7.9	36.3 ± 10.0	0.294	26.4 ± 5.7	30.5 ± 9.3	0.386
Milk fat (kg·day^{-1}) [1]	1.40 ± 0.34	1.46 ± 0.51	0.661	1.08 ± 0.21	1.16 ± 0.38	0.652
Intake (kg·day^{-1}) [2]						
Dry matter (DMI)	19.3 ± 0.4	19.5 ± 0.4	0.417	20.1 ± 1.6	19.6 ± 2.0	0.665
Organic matter	17.9 ± 0.4	18.0 ± 0.4	0.754	18.7 ± 1.5	18.1 ± 1.9	0.564
Neutral detergent fiber	6.58 ± 0.14	7.14 ± 0.17	0.002	6.91 ± 0.29	7.24 ± 0.36	0.108
Acid detergent fiber	4.01 ± 0.09	4.40 ± 0.11	0.001	4.15 ± 0.16	4.39 ± 0.20	0.045
Efficiency						
ECM per BW (kg·100 kg^{-1}) [1]	4.93 ± 0.12	5.37 ± 0.14	0.312	3.55 ± 0.97	4.02 ± 1.25	0.488
ECM per BW (g·kg$^{-0.75}$) [1]	251 ± 59	273 ± 71	0.292	185 ± 48	211 ± 65	0.458
ECM [1] per DMI [2] (kg·kg^{-1})	1.72 ± 0.41	1.86 ± 0.51	0.111	1.31 ± 0.23	1.53 ± 0.32	0.204

[1] Individual observations in the loose- (2 treatments × 20 animals) and tied- (2 treatments × 6 animals) housing system. [2] Group observations (two treatments × 4 days) in the loose housing and individual observations (2 treatments × 6 animals) in the tied-housing system.

The L cows spent more time eating (min d^{-1}, $p < 0.05$; min kg^{-1} DMI, $p < 0.10$) in the loose housing but the same time eating in the tied-housing system as the C cows (Table 3). The rumination time was longer for L cows than C cows (min·d^{-1}, $p < 0.10$; min·kg^{-1} DMI and chews per kg^{-1} DMI, $p < 0.01$) in the tied-housing system but similarly long for L cows in the loose-housing system. The daily eating time and number of eating chews (per unit DMI) were on average lower by about 4% and the rumination time and number of ruminating chews (per unit DMI) were lower by about 18% in the tied-stall system than in the loose-housing system. Considering the eating behavior exclusively from the 2 days when cows were in the respiration chambers, the cows spent an average of 5% less time eating and ruminating, and made 9% less chews than in the entire 7-day sampling period in the tied-housing system (data not shown), and there was no difference between the two diet treatments ($p > 0.10$). The digestibility of OM, NDF, and ADF did not differ between the L and C cows at the individual level. Similarly, diet did not affect the absolute and relative N balance traits and N utilization (see Appendix A, Table A1). Only the proportion of urinary N in the total manure N tended to be greater ($p < 0.10$) for the L cows than for the C cows. Energy intake (GE, digestible energy, and ME) did not differ between diets (see Appendix A, Table A2). There was no effect of linseed on energy losses, except for the energy lost as CH_4, which tended ($p < 0.10$) to be lower by 9% in L cows than in C cows. Concomitantly, the intake, retention, turnover, and utilization of energy did not differ between the two dietary treatment groups.

Table 3. Feeding behavior, digestibility, and gaseous emissions of cows from control and linseed treatments determined in the loose- and tied-housing phases (means ± standard deviation).

Housing	Loose (Group)			Tied (Individual)		
Item	Control	Linseed	*p*–Value	Control	Linseed	*p*–Value
Eating [1]						
min	406 ± 27	455 ± 15	0.019	422 ± 96	445 ± 36	0.354
min·kg^{-1} DMI [2]	21.1 ± 1.64	23.3 ± 0.95	0.051	20.4 ± 4.9	22.1 ± 3.0	0.219
Chews·kg^{-1} DMI	1439 ± 142	1598 ± 58	0.072	1386 ± 386	1536 ± 273	0.182
Rumination [1]						
min	466 ± 38	497 ± 9	0.170	411 ± 56	445 ± 34	0.078
min·kg^{-1} DMI	24.2 ± 2.2	25.4 ± 0.9	0.275	19.8 ± 2.2	22.0 ± 1.7	0.008
Chews·kg^{-1} DMI	1546 ± 146	1664 ± 66	0.162	1197 ± 172	1386 ± 188	0.006
Digestibility (%)						
Organic matter (OM)	-	-	-	76.1 ± 2.9	75.9 ± 3.1	0.935
Neutral detergent fiber (NDF)	-	-	-	61.5 ± 6.2	63.2 ± 4.1	0.580
Acid detergent fiber	-	-	-	63.2 ± 6.5	63.9 ± 4.2	0.829
Carbon dioxide (kg·LU^{-1}) [3,4]	9.18 ± 0.35	9.20 ± 0.55	0.952	9.07 ± 1.02	8.54 ± 0.60	0.295
Methane yield [4]						
g·kg^{-1} DMI	22.3 ± 1.6	21.4 ± 2.1	0.496	24.6 ± 3.1	22.9 ± 2.1	0.285
g·kg^{-1} digestible OM intake	-	-	-	34.7 ± 3.6	32.7 ± 3.6	0.361
g·kg^{-1} digestible NDF intake	-	-	-	116.3 ± 10.0	97.9 ± 8.4	0.006
kJ·MJ^{-1} gross energy intake (Y_m)	69.4 ± 5.0	64.1 ± 6.2	0.232	76.4 ± 9.7	68.6 ± 6.4	0.131
Methane emission intensity [4]						
g·kg^{-1} ECM [5]	13.6 ± 1.2	12.6 ± 1.2	0.266	19.2 ± 3.8	15.7 ± 4.4	0.165
g·LU$^{-1;4}$	321 ± 5.1	301 ± 6.7	0.292	327 ± 44	294 ± 30	0.155

[1] Individual observations in the loose (two treatments × 20 animals) and tied (two treatments × six animals) housing system. [2] Dry matter intake. [3] Livestock unit equivalent to 500 kg body weight. [4] Group observations (two treatments × 4 days) in the loose housing and individual observations (two treatments × six animals) in the tied-housing system. [5] Energy-corrected milk.

3.2. Methane and Carbon Dioxide Emissions (Group and Individual Level)

The carbon dioxide emissions (expressed per LU) did not differ between dietary treatments (L vs. C cows) but were on average 5% lower at the individual level than at the group level. In the loose (group) and tied (individual) housings, the CH_4 emissions followed the typical diurnal pattern (Figure 1), with characteristic peaks after feeding (and milking) with both diets. The average CH_4 production per animal (g·day^{-1}) was 4% lower among L cows than among C cows at the group level (414 ± 5.7 vs. 431 ± 9.0) and tended ($p < 0.10$) to be lower by 9% at the individual level (447 ± 34 vs. 491 ± 40; data not shown in table). The trends for the lower CH_4 production by L cows than the C cows were not only observed for temporal averages but also could be seen as a distinct shift in the diurnal pattern of CH_4 emission (Figure 1). There were no significant differences in the CH_4 traits related to intakes of DM, digestible OM or GE between the two dietary treatments (Table 3). However, the CH_4 yield per unit of dNDFI was lower ($p < 0.01$) in the L cows than in the C cows. Methane emission intensities did not differ between the diets at both the group and individual levels. The Y_m and DMI were 1.1-fold higher at the individual level than the group level. The methane emission intensity per unit of ECM was 1.3-fold higher at the individual level than at the group level, whereas the CH_4 production per LU was similar.

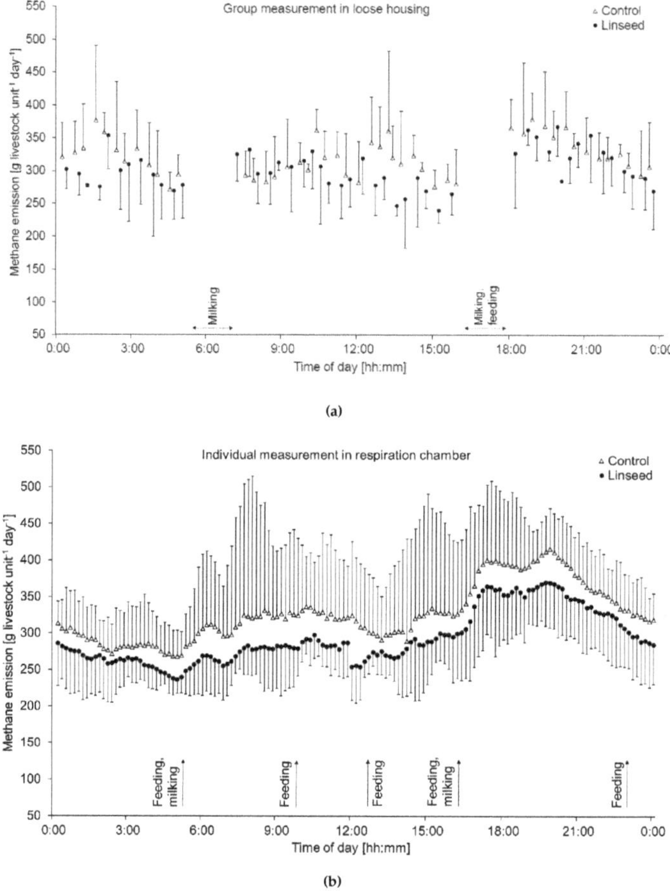

Figure 1. Diurnal pattern of methane emission per livestock unit (i.e., adjusted to 500 kg of body weight) of lactating dairy cows fed either a control or a linseed-supplemented diet measured (**a**) at the group level (mean of 4-day measurements in a group of 20 animals per dietary treatment) and (**b**) at the individual animal level (mean of 2-day measurements of six animals per diet). The whiskers in the graph extend to minimum (linseed) and maximum (control). The arrows indicate feeding and milking times.

3.3. Milk Composition (Group and Individual Level)

The milk fat and protein contents were lower ($p < 0.01$) in L cows than in the C cows both at the group (bulk milk) and individual levels (Table 4). At the group level, L cows exhibited a higher ($p < 0.01$) milk lactose content than C cows. The milk fat content of the cows selected for the tied stall system was on average slightly higher than that of all cows in the loose-housing system, but the protein and lactose contents were similar at both the individual and group levels. Enzymatic analysis revealed that MUN was significantly higher in the L cows ($p < 0.05$) than in the C cows at both the individual and group levels. The MUN analyzed by FTIR did not reveal these differences. Independent of the type of analysis, the MUN was slightly greater in the cows kept in the tied stall system than in all cows in the loose-housing system.

Table 4. Milk composition of cows from the control and linseed treatment groups measured in the loose[1]- and tied[2]-housing phases (means ± standard deviation).

Housing	Loose (Group) [1]			Tied (Individual) [2]		
Item	Control	Linseed	*p*–Value	Control	Linseed	*p*–Value
Fat (g·kg^{-1} milk)	42.9 ± 2.5	34.3 ± 2.6	0.003	45.6 ± 6.2	36.5 ± 3.1	0.009
Protein (g·kg^{-1} milk)	37.7 ± 0.4	33.7 ± 0.3	<0.001	38.8 ± 3.5	33.1 ± 1.8	0.006
Lactose (g·kg^{-1} milk)	47.6 ± 0.4	49.1 ± 0.2	<0.001	48.1 ± 0.9	48.7 ± 0.9	0.248
MUN (mg·100^{-1} mL milk)						
Enzymatic	10.9 ± 0.5	13.5 ± 2.0	0.041	12.1 ± 2.0	15.6 ± 2.4	0.022
FTIR[3]	10.8 ± 1.0	10.5 ± 1.7	0.777	12.0 ± 2.1	12.8 ± 2.0	0.521
Fatty acids (FA; g 100 g^{-1} of total FA)						
C18:0	7.24 ± 0.07	8.89 ± 0.19	<0.001	8.99 ± 1.64	11.02 ± 1.19	0.034
C18:1 *trans*-6, *trans*-8	0.25 ± 0.01	0.56 ± 0.05	<0.001	0.34 ± 0.05	0.67 ± 0.11	<0.001
C18:1 *trans*-9	0.19 ± 0.01	0.44 ± 0.03	<0.001	0.18 ± 0.02	0.44 ± 0.06	<0.001
C18:1 *trans*-10	0.34 ± 0.00	0.75 ± 0.02	<0.001	0.42 ± 0.06	1.05 ± 0.55	0.018
C18:1 *trans*-11	0.67 ± 0.01	4.39 ± 0.27	<0.001	0.80 ± 0.29	4.37 ± 1.37	<0.001
C18:1 *trans*-12	0.045 ± 0.004	0.105 ± 0.004	<0.001	0.043 ± 0.013	0.097 ± 0.044	0.016
C18:1 *trans*-13, *trans*-14, cis-6, cis-8	0.31 ± 0.00	0.72 ± 0.04	<0.001	0.36 ± 0.04	0.83 ± 0.06	<0.001
C18:1 cis-9	16.6 ± 0.6	19.4 ± 0.4	<0.001	17.4 ± 2.1	24.0 ± 3.3	0.002
C18:1 *cis*-10	0.105 ± 0.025	0.263 ± 0.146	0.076	0.078 ± 0.017	0.137 ± 0.037	0.005
C18:1 *cis*-11	0.67 ± 0.03	1.07 ± 0.08	<0.001	0.74 ± 0.13	1.28 ± 0.12	<0.001
C18:1 *cis*-12	0.208 ± 0.010	0.514 ± 0.028	<0.001	0.259 ± 0.035	0.490 ± 0.072	<0.001
C18:1 *cis*-13	0.077 ± 0.002	0.185 ± 0.019	<0.001	0.092 ± 0.025	0.212 ± 0.019	<0.001
C18:1 *cis*-14, *trans*-16	0.17 ± 0.01	0.33 ± 0.04	<0.001	0.23 ± 0.03	0.54 ± 0.16	<0.001
C18:2 n-6 *trans*	0.11 ± 0.01	0.54 ± 0.07	<0.001	0.12 ± 0.03	0.51 ± 0.16	<0.001
C18:2 *cis*-9, *trans*-13, *trans*-8, *cis*-12	0.22 ± 0.02	0.81 ± 0.08	<0.001	0.26 ± 0.04	1.21 ± 0.30	<0.001
C18:2 *cis*-9, *trans*-12	0.068 ± 0.013	0.178 ± 0.010	<0.001	0.079 ± 0.010	0.257 ± 0.077	<0.001
C18:2 *trans*-11, *cis*-15, *trans*-9, *cis*-12	0.080 ± 0.013	1.282 ± 0.109	<0.001	0.177 ± 0.075	1.482 ± 0.349	<0.001
C18:2 n-6 *cis*	1.95 ± 0.05	1.93 ± 0.05	0.599	2.05 ± 0.23	1.92 ± 0.14	0.278
C18:2 *cis*-9, *cis*-15	0.11 ± 0.01	0.18 ± 0.01	<0.001	0.14 ± 0.02	0.20 ± 0.20	<0.001
C18:2 *cis*-9, *trans*-11	0.21 ± 0.02	0.32 ± 0.07	0.022	0.33 ± 0.07	0.36 ± 0.09	0.523
C18:2 *cis*-9, *cis*-11	0.054 ± 0.003	0.241 ± 0.016	<0.001	0.064 ± 0.014	0.250 ± 0.047	<0.001
C18:2 *trans*-9, *trans*-11	0.015 ± 0.001	0.148 ± 0.009	<0.001	0.015 ± 0.007	0.158 ± 0.046	<0.001
C18:3 n-6	0.062 ± 0.004	0.087 ± 0.012	0.008	0.065 ± 0.017	0.048 ± 0.022	0.181
C18:3 n-3	0.45 ± 0.01	1.10 ± 0.05	<0.001	0.49 ± 0.09	1.23 ± 0.06	<0.001
Σ C18 FA	30.2 ± 0.6	44.4 ± 0.8	<0.001	33.7 ± 4.0	52.8 ± 4.9	<0.001
Σ SFA	72.2 ± 0.7	60.4 ± 0.7	<0.001	70.8 ± 3.3	54.5 ± 4.7	<0.001
Σ MUFA	24.0 ± 0.6	32.4 ± 0.7	<0.001	25.0 ± 2.7	37.6 ± 4.1	<0.001
Σ PUFA	3.80 ± 0.11	7.24 ± 0.24	<0.001	4.26 ± 0.56	7.97 ± 0.79	<0.001
Σ n-6 FA	2.42 ± 0.06	2.79 ± 0.08	<0.001	2.55 ± 0.30	2.68 ± 0.14	0.368
Σ n-3 FA	0.57 ± 0.01	1.24 ± 0.04	<0.001	0.60 ± 0.11	1.35 ± 0.06	<0.001
n-6:n-3 FA	4.28 ± 0.08	2.25 ± 0.13	<0.001	4.29 ± 0.34	1.99 ± 0.07	<0.001

[1] Group observations (two treatments × 4 days) in the loose-housing system. [2] Individual observations (two treatments × 6 animals) in the tied-housing system. [3] Fourier-transform infrared spectroscopy.

The milk FA composition, particularly the proportions of individual C18 FA, was substantially influenced by linseed supplementation to the diet, and this was found at both the group and individual levels. The total C18 FA proportion was higher in the L cows than in the C cows ($p < 0.001$). In particular, the C18:3 n-3 (α-linolenic acid, ALA) proportion was higher in the L cows when compared with the C cows ($p < 0.001$). The total n-3 FA proportion was higher ($p < 0.001$) and the n-6 to n-3 FA ratio was lower ($p < 0.001$) in the milk fat of L cows than that of the C cows. The milk fat of L cows had higher (mostly $p < 0.001$) proportions of the major biohydrogenation intermediates (C18:1 *trans*-11 and C18:2 *cis*-9, trans-11) and terminal products (C18:0 and C18:1 cis-9) than that of C cows. In the milk of the L than the C cows, the proportion of saturated fatty acids (SFAs) was lower and those of monounsaturated fatty acids (MUFAs) and polyunsaturated fatty acids (PUFAs) were higher ($p < 0.001$). In general, the effects of linseed feeding on the FA profile were recovered at both the group and individual levels.

4. Discussion

The main purpose of the present study was to investigate the extent to which a known methane-mitigating feed supplement can reduce methane emissions at both the individual level in respiration chambers and at the group level in naturally ventilated dairy housing.

4.1. Effects of Linseed on the Intake Pattern and Milk Yield in the Loose- and Tied-Housing Systems

In the present study, extruded linseed did not affect feed intake. The cows investigated at the individual level ate more, but this was because individual cows were the exclusively heavier multiparous cows, whereas at the group level, primiparous cows were also included. The addition of extruded linseed products in an iso-energetic manner resulted in a higher dietary content and intake of fiber. This extra fiber might explain the observed longer eating times and higher number of chews per DMI in L cows in the loose housing. In the tied-housing system, linseed also increased chews per DMI; in addition, the rumination time increased, indicating that linseed affected feeding behavior in both the loose- and tied-housing systems. Overall, rumination times were shorter in the tied-housing system than in the loose-housing system. Tied cows, which were also subjected to various experimental procedures, might have expressed less natural behavior than loose-housed cows [14,15]. The numerical increase observed in ECM in the linseed-fed cows in the loose- and tied-housing systems is in contrast with the significant increase reported by Martin et al. [6]. Despite a higher DMI, the overall ECM yield of the cows in the tied-housing system was lower than that of the cows in the loose-housing system, which was probably the result of progressing lactation, repeated transport and change in the environment (tied-housing and respiration chambers).

4.2. Effects of Linseed on Digestibility as well as Nitrogen and Energy Utilization (Individual Level)

At the moderate level of linseed supplementation studied, we expected only marginal effects on digestibility and utilization. We assumed that the high fiber content or the part of the lipids that remained unprotected in the linseed product could adversely affect ruminal nutrient degradation and microbial protein synthesis. In line with previous reports by Martin et al. [6] and Focant et al. [11], using 36 and 26 g linseed lipids·kg^{-1} feed DM, respectively, extruded linseed supplemented at 29 g lipids·kg^{-1} feed DM in the present study did not reduce the digestibility of fiber and organic matter. In another study, it was found that even feeding 20 to 40 g pure linseed oil·kg^{-1} DM did not affect digestibility in dairy cows [12]. Consistent with this, energy and protein retention were not affected either. Thus far, the energy and N balance of cattle fed extruded linseed have been reported for growing cattle, supplemented with only 16 g of additional linseed lipids per kg diet [7], reporting also no effect on energy and N utilization. The observed tendency towards a greater proportion of urinary N in the total manure N for the L cows than the C cows is in accordance with findings by Focant et al. [11] who reported higher urinary N proportion excreted by lactating Holstein cows fed similar linseed levels (26 g linseed lipids·kg^{-1} feed DM). The tendency to lose less energy via CH_4 in L cows than in C cows did not lead to greater energy retention, but in the latter variable, the variation between cows was very high. The results of the present study indicate that the linseed product was indeed iso-energetic to corn flour in the present experiment.

4.3. Effects of Linseed on Methane Emission at Group and Individual Level

In our study, we supplemented a grass silage based diet with a moderate level of 29 g linseed lipids per kg diet at a forage-to-concentrate ratio of 63:37. This lipid supplementation did not cause more than a trend for a CH_4 mitigation effect, although clear effects could have been expected based on the results reported in earlier studies. For example, Martin et al. [6]), Engelke et al. [10], and Focant et al. [11] supplemented extruded linseed at a similar moderate level (26–36 g lipids·kg^{-1} diet DM) and found CH_4 suppression by approximately 10–20%. This difference in response to extruded linseed could have resulted from the different forage type (corn vs. grass silage) and the forage-to-concentrate ratio used in the experiments. Accordingly, the effect of moderate levels of linseed on CH_4 were weaker in the studies of Engelke et al. [10] and Martin et al. [6], when exchanging corn silage with grass silage and when reducing the concentrate proportion from 50% to 40%, respectively. This is consistent with the findings of Machmüller et al. [35], who found that using a lower forage-to-concentrate proportion led to lower methane-mitigating effects of medium-chain FA. The ruminal pH is higher in forage-based diets, which may weaken or even prevent the CH_4-suppressing effect of linseed lipids [36]. In our study, linseed reduced the CH_4 yield per unit of dNDFI, which could have been simply the consequence of the higher NDF intake with the linseed diet, assuming that this extra fiber from the linseed hulls and bran was not easily digestible. However, the NDF digestibility seemed to increase in linseed-fed cows, possibly due to the longer rumination time, in response to higher levels of dietary fiber [37]. When calculating correlations between the variables involved (Table 5), no relationship was found between daily rumination time and daily CH_4 production as in Watt et al. [38]. It seems that rumination has no independent direct effect on CH_4 production, although they are both influenced by the same factors, such as intake and fiber level [37–39]. Accordingly, the dry matter intake of cows was positively correlated with rumination time and ECM in the tied-housing phase. Conversely, no relationship was found between these parameters for the cows in the loose-housing phase, most likely because individual feed intake had to be calculated from group level measurements which leveled out individual differences among cows. It is noteworthy that the multiparous cows in the loose-housing phase spent more time ruminating when eating longer, but the rumination time of the same cows in the tied housing was not related to eating time. As mentioned in Section 4.1, tied cows, which were also subjected to various experimental procedures, might have expressed less natural behavior than loose-housed cows [14,15]. In addition, the observations might be explained by a compensatory relationship between eating and ruminating time [39].

Table 5. Correlation coefficients between performance, ruminating behavior, and methane emission of cows determined in the loose- and tied-housing phases (entire sampling period).

Trait Housing System	ECM [1] (kg·day^{-1})	Methane [2] (g·day^{-1})	Rumination [1] (min·day^{-1})	Eating [1] (min·day^{-1})
DMI (kg·day^{-1}) [2]				
Loose	0.494	−0.538	−0.211	−0.107
Tied	0.777 **	0.156	0.610*	−0.156
ECM (kg·day^{-1}) [1]				
Loose		−0.387	0.143	0.397
Tied		0.151	0.466	0.156
Methane (g·day^{-1}) [2]				
Loose			−0.322	−0.184
Tied			0.198	0.254
Rumination (min·day^{-1}) [1]				
Loose				0.828*
Tied				−0.296

** $p < 0.01$; * $p < 0.05$. [1] Individual observations in the loose (two treatments × 20 animals) and tied (two treatments × six animals) housing system. [2] Group observations (two treatments × 4 days) in the loose housing and individual observations (two treatments × six animals) in the tied-housing system.

4.4. Effects of Linseed Supplementation on Milk Composition at the Group and Individual Levels

As is known from other dietary oils, feeding cows with linseed oil in the form of extruded linseed supplementation at sufficiently high levels may decrease the milk fat and protein contents in dairy milk (e.g., Martin et al. [5]). This is caused by adverse effects on fiber digestibility mediated by the lower production of acetate, which is the major precursor of milk fat and microbial protein. Accordingly, in loose- and tied-housing systems, the milk fat and protein contents were suppressed by supplementing the cows' diets with extruded linseed. Feeding cows extruded linseed increased the enzymatically determined MUN contents at both the group and individual levels, which was probably due to the higher crude protein content of the extruded linseed-containing diet. In the study by van Zijderveld et al. [40], a dietary mixture with extruded linseed containing less crude protein was used, which led to a slight decrease in the MUN content (estimated with a pH difference technique). Our data also show that the FTIR analysis did not result in sufficiently accurate MUN values to differentiate between dietary treatments. In this context, it is puzzling that the variation among individual animals was not higher than in the values determined enzymatically. It could be speculated that there were matrix effects that affected the FTIR measurements, such as changes in the milk fat composition of linseed-fed cows.

Linseed oil is particularly rich in the dietetically valuable C18:3 n-3. However, when fed as pure oil, a large proportion of the C18:3 n-3 is biohydrogenated by the ruminal microorganisms. This is why linseed is often provided in crushed or extruded form, as the oil in this form is at least partially rumen-protected but still able to exert some inhibiting effects on ruminal microorganisms [41]. Our results showed that part of the C18:3 n-3 was transferred intact to the milk, as its proportion was substantially elevated by linseed supplementation. However, ruminal biohydrogenation also occurred, as evidenced by elevated proportions of major biohydrogenation products such as C18:1 *trans*-11 (vaccenic acid) and C18:2 *cis*-9, trans-11 (rumenic acid; the most important conjugated linoleic acid), and C18:0 [42]. The two intermediates mentioned, but not the terminal product C18:0, are considered valuable to human health. Linseed has been reported to enhance both the n-3 FA and biohydrogenation intermediates [8,10,11]. The effects of linseed on the FA profile of milk fat in our study were similar at both the individual and group levels (bulk milk), indicating that changes can be clearly recovered in a few bulk milk samples obtained from linseed-fed herds of dairy cattle. Therefore, this method may be implemented as a control instrument by retailers who pay higher prices for milk produced from cows fed linseed.

4.5. Comparability of Methane Emission Measurements in an Experimental Housing System and Individually in Respiration Chambers

The methane-mitigation efficacy of extruded linseed was limited, both at the group and at individual levels. Systematic differences in CH_4 emissions between measurements at the group and individual levels are likely at least partly attributable to indirect effects, as the loose-housed cows were exposed to transport, being tied in stalls, respiration chambers, and further experimental procedures. In addition, the subpopulation of cows that underwent measurements in respiration chambers were exclusively multiparous cows. A system-dependent difference was that CH_4 emissions in the experimental dairy housing at the group level included housing-based sources, such as the floor's soiling. Integrated animal and housing emissions are highly relevant from an environmental point of view [16]; however, they may mask differences between feeding treatments. In particular, Hassanat et al. [43] recently showed that CH_4 emissions from the manure of linseed-fed cows were higher than those from the manure of cows fed a control diet devoid of linseed. In addition, certain differences in linseed intake between different group members cannot be ruled out, although this should have been minimized by fixing the cows for a certain period after the linseed supplement was provided. Simultaneous measurements performed by Schmithausen et al. [16] in a naturally ventilated experimental housing in two separate sections, similar to the setting of the present experiment, also revealed slight numerical reductions in the CH_4 emissions in lactating cows fed a CH_4-mitigating supplement (30 g of condensed tannins·kg^{-1} diet DM; effect reviewed in Beauchemin et al. [4]). This report illustrated that it may be possible to detect the CH_4 reduction potential of a feed supplement in a group-level assessment. However, this is analytically more challenging, the experimental design is more complex and detecting differences in CH_4 emissions at the group level may require a higher level of supplementation than at the individual level. In contrast to CH_4 emissions, CO_2 emissions were comparable at the group and individual levels. Unlike CH_4, the majority of CO_2 originates from nutrient oxidation in the animal's metabolism, which is less variable than ruminal fermentation. The good comparability of CO_2 emissions indicates that both concepts, i.e., respiration chambers and the tracer ratio technique, provide reliable results (if source strengths are comparable), whereas CH_4 emissions at the group and individual levels might differ to some extent.

5. Conclusions

This is the first study to assess the CH_4-mitigating effect of a feeding strategy evaluating measurements at the individual level in respiration chambers and measurements at the group level conducted in a naturally ventilated dairy loose-housing system. The moderate level of extruded linseed chosen only resulted in a slightly lower diurnal course of CH_4 emission intensity than the control. A significant emission reduction was solely observed for CH_4 per unit NDF digested. Additionally, the favorable effect of linseed on milk FA composition at both the group and individual levels suggests that no individual feed allocation and recording is necessary to control the use of linseed in diet. Overall, the results indicate that group-level measurements are more challenging with regard to the experimental conditions in the loose-housing system, but are likely suitable to detect the CH_4 reduction potential of feed supplements provided at higher concentrations. However, appropriate analytical approaches and experimental designs are needed. Due to the higher number of influencing factors, practical measurements are relevant for estimating the effective mitigation potential in a complex barn-manure storage and removal environment.

Author Contributions: Conceptualization, S.S., J.M., J.O.Z., and A.S.; methodology, J.P., S.S., K.Z., J.M., M.Z. and A.S.; formal analysis, J.P., A.S., M.Z.; investigation, J.P., K.Z. and M.Z..; resources, S.S., J.M. and M.K.; writing—original draft preparation, J.P.; writing—review and editing, S.S., J.M, M.Z., J.O.Z., M.K. and A.S.; visualization, J.P.; supervision, S.S., M.K. and A.S.; project administration, S.S.; funding acquisition, S.S., J.M. and A.S. All authors have read and agreed to the published version of the manuscript.

Funding: This research was funded by the Swiss National Science Foundation within the National Research Programme 69 'Healthy Nutrition and Sustainable Food Production' (No. 406940-145144) and the Air Pollution Control and Chemicals Division of the Federal Office for the Environment within contract numbers 11.0042.PJ/O075-1569, 11.0042.PJ/K431-4540, 11.0042.PJ/M443-0752, and 11.0042.PJ/O282-0739.

Acknowledgments: We are grateful to the expert advice, assistance, and help of numerous persons involved in the experiment from Agroscope, ETH Zurich and Strickhof (C. Bühler, F. Ferrari, F. Grandl, M. Hunziker, S. Ineichen, P. Jenni, G. Jöhl, M. Keck, M. Keller, C. Kunz, B. Kürsteiner, H. Lüthi-Wettstein, S. Mathis, M. Mergani, S. Meese, D. Näf, H.R. Ott, M. Schlatter, P. Stirnemann, R. Stoz, M. Terranova, S. Wang, and M. Wymann). We gratefully acknowledge the provision of extruded linseed by Trinova AG (Wangen, Switzerland).

Appendix A

Table A1. Nitrogen balance of cows from the control and linseed treatment groups determined individually in the tied-housing phase (six cows per treatment; means ± standard deviation).

Item	Control	Linseed	*p*-Value
N balance (g·day^{-1})			
Intake	526 ± 49	552 ± 59	0.419
Feces	176 ± 26	167 ± 26	0.561
Urine	167 ± 67	214 ± 68	0.250
Milk	125 ± 33	156 ± 53	0.246
Body	58 ± 66	15 ± 59	0.255
N balance (g·kg^{-1})			
Feces	334 ± 29	303 ± 45	0.187
Urine	312 ± 97	383 ± 89	0.211
Milk	235 ± 45	277 ± 65	0.222
Body	119 ± 122	36 ± 105	0.234
N utilization			
Apparent N digestibility (%)	64.6 ± 11.2	68.7 ± 7.2	0.468
N intake (g·kg^{-1} ECM [1])	20.5 ± 4.1	19.1 ± 4.0	0.551
Feces N (g·kg^{-1} ECM)	6.9 ± 1.9	5.8 ± 1.5	0.266
Manure N traits			
Manure N (g·kg^{-1} of N intake)	250 ± 16	232 ± 26	0.174
Urinary N (g·kg^{-1} of manure N)	473 ± 73	554 ± 78	0.094

[1] Energy-corrected milk.

Table A2. Energy balance of cows from control and linseed treatments determined individually in the tied-housing phase (six cows per treatment; means ± standard deviation).

Item	Control	Linseed	p-Value
Energy intake (MJ/day)			
Gross energy (GE)	357.8 ± 30.8	361.9 ± 37.6	0.840
Digestible energy	273.5 ± 14.3	276.5 ± 33.8	0.843
Metabolizable energy (ME)	236.5 ± 14.8	241.5 ± 34.3	0.750
Energy loss (MJ/day)			
Feces	84.3 ± 17.1	85.3 ± 13.3	0.906
Urine	9.9 ± 0.9	10.4 ± 3.1	0.717
Methane	27.1 ± 2.2	24.7 ± 1.9	0.063
Heat	119.1 ± 10.7	118.2 ± 6.2	0.858
Total energy loss	240.4 ± 25.5	238.6 ± 16.7	0.887
Energy retention (MJ/day)			
Milk	83.1 ± 17.7	95.7 ± 29.2	0.386
Body	34.3 ± 12.8	27.6 ± 17.1	0.458
Energy turnover (kJ·MJ^{-1} of GE intake)			
Feces	234 ± 26	236 ± 34	0.885
Urine	28 ± 4	29 ± 10	0.750
Methane	76 ± 10	69 ± 6	0.131
Heat	333 ± 17	328 ± 24	0.700
Total energy loss	671 ± 17	663 ± 52	0.712
Milk	231 ± 40	260 ± 54	0.313
Body	98 ± 41	77 ± 47	0.440
Heat (kJ·MJ^{-1} of ME intake)	503 ± 24	495 ± 54	0.743
Energy utilization (%)			
Apparent digestibility	76.6 ± 2.6	76.4 ± 3.4	0.885
Metabolizability (q)	66.2 ± 1.6	66.6 ± 3.7	0.822
k_L[1] (estimated from q)	62.2 ± 0.4	62.3 ± 0.9	0.882
k_L[1] (direct calculation)	57.0 ± 2.6	58.1 ± 2.4	0.493
Supply over estimated requirements			
Net energy for lactation (MJ)	21.8 ± 12.6	12.5 ± 16.6	0.299
% deviation from requirements	18.8 ± 14.1	10.7 ± 13.6	0.335
ME (MJ)	32.5 ± 19.9	17.3 ± 26.0	0.282
% deviation from requirements	17.3 ± 13.5	9.3 ± 13.0	0.327

[1] Proportionate utilization of ME for milk energy formation.

References

1. Hristov, A.N.; Hanigan, M.; Cole, A.; Todd, R.; McAllister, T.A.; Ndegwa, P.M.; Rotz, A. Ammonia emissions from dairy farms and beef feedlots. *Can. J. Anim. Sci.* **2011**, *91*, 1–35. [CrossRef]
2. Knapp, J.R.; Laur, G.L.; Vadas, P.A.; Weiss, W.P.; Tricarico, J.M. Enteric methane in dairy cattle production: Quantifying the opportunities and impact of reducing emissions. *J. Dairy Sci.* **2014**, *97*, 3231–3261. [CrossRef] [PubMed]
3. Cottle, D.J.; Nolan, J.V.; Wiedemann, S.G. Ruminant enteric methane mitigation: A review. *Anim. Prod. Sci.* **2011**, *51*, 491–514. [CrossRef]
4. Beauchemin, K.A.; Kreuzer, M.; O'Mara, F.; McAllister, T.A. Nutritional management for enteric methane abatement: A review. *Aust. J. Exp. Agr.* **2008**, *48*, 21–27. [CrossRef]
5. Martin, C.; Rouel, J.; Jouany, J.P.; Doreau, M.; Chilliard, Y. Methane output and diet digestibility in response to feeding dairy cows crude linseed, extruded linseed, or linseed oil. *J. Anim. Sci.* **2008**, *86*, 2642–2650. [CrossRef] [PubMed]
6. Martin, C.; Ferlay, A.; Mosoni, P.; Rochette, Y.; Chilliard, Y.; Doreau, M. Increasing linseed supply in dairy cow diets based on hay or corn silage: Effect on enteric methane emission, rumen microbial fermentation, and digestion. *J. Dairy Sci.* **2016**, *99*, 3445–3456. [CrossRef]

7. Hammond, K.J.; Humphries, D.J.; Crompton, L.A.; Kirton, P.; Reynolds, C.K. Effects of forage source and extruded linseed supplementation on methane emissions from growing dairy cattle of differing body weights. *J. Dairy Sci.* **2015**, *98*, 8066–8077. [CrossRef]

8. Livingstone, K.M.; Humphries, D.J.; Kirton, P.; Kliem, K.E.; Givens, D.I.; Reynolds, C.K. Effects of forage type and extruded linseed supplementation on methane production and milk fatty acid composition of lactating dairy cows. *J. Dairy Sci.* **2015**, *98*, 4000–4011. [CrossRef]

9. Judy, J.V.; Bachman, G.C.; Brown-Brandl, T.M.; Fernando, S.C.; Hales, K.E.; Harvatine, K.J.; Miller, P.S.; Kononoff, P.J. Increasing the concentration of linolenic acid in diets fed to Jersey cows in late lactation does not affect methane production. *J. Dairy Sci.* **2019**, *102*, 2085–2093. [CrossRef]

10. Engelke, S.W.; Daş, G.; Derno, M.; Tuchscherer, A.; Wimmers, K.; Rychlik, M.; Kienberger, H.; Berg, W.; Kuhla, B.; Metges, C.C. Methane prediction based on individual or groups of milk fatty acids for dairy cows fed rations with or without linseed. *J. Dairy. Sci.* **2019**, *102*, 1788–1802. [CrossRef]

11. Focant, M.; Froidmont, E.; Archambeau, Q.; Dang Van, Q.C.; Larondelle, Y. The effect of oak tannin (*Quercus robur*) and hops (*Humulus lupulus*) on dietary nitrogen efficiency, methane emission, and milk fatty acid composition of dairy cows fed a low-protein diet including linseed. *J. Dairy Sci.* **2019**, *102*, 1144–1159. [CrossRef] [PubMed]

12. Benchaar, C.; Romero-Pérez, G.A.; Chouinard, P.Y.; Hassanat, F.; Eugene, M.; Petit, H.V.; Côrtes, C. Supplementation of increasing amounts of linseed oil to dairy cows fed total mixed rations: Effects on digestion, ruminal fermentation characteristics, protozoal populations, and milk fatty acid composition. *J. Dairy. Sci.* **2012**, *95*, 4578–4590. [CrossRef] [PubMed]

13. Troy, S.H.; Duthie, C.-A.; Ross, D.W.; Hyslop, J.J.; Roehe, R.; Waterhouse, A.; Rooke, J.A. A comparison of methane emissions from beef cattle measured using methane hoods with those measured using respiration chambers. *Anim. Feed. Sci. Technol.* **2016**, *211*, 227–240. [CrossRef]

14. Storm, I.M.; Hellwing, A.L.F.; Nielsen, N.I.; Madsen, J. Methods for measuring and estimating methane emission from ruminants. *Animals* **2012**, *2*, 160–183. [CrossRef] [PubMed]

15. Huhtanen, P.; Cabezas-Garcia, E.H.; Utsumi, S.; Zimmerman, S. Comparison of methods to determine methane emissions from dairy cows in farm conditions. *J. Dairy Sci.* **2015**, *98*, 3394–3409. [CrossRef]

16. Schmithausen, A.J.; Schiefler, I.; Trimborn, M.; Gerlach, K.; Südekum, K.-H.; Pries, M.; Büscher, W. Quantification of methane and ammonia emissions in a naturally ventilated barn by using defined criteria to calculate emission rates. *Animals* **2018**, *8*, 75. [CrossRef]

17. Grainger, C.; Clarke, T.; McGinn, S.M.; Auldist, M.J.; Beauchemin, K.A.; Hannah, M.C.; Waghorn, G.C.; Clark, H.; Eckard, R.J. Methane emissions from dairy cows measured using the sulfur hexafluoride (SF_6) tracer and chamber techniques. *J. Dairy Sci.* **2007**, *90*, 2755–2766. [CrossRef]

18. Hammond, K.J.; Humphries, D.J.; Crompton, L.A.; Green, C.; Reynolds, C.K. Methane emissions from cattle: Estimates from short-term measurements using a GreenFeed system compared with measurements obtained using respiration chambers or sulphur hexafluoride tracer. *Anim. Feed. Sci. Technol.* **2015**, *203*, 41–52. [CrossRef]

19. Agroscope. Feeding Recommendations and Nutrient Tables for Ruminants. 2020. Available online: https://www.db-agroscope.ch/member/fmdb/7Fuumltterungsempfehlungenfuum_113.pdf (accessed on 11 January 2020). (In German).

20. Mohn, J.; Zeyer, K.; Keck, M.; Keller, M.; Zähner, M.; Poteko, J.; Emmenegger, L.; Schrade, S. A dual tracer ratio method for comparative emission measurements in an experimental dairy housing. *Atmos. Environ.* **2018**, *179*, 12–22. [CrossRef]

21. Schrade, S.; Zeyer, K.; Gygax, L.; Emmenegger, L.; Hartung, E.; Keck, M. Ammonia emissions and emission factors of naturally ventilated dairy housing with solid floors and an outdoor exercise area in Switzerland. *Atmos. Environ.* **2012**, *47*, 183–194. [CrossRef]

22. Buehler, K.; Wanner, M. Chapter 6: Metabolic Centre of the University of Zurich and ETH Zurich (under construction). In *Technical Manual on Respiration Chamber Designs*; Leahy, S., Parlane, K., Eds.; Ministry of Agriculture and Forestry: Wellington, New Zealand, 2018; pp. 89–106. Available online: https://globalresearchalliance.org/wp-content/uploads/2018/02/LRG-Manual-Facility-BestPract-Sept-2018.pdf (accessed on 23 April 2020).

23. Werner, J.; Leso, L.; Umstatter, C.; Niederhauser, J.; Kennedy, E.; Geoghegan, A.; Shalloo, L.; Schick, M.; O'Brien, B. Evaluation of the RumiWatchSystem for measuring grazing behaviour of cows. *J. Neurosci. Meth.* **2018**, *300*, 138–146. [CrossRef] [PubMed]

24. AOAC. *Official Methods of Analysis*; Association of Official Analytical Chemists: Arlington, VA, USA, 1995.

25. Van Soest, P.V.; Robertson, J.B.; Lewis, B.A. Methods for dietary fiber, neutral detergent fiber, and nonstarch polysaccharides in relation to animal nutrition. *J. Dairy Sci.* **1991**, *74*, 3583–3597. [CrossRef]

26. Grandl, F.; Zeitz, J.O.; Clauss, M.; Furger, M.; Kreuzer, M.; Schwarm, A. Evidence for increasing digestive and metabolic efficiency of energy utilization with age of dairy cattle as determined in two feeding regimes. *Animal* **2018**, *12*, 515–527. [CrossRef] [PubMed]

27. ISO (International Organization for Standardization). *ISO 14637: Milk—Determination of Urea Content-Enzymatic Method Using Difference in pH (Reference Method)*; ISO: Geneva, Switzerland, 2004.

28. IUPAC (International Union of Pure and Applied Chemistry). *Standard Methods for the Analysis of Oils, Fats and Derivates*, 7th ed.; Blackwell Scientific Publications: Oxford, UK, 1987.

29. Wettstein, H.-R.; Scheeder, M.R.L.; Sutter, F.; Kreuzer, M. Effect of lecithins partly replacing rumen-protected fat on fatty acid digestion and composition of cow milk. *Eur. J. Lipid. Sci. Tech.* **2001**, *103*, 12–22. [CrossRef]

30. Suter, B.; Grob, K.; Pacciarelli, B. Determination of fat content and fatty acid composition through 1-min transesterification in the food sample; principles. *Z. Lebensmitteluntersuchungs- Und -Forschung A* **1997**, *204*, 252–258. [CrossRef]

31. Collomb, M.; Bühler, T. Analysis of the fatty acid composition of the milk fat. I. Optimisation and validation of a general method adapted to the research. *Mitt. Lebensm. Hyg.* **2000**, *91*, 306–332.

32. KTBL (Kuratorium für Technik und Bauwesen in der Landwirtschaft), 2019. Grossvieheinheitenrechner 2.1. Available online: http://daten.ktbl.de/gvrechner/gvHome.do;jsessionid=C389DE28C48F29CD7A3EEA53225A478E#start (accessed on 23 April 2020).

33. Kinsman, R.; Sauer, F.D.; Jackson, H.A.; Wolynetz, M.S. Methane and Carbon Dioxide Emissions from Dairy Cows in Full Lactation Monitored over a Six-Month Period. *J. Dairy Sci.* **1995**, *78*, 2760–2766. [CrossRef]

34. Sauer, F.D.; Fellner, V.; Kinsman, R.; Kramer, J.K.G.; Jackson, H.A.; Lee, A.J.; Chen, S. Methane output and lactation response in Holstein cattle with monensin or unsaturated fat added to the diet. *J. Anim. Sci.* **1998**, *76*, 906–914. [CrossRef]

35. Machmüller, A.; Soliva, C.R.; Kreuzer, M. Methane-suppressing effect of myristic acid in sheep as affected by dietary calcium and forage proportion. *Br. J. Nutr.* **2003**, *90*, 529–540. [CrossRef]

36. Zhou, X.; Zeitz, J.O.; Meile, L.; Kreuzer, M.; Schwarm, A. Influence of pH and the degree of protonation on the inhibitory effect of fatty acids in the ruminal methanogen *Methanobrevibacter ruminantium* strain M1. *J. Appl. Microbiol.* **2015**, *119*, 1482–1493. [CrossRef]

37. Zetouni, L.; Difford, G.F.; Lassen, J.; Byskov, M.V.; Norberg, E.; Løvendahl, P. Is rumination time an indicator of methane production in dairy cows? *J. Dairy Sci.* **2018**, *101*, 11074–11085. [CrossRef] [PubMed]

38. Watt, L.J.; Clark, C.E.F.; Krebs, G.L.; Petzel, C.E.; Nielsen, S.; Utsumi, S.A. Differential rumination, intake, and enteric methane production of dairy cows in a pasture-based automatic milking system. *J. Dairy Sci.* **2015**, *98*, 7248–7263. [CrossRef] [PubMed]

39. Beauchemin, K.A. Current perspectives on eating and rumination activity in dairy cows. *J. Dairy Sci.* **2018**, *101*, 4762–4784. [CrossRef]

40. Van Zijderveld, S.M.; Dijkstra, J.; Perdok, H.B.; Newbold, J.R.; Gerrits, W.J.J. Dietary inclusion of diallyl disulfide, yucca powder, calcium fumarate, an extruded linseed product, or medium-chain fatty acids does not affect methane production in lactating dairy cows. *J. Dairy Sci.* **2011**, *94*, 3094–3104. [CrossRef] [PubMed]

41. Doreau, M.M.; Ferlay, A. Linseed: A valuable feedstuff for ruminants. *Oilseeds. Fats. Crops. Lipids.* **2015**, *22*, 1–9. [CrossRef]

42. Chilliard, Y.; Glasser, F.; Ferlay, A.; Bernard, L.; Rouel, J.; Doreau, M. Diet, rumen biohydrogenation and nutritional quality of cow and goat milk fat. *Eur. J. Lipid. Sci. Tech.* **2007**, *109*, 828–855. [CrossRef]

43. Hassanat, F.; Benchaar, C. Methane emissions of manure from dairy cows fed red clover or corn silage-based diets supplemented with linseed oil. *J. Dairy Sci.* **2019**, *102*, 11766–11776. [CrossRef]

Article

Effects of Partial Replacment of Dietary Forage Using Kelp Powder (*Thallus laminariae*) on Ruminal Fermentation and Lactation Performances of Dairy Cows

Fuguang Xue [1,†], **Fuyu Sun** [1,2,†], **Linshu Jiang** [3], **Dengke Hua** [1], **Yue Wang** [1], **Xuemei Nan** [1], **Yiguang Zhao** [1,*] and **Benhai Xiong** [1,*]

[1] State Key Laboratory of Animal Nutrition, Institute of Animal Science, Chinese Academy of Agricultural Sciences, Beijing 100193, China; xuefuguang123@163.com (F.X.); sfycaas@163.com (F.S.); dengke_h@163.com (D.H.); wangyue9313@163.com (Y.W.); xuemeinan@126.com (X.N.)
[2] College of Animal Science and Technology, China Agricultural University, Beijing 100193, China
[3] Beijing Key Laboratory for Dairy Cow Nutrition, Beijing University of Agriculture, Beijing 102206, China; jls@bua.edu.cn
* Correspondence: zhaoyiguang@caas.cn (Y.Z.); xiongbenhai@caas.cn (B.X.)
† These authors contributed equally to this work.

Received: 30 September 2019; Accepted: 9 October 2019; Published: 22 October 2019

Simple Summary: Kelp powder has been widely used as feed ingredient or additives in monogastric animal production and presented positive effects on production performance and intestinal microbiota because of its abundance in biomass output, novel oligosaccharides, and iodine. However, little information is available for the nutritional effects of kelp on ruminants. Therefore, the objective of the present study was to investigate the effects of kelp partial replacing dietary forage on rumen fermentation and production performance of dairy cows. Results indicated that, a substitution of 5% kelp powder for forage in the diet reduced the fiber content in the feeding diet and then regulated the proliferation of ruminal microbiota, which lead to a significant reduction of NH_3-N while increasing acetate, propionate, and total volatile fatty acids (TVFA) concentrations in rumen, and finally partly increasing milk production and milk fat content. The present study may provide more comprehensive cognition of the nutritional value of kelp powder as a dietary raw material. It may further promote the utilization of kelp powder and provide a useful dietary raw material to the husbandry of dairy cows.

Abstract: Background: Kelp powder, which was rich in novel oligosaccharides and iodine might be utilized by the rumen microbiome, promoted the ruminal fermentation and finally enhanced the lactation performance of dairy cows. Therefore, the purpose of this study was to investigate the effects of kelp powder partially replacing dietary forage on rumen fermentation and lactation performance of dairy cows. (2) Methods: In the present study, 20 Chinese Holstein dairy cows were randomly divided into two treatments, a control diet (CON) and a kelp powder replacing diet (Kelp) for a 35-d long trial. Dry matter intake (DMI), milk production, milk quality, ruminal fermentable parameters, and rumen microbiota were measured to investigate the effects of kelp powder feeding on dairy cows. (3) Results: On the lactation performance, kelp significantly increased milk iodine content and effectively enhanced milk production and milk fat content. On the fermentable aspects, kelp significantly raised TVFA while reducing the ammonia-N content. On the rumen microbial aspect, kelp feeding significantly promoted the proliferation of *Firmicutes and Proteobacteria* while suppressing *Bacteroidetes*. (4) Conclusion: kelp powder as an ingredient of feedstuff might promote the rumen fermentation ability and effectively increase milk fat and iodine content, and consequently improve the milk nutritional value.

Keywords: dairy cows; metagenomic sequencing; kelp powder; rumen fermentation

---

1. Introduction

Kelp powder (*Thallus laminariae*), attracts more attentions in recent years in the husbandry because of its high-yielding biomass and abundant nutritional content such as crude protein, amino acids, polyunsaturated fatty acids, and oligosaccharides [1,2]. Kelp has been used as the feed ingredient or feed additive in the production of monogastric animals such as layers, broilers, and pigs, and expressed positive effects on animal performance, such as reducing the incidence of diseases and promoting production performance [3–6]. Whereas little information is available on kelp nutritional effects on ruminants, while it has been found that that kelp may potentially lower somatic cell count and improve body condition of dairy cattle [7].

Ruminants have high digestibility of feeding ingredients for the ruminal microbial ecosystem comprised of an immense variety of microbiota which execute the complex anaerobic metabolism [8]. Bacteria in the rumen contribute more than 95% of the whole biomass and perform the main function in rumen fermentation [9,10]. For this purpose, oligosaccharides which were not degraded in monogastric creatures could be degraded in the rumen and hydrolyzed into mono- or di- saccharides and produced formate, acetate, and lactate as the end products [11]. Moreover, ruminal microbiota might be altered for the increasing ruminal end products and so might the fermentation procedure. Finally, the production performance might be influenced.

Therefore, we hypothesized that oligosaccharides in kelp will be digested in rumen, increase the volatile fatty acids (VFAs) content, and finally promote the lactation performance of ruminants. The objective of the present study was to investigate the effects of partially replacing dietary forage using kelp powder on ruminal microbiota, ruminal fermentation parameters, and the lactation performance of dairy cows.

2. Materials and Methods

All experimental protocols performed in this study were approved by the Animal Ethics Committee of the Chinese Academy of Agricultural Sciences (Beijing, China). The experimental procedures used in this study were in accordance with the recommendations of the academy's guidelines for animal research.

2.1. Kelp Powder

The kelp powder (brown algae, purity 100%, Wanfang Biological Co., Ltd., Xi'an, China) used in the current study was harvested in Weihai city, Shandong province, China (37°16′ N, 122°41′ E) in July 2017. It was then dried and milled as a gray-green powder. The chemical composition of the kelp powder is shown in Table S1.

2.2. Animals and Experimental Design

A completely randomized design was applied in the present study. Twenty Chinese Holstein lactating dairy cows (589 ± 19.9 kg BW; 160 ± 18 DIM, milk yield (22 ± 2.3 kg/day)) were randomly divided into two treatments. Treatments included a control diet (CON) and a kelp powder replacing diet (Kelp). In the kelp treatment, kelp was used partly replace corn silage, oat hay and fatty acid calcium which resulted in kelp content came to 5% (dry matter (DM) basis). The diets were formulated according to NRC (2001) to meet or exceed the energy requirements of Holstein dairy cows yielding 20 kg of milk/day with 3.5% milk fat and 3.0% true protein. Details of ingredient analysis and chemical composition of diets were shown in Table 1.

Table 1. Ingredients and chemical composition of the diets (DM basis) [A,B].

Items	Kelp	Control
Ingredients (%)		
Corn silage	27.0	29.1
Oat hay	20.4	22.8
Kelp powder	5.0	—
Alfalfa haylage	7.2	7.2
Ground corn	3.6	3.6
Extruded soybeans	2.6	2.6
Soybean meal	9.9	9.9
Rapeseed meal	4.5	4.5
Cottonseed meal	2.6	2.6
Pressed corn	2.2	2.2
Cottonseed	2.7	2.7
Corn hull	9.6	9.6
Fat powder	1.2	1.2
Fatty acid calcium	1.2	1.7
Vitamin/mineral premix	0.3	0.3
Nutrient composition (%, unless otherwise stated)		
NE_L (Mcal/kg)	1.40	1.41
CP	12.81	12.96
EE	3.42	3.57
NDF	36.78	38.14
ADF	20.69	22.15

[A] Nutrient composition of the experimental diets were calculated according to NRC (2001); 0% Kelp = no kelp powder in the diet; 5% Kelp = 5% kelp powder in the diet; EE = ether extract; NE_L = net energy for lactation and calculated according to NRC (2001). [B] Vitamin/mineral premix contained (DM basis): 15.7% of Ca, 4.1% of P, 1600 mg/kg of Fe, 700 mg/kg of Cu, 3500 mg/kg of Mn, 7500 mg/kg of Zn, 80 mg/kg of Se, 400 mg/kg of I, 50 mg/kg of Co, 190,000 IU/kg of vitamin A, 55,000 IU/kg of vitamin D and 1900 IU/kg of vitamin E. NDF = neutral detergent fiber, ADF = acid detergent fiber.

2.3. Feeding Management

All cows were raised in individual stalls of a 35 d-long period feeding procedure. The cows were fed three times a day (07:00, 13:00 and 18:00 h, respectively) ad libitum in order to ensure that the total mixed ration (TMR) were fresh and available for the cows for at least 20 h a day. Cows were milked three times a day (09:00, 15:00, and 20:00 h, respectively) and were managed with natural lighting. All cows were dewormed before the commencement of this study.

2.4. Sampling and Parameters Measurement

During the experimental period, automatic feeding equipment (made by Institute of Animal Science Chinese Academy of Agricultural Sciences, Beijing, China and NanShang Husbandry Science and Technology Ltd. Henan, China) was used to record dry matter intake. Milking facilities (90 Side-by-Side Parallel Stall Construction, Afimilk, Israel) were applied to record milk production of each cow.

On the last day of the experiment, a gastric rumen sampler was used to collect rumen fluid samples through the esophagus at 3 h after the morning feeding. Collected samples were strained through four layers of cheesecloth to obtain rumen fluid. Rumen fluid was then divided into two parts. One part was processed to analyze the pH value, rumen volatile fatty acid (VFAs) and ammonia-N (NH_3-N). The other part was put into the liquid nitrogen immediately after adding stabilizer and then stored at −80 °C for DNA extraction. Rumen contents were strained through four layers of cheesecloth with a mesh size of 250 μm. The pH of each rumen fluid sample was measured immediately using a portable pH meter (Testo 205, Testo AG, Lenzkirch, Germany). Individual and total VFAs (TVFA) in the aliquots of ruminal fluid were determined by gas chromatograph (GC-2010, Shimadzu, Kyoto,

Japan). Concentration of NH_3-N was determined by indophenol method and the absorbance value was measured through UV-2600 ultraviolet spectrophotometer (Tianmei Ltd., China).

Milk samples were collected from individual cows during the last four consecutive days in 100-mL vials in each milking period. Samples were preserved with 2-bromo-2-nitropropan-1,3-diol and stored at 4 °C, before sent to the Milk and Dairy Products Quality Supervision and Testing Center, Ministry of Agriculture (Beijing, China) for analyses of milk protein, fat, lactose, somatic cell count and milk iodine content by a mid-infrared spectroscopy (Fossomatic 4000, Foss Electric A/S, Hillerød, Denmark).

2.5. DNA Extraction and Sequencing Process

DNA for metagenomics sequencing was extracted from the rumen fluid samples by using the QIAamp DNA Stool Mini Kit (Qiagen, Hilden, Germany) according to manufacturer's protocols. The DNA concentration and purity were quantified with TBS-380 and NanoDrop2000, respectively. DNA quality was examined with the 1% agarose gels electrophoresis system.

The bacteria 16S rRNA gene was amplified using the barcoded universal primers 338F (5′-barcorde-ACTCCTRCGGGAGGCAGCAG-3) and 806R (5′-GGACTACCVGGGTATCTAAT-3′) spanning the V3–V4 hyper variable region. The 16s RNA sequencing procedure was conducted through Illumina HiSeq 4000 platform (Illumina Inc., San Diego, CA, USA). Quality filtering on the raw tags were performed under specific filtering conditions to obtain the high-quality clean tags according to the QIIME (V1.7.0) quality-controlled process. The tags were compared with the reference database using UCHIME algorithm to detect chimera sequences, and then the chimera sequences were removed. Then the effective tags were finally obtained. Sequences analysis were performed by Uparse software (Uparse v7.0.1001, Tiburon, California, USA). Sequences with >97% similarity were assigned to the same OTUs. A representative sequence for each OTU was screened for further annotation. For each representative sequence, the GreenGene Database was used based on RDP classifier algorithm to annotate taxonomic information. Abundance information of OTUs was normalized using a standard of sequence number corresponding to the sample with the least sequences. Subsequent analysis of alpha diversity and beta diversity were all performed basing on this output normalized data. Alpha diversity is applied in analyzing complexity of species diversity for a sample through six indices, including, Chao1, Shannon, Simpson, ACE. All indices in our samples were calculated with QIIME (Version 1.7.0) and displayed with R software (Version 3.3.1, R Core Team, Vienna, Austria)). Beta diversity analysis was used to evaluate differences of samples in species complexity. Beta diversity on unweighted unifrac were calculated by QIIME software (Version 1.7.0, San Diego, CA, USA). Cluster analysis was preceded by principal coordinates analysis (PCoA), which was applied to reduce the dimension of the original variables. PCoA analysis was displayed by WGCNA package, stat packages, and ggplot2 package in R software. Unweighted Pair-Group Method with Arithmetic Means (UPGMA) Clustering was performed as a type of hierarchical clustering method to interpret the distance matrix using average linkage and was conducted by QIIME.

2.6. Statistical Analysis

Results were presented as means ± SEM. A normal distribution test of ruminal pH, VFAs, and NH_3-N were first conducted using SAS procedure "proc univariate data = test_normal" and then Student's t-test of SAS 9.2 was applied to analyze the differences of parameters between CON and Kelp treatments. p-value < 0.05 was considered to be significance and $0.05 \leq p < 0.10$ was considered as a tendency. Barplot, principal coordinate analysis (PCoA), hierarchical clustering analysis (HCA) for different rumen bacteria were conducted using R package. Spearman correlations between bacteria communities and ruminal fermentation variables were assessed using the PROC CORR procedure of SAS 9.2 (SAS Institute, Inc., Cary, NC, USA). Relative abundance of all phyla of bacteria were chosen to conduct the correlation analysis. A correlation matrix was created and visualized in a heatmap format using R package version 3.3.1.

3. Results

3.1. Animal Production Performance and Rumen Fermentation Parameters

The animal production performance was the most concerned in the dairy cattle production. In the present study, as shown in Table 2, the DMI, milk production, milk fat increased after kelp feeding, however, not significantly. Particularly, the milk iodine content significantly increased in the kelp treatment compared with CON treatment ($p < 0.05$). On the aspect of fermentation parameters, the kelp significantly raised ruminal total VFAs content while reduced the ammonia-N content ($p < 0.05$). For the individual VFA, kelp feeding significantly increased ruminal acetate, propionate and isobutyrate content.

Table 2. Effects of kelp powder on animal productivity and ruminal fermentation.

Items	CON	Kelp	SEM	*p*-Value
DMI (kg/day)	16.2	16.6	0.468	0.114
Milk production (kg/day)	21.1	22.7	1.894	0.235
Milk fat (%)	3.8	3.9	0.126	0.146
Milk protein(%)	3.41	3.42	0.061	0.106
Milk iodine(mg/L)	0.07 [b]	0.12 [a]	0.020	0.029
Ruminal pH	6.45	6.38	0.194	0.216
Ammonia-N (mg/dL)	17.8 [a]	13.1 [b]	1.933	<0.001
Acetate (mmol/L)	65.7 [b]	67.1 [a]	0.773	0.038
Propionate (mmol/L)	17.2 [b]	19.2 [a]	0.632	0.027
A/P	3.83	3.57	0.098	0.001
Butyrate (mmol/L)	16.3	16.1	0.137	0.356
Isobutyrate (mmol/L)	1.6 [b]	3.9 [a]	0.115	0.041
Valerate(mmol/L)	2.21	2.14	0.064	0.241
Isovalerate (mmol/L)	2.31	2.50	0.160	0.216
TVFA (mmol/L)	104.5 [b]	110.8 [a]	1.836	0.016

[a,b] means within a row with different letters differed significantly ($p < 0.05$); SEM = standard error of the mean. A/P = acetate/propionate.

3.2. Sequencing Information

The bacteria 16S rRNA gene was amplified using the barcoded universal primers 338F and 806R spanning the V3–V4 hyper variable region. The sequencing reads number of each samples were between 30,000–45,000 and the mean length of each reads were more than 430 nt. All sequencing information has been shown in Additional File 2. Sequences with > 97% similarity were assigned to the same OTUs and then aligned to GreenGene Database based on RDP classifier algorithm to annotate taxonomic information. Totally, 1338 OTUs, 17 phyla and more than 220 genera were identified in the present study, and all the taxonomy information is displayed in Additional File 3.

3.3. Effects of Kelp Replacing Diet on Ruminal Bacteria

3.3.1. α-Diversity

All identified bacteria were chosen for further analysis to investigate the effects of kelp replacing diet on the community of ruminal bacteria. Alpha diversity is applied in analyzing complexity of species diversity for a sample through Chao1, Shannon, Simpson, ACE. All these indices of α-diversity were displayed in Table 3. Based on the results, no significant changes were found in the kelp treatment compared with CON.

Table 3. Effects of kelp powder on α-abundance of ruminal microbiota.

Items	CON	Kelp	FC	Log 2 FC	SE	*p*-Value
ACE	1182.5	1170.2	0.99	−0.02	0.204	0.853
Chao1	1192.4	1189.9	1.00	0.00	0.076	0.97
Shannon	5.42	5.31	0.98	−0.03	0.08	0.398
Simpson	0.014	0.018	1.29	0.36	0.009	0.216
SOBs	1035.3	1016	0.98	−0.03	0.12	0.759

Note, FC = fold change, SE = standard error of the mean.

3.3.2. β-Diversity

Beta diversity analysis was then conducted to evaluate differences of samples in species complexity. PCoA based on unweighted UniFrac distance metrics was conducted to compare bacterial profile among the three treatments. As shown in Figure 1, PCoA axes 1 and 2 accounted for 29.77% and 11.52% of the total variation, respectively. Based on the results, bacteria community in kelp treatment could be separated from those in the CON by PCo1 except Kelp3.

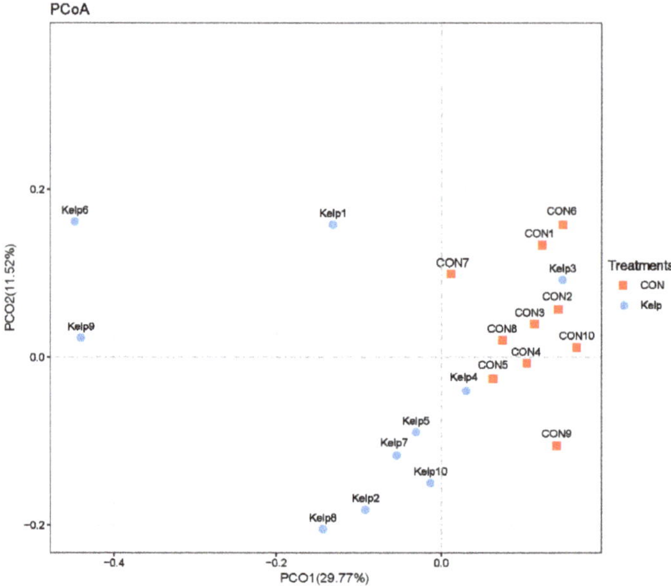

Figure 1. Principal coordinate analysis (PCoA) of bacteria community structures in CON and kelp treatment. CON = control diet; Kelp = kelp powder replacing diet.

Since the kelp made a difference to the whole bacteria community, a differential analysis on ruminal bacteria in different levels was then conducted to investigate the effects of kelp on the abundances of ruminal bacteria. Results are shown in Tables 4 and 5, in the level of phyla and genera, respectively. Based on the Table 4, *Bacteroidetes*, *Firmicutes*, and *Proteobacteria* contribute the three most abundant phyla of ruminal bacteria and the kelp feeding significantly promoted the proliferation of *Firmicutes* and *Proteobacteria* while suppressing *Bacteroidetes*. No significant effects were found of kelp feeding on other phyla. At the genera level, as shown in Table 5, *Prevotella*, *Ruminococcaceae*, *Ruminococcus*, *Bacteroidales*, *Prevotellaceae*, *Succinivibrionaceae*, *Lachnospiraceae*, and *Succiniclasticum* were the most abundant genera in the present study. *Prevotella*, *Bacteroidale*, *Rikenellacea*, *Selenomonas*, and *Saccharofermentans* were significantly decreased while *Ruminococcaceae*, *Ruminococcus*, *Lachnospiraceae*, and *Sharpea* were significantly increased after kelp feeding. There were no significant changes of other genera.

Table 4. Effects of kelp powder on the abundances of ruminal microbiota in phyla level.

Items	CON	Kelp	FC	log2 FC	SE	*p*-Value
p__Bacteroidetes	14570 [a]	10110 [b]	0.69	−0.53	0.10	<0.001
p__Firmicutes	7007 [b]	12098 [a]	1.73	0.79	0.13	0.008
p__Fibrobacteres	91.4	36.8	0.40	−1.31	0.46	0.088
p__Proteobacteria	424.6 [b]	1787.3 [a]	4.21	2.07	0.53	0.024
p__Spirochaetae	179.3	242.1	1.35	0.43	0.32	0.408
p__Saccharibacteria	133.5	143	1.07	0.099	0.32	0.765
p__Tenericutes	264	214.4	0.81	−0.30	0.52	0.462
p__Verrucomicrobia	33.8	19	0.56	−0.83	0.31	0.277
others	39.4	54.2	1.38	0.46	0.27	0.286

[a,b] means within a row, different letters differed significantly ($p < 0.05$); SE = standard error.

Table 5. Effects of kelp powder on the abundances of ruminal microbiota in genera level.

Items	CON	Kelp	FC	Log2 FC	SE	*p*-Value
g__Prevotella	9897.2 [a]	6376.8 [b]	0.64	−0.63	0.08	<0.001
g__Ruminococcaceae	1470.3 [b]	3111.1 [a]	2.12	1.08	0.17	0.021
g__Ruminococcus	539.8 [b]	2603.8 [a]	4.82	2.27	0.25	0.004
g__Bacteroidales	1969.7 [a]	1064.1 [b]	0.54	−0.89	0.15	0.002
g__Prevotellaceae	1472	1199.2	0.81	−0.30	0.15	0.185
g__Succinivibrionaceae	636.3	1238.8	1.95	0.96	0.56	0.247
g__Lachnospiraceae	566.4 [b]	1069.2 [a]	1.89	0.92	0.15	0.008
g__Succiniclasticum	626.6	653.8	1.04	0.06	0.33	0.876
g__Eubacterium	502.7	521.8	1.04	0.05	0.27	0.767
g__Rikenellaceae	600.5 [a]	315.6 [b]	0.53	−0.93	0.13	0.016
g__Christensenellaceae	430.8	428.6	0.99	−0.01	0.28	0.986
g__Selenomonas	437.1	240.3	0.55	−0.86	0.28	0.012
g__Acetitomaculum	171.7	446.2	2.60	1.38	0.18	0.121
g__Butyrivibrio	177.1	255.1	1.44	0.53	0.31	0.615
g__Treponema	285.8 [a]	124 [b]	0.43	−1.20	0.42	0.022
g__Erysipelotrichaceae	140.8	204.2	1.45	0.54	0.51	0.539
g__Anaeroplasma	214.9 [a]	113.3 [b]	0.53	−0.92	0.35	0.003
g__Lachnobacterium	170.3	143.4	0.84	−0.25	0.18	0.453
g__Candidatus Saccharimonas	169.6	111	0.65	−0.61	0.22	0.05
g__Saccharofermentans	146.9 [a]	81.2 [b]	0.55	−0.86	0.31	0.016
g__Sharpea	31.2 [b]	189.7 [a]	6.08	2.60	0.22	0.016
g__Roseburia	84.6	123.2	1.46	0.54	0.6	0.535
g__Ruminiclostridium	50.5	150.9	2.99	1.58	0.49	0.159
g__Schwartzia	93.5	77.5	0.83	−0.27	0.33	0.642
g__Lachnospiraceae	79.8	81.7	1.02	0.03	0.41	0.868
g__Fibrobacter	77.3	52.9	0.68	−0.55	0.14	0.654
g__Mogibacterium	58	64.9	1.12	0.16	0.53	0.768
g__Succinivibrio	60	34	0.57	−0.82	0.28	0.338
others	3665.1	3447.5	0.94	−0.088	0.48	0.471

[a,b] means within a row, different letters differed significantly ($p < 0.05$); SE = standard error.

Hierarchical clustering analysis (HCA) and heat map analysis was conducted to further understand the effects of kelp feeding on ruminal bacteria profile. Results of HCA analysis are displayed in Figure 2. Bacteria which belongs to CON and kelp treatments could separate into two big clusters clearly except CON7. Ruminal bacteria in genera level gathered into two big clusters. The genera in upper half part were increased after kelp feeding which consists of *g__Eubacterium*, *g__Christensenellaceae*, *g__Ruminococcaceae*, *g__Acetitomaculum*, *g__Butyrivibrio*, and *g__Mogibacterium*. Lower half part consists of those genera which were decreased after kelp feeding, such as *g__Succinivibrionaceae*, *g__Fibrobacter*, *g__Anaeroplasma*, *g__Rikenellaceae*, and *g__Saccharofermentans*.

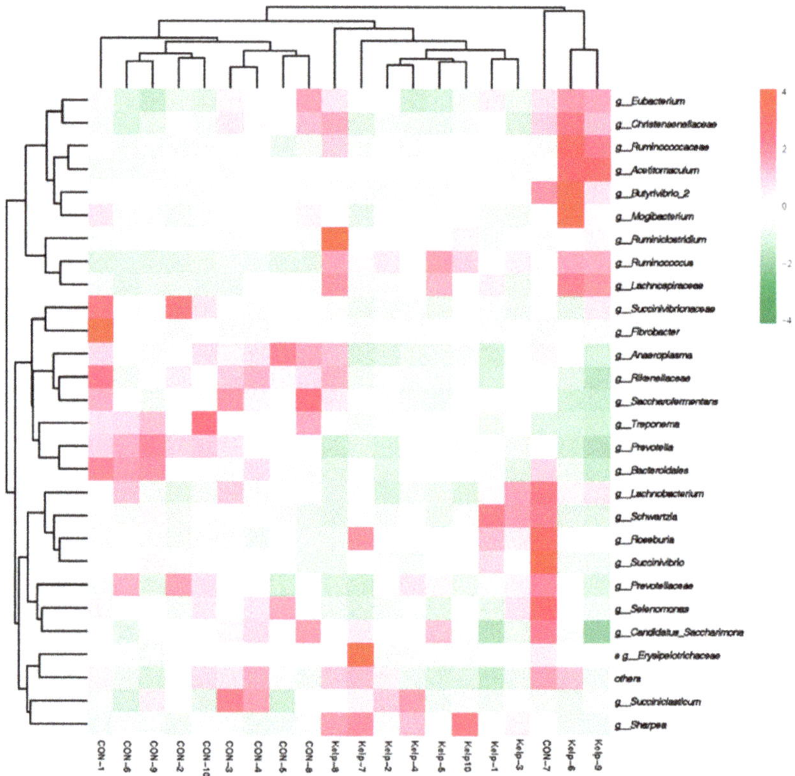

Figure 2. Hierarchical clustering analysis (HCA) and heat map analysis on abundances of ruminal bacteria content between CON and kelp treatment on the level of genera. Rows represent genera of bacteria and columns represent samples. Cells were colored based on the relative abundance of bacteria measured in rumen, red represents high rumen levels while green represents low signal intensity and white cells showing the intermediate level.

3.4. Correlations between Bacteria Communities and Ruminal Fermentation Parameters

At last, all phyla and most abundant genera were selected to conduct the correlation analysis between bacteria and fermentation parameters for the purpose to investigate the effects of ruminal bacteria on ruminal fermentation. Results are shown in Figures 3 and 4, respectively. All phyla could be separated into two parts based on the correlations with ruminal fermentation parameters. One was positively correlated with ruminal pH and VFA content, while negatively correlated with NH_3-N, this part was mainly comprised by *Firmicutes* and *Proteobacteria*.

The other part was just conversed, which was positive with NH_3-N while negative correlated with ruminal VFAs content. This part mainly consisted of *Bacteroidetes*, *Fibrobacteres*, and *Saccharibacteria*. As to the genera level, bacteria could also separate into two big clusters. The first part, which consists of *Ruminiclostridium*, *Acetitomaculum*, *Butyrivibrio*, *Ruminococcaceae*, *Ruminococcus*, *Lachnospiraceae*, and *Succiniclasticum* was positively correlated with pH, TVFAs, acetate, propionate, and Isobutyrate while being negatively correlated with NH_3-N. The other part was mainly comprised by *Prevotella*,

Saccharimonas, *Saccharofermentans*, *Bacteroidales*, *Prevotellaceae*, *Succinivibrionaceae*, and *Selenomonas* which performed conversed correlation with the fermentation parameters compared with the first part.

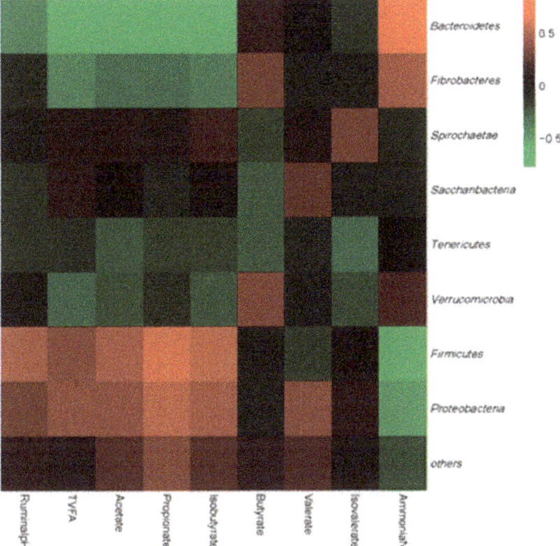

Figure 3. Correlation analyses between abundances of ruminal bacteria and DMI, milk quality parameters, and ruminal fermentation parameters on the level of phyla. The red color represents a positive correlation while the green color represents a negative correlation.

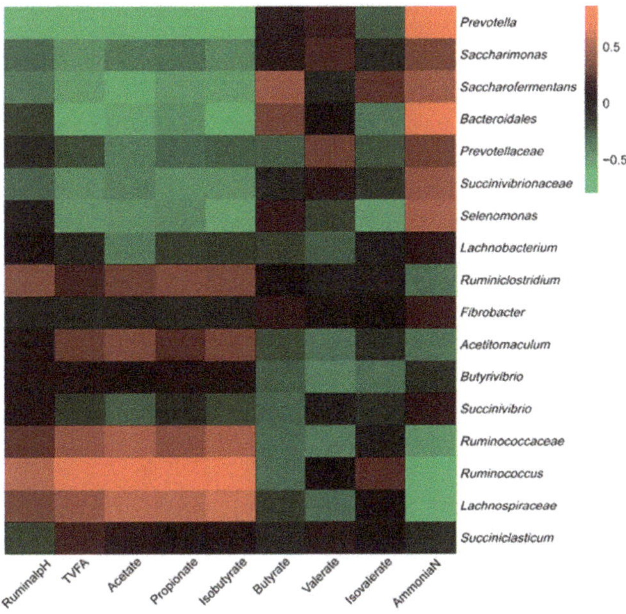

Figure 4. Correlation analyses between the most abundant genera of ruminal bacteria and DMI, milk quality parameters, and ruminal fermentation parameters. The red color represents a positive correlation while the green color represents a negative correlation.

4. Discussion

4.1. Effects of Kelp Feeding on Ruminal Carbohydrate Metabolism

In the present study, ruminal TVFA, acetate and propionate significantly increased after kelp powder diet feeding, which indicated that kelp promoted ruminal carbohydrates metabolism. In ruminal conditions, carbohydrates were always degraded into mono- or di-saccharides, and then hydrolyzed into VFAs [12]. The kelp was rich in oligosaccharides, which are always undegradable in monogastrics, which could be fermented by ruminal microbiota [11]. Because of the lower molecular weight and the simpler structure [13], oligosaccharides might be easier digestible than the polysaccharides such as cellulose. When replacing dietary forage with kelp powder, more carbohydrates were digested in the kelp feeding diets, and therefore, more VFAs were synthesized.

Kelp treatment significantly decreased the ratio of acetate/propionate. In ruminal conditions, when dietary high-fermentable content increased, it always decreased the ratio of acetate/propionate [14]. In the present study, kelp feeding diets partially decreased the fiber content, which might lead to the reduction of the mainly fiber-degrading bacteria such as *Bacteroidetes* and *Fibrobacteres* [15,16]. Besides, the kelp was added through powder form, which significantly reduced the particle size compared with their substitution, decreasing the retention time in rumen and further reducing the reaction time between fiber-degrading bacteria and dietary fiber [17].

Inversely, the easier digestible oligosaccharides help increase the abundances of *Firmicutes* and *Proteobacteria* which mainly degrade the highly-fermentable carbohydrates in ruminal conditions [18,19]. For the reasons given above, kelp feeding diets contain more highly-fermentable carbohydrate and might lead to accumulation of glucose. In ruminal conditions, glucose digested into pyruvate through glycolytic pathway [20], and then hydrolyzed into acetyl-CoA, which was the precursor of acetate with the help of pyruvate formate-lyase (PFL) [21]. When provided more high-fermentable carbohydrate, pyruvate accumulated and then turn to form more lactate, oxaloacetic acid [22], and further developed into propionate. This might contribute to the significant decrease of acetate/propionate.

4.2. Effects of Kelp Feeding on Ruminal Nitrogen Metabolism

Apart from the carbohydrate metabolism, the average rumen NH_3-N concentration in the kelp feeding treatment was significantly decreased compared with that in the control treatment, which indicated that kelp might affect the ruminal nitrogen(N) metabolism. The ruminal NH_3-N concentration could be influenced by absorption of rumen wall, ruminal outflow rate, and rumen microbial community [23]. Previous study reported that the ruminal outflow rate was negatively associated with forage length in the diet [24], the longer forage particles reduced the contact surface area with the microorganisms, resulting in a lower synthesis efficiency of rumen microbial protein and consequently increased the NH_3-N concentration [25]. In the present study, kelp was added through powder form which help increase attached area for ruminal microbiota, thereafter leading to the decrease of ruminal NH_3-N concentration in kelp feeding.

Pisulewski et al. (2010) obtained best results of ruminal NH_3-N utilization content with 88–133 mg/L [26], which opportunely included the content in kelp treatment. In ruminal conditions, ruminal bacteria could utilize NH_3-N to synthesize amino acids, and then form bacteria proteins. For synthesis of some amino acids, specific carbon sources are required, such as isobutyrate for valine [27]. In the present study, we observed that isobutyrate was significantly increased after kelp feeding. The increased isobutyrate might enhance the activity of ruminal bacteria to utilize more NH_3-N. Therefore, NH_3-N significantly decreased after kelp feeding.

In addition, utilization of ruminal NH_3-N into the bacterial protein in the rumen is energy dependent, providing adequate ruminal available energy is requirement for ruminal NH_3-N utilization [28]. In ruminal conditions, available energy was mainly provided by VFAs [29], which were found to significantly increase in kelp treatment. Therefore, the activity of rumen microorganisms will be enhanced by the provided energy, and more NH_3-N could be incorpoarated into bacteria protein.

4.3. Effect of Kelp Feeding on Milk Performance

Although not significantly, the milk production and milk fat percentage were increased after kelp feeding to some extent in the present study. Milk fat was a necessary parameter of milk quality and benefits to the human health [30]. In the current study, kelp feeding treatment could effectively promote the milk fat percentage through improving ruminal acetate content [31]. Increasing the replacement percentage of dietary forage with kelp might lead to increasing milk fat, consequenty improving the benefits of milk to human health.

Milk iodine content significantly increased in the present study, which was in line with Kaufmann [32]. For ruminants, iodine deficiency would result in low thyroid hormone concentrations and prolonged gestation [33]. The abundant iodine content in kelp might contribute to the partial increase of milk production.

5. Conclusions

In summary, a substitution of 5% kelp powder for forage in the diet reduced the fiber content in the feeding diet and then regulated the proliferation of ruminal microbiota, which lead to a significantly reduced NH_3-N content while increasing acetate and TVFA concentrations in the rumen. This result indicated that kelp powder as a dietary raw material might promoted the rumen fermentation ability and then benefit the lactation performance of dairy cows. Moreover, it could effectively increase milk fat and iodine content and consequently improve milk's nutritional value.

Supplementary Materials: The following are available online at http://www.mdpi.com/2076-2615/9/10/852/s1, Table S1: Chemical composition of kelp powder; Table S2: Sequencing information of each samples; Table S3: Taxonomy results of ruminal bacteria of each samples.

Author Contributions: Designed the study, F.X., Y.Z., and B.X.; Supervison and revised the manuscript, F.X., L.J., and B.X.; Conducted the experiment, F.X., F.S.; Analyzed the data, F.X., Y.W., D.H., and X.N.; Wrote the manuscript, F.X.; English editing, Y.Z.

Funding: National Key Research and Development Plan (2016YFD0700205).

Acknowledgments: This work was funded by Beijing Dairy Industry Innovation Team (bjcystx-ny-1). The authors thank the staff in the dairy cattle breeding center of Sanyuan Group (Beijing, China) for their assistance in the animal management.

Conflicts of Interest: All authors declare that they do not have a conflict of interests.

References

1. Overton, T.R.; Yasui, T. Practical applications of trace minerals for dairy cattle. *J. Anim. Sci.* **2014**, *92*, 416–426. [CrossRef] [PubMed]

2. Yu, L.L.; Browning, J.F.; Burdette, C.Q.; Caceres, G.C.; Chieh, K.D.; Davis, W.C.; Kassim, B.L.; Long, S.E.; Murphy, K.E.; Oflaz, R.; et al. Development of a kelp powder (*Thallus laminariae*) Standard Reference Material. *Anal. Bioanal. Chem.* **2018**, *410*, 1265–1278. [CrossRef] [PubMed]

3. Abudabos, A.M.; Okab, A.B.; Aljumaah, R.S.; Samara, E.M.; Abdoun, K.A.; Alhaidary, A.A. Nutritional Value of Green Seaweed (Ulva Lactuca) for Broiler Chickens. *Ital. J. Anim. Sci.* **2013**, *12*, 612–620. [CrossRef]

4. Pereira, M.; Tala, F.; Fernandez, M.; Subida, M.D. Effects of kelp phenolic compounds on the feeding-associated mobility of the herbivore snail Tegula tridentata. *Mar. Environ. Res.* **2015**, *112*, 40–47. [CrossRef] [PubMed]

5. Garima, K.; Bruce, R.; Glenn, S.; Nikhil, T.; Franklin, E.; Alan, C.; Jeff, H.; Balakrishnan, P. Feed supplementation with red seaweeds, Chondrus crispus and Sarcodiotheca gaudichaudii, affects performance, egg quality, and gut microbiota of layer hens. *Poult. Sci.* **2014**, *93*, 2991–3001.

6. Heins, B.J.; Chester-Jones, H. Effect of feeding kelp on growth and profitability of group-fed calves in an organic production system. *Prof. Ani. Sci.* **2015**, *31*, 368–374. [CrossRef]

7. Antaya, N.T.; Soder, K.J.; Kraft, J.; Whitehouse, N.L.; Guindon, N.E.; Erickson, P.S.; Conroy, A.B.; Brito, A.F. Incremental amounts of Ascophyllum nodosum meal do not improve animal performance but do increase milk iodine output in early lactation dairy cows fed high-forage diets. *J. Dairy Sci.* **2015**, *98*, 1991–2004. [CrossRef]

8. Pitta, D.W.; Indugu, N.; Kumar, S.; Vecchiarelli, B.; Sinha, R.; Baker, L.D.; Bhukya, B.; Ferguson, J.D. Metagenomic assessment of the functional potential of the rumen microbiome in Holstein dairy cows. *Anaerobe* **2016**, *38*, 50–60. [CrossRef]

9. Hess, M.; Sczyrba, A.; Egan, R.; Kim, T.W.; Chokhawala, H.; Schroth, G.; Luo, S.; Clark, D.S.; Chen, F.; Zhang, T. Metagenomic Discovery of Biomass-Degrading Genes and Genomes from Cow Rumen. *Science* **2011**, *331*, 463–467. [CrossRef]

10. Puniya, A.K.; Singh, R.; Kamra, D.N. *Rumen Microbiology: From Evolution to Revolution*; Springer: New Delhi, India, 2015.

11. An, D.; Cai, S.; Dong, X. Actinomyces ruminicola sp. nov., isolated from cattle rumen. *Int. J. Syst. Evol. Microbiol.* **2006**, *56*, 2043–2048. [CrossRef]

12. Hall, M.B.; Larson, C.C.; Wilcox, C.J. Carbohydrate source and protein degradability alter lactation, ruminal, and blood measures. *J. Dairy Sci.* **2010**, *93*, 311–322. [CrossRef] [PubMed]

13. Li, Z.; Bai, H.; Zheng, L.; Jiang, H.; Cui, H.; Cao, Y.; Yao, J. Bioactive polysaccharides and oligosaccharides as possible feed additives to manipulate rumen fermentation in rusitec fermenters. *Int. J. Biol. Macromol.* **2018**, *109*, 1088–1094. [CrossRef] [PubMed]

14. Vyas, D.; Beauchemin, K.A.; Koenig, K.M. Using organic acids to control subacute ruminal acidosis and fermentation in feedlot cattle fed a high-grain diet. *J. Anim. Sci.* **2015**, *93*, 3950. [CrossRef]

15. Varki, A. Biological roles of oligosaccharides: All of the theories are correct. *Glycobiology* **1993**, *3*, 97–130. [CrossRef] [PubMed]

16. Ransom-Jones, E.; Jones, D.L.; McCarthy, A.J.; McDonald, J.E. Thefibrobacteres: An important phylum of cellulose-degrading bacteria. *Microb. Ecol.* **2012**, *63*, 267–281. [CrossRef] [PubMed]

17. Rodríguez-Prado, M.; Calsamiglia, S.; Ferret, A. Effects of Fiber Content and Particle Size of Forage on the Flow of Microbial Amino Acids from Continuous Culture Fermenters. *J. Anim. Sci.* **2004**, *87*, 1413–1424. [CrossRef]

18. Naas, A.E.; Mackenzie, A.K.; Mravec, J.; Schuckel, J.; Willats, W.G.T.; Eijsink, V.G.H.; Pope, P.B. Do rumen bacteroidetes utilize an alternative mechanism for cellulose degradation? *MBIO* **2014**, *5*, 1401–1414. [CrossRef]

19. Ludwig, W.; Schleifer, K.H.; Whitman, W.B. Revised Road Map to the Phylum Firmicutes. In *Bergey's Manual® of Systematic Bacteriology*; Springer: New York, NY, USA, 2009; pp. 1–13.

20. Pan, X.H.; Yang, L.; Xue, F.G.; Xin, H.R.; Jiang, L.S.; Xiong, B.H.; Beckers, Y. Relationship between thiamine and subacute ruminal acidosis induced by a high-grain diet in dairy cows. *J. Anim. Sci.* **2016**, *99*, 8790–8801. [CrossRef]

21. Xue, F.; Pan, X.; Jiang, L.; Guo, Y.; Xiong, B.H. GC–MS analysis of the ruminal metabolome response to thiamine supplementation during high grain feeding in dairy cows. *Metabolomics* **2018**, *14*, 67. [CrossRef]

22. Pan, X.; Nan, X.; Yang, L.; Jiang, L.; Xiong, B. Thiamine status, metabolism and application in dairy cows: A review. *Br. J. Nutr.* **2018**, *120*, 491–499. [CrossRef]

23. Whitelaw, F.G.; Milne, J.S.; Chen, X.B. The effect of a rumen microbial fermentation on urea and nitrogen metabolism of sheep nourished by intragastric infusion. *Exp. Physiol.* **1991**, *76*, 91–101. [CrossRef] [PubMed]

24. Owens, F.N.; Zinn, R.A.; Kim, Y.K. Limits to starch digestion in the ruminant small intestine. *J. Anim. Sci.* **1986**, *63*, 1634. [CrossRef] [PubMed]

25. Devant, M.; Ferret, A.; Gasa, J.; Calsamiglia, S.; Casals, R. Effects of protein concentration and degradability on performance, ruminal fermentation, and nitrogen metabolism in rapidly growing heifers fed high-concentrate diets from 100 to 230 kg body weight. *J. Anim. Sci.* **2000**, *78*, 1667–1676. [CrossRef] [PubMed]

26. Pisulewski, P.M.; Okorie, A.U.; Buttery, P.J.; Haresign, W.; Lewis, D. Ammonia concentration and protein synthesis in the rumen. *Proc. Nutr. Soc.* **2010**, *32*, 759–766. [CrossRef]

27. Allison, M.J. Biosynthesis of amino acids by ruminal microorganisms. *J. Anim. Sci.* **1969**, *29*, 797–807. [CrossRef]

28. Wallace, R.J. Ruminal microbiology, biotechnology, and ruminant nutrition: Progress and problems. *J. Anim. Sci.* **1994**, *72*, 2992–3003. [CrossRef]

29. Ryle, M.; Ørskov, E.R. *Energy Nutrition of Rumen Micro-Organisms*; Springer: Dordrecht, The Netherlands, 1990; pp. 10–27.

30. Lock, A.L.; Bauman, D.E. Modifying milk fat composition of dairy cows to enhance fatty acids beneficial to human health. *Lipids* **2004**, *39*, 1197–1206. [CrossRef]

31. Storry, J.E.; Rook, J.A.F. The relationship in the cow between milk-fat secretion and ruminal volatile fatty acids. *Br. J. Nutr.* **1966**, *20*, 217–228. [CrossRef]

32. Kaufmann, S.; Rambeck, W.A. Iodine supplementation in chicken, pig and cow feed. *J. Anim. Physiol. Anim. Nutr.* **2010**, *80*, 147–152.

33. Davis, S.R.; Collier, R.J.; McNamara, J.P.; Head, H.H.; Sussman, W. Effects of thyroxine and growth hormone treatment of dairy cows on milk yield, cardiac output and mammary blood flow. *J. Anim. Sci.* **1988**, *66*, 70. [CrossRef]

Review

Effects of Propylene Glycol on Negative Energy Balance of Postpartum Dairy Cows

Fan Zhang [1,2], Xuemei Nan [1], Hui Wang [1], Yiguang Zhao [1], Yuming Guo [2] and Benhai Xiong [1,*]

[1] State Key Laboratory of Animal Nutrition, Institute of Animal Science, Chinese Academy of Agricultural Sciences, Beijing 100193, China; 18813015831@139.com (F.Z.); xuemeinan@126.com (X.N.); wanghui_lunwen@163.com (H.W.); zhaoyiguang@caas.cn (Y.Z.)

[2] State Key Laboratory of Animal Nutrition, College of Animal Science and Technology, China Agricultural University, Beijing 100193, China; guoyum@cau.edu.cn

* Correspondence: xiongbenhai@caas.cn; Tel.: +86-10-62816017; Fax: +86-10-62811680

Received: 15 July 2020; Accepted: 24 August 2020; Published: 28 August 2020

Simple Summary: After calving, the milk production of dairy cows increases rapidly, but the nutrient intake cannot meet the demand for milk production, forming a negative energy balance. Dairy cows in a negative energy balance have an increased risk of developing clinical or subclinical ketosis. The ketosis in dairy cows has a negative impact on milk production, dry matter intake, health, immunity, and reproductive performance. Propylene glycol can be used as an important gluconeogenesis in ruminants and can effectively inhibit the formation of ketones. Supplementary propylene glycol to dairy cows during perinatal is an effective method to alleviate the negative energy balance. This review summarizes the reasons and consequences of negative energy balance as well as the mechanism and effects of propylene glycol in inhibiting a negative energy balance in dairy cows. In addition, the feeding levels and methods of using propylene glycol to alleviate negative energy balance are also discussed.

Abstract: With the improvement in the intense genetic selection of dairy cows, advanced management strategies, and improved feed quality and disease control, milk production level has been greatly improved. However, the negative energy balance (NEB) is increasingly serious at the postpartum stage because the intake of nutrients cannot meet the demand of quickly improved milk production. The NEB leads to a large amount of body fat mobilization and consequently the elevated production of ketones, which causes metabolic diseases such as ketosis and fatty liver. The high milk production of dairy cows in early lactation aggravates NEB. The metabolic diseases lead to metabolic disorders, a decrease in reproductive performance, and lactation performance decline, seriously affecting the health and production of cows. Propylene glycol (PG) can alleviate NEB through gluconeogenesis and inhibit the synthesis of ketone bodies. In addition, PG improves milk yield, reproduction, and immune performance by improving plasma glucose and liver function in ketosis cows, and reduces milk fat percentage. However, a large dose of PG (above 500 g/d) has toxic and side effects in cows. The feeding method used was an oral drench. The combination of PG with some other additives can improve the effects in preventing ketosis. Overall, the present review summarizes the recent research progress in the impacts of NEB in dairy cows and the properties of PG in alleviating NEB and reducing the risk of ketosis.

Keywords: propylene glycol; dairy cows; postpartum; negative energy balance; ketosis

---

1. Introduction

The transition period from late pregnancy to early lactation is well known as the critical time for the production of cows [1]. During this period, the higher energy demand for milk production coupled with the reduction in dry matter intake (DMI) around calving means that a large number of dairy cows are in a state of negative energy balance (NEB). To support the energy requirement, the body fat and protein of dairy cows are mobilized for hepatic gluconeogenesis, which leads to the increase of non-esterified fatty acids (NEFA), β-hydroxybutyrate (BHBA), and ammonia in plasma [2]. Invariably, the dairy cows with high-producing ability have the risk of subclinical ketosis (SCK, hyperketonemia without clinical signs) or clinical ketosis (CK, hyperketonemia with clinical signs). Cows with ketosis have a greater risk of several diseases including displaced abomasum, infections of the reproductive tract, mastitis, cystic ovarian disease, leg problems, and diseases of the digit and foot [3]. Due to the rapid increase in energy demand for milk after calving, the NEB usually has an adverse impact on health and thus decreases animal welfare, production, and profitability [4].

Earlier treatment of ketosis is important to reduce future economic losses in modern high-yield dairy farms. The goal of ketosis treatment is to stimulate gluconeogenesis, increase plasma glucose, and decrease lipolysis [5]. Propionate is the major precursor for gluconeogenesis. However, the limited feed intake during early lactation restricts ruminal propionate supply to the liver, raising the requirement for alternative gluconeogenic precursors [6]. Propylene glycol (PG) is a precursor of ruminal propionate that can be rapidly absorbed from the rumen for gluconeogenesis in the liver [7]. It has long been used as a treatment against ketosis [8]. Experimental studies have shown that an oral drench of PG can be effective in increasing glucose and decreasing BHBA and NEFA in plasma [9]. Therefore, this paper reviews the effects of NEB in dairy cows, and the research progress about the properties of PG in alleviating NEB and reducing the risk of ketosis during postpartum in dairy cows.

2. The Formation of NEB in Dairy Cows

During the transition period, dairy cows experience a dramatically increased demand for nutrients for the growing fetus and the initiation of lactation [9]. In the postpartum period, the nutrient requirements for milk yield, milk fat, milk protein, and milk lactose increase rapidly and exceed the supplies from DMI [10]. In addition, the diet of dairy cows after calving changes sharply from being mainly forage-based to concentrate-rich [11]. The sudden energy requirement for milk production and the DMI lags behind, inducing the negative nutrient balance, especially the NEB. The NEB symptoms appear in postpartum, but the dynamic changes of the physiological and metabolic status are verified from the prepartum period [12]. Responding to the NEB, the cow physiologically triggers the lipomobilization of body fat reserves, thereby amounts of NEFA are released into the blood circulation [12]. The NEFA are used as a fuel source by muscle, incorporated into milk fat, and taken up by the liver in proportion to their supply [13]. However, the excessive mobilization of body fat reserves leads to the accumulation of triglycerides in the liver or conversion to ketone bodies (i.e., BHBA, acetone, and acetoacetate) and leads to the onset of ketosis, which has adverse effects on the health, milk productivity, and reproduction in dairy cows. It is well known that dairy cows already go into a period of NEB a few days before calving and that the NEB extends to a few weeks after calving, with the nadir of NEB during the first week of lactation [14]. The feed intake increases slowly at the beginning of lactation [15] and the NEB switches to a positive range after approximately 45 d of lactation [16]. Other diseases such as rumen distention, abomasum displacement, metritis, mastitis, and so on also lead to insufficient nutrient supply and trigger secondary ketosis in the early lactation period. Therefore, methods of decreasing the release of NEFA from adipose tissue are important to alleviate the NEB of dairy cows in early lactation.

3. The Effects of NEB in Dairy Cows

3.1. Increasing the Incidence of Metabolic Diseases

During the transition period, in order to meet the energy requirement for improved milk production, the rate of lipolysis excesses the lipogenesis, which results in the increasing serum NEFA of dairy cows. When the NEFA could not be completely oxidized to carbon dioxide, it will partly be oxidized to ketone bodies or be stored in the liver as triglycerides [9]. During the period of high metabolic demands, the increased hepatic oxidation of NEFA induces satiety, decreases feed intake, and increases fat mobilization [17]. The blood BHBA level of above 1.2 mmol/L or 1.4 mmol/L is related to impaired health and performance, and is a common and costly metabolic disease, which is called hyperketonemia [18]. The high concentrations of NEFA and BHBA have negative effects on the health and productivity of dairy cows. Therefore, the physiological conditions associated with insufficient energy supply predispose dairy cows to metabolic and microbial diseases such as milk fever, endometritis, ketosis, displaced abomasum, and retained placenta [11].

The epidemiology of ketosis in dairy cows in early lactation increases the risk of displaced abomasum. Each 0.1 mmol/L increases in BHBA at the first SCK-positive test increased the risk of developing a displaced abomasum by a factor of 1.1 [95% confidence interval (CI) = 1.0 to 1.2] and increased the risk of removal from the herd by a factor of 1.4 (95% CI = 1.1 to 1.8) [19]. Raboisson et al. [20] summarized the association between SCK and displaced abomasum in 38 models from 10 publications and found that the risk (95% CI) of left displaced abomasum in cows with SCK were 3.55 (2.60–4.25). Fatty liver is also a metabolic disorder of dairy cows relating to NEB in early lactation. Fatty liver develops when the hepatic uptake of lipids exceeds the oxidation and secretion of lipids by the liver and thereby causes accumulation of triacylglycerol (TAG) in the liver [21]. The severe fatty liver causes metabolic dysfunction, which will reduce the hepatic metabolism, defense function, and insulin sensitivity [12]. The results of Fiore et al. [22] showed that fatty liver already developed before parturition, and increased from moderate to severe in 10 days after calving and then progressively disappeared. Therefore, methods of preventing hepatic lipidosis should be applied during this period.

3.2. Decreasing the Milk Productivity Performance of Dairy Cows

McArt et al. [19] concluded that each 0.1 mmol/L increase in BHBA at the first SCK-positive test was associated with a decrease in milk production by 0.5 kg/d for the first 30 days in milk (DIM). The cows with CK were lower in milk production and milk protein content, but milk fat content was higher than healthy cows [23]. A high percentage of fat and a low percentage of protein in the milk were associated with significant increases in the risk of SCK [24]. The mean fat to protein percentage ratio (FPR) and the frequency of FPR > 1.5 were higher in ketosis cows than healthy cows [25]. Therefore, the FPR of milk in early lactation is negatively correlated with energy balance and has been used as an indication of ketosis. The optimal FPR values are 1.05 to 1.18, while FPR values higher than 1.3 or 1.5 suggest a severe NEB and SCK [26]. In the cows with ketosis, the plasma NEFA from fat mobilization provided the precursor for milk fat synthesis in the mammary gland. The results [23] of in vivo and in vitro data indicated that NEFA could induce cell death-inducing DNA fragmentation factor-α-like effector A (CIDEA) expression in bovine mammary epithelial cells, leading to upregulation of de novo fatty acid synthesis enzymes (fatty acid synthase and acetyl-CoA carboxylase 1) and milk lipid secretion proteins (butyrophilin and xanthine dehydrogenase), thereby contributing to an increase in milk fat content in CK cows. The decrease in milk protein percentage might be related to the increased amino acid requirements for gluconeogenesis in ketosis cows, and the spared would be limited for protein synthesis in mammary gland.

3.3. Decreasing the Reproductive Performance

The reproductive performance of cows is one of the most important factors affecting the economic benefits of dairy production. The duration and severity of dairy cows' NEB in early postpartum are also related to reproductive performance. Extensive mobilization of fat has detrimental effects on liver function due to the accumulation of TAG, impairing the detoxification of ammonia into urea [27]. The NEB of dairy cows in early lactation will also increase the mobilization of protein, which will increase the metabolic residues of ammonia and urea. Ammonia is believed to play a role in starting before ovulation, whereas urea mainly interferes negatively after fertilization [28]. The first ovulation of a dairy cow is retarded by decreasing the luteinizing hormone (LH) pulsatility because of the low blood glucose [29].

The calving-to-first-service interval was 8 d longer and the calving-to-conception interval was 16 to 22 d longer in cows with SCK than in healthy cows [20]. Rutherford et al. [30] established that the SCK cows prolonged calving to the first estrus, calving to first insemination and calving to pregnancy intervals, and the first insemination was 4.3 times less likely to be successful compared to non-SCK cows. The high BHBA values, before, after, or before and after artificial insemination were reported associated with a six to 14% reduction in the pregnancy per artificial insemination compared with cows with low BHBA values [31]. Najm et al. [32] also showed the activity of healthy cows exceeded the ketosis cows by an average of 52.6% in 4–70 DIM and the mean motion activity on the day of estrus was also higher in healthy cows. The activity level of the cow will also affect the effective monitoring of estrus, which may be a factor decreasing the reproductive performance, especially detecting estrus by automated surveillance systems. The uterine inflammation could also be exacerbated by the elevated circulating concentration of BHBA or NEFA in early postpartum [33], which would delay uterine involution and successful conception. Therefore, severe NEB will reduce the reproductive performance of cows by delaying uterine recovery, prolonging calving to the first estrus, and reducing estrus activity and successful conception rate.

3.4. Inducing Immunosuppression

During the period from late pregnancy to early lactation, the NEB of dairy cows increases the risk of metabolic and infectious diseases. The metabolic status of early-lactating cows is known to affect the immune response to pathogens and impose immune challenges [34]. In this period, the NEB decreases the efficiency in pathogen clearance and increases the magnitude and duration of inflammation [35]. As a consequence, cows are more susceptible to several economically important disease such as metritis and mastitis [36]. In ketosis cows, the inflammation biomarkers of serum amyloid A, haptoglobin, and lipopolysaccharide binding protein are increased when compared with healthy counterparts [37]. The increase in circulating NEFA impairs peripheral blood mononuclear cells and polymorphonuclear leukocytes function, along with a weakening of those cells' phagocytosis capacity and a decrease in their ability to fight bacteria [38]. The inflammatory state in early lactation may disrupt normal nutrient partitioning and decrease the productivity of dairy cows [39].

Greater concentrations of both NEFA and BHBA have been associated with impaired immune functions and mastitis in dairy cows [40]. Glucose is considered as the preferred substrate for the immune system [41], and the activation of an immune response requires energy [42]. Serum glucose levels in cows with severe NEB are significantly reduced during early lactation, affecting the energy supply of the immune system. The increase in lipid infiltration in the liver also decreases the immune response. Additionally, cows with ketosis (blood BHBA > 3 mmol/L) have higher serum concentrations of proinflammatory cytokines interleukin (IL) 18, tumor necrosis factor (TNF)-α, and IL1B, and lower concentration of anti-inflammatory cytokine IL-10 [43]. The somatic cell count (SCC) in milk is closely related to the immune status of dairy cows. Van Straten et al. [44] concluded that the odds of an event of SCC > 250,000 cells/mL or SCC > 400,000 cells/mL were 44% and 33% greater for cows with ketosis when compared with cows without, respectively. In the results of Abuajamieh et al. [37] ketosis cows had increased circulating markers (serum amyloid A, haptoglobin, and lipopolysaccharide binding protein) of inflammation pre- and post-calving and before the clinical signs of ketosis. The higher prepartum NEFA increases the risk for metritis [45]. The infection and inflammation noticeably redirect resources toward the immune system and away from the utilization and synthesis of economically relevant products [41]. Therefore, severe NEB will lead to fatty liver and high serum NEFA in cows, which also contributes to immunosuppression and increases the risk of infections in the postpartum period. Inflammation postpartum upregulated immune gene expression and mitochondrial uncoupling further increase energy requirements, which exacerbates severe NEB status in cows [11].

4. The Anti-Ketogenic Properties of PG and the Mechanism of Inhibiting NEB

During early lactation, glucose synthesis should be increased to accommodate mammary demands [46]. To avoid the occurrence of dairy cow ketosis, it is important to provide extra gluconeogenesis for dairy cows. In 1954, PG was observed to be an effective treatment of ketosis in dairy cows [8]. PG supplementation appears to increase milk yield with a slight decrease in milk fat and an increase in milk lactose percentage [47]. Propylene glycol (1, 2-propanediol; $C_3H_8O_2$) is a sweet, hygroscopic, viscous liquid that has a gluconeogenic property and is routinely used because of its therapeutic effects on cows suffering from ketosis, based on the premise that it rapidly increases blood glucose [48]. As gluconeogenic precursors, it has been proven that PG is more effective at increasing plasma glucose concentration than glycerol, since 300 mL PG is at least as effective as 600 mL of glycerol [49]. Plasma concentrations of glucose and insulin are known to increase in response to dietary PG [50,51]. Propionate is the main product of PG fermentation, which can be rapidly metabolized with short lag time [52]. This is beneficial for cows to alleviate the NEB and anti-ketogenic. After oral administration, the majority of PG escapes from the rumen wall or gastrointestinal tract and is converted to glucose by the liver [53]. However, the other mechanism of the effects of PG involves the successive production of propionate together with propanal and with the latter being converted to propanol in the rumen, which in turn is converted to propionate and thereafter glucose in the liver [54]. The main effect of PG is to increase the glucogenic status, and as a consequence, the concentration of plasma BHBA is reduced and the cows have decreased risk of developing ketosis [55]. PG is metabolized to lactate, acetate, and pyruvate in the liver. Lactate enters gluconeogenesis via pyruvate, which can be converted to oxaloacetate. The concentration of oxaloacetate is the key metabolite in determining if the acetyl-CoA enters the tricarboxylic acid (TCA)-cycle or ketogenesis [7]. When the oxaloacetate is insufficient for citrate synthase to combine with acetyl-CoA, the excessive acetyl-CoA is then partitioned toward ketone synthesis [37]. The anti-ketogenic properties of PG are partly due to increasing the oxidation of acetyl-CoA into the TCA-cycle and the supply of gluconeogenic glucose [7]. The detailed anti-ketogenic pathways of PG are shown in Figure 1.

In addition, insulin resistance is an adaptation to the very high glucose requirements for lactation, thereby conserving glucose for lactation by limiting its use by insulin-sensitive tissues (muscle, adipose tissue etc.) [56]. The insulin resistance can hence be attributed to a decrease in insulin responsiveness and a decrease in insulin sensitivity [57]. The greater extent of insulin resistance in peripartal dairy cows can contribute to excessive adipose tissue lipolysis and thus greater metabolic disease risk [58]. Chalmeh et al. [59] confirmed that the supplementary feeding with PG reduced the insulin resistance in dairy cows during the transition period by the intravenous glucose tolerance test. The decrease in insulin resistance will inhibit lipolysis and decrease the metabolic disease risk in periparturient dairy cattle. Some researchers have found a negative effect of NEFA in the insulin sensitivity of dairy cattle [57]. Therefore, the effect of PG in decreasing the insulin resistance may be related to the property of PG as a main precursor of glucose and decreasing circulating NEFA.

The energy value of PG is 5.66 Mcal/kg, and according to the assumed PG metabolizable energy utilization efficiency for lactation (80%) of Miyoshi et al. [50], the net energy for lactation (NE_L) of PG was calculated to be 4.53 Mcal/kg. Due to the higher NE_L of PG, it can supply more energy intake than other concentrates for dairy cows in early lactation and reduce the incidence of ketosis.

Therefore, the effects of PG on alleviating NEB in dairy cows are mainly by improving the precursor for hepatic gluconeogenesis and increasing the oxidation of acetyl-CoA into the TCA-cycle. The high energy content of PG can increase the energy density of the diet for dairy cows. The fatty liver and ketone bodies of dairy cows will be inhibited with the increase in liver glucose synthesis.

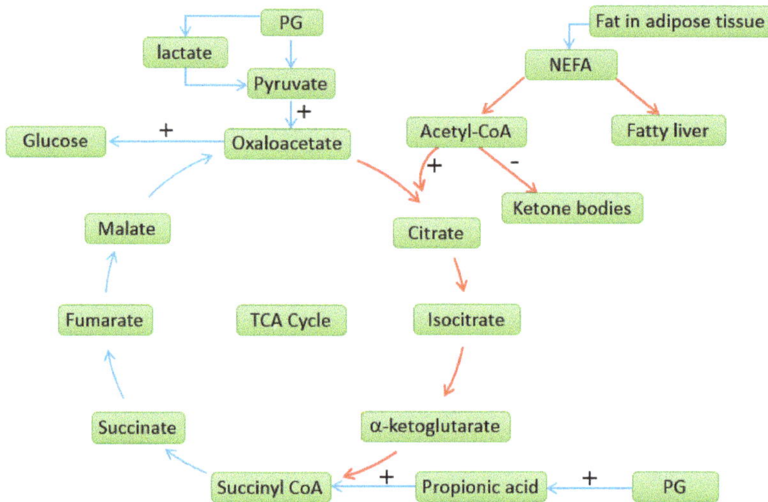

Figure 1. The anti-ketogenic pathways of propylene glycol (PG) in dairy cows [7]. The blue lines are to describe the gluconeogenic pathways of increasing the glucose to preventing ketosis. The red lines are the ways of increasing the oxidation of acetyl-CoA (coenzyme A) in the tricarboxylic acid cycle (TCA-cycle) and the supply of glucose by increasing the production of oxaloacetate, which will prevent acetyl-CoA convert to ketone bodies (β-hydroxybutyrate, acetone, and acetoacetate). PG can also reduce the triacylglycerol (TG) accumulation in liver.

5. Effects of PG on Alleviating NEB in Dairy Cows

5.1. The Effects of PG on DMI and Rumen Fermentation

The DMI at the postparturient stage has an essential effect on the NEB metabolism status of cows. The DMI of cows with CK is lower than healthy cows [23]. PG is considered unpalatable and usually reduces feed intake if not mixed thoroughly with other feed components or drenched [7]. Moallem et al. [60] reported the daily average DMI and NE_L intake from calving until 100 DIM was higher for cows supplemented with 500 g/d PG per cow until 21 DIM than the control group. The rumen fill score and body condition score (BCS) are also direct tools to evaluate the feed intake and energy balance status. The PG treatment improved the rumen fill score and lowed BCS loss in the dairy cows, which were diagnosed with ketosis in the results of Jeong et al. [61]. The increasing DMI and rumen fill scores are beneficial for decreasing the adverse impact of NEB.

The results of Kristensen and Raun [54] showed that infusion of PG did not affect ruminal pH or the total concentration of ruminal volatile fatty acids (VFA), but decreased the molar proportion of ruminal acetate and increased ruminal concentrations of PG, propanol, and propanal as well as the molar proportion of propionate. Chung et al. [62] also found the PG administration appeared to shift ruminal VFA patterns by producing more glucogenic VFA such as propionate and valerate at the expense of lipogenic VFA such as acetate. The increase of propionate and valerate can provide carbon sources for glucose biosynthesis, which is beneficial for dairy cows to alleviate NEB in early lactation. Acetate is the major source for milk fat synthesis in dairy cows [63]. Therefore, the decrease in acetate concentration may explain the decreased milk fat with PG supplementation.

5.2. The Effect of PG on Metabolic Index

It is widely accepted that PG has a glucogenic effect. The glucogenic status of the cows have effects on the liver metabolism of NEFA and, thereby, the regulation of ketogenesis [55]. The effects of PG treatment on SCK or CK have been explained by reduced adipose tissue mobilization, which leads to the decrease of NEFA in the liver and the reduction in the formation of ketone bodies [7]. As Sun et al. [12] reported, the dynamic changes of the physiological and metabolic status were from the prepartum period, so feeding of PG from prepartim is also a good method to alleviate NEB of dairy cows in postpartum. Juchem et al. [64] pointed out that prepartum PG administration increased concentrations of glucose and insulin, and decreased BHBA and NEFA in plasma. It has been validated that the dairy cows suffered deficiency of energy before calving, so the nutritional strategies should be implemented at the start of the prepartum period [12]. The supplement of PG to dairy cows before calving is effective in inhibiting the occurrence of cow ketosis. Therefore, prepartum PG administration has a glucogenic effect for dairy cows in postpartum.

Supplement PG to dairy cows in early lactation is also an important way to avoid energy metabolic diseases. The study of Butler et al. [48] found that drenching 500 mL PG to the dairy cow diet had significant beneficial effect on energy balance and increased plasma insulin and glucose, while the plasma NEFA was decreased. Kristensen and Raun [54] confirmed that when cows were dosed with PG, the plasma concentrations of PG, ethanol, propanol, propanal, glucose, L-lactate, propionate, and insulin were increased. Therefore, PG regulates the metabolism of cows by increasing the supply of L-lactate and propionate to gluconeogenesis and reducing insulin resistance. Insulin is a key hormone in the regulation of lipolysis in adipocytes. The increase of insulin is also useful for alleviating NEB for dairy cows. Bjerre-Harpøth et al. [55] observed 4-week postpartum PG allocation enhanced glucogenic status, which decreased plasma concentration of BHBA and increased plasma concentration of glucose, but had limited effect on adipose tissue mobilization. Although the metabolic changes in Simmental cows in the periparturient period were not as significant as in the case of Holstein-Friesian cows, the application of PG also resulted in higher milk yield, BCS, and serum glucose content [65]. Therefore, PG can enhance glucogenic status, and decrease the plasma NEFA and BHBA concentrations.

Displaced abomasum, fatty liver, and ketosis are common nutritional metabolic diseases of cows in the postpartum period. PG, as a glucogenic precursor of ruminants, plays an important role in inhibiting metabolic diseases caused by NEB in dairy cows. The results of McArt et al. [66] showed the cows with SCK were 1.6 times more likely to develop displaced abomasum and 2.1 times more likely to be dead or sold than SCK cows treated with PG within the first 30 DIM. The reasons for PG administration decreasing displaced abomasum and the ratio to be removed from the herd are that PG contributes to prevent ketosis and improve milk production. The results of Rukkwamsuk et al. [14] indicated dairy cows drenched with PG from seven days prepartum to seven days postpartum could reduce the risk of fatty liver. This is in accordance with PG decreasing the NEFA in plasma, which will subsequently reduce the TAG accumulation in the liver. Fatty liver is a major metabolic disease of dairy cows in early lactation. The main indicators of hepatic lesions and alterations of its function are the enzymes aspartate transaminase (AST), gamma-glutamyl transferase (GGT), and the blood metabolites glucose, cholesterol, and albumin [67]. The study of Hussein et al. [68] found that PG supplementation has the ability to reduce the enzyme activities of AST and GGT and improve serum glucose, but had no effect on the serum concentrations of total cholesterol and albumin. The PG treatment can thus reduce liver lesions. Stokes and Goff [69] reported offering PG at calving had the effects of lowering the health disorder risk (retained placenta, ketosis, hypocalcemia, displaced abomasum, and metritis etc.) in dairy cows. Feeding PG to SCK cows can effectively prevent the formation of ketone bodies, which will prevent SCK cows developing into CK cows. McArt et al. [70] showed that 300 mL/d of PG treated cows were 1.50 times more likely to resolve their SCK ($1.2 \leq$ BHBA < 3.0 mM/L) and 0.54 times less likely to develop CK (BHBA ≥ 3.0 mM/L) than the control cows. The reduction in plasma NEFA and increase in plasma glucose are related to the anti-ketogenic property of PG. The use of PG is likely to produce more propionate as the main precursor of glucose; therefore, it can reduce the NEB and insulin resistance [59]. Therefore, the PG supplement to dairy cows can decrease nutritional metabolic diseases in early lactation. The cows drenched with PG could improve the molar proportion of ruminal propionate and hepatic gluconeogenesis, which results in an elevation in serum glucose and a decrease in serum NEFA and BHBA. Drenching of PG during the transition period is therefore beneficial for dairy cows to alleviate NEB in the postpartum period.

5.3. The Effects of PG on Milk Production

In the study of Lomander et al. [9], cows supplemented with 300 g of liquid PG daily in the first 21 DIM trended to yield more milk (0.94 kg, 95% CI = −0.03–1.91) compared with control cows during the first 90 DIM, but no difference was found in energy-corrected milk. In the trial of Østergaard et al. [71] based on milk spectra analyses, the results of the treatment with PG (500 mL for 5 d) showed only few benefits in early lactation for physiologically imbalanced cows. The study of Juchem et al. [64] showed that prepartum PG administration had no effect on milk production during the first nine-weeks postpartum. Butler et al. [48] observed that when multiparous Holstein cows received 500 mL oral drench PG from d 10 before expected parturition to d 25 postpartum could increase milk lactose and tended to reduce milk fat content, but there was no difference in milk yield and milk protein percentage. McArt et al. [70] concluded that an oral dose of PG improved milk yield during early lactation in cows with SCK. Stokes and Goff [69] determined that the cows received PG within 4 h of calving and again 24 h post-calving had 3.1 kg/d greater milk production. Therefore, it is conducive to the improvement of postpartum milk performance when cows received PG after calving as soon as possible. PG can provide enough energy to support the increase of milk yield, especially to SCK cows. However, some reports showed there was no difference in milk yield. This may be because the dosages used of PG reduced the feed intake or it was used in the cows without ketosis. The reduced milk fat content affected by PG could be due to the decrease of plasma NEFA and the lower acetate in the rumen [62]. PG had no effect in milk protein possibly because there was no shortage of amino acids for milk protein synthesis. Glucose is necessary for dairy cows to synthesize milk lactose. When PG can supply enough energy and be converted to enough glucose, milk lactose will increase. Therefore, PG tends to increase milk yield and milk lactose, reduce the milk fat of ketosis cows, but has little effect on milk protein.

5.4. The Effects of PG on Reproductive Performance

The insufficient energy intake can result in poor reproductive performance such as prolonging postpartum anestrus, decreasing progesterone production by the corpus luteum, and reducing rates of conception [50]. As PG can alleviate the NEB, it therefore also effectively prevents the degradation of reproductive performance.

The results of Gamarra et al. [56] indicated that short-term dietary PG supplementation affected circulating concentrations of metabolites and metabolic hormones, increased progesterone concentrations, and the number of small follicles. The embryo losses were related to the reduced progesterone and the increase in progesterone stimulates and sustains endometrial functions essential for embryonic survival, implantation, and growth [72]. The increase in the number of small follicles is beneficial for early estrus and conception. McArt et al. [66] confirmed that oral administration of PG to SCK cows were 1.3 times more likely to conceive at first insemination than untreated cows. Insulin is necessary for maximal steroidogenesis in both follicular and luteal cells. Miyoshi et al. [50] suggested 500 mL/d of PG administration to NEB dairy cows was able to improve ovarian function in early lactation, which is due to PG induced insulin spike. PG can improve plasma glucose and stimulate insulin secretion. Thus, the increased plasma insulin in PG treated cows has effects on follicular development and LH secretion, leading to earlier ovulation [50]. However, the results of Castañeda-Gutiérrez et al. [73] showed that there was no difference in days to first ovulation in multiparous cows after daily topdressing with PG from last 21 d before expected calving to 21 DIM. Butler et al. [48] observed drenching of PG had no effect on the number of cows with follicles ovulating, undergoing atresia, or becoming cystic, but the day of maximum follicle diameter was earlier for PG treatment. PG treatment advanced the day of maximum follicle diameter, indicating that it promotes follicular growth and is conducive to early estrus.

From the above reports, it can be found that the results for the PG treatment are inconsistent. This may be due to some studies not determining the ketosis status of dairy cows or feeding the PG in diet instead of oral drench. The transient elevations in insulin and glucose, decreases in NEFA, and modest improvement in energy balance are insufficient to adequately stimulate the hypothalamic–pituitary–ovarian axis [48], especially to the dairy cows that do not have a ketosis status. However, to those cows with ketosis, PG can improve plasma glucose and decrease NEFA and therefore effectively improve reproductive performance.

5.5. The Effects of PG on Immune Performance

During the peripartum period, dairy cows experience the state of ketosis and fatty liver, which reduces the liver function coupled with increased inflammation and oxidative stress [74]. The cows drenched with PG have a remarkable reduction in TAG accumulation in the liver [14]. There are few direct studies of PG treatment on cow immunity in early lactation. However, metritis and milk SCC can indirectly reflect the immune status of cows. The cows that received PG at calving had significantly lower incidence of metritis [69]. Formigoni et al. [75] observed the mean linear SCC in the first 13 weeks of lactation period was reduced by PG administration (300 g/d from 10 d prior to expected calving until parturition and 300 g/d on days 0, 3, 6, 9, 12 d postpartum). PG is beneficial to increase the serum glucose concentration of postpartum cows and effectively inhibit the risk of fatty liver through gluconeogenesis. Therefore, feeding PG to high-ketone cows can improve the liver function by reducing the accumulation of liver fat, thus improving the immune function of cows.

6. The Toxicity and Side Effects of PG

Although PG can prevent ketosis, large doses (>500 g/d) have toxic and side effects on dairy cows due to the toxic compounds of PG during metabolism processes [12]. The clinical signs of PG in toxic doses include depression, ataxia, and excessive salivation as well as abnormal, malodorous, and foul breath and feces in dairy cows [76]. Farmers and veterinarians in Denmark found that some cows expressed the toxicity and side effects of PG [7]. PG toxicity causes oxygen saturation of arterial blood hemoglobin and the oxygen pressure in arterial blood decreases, along with the appearance of dyspnea and ruminal atony upon intake of concentrate containing PG [77]. The sulfur-containing gases produced during PG fermentation in rumen contribute to the toxic effects in rumen when high doses are administered for therapeutic purposes [76]. Hydrogen sulfide is an important signaling substance in hypoxic vasoconstriction, which can explain the link between PG application to the rumen and the dyspnea [77]. The toxicity and side effects of PG limit the maximum dosage in dairy cows to reduce the risk of ketosis. However, the side effects of PG are related to individual cow susceptibility, so it is important to consider the signs of toxicity in the administering of PG, especially at dosages above 500 g per day [7].

7. The Feeding Level and Method of PG

The cows drenched with PG could improve the molar proportion of ruminal propionate and hepatic gluconeogenesis, which results in an elevation of serum glucose and decrease in serum NEFA and BHBA. Therefore, the drenching of PG during the transition period is beneficial for the dairy cows to alleviate NEB postpartum. However, the feeding level and method may affect the effects of PG administration.

Gordon et al. [5] pointed out that feeding 300 g of PG daily to ketonic animals should be considered as the base of ketosis treatment. Due to the toxicity and side effects of PG, the maximum feeding level of PG is 500 g per day. The different parity cows have different milk production abilities, so the risk of ketosis is different, which will also affect the PG application strategy. The first lactation heifers had a 47% reduction in the risk of excessive NEB compared to the older cows because of the lower milk yields and lighter bodyweights [78]. Therefore, the use of PG to alleviate NEB should be primarily applied to multiple parity dairy cows.

PG usually exists in liquid form, which is not convenient for feeding in routine by oral drenching to dairy cows. Chung et al. [79] verified feeding PG as a dry product (65% PG and 35% silicon dioxide as the carrier) in total mixed rations (TMR) also reduced plasma BHBA concentration. However, the same amount of PG as a top dress (500 g/d of cookie meal (dried bakery by-product) mixed with dry PG) was more efficient than incorporating it into TMR. The administration method seems to be of importance for the metabolic response of PG in cows because the response of allocating PG as an oral drench or in a separately fed concentrate, is better than mixing into TMR [55]. When PG is added to TMR, the chronic delivery of PG alters the environment in the rumen and inhibits more propionate production, which would decrease the feed intake, increase fat mobilization, and perpetuate the problem of ketosis [5]. Thus, PG is best administered as an oral drench. Meanwhile, feeding PG (mixed with other carriers) in dry product is beneficial to decreasing labor.

Identifying the minimum effective durations for the treatment of ketosis is important for giving PG orally. Gordon et al. [80] observed that extended PG treatment from 3 d to 5 d increased milk production by 3.4 kg/d among the dairy cows that had low blood glucose (<2.2 mmol/L) and 1.7 times more likely to cure with blood BHBA > 2.4 mmol/L, but had no significant effect on milk production in blood glucose $\geq$ 2.2 mmol/L and cure risk with blood BHBA between 1.2 to 2.4 mmol/L. The additional treatment times to the lower blood BHBA concentration cows might have had little benefit with the stress of increasing labor. McArt et al. [81] confirmed that testing cows 2 days per week from 3 to 9 DIM and treating all positive cows with 300 mL of oral PG for 5 d were the most cost-effective strategy for herds with hyperketonemia incidences between 15% and 50%. For those above 50%, treating all fresh cows with 5 d of PG was the most cost-effective strategy. Therefore, the incidence and degree of hyperketonemia also influence the PG administration durations for treatment. As 75% of cows that developed SCK could be tested as positive within 1 week postpartum (with a peak at 5 DIM) [19], the test of ketones, and treatment of PG should be mainly done in this period. The economic benefit is also a factor influencing the application of PG in preventing ketosis in dairy cows. EI-Kasrawy et al. [82] found that continuous drenching of PG with 300 mL/d for long durations during the transition of dairy cows had a higher net return (1908.52 US$/cow) than drenching for a short duration in 400 mL/d (1171.34 US$/cow) and the control group (1440.21 US$/cow). The economic benefit is affected by the improved production and reproduction performance and the decreased treatment cost of diseases and incidence of early removal from herd. In general, 300 g of PG daily is the basic treatment and the maximum level is 500 g per day. The best feeding method of PG was administered as an oral drench from within one week postpartum and primarily applied to the multiple parities. However, the durations for the treatment of PG for ketosis need to consider the degree of hyperketonemia.

8. The Research of Combination Therapy

PG plays an important role in the treatment of ketosis in cows. With 5 d of PG therapy, the rate of cure from hyperketonemia was improved. However, approximately 40% of cows still remained hyperketonemia [66]. The feeding level of PG is restricted due to its potentially toxic effects [8]. Therefore, the use of PG alone still has a poor effect on the treatment of ketosis in some cows. So, the combination of PG with other additives may have a better effect in inhibiting ketosis in dairy cows. There has been a lot of research in this field, but different additives have different effects. The addition of dexamethasone [18] and glycerol [49] to PG showed no additional benefits and the fat appeared to blunt the metabolic response to PG administration of cows [13]. The combination of butaphosphan-cyanocobalamin [80], glucocorticoids [83], L-carnitine-methionine [61], and glucose [84] with PG to the cows of hyperketonemia are beneficial in improving the chances of the resolution of ketosis compared to PG only.

9. Conclusions

The NEB in early lactation reduces cow productivity and reproductive performance, and induces immunosuppression, increasing the chance of dairy cows being eliminated. As a precursor of gluconeogenesis, the addition of PG can provide energy and glucose for cows, thus preventing metabolic diseases such as ketosis and fatty liver as well as increasing milk yield and reducing milk fat percentage. PG can also increase the reproductive performance and immune function of cows due to glucose enhancement. However, due to the toxicity and side effects, PG is used in doses within 500 g/d per cow and offered in oral drench. To improve the effects in preventing ketosis, PG is better used with other additives.

Author Contributions: F.Z. wrote the manuscript; X.N., H.W., and Y.Z. amended the English grammar and designed the figures; Y.G. and B.X. discussed and revised the draft. All authors have read and agreed to the published version of the manuscript.

Funding: This research was funded by the National Key R&D Program of China (2017YFD0701604), the Central Public-Interest Scientific Institution Basal Research Fund (No. 2019-YWF-YTS-10, 2019-YWF-YB-04), and the Key R&D Program of Guangdong Province (2019B020215002-3).

Acknowledgments: We thank the National Agricultural Information System, Chinese Academy of Agricultural Sciences (CAAS), Beijing, 100193, China for providing access to the literature during the COVID-19 pandemic.

Conflicts of Interest: The authors declare no conflict of interest.

References

1. Macrae, A.I.; Burrough, E.; Forrest, J.; Corbishley, A.; Russell, G.; Shaw, D.J. Prevalence of excessive negative energy balance in commercial united kingdom dairy herds. *Vet. J.* **2019**, *248*, 51–57. [CrossRef] [PubMed]

2. Moore, S.M.; DeVries, T.J. Effect of diet-induced negative energy balance on the feeding behavior of dairy cows. *J. Dairy Sci.* **2020**, *103*, 7288–7301. [CrossRef]

3. Lean, I.J. Non-infectious disease: Ketosis. In *Reference Module in Food Science*; Elsevier: Amsterdam, The Netherlands, 2020. [CrossRef]

4. Rousing, T.; Holm, J.R.; Krogh, M.A.; Ostergaard, S.; Gplus, E.C. Expert-based development of a generic haccp-based risk management system to prevent critical negative energy balance in dairy herds. *Prev. Vet. Med.* **2020**, *175*, 104849. [CrossRef] [PubMed]

5. Gordon, J.L.; LeBlanc, S.J.; Duffield, T.F. Ketosis treatment in lactating dairy cattle. *Vet. Clin. N. Am. Food Anim. Pract.* **2013**, *29*, 433–445. [CrossRef] [PubMed]

6. Chibisa, G.E.; Gozho, G.N.; Van Kessel, A.G.; Olkowski, A.A.; Mutsvangwa, T. Effects of peripartum propylene glycol supplementation on nitrogen metabolism, body composition, and gene expression for the major protein degradation pathways in skeletal muscle in dairy cows. *J. Dairy Sci.* **2008**, *91*, 3512–3527. [CrossRef]

7. Nielsen, N.I.; Ingvartsen, K.L. Propylene glycol for dairy cows—A review of the metabolism of propylene glycol and its effects on physiological parameters, feed intake, milk production and risk of ketosis. *Anim. Feed Sci. Technol.* **2004**, *115*, 191–213. [CrossRef]

8. Johnson, R.B. The treatment of ketosis with glycerol and propylene glycol. *Cornell Vet.* **1954**, *44*, 6–21.

9. Lomander, H.; Frossling, J.; Ingvartsen, K.L.; Gustafsson, H.; Svensson, C. Supplemental feeding with glycerol or propylene glycol of dairy cows in early lactation-effects on metabolic status, body condition, and milk yield. *J. Dairy Sci.* **2012**, *95*, 2397–2408. [CrossRef]

10. Bertoni, G.; Trevisi, E.; Lombardelli, R. Some new aspects of nutrition, health conditions and fertility of intensively reared dairy cows. *Ital. J. Anim. Sci.* **2009**, *8*, 491–518. [CrossRef]

11. Esposito, G.; Irons, P.C.; Webb, E.C.; Chapwanya, A. Interactions between negative energy balance, metabolic diseases, uterine health and immune response in transition dairy cows. *Anim. Reprod. Sci.* **2014**, *144*, 60–71. [CrossRef]

12. Sun, B.F.; Cao, Y.C.; Cai, C.J.; Yu, C.; Li, S.X.; Yao, J.H. Temporal dynamics of nutrient balance, plasma biochemical and immune traits, and liver function in transition dairy cows. *J. Integr. Agric.* **2020**, *19*, 820–837. [CrossRef]

13. Pickett, M.M.; Piepenbrink, M.S.; Overton, T.R. Effects of propylene glycol or fat drench on plasma metabolites, liver composition, and production of dairy cows during the periparturient period1. *J. Dairy Sci.* **2003**, *86*, 2113–2121. [CrossRef]

14. Rukkwamsuk, T.; Rungruang, S.; Choothesa, A.; Wensing, T. Effect of propylene glycol on fatty liver development and hepatic fructose 1,6 bisphosphatase activity in periparturient dairy cows. *Livest. Prod. Sci.* **2005**, *95*, 95–102. [CrossRef]

15. Harder, I.; Stamer, E.; Junge, W.; Thaller, G. Lactation curves and model evaluation for feed intake and energy balance in dairy cows. *J. Dairy Sci.* **2019**, *102*, 7204–7216. [CrossRef] [PubMed]

16. Grummer, R.R.; Rastani, R.R. When should lactating dairy cows reach positive energy balance? *Prof. Anim. Sci.* **2003**, *19*, 197–203. [CrossRef]

17. Allen, M.S.; Bradford, B.J.; Oba, M. Board invited review: The hepatic oxidation theory of the control of feed intake and its application to ruminants. *J. Anim. Sci.* **2009**, *87*, 3317–3334. [CrossRef]

18. Tatone, E.H.; Duffield, T.F.; Capel, M.B.; DeVries, T.J.; LeBlanc, S.J.; Gordon, J.L. A randomized controlled trial of dexamethasone as an adjunctive therapy to propylene glycol for treatment of hyperketonemia in postpartum dairy cattle. *J. Dairy Sci.* **2016**, *99*, 8991–9000. [CrossRef]

19. McArt, J.A.A.; Nydam, D.V.; Oetzel, G.R. Epidemiology of subclinical ketosis in early lactation dairy cattle. *J. Dairy Sci.* **2012**, *95*, 5056–5066. [CrossRef]

20. Raboisson, D.; Mounié, M.; Maigné, E. Diseases, reproductive performance, and changes in milk production associated with subclinical ketosis in dairy cows: A meta-analysis and review. *J. Dairy Sci.* **2014**, *97*, 7547–7563. [CrossRef]

21. Bobe, G.; Young, J.W.; Beitz, D.C. Invited review: Pathology, etiology, prevention, and treatment of fatty liver in dairy cows. *J. Dairy Sci.* **2004**, *87*, 3105–3124. [CrossRef]

22. Fiore, E.; Piccione, G.; Perillo, L.; Barberio, A.; Manuali, E.; Morgante, M.; Gianesella, M. Hepatic lipidosis in high-yielding dairy cows during the transition period: Haematochemical and histopathological findings. *Anim. Prod. Sci.* **2017**, *57*, 74–80. [CrossRef]

23. Sun, X.; Wang, Y.; Loor, J.J.; Bucktrout, R.; Shu, X.; Jia, H.; Dong, J.; Zuo, R.; Liu, G.; Li, X.; et al. High expression of cell death-inducing dffa-like effector a (cidea) promotes milk fat content in dairy cows with clinical ketosis. *J. Dairy Sci.* **2019**, *102*, 1682–1692. [CrossRef] [PubMed]

24. Steensels, M.; Maltz, E.; Bahr, C.; Berckmans, D.; Antler, A.; Halachmi, I. Towards practical application of sensors for monitoring animal health; design and validation of a model to detect ketosis. *J. Dairy Res.* **2017**, *84*, 139–145. [CrossRef] [PubMed]

25. Koeck, A.; Miglior, F.; Jamrozik, J.; Kelton, D.F.; Schenkel, F.S. Genetic associations of ketosis and displaced abomasum with milk production traits in early first lactation of canadian holsteins. *J. Dairy Sci.* **2013**, *96*, 4688–4696. [CrossRef] [PubMed]

26. Duričić, D.; Ljubić, B.B.; Vince, S.; Turk, R.; Valpotić, H.; Žaja, I.Ž.; Maćešić, N.; Benić, M.; Getz, I.; Samardžija, M. Effects of dietary clinoptilolite supplementation on β-hydroxybutirate serum level and milk fat to protein ratio during early lactation in holstein-friesian cows. *Microporous Mesoporous Mater.* **2020**, *292*, 109766. [CrossRef]

27. Tamminga, S. The effect of the supply of rumen degradable protein and metabolisable protein on negative energy balance and fertility in dairy cows. *Anim. Reprod. Sci.* **2006**, *96*, 227–239. [CrossRef]

28. Albaaj, A.; Foucras, G.; Raboisson, D. Changes in milk urea around insemination are negatively associated with conception success in dairy cows. *J. Dairy Sci.* **2017**, *100*, 3257–3265. [CrossRef]

29. Jorritsma, R.; Wensing, T.; Kruip, T.A.M.; Vos, P.L.A.M.; Noordhuizen, J.P.T.M. Metabolic changes in early lactation and impaired reproductive performance in dairy cows. *Vet. Res.* **2003**, *34*, 11–26. [CrossRef]

30. Rutherford, A.J.; Oikonomou, G.; Smith, R.F. The effect of subclinical ketosis on activity at estrus and reproductive performance in dairy cattle. *J. Dairy Sci.* **2016**, *99*, 4808–4815. [CrossRef]

31. Albaaj, A.; Jattiot, M.; Manciaux, L.; Saille, S.; Julien, C.; Foucras, G.; Raboisson, D. Hyperketolactia occurrence before or after artificial insemination is associated with a decreased pregnancy per artificial insemination in dairy cows. *J. Dairy Sci.* **2019**, *102*, 8527–8536. [CrossRef]

32. Najm, N.-A.; Zimmermann, L.; Dietrich, O.; Rieger, A.; Martin, R.; Zerbe, H. Associations between motion activity, ketosis risk and estrus behavior in dairy cattle. *Prev. Vet. Med.* **2020**, *175*, 104857. [CrossRef] [PubMed]

33. Pascottini, O.B.; LeBlanc, S.J. Modulation of immune function in the bovine uterus peripartum. *Theriogenology* **2020**, *150*, 193–200. [CrossRef] [PubMed]

34. Gross, J.J.; Grossen-Rösti, L.; Wall, S.K.; Wellnitz, O.; Bruckmaier, R.M. Metabolic status is associated with the recovery of milk somatic cell count and milk secretion after lipopolysaccharide-induced mastitis in dairy cows. *J. Dairy Sci.* **2020**, *103*, 5604–5615. [CrossRef]
35. Mann, S.; Sipka, A.; Leal Yepes, F.A.; Nydam, D.V.; Overton, T.R.; Wakshlag, J.J. Nutrient-sensing kinase signaling in bovine immune cells is altered during the postpartum nutrient deficit: A possible role in transition cow inflammatory response. *J. Dairy Sci.* **2018**, *101*, 9360–9370. [CrossRef] [PubMed]
36. Sordillo, L.M.; Contreras, G.A.; Aitken, S.L. Metabolic factors affecting the inflammatory response of periparturient dairy cows. *Anim. Health Res. Rev.* **2009**, *10*, 53–63. [CrossRef]
37. Abuajamieh, M.; Kvidera, S.K.; Fernandez, M.V.S.; Nayeri, A.; Upah, N.C.; Nolan, E.A.; Lei, S.M.; DeFrain, J.M.; Green, H.B.; Schoenberg, K.M.; et al. Inflammatory biomarkers are associated with ketosis in periparturient holstein cows. *Res. Vet. Sci.* **2016**, *109*, 81–85. [CrossRef]
38. Lacasse, P.; Vanacker, N.; Ollier, S.; Ster, C. Innovative dairy cow management to improve resistance to metabolic and infectious diseases during the transition period. *Res. Vet. Sci.* **2018**, *116*, 40–46. [CrossRef]
39. Bertoni, G.; Trevisi, E.; Han, X.; Bionaz, M. Effects of inflammatory conditions on liver activity in puerperium period and consequences for performance in dairy cows. *J. Dairy Sci.* **2008**, *91*, 3300–3310. [CrossRef]
40. Burvenich, C.; Bannerman, D.D.; Lippolis, J.D.; Peelman, L.; Nonnecke, B.J.; Kehrli, M.E.; Paape, M.J. Cumulative physiological events influence the inflammatory response of the bovine udder to escherichia coli infections during the transition period1. *J. Dairy Sci.* **2007**, *90*, E39–E54. [CrossRef]
41. Kvidera, S.K.; Horst, E.A.; Abuajamieh, M.; Mayorga, E.J.; Fernandez, M.V.S.; Baumgard, L.H. Glucose requirements of an activated immune system in lactating holstein cows. *J. Dairy Sci.* **2017**, *100*, 2360–2374. [CrossRef]
42. Ingvartsen, K.L.; Moyes, K.M. Factors contributing to immunosuppression in the dairy cow during the periparturient period. *Jpn. J. Vet. Res.* **2015**, *63*, S15–S24. [PubMed]
43. Shen, T.; Li, X.; Loor, J.J.; Zhu, Y.; Du, X.; Wang, X.; Xing, D.; Shi, Z.; Fang, Z.; Li, X.; et al. Hepatic nuclear factor kappa b signaling pathway and nlr family pyrin domain containing 3 inflammasome is over-activated in ketotic dairy cows. *J. Dairy Sci.* **2019**, *102*, 10554–10563. [CrossRef] [PubMed]
44. Van Straten, M.; Friger, M.; Shpigel, N.Y. Events of elevated somatic cell counts in high-producing dairy cows are associated with daily body weight loss in early lactation. *J. Dairy Sci.* **2009**, *92*, 4386–4394. [CrossRef]
45. Giuliodori, M.J.; Magnasco, R.P.; Becu-Villalobos, D.; Lacau-Mengido, I.M.; Risco, C.A.; de la Sota, R.L. Metritis in dairy cows: Risk factors and reproductive performance. *J. Dairy Sci.* **2013**, *96*, 3621–3631. [CrossRef] [PubMed]
46. Fiore, E.; Perillo, L.; Piccione, G.; Gianesella, M.; Bedin, S.; Armato, L.; Giudice, E.; Morgante, M. Effect of combined acetylmethionine, cyanocobalamin and alpha-lipoic acid on hepatic metabolism in high-yielding dairy cow. *J. Dairy Res.* **2016**, *83*, 438–441. [CrossRef] [PubMed]
47. Fisher, L.J.; Erfle, J.D.; Lodge, G.A.; Sauer, F.D. Effects of propylene glycol or glycerol supplementation of the diet of dairy cows on feed intake, milk yield and composition, and incidence of ketosis. *Can. J. Anim. Sci.* **1973**, *53*, 289–296. [CrossRef]
48. Butler, S.T.; Pelton, S.H.; Butler, W.R. Energy balance, metabolic status, and the first postpartum ovarian follicle wave in cows administered propylene glycol. *J. Dairy Sci.* **2006**, *89*, 2938–2951. [CrossRef]
49. Piantoni, P.; Allen, M.S. Evaluation of propylene glycol and glycerol infusions as treatments for ketosis in dairy cows. *J. Dairy Sci.* **2015**, *98*, 5429–5439. [CrossRef]
50. Miyoshi, S.; Pate, J.L.; Palmquist, D.L. Effects of propylene glycol drenching on energy balance, plasma glucose, plasma insulin, ovarian function and conception in dairy cows. *Anim. Reprod. Sci.* **2001**, *68*, 29–43. [CrossRef]
51. Studer, V.A.; Grummer, R.R.; Bertics, S.J.; Reynolds, C.K. Effect of prepartum propylene glycol administration on periparturient fatty liver in dairy cows. *J. Dairy Sci.* **1993**, *76*, 2931–2939. [CrossRef]
52. Ferraro, S.M.; Mendoza, G.D.; Miranda, L.A.; Gutiérrez, C.G. In Vitro gas production and ruminal fermentation of glycerol, propylene glycol and molasses. *Anim. Feed Sci. Technol.* **2009**, *154*, 112–118. [CrossRef]
53. Rizos, D.; Kenny, D.A.; Griffin, W.; Quinn, K.M.; Duffy, P.; Mulligan, F.J.; Roche, J.F.; Boland, M.P.; Lonergan, P. The effect of feeding propylene glycol to dairy cows during the early postpartum period on follicular dynamics and on metabolic parameters related to fertility. *Theriogenology* **2008**, *69*, 688–699. [CrossRef] [PubMed]

54. Kristensen, N.B.; Raun, B.M. Ruminal and intermediary metabolism of propylene glycol in lactating holstein cows. *J. Dairy Sci.* **2007**, *90*, 4707–4717. [CrossRef] [PubMed]

55. Bjerre-Harpøth, V.; Storm, A.C.; Eslamizad, M.; Kuhla, B.; Larsen, M. Effect of propylene glycol on adipose tissue mobilization in postpartum over-conditioned holstein cows. *J. Dairy Sci.* **2015**, *98*, 8581–8596. [CrossRef] [PubMed]

56. Gamarra, G.; Ponsart, C.; Lacaze, S.; Le Guienne, B.; Deloche, M.C.; Monniaux, D.; Ponter, A.A. Short term dietary propylene glycol supplementation affects circulating metabolic hormones, progesterone concentrations and follicular growth in dairy heifers. *Livest. Sci.* **2014**, *162*, 240–251. [CrossRef]

57. De Koster, J.D.; Opsomer, G. Insulin resistance in dairy cows. *Vet. Clin. N. Am. Food Anim. Pract.* **2013**, *29*, 299–322. [CrossRef]

58. Rico, J.E.; Bandaru, V.V.; Dorskind, J.M.; Haughey, N.J.; McFadden, J.W. Plasma ceramides are elevated in overweight holstein dairy cows experiencing greater lipolysis and insulin resistance during the transition from late pregnancy to early lactation. *J. Dairy Sci.* **2015**, *98*, 7757–7770. [CrossRef]

59. Chalmeh, A.; Pourjafar, M.; Badiei, K.; Jalali, M.; Sebdani, M.M. The comparative effects of dietary monensin and propylene glycol on insulin resistance of transition dairy cows. *Trop. Anim. Health Prod.* **2020**, *52*, 1573–1582. [CrossRef]

60. Moallem, U.; Katz, M.; Arieli, A.; Lehrer, H. Effects of peripartum propylene glycol or fats differing in fatty acid profiles on feed intake, production, and plasma metabolites in dairy cows. *J. Dairy Sci.* **2007**, *90*, 3846–3856. [CrossRef]

61. Jeong, J.-K.; Choi, I.-S.; Moon, S.-H.; Lee, S.-C.; Kang, H.-G.; Jung, Y.-H.; Park, S.-B.; Kim, I.-H. Effect of two treatment protocols for ketosis on the resolution, postpartum health, milk yield, and reproductive outcomes of dairy cows. *Theriogenology* **2018**, *106*, 53–59. [CrossRef]

62. Chung, Y.H.; Martinez, C.M.; Brown, N.E.; Cassidy, T.W.; Varga, G.A. Ruminal and blood responses to propylene glycol during frequent feeding. *J. Dairy Sci.* **2009**, *92*, 4555–4564. [CrossRef] [PubMed]

63. Urrutia, N.; Bomberger, R.; Matamoros, C.; Harvatine, K.J. Effect of dietary supplementation of sodium acetate and calcium butyrate on milk fat synthesis in lactating dairy cows. *J. Dairy Sci.* **2019**, *102*, 5172–5181. [CrossRef] [PubMed]

64. Juchem, S.O.; Santos, F.A.P.; Imaizumi, H.; Pires, A.V.; Barnabé, E.C. Production and blood parameters of holstein cows treated prepartum with sodium monensin or propylene glycol1. *J. Dairy Sci.* **2004**, *87*, 680–689. [CrossRef]

65. Adamski, M.; Kupczynski, R.; Chladek, G.; Falta, D. Influence of propylene glycol and glycerin in simmental cows in periparturient period on milk yield and metabolic changes. *Arch. Tierz.* **2011**, *54*, 238–248. [CrossRef]

66. McArt, J.A.A.; Nydam, D.V.; Oetzel, G.R. A field trial on the effect of propylene glycol on displaced abomasum, removal from herd, and reproduction in fresh cows diagnosed with subclinical ketosis. *J. Dairy Sci.* **2012**, *95*, 2505–2512. [CrossRef]

67. Gonzalez, F.D.; Muino, R.; Pereira, V.; Campos, R.; Benedito, J.L. Relationship among blood indicators of lipomobilization and hepatic function during early lactation in high-yielding dairy cows. *J. Vet. Sci.* **2011**, *12*, 251–255. [CrossRef]

68. Hussein, H.A.; Abdel-Raheem, S.M.; Abd-Allah, M.; Senosy, W. Effects of propylene glycol on the metabolic status and milk production of dairy buffaloes. *Tierärztliche Prax. Großtiere* **2015**, *43*, 25–34.

69. Stokes, S.R.; Goff, J.P. Evaluation of calcium propionate and propylene glycol administered into the esophagus of dairy cattle at calving11supported in part by a gift from kemin industries, inc., des moines, ia 50301-0070. *Prof. Anim. Sci.* **2001**, *17*, 115–122. [CrossRef]

70. McArt, J.A.A.; Nydam, D.V.; Ospina, P.A.; Oetzel, G.R. A field trial on the effect of propylene glycol on milk yield and resolution of ketosis in fresh cows diagnosed with subclinical ketosis. *J. Dairy Sci.* **2011**, *94*, 6011–6020. [CrossRef]

71. Østergaard, S.; Krogh, M.A.; Oliveira, V.H.S.; Larsen, T.; Otten, N.D. Only few benefits from propylene glycol drench in early lactation for cows identified as physiologically imbalanced based on milk spectra analyses. *J. Dairy Sci.* **2020**, *103*, 1831–1842. [CrossRef]

72. Madoz, L.V.; Rabaglino, M.B.; Migliorisi, A.L.; Jaureguiberry, M.; Perez Wallace, S.; Lorenti, N.; Domínguez, G.; Giuliodori, M.J.; de la Sota, R.L. Association between progesterone concentration and endometrial gene expression in dairy cows. *Domest. Anim. Endocrinol.* **2020**, *74*, 106481. [CrossRef] [PubMed]

73. Castañeda-Gutiérrez, E.; Pelton, S.H.; Gilbert, R.O.; Butler, W.R. Effect of peripartum dietary energy supplementation of dairy cows on metabolites, liver function and reproductive variables. *Anim. Reprod. Sci.* **2009**, *112*, 301–315. [CrossRef] [PubMed]

74. Osorio, J.S.; Trevisi, E.; Ji, P.; Drackley, J.K.; Luchini, D.; Bertoni, G.; Loor, J.J. Biomarkers of inflammation, metabolism, and oxidative stress in blood, liver, and milk reveal a better immunometabolic status in peripartal cows supplemented with smartamine m or metasmart. *J. Dairy Sci.* **2014**, *97*, 7437–7450. [CrossRef] [PubMed]

75. Formigoni, A.; Cornil, M.C.; Prandi, A.; Mordenti, A.; Rossi, A.; Portetelle, D.; Renaville, R. Effect of propylene glycol supplementation around parturition on milk yield, reproduction performance and some hormonal and metabolic characteristics in dairy cows. *J. Dairy Res.* **1996**, *63*, 11–24. [CrossRef] [PubMed]

76. Trabue, S.; Scoggin, K.; Tjandrakusuma, S.; Rasmussen, M.A.; Reilly, P.J. Ruminal fermentation of propylene glycol and glycerol. *J. Agric. Food Chem.* **2007**, *55*, 7043–7051. [CrossRef] [PubMed]

77. Bertram, H.C.; Petersen, B.O.; Duus, J.O.; Larsen, M.; Raun, B.M.L.; Kristensen, N.B. Proton nuclear magnetic resonance spectroscopy based investigation on propylene glycol toxicosis in a holstein cow. *Acta Vet. Scand.* **2009**, *51*, 1–9. [CrossRef]

78. Macrae, A.I.; Burrough, E.; Forrest, J.; Corbishley, A.; Russell, G.; Shaw, D.J. Risk factors associated with excessive negative energy balance in commercial united kingdom dairy herds. *Vet. J.* **2019**, *250*, 15–23. [CrossRef]

79. Chung, Y.H.; Brown, N.E.; Martinez, C.M.; Cassidy, T.W.; Varga, G.A. Effects of rumen-protected choline and dry propylene glycol on feed intake and blood parameters for holstein dairy cows in early lactation. *J. Dairy Sci.* **2009**, *92*, 2729–2736. [CrossRef]

80. Gordon, J.L.; LeBlanc, S.J.; Kelton, D.F.; Herdt, T.H.; Neuder, L.; Duffield, T.F. Randomized clinical field trial on the effects of butaphosphan-cyanocobalamin and propylene glycol on ketosis resolution and milk production. *J. Dairy Sci.* **2017**, *100*, 3912–3921. [CrossRef]

81. McArt, J.A.; Nydam, D.V.; Oetzel, G.R.; Guard, C.L. An economic analysis of hyperketonemia testing and propylene glycol treatment strategies in early lactation dairy cattle. *Prev. Vet. Med.* **2014**, *117*, 170–179. [CrossRef]

82. El-Kasrawy, N.I.; Swelum, A.A.; Abdel-Latif, M.A.; Alsenosy, A.E.A.; Beder, N.A.; Alkahtani, S.; Abdel-Daim, M.M.; Abd El-Aziz, A.H. Efficacy of different drenching regimens of gluconeogenic precursors during transition period on body condition score, production, reproductive performance, subclinical ketosis and economics of dairy cows. *Animals* **2020**, *10*, 937. [CrossRef] [PubMed]

83. Van der Drift, S.G.A.; Houweling, M.; Bouman, M.; Koets, A.P.; Tielens, A.G.M.; Nielen, M.; Jorritsma, R. Effects of a single glucocorticoid injection on propylene glycol-treated cows with clinical ketosis. *Vet. J.* **2015**, *204*, 144–149. [CrossRef] [PubMed]

84. Mann, S.; Yepes, F.A.L.; Behling-Kelly, E.; McArt, J.A.A. The effect of different treatments for early-lactation hyperketonemia on blood β-hydroxybutyrate, plasma nonesterified fatty acids, glucose, insulin, and glucagon in dairy cattle. *J. Dairy Sci.* **2017**, *100*, 6470–6482. [CrossRef] [PubMed]

Article

Dietary Supplementation of Lysophospholipids Affects Feed Digestion in Lambs

Qin Huo [1], Bo Li [1], Long Cheng [2], Tingting Wu [1], Peihua You [3], Shuanghua Shen [4], Yiyong Li [3], Yuhua He [1], Wannian Tian [1], Rongquan Li [1], Changsheng Li [1], Jianping Li [1], Baijun Song [1], Chunqing Wang [1] and Xuezhao Sun [1,*]

[1] The Innovation Centre of Ruminant Precision Nutrition and Smart and Ecological Farming, Jilin Agricultural Science and Technology University, Jilin 132109, China; huoqin0605@163.com (Q.H.); ma_lee@163.com (B.L.); ttw528@126.com (T.W.); yhyttt0235@sina.com (Y.H.); wannian2000@hotmail.com (W.T.); 16643429184@sohu.com (R.L.); csli-64@163.com (C.L.); sxljp2004@126.com (J.L.); jlsbj2005@126.com (B.S.); chunqingw@sina.com (C.W.)

[2] Faculty of Veterinary and Agricultural Sciences, Dookie Campus, The University of Melbourne, Victoria 3647, Australia; long.cheng@unimelb.edu.au

[3] Portal Agri-Industries Co., Ltd., Nanjing 211803, China; ph-u@163.com (P.Y.); liyiyongfeed@163.com (Y.L.)

[4] Floor 15, Building B, 650 Xinzhuang Avenue, Xinqiao Town, Songjiang District, Shanghai 201612, China; sshqz@sina.com

* Correspondence: xuezhaos@hotmail.com; Tel.: +86-187-4327-5745

Received: 27 September 2019; Accepted: 12 October 2019; Published: 15 October 2019

Simple Summary: Previous works showed that supplementation of lysophospholipid as a feed additive improves animal growth and milk yield in beef and dairy cattle production. However, its effects on fattening lambs have not been reported before. In this study, we fed fattening lambs a diet with no or 0.5 g lysophospholipid in a kilogram of diet. We found that lysophospholipid did not or slightly improved the growth of fattening lambs. Feed digestibility, ruminal fermentation parameters and rumen bacterial community were altered, which may be associated with decreased fiber digestion. However, lipase concentration in serum was decreased, which might enhance fat deposition in muscle and thus may increase meat quality. Effects of lysophospholipid on sheep observed in this study are different from those on cattle, which warrants further study.

Abstract: Five experiments were conducted to examine effects of lysophospholipids (LPL) on live weight gain, nutrient digestibility, ruminal fermentation parameters, serum biochemical parameters and rumen bacterial community profile in fattening lambs. Two dietary treatments (pelleted complete feed supplemented without (control diet; CON) or with 0.05% LPL on dry matter basis) were tested in these experiments. Feed and water were provided *ad libitum* to lambs. The results showed that average daily gain (ADG) tended to increase or was not affected by LPL supplementation. Compared with CON, the supplementation of LPL resulted in an increase in dry matter, crude protein and organic matter digestibilities, and a decrease in neutral detergent fiber and acid detergent fiber digestibilities. Ruminal pH values did not change with LPL supplementation, but the concentrations of ammonia and total short chain fatty acids (SCFAs) were increased. The molar proportion of major individual SCFAs and the ratio of acetate to propionate were not affected by LPL supplementation. While the activity of lipase was decreased with LPL supplementation, all other serum biochemical parameters did not change. Rumen bacterial community was altered by LPL supplementation with the relative abundance of fibrolytic bacteria in the total bacterial population, such as *Prevotella*, decreased. In conclusion, LPL supplementation can alter feed digestion, but may not result in consistent positive responses in animal growth performance.

Keywords: lysophospholipid; growth performance; digestibility; rumen function; blood metabolite; sheep; pelleted complete feed

1. Introduction

Pelleted complete feed has been increasingly applied to sheep farming in recent years [1,2]. The indoor sheep production system based on pelleted complete feed provides us with more opportunities than the grazing system, as it allows better manipulation of rumen fermentation, as well as increased feed efficiency and animal growth performance. A wide range of feed additives has been studied and applied for rumen fermentation manipulation in sheep production [3–6]. In addition to directly providing nutrients, such as rumen-protected lysine, the supplementation of microbial additives and antibiotics has been the major nutritional approach to altering rumen function and animal performance [7,8]. However, microbial additives are normally unstable in their quality and consequently in the response of animal performance [7]. Antibiotics, such as monesin, have been banned in some countries due to concerns over the formation of antibiotic-resistant microbial mutants [8,9]. This leads animal scientists to keep seeking new feed additives to increase animal production.

Lysophospholipids (LPL) are produced from the enzymatic hydrolysis of phospholipid with a chain of fatty acids removed, mainly containing lysophosphatidylcholine, lysophosphatidic acid, lysophosphatidylethanolamine, and lysophosphatidylinositol [10]. The LPL can selectively inhibit the growth of Gram-positive but not Gram-negative bacteria [11]. They also could emulsify dietary fat [12] and upregulate genes in the intestinal epithelium [13] to increase fat absorption. This group of compounds has been used as feed additives in pig and poultry for improved feed efficiency and increased growth rate [14–17].

To our best knowledge, limited study has been conducted with LPL supplementation to sheep, though a few trials were performed on dairy cows [18–20] and beef cattle [21]. Lee et al. [19] found that adding 0.05–0.075% LPL (dry matter (DM) basis) to the diet of lactating dairy cows can increase milk production. Our preliminary trials showed that adding 0.5 mg LPL/kg DM to the diet for lactating cows increased milk yield from 28.0 to 29.5 L/day/cow. The positive animal performance might result from the emulsifying effect of LPL as emulsifying agents could increase enzymatic activities in the rumen [22] and also from the selective inhibition of Gram-positive bacteria, such as *Roseburia* spp. in the rumen [11].

The hybrid of thin-tailed sheep and Northeast fine-wool sheep was the most commonly farmed breed in Northeast China [23]. This breed has a relatively high growth performance and is well accommodated to the local environment.

We hypothesized that the addition of LPL to pelleted formulated complete feed for fattening lambs may result in more efficient feed digestion and rumen fermentation towards more propionate and less acetate formation, which consequently may lead to more nutrients available for growth. This study aimed to investigate the effect of the addition of LPL to the pelleted diet of fattening lambs on animal growth performance, blood parameters, feed digestibility, ruminal fermentation parameters and rumen bacterial community.

2. Materials and Methods

The use of animals and all animal handling and management in this study were approved in advance by the Committee of Animal Ethics and Welfare of Jilin Agricultural Science and Technology University (Approval number 2018001).

2.1. Experimental Design and Animals

The study included two dietary treatments: a control diet (CON) and a control diet supplemented with 0.5 g/kg DM of LPL. It was conducted in five experiments at three experimental sites. Experiment

1 (Exp 1) was for the determination of growth performance, digestibility, blood parameters, ruminal fermentation parameters and rumen bacterial community, while the other experiments (Exps 2–5) were for the determination of growth performance only.

Healthy hybrids of thin-tailed sheep and Northeast fine-wool sheep with similar liveweight were used for the study. All sheep were negative in the rose bengal plate test for brucellosis (*Bruceila* spp.) [24].The slow-growing sheep with daily liveweight gain less than 100 g/day during the adaptation period were excluded from the study. Sheep were stratified by liveweights measured at the end of the adaptation period and randomly allocated to one of the two dietary treatments. During the formal experimental period, sick lambs or pregnant female lambs were excluded from the study.

Exp 1 was conducted at the Animal Experimental Station of Jilin Agricultural Science and Technology University, Jilin city, Jilin province, China. Eighteen three-month-old ram lambs with a liveweight of 24.7 ± 1.9 (mean ± SD) kg were adapted for 10 days and then 14 of them with the best growth performance were randomly allocated to one of two groups with eight lambs each. After three more days for further adaptation, the formal experimental period started and lasted for 68 days (from 19 August to 26 October 2018). Sheep were fed in a half-opened feedlot and transferred to individual metabolic cages on day 15 (3 September 2018) for digestibility measurements. They were adapted to the cage housing conditions for 4 days and then the digestibility sample collection period lasted for 7 days. After the digestibility trial, animals were transferred back to the feedlot. Rumen samples were taken on days 29 and 68 (17 September and 26 October 2018) and blood samples were taken on day 42 (30 September 2018).

Exp 2 had the same experimental design as the Exp 1 and was conducted at the same site as the Exp 1, but the ram lambs were heavier and older than those in the Exp 1. The initial liveweight was at 43.8 ± 4.0 kg and aged at five months. After two weeks of isolation for the detection of brucellosis, 16 lambs were allocated to one of the two dietary treatments with seven each, housed in a pen for each treatment, and adapted to their assigned diets. The formal experimental period lasted for 28 d (from 27 September to 25 October 2018).

Exp 3 was conducted at Jiaogeermiao sheep farm in Tongyu county, Jilin province, China. Thirty six ewe lambs with a liveweight of 25.0 ± 3.1 kg were adapted to pelleted complete feed for 10 days from 11 to 21 August 2018 and then assigned to one of the two treatment groups. The formal experimental period lasted for 29 days and ended on 19 September 2018.

Exp 4 and Exp 5 were conducted at the Portal Chifeng Experimental Station, Inner Mongolia, China. Exp 4 had ram lambs with a liveweight of 29.9 ± 5.4 kg and in Exp 5 ewe lambs had a liveweight of 27.0 ± 3.5 kg. Lambs were adapted to pelleted complete feed for 14 days and then were weighed and grouped by weight. Lambs from each treatment were fed in two pens with the equal number of animals. The formal experimental period lasted for 62 days from 7 November 2018 to 8 January 2019. Animals were weighed on days 0, 24 and 62. At the end of the experiment, all lambs were slaughtered and hot carcasses were weighed.

2.2. Diets and Feeding Management

The experimental diets were formulated according to the Chinese Feeding Standard for Lamb Finishing (NY/T 816-2004). The ingredients and nutrient contents of the diets are listed in Table 1. Sorghum husk was ground using a 2 mm sieve and maize, maize germ meal, sunflower seed meal, peanut shell, rice hull, cottonseed meal, and barley malt root using a 4/6 mm sieve (half sieve with holes having a diameter of 4 mm and another half 6 mm).

Table 1. Ingredients and nutrient contents of experimental diets supplemented without (CON) or with (LPL) lysophospholipids.

Item	Diet	
	CON	LPL
Ingredient (kg/t of fresh weight)		
Maize	350	350
Maize germ meal	120	120
Sunflower seed meal	120	120
Peanut shell	113	113
Rice hull	70	70
Cottonseed meal	30	30
Bentonite	20	20
Barley malt root	100	100
Limestone	14	14
Sorghum husk	10	10
Calcium hydrogen phosphate	7	7
Soybean powder	20	20
Sodium chloride	6	6
Premix (Trace mineral salt and vitamins) [1]	20	20
Lysophospholipids		0.5
Nutrient contents [2] (g/kg of DM)		
Dry matter (DM) (g/kg of fresh weight)	880	874
Organic matter (OM)	903	905
Crude protein (CP)	158	162
Neutral detergent fiber (NDF)	427	435
Acid detergent fiber (ADF)	174	178
Ether extract (EE)	16	16
Metabolizable energy (MJ/kg of DM) [3]	11.7	11.8

[1] Premix per kg contained 24,000 IU vitamin A, 4800 IU vitamin E, 120 mg Fe, and 24 mg Cu; [2] The nutrient contents were measured values. [3] Metabolizable energy was estimated from NRC (2007) [25].

All ingredients were thoroughly mixed, conditioned at 90 °C for 45 s and pelletized at 85 °C using a pelleting machine with the compression ratio of ring to die at 1:7 (model YPM508E; Jiangsu Yongli Machinery Co., Ltd., Liyang City, Jiangsu Province, China). The pellets were air-cooled. Feed pellets were 5 mm in diameter and 8–10 mm in length. All feed used in the whole period of the experiments was pelleted in one batch at the Chifeng Subsidiary Company of Jiangsu Portal Agri-Industries Co., Ltd.

During the adaptation period, feed provided to lambs was gradually changed from hay to pellets by an increase of 100 g of pellets a day. Pellets provided in the first week contained an antiparasitic agent (triclabendazole, 250 mg/kg feed) and then experimental pellets were provided. Feed was provided twice a day with equal portions at 8:00 and 17:00. Once animals were completely on pelleted feed, feed allowance was set to allow leftover at ca. 10% of total feed provided on the previous day to achieve feeding *ad libitum*. Leftovers were cleaned up and weighed before morning feeding for intake calculation. Drinking water was available all the time. Weather, temperature and humidity were recorded daily, and animal behaviors were observed.

2.3. Liveweight Measurement and Sampling

Liveweight was measured before morning feeding after 16 h fasting with an accuracy of 0.05 kg using an electronic scale for living bodies (YongHuang, Jinhua, Zhejiang, China) at the beginning and the end of the formal experimental period, and in the middle of the experiment if the experimental period was longer than one month. Average daily gain (ADG) was calculated as the slope of liveweight against date.

Rumen sample (50 mL) was taken from each lamb in the Exp 1 via mouth using stomach tubing (1500 mm in length, 8 mm in inner diameter, 6 openings with a diameter of 2 mm in the end to the rumen, made with polyurethane material) at 0 and 3 h after morning feeding. Ruminal pH value was measured immediately after sampling using a pH meter with the accuracy of 0.01 (LICHEN pH-100A, Shanghai Lichen Scientific Laboratory Instrument Ltd., Shanghai, China). The rumen sample was placed in a 2-mL cryogenic vial with the circular bottom (Corning Inc., New York, USA) and then stored at $-20\ ^{\circ}$C for the analysis of short chain fatty acid (SCFA) and ammonia concentrations. The remaining samples (3 h after morning feeding samples only) were stored at $-80\ ^{\circ}$C for the characterization of rumen microbial profile.

Blood sample (5 mL) was collected into a coagulation promoting tube with separating gel (Sanli Industrial Co., Ltd., Huizhou, China) from the jugular vein of each lamb in the Exp 1 before morning feeding.

2.4. Digestibility Trial

Digestibility measurement was conducted in metabolic cages using the total feces collection technique. The digestibility trial lasted for 10 days, including a 3-day adaptation and 7-day collection periods. Feed allowance was adjusted to allow 5–10% refused and feed provided at 8:00 and 15:00 with equal portions. Feed sample (50 g) was collected daily. Refusals, feces and urine were collected and quantified daily. Refusals were all kept and feces and urine subsampled. One fifth of collected feces was stored at $-20\ ^{\circ}$C, with half acidified with 10% H_2SO_4 at a ratio of 1:10 for crude protein (CP) determination and another half for the determination of other nutrients. Total urine was collected in a tank with 100 mL of 10% H_2SO_4 added and diluted to 3000 mL, from which 30 mL was taken and kept at $-20\ ^{\circ}$C. At the end of the experiment, samples were pooled over individual animals. Feed, refusal and feces samples were dried at 65 $^{\circ}$C for 48 h and stored in cold for later analysis. Data from one sheep in the LPL group were discarded due to health problem of the sheep.

2.5. Laboratory Analysis

Feed samples were analyzed for DM, organic matter (OM), CP, neutral detergent fiber (NDF), acid detergent fiber (ADF), ether extract (EE), Ca and P, refusal and feces samples for DM, OM, CP, NDF and ADF, and urine samples for CP. Crude protein was determined in the modified method of Kjeldahl (method GB/T 6432-2018) using an automatic nitrogen determination apparatus (model K9860; Haineng Instruments Ltd., Jinan, Shandong, China). Dry matter was determined by the loss of water during drying at 105 $^{\circ}$C for 3 h (method GB/T 6435-2014). Ash was determined by samples heated until no smoking and then in a muffle at 550 $^{\circ}$C for 4 h (GB/T 6438-2007). Ether extract was determined as dry matter loss after extraction in petroleum ether for 24 h at 30–60 $^{\circ}$C (GB/T 6433-2006). NDF and ADF were consecutively determined after boiling in 3% neutral detergent for 1 h (GB/T 20806-2006) and 2% acid detergent for 1 h (NY/T 1459-2007), respectively.

A cryogenic vial with a rumen sample was thawed in tap water and centrifuged at 2000× g for 10 min. The supernatant was transferred to a 5 mL centrifugation tube, mixed with 25% metaphosphoric acid at a 4:1 ratio, vortexed and centrifuged at 4 $^{\circ}$C in 8000× g for 10 min. The supernatant from this centrifugation was filtered by a 0.22 μm filter for water solution (Shanghai Rebus Biotechnology Co., Ltd., Shanghai, China) and transferred to a new 5 mL centrifugal tube for the determination of SCFA and ammonia concentrations.

Short chain fatty acids were identified and quantified using a gas chromatograph system (GC9790, Fuli Instruments Ltd., Wenling, Zhenjiang, China) fitted with a flame ionisation detector (FID). The GC was fitted with a FFAP 30 m × 3 mm × 0.25 μm capillary polar column. The column temperature was set at 140 $^{\circ}$C, the FID set at 250 $^{\circ}$C and vaporization chamber temperature at 250 $^{\circ}$C with N_2 as a carrier gas and shunt ratio at 1:50. The injection volume was 1 μL.

The above filtered rumen sample was diluted with double evaporated water in a ratio of 1:7 and vortexed in a 5 mL centrifugal tube. Ammonia concentration was determined using the Indigo phenol blue-spectrophotometry method [26] modified by Feng and Gao [27]. Briefly, the sample solution was mixed with a color developer 37 °C for 20 min, cooled down with tap water to room temperature, and measured for absorbance at 637 nm.

The collected blood sample was centrifuged at 1000× g for 5 min (Model TDL-80-2B; Anting Scientific Instrument Factory, Shanghai, China) and the serum analyzed for blood biochemical parameters using an automatic biochemical analyzer (Model 7160; Hitachi Ltd., Tokyo, Japan). The reagents for the analysis were purchased from Mairui Biomedical Electronics Co., Ltd. (Shenzhen, China).

Rumen bacterial populations were analyzed at Shanghai Biozeron Biotechnology Co., Ltd. (Shanghai, China). Total genome DNA was extracted using the CTAB/SDS method, monitored for DNA concentration and purity on 1% agarose gels and then diluted to 1 ng/μL. The 16S rRNA genes were amplified using hypervariable V3-V4 region PCR primers (341F: 5′- CCTAYGGGRBGCASCAG-3′; 806R: 5′- GGACTACNNGGGTATCTAAT-3′) with the barcode. All PCR reactions were carried out in 30 μL reactions with 15 μL of Phusion®High-Fidelity PCR Master Mix (New England Biolabs, Ipswich, MA, USA), 0.2 μM of forward and reverse primers and about 10 ng template DNA. Thermal cycling consisted of initial denaturation at 98 °C for 1 min, followed by 30 cycles of denaturation at 98 °C for 10 s, annealing at 50 °C for 30 s, elongation at 72 °C for 60 s and finalization at 72 °C for 5 min. After amplification, PCR products were mixed with the same volume of SYBR green (QIAGEN Inc., Alameda, CA, USA) containing buffer and then operated electrophoresis on 2% agarose gel. Samples with bright main strip between 400–450 bp were purified with GeneJET Gel Extraction Kit (Thermo Fisher Scientific (China) Ltd., Shanghai, China). Sequencing libraries were generated using NEB Next®Ultra™DNA Library Prep Kit for Illumina (New England Biolabs, Ipswich, MA, USA) following the manufacturer's recommendations. The library quality was assessed on the Qubit@ 2.0 Fluorometer (Thermo Fisher Scientific (China) Ltd., Shanghai, China) and Agilent Bioanalyzer 2100 system (Agilent Technologies, Inc., Santa Clara, CA, USA). At last, the library was sequenced on an Illumina MiSeq platform (Illumina, Inc., San Diego, CA, USA) and 250 bp/300 bp paired-end reads were generated. Paired-end reads from the original DNA fragments were merged using FLASH [28], which was designed to merge paired-end reads when at least some of the reads overlap the read generated from the opposite end of the same DNA fragment. Sequences analysis were performed by UPARSE software package using the UPARSE-OTU and UPARSE-OTU ref algorithms [29]. In-house Perl scripts were used to analyze alpha (within samples) and beta (among samples) diversity. Sequences with ≥97% similarity were assigned to the same OTUs and the RDP classifier was used to annotate taxonomic information for the representative sequence in each OTU according to the database Silva (www.arb-silva.de) updated for the taxonomic assignment of ruminal bacteria [30]. Details in bioinformatics analysis are described in Appendix A.

2.6. Statistical Analysis

The statistical analysis was conducted using GenStat 19th edition (VSN International, Hemel Hempstead, UK, 2017) and significance was declared at $p < 0.05$. The data of liveweight, ADG, carcass weight, dressing percentage, feed intake, digestibility and blood biochemical parameters were analyzed with a one-way ANOVA. Rumen fermentation parameters were analyzed using a mixed model (REML) with repeated measurements. Treatment, sampling time and the interaction of treatment and sampling time were defined to have fixed effects in the model, sampling date, and animal as random effects. Parameters obtained from each lamb at different sampling times were treated as repeated measures. A meta-analysis was conducted for ADG with experimental site using block and dietary treatment, sex, and their interaction as fixed effects. As the interaction was not significant, it was omitted from the model for analysis. For the data of ruminal bacterial community, only when average relative abundance of a species in any treatment group was over 0.5%, the species and its corresponding genus and phylum were statistically analyzed with a one-way ANOVA.

3. Results

3.1. Growth Performance

In Exp 1, the initial liveweight was similar for the two treatments ($p = 0.826$), but after 68 days the final liveweight tended to be higher for the LPL treatment than for the control ($p = 0.084$; Table 2). The LPL treatment was 12% higher in ADG than the control, however, the difference did not reach the level of significance ($p = 0.165$). In Exp 2, the supplementation of LPL tended to increase ADG ($p = 0.099$). In other experiments, a significant change in ADG with LPL supplementation was not observed ($p \geq 0.601$). The meta-analysis with all ADG data obtained in this study showed that the supplementation of LPL ($p = 0.471$) or the interaction of LPL and animal sex ($p = 0.400$) did not significantly change ADG, although male lambs (274 g/day) had higher ($p < 0.001$) ADG than female lambs (200 g/day).

Table 2. Growth performance of fattening lambs fed experimental diets supplemented without (CON) or with (LPL) lysophospholipids.

Experiment No.	Site	Sex	Item	Diet		*p* Value
				CON	LPL	
Exp 1	Jilin	Male	n	7	6	
			Days	68	68	
			Initial liveweight (kg)	24.6 ± 0.56	24.8 ± 0.58	0.826
			Final liveweight (kg)	44.6 ± 1.12	46.9 ± 0.86	0.084
			Average daily gain (ADG) (g/day)	294 ± 17.3	330 ± 8.1	0.165
Exp 2	Jilin	Male	n	8	8	
			Days	28	28	
			Initial liveweight (kg)	43.6 ± 1.34	44.7 ± 1.38	0.568
			Final liveweight (kg)	48.9 ± 1.65	52.4 ± 1.74	0.158
			Average daily gain (ADG) (g/day)	198 ± 38.9	282 ± 24.3	0.099
Exp 3	Tongyu	Female	n	18	17	
			Days	29	29	
			Initial liveweight (kg)	24.9 ± 0.75	25.0 ± 0.77	0.972
			Final liveweight (kg)	30.9 ± 0.93	30.7 ± 0.96	0.851
			Average daily gain (ADG) (g/day)	207 ± 13.2	197 ± 13.5	0.601
Exp 4	Chefeng	Male	n	23	26	
			Days	62	62	
			Initial liveweight (kg)	30.4 ± 1.17	31.6 ± 0.76	0.378
			Final liveweight (kg)	47.5 ± 1.56	48.5 ± 0.92	0.569
			Average daily gain (ADG) (g/day)	276 ± 12.3	275 ± 9.6	0.982
			Carcass (kg)	20.3 ± 0.64	20.2 ± 0.47	0.815
			Dressing percentage (%)	42.9 ± 0.38	41.5 ± 0.38	0.012
Exp 5	Chefeng	Female	Number of animals	18	27	
			Days	62	62	
			Initial liveweight (kg)	28.7 ± 0.76	29.8 ± 0.65	0.286
			Final liveweight (kg)	41.0 ± 1.14	42.4 ± 0.90	0.339
			Average daily gain (ADG) (g/day)	198 ± 13.7	203 ± 8.5	0.764
			Carcass (kg)	17.4 ± 0.48	18.1 ± 0.43	0.276
			Dressing percentage (%)	42.5 ± 0.38	42.8 ± 0.34	0.621

At the end of Exp 3 and Exp 4, all lambs were slaughtered. The supplementation of LPL did not significantly ($p \geq 0.276$) affect hot carcass weights for either male or female lambs. Dressing percentage was similar for the two treatments in female lambs. However, male lambs had a lower dressing percentage ($p = 0.012$) by 1.4 unit when the diet was supplemented with LPL than with the control.

3.2. Total Tract Nutrient Apparent Digestibility

During the period of digestibility trial, dry matter intake and other nutrient intakes were similar for the treatment and control animals ($p \geq 0.363$; Table 3). The supplementation of LPL slightly (by 2%), but statistically significantly ($p < 0.05$) increased DM and OM digestibilities. The increase in CP digestibility resulted from the supplementation of LPL was 4.3% ($p = 0.024$). However, NDF and ADF digestibilities had both statistically ($p = 0.012$ for NDF and $p = 0.001$ for ADF, respectively) and biologically (by -8% for NDF and -35% for ADF, respectively) significant decreases. The digestibility of EE did not significantly ($p = 0.263$) change with the supplementation of LPL.

Table 3. Intake and total tract apparent nutrient digestibility of fattening lambs offered a diet supplemented without (CON) or with (LPL) lysophospholipids.

Index	Diet		*p* Value
	CON (n = 7)	**LPL (n = 6)**	
Intake (g/day)			
Dry matter intake (DMI)	1328 ± 53.2	1382 ± 97.3	0.620
Organic matter intake (OMI)	1199 ± 48.1	1256 ± 88.3	0.566
Crude protein intake (CP intake)	207 ± 8.4	224 ± 15.7	0.363
Neutral detergent fiber intake (NDF intake)	562 ± 23.1	560 ± 46.9	0.969
Acid detergent fiber intake (ADF intake)	233 ± 9.3	240 ± 18.7	0.730
Ether extract intake (EE intake)	21.2 ± 1.83	22.2 ± 2.00	0.710
Digestibility (g/kg of DM)			
Dry matter (DM)	617 ± 3.3	629 ± 4.4	0.047
Organic matter (OM)	655 ± 3.1	669 ± 4.9	0.039
Crude protein (CP)	694 ± 7.0	723 ± 9.3	0.024
Neutral detergent fiber (NDF)	503 ± 8.3	463 ± 10.3	0.012
Acid detergent fiber (ADF)	285 ± 13.6	186 ± 17.3	0.001
Ether extract (EE)	941 ± 10.8	928 ± 1.3	0.263

3.3. Rumen Fermentation Parameters

Although ruminal pH value was higher before morning feeding than 3 h after morning feeding ($p < 0.001$), the supplementation of LPL did not significantly affect ruminal pH value ($p = 0.235$; Table 4). Both LPL supplementation ($p = 0.011$) and sampling time ($p < 0.001$) significantly affected ammonia-N concentration in the rumen and there was no significant interaction between them. The supplementation of LPL resulted in 37% higher ammonia-N concentration compared with the control. Supplementation, sampling time and their interaction significantly affected or tended to affect total SCFA concentration ($p \leq 0.083$). Although the total SCFA concentration at 3 h after morning feeding was similar for both LPL supplemented and unsupplemented lambs, the supplementation of LPL increased the total SCFA concentration before morning feeding by 61% ($p < 0.05$). In term of the molar proportion of individual SCFAs and the ratio of acetate to propionate, LPL supplementation and its interaction with sampling time did not have significant effects ($p \geq 0.172$) although sampling time did significantly affect the molar proportions of acetate and propionate and their ratio ($p \leq 0.026$).

Table 4. Ruminal fermentation parameters of fattening lambs (n = 7) fed experimental diets supplemented with (LPL) or without (CON) lysophospholipids.

Index	Diet				p Value		
	LPL (n = 7)		CON (n = 7)		Treatment (T)	Sampling Time (S)	T × S
	Before [1]	After [1]	Before	After			
Ruminal pH	6.75 ± 0.127 [a2]	5.94 ± 0.159 [b]	7.01 ± 0.048 [a]	5.98 ± 0.137 [b]	0.235	<0.001	0.372
Ammonia N (mmol/L)	8.92 ± 1.450 [b]	13.13 ± 1.328 [a]	6.92 ± 0.721 [b]	9.19 ± 0.911 [b]	0.011	0.006	0.384
Total SCFA [3] (mmol/L)	44.4 ± 6.93 [b]	65.9 ± 4.97 [a]	27.5 ± 2.28 [c]	66.0 ± 3.86 [a]	0.052	<0.001	0.083
Individual SCFA molar proportion (mol/100 mol)							
Acetate	60.5 ± 0.93 [a]	58.0 ± 1.47 [ab]	61.6 ± 1.86 [a]	54.9 ± 1.35 [b]	0.627	0.003	0.172
Propionate	22.3 ± 2.35 [b]	25.6 ± 2.35 [ab]	26.0 ± 2.63 [a]	31.8 ± 2.40 [a]	0.194	0.026	0.674
Butyrate	17.2 ± 1.88	16.4 ± 2.09	12.4 ± 1.37	13.3 ± 1.86	0.151	0.930	0.340
Acetate: Propionate	2.96 ± 0.322 [a]	2.50 ± 0.262 [ab]	2.77 ± 0.359 [a]	1.93 ± 0.216 [b]	0.443	0.011	0.447

[1] Before, rumen samples taken before morning feeding; After, rumen samples taken 3 h after morning feeding; [2] a, b, c: Different letters on the shoulder of values in a row mean a significant difference ($p < 0.05$); [3] SCFA, short chain fatty acid.

3.4. Blood Metabolites

Among serum biochemical parameters measured in this study (Table 5), only lipase activity was significantly ($p = 0.034$) altered by the supplementation of LPL and the activity of the enzyme decreased by 71% compared with the control.

Table 5. Serum biochemical parameters of fattening lambs fed experimental diets supplemented without (CON) or with (LPL) lysophospholipids.

Item	Diet		Normal Range	p Value
	CON (n = 7)	LPL (n = 7)		
α-Amylase (AMYL) (U/L)	21.7 ± 4.54	26.8 ± 4.50	1–30	0.439
Albumin (ALB) (g/L)	27.5 ± 1.31	27.2 ± 1.31	24–37	0.854
Alkaline phosphatase (ALKP) (U/L)	214 ± 18.0	186 ± 14.0	50–228	0.249
Alkaline transaminase (ALT) (U/L)	6.93 ± 0.913	8.84 ± 1.503	5–17	0.299
Aspartate transaminase (AST) (U/L)	54.4 ± 5.71	62.9 ± 8.01	40–96	0.403
Low density lipoprotein cholesterol (LDL) (mmol/L)	0.573 ± 0.0464	0.497 ± 0.0785	–	0.423
High density lipoprotein cholesterol (HDL) (mmol/L)	0.471 ± 0.0522	0.400 ± 0.0724	–	0.439
Total cholesterol (TC) (mmol/L)	1.29 ± 0.156	1.22 ± 0.095	1.14–2.12	0.725
Urea N (UREA) (mg/dL)	9.13 ± 0.861	8.66 ± 1.180	–	0.753
Creatinine (CREA) (μmol/L)	88.0 ± 18.87	53.5 ± 14.44	53–133	0.186
Triglyceride (TG) (mmol/L)	0.223 ± 0.0231	0.210 ± 0.0225	0.10–0.34	0.700
Total protein (TP) (g/L)	75.8 ± 1.60	74.5 ± 1.58	56–78	0.565
Glucose (GLU) (mmol/L)	4.10 ± 0.210	4.19 ± 0.238	2.78–4.45	0.778
Lipase (LIP) (U/L)	61.4 ± 16.29	17.8 ± 3.24	1–71	0.034
Globulin (GLOB) (g/L)	24.2 ± 1.01	25.6 ± 2.38	32–41	0.585

3.5. Rumen Bacterial Community

The supplementation of LPL altered the composition of bacterial community in the rumen (Figure 1). At the phylum level, the dominant phyla were *Bacteroidetes*, *Firmicutes*, and *Proteobacteria* in the rumen of lambs fed a diet containing either LPL or no LPL, accounting for 95.4% of total bacterial relative abundance (Figure S1). The relative abundance of these three phyla were 51.5%, 41.0%, and 2.8% in LPL treatment, 58.3%, 20.0%, and 17.1%, respectively. *Firmicutes* (CON 20.0% vs. LPL 41.0%; $p = 0.055$) and *Fibrobacteres* (CON 1.87% vs. LPL 0.36%; $p = 0.011$) were significantly different between the two treatments (Figure S1; Table S1). On the average the relative abundance of *Proteobacteria* dropped over 83.5% when LPL was supplemented, although the difference was insignificant due to a large variation within treatment groups. At the genus level (Figure 2), *Christensenellaceae* (*p* = 0.012), *Lachnospiraceae* (*p* = 0.090), and *Saccharofermentans* (*p* = 0.058) were increased and *Fibrobacter* (*p* = 0.011), *Prevotella* 7 (*p* = 0.037), *Prevotella* 9 (*p* = 0.031), unclassified *Prevotellaceae* (*p* = 0.011), and *Syntrophococcus* (*p* = 0.050) were decreased with the supplementation of LPL (Supplementary Table S1). At the species level, the *Christensenellaceae* R-7 group increased from 0.6% to 2.4% (*p* = 0.012), *Erysipelotrichaceae* from 0.4% to 2.2% (*p* = 0.078), *Lachnospiraceae* from 2.7% to 11.9% (*p* = 0.089), and uncultured *Oribacterium* from 0% to 1.5% (*p* = 0.088) when LPL was supplemented, while *Fibrobacter* was decreased from 1.8% to 0.3% (*p* = 0.013) and uncultured *Syntrophococcus* from 0.5% to 0.1% (*p* = 0.036). The supplementation of LPL reduced the relative abundance of Gram-positive bacteria *Roseburia* spp. 0.11% to 0.02%.

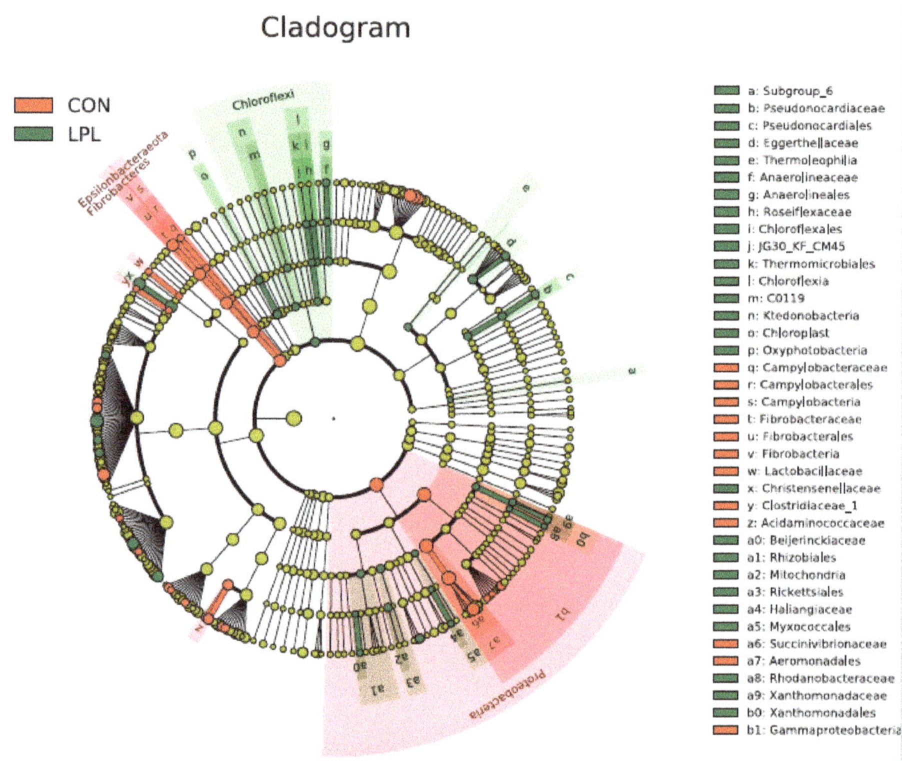

Figure 1. LEfSe cladogram of the composition of bacterial community in the rumen fattening lambs fed a diet supplemented without (CON) or with (LPL) 0.05% lysophospholipid. Differences are represented in the color of the group, where taxa are most abundant. Red: taxa abundant in CON; Green: taxa abundant in LPL.

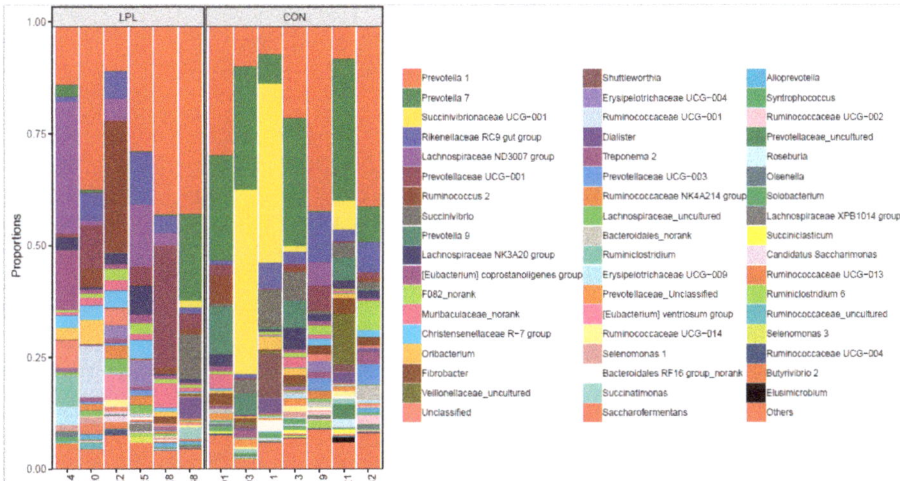

Figure 2. Distribution of the most dominant genera in the rumen of fattening lambs fed a diet supplemented without (CON) or with (LPL) 0.05% lysophospholipid (those with a relative abundance of less than 1% were combined as other).

4. Discussion

4.1. Animal Performance

In this study, the response of lambs to the supplementation of LPL at 0.05% of dietary DM was inconsistent. The ADG was increased in two experiments, but the difference did not reach the significance level (Exp 1) or just reached the tendency level (Exp 2). In the other three experiments, ADG did not change with the supplementation of LPL. There are no reported studies with LPL supplemented to sheep diet in the literature for us to directly compare with. In other ruminant species, the response of animal production to LPL supplementation was inconsistent as well. Song et al. [21] found that the supplementation of LPL at 0.3–0.5% of dietary DM to the diet of beef cattle did not improve growth performance. However, our preliminary trial showed that eleven 15.5-month-old Hanwoo beef cattle supplemented with LPL at 0.1% of the diet for 3 months had an average weight gain of 123.4 kg, being higher than the average weight gain of 117.5 kg from the control animals. Rico et al. [18] supplemented LPL at 10 g/day (approximately 0.035% of dietary DM) to lactating dairy cows for 10 days and did not observe an increase of milk yield. In contrast, a short-term supplementation of LPL at 0.05–0.075% of dietary DM did increase milk yield by 5.5%–6.5% [19].

The inconsistency in animal production in ruminants is contradictory to the studies on non-ruminants, in which LPL consistently increased growth and lactation performance and feed efficiency [14–16]. The inconsistent results among different ruminant studies might result from the variation in source of LPL (Sontakke et al. [20] vs. Lee et al. [19]), the dose of LPL (this study vs. Song et al. [21]), the duration of supplementation (e.g., this study vs. Lee et al. [19]), active compounds in LPL, the degradation of LPL in the rumen, etc. The degree of the removal of fatty acid chain from phospholipid during enzymatic hydrolysis affects the bioactivity of LPL [31]. The studies conducted previously might have used LPL with lower activity than the recent studies, which is expected to have different responses. It is well known that feed is subjected to degradation in the rumen before further digestion in a way similar to nonruminants. Phospholipids, from which LPL is produced, can be degraded in the rumen, but a considerable amount of LPL was able to bypass the rumen [32]. The degree of degradation may vary with the source of LPL and rumen environment which is moderated by diets and other factors [33]. We are not sure how much LPL can escape from the rumen. The discrepancy in growth response among experiments within this study may result from the age of animals, seasons and environment which all might affect rumen condition for degradation. In the future, studies are needed to determine the degradation of LPL in the rumen and to explore rumen-protected LPL.

4.2. Digestibility

To the best of our knowledge, there are only two studies in the literature measuring digestibility when LPL is supplemented to a diet for ruminants (Song et al. [21] in beef cattle and Lee et al. [19] in dairy cows). We found that LPL supplementation increased DM digestibility, which is consistent with the findings of Song et al. [21] who supplemented LPL to beef cattle diet at 0.3%–0.5% of dietary DM, but is contrasted with the results obtained from lactating dairy cows where DM digestibility slightly decreased when LPL was supplemented at 0.05–0.075% in the diet [19]. Although LPL supplementation resulted in an increase (this study) or a decrease [19] in DM digestibility, the difference between LPL and CON was only 12–14 g/kg. This small difference may not be biologically significant. However, in the study by Song et al. [21], the difference was small, only 8–17 g/kg on day 30, but increased with time. After 90 days, the difference in DM digestibility increased by 64 g/kg for LPL supplementation at 0.5% of the diet. Our digestibility trial started on day 15 of the experimental period. The measurement should be taken after LPL is supplemented for a longer time as well for better understanding in the effect of LPL on digestibility.

Song et al. [21] measured DM digestibility only. Lee et al. [19] is only a reference for comparison in nutrient digestibility. In our study, OM digestibility was increased, but decreased in the study conducted by Lee et al. [19]. Being similar to DM digestibility, the difference between LPL and CON was small as well and might be not relevant biologically at least for a short-term after LPL supplementation.

Our result showed that CP digestibility was increased, while it did not change in the study by Lee et al. [19]. The increase in CP digestibility was consistent with increased ruminal ammonia-N concentration 3 h after morning feeding with LPL supplementation. How LPL affect CP digestion is not known and warrants further study.

NDF digestibility decreased by around 40 g/kg DM in both Lee et al. [19] and our study, although the decrease was not statistically significant in the study by Lee et al. [19]. Lee et al. [19] did not measure ADF digestibility. Our result showed that LPL supplementation resulted in a large decrease in ADF digestibility by 99 g/kg DM. The great decrease in fiber digestibility is conflicted with our hypothesis proposed from the finding that fiber degradation is improved by emulsifying agents due to increased cellulolytic enzyme activity [34]. Further study is warranted for the mechanisms underlying the inhibition of fiber degradation by LPL.

4.3. Rumen Fermentation Parameters

Our study was contrasted with Lee et al. [19]'s findings on rumen fermentation parameters although ruminal pH value was not affected by LPL in both studies. We found ammonia and total SCFA concentrations were increased with LPL, but Lee et al. [19] did not. There was a lack of difference in the molar proportion of major individual SCFAs and the ratio of acetate to propionate in both studies. The reason for different effects of LPL on rumen fermentation is not known, but the chemical composition of the diet, especially EE content, was greatly different between the two studies, which might alter LPL reaction in the rumen. Our results did support the finding that LPL enhances feed digestion, but did not support the hypothesis that LPL results in rumen fermentation towards more propionate and less acetate formation that we proposed before.

4.4. Blood Biochemical Parameters

The major change observed in blood biochemical parameters was a decrease in lipase activity associated with the supplementation of LPL. Rico et al. [18] found that milk fat concentration was increased 5 days after dairy cows were supplemented to a higher fiber and lower unsaturated fatty acid diet. These authors attributed the increase in milk fat concentration to increased acetate supply. We suspect their findings might be associated with the decrease in lipase activity. How LPL affects lipase activity warrants further mechanistic investigation. We suspect LPL that escaped from the rumen is absorbed in the small intestines and stimulates the body to depress the expression of genes encoding lipase. From another point of view, the decrease in lipase activity might enhance fat deposition in muscle, which could be useful for the production of quality meat.

Most blood biochemical parameters tested did not alter with the supplementation of LPL, suggesting the status of energy and nitrogen in the body has not been improved, which supports the outcome of no or a slight difference in ADG between the two treatment groups.

4.5. Rumen Bacterial Community Profile

A significant change in bacterial community profile was observed when LPL was supplemented to the diet for lambs. Of the most abundant genera found in the study, *Prevotella* was most significantly decreased with LPL supplementation based on LDA effect size factors. *Prevotella* is well-known hemicellulose degrading bacteria [35]. The decreased abundance of *Prevotella* was consistent with our finding that fiber digestibility was decreased with LPL. It is worthy to mention that among the *Prevotella* bacteria, *Prevotella* 1 was increased, while *Prevotella* 7 and *Prevotella* 9 decreased with LPL, suggesting a wide diversity of *Prevotella* bacteria in their biochemical characteristics.

Lachnospiraceae ND3007 group was increased, having the largest LDA effect size factor, with LPL supplementation in the diet. The bacteria in this genus was found to be decreased with an increase in temperature and humidity index [36]. The genus has not been well described in the literature, but is a butyric acid producing strain and considered as a probiotics [36]. The increase of the genus might promote a healthy rumen environment.

To date, the six culturable lipolytic bacteria are *Anaerovibrio lipolytica* and some species in the genera *Butyrivibrio, Clostridium,* and *Propionibacterium* [37]. The populations of these bacteria were quite small in the rumen of lambs fed the two diets in this study. This reflected the low content of ether extract (1.6 g/kg) in the diet. The content is at the low end of the lipid content range of 20 to 100 g/kg in the diet for ruminants [38]. The abundance of these bacteria was not changed with LPL supplementation, suggesting lipid metabolism in the rumen is not affected. The decreased activity of lipase in serum in the lambs fed a diet supplemented with LPL might just result from the absorption of LPL into the blood and was not related to lipid metabolism in the rumen.

Roseburia spp., Gram-positive bacteria, can be selectively inhibited by LPL in vitro [11]. This is confirmed true in vivo as the relative abundance of the genus in the rumen dropped from 0.11% to 0.02% with the supplementation of LPL in this study although the genus might be not biologically important for fattening lambs due to the low abundance. However, the phylum *Firmicutes* belongs to Gram-positive bacteria, increasing from 20% to 41% in the relative abundance. It might be because LPL may not inhibit all Gram-positive bacteria or the LPL concentration is not high enough to have an inhibitory effect.

5. Conclusions

The supplementation of lysophospholipid did not consistently result in positive animal performance in fattening lambs in this study. Further studies on a large scale are warranted to confirm the effect of lysophospholipid in animal performance. Feed digestion and rumen fermentation were altered with lysophospholipid supplementation, which was in the opposite direction to the results reported in other ruminant species and contradictory with our hypothesis that lysophospholipid enhances fiber degradation. Mechanism studies are needed to explore the effects of lysophospholipid on rumen fermentation, ruminal degradation and bypass of lysophospholipid, and body metabolisms, especially fat metabolism.

Supplementary Materials: The following are available online at http://www.mdpi.com/2076-2615/9/10/805/s1, Figure S1: Distribution of the most dominant phyla in the rumen of fattening lambs fed a diet supplemented without (CON) or with (LPL) 0.05% lysophospholipid (those with a relative abundance of less than 2% were combined as other), Figure S2: Linear discriminant analysis effect size of rumen bacterial populations of fattening lambs fed a diet supplemented without (CON) or with (LPL) 0.05% lysophospholipid. This plot represents the most differentially abundant taxa according to the logarithmic linear discriminant analysis (LDA) score cutoff of at 4.0. Table S1: Ruminal bacterial composition (% of total sequence reads) in fattening lambs fed experimental diets supplemented with (LPL) or without (CON) lysophospholipids.

Author Contributions: Conceptualization, P.Y. and S.S.; methodology, X.S.; formal analysis, X.S.; investigation, Q.H., B.L., T.W., Y.H., W.T., R.L., C.L., J.L., B.S., C.W. and X.S.; resources, P.Y. and Y.L.; data curation, X.S.; writing—original draft preparation, X.S.; writing—review and editing, X.S. and L.C.; supervision, X.S.; project administration, X.S.; funding acquisition, X.S. and S.S.

Funding: This research was funded by Jilin Agricultural Science and Technology University (Start-up fund number 2018:5001) and Development of Science and Technology of Jilin Province (Grant number 20180201041NY).

Acknowledgments: We thank Portal Agri-Industries Co., Ltd. for manufacturing pelleted feeds and Yue Yu for convenience during the experiment conducted at the Animal Experimental Station of Jilin Agricultural Science and Technology University.

Conflicts of Interest: The authors declare no conflict of interest. The funders had no role in the design of the study; in the collection, analyses, or interpretation of data; in the writing of the manuscript, or in the decision to publish the results.

Appendix A

Graphical representation of the relative abundance of bacterial diversity from phylum to species was visualized using Krona chart. Cluster analysis was preceded by principal component analysis (PCA) to reduce the dimension of the original variables using the QIIME software package [39]. QIIME calculates both weighted and unweighted unifrac distance, which are phylogenetic measures of beta diversity. We used unweighted unifrac distance for Principal Coordinate Analysis (PCoA) and Unweighted Pair Group Method with Arithmetic mean (UPGMA) Clustering.

To confirm differences in the abundances of individual taxonomy between the two groups, Metastats software was utilized. LEfSe was used for the quantitative analysis of biomarkers within different groups. This method was designed to analyze data in which the number of species is much higher than the number of samples and to provide biological class explanations to establish statistical significance, biological consistency, and effect-size estimation of predicted biomarkers. To identify differences of microbial communities between the two groups, ANOSIM [40] and MRPP (multi-response permutation procedure) [41] were performed based on the Bray-Curtis dissimilarity distance matrices.

References

1. Sun, X.; Song, B.; He, Y.; You, P. A review on pelleted complete feed for sheep and goats. *Mod. J. Anim. Husb. Vet. Med.* **2017**, *46*, 162–165. (In Chinese)
2. Toteda, F.; Facciolongo, A.M.; Ragni, M.; Vicenti, A. Effect of suckling type and PUFA use on productive performances, quanti-qualitative characteristics of meat and fatty acid profile in lamb. *Prog. Nutr.* **2011**, *13*, 125–134.
3. Salem, A.F.Z.M. *Nutritional Strategies of Animal Feed Additives*; Nova Science Publishers, Inc.: Hauppauge, NY, USA, 2013; pp. 1–221.
4. Guo, X.; Cheng, L.; Liu, J.; Zhang, S.; Sun, X.; Al-Marashdeh, O. Effects of licorice extract supplementation on feed intake, digestion, rumen function, blood indices and live weight gain of Karakul sheep. *Animals* **2019**, *9*, 279. [CrossRef] [PubMed]
5. Zhong, R.; Xiang, H.; Cheng, L.; Zhao, C.; Wang, F.; Zhao, X.; Fang, Y. Effects of feeding garlic powder on growth performance, rumen fermentation, and the health status of lambs infected by gastrointestinal nematodes. *Animals* **2019**, *9*, 102. [CrossRef]
6. Facciolongo, A.M.; De Marzo, D.; Ragni, M.; Lestingi, A.; Toteda, F. Use of alternative protein sources for finishing lambs. 2. Effects on chemical and physical characteristics and fatty acid composition of meat. *Prog. Nutr.* **2015**, *17*, 165–173.

7. Nagpal, R.; Shrivastava, B.; Kumar, N.; Dhewa, T.; Sahay, H. Microbial feed additives. In *Rumen Microbiology: From Evolution to Revolution*; Puniya, A.K., Singh, R., Kamra, D.N., Eds.; Springer: Berlin, Germany, 2015; pp. 161–175.

8. Bernal-Barragán, H.; Cerrillo-Soto, M.A.; García-Mazcorro, J.F.; Juárez-Reyes, A.S.; Salem, A.Z.M. Antibiotics in animal nutrition. In *Nutritional Strategies of Animal Feed Additives*; Salem, A.F.Z.M., Ed.; Nova Science Publishers, Inc.: Hauppauge, NY, USA, 2013; pp. 25–46.

9. Maron, D.F.; Smith, T.J.S.; Nachman, K.E. Restrictions on antimicrobial use in food animal production: An international regulatory and economic survey. *Glob. Health* **2013**, *9*. [CrossRef]

10. Mnasri, T.; Hérault, J.; Gauvry, L.; Loiseau, C.; Poisson, L.; Ergan, F.; Pencréac'H, G. Lipase-catalyzed production of lysophospholipids. *OCL Oilseeds Fats Crop. Lipids* **2017**, *71*. [CrossRef]

11. Van Rensburg, C.E.J.; Joone, G.K.; O'Sullivan, J.F.; Anderson, R. Antimicrobial activities of clofazimine and B669 are mediated by lysophospholipids. *Antimicrob. Agents Chemother.* **1992**, *36*, 2729–2735. [CrossRef]

12. Zhao, P.Y.; Li, H.L.; Hossain, M.M.; Kim, I.H. Effect of emulsifier (lysophospholipids) on growth performance, nutrient digestibility and blood profile in weanling pigs. *Anim. Feed Sci. Technol.* **2015**, *207*, 190–195. [CrossRef]

13. Brautigan, D.L.; Li, R.; Kubicka, E.; Turner, S.D.; Garcia, J.S.; Weintraut, M.L.; Wong, E.A. Lysolecithin as feed additive enhances collagen expression and villus length in the jejunum of broiler chickens. *Poult. Sci.* **2017**, *96*, 2889–2898. [CrossRef]

14. Zampiga, M.; Meluzzi, A.; Sirri, F. Effect of dietary supplementation of lysophospholipids on productive performance, nutrient digestibility and carcass quality traits of broiler chickens. *Ital. J. Anim. Sci.* **2016**, *15*, 521–528. [CrossRef]

15. Zhao, P.Y.; Kim, I.H. Effect of diets with different energy and lysophospholipids levels on performance, nutrient metabolism, and body composition in broilers. *Poult. Sci.* **2017**, *96*, 1341–1347. [CrossRef] [PubMed]

16. Zhao, P.Y.; Zhang, Z.F.; Lan, R.X.; Liu, W.C.; Kim, I.H. Effect of lysophospholipids in diets differing in fat contents on growth performance, nutrient digestibility, milk composition and litter performance of lactating sows. *Animal* **2017**, *11*, 984–990. [CrossRef] [PubMed]

17. Wang, Q.Q.; Long, S.F.; Hu, J.X.; Li, M.; Pan, L.; Piao, X.S. Effects of dietary lysophospholipid complex supplementation on lactation performance, and nutrient digestibility in lactating sows. *Anim. Feed Sci. Technol.* **2019**, *251*, 56–63. [CrossRef]

18. Rico, D.E.; Ying, Y.; Harvatine, K.J. Short communication: Effects of lysolecithin on milk fat synthesis and milk fatty acid profile of cows fed diets differing in fiber and unsaturated fatty acid concentration. *J. Dairy Sci.* **2017**, *100*, 9042–9047. [CrossRef]

19. Lee, C.; Morris, D.L.; Copelin, J.E.; Hettick, J.M.; Kwon, I.H. Effects of lysophospholipids on short-term production, nitrogen utilization, and rumen fermentation and bacterial population in lactating dairy cows. *J. Dairy Sci.* **2019**, *102*, 3110–3120. [CrossRef]

20. Sontakke, U.B.; Kaur, H.; Tyagi, A.K.; Kumar, M.; Hossain, S.A. Effect of feeding rice bran lyso-phospholipids and rumen protected fat on feed intake, nutrient utilization and milk yield in crossbred cows. *Indian J. Anim. Sci.* **2014**, *84*, 998–1003.

21. Song, W.-S.; Yang, J.; Hwang, I.H.; Cho, S.; Choi, N.-J. Effect of dietary lysophospholipid (LIPIDOL^{TM}) supplementation on the improvement of forage usage and growth performance in Hanwoo heifer. *J. Korean Soc. Grassl. Forage Sci.* **2015**, *35*, 232–237. [CrossRef]

22. Kamande, G.M.; Baah, J.; Cheng, K.J.; McAllister, T.A.; Shelford, J.A. Effects of tween 60 and tween 80 on protease activity, thiol group reactivity, protein adsorption, and cellulose degradation by rumen microbial enzymes. *J. Dairy Sci.* **2000**, *83*, 536–542. [CrossRef]

23. Ma, N. Introduction of breeding stock and protection and utilization of genetic resources of sheep. *J. Jilin Agric. Univ.* **2004**, *26*, 77–82. (In Chinese)

24. Alton, G.G.; Jones, L.M.; Pietz, D.E. *Laboratory Techniques in Brucellosis*; Monograph Series; World Health Organization: Geneva, Switzerland, 1975; pp. 1–163.

25. National Research Council. *Nutrient Requirements of Small Ruminants: Sheep, Goats, Cervids and New World Camelids*, 6th ed.; National Academy Press: Washington, DC, USA, 2007; p. 384.

26. Liang, J.; Zhu, L.; Xu, Z. Study on the determination of NH_4^+-N content in microbial fermentation liquor by indophenol blue spectrophotometric method. *Food Ferment. Ind.* **2006**, *32*, 134–137. (In Chinese)

27. Feng, Z.; Gao, M. A modified spectrophotometric method for the determination of ammonia concentration in ruminal liquor. *Anim. Husb. Feed Sci.* **2010**, *31*, 37. (In Chinese)

28. Magoč, T.; Salzberg, S.L. FLASH: Fast length adjustment of short reads to improve genome assemblies. *Bioinformatics* **2011**, *27*, 2957–2963. [CrossRef] [PubMed]

29. Edgar, R.C. UPARSE: Highly accurate OTU sequences from microbial amplicon reads. *Nat. Methods* **2013**, *10*, 996–998. [CrossRef] [PubMed]

30. Henderson, G.; Yilmaz, P.; Kumar, S.; Forster, R.J.; Kelly, W.J.; Leahy, S.C.; Guan, L.L.; Janssen, P.H. Improved taxonomic assignment of rumen bacterial 16S rRNA sequences using a revised SILVA taxonomic framework. *PeerJ* **2019**, *2019*. [CrossRef]

31. Fan, K.; Yi, Y.; Liu, Y.; Li, M.; Wang, J.; Wang, Z.; Zhang, H.; Gu, K. Preparation of soybean lysophospholipids and its biosafety analysis. *China Oils Fats* **2019**, *44*, 124–127, 132. (In Chinese)

32. Jenkins, T.C.; Gimenez, T.; Cross, D.L. Influence of phospholipids on ruminal fermentation in vitro and on nutrient digestion and serum lipids in sheep. *J. Anim. Sci.* **1989**, *67*, 529–537. [CrossRef]

33. Russell, J.B.; Rychlik, J.L. Factors that alter rumen microbial ecology. *Science* **2001**, *292*, 1119–1122. [CrossRef]

34. Hwang, I.H.; Lee, C.H.; Kim, S.W.; Sung, H.G.; Lee, S.Y.; Lee, S.S.; Hong, H.; Kwak, Y.C.; Ha, J.K. Effects of mixtures of Tween80 and cellulolytic enzymes on nutrient digestion and cellulolytic bacterial adhesion. *Asian Australas. J. Anim. Sci.* **2008**, *21*, 1604–1609. [CrossRef]

35. Rubino, F.; Carberry, C.; Waters, S.M.; Kenny, D.; Mccabe, M.S.; Creevey, C.J. Divergent functional isoforms drive niche specialisation for nutrient acquisition and use in rumen microbiome. *ISME J.* **2017**, *11*, 932–944. [CrossRef]

36. Zhong, S.; Ding, Y.; Wang, Y.; Zhou, G.; Guo, H.; Chen, Y.; Yang, Y. Temperature and humidity index (THI)-induced rumen bacterial community changes in goats. *Appl. Microbiol. Biotechnol.* **2019**, *103*, 3193–3203. [CrossRef] [PubMed]

37. Privé, F.; Newbold, C.J.; Kaderbhai, N.N.; Girdwood, S.G.; Golyshina, O.V.; Golyshin, P.N.; Scollan, N.D.; Huws, S.A. Isolation and characterization of novel lipases/esterases from a bovine rumen metagenome. *Appl. Microbiol. Biotechnol.* **2015**, *99*, 5475–5485. [CrossRef] [PubMed]

38. Harfoot, C.G.; Hazlewood, G.P. Lipid metabolism in the rumen. In *The Rumen Microbial Ecosystem*; Hobson, P.N., Stewart, C.S., Eds.; Blackie Academic and Professional Publishers: London, UK, 1997; pp. 382–426.

39. Caporaso, J.G.; Kuczynski, J.; Stombaugh, J.; Bittinger, K.; Bushman, F.D.; Costello, E.K.; Fierer, N.; Peña, A.G.; Goodrich, J.K.; Gordon, J.I. QIIME allows analysis of high-throughput community sequencing data. *Nat. Methods* **2010**, *7*, 335–336. [CrossRef] [PubMed]

40. Philippot, L.; Spor, A.; Hénault, C.; Bru, D.; Bizouard, F.; Jones, C.M.; Sarr, A.; Maron, P.A. Loss in microbial diversity affects nitrogen cycling in soil. *ISME J.* **2013**, *7*, 1609–1619. [CrossRef] [PubMed]

41. Price, M.N.; Dehal, P.S.; Arkin, A.P. FastTree 2—Approximately maximum-likelihood trees for large alignments. *PLoS ONE* **2010**, *5*, e9490. [CrossRef] [PubMed]

Article

Pre-Exposure of Early-Weaned Lambs to a Herb-Clover Mix Does Not Improve Their Subsequent Growth

Lukshman Jay. Ekanayake [1],*, Rene Anne Corner-Thomas [2], Lydia Margaret Cranston [2], Paul Richard Kenyon [2], Stephen Todd Morris [2] and Sarah Jean Pain [2]

[1] Department of Animal Science, Faculty of Agriculture, University of Peradeniya, Peradeniya 20000, Sri Lanka
[2] School of Agriculture and Environment, Private Bag 11-222, Massey University, Palmerston North 4442, New Zealand; R.Corner@massey.ac.nz (R.A.C.-T.); L.Cranston@massey.ac.nz (L.M.C.); P.R.Kenyon@massey.ac.nz (P.R.K.); S.T.Morris@massey.ac.nz (S.T.M.); S.J.Pain@massey.ac.nz (S.J.P.)
* Correspondence: jayampathiekn@gmail.com; Tel.: +94-71-071-8538; Fax: +94-81-239-5012

Received: 10 June 2020; Accepted: 3 August 2020; Published: 5 August 2020

Simple Summary: Exposure of lambs to herbage-based diets prior to weaning may facilitate the development of the rumen which may subsequently increase animal performance after early weaning. The aim of this study was to estimate the effects of varying durations of exposure of lambs to a herb–clover mix containing chicory, plantain, red clover, and white clover prior to early weaning (at ~45 days of age) on their subsequent growth and rumen development at conventional weaning age. Prolonged exposure of lambs to the herb–clover mix prior to early weaning had no impact on lamb growth or rumen development, suggesting that using this management option will not improve performance of lambs after early weaning.

Abstract: Twin sets of lambs were randomly allocated to one of six treatments: (1) lambs born and managed on ryegrass–clover-based pasture until conventional weaning approximately at 99 days of age (Grass–Grass$_{CW}$); (2) lambs born on ryegrass–clover-based pasture and early weaned onto a herb–clover mix at ~45 days of age (Grass–Herb$_{EW}$); (3) lambs born on ryegrass–clover-based pasture, transferred with their dam onto a herb–clover mix at ~45 days of age until conventional weaning (Grass–Herb$_{CW}$); (4) lambs born on ryegrass–clover-based pasture, transferred with their dam onto a herb–clover mix at ~15 days of age and early weaned onto a herb–clover mix at ~45 days of age (Grass–Herb$_{D15EW}$); (5) lambs born and managed on herb–clover mix until conventional weaning (Herb–Herb$_{CW}$); (6) lambs born on herb–clover mix and weaned early onto a herb–clover mix at ~45 days of age (Herb–Herb$_{EW}$). In both years, Herb–Herb$_{CW}$ lambs had greater ($p < 0.05$) growth rates than lambs in other treatments. The liveweight gains and rumen papillae development of Herb–Herb$_{EW}$, Grass–Herb$_{D15EW}$ and Grass–Herb$_{EW}$ lambs did not differ ($p > 0.05$). The weight of the empty digestive tract components at either early weaning or conventional weaning did not differ ($p > 0.05$) between treatments. Exposing early-weaned lambs to the herb mix for a prolonged period, prior to early weaning, does not improve their subsequent growth.

Keywords: early weaning; lamb; rumen development; herb–clover mix

---

1. Introduction

Weaning lambs early can be a useful management option that farmers can use when herbage production is inadequate or of poor quality, to reduce the overall ewe feed demand and to support the growth of early-weaned lambs to achieve liveweight targets [1,2]. Lambs weaned early at 4–8 weeks of age onto a ryegrass–clover-based pasture have consistently displayed lower live weight gains and poorer survival compared to unweaned lambs on ryegrass–clover-based pastures, to approximately 14 weeks of age [3,4]. This is likely explained by the inadequate nutritional quality of ryegrass–clover-based pastures. In contrast, a herb–clover mix containing plantain (*Plantago lanceolata*), chicory (*Cichorium intybus*), red clover (*Trifolium pretanse*) and white clover (*Trifolium repens*) has been shown to have a greater nutritional quality compared to a ryegrass–clover-based pasture [5–7]. This suggests that a herb–clove mix can be used to support lamb growth, post-early weaning. However, the liveweight gain of lambs weaned early, at a minimum live weight of 14 to 16 kg, onto a herb–clover mix, has been inconsistent across studies, compared to unweaned lambs on ryegrass–clover-based pastures [8–11]. In contrast, the live weight of ewes whose lambs were weaned early has consistently been greater than those left unweaned on ryegrass–clover [10,11]. The inconsistency in lamb liveweight gain has not yet been explained. The reasons for the variation in lamb growth post-early weaning on a herb–clover mix appear not to be driven by variation of metabolisable energy [9,11] or crude protein content [9–12] of the herbage. If farmers are to utilize early weaning, a consistent method for achieving suitable lamb growth post weaning is required. The impact of prior exposure to the herb–clover mix on lamb growth, post early weaning, has not been previously examined. If prior exposure to the herb–clover mix was found to result in a consistent positive effect on lamb growth, management guidelines for farmers could be developed.

Early development of the reticulo-rumen and its microbial population is important for efficient utilization of herbage-based diets after weaning [13,14]. Early exposure to herbage-based diets can help develop a strong host rumen microbiota [15,16] and familiarity with a herbage type can influence a lamb's preference for that herbage later in life [17]. Sheep avoid grazing novel herbages [18,19] for several reasons including taste aversion [20], as a strategy to maintain effective rumen micro-flora [21] and to meet their ideal protein intake [22,23]. Combined, these studies suggest that early exposure of lambs to herbage may facilitate rumen development and increase subsequent animal performance. To date the rumen development of early-weaned lambs offered herb–clover mix has not been studied. It was hypothesized that prolonged exposure of lambs to a herb–clover mix, prior to early weaning, would have a positive effect on their subsequent growth and rumen development. The aims of the present study were to test the effects of (i) weaning lambs early at approximately 45 days of age onto a herb–clover mix and (ii) early prolonged exposure of lambs to a herb–clover mix prior to early weaning at approximately 45 days of age on their subsequent growth and rumen development until conventional weaning age (~99 days of age).

2. Materials and Methods

2.1. Herbage Treatments

Eight paddocks (12.85 ha in total land area) of herb–clover mix and seven paddocks (14.85 ha in total land area) of ryegrass–clover-based pasture containing perennial ryegrass (*Lolium perenne* L.) and white clover (*Trifolium repens*) were used over two spring periods in 2017 and 2018. Herb–clover mix paddocks were sown during autumn in 2012 and 2013 with a seed mixture of chicory (*Cichorium intybus*; 6 kg/ha), plantain (*Plantago lanceolata*; 6 kg/ha), white clover (4 kg/ha) and red clover (*Trifolium pratense*; 6 kg/ha). During the experimental period (August to December in both years), a rotational grazing system was used whereby the post-grazing sward surface heights were maintained at a minimum of 5 to 7 cm for both ryegrass–clover-based pasture and herb–clover mix, respectively, in order to provide ad-libitum dry matter intakes for lambs and ewes [24,25]. The management of the herbages in this study was similar to the management applied in previous studies [8–12].

2.2. Experimental Design

The studies were conducted at Massey University's Keeble farm, 5 km southeast of Palmerston North, New Zealand (40°24′ S and 175°36′ E). All animal manipulations were approved by the

Massey University Animal Ethics Committee (MUAEC 17-40). Each year, Romney ewes, from a large commercial flock, were naturally bred during a 17-day breeding period. A random subset of ewes diagnosed as twin-bearing using transabdominal ultrasound at approximately 90 days of pregnancy, were included in each year. Throughout the gestation period, within each year, ewes were managed as a single flock (group) under commercial pastoral farming conditions on a ryegrass–clover-based pasture.

Each year, twin-bearing ewes were allocated to treatments three days prior to lambing using a stratified random sampling procedure, in order to balance the groups for ewe live weight. A summary of the experimental design can be found in Table 1. Briefly, in 2017, the study included six treatments (Grass–Grass$_{CW}$, Grass–Herb$_{EW}$, Grass–Herb$_{CW}$, Grass–Herb$_{D15EW}$, Herb–Herb$_{CW}$ and Herb–Herb$_{EW}$, Table 1), while in 2018, Grass–Grass$_{CW}$ was omitted resulting in five treatments (Grass–Herb$_{EW}$, Grass–Herb$_{CW}$, Grass–Herb$_{D15EW}$, Herb–Herb$_{CW}$ and Herb–Herb$_{EW}$). The Grass–Grass$_{CW}$ treatment had previously been examined over three years [10,11] and was determined to be unnecessary as the overall aim was to evaluate the effects of early exposure of lambs to a herb–clover mix prior to early weaning on their subsequent growth and rumen development until a conventional weaning age. Three days prior to lambing, ewes (n = 117 in 2017 and n = 161 in 2018) were allocated to Grass–Herb$_{EW}$, Grass–Herb$_{CW}$, and Grass–Herb$_{D15EW}$ and Grass–Grass$_{CW}$ (only in 2017) treatments and grazed on a ryegrass–clover-based pasture. Similarly, on the same date, ewes (n = 80 in 2017 and n = 82 in 2018) were allocated to Herb–Herb$_{CW}$ and Herb–Herb$_{EW}$ treatments begun grazing the herb–clover mix.

Table 1. Summary of the experimental design with a description of treatments, year, number of lambs in each treatment, treatment herbage description and age of weaning.

Treatment	Year	Number of Lambs	Treatment Description	Lamb Age [1]			
				0	15	47–51	87–99
Grass–Grass$_{CW}$	2017	36	Lambs born and remained on ryegrass–clover-based pasture until conventional weaning	◆	◆	◆	◆,•
Grass–Herb$_{EW}$	2017 2018	36 42	Lambs born on ryegrass–clover-based pasture and early weaned onto herb–clover mix	◆	◆	★,•	★
Grass–Herb$_{CW}$	2017 2018	36 44	Lambs born on ryegrass–clover-based pasture, transferred with dam onto herb–clover mix at early weaning and remained until conventional weaning	◆	◆	★	★,•
Grass–Herb$_{D15EW}$	2017 2018	30 42	Lambs born on ryegrass–clover-based pasture, transferred with dam onto herb–clover mix at 15 days of age and early weaned	◆	★	★,•	★
Herb–Herb$_{CW}$	2017 2018	34 36	Lambs born and remained on herb–clover mix until conventional weaning	★	★	★	★,•
Herb–Herb$_{EW}$	2017 2018	36 40	Lambs born on herb–clover mix and early weaned onto herb–clover mix	★	★	★,•	★

[1] Management of the lambs during each time period. ◆ Lambs on ryegrass–clover-based pasture; ★ Lambs on herb–clover mix; • Weaning age; Early weaned ewes were removed and grazed on ryegrass–clover-based pastures; early weaning occurred at L51 and L44 (average days of age) in 2017 and 2018, respectively; conventional weaning occurred at L99 and L87 days of age in 2017 and 2018, respectively. Grass–GrassCW, Grass–HerbEW and Grass–HerbCW lambs have been used in a previous study [11].

Lambing began on 28 and 27 August in 2017 and 2018 (L0), respectively. Lambs were born over 26- and 28-day period in 2017 and 2018 (mid-point of the lambing period each year was defined as L1, i.e., average day of age at that point was one). All lambs were weighed, ear tagged and identified to their dam within 24 h of birth. In total, n = 136 and n = 138 lambs were born on herb–clover mix in 2017 and 2018, respectively. On the ryegrass–clover-based pasture n = 224 and n = 302 lambs were born in 2017 and 2018, respectively. In accordance with veterinary advice, at L27 (average 27 days of age) in 2017 and L26 in 2018, lambs were drenched with an oral triple combination drench (Matrix, Merial Ancare, Manukau City, New Zealand) and thereafter at 28-day intervals throughout the study at a rate of 1 mL per 5 kg live weight to control internal parasites. This method has been successfully used in previous studies [8–11], therefore, lamb growth in this study was unlikely to be affected by internal parasites.

At L15 (average 15 dayss of age) in both years, ewes and lambs allocated to the Grass–Herb$_{D15EW}$ treatment were transferred from ryegrass–clover-based pasture to a herb–clover mix where they remained until early weaning (at L51 and L44 in 2017 and 2018, respectively, Table 1). Ewes and lambs allocated to the remaining treatments were maintained on their respective herbage until early weaning. At L51 in 2017 and at L44 in 2018 (early weaning), ewes rearing twin lambs ($n = 123$ and $n = 112$ in 2017 and 2018, respectively) that were both a minimum of 14 kg live weight ($n = 246$ and $n = 224$, in 2017 and 2018, respectively) were included in the study. Any ewes and lambs that did not fulfil these criteria were excluded from the remainder of the study. Lambs in each treatment ($n = 42, 44, 38, 42, 46, 34$ in Herb–Herb$_{CW}$, Herb–Herb$_{EW}$, Grass–Herb$_{D15EW}$, Grass–Herb$_{CW}$, Grass–Herb$_{EW}$, Grass–Grass$_{CW}$ in 2017 and $n = 36, 36, 42, 34, 34$ in Herb–Herb$_{CW}$, Herb–Herb$_{EW}$, Grass–Herb$_{D15EW}$, Grass–Herb$_{CW}$, Grass–Herb$_{EW}$ in 2018, respectively) remained on these treatments until conventional weaning at L99 in 2017 and L87 in 2018.

After early-weaning, ewes in the Grass–Herb$_{EW}$, Grass–Herb$_{D15EW}$ and Herb–Herb$_{EW}$ treatments (weaned ewes) were managed with Grass–Grass$_{CW}$ treatment in 2017 until L99 or with a commercial flock under pastoral farming conditions on ryegrass–clover-based pasture until L87 in 2018, receiving similar allowances in terms of pasture offered. Early-weaned lambs (Grass–Herb$_{EW}$, Grass–Herb$_{D15EW}$ and Herb–Herb$_{EW}$) were managed with the Herb–Herb$_{CW}$ ewes and lambs until L99 and L87 in 2017 and 2018, respectively. Within 1 to 2 h of birth, ewes develop an exclusive bond with their lambs and reject any alien lamb that attempts to suck for the remainder of the lactation [26]. Therefore, there was no concern regarding the early-weaned lambs suckling from unweaned ewes.

2.3. Animal Measurements

Lambs and ewes were weighed within 1 h of removal from herbage at L15, L51, L65 and L99 in 2017 and L15, L44, L61 and L87 in 2018. Ewe body condition score (BCS, scale 1–5, including half units, [27]) was assessed at each weighing by a single experienced operator.

2.4. Anatomical and Histological Examination of the Digestive System

A subset of male lambs from the complete twin sets were selected for anatomical and histological examination of the digestive system at L44 ($n = 12$) and L87 ($n = 24$) in 2018 to determine the effect of feed type prior to early weaning, and the combination of feed type and early weaning on the rumen development, respectively. Lambs were selected using stratified random sampling procedure, based on live weights, in order to obtain a representative sample of lambs from each treatment. At L44, lambs in the Grass–Herb$_{EW}$ ($n = 6$, average live weight 19.0 ± 1.5) and Herb–Herb$_{EW}$ ($n = 6$, average live weight 21.0 ± 1.5) treatments were euthanised. At L87, 24 male lambs were selected from the Grass–Herb$_{EW}$ ($n = 6$; average live weight 27.4 ± 2.0), Grass–Herb$_{CW}$ ($n = 6$; average live weight 31.0 ± 2.0), Herb–Herb$_{CW}$ ($n = 6$; average live weight 35.0 ± 2.0) and Herb–Herb$_{EW}$ ($n = 6$; average live weight 32.0 ± 2.0) treatments and euthanised. Lambs were weighed within 1 h of removal from herbage and were euthanised using captive bolt stunning and exsanguination.

Immediately after euthanasia, their skins removed, they were eviscerated, and the head was separated from the carcass. Hot carcass weight was then recorded. The digestive system was removed by severing the esophagus which was tied at the cardia of the stomach. All the organs attached to the digestive system were removed and then weighed. The digestive system was then separated into its components (reticulo-rumen, small intestine and large intestine) which were cleaned using tap water. Empty component weights were recorded to the nearest gram. The cleaned reticulo-rumen was dissected according to the procedure of [28] and the reticulo-rumen was opened and laid flat symmetrically. Two samples (1 cm^2 each) were collected from the right side of the caudal dorsal sac as [29] reported that samples from this region were representative of rumen development.

Rumen tissue samples (two from each animal) were immediately fixed in a 10% formalin saline solution for 24 h. Samples were then processed using a Excelsior ES Tissue Processor (ThermoFisher$^®$, Waltham, MA, USA). Dehydration was carried out by ascending graded series of ethyl alcohol (70, 95% and absolute Alcohol) for 12 h, followed by clearing in Xylene at 60 °C. The samples were then embedded and impregnated in melted Paraffin wax at 55–58 °C using a HistoStar Embedder (ThermoFisher$^®$, MA, USA). Embedded samples were cut using Rotary Microtome (MicroTec$^®$, Wetzlar, Germany) into sections (4 μm). Two sections were produced, 150 μm apart from each tissue sample resulting in 4 sections per animal (replicates). The sections were floated in a Tissue Bath (ThermoFisher$^®$,MA, USA) at 43 °C and were mounted onto adhesive pre-cleaned slides (90° ground edges, 76 mm × 26 mm). The sections were stained using Haemytoxelin and Eosin on an Autostainer XL (Leica$^®$, Wetzlar, Germany). The stained slides, 2 slides per lamb, each contained two sections were covered using a Leica$^®$ CV5030 cover slip. All complete papillae in each section were examined microscopically to determine their length and width at the base (Figure 1).

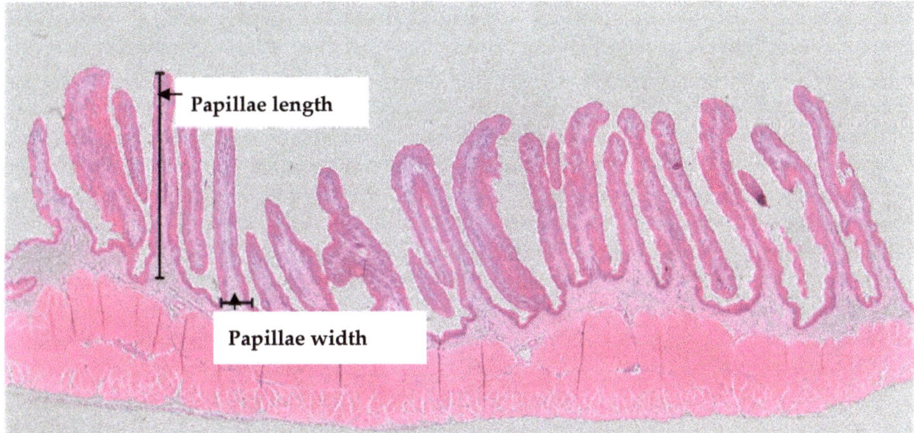

Figure 1. Example of histomorphometric measurements (papillae length and width) of the rumen of lambs measured using a computerised micrometre at 100× magnification.

2.5. Herbage Measurements

Herbage masses were measured at L15, L51 and L99 in 2017 and L15, L44 and L87 in 2018. Four random quadrat cuts (0.1 m^2 each) were taken to ground level from each herbage type at each sampling date using an electric shearing handpiece [30]. Samples were then oven dried to a constant weight to estimate herbage mass. In addition, four composite herbage samples, each containing ten grab samples per herbage type, were also collected, to mimic post grazing height, at each sampling date to determine the botanical and nutritional composition [30]. To determine the botanical composition, a subsample from each composite sample (4 per herbage type) was sorted into each species (herb–clover mix: plantain, chicory, red clover, white clover; ryegrass–clover-based pasture: ryegrass, clover; other grasses (combined), weeds, and dead matter) and then oven dried and weighed to determine the botanical composition. The remaining sample was then frozen, dried, ground, sieved (1 mm) and analysed using in vitro methods to determine the nutritional quality. These measures included dry matter digestibility (DMD, [31]), digestible organic matter digestibility (DOMD, [31]), percentage crude protein (CP; "Dumas" procedure, AOAC method 968.06 using a Leco total combustion method, LECO Corporation, St. Joseph, MI, USA). Percentage acid detergent fibre (ADF) was analysed by a Tecator Fibretec System [32]. Metabolisable energy (ME) content of herbages was calculated from the organic matter digestibility (DOMD × 0.16 MJ/Kg DM, [31]). These herbage measurements have been used in previous studies [8–12] to estimate the dry matter mass, botanical and chemical composition of both ryegrass–clover-based pasture and herb–clover mix.

2.6. Statistical Analysis

The individual animal was considered the experimental unit for these analyses. Live weight of lambs and ewes were subjected to analysis of variance for repeated measures using the MIXED procedure in SAS (Statistical Analysis System, version 9.2; SAS Institute Inc., Cary, NC, USA). Analyses were performed separately for each year due to the differences in the days on which measurements were collected, number of treatments in 2017 and 2018, differences in climate and herbage quality, and different animals used between years. The model for lamb live weight at L51 and L99 and at L44 and L87 in 2017 and 2018, respectively, included the fixed effects of weaning treatment, sex of lamb (male, female), measurement date and the two-way interactions of treatment and measurement date. The live weight of lambs at the start of the treatment period was included in the model as a covariate. The model for lamb liveweight gain from early weaning to conventional weaning included the fixed effect of weaning treatment and sex of lamb.

The models for ewe live weight included the fixed effects of weaning treatment, measurement date and two-way interactions of treatment and measurement date. The model for ewe liveweight gain from early weaning to conventional weaning included the fixed effect of weaning treatment. The live weight of the ewe at the start of the treatment period was included in the model as a covariate. Ewe body condition score was analysed using a Poisson distribution and logit transformation using the GENMOD procedure in SAS. The Poisson distribution was chosen as it is a nonlinear regression model for discrete outcomes. The model included the fixed effects of weaning treatment, measurement date and the two-way interaction of treatment and measurement date.

Papillae length and width were subjected to analysis of variance using the MIXED procedure in SAS. The models for papillae length and width included the fixed effect of weaning treatment and the random effect of lamb, sample and replicate (sections) for each sample. The live weight of the lamb at slaughter was included in the model as a covariate. The models for the anatomical measurements of the rumen included the fixed effect of weaning treatment and lamb live weight at slaughter as a covariate.

The botanical composition of herbages was subjected to an analysis of variance for repeated measures using the MIXED procedure in SAS. The model included fixed effects of plant species and measurement date. Herbage masses were analysed using a model that included herbage type and measurement date as fixed effects. The nutritional quality data were analysed using the MIXED procedure in a model that included the fixed effects of herbage type and measurement date.

3. Results

3.1. Botanical Composition, Herbage Mass and Nutritional Quality of Herbage

The percentage of chicory and clover was 16% greater ($p < 0.05$) and plantain was 23% lower ($p < 0.05$) in the herb–clover mix at L44 in 2018 than at L47 in 2017 (Figure 2). The percentage of plantain was 20% greater ($p < 0.05$) and dead matter content was 35% lower ($p < 0.05$) in the herb–clover mix at L87 in 2018 than at L99 in 2017. Other plant components in the herb–clover mix did not differ ($p > 0.05$) between years and time points. At L15, the percentage of ryegrass in the ryegrass–clover-based pasture was 22% greater ($p < 0.05$) in 2018 than 2017. Other plant components in the ryegrass–clover-based pasture did not differ ($p > 0.05$) between years and time points. In both years, herbage mass of both herbage types was above 1600 kg DM/ha at all sampling dates throughout the study (Table 2).

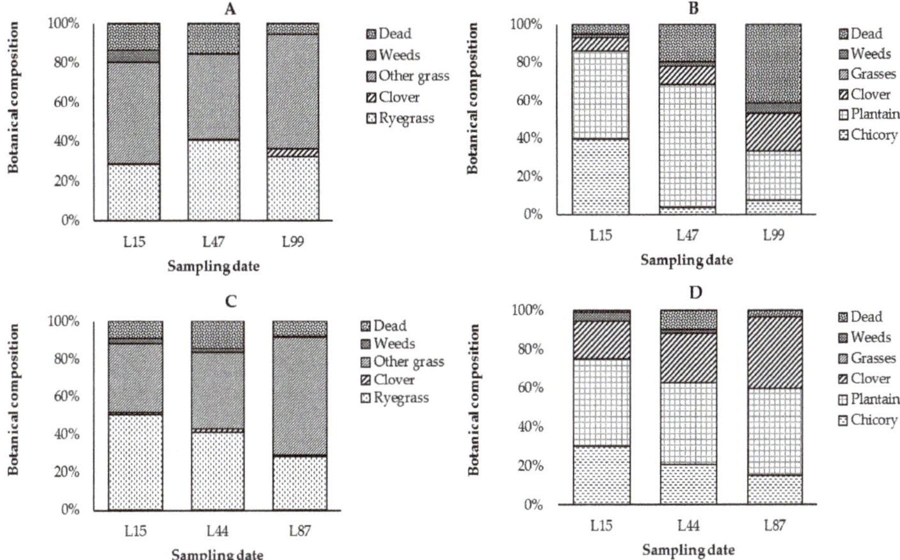

Figure 2. The botanical composition of the components of ryegrass–clover-based pasture (**A**), herb–clover mix (**B**) in 2017 and ryegrass–clover-based pasture (**C**) and herb–clover mix (**D**) in 2018 on 15, 51 and 99 days of age in 2017 and 2018 (L15, L44 and L87).

Table 2. Herbage mass (HM) offered to both lambs and ewes, crude protein (CP), neutral detergent fibre (NDF) acid detergent fibre (ADF), dry matter digestibility (DMD) and metabolisable energy content (ME) of herbages collected L15, L51, L99 in 2017 and 2018 (L15, L44, L87) (least-squares mean ± SEM).

Year	Herbage Type	Study Day	HM (kg DM/ha)	CP (%)	NDF (%)	ADF (%)	DMD (%)	ME (MJ/Kg)
2017	Herb–clover mix	L15	2742 [b,c] ± 260	10.9 [a] ± 1.1	50.4 [c,d] ± 1.4	35.6 [b] ± 1.3	65.7 [b] ± 0.7	9.2 [a,b] ± 0.1
		L51	3221 [c] ± 260	17.3 [b] ± 1.1	46.7 [c] ± 1.4	26.6 [a] ± 1.3	65.8 [b] ± 0.7	9.5 [b] ± 0.1
		L99	3048 [b,c] ± 260	15.6 [b] ± 1.1	53.7 [d] ± 1.4	30.5 [b] ± 1.3	63.5 [a] ± 0.7	9.1 [a] ± 0.1
	Ryegrass–clover-based pasture	L15	2313 [a,b] ± 260	16.5 [b] ± 1.1	28.7 [a] ± 1.4	22.9 [a] ± 1.3	75.5 [d] ± 0.7	11.0 [d] ± 0.1
		L51	1899 [a] ± 260	12.0 [a] ± 1.1	32.7 [b] ± 1.4	26.2 [a] ± 1.3	72.0 [c] ± 0.7	10.6 [c] ± 0.1
		L99	2680 [b] ± 260	13.0 [a,b] ± 1.1	31.4 [a,b] ± 1.4	23.7 [a] ± 1.3	73.1 [c] ± 0.7	10.6 [c] ± 0.1
2018	Herb–clover mix	L15	4298 [d] ± 260	14.4 [a] ± 1.1	34.0 [b] ± 1.4	24.2 [b] ± 1.3	71.4 [b,c] ± 0.7	10.5 [c,d] ± 0.1
		L44	3481 [c] ± 260	17.2 [b] ± 1.1	27.6 [a] ± 1.4	19.5 [a] ± 1.3	74.1 [d] ± 0.7	11.0 [e] ± 0.1
		L87	3857 [c] ± 260	12.1 [a] ± 1.1	31.7 [b] ± 1.4	22.4 [a,b] ± 1.3	73.2 [c,d] ± 0.7	10.8 [d,e] ± 0.1
	Ryegrass–clover-based pasture	L15	1684 [a] ± 260	17.3 [b] ± 1.1	47.5 [d] ± 1.4	26.1 [b] ± 1.3	64.7 [a] ± 0.7	9.3 [a] ± 0.1
		L44	2057 [a] ± 260	12.9 [a] ± 1.1	48.0 [d] ± 1.4	24.6 [b] ± 1.3	65.9 [a] ± 0.7	9.8 [b] ± 0.1
		L87	2068 [a] ± 260	18.4 [b] ± 1.1	42.4 [c] ± 1.4	21.8 [a,b] ± 1.3	69.5 [b] ± 0.7	10.4 [b,c] ± 0.1

L, average days of age. [a–e] Means with different superscripts within columns are significantly different across years and treatments ($p < 0.05$).

At L15 in 2017, the CP content of ryegrass–clover-based pasture was ~5% greater ($p < 0.05$) than herb–clover mix whereas the reverse was seen at L51. At L99, CP content did not differ ($p > 0.05$) between herbages. At L15 and L99, the ADF content of herb–clover mix was greater (13% and 7%, respectively, $p < 0.05$) than that of ryegrass–clover-based pasture, however, at L51 ADF did not differ ($p > 0.05$). The DMD, neutral detergent fibre (NDF) and ME of herb–clover mix were lower (~6 to 10%, 14 to 22% and 1.1 to 1.8 MJ/Kg, respectively, $p < 0.05$) than ryegrass–clover-based pasture at all sampling times.

In 2018, at L15 and L87, the CP content of herb–clover mix was lower (~3 and 6%, respectively, $p < 0.05$) than ryegrass–clover-based pasture, whereas the CP content of herb–clover mix was ~3% greater ($p < 0.05$) at L44 (Table 2). The ADF content of herb–clover mix did not differ ($p > 0.05$) from that of ryegrass–clover-based pasture at L15 and L87 but was ~5% lower ($p < 0.05$) at L44. The NDF of herb–clover mix was 10 to 20% lower ($p < 0.05$) than that of ryegrass–clover-based pasture at L15, L44 and L87. The DMD and ME of herb–clover mix were greater (~4 to 8% and 0.4 to 1.2 MJ/Kg, respectively, $p < 0.05$) than that of ryegrass–clover-based pasture at L15, L44 and L87.

3.2. Lamb Live Weight and Liveweight Gain

In 2017, at L51 (date of early weaning), the live weights of lambs did not differ ($p > 0.05$) among treatments (Table 3). At L65, the live weight of lambs in the Herb–Herb[EW], Grass–Herb[D15EW] and Grass–Herb[EW] treatments did not differ ($p > 0.05$) but were ~1 kg lighter ($p < 0.05$) than Herb–Herb[CW], Grass–Herb[CW] and Grass–Grass[CW] lambs. At L65, Herb–Herb[CW] lambs were 0.8 kg heavier ($p < 0.05$) than Grass–Grass[CW] lambs but neither differed ($p > 0.05$) from Grass–Herb[CW] lambs. At L99 (at conventional weaning), Herb–Herb[CW] and Grass–Herb[CW] lambs did not differ ($p > 0.05$) in live weight but were 4 to 5 kg heavier ($p < 0.05$) than lambs in all other treatments, which did not differ ($p > 0.05$) from one another. Lamb liveweight gains between L51 and L99 in the Herb–Herb[CW] and Grass–Herb[CW] treatments were greater ($p < 0.05$) than lambs in the Herb–Herb[EW], Grass–Herb[D15EW], Grass–Herb[EW] and Grass–Grass[CW] treatments that did not differ ($p > 0.05$).

Table 3. Impact of weaning treatment; Herb–Herb[CW], Herb–Herb[EW], Grass–Herb[D15EW], Grass–Herb[CW], Grass–Herb[EW], Grass–Grass[CW] on live weight of lambs on L51, L65 and L99 in 2017 and L44, L61 and L87 in 2018; and liveweight gain (LWG) during L51–L99 in 2017 and during L44–L87 in 2018 (least-squares mean ± SEM).

Herbage Treatment	Lamb Live Weight (kg)						LWG (g/day)
	n		*n*		*n*		
2017							
		L51		L65		L99	L51–L99
Herb–Herb[CW]	42	19.3 ± 0.2 [a]	41	23.5 ± 0.2 [d]	42	34.7 ± 0.4 [f]	317 ± 7 [b]
Herb–Herb[EW]	44	19.4 ± 0.2 [a]	43	21.9 ± 0.2 [b]	43	31.1 ± 0.4 [e]	244 ± 7 [a]
Grass–Herb[D15EW]	38	19.1 ± 0.2 [a]	38	21.7 ± 0.2 [b]	38	30.1 ± 0.4 [e]	231 ± 8 [a]
Grass–Herb[CW]	42	19.3 ± 0.2 [a]	41	23.1 ± 0.2 [c,d]	42	34.6 ± 0.4 [f]	318 ± 7 [b]
Grass–Herb[EW]	46	19.0 ± 0.2 [a]	45	21.6 ± 0.2 [b]	46	30.6 ± 0.4 [e]	239 ± 7 [a]
Grass–Grass[CW]	34	19.2 ± 0.2 [a]	33	22.7 ± 0.2 [c]	34	30.3 ± 0.4 [e]	231 ± 8 [a]
2018							
		L44		L61		L87	L44–L87
Herb–Herb[CW]	36	19.0 ± 0.1 [a]	36	25.0 ± 0.2 [c]	35	31.7 ± 0.4 [e]	289 ± 8 [b]
Herb–Herb[EW]	36	18.9 ± 0.1 [a]	34	23.0 ± 0.2 [b]	33	28.9 ± 0.4 [d]	232 ± 8 [a]
Grass–Herb[D15EW]	42	18.9 ± 0.1 [a]	41	22.7 ± 0.2 [b]	41	28.7 ± 0.3 [d]	231 ± 7 [a]
Grass–Herb[CW]	34	18.8 ± 0.1 [a]	34	24.8 ± 0.2 [c]	34	32.1 ± 0.3 [e]	321 ± 8 [c]
Grass–Herb[EW]	34	18.8 ± 0.1 [a]	34	22.6 ± 0.2 [b]	34	28.7 ± 0.4 [d]	237 ± 8 [a]

L, average days of age; Herb–Herb[CW], lambs born on herb–clover mix and conventional weaning (at ~99 days of age); Herb–Herb[EW], lambs born on herb–clover mix and early weaned onto herb–clover mix at ~45 days of age; Grass–Herb[D15EW], lambs born on ryegrass–clover-based pasture and transferred with dam onto herb–clover mix at 15 days of age and early weaned at ~45 days of age; Grass–Herb[CW], lambs born on ryegrass–clover-based pasture and transferred with dam onto herb–clover mix at ~45 days of age and conventional weaning; Grass–Herb[EW], lambs born on ryegrass–clover-based pasture and early weaned at ~45 days of age onto herb–clover mix; Grass–Grass[CW], lambs born on ryegrass–clover-based pasture and conventional weaning. [a–f] Means with different superscripts are significantly different across years and treatments.

In 2018, at L44 (date of early weaning), the live weights of lambs did not differ ($p > 0.05$) among the treatments (Table 3). At L61 and L87, Herb–Herb[CW] and Grass–Herb[CW] lambs were ~2 kg heavier ($p < 0.05$) than Herb–Herb[EW], Grass–Herb[D15EW] and Grass–Herb[EW] lambs that did not differ ($p > 0.05$). Liveweight gain of Grass–Herb[CW] lambs between L44 and L87 was greater ($p < 0.05$) than Herb–Herb[CW] lambs, which in turn was greater ($p < 0.05$) than Herb–Herb[EW], Grass–Herb[D15EW] and Grass–Herb[EW] lambs, which did not differ ($p > 0.05$).

3.3. Ewe Live Weight and Liveweight Gain

In 2017, at L51 (date of early weaning), the live weights of ewes did not differ ($p > 0.05$) between treatments (Table 4). At L65, the live weights of Herb–Herb$_{CW}$, Herb–Herb$_{CW}$, Grass–Herb$_{D15EW}$, Grass–Herb$_{CW}$ ewes did not differ ($p > 0.05$). Herb–Herb$_{EW}$ ewes were 3 kg heavier ($p < 0.05$) than Grass–Herb$_{EW}$ and Grass–Grass$_{CW}$ ewes whose live weight did not differ ($p > 0.05$). At L99 (at conventional weaning), Herb–Herb$_{CW}$, Herb–Herb$_{EW}$, Grass–Herb$_{D15EW}$, Grass–Herb$_{CW}$ ewes did not differ ($p > 0.05$) in live weight but were 2 to 5 kg heavier ($p < 0.05$) than Grass–Herb$_{EW}$ and Grass–Grass$_{CW}$ ewes. Between L51 and L99, the liveweight gain of Herb–Herb$_{CW}$, Herb–Herb$_{EW}$, Grass–Herb$_{D15EW}$, Grass–Herb$_{CW}$ and Grass–Herb$_{EW}$ ewes did not differ ($p > 0.05$) but were at least 30 g/day greater ($p < 0.05$) than Grass–Grass$_{CW}$ ewes.

Table 4. Impact of weaning treatment; Herb–Herb$_{CW}$, Herb–Herb$_{EW}$, Grass–Herb$_{D15EW}$, Grass–Herb$_{CW}$, Grass–Herb$_{EW}$, Grass–Grass$_{CW}$ on live weight of ewes on L51, L65 and L99 in 2017 and L44, L61 and L87 in 2018; and liveweight gain (LWG) during L51–L99 in 2017 and during L44–L87 in 2018 (least-squares mean ± SEM).

Weaning Treatment		Ewe Live Weight (kg)					LWG (g/day)
	n		*n*		*n*		
				2017			
		L51		L65		L99	L51–L99
Herb–Herb$_{CW}$	21	68.8 ± 0.9 [a]	21	71.0 ± 0.9 [a,b]	20	76.3 ± 1.0 [b]	145 ± 18 [b]
Herb–Herb$_{EW}$	22	70.0 ± 0.8 [a]	22	73.0 ± 0.8 [b]	21	76.7 ± 0.9 [b]	129 ± 17 [b]
Grass–Herb$_{D15EW}$	18	70.1 ± 0.9 [a]	18	71.7 ± 1.0 [a,b]	18	76.3 ± 1.0 [b]	119 ± 19 [b]
Grass–Herb$_{CW}$	21	70.0 ± 0.9 [a]	20	71.1 ± 0.9 [a,b]	21	77.1 ± 0.9 [b]	136 ± 17 [b]
Grass–Herb$_{EW}$	23	68.2 ± 0.9 [a]	23	70.5 ± 0.9 [a]	22	73.9 ± 0.9 [a]	112 ± 17 [b]
Grass–Grass$_{CW}$	17	68.1 ± 0.9 [a]	16	70.4 ± 1.0 [a]	17	72.0 ± 1.0 [a]	78 ± 19 [a]
				2018			
		L44		L61		L87	L44–L87
Herb–Herb$_{CW}$	18	69.5 ± 1.0 [a]	18	72.2 ± 1.0 [a]	18	74.0 ± 1.0 [a]	113 ± 19 [a]
Herb–Herb$_{EW}$	18	71.2 ± 0.9 [a]	17	72.8 ± 1.0 [a]	17	74.7 ± 1.0 [a]	81 ± 19 [a]
Grass–Herb$_{D15EW}$	21	71.0 ± 0.9 [a]	21	72.8 ± 0.9 [a]	21	74.5 ± 0.9 [a]	76 ± 18 [a]
Grass–Herb$_{CW}$	17	69.8 ± 0.8 [a]	16	73.8 ± 0.8 [a]	16	73.9 ± 0.9 [a]	89 ± 17 [a]
Grass–Herb$_{EW}$	17	70.1 ± 0.9 [a]	17	74.3 ± 0.9 [a]	14	74.6 ± 1.0 [a]	103 ± 19 [a]

L, average days of age; Herb–Herb$_{CW}$, lambs born on herb–clover mix and conventional weaning (at ~99 days of age); Herb–Herb$_{EW}$, lambs born on herb–clover mix and early weaned onto herb–clover mix at ~45 days of age; Grass–Herb$_{D15EW}$, lambs born on ryegrass–clover-based pasture and transferred with dam onto herb–clover mix at 15 days of age and early weaned at ~45 days of age; Grass–Herb$_{CW}$, lambs born on ryegrass–clover-based pasture and transferred with dam onto herb–clover mix at ~45 days of age and conventional weaning; Grass–Herb$_{EW}$, lambs born on ryegrass–clover-based pasture and early weaned at ~45 days of age onto herb–clover mix; Grass–Grass$_{CW}$, lambs born on ryegrass–clover-based pasture and conventional weaning. [a–b] Means with different superscripts are significantly different across years and treatments.

In 2018, at L44, L61 and L87, the live weights of ewes in all five treatments did not differ ($p > 0.05$) (Table 4). Similarly, the liveweight gain of ewes between L44 and L87 in each treatment did not differ ($p > 0.05$).

3.4. Ewe Body Condition Score

In 2017, at L51 (date of early weaning), the BCS of ewes did not differ ($p > 0.05$) between treatments (Table 5). At L65, the BCS of Herb–Herb$_{CW}$, Herb–Herb$_{EW}$, Grass–Herb$_{D15EW}$ and Grass–Herb$_{EW}$ ewes did not differ ($p > 0.05$). The BCS of Grass–Herb$_{CW}$ and Grass–Grass$_{CW}$ ewes did not differ ($p > 0.05$) but were lower ($p < 0.05$) than Herb–Herb$_{EW}$, Grass–Herb$_{D15EW}$ and Grass–Herb$_{EW}$ ewes. At L99 (at conventional weaning), the BCS of Herb–Herb$_{CW}$, Herb–Herb$_{EW}$, Grass–Herb$_{D15EW}$ and Grass–Herb$_{EW}$ and Grass–Herb$_{CW}$ were greater ($p < 0.05$) than Grass–Grass$_{CW}$ ewes.

Table 5. Impact of weaning treatment; Herb–Herb$_{CW}$, Herb–Herb$_{EW}$, Grass–Herb$_{D15EW}$, Grass–Herb$_{CW}$, Grass–Herb$_{EW}$, Grass–Grass$_{CW}$ on the BCS of ewes at L51, L65 and L99 in 2017 and at L44, L61 and L87 in 2018 (results displayed as back transformed logit mean and 95% confidence interval).

Herbage Treatment	Ewe Body Condition Score					
	n		*n*		*n*	
2017						
		L51		L65		L99
Herb–Herb$_{CW}$	21	2.9 (2.5–2.9) [a]	21	3.1 (3.1–3.6) [a,b]	20	3.5 (3.1–3.4) [b]
Herb–Herb$_{EW}$	22	2.9 (2.8–3.1) [a]	22	3.3 (3.1–3.6) [b]	21	3.3 (3.1–3.6) [b]
Grass–Herb$_{D15EW}$	18	2.9 (2.7–3.1) [a]	18	3.4 (3.2–3.7) [b]	18	3.5 (3.3–3.7) [b]
Grass–Herb$_{CW}$	21	2.7 (2.5–2.9) [a]	20	2.8 (2.6–3.1) [a]	21	3.3 (3.1–3.4) [b]
Grass–Herb$_{EW}$	23	2.9 (2.8–3.1) [a]	23	3.4 (3.1–3.6) [b]	22	3.4 (3.2–3.5) [b]
Grass–Grass$_{CW}$	17	2.8 (2.6–3.0) [a]	16	2.9 (2.2–3.0) [a]	17	2.9 (2.7–3.1) [a]
2018						
		L44		L61		L87
Herb–Herb$_{CW}$	18	2.5 (2.5–2.9) [b]	18	2.8 (2.6–3.0) [a,b]	18	3.1 (2.8–3.4) [a,b]
Herb–Herb$_{EW}$	18	2.9 (2.7–3.3) [b]	17	3.0 (2.8–3.3) [b]	17	3.1 (2.8–3.6) [a,b]
Grass–Herb$_{D15EW}$	21	2.9 (2.6–3.2) [b]	21	3.0 (2.9–3.2) [b]	21	3.4 (3.2–3.7) [b]
Grass–Herb$_{CW}$	18	2.4 (2.2–2.6) [a]	16	2.6 (2.4–2.7) [a]	16	2.8 (2.6–3.1) [a]
Grass–Herb$_{EW}$	17	2.3 (2.1–2.5) [a]	17	2.9 (2.6–3.1) [a,b]	14	3.3 (3.3–3.8) [b]

L, average days of age; Herb–Herb$_{CW}$, lambs born on herb–clover mix and conventional weaning (at ~99 days of age); Herb–Herb$_{EW}$, lambs born on herb–clover mix and early weaned onto herb–clover mix at ~45 days of age; Grass–Herb$_{D15EW}$, lambs born on ryegrass–clover-based pasture and transferred with dam onto herb–clover mix at 15 days of age and early weaned at ~45 days of age; Grass–Herb$_{CW}$, lambs born on ryegrass–clover-based pasture and transferred with dam onto herb–clover mix at ~45 days of age and conventional weaning; Grass–Herb$_{EW}$, lambs born on ryegrass–clover-based pasture and early weaned at ~45 days of age onto herb–clover mix; Grass–Grass$_{CW}$, lambs born on ryegrass–clover-based pasture and conventional weaning. [a-b] Means with different superscripts are significantly different across years and treatments.

In 2018, at L44 (date of early weaning), the BCS of Herb–Herb$_{CW}$, Herb–Herb$_{EW}$ and Grass–Herb$_{D15EW}$ ewes were greater ($p < 0.05$) than the Grass–Herb$_{CW}$ and Grass–Herb$_{EW}$ ewes that did not differ ($p > 0.05$). At L61, the BCS of Grass–Herb$_{CW}$ ewes was lower ($p < 0.05$) than Herb–Herb$_{EW}$ and Grass–Herb$_{D15EW}$ ewes but was similar ($p > 0.05$) to Herb–Herb$_{CW}$ and Grass–Herb$_{EW}$ ewes. At L87 (at conventional weaning), the BCS of Herb–Herb$_{CW}$, Herb–Herb$_{EW}$, Grass–Herb$_{D15EW}$ and Grass–Herb$_{EW}$ ewes did not differ ($p > 0.05$). The BCS of Grass–Herb$_{CW}$ ewes was lower ($p < 0.05$) than Grass–Herb$_{D15EW}$ and Grass–Herb$_{EW}$ ewes but was similar ($p > 0.05$) to that of Herb–Herb$_{CW}$ and Herb–Herb$_{EW}$ ewes.

3.5. Digestive Tract Components and Rumen Papillae Dimensions of Lambs in 2018

The weight of the empty digestive tract components of lambs between treatments at either L44 (at early weaning) or L87 (at conventional weaning) did not differ ($p > 0.05$; Table 6). At L44, papillae length and width of Grass–Herb$_{EW}$ and Herb–Herb$_{EW}$ lambs did not differ ($p > 0.05$; Table 6). At L87, Grass–Herb$_{EW}$ and Herb–Herb$_{EW}$ lambs had ~200 µm longer ($p < 0.05$) papillae than Grass–Herb$_{CW}$ and Herb–Herb$_{CW}$ lambs, however, the width of ruminal papillae did not differ ($p > 0.05$) between any of the treatments.

Table 6. Mean weight of hot carcass, total digestive tract, reticulo-rumen, empty small intestine and empty large intestine and mean length and width of rumen papillae of Herb–Herb$_{EW}$ and Grass–Herb$_{EW}$ lambs at L44; and Grass–Herb$_{CW}$, Herb–Herb$_{CW}$, Herb–Herb$_{EW}$ and Grass–Herb$_{EW}$ lambs at L87 in 2018 (least-squares mean ± SEM).

Treatment		Hot Carcass Weight (kg)	Total Digestive Tract Weight (kg)	Reticulo-Rumen Weight (g)	Empty Small Intestine Weight (g)	Empty Large Intestine Weight (g)	Papillae Length (µm)	Papillae Width (µm)
Herb–Herb$_{EW}$	L44	8.9 ± 0.3	3.5 ± 0.3	458 ± 40	458 ± 40	296 ± 51	809 ± 136	268 ± 16
Grass–Herb$_{EW}$		9.6 ± 0.3	3.3 ± 0.3	410 ± 40	410 ± 40	234 ± 51	732 ± 148	260 ± 17
Grass–Herb$_{CW}$	L87	14.7 ± 0.3	10.1 ± 0.3	951 ± 50	1353 ± 85	483 ± 80	892 ± 135 [a]	221 ± 16
Herb–Herb$_{CW}$		14.6 ± 0.3	10.3 ± 0.3	884 ± 53	1275 ± 90	425 ± 85	1090 ± 152 [a]	211 ± 18
Herb–Herb$_{EW}$		14.3 ± 0.3	10.8 ± 0.3	1013 ± 51	1119 ± 86	611 ± 80	1225 ± 137 [b]	242 ± 16
Grass–Herb$_{EW}$		14.0 ± 0.3	10.9 ± 0.3	970 ± 55	1280 ± 93	463 ± 87	1333 ± 122 [b]	245 ± 14

L, average days of age; Herb–Herb$_{EW}$, lambs born on herb–clover mix and early weaned onto herb–clover mix at ~45 days of age; Grass–Herb$_{EW}$, lambs born on ryegrass–clover-based pasture and early weaned at ~45 days of age onto herb–clover mix; Grass–Herb$_{CW}$, lambs born on ryegrass–clover-based pasture and transferred with dam onto herb–clover mix at ~45 days of age and conventional weaning; Herb–Herb$_{CW}$, lambs born on herb–clover mix and conventional weaning (at ~99 days of age). [a,b] Means with different superscripts are significantly different across treatments.

4. Discussion

The duration of the exposure of lambs to a herb–clover mix prior to early weaning, either from birth to early weaning, or from 15 days of age (L15) to early weaning, had no effect on lamb growth rates post-early weaning or their live weights at conventional weaning age, in either year of the current study. This suggests that this management option is not needed to improve the growth of lambs after early weaning onto herb–clover mix. This finding is supported by the limited differences in rumen measurements observed in 2018. In early life, lambs rely primarily on milk to fulfil their nutritional requirements and herbage consumption is low [33]. In the present study, lambs had ad-libitum access to their dams' milk prior to early weaning, potentially limiting their need to consume herbage to fulfil their nutritional requirements thereby mitigating against any potential advantage from prolonged early exposure to the herb–clover mix. Combined, these results, and those from previous studies [9–11], indicate that a four-day adaptation period to the herb–clover mix, before early weaning, is all that is required to ensure adequate growth to a traditional weaning age.

Lambs weaned early onto a herb–clover mix grew at a similar rate, and were as heavy at conventional weaning, as unweaned lambs on a ryegrass–clover-based pasture in 2017. This was the only year that this later treatment was utilised. The growth rates achieved in the current study for early-weaned lambs on a herb–clover mix, were similar to those previously reported for commercially-reared-twin lambs on ryegrass white clover in New Zealand [34]. This suggests that lambs weaned at ~45 days of age and at a minimum live weight of 14 kg onto a herb–clover mix, have the potential to achieve liveweight gains similar to unweaned lambs reared on ryegrass–clover-based pasture. This adequate level of growth of lambs weaned early onto a herb–clover mix was likely driven by two potential factors. Firstly, increased herbage intake, due to herbs and clovers having lower fiber content and faster rumen flow through rates, than ryegrass [35]. Secondly, through greater preferential selection of different plant species [33] as lambs prefer herbs and clovers over ryegrass [36], or a combination of these variables. These results, in combination with previous studies [9–11] indicate early-weaning, onto a herb–clover mix is a management tool that can be used by farmers without risk of poor live weights at ~100 days of age.

In both years of the current study, lambs weaned early onto a herb–clover mix had lower liveweight gains to a conventional weaning age, than unweaned lambs on a herb–clover mix. This was likely due to unweaned lambs having greater total nutrient intake from both milk and herbage, compared to early-weaned lambs that had access only to herbage post-early weaning. Interestingly, the papillae length of early-weaned lambs was 200 µm greater than unweaned lambs on the herb–clover mix at conventional weaning age in 2018. The longer papillae of early-weaned lambs were likely driven by the need for greater dry matter intake to fulfil their nutritional requirements. Dry matter supplementation early in life is known to improve dry matter intake, rumination, the establishment of rumen microflora and alter feeding behaviors of neonatal ruminates [37]. Previously it has been shown that the development of the rumen, either reticulo-rumen capacity or the papillae length, of lambs did not differ at 90 days of age, even under conditions when differences in lamb growth and dry matter intake occurred [38]. That finding, combined with the present finding in regards to papillae length, may suggest that prior to day 90, any potential positive effects of rumen development on lamb growth may not occur and that studies should consider examining lambs for longer term impacts, post a traditional weaning age.

Lambs left unweaned on the herb–clover mix were 4 kg heavier at conventional weaning than those on ryegrass–clover-based pastures in 2017, supporting previous findings [9–11]. In both years of the study, their variations in herbage availability and quality were observed, although not at levels which would have restricted lamb intake [24,25]. Similarly, herbage quality and, in particular, ME and NDF were not inadequate to meet the requirements for lamb growth [39]. These parameters, therefore, do not explain why unweaned lamb growth was greater in the herb–clover mix treatment than unweaned lambs on ryegrass–clover-based pastures. As eluded to earlier, the likely explanation for the greater growth rate of unweaned lambs on herb–clover mix is the greater milk production of their dams, compared to those on ryegrass white clover [40], although the milk and herbage intake of lambs was not measured in this study. The results of this study, and previous studies [9–11] indicate grazing unweaned lambs on herb–clover mix is a management tool that farmers can utilize to increase lamb live weight at the conventional weaning age.

This study also allowed for the examination of the impact of moving ewes and lambs from a ryegrass pasture onto herb–clover mix at approximately 45 days of age without early weaning. The results in 2017 indicated that lambs were heavier and grew faster to conventional weaning than unweaned lambs that remained on ryegrass–clover-based pastures. Farmers, therefore, could consider moving ewes and lambs onto the herb–clover mix as an alternative to lambing ewes on a herb–clover mix. Further, unweaned lambs moved to a herb–clover mix at approximately 45 days of age grew at similar rates in 2017 and faster in 2018 than lambs reared with their dams on herb–clover mix from birth to conventional weaning. This suggests that farmers could use ryegrass–clover-based pasture for the lambing period and then move ewes and lambs to a herb–clover mix at 45 days of age to improve lamb weaning weights.

Ewes whose lambs were weaned early and offered ryegrass–clover-based pasture until conventional weaning had greater liveweight gains and BCS than unweaned ewes on ryegrass–clover-based pasture. This was likely due to the cessation of lactation, allowing ewes to partition nutrients to gaining live weight and was consistent with previous studies [10,11]. In both years, unweaned ewes on the herb–clover mix also had similar liveweight gains from early weaning to conventional weaning compared to ewes whose lambs were weaned early and offered ryegrass–clover-based pasture, matching previous study results [11]. Combined, these results suggest that ewes can either be weaned early onto ryegrass–clover-based pasture or left unweaned on herb–clover mix and achieve greater live weights than ewes on ryegrass–clover-based pasture with their lambs. Greater live weights at weaning can lead to greater live weights at mating which has been reported to have a positive impact on ewe reproductive performance [41]. Early weaning can be used as a technique to improve future ewe live weights and reproductive performance.

5. Conclusions

Early-weaned lambs on a herb–clover mix had similar liveweight gains to conventional weaning as lambs left unweaned on a ryegrass–clover-based pasture. Prolonged exposure of lambs to the herb–clover mix prior to early weaning, however, had no impact on their rumen development or subsequent growth, therefore, this management option cannot be used to improve performance post-early weaning. Ewes whose lambs were weaned early gained greater live weights and BCS compared to ewes which remained with their lambs on a ryegrass–clover-based pasture. Early weaning of lambs with a herb–clover mix is a management tool that farmers can use to achieve adequate growth in their lambs to a conventional weaning age whilst allowing their ewes to gain additional live weight. Greater lamb performance to conventional weaning, however, was achieved when lambs were left unweaned on the herb–clover mix.

Author Contributions: Conceptualization, L.J.E.; investigation and supervision, R.A.C.-T., L.M.C., P.R.K., S.T.M. and S.J.P.; writing—review and editing, L.J.E., R.A.C.-T., L.M.C., P.R.K. and S.T.M. All authors have read and agreed to the published version of the manuscript.

Funding: This research received funding from Beef + Lamb New Zealand, and Massey University.

Acknowledgments: The authors wish to thank D. Burnham and Keeble farm staff for their technical assistance.

Conflicts of Interest: The authors have no conflict of interest with any financial organization regarding the material discussed in the manuscript.

References

1. Kenyon, P.R.; Webby, R.W. Pastures and supplements in sheep production systems. In *Pasture and Supplements for Grazing Animals*; Rattray, P., Brookes, I.M., Nicol, A., Eds.; New Zealand Society of Animal Production: Hamilton, New Zealand, 2007; pp. 255–274.

2. Muir, P.D.; Smith, N.B.; Wallace, G.J.; Fugle, C.J.; Bown, M.D. Maximizing lamb growth rates. *Proc. N. Z. Soc. Anim. Prod.* **2000**, *62*, 55–58.

3. Mulvaney, F.J.; Morris, S.T.; Kenyon, P.R.; West, D.M.; Morel, P.C.H. The effect of weaning at 10 or 14 weeks of age on liveweight changes in the hogget and her lambs. *Proc. N. Z. Soc. Anim. Prod.* **2009**, *69*, 68–70.

4. Mulvaney, F.J.; Morris, S.T.; Kenyon, P.R.; Morel, P.C.H.; West, D.M. Is there any advantage of early weaning of twin lambs born to yearlings? *Proc. N. Z. Soc. Anim. Prod.* **2011**, *71*, 79–82.

5. Somasiri, S.C. Effect of Herb-Clover Mixes on Weaned Lamb Growth. Ph.D. Thesis, Massey University, Palmerston North, New Zealand, 2014.

6. Cranston, L.M.; Kenyon, P.R.; Morris, S.T.; Kemp, P.D. A review of the use of chicory, plantain, red clover and white clover in a sward mix for increased sheep and beef production. *J. N. Z. Grassl.* **2015**, *77*, 89–94. [CrossRef]

7. Somasiri, S.C.; Kenyon, P.R.; Kemp, P.D.; Morel, P.C.H.; Morris, S.T. Effect of herb-clover mixes of plantain and chicory on yearling lamb production in the early spring period. *Anim. Prod. Sci.* **2016**, *56*, 1662–1668. [CrossRef]

8. Corner-Thomas, R.; Cranston, L.M.; Kemp, P.; Morris, S.T.; Kenyon, P. Can herb-clover mixes compensate for the lack of milk in the diet of early-weaned lambs? *N. Z. J. Agric. Res.* **2020**, *63*, 233–245. [CrossRef]

9. Ekanayake, W.E.M.L.J.; Corner-Thomas, R.A.; Cranston, L.M.; Kenyon, P.R.; Morris, S.T. The effect of live weight at weaning on liveweight gain of early weaned lambs onto a herb-clover mixed sward. *Proc. N. Z. Soc. Anim. Prod.* **2017**, *77*, 37–42.

10. Ekanayake, W.E.M.L.J.; Corner-Thomas, R.A.; Cranston, L.M.; Kenyon, P.R.; Morris, S.T. A comparison of liveweight gain of lambs weaned early onto a herb-clover mixed sward and weaned conventionally onto a ryegrass-clover pasture and herb-clover mixed sward. *Asian-Australas. J. Anim. Sci.* **2019**, *32*, 201–208. [CrossRef]

11. Ekanayake, L.J.; Corner-Thomas, R.A.; Cranston, L.M.; Kenyon, P.R.; Morris, S.T. Lambs Weaned Early onto a Herb-Clover Mix Have the Potential to Grow at a Similar Rate to Unweaned Lambs on a Grass-Predominant Pasture. *Animals* **2020**, *10*, 613. [CrossRef] [PubMed]

12. Ekanayake, L.J.; Corner-Thomas, R.A.; Cranston, L.M.; Hickson, R.E.; Kenyon, P.R.; Morris, S.T. Characterisation of the nutritional composition of plant components of a herb-clover mix during November to May in New Zealand. *N. Z. J. Anim. Sci. Prod.* **2019**, *79*, 162–167.

13. Jiao, J.; Li, X.; Beauchemin, K.A.; Tan, Z.; Tang, S.; Zhou, C. Rumen development process in goats as affected by supplemental feeding v. grazing: Age-related anatomic development, functional achievement and microbial colonisation. *Br. J. Nutr.* **2015**, *113*, 888–900. [CrossRef] [PubMed]

14. Yanez-Ruiz, D.R.; Abecia, L.; Newbold, C.J. Manipulating rumen microbiome and fermentation through interventions during early life: A review. *Front. Microbiol.* **2015**, *6*, 1–12. [CrossRef] [PubMed]

15. Distel, R.A.; Villalba, J.J.; Laborde, H.E. Effects of early experience on voluntary intake of low-quality roughage by sheep. *J. Anim. Sci.* **1994**, *2*, 1191–1195. [CrossRef] [PubMed]

16. Kittelmann, S.; Pinares-Patiño, C.S.; Seedorf, H.; Kirk, M.R.; Ganesh, S.; McEwan, J.C.; Janssen, P.H. Two different bacterial community types are linked with the low-methane emission trait in sheep. *PLoS ONE* **2014**, *9*, 1–9. [CrossRef] [PubMed]

17. Provenza, F.D.; Balph, D.F. Applicability of five diet selection models to various foraging challenges ruminants encounter. In *Behavioural Mechanisms of Food Selection*; Hughes, R.N., Ed.; Springer: Berlin, Germany, 1990; pp. 423–459.

18. Ramos, A.; Tennessen, T. Effect of previous grazing experience on the grazing behaviour of lambs. *Appl. Anim. Behav. Sci.* **1992**, *33*, 43–52. [CrossRef]

19. Parsons, A.J.; Newman, J.A.; Penning, P.D.; Harvey, A.; Orr, R.J. Diet preference of sheep: Effects of recent diet, physiological state and species abundance. *J. Anim. Ecol.* **1994**, *63*, 465–478. [CrossRef]

20. Provenza, F.D. Post-ingestive feedback as an elementary determinant of food preference and intake in ruminants. *J. Range Manag.* **1995**, *48*, 2–17. [CrossRef]

21. Rutter, S.M.; Orr, R.J.; Rook, A.J. Dietary preference for grass and white clover in sheep and cattle: An overview. In *Grazing Management: The Principles and Practice of Grazing, for Profit and Environmental Gain, Within Temperate Grassland Systems*; Rook, A.J., Penning, P.D., Eds.; British Grassland Society Occasional Symposium No. 34; British Grassland Society: Harrogate, UK, 2000; pp. 73–78.

22. Kyriazakis, I.; Oldham, J.D. Diet selection in sheep: The ability of growing lambs to select a diet that meets their crude protein requirements. *Br. J. Nutr.* **1993**, *69*, 617–629. [CrossRef]

23. Kyriazakis, I.; Oldham, J.D.; Coop, R.L.; Jackson, F. The effect of sub-clinical intestinal nematode infection on the diet selection of growing sheep. *Br. J. Nutr.* **1994**, *72*, 665–677. [CrossRef]

24. Morris, S.T.; Kenyon, P.R. The effect of litter size and sward height on ewe and lamb performance. *N. Z. J. Agric. Res.* **2004**, *47*, 275–286. [CrossRef]

25. Cranston, L.M.; Kenyon, P.R.; Morris, S.T.; Lopez Villalobos, N.; Kemp, P.D. Effect of post grazing height on the productivity, population and morphology of a herb and legume mix. *N. Z. J. Agric. Res.* **2015**, *58*, 397–411. [CrossRef]

26. Nowak, R.; Poindron, P. From birth to colostrum: Early steps leading to lamb survival. *Reprod. Nutr. Dev.* **2006**, *46*, 431–446. [CrossRef] [PubMed]

27. Jefferies, B.C. Body condition scoring and its use in management. *Tasman. J. Agric.* **1961**, *32*, 19–21.

28. McGavin, M.D.; Morrill, J.L. Dissection technique for examination of the bovine rumino-reticulum. *J. Anim. Sci.* **1976**, *42*, 535–538. [CrossRef] [PubMed]

29. Lesmeister, K.E.; Tozer, P.R.; Heinrichs, A.J. Development and analysis of a rumen tissue sampling procedure. *J. Dairy Sci.* **2004**, *87*, 1336–1344. [CrossRef]

30. Frame, J. Herbage Mass. In *Sward Measurement Handbook*; Hodgson, J., Baker, R., Davies, D.A., Laidlaw, A.S., Leaver, J., Eds.; British Grassland Society: Kenilworth, UK, 1993; pp. 39–69.

31. Roughan, P.; Holland, R. Predicting in-vivo digestibility of herbages by exhaustive enzymic hydrolysis of cell walls. *J. Sci. Food Agric.* **1977**, *28*, 1057–1064. [CrossRef]

32. Robertson, J.; Van Soest, P. The detergent system of analysis and its application to human foods. In *The Analysis of Dietary Fiber in Food*; James, W.P.T., Theander, O., Eds.; Marcel Dekker Inc.: New York, NY, USA, 1981; pp. 123–158.

33. Kerr, P. *400 Plus—A Guide to Improved Lamb Growth*; New Zealand Sheep Council: Wellington, New Zealand, 2000; p. 160.

34. Litherland, A.J.; Lambert, M.G. Herbage quality and growth rate of single and twin lambs at foot. *Proc. N. Z. Soc. Anim. Prod.* **2000**, *60*, 55–57.

35. Golding, K.P.; Kemp, P.D.; Kenyon, P.R.; Morris, S.T. High weaned lamb live weight gains on herbs. *Agron. N. Z.* **2008**, *38*, 33–39.

36. Pain, S.J.; Hutton, P.G.; Kenyon, P.R.; Morris, S.T.; Kemp, P.D. Preference of lambs for novel pasture herbs. *Proc. N. Z. Soc. Anim. Prod.* **2010**, *70*, 258–287.

37. Xiao, J.; Alugongo, G.M.; Li, J.; Wang, Y.; Li, S.; Cao, Z. Review: How Forage Feeding Early in Life Influences the Growth Rate, Ruminal Environment, and the Establishment of Feeding Behavior in Pre-Weaned Calves. *Animals* **2020**, *10*, 188. [CrossRef]

38. Xie, B.; Huang, W.; Zhang, C.; Diao, Q.; Cui, J.; Wang, S.; Lv, X.; Zhang, N. Influences of starter NDF level on growth performance and rumen development in lambs fed isocaloric and isonitrogenous diets. *J. Anim. Sci.* **2020**, *98*, 1–8. [CrossRef] [PubMed]

39. Hodgson, J.; Brookes, I.M. Nutrition of grazing animals. In *New Zealand Pasture and Crop Science*; Hodgson, J., White, J., Eds.; Oxford University Press: Auckland, New Zealand, 2002; pp. 133–153.

40. Hutton, P.G.; Kenyon, P.R.; Bedi, M.K.; Kemp, P.D.; Stafford, K.J.; West, D.M.; Morris, S.T. A herb and legume sward mix increased milk production and ewe and lamb live weight gain to weaning compared to a ryegrass dominant sward. *Anim. Feed Sci. Technol.* **2011**, *164*, 1–7. [CrossRef]

41. Corner-Thomas, R.A.; Ridler, A.L.; Morris, S.T.; Kenyon, P.R. Ewe lamb live weight and body condition scores affect reproductive rates in commercial flocks. *N. Z. J. Agric. Res.* **2015**, *58*, 26–34. [CrossRef]

Article

Effect of Mineral Supplementation on the Macromineral Concentration in Blood in Pre- and Postpartum Blackbelly Sheep

Juan Carlos Moyano Tapia [1,2,*], **Simon Alexander Leib** [1], **Pablo Roberto Marini** [2,3,4] **and Maria Laura Fischman** [2,5,6]

1   Facultad de Ciencias de la Vida/Centro de Investigación, Posgrado y Conservación Amazónica, Universidad Estatal Amazónica, Puyo 160150, Ecuador; simonleib@gmx.de
2   Centro Latinoamericano de Estudios de Problemáticas Lecheras (CLEPL), Casilda 2170, Argentina; pmarini@unr.edu.ar (P.R.M.); fischman@fvet.uba.ar (M.L.F.)
3   Facultad de Ciencias Veterinarias, Universidad Nacional de Rosario, Casilda 2170, Argentina
4   Consejo de Investigaciones (CIC-UNR), Rosario 2000, Argentina
5   Facultad de Ciencias Veterinarias, Universidad de Buenos Aires, CABA 1427, Argentina
6   Instituto de Investigación y Tecnología en Reproducción Animal (INITRA), Universidad de Buenos Aires, CABA 1427, Argentina
*   Correspondence: jmoyano@uea.edu.ec

Received: 27 March 2020; Accepted: 12 May 2020; Published: 16 July 2020

Simple Summary: The objective of this study was to determine the effect of mineral supplementation on the serum concentration of calcium, phosphorus, and magnesium in pre- and postpartum Blackbelly sheep throughout three successive lambing periods under free grazing conditions in the Ecuadorian Amazon Region. The field work was carried out between January 2015 and February 2018 using 20 Blackbelly females. The flock was randomly divided into two groups: Group 1 (G1), who were fed forage plus a supplementation (Pecutrin® Mineral supplement plus vitamins A, D3, and E. Bayer HealthCare); and Group 2 (G2), who were fed forage only without mineral supplementation. In this study, we showed that Blackbelly sheep raised under free grazing conditions (G2) had very low serum calcium values, and supplementation was unable to improve them. Meanwhile, phosphorus and magnesium values were below the required levels, but after supplementation (G1), they exceeded the minimum threshold. Mineral supplementation in the rearing of sheep in grazing systems is necessary during the entire production cycle, but it must be done taking into account the soil–plant–animal relationship, specifically for the Amazonian Region systems.

Abstract: The objective of this study was to determine the effect of mineral supplementation on the serum concentration of calcium, phosphorus, and magnesium in pre- and postpartum Blackbelly sheep throughout three successive lambing periods under free grazing conditions in the Ecuadorian Amazon Region. The field work was carried out between January 2015 and February 2018 using 20 Blackbelly sheep belonging to the Centre for Research, Postgraduate Studies and Conservation of Amazon Biodiversity, Ecuador. The flock was randomly divided into two groups: Group 1 (G1) was fed with forage plus a supplementation (Pecutrin® Mineral supplement plus vitamins A, D3, and E. Bayer HealthCare) and Group 2 (G2) was fed only with forage without mineral supplementation. Three blood samples from the coccygeal vein were taken from each sheep 30 days before lambing, 30 days after, and 60 days after lambing. Concerning the average of calcium, significant differences were found at different times inside each group and also between them ($p < 0.0001$ in both cases). As for the phosphorus, significant differences were found between the means of the groups for all times from 30 days after the second lambing season ($p < 0.05$). It was observed that the groups differed significantly in terms on the average of magnesium (considering a significance level of 0.05) 30 days before the first lambing and at all times measured from the 30 days after the second lambing ($p < 0.005$). In this study, we showed that Blackbelly sheep raised under free grazing conditions in the Ecuadorian Amazon Region had very low serum calcium values, and supplementation was unable to improve them. Meanwhile, phosphorus and magnesium levels were below the required values, but after supplementation, they exceeded the minimum threshold. Mineral supplementation in the rearing of sheep in grazing systems is necessary during the entire production cycle, but it must be done taking into account the soil–plant–animal relationship specifically for the Amazonian Region systems.

Keywords: minerals; sheep; pre- and postpartum; grazing system

1. Introduction

The Blackbelly breed is of utmost relevance for meat production in the Ecuadorian Amazon Region, as it can adapt to extreme conditions. This fact is reflected in its low mortality, precocity and high reproductive fertility (multiple births), medium size, and productive longevity [1].

Many authors have shown that one of the problems regarding grazing ruminants is that pasture does not cover their protein, energy, or mineral needs, affecting the normal development of their metabolic processes [2], because their requirements depend exclusively on the forage's composition [3]. The soils of the Amazonian Region, with a pH lower than 5.5 during most of the year, generate restrictions on plants. The soil acidity modifies physical and chemical properties. These negative effects are reflected by an excess of certain minerals—such as aluminum—and the deficiency of others, for instance, molybdenum. It should also be considered that forage resources have fluctuations throughout the year, both in quantity and in quality [4].

The intake of minerals in ruminants reared in grazing systems depends on the composition and the total consumption of forage, the consumption and the minerals content of water, and the composition of the soil [5]. Adequate amounts of essential minerals are critical to maximize the productivity and the health of livestock [6]. Furthermore, appropriate mineral levels improve the interaction between production and reproduction [7,8]. A widely used indicator to assess nutritional status, health, and well-being of sheep is the Body Condition Index, as indicated in the Animal Welfare Indicators (AWIN) protocols [9].

A suitable quantity and proportion of essential minerals prevents diseases by activating the immune system, improving growth, production, viability, and fertility [10]. Previously, some authors described the need for supplementation [11,12] and the importance of using concentrated mixtures for sheep according to age groups [13–15]. However, for many years, the supplementation under grazing conditions was not considered a cost-effective alternative. Preliminary studies showed that supplementation is profitable if kilograms of weaned lamb per female per year were analyzed. This variable involves the factors that most influence the profitability of the herd, since it considers prolificity of the female, milk production, maternal instinct, mortality, and weight gain of lambs [16].

The objective of this study was to determine the effect of mineral supplementation on the serum concentration of calcium, phosphorus, and magnesium in pre- and postpartum Blackbelly sheep throughout three successive lambing periods under free grazing conditions in the Ecuadorian Amazon Region.

2. Materials and Methods

The field work was carried out in the Centre for Research, Postgraduate Studies and Conservation of Amazon Biodiversity (CIPCA). It is located in the Arosemena Tola canton, province of Napo, Ecuador (coordinates: 01°14.325′ S; 077°53.134′ W). The altitude varies between 580 and 990 m above sea level. The environment is tropical with an average annual rainfall of 4000 mm, an average relative humidity of 80%, and temperatures ranging between 15 and 25 °C.

The study was performed between January 2015 and February 2018. This interval included three lambing periods by the sheep under study. The same 20 Blackbelly females aged 24 to 32 months and with an average weight and standard deviation of 34 ± 4 kg were used.

The sheep were free to graze from 07:00 to 18:00, then housed overnight with water consumption ad libitum. They were randomly divided into two groups. Group 1 (G1) was fed with forage plus a supplement at a dose of 0.005 kg per animal per day every day from when the firstborn lamb was weaned until the end of the trial (Pecutrin® Mineral supplement plus vitamins A, D3, and E. Bayer HealthCare; parts of the formula components: calcium min. 17% max. 20%; phosphorus min. 18%; magnesium min. 3.0%; relationship calcium–phosphorus 1.3:1) The mineral supplement dose was the minimum recommended by the manufacturer; each sheep had its own feeder space. Group 2 (G2) was fed only with forage without mineral supplementation. Table 1 shows the chemical composition of the predominant species in the pasturage used. The presence of aluminum and hydrogen ions determined by intense washing phenomena as a consequence of high rainfall generates acid soils with a pH of 5.2 [17]. This reduction in the pH affects chemical and biological characteristics of the soil, reducing the growth of plants. It also decreases the availability of nutrients such as calcium, magnesium, phosphorus, and potassium. In turn, acidic soils tend toward the accumulation of toxic elements for plants, such as aluminum and manganese [18].

Table 1. Chemical composition of the pasture.

Pasturage	DM (kg/ha/yr)	Protein (%)	Calcium (%)	Phosphorus (%)	Magnesium (%)	IVD (%)
Brachiaria decumbens	17.585	10.6	0.20	0.18	0.15	44.4
Arachis Pintoi	6.212	19.4	1.7	0.21	0.63	59.2

DM: dry matter; IVD: in vitro digestibility. References: [19,20].

The sanitary management routinely used for the sheep flock of CIPCA was applied. This includes deworming, baths to combat ticks and flies, vaccines for foot and mouth disease, alongside antifungal and antibacterial medicines.

Through the fodder mixture, the animals consumed 0.612 kg/day of dry matter (1.8% of their average weight). This percentage was estimated based on pastures advanced in their physiological maturity [21]. The contribution of Ca^{2+}, P^{3-}, and Mg^{2+} was 2.5 g/day, 0.5 g/day, and 1.8 g/day, respectively. Except for Mg^{2+}, the mineral levels were below the daily minimum allowance recommended, given that the animals should consume 3.1 g/day, 2.9 g/day, and 1.0 g/day of Ca^{2+}, P^{3-}, and Mg^{2+}, respectively [21]. Pecutrin® (Bayer, Leverkusen, Alemania)., at the dose offered, contributed with a total of 0.6 g/day of $Ca^{2+,}$ 0.5 g/day of P^{3-} and 0.9 g/day of Mg^{2+}.

Blood samples were only taken from the female adult sheep. Three blood samples from the coccygeal vein were taken 30 days before lambing as well as 30 days and 60 days post lambing. Ten milliliters of blood was obtained from each animal in a vacuum tube without anticoagulant (BD Vacutainer® red cap, Franklin Lakes, NJ, USA). The blood was centrifuged (3000 rpm × 15–30 min), and the separated serum was stored at −20 °C until analyzed.

Serum minerals were determined, including phosphorus (P^{3-}) and magnesium (Mg^{2+}). These two tests were performed by means of molecular spectrophotometry with the GENESYS 10 UV Series Thermo Scientific Spectrophotometer (Waltham, MA, USA). The reagents used were HUMAN (Germany), and the equipment was used according to its technical specifications for each of the samples. For calcium (Ca^{2+}), we employed the AUDICOM, AC 9801 Electrolytic Analyzer (Maharashtra, India) with specific AUDICOM reagents according to the equipment's technical specifications.

A descriptive statistical analysis was performed by calculating means. A linear regression model for correlated data was adjusted for inferential statistical analysis. This model took into account the correlation between the observations measured at the same unit over time by incorporating an intra-unit correlation structure. First, the correlation structure that best fit the data was chosen using the

Akaike Penalized Likelihood Criterion. Then, we proceeded to interpret the results of the hypothesis tests for the model's terms. If interaction between group and time was significant, contrasts were performed to compare if there were differences on average between the groups for each time. Statistical program R 3.6.3 was used.

3. Results

3.1. Calcium

There was not significant interaction between group and time for the calcium variable ($p = 0.1469$). There were significant differences between the average of calcium between times, and there were also significant differences between groups regarding the average of calcium ($p < 0.0001$ in both cases) (Table 2). Even though the model detected differences in the average calcium between groups, and the means were 2.44 mg/dL and 2.25 mg/dL, they were statistically different but not biologically.

Table 2. Serum calcium levels throughout the three successive lambing seasons.

Ca^{2+} (mg/dL)	Lambing Season								
	First Lambing			Second Lambing			Third Lambing		
Days before or after	−30	+30	+60	−30	+30	+60	−30	+30	+60
Supplemented group	2.17	2.44	2.24	2.48	2.37	2.40	2.66	2.61	2.56
Not supplemented group	2.20	2.20	2.20	2.22	2.28	2.29	2.28	2.36	2.41

Results are expressed as mean values of calcium per days before or after lambing. No significant differences were observed between groups.

3.2. Phosphorus

There was an interaction between group and time for the variable phosphorus ($p = 0.0013$). When performing contrasts to compare the means of the groups at each moment, significant differences were found between groups for all times from 30 days after the second lambing season ($p < 0.05$) (Table 3).

Table 3. Serum phosphorus levels throughout the three successive lambing seasons.

P^{3-} (mg/dL)	Lambing Season								
	First Lambing			Second Lambing			Third Lambing		
Days before or after	−30	+30	+60	−30	+30	+60	−30	+30	+60
Supplemented group	4.50	3.31	4.29	4.67	5.71	5.70	6.80	6.39	6.96
Not supplemented group	3.90	4.24	4.30	4.77	4.16	4.31	5.61	4.37	4.54
	ns	ns	ns	ns	*	ns	ns	*	*

Results are expressed as mean values of phosphorus per days before or after lambing. (*): significant differences ($p < 0.001$); (ns): no significant differences were observed between groups.

3.3. Magnesium

When adjusting the model, it was observed that there was interaction between group and time ($p < 0.0001$). The mean trajectory in time was not the same for both groups with respect to the variable Mg^{2+} (Table 4). When performing contrasts, it was observed that the groups differed significantly in terms of the average of magnesium (considering a significance level of 0.05) 30 days before the first lambing and at all times measured from the 30 days after the second lambing ($p < 0.005$).

Table 4. Serum magnesium levels throughout the three successive lambing seasons.

Mg^{2+} (mg/dL)	Lambing Season								
	First Lambing			Second Lambing			Third Lambing		
Days before or after	−30	+30	+60	−30	+30	+60	−30	+30	+60
Supplemented group	2.18	2.42	2.25	2.70	3.45	4.50	3.49	3.95	3.86
Not supplemented group	2.72	2.90	2.54	2.17	2.39	2.29	2.36	2.50	2.73
	ns	ns	ns	ns	ns	*	*	*	*

Results are expressed as mean values magnesium per days before or after lambing. (*): significant differences ($p < 0.001$); (ns): no significant differences were observed between groups.

4. Discussion

Few studies exist concerning feeding sheep raised in an Amazonian environment. Most regions have adapted their production systems according to knowledge obtained from production systems in temperate climates [22]. In the conditions of the Amazon Region, sheep grow consuming poor quality forage (that has excessive fiber, low protein, and low energy). The low quality of plants in acidic soils is due to the combination of toxicities of aluminum and manganese as well as deficiencies of sodium, phosphorus, potassium, calcium, magnesium, and some micronutrients, such as iron and zinc. Plants are subject to varying degrees of abiotic stress (soil acidity, mineral deficiencies and/or toxicities, droughts/floods, light/shadow quality, and extreme temperatures) and biotic stress (pests, diseases, and weeds). These tensions have an effect on growth and development, causing less absorption, a reduced utilization of absorbed nutrients and, consequently, a reduction in the efficient use of nutrients [23,24].

These characteristics could affect the normal development of metabolic and mineral processes in sheep. Blood is the most important bio-substrate for the estimation of an animal's mineral state [25,26].

Minerals are essential nutrients for small ruminants and their concentrations in blood must oscillate within bounded intervals in order to maintain their health and well-being. Adequate concentrations of macronutrients allow structural, physiological, catalytic, and regulatory functions of the organism to be rightly developed [27,28].

Normal calcemic values in sheep are 11.5–12.8 mg/dL [29]. In this study, calcium levels in the ewes ranged between 2.6 and 2.8 mg/dL, well below the normal range. Quintero-Moreno [30] found values in sheep without supplementation of 8.73 mg/dL, while in supplemented sheep, they obtained values of 15.2 mg/dL. Underwood and Suttle [31] found that normal blood serum calcium values ranged from 7 to 8 mg/dL in young sheep, while Norton [32] described that calcium rarely is a limitation in fodder diets. The lower Ca^{2+} content in Amazonian pastures could be due to the natural dilution of the process by which the production of dry matter exceeds mineral uptake [33]. In this case, supplementation did not help to raise the calcium level; the calcium level in G1 increased (2.44 mg/dL) in comparison to G2 (2.25 mg/dL), thus they were statistically different but not biologically. This result could be explained in part because a minimum dose of 5 g/animal/day was used. However, the interactions with toxic minerals, which could interfere and decrease absorption, should also be considered. No placental retention problems were observed, nor ewes that fell after lambing, spontaneous fractures, or other related pathologies, although the values suggest acute subclinical hypocalcaemia. Herdt [34] suggested that animals would adapt to this condition and that such adaptation would be mainly based on glucose, non-esterified fatty acids (NEFA), and ketone body availability and supply. The improvement in Ca^{2+} levels could be reflected in the general state and surely in higher and better milk production, which should be seen in the weight of lamb weaned. Abarghani [35] found that Ca^{2+} and P^{3-} were deficient during different seasons in grazing animals in all regions and for all sheep categories. These authors concluded that, for sheep in continuous grazing, it was not possible to prevent deficiencies in the concentration of Ca^{2+} and P^{3-} in serum or plasma. They concluded that supplementing these small ruminants with a bioavailable mineral mixture could increase the blood level of these minerals. However, the authors stressed the importance of conducting more studies to determine requirements and economic benefits of mineral supplements.

Normal values of phosphatemia in sheep are 5–7.3 mg/dL [29]. In this study, the ewes' phosphatemia values ranged between 4.03 and 6.71 mg/dL. In the first lambing, the values were lower than normal, but after incorporating the supplementation, the G1 values were within the normal values, as was observed for the G2 in the third lambing. Stojkovic [36] obtained similar results; he found that the phosphorus level in the blood serum of the sheep studied was between 3.84–4.61 mg/dL, which was below the normal limits necessary to meet the sheep's needs for this macroelement. The low phosphorus content could be due to the low presence of P^{3-} in the acidic Amazonian soil and also to the low amount of this mineral in *Arachis pintoi* [23,24,37,38]. In ruminants, the amount of mineral excreted varies according to the type of diet. The amount and the source of phosphorus used in the diet also affect the kinetics of the mineral in the body [39,40]. The age of the animal can influence the use of dietary phosphorus, since it has been observed that young animals have higher absorption efficiency [41]. In this way, there are several aspects that can influence the phosphorus homeostasis in the organism and, therefore, the mineral's kinetics. In recent years, the study of this mineral in the context of animal production has mainly focused on redefining the requirements of the mineral, seeking to minimize excretion without affecting animal performance [42]. In addition, the absorptions of calcium and phosphorus are independent of each other, which allow the sheep to adjust to different physiological demands [43]. The results showed an inverse Ca^{2+}: P^{3-} ratio. For optimal performance in ruminants, it must be between 1.5:1 and 2.5:1 [26]. In our study, it was 1:1.7 at the beginning and 1:2.57 at the end, showing a clear mineral imbalance. However, in ruminants, the ratio of Ca^{2+} to P^{3-} only affects the absorption of these minerals if the diet is inadequate in said minerals [44]. The metabolism of Ca^{2+} is closely related to that of P^{3-}. The excess or the deficit of one mineral can affect the use of the other [44].

In this study, the serum levels of Mg^{2+} found in the sheep studied ranged between 2.5 and 2.7 mg/dL in the first lambing, even when the sheep of both groups were not supplemented. Supplementation allowed Mg^{2+} in blood to reach normal values (2.5–3.5 mg/dL) and to exceed it in the third lactation. The normal values of magnesemia in sheep are 5–7.3 mg/dL [29]. Our results coincided with those reported by Abarghani [35], with values that ranged between 2.9 and 3.7 mg/dL. The organic deficit of an important element such as Mg^{2+} results in less growth and development in the animal. McDowell [2] stated that magnesium is an enzymatic activator involved in the metabolism of carbohydrates and lipids, since it is a catalyst for a wide variety of enzymes. It is also part of protein synthesis through its action in ribosomal aggregation.

5. Conclusions

In this study, we showed that Blackbelly sheep raised in free grazing conditions in the Ecuadorian Amazon Region had very low calcium values in serum, and supplementation failed to improve them. Phosphorus and magnesium values were below the required values, but after supplementation, they exceeded the minimum threshold. Mineral supplementation in the rearing of sheep in grazing systems is necessary throughout the production cycle, but adjustments are essential upon taking into account the soil–plant–animal relationship for systems in the Amazon region.

Author Contributions: Conceptualization, M.L.F. and P.R.M..; methodology, J.C.M.T. and S.A.L.; software, P.R.M.; validation, J.C.M.T., S.A.L., P.R.M. and M.L.F.; formal analysis, P.R.M. and M.L.F.; investigation, J.C.M.T., P.R.M., S.A.L.; resources, S.A.L. and J.C.M.T.; data curation, P.R.M.; writing—original draft preparation, P.R.M. and M.L.F.; writing—review and editing, P.R.M. and M.L.F.; visualization, S.A.L. and J.C.M.T.; supervision, M.L.F. and P.R.M.; project administration, S.A.L. and J.C.M.T.; funding acquisition, J.C.M.T. All authors have read and agreed to the published version of the manuscript.

Funding: This research received no external funding.

Conflicts of Interest: The authors declare no conflict of interest.

References

1. Mendives, J.A. Importancia de los ovinos tropicales introducidos al país: Características productivas y reproductivas. *Latinoam. Prod. Anim.* **2007**, *15*, 310–315. Available online: http://www.bioline.org.br/pdf?la07068 (accessed on 13 May 2020).
2. McDowell, L.R. *Minerals in Animal and Human Nutrition*, 2nd ed.; Elsevier: Amsterdam, The Netherlands, 2003; p. 644.
3. Morales, A.E.; Domínguez, V.I.; González Ronquillo, M.; Jaramillo, E.G.; Castelán, O.O.; Pescador, S.N.; Huerta, B.M. Diagnóstico mineral en forraje y suero sanguíneo de bovinos lecheros en dos épocas en el valle central de México. *Técnica Pecu. México* **2007**, *45*, 329–344.
4. Organización de las Naciones Unidas Para la Alimentación y la Agricultura (FAO). Suelos Ácidos. 2019. Available online: http://www.fao.org/soils-portal/soil-management/manejo-de-suelos-problematicos/suelos-acidos/es/ (accessed on 13 May 2020).
5. Rodríguez, A.; Banchero, G. Deficiencia de minerals en ruminates. *Revista INIA* **2007**, *13*, 11–15.
6. Wang, H.; Liu, Z.; Huang, M.; Wang, S.; Cui, D.; Dong, S.; Li, S.; Qi, Z.; Lin, Y. Effects of Long-Term Mineral Block Supplementation on Antioxidants, Immunity, and Health of Tibetan Sheep. *Biol. Trace Elem. Res.* **2016**, *172*, 326–335. [CrossRef] [PubMed]
7. Almeida, A.M.; Schwalbach, L.M.J.; Cardoso, L.A.; Greyling, J.P.C. Scrotal, testicular and semen characteristics of young Boer bucks fed winter veld hay: The effect of nutritional supplementation. *Small Rumin. Res.* **2007**, *73*, 216–220. [CrossRef]
8. Griffiths, L.M.; Loeffler, S.H.; Socha, M.T.; Tomlinson, D.J.; Johnson, A.B. Effects of supplementing complexed zinc, manganese, copper and cobalt on lactation and reproductive performance of intensively grazed lactating dairy cattle on the South Island of New Zealand. *Anim. Feed Sci. Technol.* **2007**, *137*, 69–83. [CrossRef]
9. Animal Welfare Indicators (AWIN). AWIN Welfare Assessment Protocol for Sheep. 2015. Available online: https://www.researchgate.net/publication/275887069_AWIN_Welfare_Assessment_Protocol_for_Sheep (accessed on 13 May 2020). [CrossRef]
10. Underwood, E.J. *The Mineral Nutrition of Livestock*, 2nd ed.; Commonwealth Agriculture Bureau: Slough, UK, 1981.
11. Yadav, K.K.; Mandokhot, V.M. Effect of sulfur supplementation on the performance of stall-fed Nali lambs, growth responses, nutrient and minerals. *Indian J. Anim. Sci.* **1988**, *58*, 843–848.
12. Samanta, A.; Samanta, G. Mineral profile of different feed and fodders and their effect on plasma profile in ruminants of West Bengal. *Indian J. Anim. Nutr.* **2002**, *19*, 278–281.
13. Karim, S.A.; Verma, D.L. Growth performance and carcass characteristics of finisher lambs maintained on intensive feeding or grazing with supplementation. *Indian J. Anim. Sci.* **2001**, *71*, 959–961.
14. Karim, S.A.; Santra, A.; Sharma, V.K. Milk yield, its composition and performance of Malpura ewews under protocol of grazing and supplementation. *Indian J. Anim. Sci.* **2001**, *71*, 258–260.
15. Santra, A.; Karim, S.A.; Chaturvedi, O.H. Effect of concentrate supplementation on nutrient intake and performance of lambs of two genotypes grazing a semiarid rangeland. *Small Rumin. Res.* **2002**, *44*, 37–45. [CrossRef]
16. Moyano, J.C.; Marini, P.R.; Fischman, M.L. Biological Efficiency in Hair Sheep Reared in a Sustainable Farming System in the Ecuadorian Amazon Region. *Dairy Vet. Sci. J.* **2019**, *11*. [CrossRef]
17. Calva, C.; Espinosa, J. Effect of four liming materials on acidity control of a soil from Loreto, Orellana. *Siembra* **2017**, *4*, 110–120. [CrossRef]
18. Sumner, M.; Pavan, M. Alleviating Soil Acidity through Organic Matter Management. 2005. Available online: http://brasil.ipni.net/ipniweb/region/brasil.nsf/e0f085ed5f091b1b852579000057902e/5dfb8bcc4691864983257b0c004a898f/$FILE/Palestra%20Malcolm%20Sumner.pdf (accessed on 13 May 2020).
19. Instituto Nacional Autónomo de Investigaciones Agropecuarias (INIAP). *Manual de Pastos Tropicales Para la Amazonia Ecuatoriana*; Manual N33, pp. 66; Napo-Ecuador; 1997; Available online: http://repositorio.iniap.gob.ec/bitstream/41000/1622/1/Manual%20n%C2%BA%2011%20de%20pastos%20tropicales%20reducido%20ultimo.pdf (accessed on 13 May 2020).
20. Leonard, I. Recursos forrajeros autóctonos y promisorios para la ganadería en la provincia de Pastaza. In *Retos y Posibilidades para una Ganadería Sostenible en la Provincia de Pastaza de la Amazonia Ecuatoriana*; Universidad Estatal Amazónica: Puyo, Ecuador, 2015; Chapter IV; pp. 46–69.

21. Romero Yáñez, O.; Bravo Marchan, S. Fundamentos de la producción ovina en la Región de La Araucanía. Ministerio de Agricultura Instituto de Investigaciones Agropecuarias Centro Regional de Investigación Carillanca. *BOLETÍN INIA* **2012**, *245*, 204.

22. Pereira, E.S.; Fontenele, R.M.; Medeiros, A.N.; Lopes, R.O.; Campos, A.C.N.; Heinzen, E.L.; Bezerra, L.R. Requirements of protein for maintenance and growth in ram hair lambs. *Trop. Anim. Health Prod.* **2016**, *48*, 1323–1330. [CrossRef] [PubMed]

23. Fageria, N.K.; Baligar, V.C.; Li, Y.C. The role of nutrient efficient plants in improving crop yields in twenty first century. *J. Plant Nutr.* **2008**, *31*, 1121–1157. [CrossRef]

24. Lyda, S.D. Alleviating pathogen stress. In *Modifying the Root Environment to Reduce Crop Stress*; ASAE Monograph No. 4; Arkin, G.F., Taylor, H.M., Eds.; American Society of Agricultural Engineers: St. Joseph, MI, USA, 1981; pp. 195–214.

25. Khan, Z.I.; Ashraf, M.; Ahmad, K.; Valeem, E.E.; McDowell, L.R. Mineral Status of Forage and Its Relationship with that of Plasma of Farm Animals in Southern Punjab; Pakistan. *Pak. J. Bot.* **2009**, *36*, 851–856.

26. Khan, Z.I.; Hussain, A.M.; Ashraf, M.; McDowell, L.R. Mineral Status of Soils and Forages in Southwestern Punjab; Pakistan: Micro Minerals. *Asian-Aust. J. Anim. Sci.* **2006**, *19*, 1139–1147. [CrossRef]

27. *NRC: Nutrient Requirements of Small Ruminants: Sheep; Goats; Cervids and New World Camelids*, 1st ed.; National Research Council, National Academy Press: Washington, DC, USA, 2007.

28. Suttle, N.F. *Mineral Nutrition of Livestock*, 4th ed.; CABI Publishing: Wallingford, UK, 2010; p. 595.

29. Kaneko, J.J.; Harvey, J.W.; Bruss, M.L. Appendix. In *Clinical Biochemistry of Domestic Animals*, 5th ed.; Academics Press: Cambridge, MA, USA, 1997; pp. 885–905.

30. Quintero-Moreno, A.; Miranda, S.; López, R.; Dean, D.; Rojas, N.; González, A.; Palomares, R.; Boscan, J. Crecimiento, niveles de calcio; fósforo y magnesio y perfiles séricos de progesterona en corderas prepúberes mestizas west african suplementadas con tres fuentes de minerales. *Rev. Científica FCV-LUZ* **2000**, *3*, 205–211.

31. Underwood, E.J.; Suttle, N.F. *Los Minerales en la Nutrición del Ganado*, 3rd ed.; Editorial Acribia, S.A.: Zaragoza, Spain, 2003; p. 637.

32. Norton, B.W. The nutritive value of tree legumes. In *Forage Tree Legumes in Tropical Agricultura*; Gutteridge, R.C., Shelton, H.M., Eds.; CAB International: Oxon, UK, 1994; pp. 177–191.

33. Fleming, G.A. Mineral composition of herbage. In *Chemistry and Biochemistry of Herbage*; Butler, G.W., Bailey, W., Eds.; Academic Press: London, UK, 1973; pp. 529–566.

34. Herdt, T.H. Ruminant adaptation to negative energy balance. Influences on the etiology of ketosis and fatty liver. *Vet. Clin. N. Am. Food Anim. Pract.* **2000**, *16*, 215–230. [CrossRef]

35. Abarghani, A.; Mostafaei, M.; Alamisaed, K.; Ghanbari, A.; Sahraee, M.; Ebrahimi, R. Investigation of Calcium, Phosphorous and Magnesium Status of Grazing Sheep in Sabalan Region. *Iran. J. Agric. Sci. Tech.* **2013**, *15*, 65–76.

36. Stojković, J. The contents of some mineral elements in the blood serum of sheep in different physiological cycles and with the different level of minerals in a meal. *Biotechnol. Anim. Husb.* **2009**, *25*, 979–984.

37. Masters, D.G.; Purser, D.B.; Yu, S.X.; Wang, Z.S.; Yang, R.Z.; Liu, N.; Lu, D.X.; Wu, L.H.; Ren, J.K.; Li, G.H. Mineral nutrition of grazing sheep in Northern China. 1 Macro-minerals on pasture feed supplement and sheep. *Asian-Aust. J. Anim. Sci.* **1993**, *6*, 99–105. [CrossRef]

38. Alvaro Rincón, C. Maní forrajero (*Arachis pintoi*), la leguminosa para sistemas sostenibles en producción agropecuaria. *Corpoica Inf. Técnica Año* **1999**, *3*, 1–8.

39. Vitti, D.M.S.S.; Kebreab, E.; Lopes, J.B.; Abdalla, A.L.; De Carvalhos, F.F.R.; De Resende, K.T.; Crompton, L.A.; France, J. A kinetic model of phosphorus metabolism in growing goats. *J. Anim. Sci.* **2000**, *78*, 2706–2712. [CrossRef]

40. Dias, R.S.; Lopes, S.; Silva, T.; Pardo, R.M.P.; Silva Filho, J.C.; Vitti, D.M.S.S.; Kebreb, E.; France, J. Rumen phosphorus metabolism in sheep. *J. Agric. Sci.* **2009**, *147*, 391–398. [CrossRef]

41. Braithwaite, G.D. Studies on the absorption and retention of calcium and phosphorus by young and mature Ca-deficient sheep. *Br. J. Nutr.* **1975**, *34*, 311–324. [CrossRef] [PubMed]

42. Pinares-Patiño, C.S.; McEwan, J.C.; Dodds, K.G.; Cárdenas, E.A.; Hegarty, R.S.; Koolaard, J.P.; Clark, H. Repeatability of methane emissions from sheep. *Anim. Feed Sci. Technol.* **2011**, *166*, 210–218.

43. Pfeffer, E.; Beede, D.K.; Valk, H. Phosphorus metablolism in ruminants and requirements of cattle. In *Nitrogen and Phosphorus Nutrition of Cattle and the Environment*; Pfeffer, E., Hristov, A., Eds.; CAB International: Wallingford, UK, 2005; pp. 195–231.
44. Veum, T.L. Phosphorus and calcium nutrition and metabolism. In *Phosphorus and Calcium Utilization and Requirements in Farm Animals*; Kebreab, E., Vitti, D.M.S.S., Eds.; CAB International: Wallingford, UK, 2010; pp. 94–111.

Article

The Potential Use of Layer Litter in Awassi Lamb Diet: Its Effects on Carcass Characteristics and Meat Quality

Belal S. Obeidat [1,*], Mohammad A. Mayyas [1], Abdullah Y. Abdullah [1], Mofleh S. Awawdeh [2], Rasha I. Qudsieh [3], Mohammad D. Obeidat [1], Basheer M. Nusairat [1], Kamel Z. Mahmoud [1], Serhan G. Haddad [1], Fatima A. Al-Lataifeh [4], Mysaa Ata [4], Majdi A. Abu Ishmais [1] and Ahmed E. Aljamal [1]

[1] Department of Animal Production, Faculty of Agriculture, Jordan University of Science and Technology, Irbid 22110, Jordan
[2] Department of Veterinary Pathology and Public Health, Faculty of Veterinary Medicine, Jordan University of Science and Technology, Irbid 22110, Jordan
[3] Prestage Department of Poultry Science, North Carolina State University, Raleigh, NC 27695-7608, USA
[4] Department of Animal Production and Protection, Faculty of Agriculture, Jerash University, Jerash 26150, Jordan
* Correspondence: bobeidat@just.edu.jo; Tel.: +962-2-7201000 (ext. 22214); Fax: +962-2-7201078

Received: 19 August 2019; Accepted: 7 October 2019; Published: 11 October 2019

Simple Summary: Inclusion of local agro–industrial by-products as alternatives to cereal-based concentrates is a promising solution with an increased usage in the area. In this study layer litter is included at 0, 150, or 300 g/kg in the diets of growing lambs. Except with minor effects on carcass characteristics for lambs fed layer liter at 150 g/kg, the inclusion of layer liter did not affect carcass characteristics and meat quality.

Abstract: Carcass parameters and meat quality in lambs that consumed diets having layer hen litter (LL) were evaluated in a complete randomized study. Forty-two lambs were allocated equally (14 lambs/treatment diet) into one of three iso-nitrogenous diets for 75 days. To partially replace soybean meal and barley, LL was given at 0 (LL0), 150 (LL150), or 300 g/kg (LL300) of dietary dry matter (DM). At the termination of the trial, the characteristics of carcasses (hot and cold carcass weight, dressing percentage, and carcass cuts) and meat quality (*Musculus longissimus linear dimensions*, ultimate pH, cooking loss, water holding capacity (WHC), shear force (SF), color coordinates) were measured after slaughtering all lambs. *Longissimus* muscle weight was greatest ($p < 0.05$) for the LL150. For the dissected loin, intermuscular fat content was lowest for the LL0 diet. However, subcutaneous fat content was lower ($p < 0.05$) in the LL300 diet than LL0 and LL150 diets. Rib fat depth and *Musculus longissimus* area were greater ($p < 0.05$) for LL150 than L0. No differences were found in meat pH or color parameters among treatments but WHC and SF were lower in L0 lambs than in lambs fed LL containing diets. Cooking loss was greater for the LL300 diet than the LL0 diet. In summary, quality of meat and carcasses data indicate the possibility of inclusion of LL up to 300 g/kg DM to growing Awassi lambs.

Keywords: awassi lambs; carcass characteristics; meat quality; layer litter

---

1. Introduction

A good approach to measure the productivity of slaughtering animals is by estimating the efficiency of production. Almost 70% of the cost of red meat production is associated with diet ingredients, which alerts the necessity to use unconventional feed ingredients in order to mitigate the production cost [1]. Furthermore, feed sources available for ruminants during the dry season are of low quality and have low protein content in Jordan and worldwide [2]. Therefore, layer hen litter (LL) is a possible feed to be used as an alternative feed ingredient to provide protein for ruminants [3]. Noland et al. [4] were among the first to investigate the potential for using broiler litter as a nitrogen source for sheep. Sheep have the ability to digest this cheap nitrogen-rich LL and, at the same time, poultry producers would have a safe way for disposing LL. Several trials have reported successful performance with inclusion of LL in sheep rations [5–7]. For example, Obeidat et al. [6] reported that growth performance, carcass characteristics, and meat quality was not affected when broiler litter was included in the diets of lambs at 0, 100, or 200 g/kg dry matter (DM). However, the effects of feeding LL on characteristics of carcasses and quality of meat are still essential to investigate. Azizi-Shotorkhoft et al. [8] reported that feeding up to 210 g/kg dry matter of heat-processed broiler litter to fat-tailed Moghani lambs reduced loin fat weights, and had no effects on lean, bone, and fat weights in carcasses.

Our hypothesis was that feeding LL to Awassi lambs during the fattening period would not impact carcass characteristics and meat quality. Therefore, this experiment was conducted to evaluate the influence of the inclusion of different proportions of LL on carcass and meat quality parameters of growing Awassi lambs.

2. Materials and Methods

2.1. Experimental Procedures

The Institutional Animal Care and Use Committee approved all methods and procedures used in the current study at Jordan University of Science and Technology (JUST). Procedures and data of nutrient intake and growth performance were previously described in Obeidat et al. [9]. In brief, forty-two Awassi lambs (20.5 ± 0.88 kg initial body weight (BW), 70 ± 2.02 days of age) were assigned randomly into one of three iso-nitrogenous [174 g/kg crude protein (CP); dry matter (DM) basis] diets. To partially replace soybean meal and barley, LL was given at 0 (LL0), 150 (LL150), or 300 g/kg (LL300) of dietary DM. Lambs were individually housed (1.5 × 0.75 m) and the experimental diets were offered twice daily ad libitum during the whole experimental period (75 days). The feed intake was calculated by subtracting the collected refusal from the offered feed. In brief, DM intake, final body weight, and average daily gain were greater in lambs fed diets containing LL at 150 g/kg DM compared with the lambs fed LL at 0 g/kg DM, whereas in lambs fed LL at 300 g/kg DM, results were intermediate. Dry matter intake was 894, 1006, and 940 g/d for LL0, LL150, and LL300 g/kg, respectively. Layer litter was obtained from a local floor-reared laying hen farm. Before mixing the diets, LL was placed in plastic bags and autoclaved at 121 °C for 20 min to kill litter microflora. After autoclaving, LL was ground to pass a 3 mm screen in order to facilitate its mixing with the other dietary ingredients (Table 1). The chemical composition of the LL and the diet ingredients was analyzed following the procedures of AOAC [10]. Composition of the LL and experimental diets is shown in Table 1.

Table 1. Ingredients and chemical composition of diets and layer litter (LL) fed to Awassi lambs.

Item	Diets [1]			
	LL0	**LL150**	**LL300**	**LL**
Ingredients (g/kg DM)				
Barley grain	629	519	409	
Soybean meal (440 g/kg CP [2] (solvent))	150	110	70	
Layer litter	0	150	300	
Wheat straw	200	200	200	
Soybean oil	5	5	5	
Salt	7.5	7.5	7.5	
Limestone	7.5	7.5	7.5	
Mineral vitamin premix [3]	1.0	1.0	1.0	
Nutrients				
Dry matter (g/kg DM)	909	907	916	895
Organic matter (g/kg DM)	926	890	878	725
Crude protein (g/kg DM)	173	174	174	269
Neutral detergent fiber (g/kg DM)	226	254	274	236
Acid detergent fiber (g/kg DM)	117	142	154	97
Ether extract (g/kg DM)	69	66	65	25
Copper (µg/g)	4.86	8.76	14.18	41.6

[1] Diets were: LL included in the diets at 0 (LL0), 150 (LL150), and 300 g/kg (LL300) of dietary dry matter (DM). [2] CP: crude protein, [3] Composition per kg use (vitamin A, 2,000,000 IU; vitamin D_3, 40,000 IU; vitamin E, 400 mg, Mn, 12.80 g; Zn, 9.00 g; I, 1.56 g; Fe, 6.42 g; Cu, 1.60 g; Co, 50 mg; Se, 32 mg).

2.2. Slaughtering Procedures and Meat Quality Measurements

At the end of the experimental period, lambs were transported to a slaughterhouse. After 18 h of fasting, live BW was recorded and animals were slaughtered following the procedure described by Abdullah et al. [11]. Non-carcass components (spleen, liver, kidneys, lungs, and trachea) were removed and weighed and carcass weight was recorded immediately after slaughter (HCW) and after 24 h of refrigeration at 4 °C (CCW) to calculate chilling losses and dressing percentage. Upon cutting, the right leg and loin cut were dissected and vacuum-packed immediately and stored at −20 °C for further measurements. Then, the *longissimus* muscle was removed and kept frozen at −20 °C for further analysis.

To measure meat quality parameters of the *longissimus* muscle, frozen muscle was thawed over night at 4 °C using the procedure described by Obeidat et al. [12]. The measured parameters were water holding capacity (WHC), meat color, pH, shear force, and cooking loss. In brief, each thawed muscle was divided into slices of 15 mm thick to be used to measure color coordinates (CIE *L*a*b**). To allow oxygenation of the sample, 2 h prior to the measurement, color slices were covered with permeable film and kept at 4 °C. Then, color coordinates were measured in triplicates by using a handheld colorimeter device (12 mm Aperture U 59730-30, Cole-Parameter International, Accuracy Microsensors Inc., Pittsford, NY, USA). Cooking loss was measured using duplicate-slices (25 mm thickness) after weighing raw slices and placing and cooking them in plastic bags in a water bath at 75 °C for 90 min (until reaching an internal temperature of 72 °C). After bathing the slices, weight was measured to calculate water lost percentage. To measure the shear force (i.e., tenderness measurement) cooked slices were then stored in the chiller for 24 h at 4 °C. By using a Warner–Bratzler (WB) shear blade (Warner-Bratzler meat shear, GR manufacturing Co. 1317 Collins LN, Manhattan, Kansas, 66502, USA) with the triangular slot cutting edge mounted on a Salter Model 235, a total of 6 cores, with a size of 1 cm^3, were cut from the slices and sheared in a perpendicular direction of muscle fiber. Peak force (kg) required to shear the cores was measured as an indicator of meat tenderness. Muscle pH was evaluated on a homogenate of 2 g of raw meat and 10 mL of neutralized 5 mM iodoacetate reagent using a pH spear (pH Spear, Eutech Instrument, USA). Water holding capacity was evaluated using methods described by Grau and Hamm [13].

2.3. Statistical Analysis

Data were subjected to analysis of variance using the mixed procedure of SAS (Version 8.1, 2000, SAS Inst. Inc., Cary, NC, USA). [14]. Diet was included as fixed effect in the statistical model, the animal nested to diet being the residual error. Mean separation for all traits was conducted using pair-wise *t*-tests and significance level was determined at $p < 0.05$.

3. Results

Fasting weight, HCW, CCW, dressing percentage, and non-carcass components were not significantly different ($p > 0.05$) among treatment diets (Table 2). However, carcass cut weights were greater ($p < 0.05$) in lambs fed the LL150 diet than the LL0 and LL300 diets. Fat tail and mesenteric fat were similar ($p > 0.05$) among the diets. The weight of kidney fat was greater ($p < 0.05$) in the LL150 diet than the LL0 diet, whereas the LL300 was intermediate.

Table 2. Carcass components, dissected loin and leg carcass cut weights, and percentages and fat depth of Awassi lambs fed diets containing layer litter.

Item	Diets [1]			
	LL0	LL150	LL300	SEM [2]
Fasting live weight (kg)	35.7	38.4	36.6	1.32
Hot carcass weight (kg)	16.7	18.2	16.8	0.72
Cold carcass weight (kg)	16.1	17.6	16.1	0.70
Dressing percentage (%)	45.7	45.8	44.3	1.19
Non-carcass components (kg [3])	1.42	1.41	1.33	0.06
Carcass cut weights (kg [4])	14.4 [a]	16.3 [b]	15.3 [a]	0.46
Fat tail (kg)	1.60	1.71	1.42	0.16
Mesenteric fat (g)	348	407	327	45.0
Kidney fat (g)	157 [a]	228 [b]	185 [ab]	20.1
Loin weight (g)	807 [a]	976 [b]	879 [ab]	39.8
Longissimus muscle (g)	200.7 [a]	252.1 [b]	237.9 [b]	8.01
Intermuscular fat (g/100 g)	5.70 [a]	7.13 [b]	6.71 [ab]	0.53
Subcutaneous fat (g/100 g)	16.9 [b]	16.7 [b]	12.8 [a]	1.29
Total fat (g/100 g)	22.6	23.9	19.5	1.64
Total lean (g/100 g)	51.1	49.7	50.9	1.30
Total bone (g/100 g)	17.7 [a]	17.9 [a]	22.4 [b]	0.77
Meat to bone ratio	2.93 [b]	2.87 [b]	2.30 [a]	0.10
Meat to fat ratio	2.85	2.22	2.89	0.272
Leg weight (g)	2675	2937	2891	108.3
Intermuscular fat (g/100 g)	3.21 [ab]	3.50 [b]	3.00 [a]	0.17
Subcutaneous fat (g/100 g)	13.9 [a]	16.6 [b]	13.7 [a]	0.62
Total fat (g/100 g)	17.0 [a]	20.1 [b]	16.6 [a]	0.66
Total lean (g/100 g)	58.3	59.3	57.9	1.65
Total bone (g/100 g)	19.2	19.3	19.7	0.62
Meat to bone ratio	3.05 [ab]	3.10 [b]	2.95 [a]	0.050
Meat to fat ratio	3.57 [b]	3.02 [a]	3.52 [b]	0.15
Tissue depth (GR) (mm)	14.9	16.2	14.3	0.91
Rib fat depth (J) (mm)	5.90 [a]	8.62 [b]	5.81 [a]	0.64
Shoulder fat depth (C) (mm)	3.92	4.71	3.60	0.61

[1] Diets were: layer litter (LL) included in the diets at 0 (LL0), 150 (LL150), and 300 g/kg (LL300) of dietary dry matter. [2] SEM: Standard Error of the Mean, [3] Non-carcass components (heart, liver, spleen, kidney, and lungs and trachea). [4] Carcass cut (shoulder, racks, loins, and legs). [ab] Within a row, means without a common superscript differ ($p < 0.05$).

Results showed that lambs fed LL150 diet had greater ($p < 0.05$) weight of loin cut compared with lambs fed the LL0 diet (Table 2). *Longissimus* muscle was greater ($p < 0.05$) for the LL diets than the control group (LL0). The content of intermuscular fat was lower ($p < 0.05$) for lambs fed the LL0 than for lambs fed the LL150. Subcutaneous fat content was lower ($p < 0.05$) in the LL300 than the LL0 and LL150 diets. Content of total lean and fat and ratio of meat to fat did not differ ($p > 0.05$) among diets.

However, total bone content was higher ($p < 0.05$) in lambs fed LL300 than LL0 or LL150 diets. As a result, the ratio of meat to bone was lower ($p < 0.05$) in LL300 diets than the LL0 and LL150 diets.

Intermuscular fat content was lower ($p < 0.05$) in the LL300 diet than the LL150 diet, while the LL0 diet was similar compared with LL150 and LL300 diets in the dissected leg (Table 2). However, the content of total fat and subcutaneous fat were greater significantly in LL150 compared with the LL0 and LL300 diets. However, the content of total bone and lean was comparable among different diets. The ratio of meat to bone in the loin was lower ($p < 0.05$) in lambs fed the LL300 than lambs fed the LL150 diet. However, the ratio of meat to fat in the leg was lower ($p < 0.05$) in the LL150 diet compared to the LL0 and LL300 diets. Meat to bone ratio in the leg was lower ($p < 0.05$) in the LL150 diet than the LL0 and LL300 diets. Tissue depth did not differ among diets. However, the depth of rib fat was greater ($p < 0.05$) in lambs fed the LL150 diet compared to lambs fed the LL0 and LL300 diets. No significant difference was observed for the shoulder fat depth between the diets.

Results of meat quality are shown in Table 3. Ultimate pH did not differ ($p > 0.05$) among diets. Cooking loss was greater ($p < 0.05$) in lambs fed the LL300 diet compared to the LL0 diet, whereas cooking loss for the LL150 diet was intermediate. However, greater water holding capacity and shear forces were observed in lambs fed LL-containing diets compared with the control diet. Meat color (i.e., $L*$, $a*$, and $b*$) did not differ among diets. *M. longissimus* width was greater ($p < 0.05$) in the LL300 diet than the LL0 diet; whereas the LL150 diet was not different from the other two diets. *M. longissimus* depth was lower ($p < 0.05$) for the LL0 diet than for the LL150 and LL300 diets. *M. longissimus* area was greater ($p < 0.05$) for the LL150 diet than for the LL0 diet.

Table 3. Meat quality characteristics and *M. longissimus* linear dimensions of Awassi lambs fed finishing diets containing layer litter.

Item	Diets [1]			
	LL0	LL150	LL300	SEM [2]
Ultimate pH [3]	5.87	5.89	5.84	0.019
Cooking loss (g/100 g)	41.6 [a]	42.6 [ab]	43.2 [b]	0.61
Water holding capacity (g/100)	21.6 [a]	24.6 [b]	24.2 [b]	0.73
Shear force (kg/cm^2)	5.8 [a]	8.8 [b]	7.7 [b]	0.50
Color coordinates				
$L*$ (whiteness)	38.9	38.3	39.4	0.50
$a*$ (redness)	3.0	2.8	3.7	0.54
$b*$ (yellowness)	19.0	18.8	19.3	0.63
M. longissimus width (A) (mm)	26.1 [a]	27.3 [ab]	28.2 [b]	0.70
M. longissimus depth (B) (mm)	55.1 [a]	59.9 [b]	58.1 [b]	0.81
M. longissimus area (cm^2)	13.4 [a]	15.3 [b]	14.3 [ab]	0.35

[1] Diets were: layer litter (LL) included in the diets at 0 (LL0), 150 (LL150), and 300 g/kg (LL300) of dietary dry matter.
[2] SEM: Standard Error of the Mean, [3] pH measured after thawing. [ab] Within a row, means without common letters differ ($p < 0.05$).

4. Discussion

In accordance with previous studies [6,15], neither carcass weight nor dressing percentage were affected by broiler litter supplementation when fed at 100 and 200 g/kg of dietary DM fed to Awassi lambs [6] and 0, 280, 560, and 850 g/kg DM) in diets of South African Mutton Merino wethers [15], which would be in agreement with results obtained herein. Nevertheless, meat to bone ratio decreased in the highest LL supplementation. Obeidat et al. [6] did not observe any effect of meat to bone ratio in leg cuts, but they tested levels of LL lower than 200 g/kg feed. In our experiment, mineral content was not equilibrated, and it increased as the LL proportion increased. Therefore, it would be possible that a greater mineral supply affects bone development or density.

When looking at the meat to fat ratio in both loin and leg cuts, diets formulated with 150 g/kg LL had the lowest ratio in both cuts (even though differences did not reach a significant level in the loin) indicating that total fat content was highest in those lambs. It is the metabolic rate and physiological importance that play an important role in the process of separating the use of nutrients between tissues and different organs in the body. Therefore, when providing ad libitum balanced feed, the highest efficiency of feed conversion was achieved. Based on live performance data of Obeidat et al. [9], lambs fed LL150 had the highest dry matter intake, final body weight, and the best feed efficiency compared to the other treatments. Dry matter intake was 894, 1006, and 940 g/d for LL0, LL150, and LL300 g/kg, respectively.

Consistent with differences observed among treatments in carcass cuts, LL150 lambs showed significantly greater rib fat depth and kidney fat, mesenteric fat and shoulder fat depth being numerically greater as well. In a study that compared linear dimensions of *M. longissimus* and measurements of fat for lambs that consumed broiler litter at various levels (i.e., 0, 100, or 200 g/kg), Obeidat et al. [6] reported that tissue depth, *M. longissimus* depth, width, and area, fat depth, fat tail weight, and weight of kidney fat did not change among dietary treatments. On the other hand, rib fat depth or mesenteric fat weight were lower for 200 g/kg broiler litter diet compared with the control diet but 100 g/kg broiler litter diet did not differ from the other two treatments [6]. Furthermore, *M. longissimus* dimensions were also the largest in lambs fed 150 g LL/ kg feed, which is in accordance with the heaviest weight of *M. longissimus* harvested from those lambs. The discrepancies in linear dimensions that was observed among different studies could be due to the difference in the diets' compositions, the chemical content in the LL, and/or the age and the type of animals.

Values for meat quality attributes measured in the current study were within acceptable range. Consistent with our results, Obeidat et al. [6]) reported that broiler litter supplementation at 0, 100, or 200 g/kg DM did not affect the pH of meat. Ultimate pH of *M. longissimus* observed herein did not differ from other studies [16–18]. Inconsistent with results obtained herein, some of the meat quality characteristics (shear force, cooking loss, or water holding capacity) was similar when lambs fed diets containing various levels of broiler litter (0, 100, and 200 g/kg of dietary DM) [6]. In addition, shear force was not different in Holstein steers fed diets containing broiler litter [19]. In the current study, shear force values were higher for *M. longissimus* from LL-fed lambs, which relates to higher cooking loss, and thus the meat was drier. Despite that, the shear force values were greater in diets containing LL when compared to the control diet; they were still within the normal range. Color coordinates measured were comparable among the different diets. Overall, current results showed that feeding LL to Awassi lambs did not negatively affect meat quality except for the greater shear force in diets containing LL.

5. Conclusions

Results of the current study shows that the inclusion of layer litter at 0, 150, or 300 g/kg of DM did not affect fasting live weight, hot and cold carcass weight, and dressing percentage among diets. However, there are few minor changes in the meat quality parameters especially in lambs fed the 150 g/kg diet. Therefore, it is recommended to feed diet containing layer litter between 150 to 300 g/kg to lambs.

Author Contributions: Conceptualization, Methodology, and Writing—original draft, B.S.O. and M.A.M.; Writing—review and editing, A.Y.A., M.S.A., R.I.Q., M.D.O., B.M.N., K.Z.M., S.G.H., F.A.A.-L., M.A., M.A.A.I., and A.E.A.

Funding: This project is funded by Deanship of Scientific Research at Jordan University of Science and Technology (135/2010).

Acknowledgments: The authors wish to acknowledge the Deanship of Scientific Research at JUST for the financial funding of this study.

Conflicts of Interest: The authors declare no conflict of interest regarding this research.

References

1. Awawdeh, M.S. Alternative feedstuffs and their effects on performance of Awassi sheep: A review. *Trop. Anim. Health Prod.* **2011**, *43*, 1297–1309. [CrossRef] [PubMed]
2. Roothaert, R.L.; Matthewman, R.W. Poultry wastes asods for ruminants and associated aspects of animal welfare. *Asian-Australas J. Anim. Sci.* **1992**, *5*, 593–600. [CrossRef]
3. Talib, N.H.; Ahmed, F.A. Digestibility, degradability and dry matter intake of deep-stacked poultry litter by sheep and goats. *J. Anim. Vet. Adv.* **2008**, *7*, 1474–1479.
4. Noland, P.R.; Ford, B.F.; Ray, M.L. The use of ground chicken litter as a source of nitrogen for gestating lactating ewes and fattening steers. *J. Anim. Sci.* **1955**, *14*, 860–865. [CrossRef]
5. Elemam, M.B.; Fadelelseed, A.M.; Salih, A.M. Growth performance, digestibility, N-balance and rumen fermentation of lambs fed different levels of deep-stack broiler litter. *Res. J. Anim. Vet. Sci.* **2009**, *4*, 9–16.
6. Obeidat, B.S.; Awawdeh, M.S.; Abdullah, A.Y.; Muwalla, M.M.; Abu Ishmais, M.A.; Telfah, B.T.; Ayrout, A.J.; Matarneh, S.K.; Subih, H.S.; Osaili, T.O. Effect of feeding broiler litter on performance of Awassi lambs fed finishing diets. *Anim. Feed Sci. Technol.* **2011**, *165*, 15–22. [CrossRef]
7. Chaudhry, S.M.; Naseer, Z. Processing and nutritional value of broiler litter as a feed for buffalo steers. *J. Anim. Plant Sci.* **2012**, *22*, 358–364.
8. Azizi-Shotorkhoft, A.; Rezaei, J.; Papi, N.; Mirmohammadi, D.; Fazaeli, H. Effect of feeding heat-processed broiler litter in pellet-form diet on the performance of fattening lambs. *J. Appl. Anim. Res.* **2015**, *43*, 184–190. [CrossRef]
9. Obeidat, B.S.; Abdullah, Y.A.; Mayyas, M.A.; Awawdeh, M.S. The potential use of layer litter in Awassi lambs' diet: Its effects on nutrient intake, digestibility, N balance, and growth performance. *Small Rumin. Res.* **2016**, *137*, 24–27. [CrossRef]
10. AOAC. *Official Methods of Analysis*; Association of Official Analytical Chemists: Gaithersburg, MD, USA, 1997.
11. Abdullah, A.Y.; Purchas, R.W.; Davis, A.S. Patterns of change with growth for muscularity and other composition characteristics of southern rams selected for high and low back depth. *N. Z. J. Agric. Res.* **1998**, *41*, 367–376. [CrossRef]
12. Obeidat, B.S.; Abdullah, A.Y.; Al-Lataifeh, F.A. The effect of partial replacement of barley grains by Prosopis juliflora pods on growth performance, nutrient intake, digestibility, and carcass characteristics of Awassi lambs fed finishing diets. *Anim. Feed Sci. Technol.* **2008**, *146*, 42–54. [CrossRef]
13. Grau, R.; Hamm, G. Eine enfache Methode zur Bestimmung der Wasserbindung im Muskel. *Naturwissenschaften* **1953**, *40*, 29–30. [CrossRef]
14. SAS. *SAS User Guide Statistics*; Version 8.1; SAS Inst.: Cary, NC, USA, 2000.
15. Mavimbela, D.T.; Webb, E.C.; Van Ryssen, J.B.J.; Bosman, M.J.C. Sensory characteristics of meat and composition of carcass fat from sheep fed diets containing various levels of broiler litter. *S. Afr. J. Anim. Sci.* **2000**, *30*, 26–32. [CrossRef]
16. Abdullah, A.Y.; Qudsieh, R.I. Carcass characteristics of Awassi ram lambs slaughtered at different weights. *Livest. Sci.* **2008**, *117*, 165–175. [CrossRef]
17. Abdullah, A.Y.; Qudsieh, R.I. Effect of slaughter weight and aging time on the quality of meat from Awassi ram lambs. *Meat Sci.* **2009**, *82*, 309–316. [CrossRef] [PubMed]
18. Abdullah, A.Y.; Qudsieh, R.I.; Nusairat, B.M. Effect of crossbreeding with exotic breeds on meat quality of Awassi lambs. *Livest. Sci.* **2011**, *142*, 121–127. [CrossRef]
19. Jeremiah, L.E.; Gibson, L.L. The effects of postmortem product handling and aging time on beef palatability. *Food Res. Int.* **2003**, *36*, 929–941. [CrossRef]

Article

Effect of Age and Weaning on Growth Performance, Rumen Fermentation, and Serum Parameters in Lambs Fed Starter with Limited Ewe–Lamb Interaction

Shiqin Wang [1,2,†], Tao Ma [2,†], Guohong Zhao [2], Naifeng Zhang [2], Yan Tu [2], Fadi Li [1], Kai Cui [2], Yanliang Bi [2], Hongbiao Ding [2] and Qiyu Diao [1,2,*]

[1] State Key Laboratory of Grassland Agro-Ecosystems, Key Laboratory of Grassland Livestock Industry Innovation, Ministry of Agriculture and Rural Affairs, College of Pastoral Agriculture Science and Technology, Lanzhou University, Lanzhou 730020, China; wshq1988@163.com (S.W.); lifd@lzu.edu.cn (F.L.)

[2] Feed Research Institute, Chinese Academy of Agricultural Sciences/Key Laboratory of Feed Biotechnology of the Ministry of Agriculture and Rural Affairs, Beijing 100081, China; matao@caas.cn (T.M.); zhaoguoh@foxmail.com (G.Z.); zhangnaifeng@caas.cn (N.Z.); tuyan@caas.cn (Y.T.); cuikai@caas.cn (K.C.); vetbi2008@163.com (Y.B.); dinghongbiao@caas.cn (H.D.)

* Correspondence: diaoqiyu@caas.cn; Tel.: +86-10-8210-6055

† Shiqin Wang and Tao Ma contribute equally to the article.

Received: 22 September 2019; Accepted: 16 October 2019; Published: 18 October 2019

Simple Summary: Early weaning is a common practice in the modern lamb industry, which shortens the breeding cycle of ewes and improves the flock productivity by increasing the frequency of lambing. Recently, it was suggested that weaning age (21 or 35 days of age) and milk replacer feeding level had limited effect on apparent digestibility of nutrients in lambs. However, artificial rearing may increase the cost due to the use of milk replacer, labor, and feeding facilities. In this study, we compared the difference in growth performance of lambs weaned at 21 days (early weaning) and 49 days (conventional weaning) of age. All lambs were initially reared with ewes and supplemented with starter at 7 days of age. The results showed that diarrhea rate was increased when lambs were weaned at 21 days of age while average daily gain was decreased. Early weaning reduced average daily gain without compromising lambs' overall immunity. In conclusion, early weaning (21 days of age) may have a negative impact on lambs' performance based on a short-term study.

Abstract: Sixty neonatal Hu lambs were weaned at either 21 (n = 30) (early weaning, EW) or 49 days (n = 30) of age (control, CON). The starter intake and body weight (BW) of lambs was recorded weekly from birth to 63 days of age. Diarrhea rate of lambs was measured from birth to 35 days. Six randomly selected lambs from each treatment were slaughtered at 26, 35, and 63 days of age, respectively. Ruminal pH, NH_3-N, and volatile fatty acid (VFA) concentration, as well as serum parameters including immunity, antioxidant status, and inflammatory parameters from randomly selected lambs from each treatment were measured. There was no difference in BW at birth and day 21 between the two groups of lambs ($p > 0.05$). However, BW of the lambs in the EW group was significantly lower than those in the CON group ($p < 0.01$) from 28 to 49 days of age. Average daily gain (ADG) of the lambs in the EW group was significantly lower than those in the CON group ($p < 0.01$) at three weeks after early weaning. Starter intake of the lambs in the EW group was obviously higher than that in the CON group ($p < 0.01$) from day 28 to 49. In addition, the diarrhea rate was significantly higher than that in the CON group from day 5 to 14 after weaning ($p < 0.01$). The EW group had heavier carcasses ($p < 0.01$) and rumen relative to whole stomach weights ($p < 0.01$). Rumen pH was increased by age ($p < 0.01$) and was not affected by early weaning ($p > 0.05$). Early weaning decreased abomasum relative to whole stomach weight ($p < 0.01$) and increased total VFA concentrations ($p < 0.01$) at day 26. There was no difference in lambs' immunity and stress indicators

($p > 0.05$). The results indicated that lambs weaned at 21 days of age had decreased ADG and higher diarrhea rate, although the overall immunity was not compromised. Long-term study is needed to further validate the feasibility of early weaning strategy in lambs.

Keywords: early weaning; weaning stress; lamb; growth performance; diarrhea

---

1. Introduction

Weaning time is critical to the management of ewes and lambs. Although there is no 'best' age to wean, lambs can be weaned at as early as three-weeks old [1], and early weaning is a common practice in modern lamb industry. Early weaning shortens the breeding cycle of ewes, which improves the flock productivity by increasing the frequency of lambing [2,3]. More importantly, lambs utilize diet more efficiently than ewes because lambs directly convert feed to gain while ewes convert the feed to milk and then to lamb gain. Therefore, an increase in economic profits can be expected if early weaning is successful. However, early weaning can be an important stressor for lambs due to separation from dams [4], as it is suggested that suckling is a major factor in the strength of the ewe–lamb contact [5]. In addition, the transition from liquid (milk/milk replacer) to solid diets (starter or grass) during weaning requires the development of a functional rumen [6]. Consequently, early weaning may cause a drop in feed consumption and decrease in growth rate of lambs [7] if the rumen structure and function has not been fully developed before weaning [8].

To minimize the negative effect of early weaning as well as ensure economic benefits, great efforts have been made to investigate the optimal weaning strategy for lambs. Early study suggested that lambs reared under a mixed system, featured by separating lambs from their dams for 15 h while allowing them to suckle for the remaining 9 h daily, had better financial return than those exclusively suckling their dams and weaned at 30 days after birth [9]. On the other hand, it is reported that sufficient starter intake can stimulate the development of a functional rumen as well as the establishment of rumen microbiota [6]. Our previous study suggested that when starter was offered at day 15 of age, lambs separated from their dams at 20 days of age and artificially fed milk replacer had superior growth performance than their ewe-reared counterparts at the end of the feeding trial (90 days of age) [10]. Recently, it was suggested that weaning age (21 or 35 days of age) and milk replacer feeding level had limited effect on apparent digestibility of nutrients, ruminal microbiota, and fermentation of lambs at 50 days [11]. However, artificial rearing may increase the cost due to the involvement of milk replacer, labor, and feeding facilities [12].

Here, we compared the effect of limited ewe–lamb contact (18 h daily) and supplementation of starter at an early age (seven-days old) on growth performance, rumen fermentation, slaughter performance, and serum parameters of lambs weaned at either 21 or 49 days of age in a 63-day feeding trial. We hypothesized that early weaning (21 days of age) may have temporal negative effect on lambs' performance, but the overall performance of lambs may be similar between two groups.

2. Materials and Methods

The study was conducted from October to December, 2018 at Runlin sheep farm (N 36°82′, E 115°83′) in Linqing county, Liaocheng city, Shandong province, China. The experiment protocol was approved by the Animal Ethics and Humane Animal Care of the Chinese Academy of Agricultural Sciences (with protocol FRI-CAAS-20180810).

A total of 60 neonatal Hu lambs (48 male and 12 female) with similar birth weights (3.82 ± 0.46 kg) were used. Lambs were initially reared with ewes in 6 well-ventilated sheep pens (4 m × 5 m) with controlled temperature and humidity, with 10 lambs (8 male and 2 female) and their ewes in each pen. All ewes were fed two times daily according to the farm's feeding management schedule. None of the lambs in the groups had access to the ewes' feed. Ewes were fed a total mixed ration consisting of 35%

corn silage, 28% peanut straw, 7% garlic straw, 3% soybean residue, 15% corn, 6% wheat bran, and 6% soybean meal.

From day 7 of age, all lambs were separated from the ewes and offered pelleted starter for 6 h daily (08:00–10:00, 12:00–14:00, and 16:00–18:00). At other times, the lambs remained with their dams and still had free access to starter. All lambs were fed starter 1 from 7 to 35 and starter 2 from 36 to 63 days of age, respectively. The ingredient and nutritional composition of starters are listed in Table 1. Half of the lambs from three pens (n = 30) were weaned and abruptly separated from their dams at 21 days of age (early weaning, EW) while the other half of the lambs from another three pens (n = 30) were weaned and abruptly separated from their dams at 49 days of age (control, CON). All lambs and ewes had ad libitum access to water. During the experiment, two lambs from the CON group did not have enough milk intake as their dams were sick, and one lamb from the CON group died accidentally. Therefore, those three lambs were removed from further analysis.

Table 1. Ingredients and chemical composition of the starters.

Item	Starter 1 (7–35 Days)	Starter 2 (36–63 Days)
Ingredients, air dry basis, %		
Alfalfa hay		7.0
Oat grass		5.0
Corn	50.0	45.5
Soybean meal	23.5	20.0
Corn germ meal	12.0	10.0
Wheat bran	10.0	8.0
Vitamin and mineral mixture [1]	1.0	1.0
Limestone	2.0	2.0
$CaHPO_4$	0.5	0.5
Salt	0.5	0.5
$NaHCO_3$	0.5	0.5
Chemical composition [2], % of DM		
DM, air dry basis	90.2	89.1
CP	21.5	21.5
NDF	15.1	18.9
ADF	6.4	8.2
Ash	8.0	5.7
Ca	0.9	0.8
P	0.5	0.6
ME, MJ/kg	10.9	10.5

[1] Contained per kilogram of supplement: 800,000 IU of vitamin A, 300,000 IU of vitamin D3, 3000 mg of vitamin E, 4 g of Fe, 4 g of Mn, 0.8 g of Cu, 5 g of Zn, 20 mg of Se, 70 mg of I, 40 mg of Co. [2] DM, CP, NDF, Cam and P were measured values, while ME (metabolizable energy) was calculated according to National Research Council 2007 [13].

The body weight (BW) of each lamb was recorded weekly (before morning feeding) from 7 to 63 days of age. The starter was offered to each pen and the intake of starter was measured by each pen daily before morning feeding. Representative samples of starter were collected weekly and were immediately frozen at −20 °C for further analysis. Diarrhea incidence was also monitored on a daily basis from birth to 35 days of age.

Blood sampled from six male lambs per treatment was collected in heparinized tubes by jugular venipuncture before morning feeding at 26, 35m and 63 days of age, respectively. The blood sample was centrifuged at 3500× *g* for 15 min at 4 °C to separate serum, which was pipetted into 2 mL cryotubes and stored at −20 °C until subsequent analysis. After blood sampling, those six lambs from each group were slaughtered at 26, 35, and 63 days of age, respectively. Ruminal digesta sample was collected and pH value was measured immediately. The rumen digesta was then filtered through four layers of cheesecloth and a 10 mL subsample of each strained fluid was collected, stored at −20 °C for analysis of the volatile fatty acids (VFAs) and ammonia nitrogen (NH_3-N). Immediately after slaughter, the rumen, reticulum, omasum, and abomasum of each lamb was dissected and weighted after the digesta was removed. The carcass weight of each lamb was also recorded.

The starter samples were ground to pass through a 1 mm sieve and dried in an oven at 135 °C for 2 h (method 930.15; AOAC, 1990) [14] to measure the dry matter (DM) content. The ash content, nitrogen, neutral detergent fiber (NDF), acid detergent fiber (ADF), calcium, and total phosphorus was measured according to methods described by previous studies [14–17]. Crude protein (CP) was calculated as 6.25 × Nitrogen.

Feces were scored using a 1 to 4 scale classified as firm and well-formed (score 1), soft and pudding-like (score 2), runny and pancake batter-like (score 3), or liquid and splatters (score 4) as described previously [18]. If an animal presented a fecal score ≥ 3 for 3 consecutive days, it was considered diarrheic. Diarrhea rate (%) = number of lambs of diagnosed at least once for diarrhea/the total number of lambs. Diarrhea frequency (%) = the number of diarrhea lambs × diarrhea days/the total number of lambs × experimental days.

Serum concentration of immunoglobulin (IgG, IgM, and IgA), superoxide dismutase (SOD), Glutathione Peroxidase (GSH-Px), catalase (CAT), total antioxidative capacity (T-AOC), and malondialdehyde (MDA) were determined using the Hitachi 7020 autobiochemistry instrument (Hitachi, Tokyo, Japan) with corresponding commercial test kits (Nanjing Jiancheng Bioengineering Institute, Nanjing, Jiangsu, China) following the manufacturer's instructions. Interleukin 1β (IL-1β), interleukin 4 (IL-4), interleukin 6 (IL-6), cortisol, interferon-gamma (IFN-γ), and tumor necrosis factor α (TNF-α) were determined using bovine ELISA kits (Nanjing Jiancheng Bioengineering Institute, Nanjing, Jiangsu, China) following the manufacturer's instructions.

Individual and Total VFA concentrations were determined as described previously [19]. Briefly, 1 mL rumen fluid filtrate was added to 0.2 mL of metaphosphoric acid solution (250 g/L) containing 2 g/L 2-ethyl butyrate, mixed overnight, and analyzed by gas chromatography (SP-3420, Beijing Analytical Instrument Factory, Beijing, China). The concentration of NH_3-N was measured using phenol hypochlorite colorimetric method as described by [20]

Starter intake, BW, and average daily gain (ADG) of lambs from each group were pooled by week. Data for weights of carcasses and stomach compartment, blood variables, as well as rumen fermentation parameters of lambs from each group were pooled at day 21, 35, and 63, respectively. Those data were analyzed according to a two-way ANOVA using PROC GLM of SAS (SAS Inst. Inc., Cary, NC, USA) to examine the effect of treatment, age, and the interaction between treatment and age. The statistical model was as follows:

$$Yij = \mu + Ti + Aj + TAij + Eij \tag{1}$$

where Yij is the dependent variable, μ is the overall mean, Ti is the treatment effect, Aj is the age effect, TAij is the interaction of treatment and age, and Eij is the error term.

One-way ANOVA was used to analyze the effect of weaning time on diarrhea rate and diarrhea frequency. The statistical model was as follows:

$$Yi = \mu + Ti + Ei \tag{2}$$

where Yi is the dependent variable, μ is the overall mean, Ti is the Treatment effect, and Ei is the error term. Multiple comparisons of means among treatments were performed using Tukey's tests. Significance was defined as $p < 0.05$.

3. Results

3.1. Growth Performance and Diarrhea Rate

There was no difference in BW from birth until day 21 of age between the two groups of lambs ($p > 0.05$) (Figure 1a). However, BW of lambs in the EW group was significantly lower than BW in the

CON group ($p < 0.05$) after 21 days of age (Figure 1a). No difference was observed in ADG at 21 days of age between two groups of lambs ($p > 0.05$) (Figure 1b). The ADG of lambs in the EW group was significantly lower than ADG of those in the CON group at 28 ($p < 0.01$), 35 ($p < 0.001$), 42 ($p < 0.001$), and 49 ($p < 0.001$) days of age (Figure 1b).

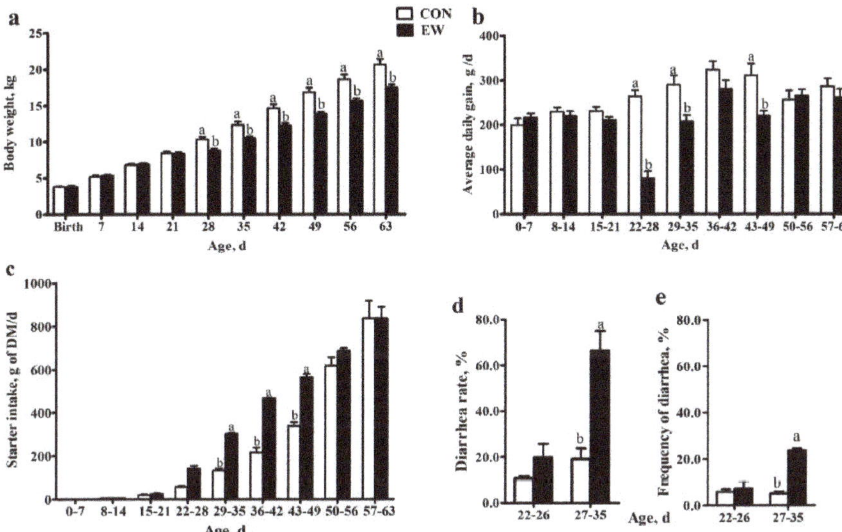

Figure 1. Body weight (**a**), average daily gain (**b**), starter intake (**c**), and diarrhea rate (**d,e**) for lambs weaned at day 21 as early weaned group (EW; black columns) or lambs weaned at day 49 as conventional weaned group (CON; white columns). Columns with different letters at a single time point indicate means that differed based on the means separation ($p < 0.05$). Error bars indicate SEM for the treatment × age interaction.

No difference was observed in starter intake from 8 to 28 days of age between the two groups of lambs ($p > 0.05$) (Figure 1c). The starter intake of lambs in EW group was significantly higher than that in CON group ($p < 0.05$) from 29 to 49 days of age (Figure 1c). No difference in starter intake between two groups of lambs was observed from 50 to 56 ($p > 0.05$) and 57 to 63 days of age ($p > 0.05$) (Figure 1c).

No diarrhea incidence was observed for any lamb from birth to 21 days of age. From 22 to 26 days of age, the diarrhea rate as well as frequency was not different between lambs in EW and CON group ($p > 0.05$) (Figure 1d). From day 27 to 35, the diarrhea rate and frequency of lambs in EW group was significantly higher than that in CON group ($p < 0.001$). No diarrhea incidence was observed after 35 days of age for lambs in both groups.

3.2. Slaughter Performance and Development of Visceral Organs

Both slaughter BW ($p = 0.010$) and hot carcass weight ($p = 0.007$) of lambs from EW group was significantly lower than that of lambs from CON group (Table 2). Dressing percentage decreased over time for both groups ($p < 0.001$) and tended to be lower in EW than CON group ($p = 0.065$). No change was found in the weight of whole stomach ($p = 0.705$), but abomasum relative to whole stomach weight ($p = 0.003$) was lower, while rumen ($p = 0.007$) and omasum ($p = 0.035$) relative to whole stomach weight was higher for lambs in EW than those in CON group. Reticulum relative to whole stomach weight was not different between two groups of lambs ($p = 0.821$). Both treatment and age interactively affected dressing percentage ($p = 0.021$) and reticulum relative to whole stomach weight ($p = 0.049$).

Table 2. Carcass weight, weight of rumen, reticulum, omasum, and abomasum relative to whole stomach weight on 26, 35, and 63 days of age [1].

Item	Treatment	Day of Age			SEM [2]	*p*-Value [2]		
		26	35	63		Treatment	Age	Treatment × Age
Slaughter BW, kg	CON	9.8 [bc]	12.3 [b]	22.2 [a]	0.779	0.010	<0.001	0.675
	EW	8.8 [c]	10.2 [bc]	20.0 [a]				
Hot carcass, kg	CON	4.7 [b]	5.9 [b]	10.2 [a]	0.443	0.007	<0.001	0.376
	EW	4.4 [b]	4.5 [b]	8.7 [a]				
Dressing percentage, %	CON	48.3 [ab]	47.6 [abc]	45.8 [bdc]	0.491	0.065	<0.001	0.021
	EW	49.8 [a]	44.4 [dc]	43.5 [d]				
Whole stomach, g	CON	135.5 [c]	237.0 [b]	566.6 [a]	21.400	0.705	<0.001	0.936
	EW	127.7 [c]	238.7 [b]	552.7 [a]				
relative to whole stomach weight, %								
Rumen	CON	45.1 [e]	55.5 [dc]	67.9 [ab]	1.784	0.007	<0.001	0.300
	EW	51.3 [de]	61.0 [bc]	68.9 [a]				
Reticulum	CON	7.8	9.1	9.9	0.502	0.821	0.258	0.049
	EW	9.1	8.7	8.7				
Omasum	CON	4.5 [bdc]	3.7 [d]	6.3 [ab]	0.434	0.035	<0.001	0.648
	EW	5.7 [abc]	4.1 [dc]	7.0 [a]				
Abomasum	CON	42.6 [a]	31.7 [b]	15.9 [c]	1.852	0.003	<0.001	0.102
	EW	33.9 [b]	26.2 [b]	15.4 [c]				

[1] Lambs weaned at day 21 as early weaned group (EW) or lambs weaned at day 49 as conventional weaned group (CON). Values are presented as means; n = 6 per group, BW= body weight. [2] Treatment = the effect of weaning time; age = the effect of age; treatment × age= the interaction between treatment and age. SEM = standard error of the mean. [a–d] Means within two rows (six values including age and treatment) with different superscripts differ (*p* < 0.05).

3.3. Rumen Fermentation Parameters

Both rumen pH (*p* = 0.007) and NH_3-N concentration (*p* < 0.001) increased over time and was not affected by treatment (*p* = 0.694 and 0.999, respectively) (Table 3). Total VFA concentration was higher in EW than CON group (*p* = 0.027). Molar proportion of acetate (*p* = 0.804), propionate (*p* = 0.154), butyrate (*p* = 0.171), isobutyrate (*p* = 0.991), isovalerate (*p* = 0.159), or acetate/propionate ratio (*p* = 0.253) was not affected by treatment. Molar proportion of valerate tended to be higher in the EW than that in the CON group (*p* = 0.098). Treatment and age interactively affected the NH_3-N concentration (*p* = 0.040) and molar proportion of isovalerate (*p* = 0.041).

Table 3. Rumen pH, ammonia nitrogen (NH_3-N), and volatile fatty acid (VFA) profiles of lambs at 26, 35, and 63 days of age [1].

Item	Treatment	Day of Age			SEM [2]	*p*-Value [2]		
		26	35	63		Treatment	Age	Treatment × Age
pH	CON	5.7 [b]	5.5 [b]	6.3 [a]	0.081	0.694	0.007	0.684
	EW	5.6 [b]	5.6 [b]	6.0 [a]				
NH_3-N, mg/100 mL	CON	10.6 [bc]	15.7 [ab]	17.5 [ab]	1.048	0.999	<0.001	0.040
	EW	8.1 [c]	12.9 [bc]	22.8 [a]				
Total VFA, mmol/L	CON	66.4 [b]	106.5 [ab]	92.9 [ab]	4.884	0.027	0.049	0.197
	EW	109.7 [ab]	122.0 [a]	97.6 [ab]				
Molar proportion of VFA, mol/100 mol								
Acetate	CON	50.9	50.1	52.8	1.234	0.804	0.676	0.340
	EW	49.3	55.8	50.2				
Propionate	CON	24.1	26.4	25.0	0.775	0.154	0.744	0.482
	EW	29.4	27.4	26.2				
Butyrate	CON	20.1	16.5	16.7	1.331	0.171	0.379	0.526
	EW	14.8	10.0	17.1				
Valerate	CON	1.9 [b]	5.3 [a]	3.1 [ab]	0.378	0.098	0.015	0.185
	EW	4.9 [ab]	5.6 [a]	3.5 [ab]				
Isobutyrate	CON	1.0 [ab]	0.8 [ab]	0.9 [ab]	0.075	0.991	0.033	0.068
	EW	0.8 [ab]	0.6 [b]	1.3 [a]				
Isovalerate	CON	1.9 [a]	0.9 [ab]	1.5 [ab]	0.135	0.159	0.01	0.041
	EW	0.7 [b]	0.7 [b]	1.8 [a]				
Acetate/Propionate ratio	CON	2.1	2.1	2.3	0.099	0.253	0.71	0.616
	EW	1.7	2.1	1.9				

[1] Lambs weaned at day 21 as early weaned group (EW) or lambs weaned at day 49 as conventional weaned group (CON). Values are presented as means; n = 6 per group. [2] Treatment = the effect of weaning time; age = the effect of age; treatment × age = the interaction between treatment and age. SEM = standard error of the mean. [a–c] Means within two rows (six values including age and treatment) with different superscripts differ (*p* < 0.05).

3.4. Serum Parameters

The concentration of IgG ($p = 0.761$), IgA ($p = 0.809$) or IgM ($p = 0.889$) was not affected by treatment (Table 4). The concentration of IgG ($p < 0.001$) was lower, while that of IgA ($p < 0.001$) and IgM ($p < 0.001$) was higher at 63 than 26 days of age for both groups. No difference was observed in activity of SOD ($p = 0.839$), GSH-Px ($p = 0.290$), CAT ($p = 0.145$), T-AOC ($p = 0.605$), or MDA ($p = 0.464$) between the two groups of lambs (Table 4). Activity of SOD ($p < 0.001$), GSH-Px ($p < 0.001$), CAT ($p < 0.001$), and T-AOC ($p < 0.001$) was higher, while that of MDA ($p < 0.001$) was lower at 35 than those at 26 or 63 days of age for both groups. There was no difference in the concentration of IL-1β ($p = 0.659$), IL-4 ($p = 0.361$), IL-6 ($p = 0.085$), cortisol ($p = 0.946$), IFN-γ ($p = 0.843$), or TNF-α ($p = 0.927$) between the two groups of lambs. The concentration of IL-1β ($p < 0.001$), IL-4 ($p < 0.001$), IL-6 ($p < 0.001$), cortisol ($p < 0.001$), IFN-γ ($p < 0.001$), and TNF-α ($p < 0.001$) were higher, while that of MDA ($p < 0.001$) was lower at 35 than those in 26 or 63 days of age for both groups. There was no interactive effect of treatment and age on any serum parameter ($p > 0.05$).

Table 4. Serum parameters of lambs at 26, 35, and 63 days of age [1].

Item [2]	Treatment	Day of Age 26	Day of Age 35	Day of Age 63	SEM [3]	p-Value [3] Treatment	p-Value [3] Age	p-Value [3] Treatment × Age
IgG, g/L	CON	24.0 [a]	22.5 [a]	19.0 [b]	0.422	0.761	<0.001	0.454
	EW	22.8 [a]	22.8 [a]	19.4 [b]				
IgA, g/L	CON	0.5 [c]	0.6 [bc]	0.7 [a]	0.014	0.809	<0.001	0.658
	EW	0.6 [c]	0.6 [c]	0.7 [ab]				
IgM, g/L	CON	1.1 [b]	1.1 [b]	1.3 [a]	0.026	0.889	<0.001	0.046
	EW	1.1 [b]	1. 1 [b]	1.3 [a]				
SOD, U/mL	CON	93.5 [b]	111.5 [a]	94.8 [b]	1.548	0.839	<0.001	0.556
	EW	96.7 [b]	110.4 [a]	93.8 [b]				
GSH-Px, μmol/L	CON	754.7 [bc]	818.6 [a]	732.4 [c]	8.645	0.290	<0.001	0.199
	EW	712.7 [c]	811.1 [ab]	743.4 [c]				
CAT, U/mL	CON	7.6 [c]	8.9 [a]	7.8 [bc]	0.118	0.145	<0.001	0.844
	EW	7.5 [c]	8.6 [ab]	7.6 [c]				
T-AOC, U/mL	CON	8.4 [c]	9.8 [a]	8.8 [bc]	0.115	0.605	<0.001	0.134
	EW	8.9 [bc]	9.5 [ab]	8.9 [bc]				
MDA, nmol/mL	CON	5.3 [a]	4.3 [b]	5.3 [a]	0.112	0.464	<0.001	0.948
	EW	5.5 [a]	4.4 [b]	5.4 [a]				
Cortisol, μg/dL	CON	1.5 [b]	2.7 [a]	1.6 [b]	0.107	0.946	<0.001	0.695
	EW	1.6 [b]	2.6 [a]	1.5 [b]				
IL-1β, pg/mL	CON	98.9 [b]	111.4 [a]	99.7 [b]	1.534	0.659	<0.001	0.471
	EW	97.1 [b]	114.0 [a]	96.1 [b]				
IL-4, pg/mL	CON	13.3 [c]	14.3 [ab]	13.2 [c]	0.124	0.361	<0.001	0.850
	EW	13.4 [bc]	14.6 [a]	13.3 [c]				
IL-6, pg/mL	CON	45.9 [c]	54.2 [ab]	45.8 [c]	1.055	0.085	<0.001	0.356
	EW	47.5 [bc]	59.2 [a]	46.2 [c]				
IFN-γ, pg/mL	CON	143.2 [b]	159.0 [a]	141.9 [b]	1.443	0.843	<0.001	0.772
	EW	144.2 [b]	157.1 [a]	141.9 [b]				
TNF-α, pg/mL	CON	54.4 [b]	67.5 [a]	55.7 [b]	1.133	0.927	<0.001	0.549
	EW	56.4 [b]	66.3 [a]	54.5 [b]				

[1] Values are presented as means; n = 6 per group. Lambs weaned at day 21 as early weaned group (EW) or lambs weaned at day 49 as conventional weaned group (CON). [2] IgG = immunoglobulin G, IgM = immunoglobulin M, IgA = immunoglobulin A, SOD = superoxide dismutase, GSH-Px = Glutathione Peroxidase, CAT = catalase, T-AOC = total antioxidative capacity, MDA = malondialdehyde IL-1β = interleukin 1β, IL-4 = interleukin 4, IL-6 = interleukin 6, IFN-γ = interferon-gamma, TNF-α = tumor necrosis factor α. [3] Treatment = the effect of weaning time; age = the effect of age; treatment × age = the interaction between treatment and age. SEM = standard error of the mean. [a–c] Means within two rows (six values including age and treatment) with different superscripts differ ($p < 0.05$).

4. Discussion

It has been generally accepted that intake of solid feed before weaning stimulates rumen development and functionality [21,22]. Early starter feeding (7 days of age) was reported to increase ruminal VFA concentration in lambs compared with late starter feeding (42 days of age) [23]. Therefore, lambs were provided starter at 7 days of age in the current study. However, a sharp decrease in ADG (from 21 to 28 days of age) of lambs weaned at 21 days of age was not prevented. This might be partly due to the short time for ewe-lamb separation (6 h daily), which was insufficient for lambs to get used to the abrupt separation from their dams at 21 days of age. The final BW of lambs weaned at 21 days of age was still lower than that of their counterparts weaned at 49 days of age, which again proved that early-life stress may have a long-lasting negative effect on the performance of lambs. For example, it is suggested that lambs subjected to nutrient restriction at 15 days of age followed by realimentation at 60 days of age still physically lag behind those continuously provided sufficient nutrients, featured by a lower BW at 90 days of age (24.1 vs. 26.1 kg) [24]. On the other hand, we found that the growth performance of lambs weaned and separated from their dams at 49 days of age (>20 kg of BW at 60 days of age) was better than those weaned and separated from their dams at 60 days of age (12.8 kg of BW at 60 days of age) [25]. This discrepancy might be explained by the difference in time of introduction of starter to lambs between current (day 7) and our previous study (day 15), as well as the higher protein and energy level of starter used in the current study than our previous study [25].

Post-weaning diarrhea is an economically important enteric disease due to financial losses [26]. In our present study, diarrhea rate and frequency were increased in the EW group due to early weaning. Similarly, this disease occurs most frequently within the two weeks after weaning and is characterized by a profuse diarrhea, dehydration, significant mortality, and loss of body weight of pigs [27]. Our previous study showed a sudden change in the feeding system due to early weaning resulted in changes of the intestinal morphology and functions [28], that may trigger reductions in digestive and absorptive capacities. Studies suggest that increasing weaning age reduces stress associated with this period and allows animals to have a more mature gastrointestinal tract and become increasingly familiar with solid feed during milk-period with an improvement in growth performance [29,30].

Ruminal pH is crucial for normal development, fermentation, and overall lamb health. Generally, rumen fluid pH is influenced by the rate of fermentation and absorption of VFA, which in turn, are affected by passage rate of digesta and the buffering capacity of the rumen contents [31]. In this experiment, rumen pH of lambs on day 26 or day 35 were lower than day 63 and not affected by weaning time, mainly due to the difference in starter intake between pre-and post-weaning. The increase in pH with age may be explained by the continuous development of rumen and starter intake. Furthermore, when the rumen epithelium is still underdeveloped, early starter feeding in lambs can increase ruminal VFA concentrations and affect the expression levels of genes involved in VFA transport and pH regulation in the ruminal epithelium [23].

Previous studies showed that ruminal VFA concentration increased with increasing solid feed intake and was affected by either dietary (e.g., nutrient fractionation) or animal factors (e.g., absorption and passage rate) [6]. Rumen development is not only driven by the presence of VFA, but also by the nutrients supplied from liquid feed [6]. In our study, weaning and feeding starter to lambs at day 21 promoted the development of the rumen, which is in accordance with the results in previous studies as providing solid feed as early as possible is beneficial to the rapid development of the rumen [32,33]. In the current study, the concentration of Total VFA, molar proportion of acetate and propionate in the EW group increased only at 5 days after weaning. This is because more starter intake in the EW group promoted rumen development, while the CON lambs were mainly fed milk. The consumption of starter in lambs has been shown to increase the ruminal concentrations of VFAs, which have positive effects on the development of ruminal epithelium [23]. This result is similar with Yang et al. [34] who showed no difference in Total VFA concentrations in lambs between pre- and post-weaning.

The effect of passive immunity on mortality, morbidity, and subsequent growth of newborn animals is receiving increasing attention [35]. Among three major immunoglobulin isotypes, namely IgG, IgM, and IgA, IgG usually has the highest concentration in serum and is the main antibody for humoral immunity. A previous study showed that serum IgG concentration was positively associated with calf health, as the morbidity and intensity of disease were the lowest in heifer calves with serum IgG concentration exceeding 10 g/L at 30 to 60 h of life [36]. In our study, the IgG concentration was higher than that of lambs (same age) reported by Hernández-Castellano et al. [37], which may due to the difference in sheep species or different test method of immunodiffusion, as a recent review suggested that the concentration of IgG may differ between methods such as ELISA or radial immunodiffusion [35]. In the present study, no change was found in the concentration of IgG, IgA, or IgM. Similarly, our previous study also showed no difference in serum IgG and IgA between lambs differing in weaning time [25]. Those results are in agreement with Hernández-Castellano et al. [37] who found no difference in the IgG concentration pre- and post-weaning, which may be due to the beginning of the endogenous immunoglobulin production by lambs and increase in the concentration of IgG over time. The authors also found the concentrations of IgG and IgM in blood of lambs decreased after birth with aging until the end of milk feeding period and then increased 30 days after weaning, and this phenomenon is not affected by milk sources (nursing, milk replacer, or cow milk) [37]. In our study, during the first few days of birth, serum immunoglobulin may typically reflect the efficacy of passive immunity transfer through adequate and timely consumption of colostrum. Later on, as feed intake increased, the systematic humoral immune function of lambs developed.

GSH-Px is an antioxidant enzyme that helps to control the levels of hydrogen peroxide and lipid peroxides under normal metabolic conditions. T-AOC indicates the oxidation resistance capacity of the whole body. MDA is one of the final products of polyunsaturated fatty acid peroxidation in cells and is considered as a marker of oxidative stress. The insignificant difference of GSH-Px, T-AOC, and MDA between lambs in different treatments suggested that the antioxidant status was not affected by weaning time. The activity of SOD, GSH-Px, CAT, and T-AOC was the highest, while that of MDA concentration was the lowest at 35 days of age for both groups, suggesting that antioxidant status improved with age. It was reported that weaning induced lower activities of antioxidant enzymes and a greater content of MDA, reflecting an imbalance of redox status in piglets [38]. However, Buchet et al. [39] reported that the observed effects are related with weaning but not with age.

In our study, no changes in the concentrations of cortisol were observed in treatment. Previous study suggested that the plasma concentration of cortisol increased in calves when they were subjected to weaning [40]. On the contrary, it was reported that the change in plasma cortisol concentration was not affected by weaning [41], and a similar result was observed in lambs [42] and piglets [43]. This may be related to the sampling time after weaning, because inflammatory response is a short-term change that only occurs within a few hours after weaning and does not last for a long time [40,44].

Several cytokines biomarkers such as IL-1β, IFN-γ, and TNF-α were used to evaluate stress and inflammatory response before and after weaning [40,42]. Early weaning can stimulate the secretion of proinflammatory cytokines such as IL-1, TNF-α, and stimulate inflammatory response [42]. Yang et al. [45] found that plasma IFN-γ declined with age during weaning period in calves. TNF-α often work synergistically with IFN-γ and IL-1β to improve immune responses [46]. Study in lambs have shown that serum TNF-α concentration was not affected by early weaning or age [25,42], and are responsive to proinflammatory cytokine regulation due to early weaning. In our study, no changes in the concentrations of IL-1β, IL-4, IL-6, IFN-γ, or TNF-α were observed in response to weaning time. Therefore, more sampling time is needed in future study to validate the findings in the current experiment.

5. Conclusions

The results indicate that early starter feeding combined with limited ewe-lamb contact promoted rumen development and increased Total VFA concentration, without compromising lambs' overall immunity weaning at 21 days of age. However, the increase in diarrhea rate and decrease in ADG in lambs weaned at 21 days of age suggested that a long-term effect of weaning on overall performance as well as development in gut microbiota and barrier function needs to be assessed.

Author Contributions: Conceptualization, Q.D. and Y.T.; methodology, S.W. and G.Z.; data curation, S.W.; software, S.W. and Y.B.; formal analysis, S.W.; investigation, S.W.; resources, S.W. and T.M.; writing—original draft, K.C. and H.D.; writing—review and editing, Q.D. and Y.T.; supervision, N.Z. and F.L.; project administration, S.W. and G.Z.; funding acquisition.

Funding: This work was supported by the earmarked fund for China Agriculture Research System (CARS-38), National Natural Science Foundation of China (31660670), The Agricultural Science and Technology Innovation Program (CAAS-ASTIP-2017-FRI-04), and Double first-class Funding-International Cooperation and Exchange Program (227000-560001).

Acknowledgments: We thank students in the ruminant nutrition laboratory of Feed Research Institute of Chinese Academy of Agricultural Sciences (CAAS) for their technical assistance. We thank Linqing Runlin Sheep Industry Co. Ltd. (Shangdong Province, China) for their assistance with the animal experiment.

Conflicts of Interest: The authors declare no conflict of interest. The funders had no role in the design of the study; in the collection, analyses, or interpretation of data; in the writing of the manuscript, or in the decision to publish the results.

References

1. Jagusch, K.T.; Clark, V.R.; Jay, N.P. Lamb production from animals weaned at 3 to 5 weeks of age on to lucerne. *New Zeal J. Agr. Res.* **1970**, *13*, 808–814. [CrossRef]
2. Cheng, C.S.; Wei, H.K.; Wang, P.; Yu, H.C.; Zhang, X.M.; Jiang, S.W.; Peng, J. Early intervention with faecal microbiota transplantation: An effective means to improve growth performance and the intestinal development of suckling piglets. *Animal* **2019**, *13*, 533–541. [CrossRef] [PubMed]
3. He, X.; Lu, Z.; Ma, B.; Zhang, L.; Li, J.; Jiang, Y.; Zhou, G.; Gao, F. Effects of chronic heat exposure on growth performance, intestinal epithelial histology, appetite-related hormones and genes expression in broilers. *J. Sci. Food Agric.* **2018**, *98*, 4471–4478. [CrossRef] [PubMed]
4. Orgeur, P.; Bernard, S.; Naciri, M.; Nowak, R.; Schaal, B.; Levy, F. Psychobiological consequences of two different weaning methods in sheep. *Reprod. Nutr. Dev.* **1999**, *39*, 231–244. [CrossRef] [PubMed]
5. Hinch, G.N.; Lynch, J.J.; Elwin, R.L.; Green, G.C. Long term associations between Merino ewes and their offspring. *Appl. Anim. Behav. Sci.* **1990**, *27*, 93–103. [CrossRef]
6. Khan, M.A.; Bach, A.; Weary, D.M.; von Keyserlingk, M.A.G. Invited review: Transitioning from milk to solid feed in dairy heifers. *J. Dairy Sci.* **2016**, *99*, 885–902. [CrossRef]
7. Demir, H. Investigations on growing Kivircik lambs weaned at different times. *Vet. Fak. Derg. Istambul* **1995**, *21*, 142–150.
8. Baldwin, R.L.; McLeod, K.R.; Klotz, J.L.; Heitmann, R.N. Rumen development, intestinal growth and hepatic metabolism in the pre- and postweaning ruminant. *J. Dairy Sci.* **2004**, *87*, E55–E65. [CrossRef]
9. McKusick, B.C.; Thomas, D.L.; Berger, Y.M. Effect of weaning system on commercial milk production and lamb growth of East Friesian dairy sheep. *J. Dairy Sci.* **2001**, *84*, 1660–1668. [CrossRef]
10. Chai, J.; Diao, Q.Y.; Wang, H.C.; Tu, Y.; Tao, X.J.; Zhang, N.F. Effects of weaning age on growth, nutrient digestibility and metabolism, and serum parameters in Hu lambs. *Anim. Nutr.* **2015**, *1*, 344–348. [CrossRef]
11. Zhang, Q.; Li, C.; Niu, X.L.; Zhang, Z.F.; Li, F.; Li, F.D. The effects of milk replacer allowance and weaning age on the performance, nutrients digestibility, and ruminal microbiota communities of lambs. *Anim. Feed. Sci. Technol.* **2019**, *257*, 1–9. [CrossRef]
12. Bimczok, D.; Röhl, F.W.; Ganter, M. Evaluation of lamb performance and costs in motherless rearing of German Grey Heath sheep under field conditions using automatic feeding systems. *Small Rumin Res.* **2005**, *60*, 255–265. [CrossRef]
13. National Research Council (NRC). *Nutrient Requirements of Small Ruminants: Sheep, Goat, Cervids, and New World Camelids*; National Academy Press: Washington, WA, USA, 2007.
14. AOAC (Association of Official Analytical Chemists). *Official Methods of Analysis*, 15th ed.; AOAC: Arlington, VA, USA, 1990.
15. Marshall, C.M.; Walker, A.F. Comparison of a short method for Kjeldahl digestion using a trace of selenium as catalyst, with other methods. *J. Sci. Food Agric.* **1978**, *29*, 940–942. [CrossRef]
16. Van Soest, P.J.; Robertson, J.B.; Lewis, B.A. Methods for dietary fiber, neutral detergent fiber, and nonstarch polysaccharides in relation to animal nutrition. *J. Dairy Sci.* **1991**, *74*, 3583–3597. [CrossRef]
17. Goering, H.G.; Van Soest, J.P. *Forage Fiber Analysis. Agricultural Handbook*; UPSDA: Washington, DC, USA, 1970.
18. Kargar, S.; Mousavi, F.; Karimi-Dehkordi, S. Effects of chromium supplementation on weight gain, feeding behaviour, health and metabolic criteria of environmentally heat-loaded Holstein dairy calves from birth to weaning. *Arch. Anim. Nutr.* **2018**, *72*, 443–457. [CrossRef]

19. Zhang, R.; Zhou, M.; Tu, Y.; Zhang, N.F.; Deng, K.D.; Ma, T.; Diao, Q.Y. Effect of oral administration of probiotics on growth performance, apparent nutrient digestibility and stress-related indicators in Holstein calves. *J. Anim. Physiol. Anim. Nutr. (Berl)* **2016**, *100*, 33–38. [CrossRef]

20. Broderick, G.; Kang, J. Automated stimultaneous determination of ammonia and total amino acids in ruminal fluid and in vitro media. *J. Dairy Sci.* **1980**, *63*, 64–75. [CrossRef]

21. Spencer, J.D.; Boyd, R.D.; Cabrera, R.; Allee, G.L. Early weaning to reduce tissue mobilization in lactating sows and milk supplementation to enhance pig weaning weight during extreme heat stress. *J. Anim. Sci.* **2003**, *81*, 2041–2052. [CrossRef]

22. Jesse, B.W. *Energy Metabolism in the Developing Rumen Epithelium*; Elsevier: London, UK, 2005; Volume 3.

23. Liu, T.; Li, F.; Wang, W.M.; Yue, X.P.; Li, F.; Li, C.; Pan, X.Y.; Mo, F.T.; Wang, F.B.; La, Y.F.; et al. Effects of lamb early starter feeding on the expression of genes involved in volatile fatty acid transport and pH regulation in rumen tissue. *Anim. Feed Sci. Technol.* **2016**, *217*, 27–35. [CrossRef]

24. Ma, T.; Wang, B.; Zhang, N.F.; Tu, Y.; Si, B.W.; Cui, K.; Qi, M.L.; Diao, Q.Y. Effect of protein restriction followed by realimentation on growth, nutrient digestibility, ruminal parameters, and transporter gene expression in lambs. *Anim. Feed Sci. Technol.* **2017**, *231*, 19–28. [CrossRef]

25. Chai, J.M.; Ma, T.; Wang, H.C.; Qi, M.L.; Tu, Y.; Diao, Q.Y.; Zhang, N.F. Effect of early weaning age on growth performance, nutrient digestibility, and serum parameters of lambs. *Anim. Prod. Sci.* **2017**, *57*, 110–115. [CrossRef]

26. Amezcua, R.; Friendship, R.M.; Dewey, C.E.; Gyles, C.; Fairbrother, J.M. Presentation of postweaning Escherichia coli diarrhea in southern Ontario, prevalence of hemolytic E-coli serogroups involved, and their antimicrobial resistance patterns. *Can. J. Vet. Res.* **2002**, *66*, 73–78. [PubMed]

27. Fairbrother, J.M.; Nadeau, É.; Gyles, C.L. Escherichia coli in postweaning diarrhea in pigs: An update on bacterial types, pathogenesis, and prevention strategies. *Anim. Health Res. Rev.* **2007**, *6*, 17–39. [CrossRef]

28. Cui, K.; Wang, B.; Zhang, N.F.; Tu, Y.; Ma, T.; Diao, Q.Y. iTRAQ-based quantitative proteomic analysis of alterations in the intestine of Hu sheep under weaning stress. *PLoS ONE* **2018**, *13*, e0200680. [CrossRef] [PubMed]

29. Main, R.G.; Dritz, S.S.; Tokach, M.D.; Goodband, R.D.; Nelssen, J.L. Increasing weaning age improves pig performance in a multisite production system. *J. Anim. Sci.* **2004**, *82*, 1499–1507. [CrossRef]

30. Campbell, J.M.; Crenshaw, J.D.; Polo, J. The biological stress of early weaned piglets. *J. Anim. Sci. Biotechnol.* **2013**, *4*, 19. [CrossRef]

31. Pazoki, A.; Ghorbani, G.R.; Kargar, S.; Sadeghi-Sefidmazgi, A.; Drackley, J.K.; Ghaffari, M.H. Growth performance, nutrient digestibility, ruminal fermentation, and rumen development of calves during transition from liquid to solid feed: Effects of physical form of starter feed and forage provision. *Anim. Feed Sci. Technol.* **2017**, *234*, 173–185. [CrossRef]

32. Liu, J.; Bian, G.; Sun, D.; Zhu, W.; Mao, S. Starter feeding altered ruminal epithelial bacterial communities and some key immune-related genes' expression before weaning in lambs. *J. Anim. Sci.* **2017**, *95*, 910–921. [CrossRef]

33. Liu, J.; Bian, G.; Sun, D.; Zhu, W.; Mao, S. Starter Feeding Supplementation Alters Colonic Mucosal Bacterial Communities and Modulates Mucosal Immune Homeostasis in Newborn Lambs. *Front. Microbiol.* **2017**, *8*, 429. [CrossRef]

34. Yang, B.; He, B.; Wang, S.S.; Liu, J.X.; Wang, J.K. Early supplementation of starter pellets with alfalfa improves the performance of pre- and postweaning Hu lambs. *J. Anim. Sci.* **2015**, *93*, 4984–4994. [CrossRef]

35. McGee, M.; Earley, B. Review: Passive immunity in beef-suckler calves. *Animal* **2019**, *13*, 810–825. [CrossRef] [PubMed]

36. Furman-Fratczak, K.; Rzasa, A.; Stefaniak, T. The influence of colostral immunoglobulin concentration in heifer calves' serum on their health and growth. *J. Dairy Sci.* **2011**, *94*, 5536–5543. [CrossRef] [PubMed]

37. Hernández-Castellano, L.E.; Moreno-Indias, I.; Morales-delaNuez, A.; Sánchez-Macías, D.; Torres, A.; Capote, J.; Argüello, A.; Castro, N. The effect of milk source on body weight and immune status of lambs. *Livest. Sci.* **2015**, *175*, 70–76. [CrossRef]

38. Cao, S.T.; Wang, C.C.; Wu, H.; Zhang, Q.H.; Jiao, L.F.; Hu, C.H. Weaning disrupts intestinal antioxidant status, impairs intestinal barrier and mitochondrial function, and triggers mitophagy in piglets. *J. Anim. Sci.* **2018**, *96*, 1073–1083. [CrossRef] [PubMed]

39. Buchet, A.; Belloc, C.; Leblanc-Maridor, M.; Merlot, E. Effects of age and weaning conditions on blood indicators of oxidative status in pigs. *PLoS ONE* **2017**, *12*, e0178487. [CrossRef]

40. O'Loughlin, A.; McGee, M.; Doyle, S.; Earley, B. Biomarker responses to weaning stress in beef calves. *Res. Vet. Sci.* **2014**, *97*, 458–463. [CrossRef]

41. Hickey, M.C.; Drennan, M.; Earley, B. The effect of abrupt weaning of suckler calves on the plasma concentrations of cortisol, catecholamines, leukocytes, acute-phase proteins and in vitro interferon-gamma production. *J. Anim. Sci.* **2003**, *81*, 2847–2855. [CrossRef]

42. Zhang, Q.; Li, C.; Niu, X.L.; Zhang, Z.A.; Li, F.; Li, F.D. An intensive milk replacer feeding program benefits immune response and intestinal microbiota of lambs during weaning. *BMC Vet. Res.* **2018**, *14*, 366. [CrossRef]

43. Turpin, D.L.; Langendijk, P.; Chen, T.Y.; Pluske, J.R. Intermittent Suckling in Combination with an Older Weaning Age Improves Growth, Feed Intake and Aspects of Gastrointestinal Tract Carbohydrate Absorption in Pigs after Weaning. *Animals* **2016**, *6*, 66. [CrossRef]

44. Hulbert, L.E.; Moisa, S.J. Stress, immunity, and the management of calves. *J. Dairy Sci.* **2016**, *99*, 3199–3216. [CrossRef]

45. Yang, C.; Zhang, L.; Cao, G.; Feng, J.; Yue, M.; Xu, Y.; Dai, B.; Han, Q.; Guo, X. Effects of dietary supplementation with essential oils and organic acids on the growth performance, immune system, fecal volatile fatty acids, and microflora community in weaned piglets. *J. Anim. Sci.* **2019**, *97*, 133–143. [CrossRef] [PubMed]

46. Schroder, K.; Hertzog, P.J.; Ravasi, T.; Hume, D.A. Interferon-gamma: An overview of signals, mechanisms and functions. *J. Leukoc. Biol.* **2004**, *75*, 163–189. [CrossRef] [PubMed]

Article

Effects of Mannan Oligosaccharides on Gas Emission, Protein and Energy Utilization, and Fasting Metabolism in Sheep

Chen Zheng [1], Junjun Ma [1], Ting Liu [1,*], Bingdong Wei [2] and Huaming Yang [2]

[1]   College of Animal Science and Technology, Gansu Agricultural University, No. 1 Yingmen Village, Anning District, Lanzhou 730070, China; zhengc@gsau.edu.cn (C.Z.); majunjungsau@gmail.com (J.M.)
[2]   Branch of Animal Husbandry, Jilin Academy of Agricultural Sciences, No.186 East Xinhua Road, Gongzhuling 136100, China; weibingdong@dlut.edu.cn (B.W.); yhmjl@163.com (H.Y.)
*   Correspondence: liuting@gsau.edu.cn; Tel.: +86-931-763-1225

Received: 16 August 2019; Accepted: 26 September 2019; Published: 28 September 2019

Simple Summary: Mannan oligosaccharides (MOS) are a promising feed additive to improve animal health, immune capacity, and antioxidation. Based on the previous studies, we carried out three experiments to investigate the effects of MOS on the gas emission, protein and energy utilization, and fasting metabolism of sheep. The results showed that 2.0% MOS supplementation led to the lowest *in vitro* CO_2 production and lower CH_4 production and decreased *in vivo* intake. However, it also decreased urine nitrogen excretion and energy released as CH_4, and then improved the utilization of crude protein and energy of sheep. There were no differences in the parameters of respiration and energy metabolism of sheep under the fasting condition. The findings indicated that MOS slightly affected the gas emission and nutrients and energy utilization of sheep.

Abstract: This study investigated the effects of mannan oligosaccharides (MOS) on *in vitro* and *in vivo* gas emission, utilization of crude protein (CP) and energy, and relative parameters of sheep under fasting metabolism conditions. *In vitro* gas productions were evaluated over 12 h in sheep diets containing different amounts of MOS (from 0% to 6.0%/kg, the increment was 0.5%). A control experiment was used to assess the gas emission, utilization of CP and energy, and fasting metabolism in control sheep and sheep treated with 2.0% MOS over 24 days (d). The results showed that 2.0% MOS supplementation led to the lowest *in vitro* CO_2 production and less CH_4 production, while also leading to decrease *in vivo* nutrients intake, CP and energy excretion, digested and retained CP, and energy released as CH_4 ($p < 0.05$). Furthermore, 2.0% MOS supplementation appeared to decrease *in vivo* O_2 consumption and CH_4 production per metabolic body weight ($BW^{0.75}$), and increase the CP retention rate of sheep ($p < 0.074$). MOS did not affect other parameters, along with the same parameters of sheep under fasting metabolism conditions ($p > 0.05$). The findings indicate MOS has only slight effects on the gas emission and nutrients and energy metabolism of sheep.

Keywords: energy; gas; mannan oligosaccharides; protein; sheep

---

1. Introduction

Prebiotics are non-digestible food ingredients that benefit the host by selectively stimulating the activity of one or a limited number of bacteria in the intestine [1,2]. The predominant prebiotics include oligosaccharides such as fructooligosaccharides (FOS), inulin, mannan oligosaccharides (MOS), and xylooligosaccharides [1]. They have been termed a nutricine, meaning that they have no direct nutritive value, but maintain intestinal digestive and absorptive functions, thus improving the health and performance of farmed animals [3].

MOS are found in the outer layer of yeast, *Saccharomyces cerevisiae*, in the cell wall. As the most bioactive compounds, they contain both mannan proteins and complex carbohydrates, including β-glucans [4], and are widely used as a dietary supplement to boost the immune system and eliminate pathogens from the intestinal tract [5,6]. For instance, MOS bind to the mannose-specific lectin of Gram-negative pathogens that express Type-1 fimbriae (e.g., *Escherichia coli*), resulting in their excretion from the intestine [7]. The limited research focusing on MOS in ruminants found that MOS improved the immunoglobulin G (IgG) concentration of lamb's blood [8], enhanced the health of the ruminal epithelium of sheep [9], improved the colostrum quantity of cows [4], and increased the antioxidant capacity of sheep [10]. In previous study, β 1-4 galacto-oligosaccharides (GOS) reduced the CH_4 production in sheep [11]; however, very few scientists have studied the effect of MOS on the energy metabolism and gas emission of ruminants. So, we hypothesized that MOS could adjust the CO_2 and CH_4 emission in the rumen and improve the energy utilization of sheep. The purpose of this study was to investigate the effects of MOS on *in vitro* and *in vivo* gas production, respiration, protein digestion, energy metabolism, and fasting metabolism in sheep.

2. Materials and Methods

2.1. Experimental Design, Animals, and Housing

All experiments in this study were carried out in accordance with the approved guidelines of the Regulation of the Standing Committee of Gansu People's Congress. All experimental protocols and the collection of samples were approved by the Ethics Committee of Gansu Agriculture University under permission no. DK-005.

A single-factor experimental design was used for the *in vitro* gas production experiments. Thirteen different doses of MOS (Bio-Mos®, Alltech, Nicholasville, KY, USA) were tested: 0, 0.5, 1.0, 1.5, 2.0, 2.5, 3.0, 3.5, 4.0, 4.5, 5.0, 5.5, and 6.0%/kg of basal diet (as fed basis). Each treatment was repeated three times. Four Chinese Northeast Merino rams with an eternal ruminal fistula and similar body weights (64.82 ± 3.17 kg) were used as ruminal fluid donors, and were fed basal diet and water ad libitum. An automatic recording device of trace gas production with six channels (Branch of Animal Husbandry, Jilin Academy of Agricultural Sciences, Gongzhuling, Jilin Province, China) equipped with gas flowmeters (CH_4 and CO_2, Sensors, Inc., Saline, MI, USA; H_2, CITY Technology, Inc., Hampshire, UK) was used to measure the gas flow. The *in vitro* tests lasted for 12 h.

A control experiment was used to assess *in vivo* CO_2, CH_4, and NH_3 production, O_2 consumption, urine nitrogen excretion, total heat output, and the apparent digestibility and retention rates of crude protein (CP) and energy in control sheep and sheep treated with 2.0% MOS (because adding 2.0% MOS to the basal diet led to the lowest CO_2 and lower CH_4 production in an *in vitro* experiment; then, this dose was chosen to apply in an *in vivo* experiment). The experimental group consisted of eight healthy rams (*Dorper ♂× Small tail Han-yang ♀*; four rams per group) with similar body weights (42.50 ± 3.24 kg). The test period included a 14-day (d) acclimation period, a 6-d digestive and metabolic experiment, a 60-hour (h) fasting acclimation period (the respiratory quotient (RQ) approximately was below 0.71 when sheep fasted for over 60 h), and a 24-h fasting metabolism experiment. All rams were housed individually in a respiratory chamber (Branch of Animal Husbandry, Jilin Academy of Agricultural Sciences, Gongzhuling, Jilin Province, China). Each chamber was equipped with a feeder and a drinker that provided ad libitum access to feed and water. Each chamber was also equipped with automatic gas flowmeters (CH_4 and CO_2, Sensors, Inc., Saline, MI, USA; O_2, Advanced Micro Instruments, Inc., Costa Mesa, CA, USA; NH_3, Industrial Health & Safety Instrumentation, Inc., St. Petersburg, FL, USA). Before rams went into the respiratory chamber, the 6.14-L calibrating CO_2 was pumped into the respiratory chamber to test the recovery rate, and repeated three times; the result of the gas recovery tests was 98.22% (Table S1). The data on gas production and O_2 consumption were calibrated by the gas recovery rate of each chamber.

2.2. Experimental Diets

The basal diet was formulated to meet or exceed the recommendations for all nutrients of ram sheep under China Agricultural Industry Standard NY/T816-2004 (Table 1).

Table 1. Ingredient and chemical composition of the basal diet.

Items	Concentrate
Ingredient (%)	
Corn	24.90
Soybean meal	12.72
Chinese wildrye	60.00
Calcium hydrophosphate	0.78
Limestone	0.80
Salt	0.40
Additive premix [1]	0.40
Chemical composition (%) [2]	
DM	85.71
DE (MJ/kg) [3]	9.25
CP	9.46
DE/CP (MJ/g)	0.10
NDF	46.06
ADF	23.75
Ca	0.65
P	0.40

[1] Additive premix includes mineral elements (mg/kg): S, 200; Fe, 25; Zn, 40; Cu, 8; I, 0.3; Mn, 40; Se, 0.2; Co, 0.1; vitamins (IU/kg): vitamin A, 940; vitamin E, 20. [2] Concentrations of dry matter (DM), crude protein (CP), Ca, and P were measured in accordance with AOAC (DM: method 930.15, CP: method 990.03, Ca: method 978.02, P: method 946.06) [12], and concentrations of neutral detergent fiber (NDF) and acid detergent fiber (ADF) were measured in accordance with Goering and Soest [13]. [3] DE = digestible energy.

2.3. In Vitro Procedures

Rumen contents from each donor sheep were obtained immediately before morning feeding by soft tube and syringe via ruminal fistula and strained through four layers of cheesecloth. Then, the fluid from each sheep was pooled and mixed with culture medium solution in a 1:2 ratio (vol/vol) at 39 °C in accordance with the description by Menke et al. [14], and the mixed solution was also constantly filled with N_2 to maintain anaerobic condition. Before incubation, in order to eliminate the metering error, 0.2843% calibrating CO_2 was pumped into the channel, and then the gas flowmeter was calibrated to 0.2843%. All *in vitro* cultures contained 2 g of experimental diets, which were carefully weighed into the fermentation channel of the automatic recording device of trace gas production, while the control channel did not receive any diet in order to calibrate the gas production from the ruminal fluid. Additionally, different doses of MOS were added into the channel in terms of experimental treatment. Then, each channel received 150 mL of mixed solution (including 50 mL of ruminal fluid and 100 mL of culture medium solution), and was maintained at 39 °C for 12 h of incubating. Since there were six channels for incubation, the culture was divided into nine periods, which meant that control, 0%, 0.5%, 1.0%, 1.5%, and 2.0% MOS treatments were carried out during the first period; control, 2.5%, 3.0%, 3.5%, 4.0%, and 4.5% MOS treatments were carried out during the second period; control, 5.0%, 5.5%, and 6.0% MOS treatments were carried out during the third period; and then, this procedure was repeated two times during the fourth to ninth periods. The productions of CO_2, CH_4, and H_2 were recorded by a gas flowmeter automatically per every 6 min. Then, the gas production was calculated per min from a total of 12 h of incubated gas production, after which the gas production per every 24 h was calculated.

2.4. In Vivo Procedures

In the acclimation period and digestive and metabolic experiments, all sheep received a basal diet or 2.0% MOS-treated diet randomly. Since they were housed in the respiratory chamber, the productions of CO_2, CH_4, NH_3, and H_2 along with the consumption of O_2 were measured using the gas flowmeter automatically per every 27 min.

The weight of feed intake and residues were recorded daily carefully. During the digestive and metabolic experiments, 10% of total diet, 10% of total feces output, and 5% of total urine output (5 mL of sulfuric acid added into total urine daily before collection to prevent nitrogen losses) were sampled for 6 sequential d and stored at −20 °C. At the end of the data collection period, the diets, fecal samples, and urine samples were thawed and pooled for each sheep. For nitrogen analysis, 3% of total feces output was sampled daily and stored in wide-mouth bottles containing 20 mL of 10% sulfuric acid for nitrogen fixation, and these samples were also pooled for 6 d individually.

After the digestive and metabolic experiments, sheep fasted approximately over 60 h until the RQ was below 0.71. Then, the sheep fasting metabolism experiment was carried out. The procedures of gas production and O_2 consumption recording, feces output sampling, and urine output sampling was as previously stated, but the recording and sampling times were 24 h.

2.5. Chemical Analysis

The diets and fecal samples were dried at 65 °C for 72 h in a forced-air oven, and then ground through a 1-mm screen using a Wiley Mill (Ogaw Seiki Co., Ltd., Tokyo, Japan). The gross energy of diet, feces, and urine samples were measured using a calorimeter (C2000, IKA®-Werke GmbH & Co. KG, Staufen, Germany). The nitrogen contents of the diet, feces containing 10% sulfuric acid, and urine samples were determined by the Kjeldahl method (AOAC, method 990.03) [12].

2.6. Calculations and Statistical Analysis

A gas flowmeter records the dynamic gas production or consumption; therefore, the total gas production or consumption is calculated using the following equations:

In vitro gas production (mL/24 h) = average production (mL/min) × 60 × 24;

In vivo gas production/consumption (L/24 h) = average production/consumption (L/min) × 60 × 24.

Parameters related to energy metabolism are calculated using the following equations [15]:

Total heat output (J) = 16.175 × O_2 (L) + 5.021 × CO_2 (L) − 2.167 × CH_4 (L) − 5.987 × Urine nitrogen (g).

Oxidation of protein (g) = Urine nitrogen (g) × 6.25.

CO_2 production from the oxidation of protein (L) = Urine nitrogen (g) × 4.754.

O_2 consumption from the oxidation of protein (L) = Urine nitrogen (g) × 5.923.

Heat output from the oxidation of protein (J) = Urine nitrogen (g) × 113.76.

Data from the *in vitro* experiment were analyzed by one-way analysis of variance (SPSS 19.0, IBM Co. Limited, Chicago, IL, USA) using the following model:

$$X_{ij} = \mu + \alpha_i + e_{ij} \tag{1}$$

where X_{ij} is the observation of the dependent variable ($i = 1$ to 13, $j = 1$ to 3), μ is the population mean, α_i is the random effect of treatment, and e_{ij} is the random error associated with the observation.

Data from the *in vivo* experiments were analyzed by an independent samples *t* test.

Significance was determined at $p \leq 0.05$ and tendency at $0.05 < p \leq 0.10$ using Tukey's multiple comparison tests.

3. Results

3.1. In Vitro CO_2, CH_4 and H_2 Production

The 24-h *in vitro* CO_2, CH_4, and H_2 production under different amounts of adding MOS are shown in Table 2. MOS influenced *in vitro* CO_2 and CH_4 productions significantly ($p < 0.05$). An MOS of 3.5% led to the highest production of CO_2 and CH_4, while 2.0% MOS resulted in the lowest amount of CO_2 emissions, and 0.5% and 5.0% MOS resulted in the lowest CH_4 emissions ($p < 0.05$). No differences in H_2 production were observed between different MOS doses ($p > 0.05$).

Table 2. *In vitro* CO_2, CH_4, and H_2 production with different amounts of adding mannan oligosaccharides (MOS).

MOS (%) [1]	CO_2 (mL/24 h)	CH_4 (mL/24 h)	H_2 (mL/24 h)
0	161.52 [cde]	27.87 [bc]	0.11
0.5	153.03 [cde]	26.21 [c]	0.06
1.0	155.61 [cde]	28.29 [bc]	0.04
1.5	145.64 [de]	28.50 [bc]	0.10
2.0	143.65 [e]	27.65 [bc]	0.10
2.5	169.92 [abc]	30.80 [abc]	0.05
3.0	169.16 [abc]	30.88 [abc]	0.06
3.5	189.27 [a]	36.58 [a]	0.08
4.0	186.53 [ab]	34.66 [ab]	0.09
4.5	172.96 [abc]	28.92 [abc]	0.11
5.0	162.40 [cde]	26.64 [c]	0.10
5.5	166.69 [bcd]	27.73 [bc]	0.09
6.0	162.60 [cde]	27.12 [bc]	0.10
SEM [2]	2.05	0.58	0.01
p-value	<0.001	0.001	0.841

[1] MOS doses were 0, 0.5, 1.0, 1.5, 2.0, 2.5, 3.0, 3.5, 4.0, 4.5, 5.0, 5.5, and 6.0% added to a 2-g basal diet. Mean results of *in vitro* gas production are shown for yield per 24 h ($n = 3$ per treatment, *in vitro* gas production (mL/24 h) = average production (mL/min) $\times$ 60 $\times$ 24). [2] SEM: standard error of mean. Values within a column with different superscripts differ significantly at $p < 0.05$.

3.2. In Vivo Gas Production, CP, and Energy Digestion and Retention of Sheep

The CO_2, CH_4, and NH_3 production, O_2 consumption, urine nitrogen excretion, total heat output, and relative parameters regarding the energy metabolism of the control and 2.0% MOS-treated sheep are shown in Table 3. MOS did not affect RQ, CO_2, CH_4, and NH_3 production, O_2 consumption, and total heat output ($p > 0.05$). However, MOS decreased the urine nitrogen excretion and parameters of protein oxidation related to the energy metabolism of sheep ($p < 0.05$). Furthermore, MOS tended to decrease O_2 consumption and CH_4 production per metabolic body weight ($BW^{0.75}$), along with the ratio of heat output from protein oxidation to total heat output ($p < 0.074$).

The effects of MOS on sheep dry matter intake (DMI), CP, and energy digestion and retention are shown in Table 4. The supplementation of MOS decreased dry matter (DM), CP, and energy intake, CP and energy excreted in feces and urine, digested and retained CP, and the energy released as CH_4 in sheep ($p < 0.05$). The addition of MOS did not affect the apparent digestibility of CP and energy, retention of energy, digestible energy (DE), and metabolizable energy (ME) ($p > 0.05$). However, MOS tended to increase the retention rate of CP ($p < 0.054$).

Table 3. *In vivo* gas production and energy metabolism of control and MOS-treated sheep.

Item [1]		Control	MOS-Treated	SEM [9]	*p*-Value
RQ [2]		0.86	0.87	0.002	0.328
CO_2 production	(L/24 h)	375.05	372.29	4.04	0.111
	(L/24 h/kg $BW^{0.75}$) [3]	20.77	20.58	0.20	0.641
CH_4 production	(L/24 h)	21.64	20.22	0.40	0.600
	(L/24 h/kg $BW^{0.75}$)	1.20	1.12	0.02	0.072
NH_3 production	(L/24 h)	32.58	33.43	0.61	0.923
	(L/24 h/kg $BW^{0.75}$)	1.80	1.85	0.03	0.507
O_2 consumption	(L/24 h)	438.37	430.69	4.96	0.074
	(L/24 h/kg $BW^{0.75}$)	24.27	23.81	0.25	0.358
Urine nitrogen excretion	(g/24 h)	14.87 [a]	13.56 [b]	0.30	0.028
	(g/24 h/kg $BW^{0.75}$)	0.82 [a]	0.75 [b]	0.02	0.035
Total heat output [4]	(MJ/24 h)	8.65	8.57	0.10	0.694
	(MJ/24 h/kg $BW^{0.75}$)	0.48	0.47	0.01	0.553
Oxidation of protein [5]	(g/24 h)	92.93 [s]	84.78 [b]	1.88	0.028
	(g/24 h/kg $BW^{0.75}$)	5.15 [a]	4.70 [b]	0.11	0.035
CO_2 production from the oxidation of protein [6]	(L/24 h)	70.69 [a]	64.48 [b]	1.43	0.028
	(L/24 h/kg $BW^{0.75}$)	3.92 [a]	3.57 [b]	0.08	0.035
O_2 consumption from the oxidation of protein [7]	(L/24 h)	88.07 [a]	80.34 [b]	1.78	0.028
	(L/24 h/kg $BW^{0.75}$)	4.88 [a]	4.45 [b]	0.10	0.035
Heat output form the oxidation of protein [8]	(MJ/24 h)	1.69 [a]	1.54 [b]	0.03	0.028
	(MJ/24 h/kg $BW^{0.75}$)	0.094 [a]	0.086 [b]	0.002	0.035
The ratio of heat output from the oxidation of protein to total heat output (%)		19.55	18.12	0.39	0.069

[1] Sheep were fed diets containing 0% and 2.0% MOS ($n = 4$ per treatment). Mean results of CO_2, CH_4, and NH_3 production, O_2 consumption, urine nitrogen excretion, total heat output, and relative parameters regarding energy metabolism are shown for the 6-d collection phase of the study for each treatment. The *in vivo* gas production or consumption was calculated using the following equation: *in vivo* gas production/consumption (L/24 h) = average production/consumption (L/min) × 60 × 24. [2] RQ = respiratory quotient. [3] $BW^{0.75}$ = Metabolic body weight. [4] Total heat output (J) = 16.175 × O_2 (L) + 5.021 × CO_2 (L) − 2.167 × CH_4 (L) − 5.987 × Urine nitrogen (g). [5] Oxidation of protein (g) = Urine nitrogen (g) × 6.25. [6] CO_2 production from the oxidation of protein (L) = Urine nitrogen (g) × 4.754. [7] O_2 consumption from the oxidation of protein (L) = Urine nitrogen (g) × 5.923. [8] Heat output from the oxidation of protein (J) = Urine nitrogen (g) × 113.76. [9] SEM: standard error of the mean. Values within a row with different superscripts differ significantly at $p < 0.05$.

Table 4. *In vivo* nutrients and energy digestion and retention of control and MOS-treated sheep.

Item [1]		Control	MOS-Treated	SEM [11]	*p*-Value
Dry matter intake (DMI, kg/24 h)		1.66 [a]	1.48 [b]	0.02	<0.001
Crude protein (CP)					
CP intake	(g/24 h)	182.96 [a]	163.56 [b]	2.27	<0.001
	(g/24 h/kg BW [0.75]) [2]	10.07 [a]	9.22 [b]	0.13	<0.001
CP in feces	(g/24 h)	46.90 [a]	38.78 [b]	2.03	0.044
	(g/24 h/kg BW [0.75])	2.51	2.12	0.12	0.107
CP in urine	(g/24 h)	92.93 [a]	84.78 [b]	1.88	0.028
	(g/24 h/kg BW [0.75])	9.40	9.41	0.19	0.976
Digested CP [3]	(g/24 h)	136.05 [a]	124.78 [b]	2.14	0.007
	(g/24 h/kg BW [0.75])	7.56	7.09	0.12	0.058
Retained CP [4]	(g/24 h)	43.12 [a]	40.00 [b]	0.66	0.016
	(g/24 h/kg BW [0.75])	2.34	2.28	0.03	0.339
Apparent digestibility (%) [5]		74.41	76.40	1.02	0.336
Retention rate (%) [6]		23.56	24.42	0.22	0.054

Table 4. *Cont.*

Item [1]		Control	MOS-Treated	SEM [11]	*p*-Value
Energy					
Energy intake	(MJ/24 h)	35.39 [a]	31.64 [b]	0.44	< 0.001
	(MJ/d/kg BW $^{0.75}$)	1.96 [a]	1.73 [b]	0.03	< 0.001
Energy in feces	(MJ/24 h)	12.69 [a]	10.05 [b]	0.61	0.028
	(MJ/d/kg BW $^{0.75}$)	0.74 [a]	0.48 [b]	0.03	<0.001
Energy in urine	(MJ/24 h)	0.46 [a]	0.42 [b]	0.01	0.028
	(MJ/d/kg BW $^{0.75}$)	0.03	0.02	0.001	0.301
Energy in CH_4	(MJ/24 h)	0.86 [a]	0.80 [b]	0.01	0.015
	(MJ/d/kg BW $^{0.75}$)	0.05	0.04	0.001	0.094
Digestible energy (DE) [7]	(MJ/24 h)	22.70	21.59	0.46	0.231
	(MJ/d/kg BW $^{0.75}$)	1.22	1.25	0.02	0.549
Metabolizable energy (ME) [8]	(MJ/24 h)	21.40	20.39	0.45	0.270
	(MJ/d/kg BW $^{0.75}$)	1.15	1.18	0.02	0.470
Apparent digestibility (%) [9]		64.34	68.59	1.52	0.165
Retention rate (%) [10]		60.63	64.32	1.48	0.216

[1] Sheep were fed 0% and 2.0% MOS (*n* = 4 per treatment). Mean results of DMI, CP, and energy digestion and retention are shown for the 6-d collection phase of the study for each treatment. [2] BW $^{0.75}$ = Metabolic body weight. [3] Digested CP (g) = CP intake − CP in feces. [4] Retained CP (g) = CP intake − CP in feces − CP in urine. [5] CP apparent digestibility (%) = (CP intake − CP in feces)/(CP intake). [6] CP retention rate (%) = (CP intake − CP in feces − CP in urine)/(CP intake). [7] Digestible energy (MJ) = energy intake − energy in feces. [8] Metabolizable energy (MJ) = energy intake − energy in feces − energy in urine − energy in CH_4. [9] Energy apparent digestibility (%) = (energy intake − energy in feces)/(energy intake). [10] Energy retention rate (%) = (energy intake − energy in feces − energy in urine − energy in CH_4) / (energy intake). [11] SEM: standard error of the mean. Values within a row with different superscripts differ significantly at *p* < 0.05.

3.3. In Vivo Gas Production and Energy Metabolism of Sheep under Fasting Metabolism Conditions

There were no differences in RQ, CO_2 production, O_2 consumption, urine nitrogen excretion, total heat output, and relative parameters regarding energy metabolism between the control and MOS-treated sheep under fasting metabolism conditions (*p* > 0.05, Table 5).

Table 5. *In vivo* gas production and energy metabolism of control and MOS-treated sheep under fasting metabolism conditions.

Item [1]		Control	MOS-treated	SEM [9]	*p*-Value
RQ [2]		0.66	0.69	0.006	0.267
CO_2 production	(L/24 h)	69.00	75.03	3.93	0.849
	(L/24 h/kg BW$^{0.75}$) [3]	3.83	4.15	0.22	0.494
O_2 consumption	(L/24 h)	103.45	109.57	5.24	0.921
	(L/24 h/kg BW$^{0.75}$)	5.74	6.06	0.29	0.617
Urine nitrogen excretion	(g/24 h)	3.28	2.95	1.52	0.828
	(g/24 h/kg BW$^{0.75}$)	0.18	0.35	0.08	0.490
Total heat output [4]	(MJ/24 h)	2.03	1.99	0.18	0.783
	(MJ/24 h/kg BW$^{0.75}$)	0.11	0.11	0.004	0.788
Oxidation of protein [5]	(g/24 h)	20.53	18.42	3.87	0.828
	(g/24 h/kg BW$^{0.75}$)	1.26	1.11	0.26	0.824
CO_2 production from the oxidation of protein [6]	(L/24 h)	15.61	14.01	2.94	0.828
	(L/24 h/kg BW$^{0.75}$)	0.96	0.85	0.20	0.824
O_2 consumption from the oxidation of protein [7]	(L/24 h)	19.45	17.45	3.67	0.828
	(L/24 h/kg BW$^{0.75}$)	1.19	1.05	0.24	0.824
Heat output form the oxidation of protein [8]	(MJ/24 h)	0.37	0.34	0.07	0.828
	(MJ/24 h/kg BW$^{0.75}$)	0.02	0.02	0.005	0.824
The ratio of heat output from the oxidation of protein to total heat output (%)		19.84	19.09	4.95	0.953

[1] Sheep were under the fasting metabolism conditions (*n* = 4 per treatment). Mean results of CO_2 production, O_2 consumption, urine nitrogen excretion, total heat output, and relative parameters regarding the energy metabolism of control and MOS-treated sheep are shown for the 24-h collection phase of the study for each treatment. The *in vivo* gas production or consumption was calculated using following equation: *in vivo* gas production/consumption (L/24 h) = average production/consumption (L/min) × 60 × 24. [2] RQ = respiratory quotient. [3] BW$^{0.75}$ = Metabolic body weight. [4] Total heat (J) = 16.175 × O_2 (L) + 5.021 × CO_2 (L) − 5.987 × Urine N (g). [5] Oxidation of protein (g) = Urine nitrogen (g) × 6.25. [6] CO_2 production from the oxidation of protein (L) = Urine nitrogen (g) × 4.754. [7] O_2 consumption from the oxidation of protein (L) = Urine nitrogen (g) × 5.923. [8] Heat output from the oxidation of protein (J) = Urine nitrogen (g) × 113.76. [9] SEM: standard error of the mean.

4. Discussion

4.1. In Vitro CO_2, CH_4 and H_2 Production

In this study, the MOS addition range from 0.5% to 2.0% decreased *in vitro* CO_2 and CH_4 production. Basically, carbohydrate was degraded by microbes in the rumen to produce approximately 65.5% CO_2, 28.8% CH_4, and small quantities of N_2, O_2, and H_2 [16]. Although some bicarbonates from saliva through the ruminal wall produce CO_2, the degradation of carbohydrates by ruminal microbes is the major pathway for CO_2 production [16]. In the rumen, methanogenesis evolves from carbon oxide, acetic acids, and methanol by methanogens. For most methanogens, methanogenesis from CO_2 and H_2 is the sole energy source. CH_4 production is considerably enhanced by the CO_2 present in the headspace, and there is an equilibrium established between the dissolved CO_2 in the media and the partial pressure of CO_2 in headspace gas. Meanwhile, the higher CO_2 concentration in the headspace would result in a greater dissolved CO_2 concentration in the media [17,18]. The result of the current study was in accordance with the previous conclusion that a positive relationship existed between CO_2 and CH_4; however, some scientists believed that there was a negative relationship, because CO_2 and H_2 are in general the precursors for CH_4 formation in the rumen [19,20]. This experiment illustrated that in an *in vitro* condition, MOS as an additive had only slight effects on ruminal fermentation, the partial pressure of CO_2, and methanogens and methanogenesis, and adding 0.5–2.0% MOS decreased CO_2 production and led to lower CH_4 production. It might be that MOS doses ranging from 0.5% to 2.0% were appropriate doses for ruminal microbes' fermentation pathway, and most of the produced CO_2 was potentially absorbed by the rumen wall to synthesize some nutrients, such as aspartic acid and isoleucine, and a lower partial pressure of CO_2 and dissolved CO_2 led to lower methanogenesis. Furthermore, higher doses of MOS meant higher oligosaccharides transferred into rumen as a substrate for microbes' fermentation, so, more substrates produced more CO_2, and instead promoted methanogenesis. However, MOS doses ranging from 2.5% to 4.5% generated the highest CO_2 and CH_4 emissions, but MOS doses ranging 5.0% to 6.0% resulted in lower CO_2 and CH_4 production. Further investigation is needed to confirm whether moderate doses of MOS improve carbohydrate fermentation in the rumen and much higher doses of MOS inhibit ruminal fermentation because more soluble carbohydrates led to lower pH and were harmful to microbes [20]. In addition, the fate of CO_2 is more complicated because of the pooling and recycling of animal metabolic carbon as urea and bicarbonates in saliva with that produced by the rumen organisms [21]. Furthermore, the process of methanogenesis is affected by many environmental factors, including the carbohydrate type and digestion passage rate. Feed additives such as yeast have the ability to shift H_2 utilization from methanogenesis to reductive acetogenesis through the homoacetogenic bacteria that can produce acetate from CO_2 and H_2 [22], along with internal factors [23]. So, more research about MOS on ruminal fermentation and the microbial population need to be undertaken in the future.

4.2. In Vivo Gas Production, CP, and Energy Digestion and Retention of Sheep

In the present study, the supplementation of MOS decreased ingestion, CP, and energy excreted in feces and urine, digested and retained CP, energy released as CH_4, along with the same parameters per $BW^{0.75}$, but MOS did not affect the apparent digestibility of CP and energy. Surprisingly, MOS decreased the ingestion of DM in sheep; the reason may be that there were low replicates in each treatment, indicating that individual differences may be a key factor in the results. Additional experiments with more sheep are needed to verify these effects of MOS.

Although MOS decreased the DMI of sheep, it decreased the CP and energy excretion in feces and urine at the same time; as a result, no differences were observed on the digestion and retention of CP and energy, even though higher digestibility and retention rates occurred in MOS-treated sheep. It indicates that MOS negatively influences the nutrients and energy intake of sheep; however, it makes sheep improve the utilization of nutrients and energy. Accordingly, because of improving the utilization of energy both in the rumen and the whole body of sheep, the methanogenesis in rumen was restricted, and resulted in less energy released as CH_4. A previous study found that GOS significantly increased the DE of sheep fed with Italian ryegrass hay and concentrate (3:2, on a DM basis), but did not affect nitrogen digestion and retention [24]. Other researchers reported that MOS did not affect the apparent digestibility and retention rate of nutrients in cattle and sheep [4,10]; these findings are in accordance with that of Goiri et al. [25], who pointed out that chitosan did not influence the nutrients' apparent digestibility. However, another study indicated that MOS enhanced the health of the ruminal epithelium of sheep by reducing the thickness of the stratum corneum, and it might have increased the nutrients digestion [9]. In addition, similar to this study, Sharma et al. [26] also reported that MOS increased the apparent digestibility of CP in Murrah buffalo calves. These findings indicated that MOS improved the CP utilization via decreased urine nitrogen excretion. So, more research about the effects of MOS on nutrients and energy metabolism in rumen also need to be undertaken in the future.

In the current study, treatment with 2.0% MOS resulted in less DM ingestion compared to the control. This led to less O_2 consumption and CO_2 production, because the digestion of feed and metabolism of nutrients in MOS-treated sheep consumed less O_2 and produced less CO_2. In *in vivo* conditions, CO_2 production from respiration is much larger than ruminal fermentation, and MOS only adjusts ruminal fermentation slightly; so, the decreasing tendency of CH_4 is observed from sheep in the respiratory chamber, not CO_2 production. Thus, it is possible that the basic vital activities of sheep were not affected by MOS. Respiration is a kind of rhythmic vital activity; its frequency and amplitude are determined by the body condition, which depends on nervous humoral regulation. Normally, feed intake cannot change the basic respiration frequency and amplitude, not to mention that there was only a small quantity of MOS; the results of the current study are also in consonance with the physiological theory.

Food ingestion affects the production of body heat in animals; however, for example, animals continuously produce heat and lose it to their surroundings, either directly by radiation, conduction, and convection, or indirectly by the evaporation of water [27]. In the current study, MOS decreased the urine nitrogen excretion and relative protein oxidation in sheep; however, the ratio of heat output from protein oxidation in the body to total heat output was decreased at the same time, indicated that the basal metabolism process is hardly influenced by MOS.

4.3. In Vivo Gas Production and Energy Metabolism of Sheep under Fasting Metabolism Conditions

MOS did not affect the CO_2 production, O_2 consumption, urine nitrogen excretion, total heat output, and relative parameters regarding the energy metabolism of sheep under fasting metabolism conditions in the current study. In a fasting animal, the quantity of heat produced is equal to the energy of the tissue catabolized. When measured under specific conditions, the energy is known as the animal's basal metabolism. A fasting animal must oxidize reserves of nutrients to provide the energy needed for essential processes such as respiration and circulation of the blood [27]. MOS cannot modify primary vital activities; no differences were observed in CO_2 production, O_2 consumption, urine nitrogen excretion, total heat output, and relative parameters regarding energy metabolism in 2.0% MOS-treated sheep and non-MOS fed sheep under fasting metabolism conditions.

5. Conclusions

The supplementation of MOS did not affect *in vitro* H_2 production, *in vivo* CO_2, CH_4, and NH_3 production, O_2 consumption, total heat output, the apparent digestibility of CP and energy, the retention rates of energy of sheep, and the respiration and energy metabolism of sheep under fasting metabolism conditions. However, the addition of 2.0% MOS to the sheep diet led to the lowest *in vitro* CO_2 production and less CH_4 production. Furthermore, treatment with 2.0% MOS decreased the intake of DM, CP, and energy, along with the CP and energy in feces and urine, digested and retained CP, and released the energy as CH_4. Treatment with 2.0% MOS appeared to increase the retention rate of CP in sheep. The results suggest that MOS only slightly affects ruminal fermentation and metabolism in sheep, and the effects were not strong enough to lead to substantial changes.

Supplementary Materials: The following are available online at http://www.mdpi.com/2076-2615/9/10/741/s1, Table S1: Gas recovery rate by calibrating CO_2.

Author Contributions: Conceptualization, C.Z., B.W., and H.Y.; methodology, C.Z. and B.W.; investigation, J.M.; data curation, C.Z.; writing—original draft preparation, C.Z.; writing—review and editing, T.L.; supervision, J.M., B.W. and H.Y.

Funding: This research was funded by Discipline Construction Fund Project of Gansu Agricultural University (grant number GSAU-XKJS-2018-036).

Acknowledgments: We thank all the people who participated in the experiments. We also thank Zhiyuan Ma from the Institute of Subtropical Agriculture, Chinese Academy of Science for his comments on this paper along with the native English-speaking scientists of Elixigen Company (Huntington Beach, California) for editing this manuscript.

Conflicts of Interest: All authors declare no conflicts of interest.

References

1. Adhikari, P.A.; Kim, W.K. Overview of prebiotics and probiotics: Focus on performance, gut health and immunity—a review. *Ann. Anim. Sci.* **2017**, *17*, 949–966. [CrossRef]
2. Gibson, G.R.; Roberfroid, M.B. Dietary modulation of the human colonic microbiota: Introducing the concept of prebiotics. *J. Nutr.* **1995**, *125*, 1401–1412. [CrossRef] [PubMed]
3. Halas, V.; Nochta, I. Mannan oligosaccharides in nursery pig nutrition and their potential mode of action. *Animals* **2012**, *2*, 261–274. [CrossRef] [PubMed]
4. Westland, A.; Martin, R.; White, R.; Martin, J.H. Mannan oligosaccharide prepartum supplementation: Effects on dairy cow colostrum quality and quantity. *Animal* **2017**, *11*, 1779–1782. [CrossRef] [PubMed]
5. Najdegerami, E.H.; Tokmachi, A.; Bakhshi, F. Evaluating the Effects of Dietary Prebiotic Mixture of Mannan Oligosaccharide and Poly-β-Hydroxybutyrate on the Growth Performance, Immunity, and Survival of Rainbow Trout, *Oncorhynchus mykiss* (Walbaum 1792), Fingerlings. *J. World Aquacult. Soc.* **2017**, *48*, 415–425. [CrossRef]
6. Fernandez, F.; Hinton, M.; van Gils, B. Dietary mannan-oligosaccharides and their effect on chicken caecal microflora in relation to Salmonella Enteritidis colonization. *Avian Pathol.* **2002**, *31*, 49–58. [CrossRef] [PubMed]
7. Thomas, W.E.; Nilsson, L.M.; Forero, M.; Sokurenko, E.V.; Vogel, V. Shear-dependent 'stick-and-roll' adhesion of type 1 fimbriated *Escherichia coli*. *Mol. Microbiol.* **2004**, *53*, 1545–1557. [CrossRef] [PubMed]
8. Demirel, G.; Turan, N.; Tanor, A.; Kocabagli, N.; Alp, M.; Hasoksuz, M.; Yilmaz, H. Effects of dietary mannanoligosaccharide on performance, some blood parameters, IgG levels and antibody response of lambs to parenterally administered *E. coli* O157:H7. *Arch. Anim. Nutr.* **2007**, *61*, 126–134. [CrossRef] [PubMed]
9. Diaz, T.G.; Branco, A.F.; Jacovaci, F.A.; Jobim, C.C.; Bolson, D.C.; Daniel, J.L.P. Inclusion of live yeast and mannan-oligosaccharides in high grain-based diets for sheep: Ruminal parameters, inflammatory response and rumen morphology. *PLoS ONE* **2018**, *13*, e0193313.
10. Zheng, C.; Li, F.D.; Hao, Z.L.; Liu, T. Effects of adding mannan oligosaccharides on digestibility and metabolism of nutrients, ruminal fermentation parameters, immunity, and antioxidant capacity of sheep. *J. Anim. Sci.* **2018**, *96*, 284–292. [CrossRef]
11. Mwenya, B.; Santoso, B.; Sar, C.; Gamo, Y.; Kobayashi, T.; Arai, I.; Takahashi, J. Effects of including β-1,4 galacto-oligosaccharides, lactic acid bacteria or yeast culture on methanogenesis as well as energy and nitrogen metabolism in sheep. *Anim. Feed Sci. Tech.* **2004**, *115*, 313–326. [CrossRef]
12. AOAC. *Official Methods of Analysis*, 17th ed.; Association of Offical Analytical Chemists: Arlington, VA, USA, 2000.
13. Goering, H.R.; Soest, P.J.V. *Forage Fiber Analysis. Agricultural Handbook No. 379*; Agricultural Research Service-US Department of Agriculture: Washington, DC, USA, 1970.
14. Menke, K.H.; Raab, L.; Salewski, A.; Steingass, H.; Fritz, D.; Schneider, W. The estimation of the digestibility and metabolizable energy content of ruminant feedingstuffs from the gas production when they are incubated with rumen liquor *in vitro*. *J. Agr. Sci.* **1979**, *93*, 217–222. [CrossRef]
15. Brouwer, E. Report of Subcommittee on Constants and Factors. In *Proceedings of the 3rd Symposium on Energy Metabolism, Troon, Scotland, May 1964*; EAAP Publ.: Academic Press: London, UK, 1965; pp. 441–443.
16. Aschenbach, J.R.; Penner, G.B.; Stumpff, F.; Gäbel, G. Ruminant Nutrition Symposium: Role of fermentation acid absorption in the regulation of ruminal pH. *J. Anim. Sci.* **2011**, *89*, 1092–1107. [CrossRef] [PubMed]

17. Patra, A.K.; Yu, Z.T. Effects of gas composition in headspace and bicarbonate concentrations in media on gas and methane production, degradability, and rumen fermentation using *in vitro* gas production techniques. *J. Dairy Sci.* **2013**, *96*, 4592–4600. [CrossRef] [PubMed]

18. Alford, J.S., Jr. Measurement of dissolved carbon dioxide. *Can. J. Microbiol.* **1976**, *22*, 52–56. [CrossRef]

19. Sirohi, S.K.; Pandey, N.; Singh, B.; Puniya, A.K. Rumen methanogens: A review. *Indian J. Microbiol.* **2010**, *50*, 253–262. [CrossRef] [PubMed]

20. Parra-Garcia, A.; Elghandour, M.M.M.Y.; Greiner, R.; Barbabosa-Pilego, A.; Camacho-Diaz, L.M.; Salem, A.Z.M. Effects of *Moringa oleifera* leaf extract on ruminal methane and carbon dioxide production and fermentation kinetics in a steer model. *Environ. Sci. Pollut. Res.* **2019**, *26*, 15333–15344. [CrossRef] [PubMed]

21. Soest, P.V. *Nutritional Ecology of the Ruminant*, 2nd ed.; Cornell University Press: New York, NY, USA, 1994.

22. Elghandour, M.M.Y.; Kholif, A.E.; López, S.; Mendoza, G.D.; Odongo, N.E.; Salem, A.Z.M. *In vitro* Gas, Methane, and Carbon Dioxide Productions of High Fibrous Diet Incubated With Fecal Inocula From Horses in Response to the Supplementation With Different Live Yeast Additives. *J. Equine Vet. Sci.* **2016**, *38*, 64–71. [CrossRef]

23. Zhao, L.P.; Meng, Q.X.; Li, Y.; Wu, H.; Huo, Y.L.; Zhang, X.Z.; Zhou, Z.M. Nitrate decreases ruminal methane production with slight changes to ruminal methanogen composition of nitrate-adapted steers. *BMC Microbiol.* **2018**, *18*, 21. [CrossRef]

24. Pen, B.; Takaura, K.; Yamaguchi, S.; Asa, R.; Takahashi, J. Effects of *Yucca schidigera* and *Quillaja saponaria* with or without β1–4 galacto-oligosaccharides on ruminal fermentation, methane production and nitrogen utilization in sheep. *Anim. Feed Sci. Tech.* **2007**, *138*, 75–88. [CrossRef]

25. Goiri, I.; Oregui, L.M.; Garcia-Rodriguez, A. Use of chitosans to modulate ruminal fermentation of a 50:50 forage-to-concentrate diet in sheep. *J. Anim. Sci.* **2010**, *88*, 749–755. [CrossRef] [PubMed]

26. Sharma, A.N.; Kumar, S.; Tyagi, A.K. Effects of mannan-oligosaccharides and *Lactobacillus acidophilus* supplementation on growth performance, nutrient utilization and faecal characteristics in Murrah buffalo calves. *J. Anim. Physiol. Anim. Nutr.* **2018**, *102*, 679–689. [CrossRef] [PubMed]

27. McDonald, P.; Edwards, R.A.; Greenhalgh, J.F.D.; Morgan, C.A. *Animal Nutrition*; Pearson Education Limited: Upper Saddle River, NJ, USA, 2002.

Article

Effects of Increasing Doses of Lactobacillus Pre-Fermented Rapeseed Product with or without Inclusion of Macroalgae Product on Weaner Piglet Performance and Intestinal Development

Gizaw Dabessa Satessa [1,*], Paulina Tamez-Hidalgo [2], Søren Kjærulff [2], Einar Vargas-Bello-Pérez [1], Rajan Dhakal [1] and Mette Olaf Nielsen [3,*]

[1] Department of Veterinary and Animal Sciences, University of Copenhagen, Grønnegårdsvej 3, DK-1870 Frederiksberg C, Denmark; evargasb@sund.ku.dk (E.V.-B.-P.); dhakal@sund.ku.dk (R.D.)

[2] Fermentationexperts A/S, Vorbassevej 12, DK-6622 Copenhagen, Denmark; pat@fexp.eu (P.T.-H); skj@fermbiotics.com (S.K.)

[3] Department of Animal Sciences, Faculty of Technical Sciences, Aarhus University, Blichers Allé 20, 8830 Tjele, Denmark

* Correspondence: gizaw.satessa@sund.ku.dk (G.D.S.); mon@anis.au.dk (M.O.N.)

Received: 15 February 2020; Accepted: 25 March 2020; Published: 27 March 2020

Simple Summary: This study investigated the effects of dietary addition of increasing amounts (8, 10, 12, 15 and 25%) of pre-fermented rapeseed meal (FRM) or 10% FRM with the inclusion of 0.6% or 1.0% *Ascophyllum nodosum* (AN), a brown macroalgae, on performance and gut health as compared to either a non-supplemented diet (negative control, NC) or a diet supplemented with 2500 ppm ZnO (PC) in piglets weaned at 28 days of age. At any of the amounts supplemented, FRM sustained growth performance similar to the PC group during the first 10 days before weaning (18–27 days of age) and improved performance better than the PC from 28–41 days of age when fed at 8%. Inclusion of AN (0.6% or 1.0%) on top of 10% FRM did not affect growth performance. The percent of piglets that completed the experiment was increased at all levels of FRM (maximum of 91% at 8% FRM) or a combination of 10% FRM with AN (maximum of 90% at 10% FRM + 0.6% AN). Maximum intestinal development (villus, crypts, enterocytes) was observed at lower levels (8–10%) of FRM supplementation, but this was abolished by inclusion of AN. Feeding of FRM with or without AN increased some hematological parameters at all doses and enhanced immunoglobulin and interleukin-6 titers at lower doses (8% or 10%). In conclusion, FRM sustained piglet growth performance and intestinal development similar to ZnO with an optimal dietary inclusion level at 8–10% of dietary DM.

Abstract: This study evaluated the effects of increasing doses of pre-fermented rapeseed meal (FRM) without or with inclusion of the brown macroalgae *Ascophyllum nodosum* (AN) on weaner piglets' performance and gut development. Ten days pre-weaning, standardized litters were randomly assigned to one of nine isoenergetic and isoproteic diets comprising (on DM basis): no supplement (negative control, NC), 2500 ppm ZnO (positive control, PC), 8, 10, 12, 15 or 25% FRM, and 10% FRM plus 0.6 or 1.0% AN. Fifty piglets receiving the same pre-weaning diets were weaned at 28 days of age and transferred to one pen, where they continued on the pre-weaning diet until day 92. At 41 days, six piglets per treatment were sacrificed for blood and intestinal samplings. The average daily gain was at least sustained at any dose of FRM (increased at 8% FRM, 28–41 days) from 18–41 days similar to PC but unaffected by inclusion of AN. The percentage of piglets that completed the experiment was increased by FRM compared to NC, despite detection of diarrhea symptoms. FRM showed quadratic dose-response effects on colon and mid-jejunum crypts depth, and enterocyte and mid-jejunum villus heights with optimum development at 8% or 10% FRM, respectively, but this was abolished when AN was also added. In conclusion, FRM sustained piglet growth performance and intestinal development similar to ZnO with an optimum inclusion level of 8–10% of dietary DM.

Keywords: fermented feed; *Ascophyllum nodossum*; villi development; gut health

1. Introduction

Weaning is the most challenging event in modern pig production, which often results in reduced feed intake, decreased growth, increased risks of post-weaning diarrhea (PWD) and mortality in newly weaned piglets [1–3]. Thus, pig producers in Europe, particularly in Denmark, heavily depend on dietary supplementation of pharmacological doses (2500 to 3000 ppm) of zinc oxide (ZnO) to avert weaning related diarrhea and associated piglet mortality. However, due to environmental concerns [4] and increased risk of antibiotic-resistant bacteria [5,6], the European Union passed a new law in July 2017, which bans the use of in-feed ZnO at pharmacological doses beyond 2022 [7]. As a consequence, pig producers will face huge challenges to sustain piglet performance and prevent PWD and associated mortality unless effective alternative feeding strategies are developed.

Dietary inclusion of feeds pre-fermented with probiotic bacteria could be a promising dietary strategy to replace in-feed ZnO in weaner piglet diets. A meta-analysis-based investigation revealed that fermented feeds increased growth performance irrespective of age when fed to pigs by reducing intestinal coliforms, increasing villi development, and hence nutrient absorption, and enhancing immune responses [8]. Gut health has also been improved during weaning when pre-fermented diets have been fed to newly weaned piglets due to increased abundance of beneficial bacteria, such as lactic acid bacteria, and decreased numbers of enteropathogenic bacteria such as *Eschdurierichia coli* and overall coliforms throughout the gastrointestinal tract [9–11]. Unlike non-fermented products, fermented feed has been shown to lower fecal excretion of pathogenic coliforms when fed to sows [12].

Recently, rapeseed, particularly in its fermented form, has attracted interest in monogastric animals such as poultry and pigs as an alternative protein source as well as for manipulation of gut health. Feeding *Lactobacillus* pre-fermented rapeseed meals (FRM) to broiler chicken has been shown to increase feed digestibility and utilization (by reducing impact of anti-nutritional factors), gut morphology and antioxidant capacity in the body [13]. We have also previously tested FRM or rapeseed meal co-fermented with the brown macroalgae *Ascophyllum nodosum* (AN) and or *Saccharina latissima* (SL) in weaned piglets where 10% FRM inclusion improved growth performance, intestinal health and colon microbiota. However, the inclusion of either AN or AN plus SL in FRM resulted in inconsistent and contradictory findings between growth performance and gut health indices [14].

Nevertheless, brown macroalgae are currently viewed as another promising dietary strategy to improve growth performance and gut health in weaned piglets due to their wide range of bioactive components [15]. The bioactive carbohydrate components from brown macroalgae, such as laminarin and fucoidan, are unique in that they are not found in terrestrial plants and possess antimicrobial [16,17], immunomodulatory [18] and antioxidant [19] properties. Inclusion of these purified extracts in weaner piglets' diets also improved overall growth performance [20,21], but are relatively expensive compared to the whole plant due to the cost of extraction. Available studies involving intact macroalgae, which are limited in contrast to purified extracts, have reported contradicting findings on the growth performance and gut health in pigs [22–24]. These variations could be attributed to, among others, the amount and species of macroalgae administered. In our previous study, we observed that the positive effects of FRM were constrained following inclusion of either AN or AN plus SL. However, the optimum doses for both FRM and the macroalgae products were generally unknown.

Thus, we hypothesized that: (1) At optimal dietary inclusion levels, FRM can sustain weaner piglet performance, promote gut development and prevent PWD as efficiently as dietary addition of pharmacological doses of ZnO (2500 ppm), (2) inclusion of the brown macroalgae AN can act synergistically with FRM at a lower but optimal dose to promote weaner piglet performance and health.

Therefore, the objectives of this pilot study were (1) to investigate the effects of increasing doses of FRM on weaner piglet performance and intestinal development as compared to medicinal ZnO or no supplement, and (2) to investigate the potential synergistic effects between AN and FRM when AN was included at lower doses.

2. Materials and Methods

This study was carried out at a commercial pig farm (Chotycze Farm, Łosice, Poland) from September 2017 to March 2018. The experiment was approved by the local ethics committee for animal experiments of the University of Lublin with authorization number 67/2017.

2.1. Experimental Design, Animals and Feeding

A total of 450 piglets (Landrace & Yorkshire hybrid [LY] × Duroc & Pietrain hybrid [DP] crossbreed) were recruited for the study from 45 sows (LY Danbred C22 line, DanBred, Denmark) in their second or third parity. The piglets came from litters that were standardized to 14 piglets per sow. Standardization of litters to 14 piglets per sow was made by transferring piglets among sows until 14 days of age. Thereafter, no movement of piglets among sows was allowed. At 18 days of age (10 days before weaning), piglets were introduced to one of nine experimental diets comprising a basal diet (pre-starter, 18 to 64 days of age or starter diet, 65 to 92 days age) with: no supplement (NC), 2500 ppm (PC), increasing levels of FRM (8, 10, 12, 15 or 25% of dietary DM), and a combination of 10% FRM with either 0.6 or 1.0% AN of dietary DM while still with the sows, with piglets from five sows receiving the same experimental diet. At 28 days of age, the piglets were weaned and 50 of the piglets already receiving the same dietary treatment, excluding the runts, were then transferred from five sows to one slatted floor pen in the weaner unit per treatment and continued on the same dietary treatment until they exited the weaner unit at 92 days of age. All diets were based on the same basal diet, where wheat and soybean products were replaced with increasing amounts of FRM and AN.

All sows received the same standard lactation diet commonly used at the farm. Piglets were fed pre-starter diets from 18 (10 days pre-weaning) to 64 days of age, and a starter diet from 65 to 91 days of age, i.e., until the end of the experiment, ad libitum from open troughs along with ad libitum access to water from drinking nipples during the entire trial.

2.2. Preparation of Experimental Diets

FRM (commercial name EP100i) was obtained from European Protein (Baekke, Denmark). It was prepared with solid-state fermentation using inocula from three lactic-acid fermentative bacteria, namely, *Pediococcus acidilactici* (DSM 16243), *Pediococcus pentosaceus* (DSM 12834) and *Lactobacillus plantarum* (DSM 12837) where ground rapeseed meal (80% on DM basis) was mixed with the inoculant broth in a one-step fermentation process with constant stirring, followed by decanting and incubation at 38 °C for four days.

The mixing of the final experimental diets was conducted by a certified feed company where no heat treatment was involved after the addition of *Lactobacillus* spp. containing FRM product. Formulation of the diets was performed by AgroSoft®(Tørring, Denmark) according to the Danish nutritional recommendations for piglets [25]. Analyses for energy contents and nutritional composition of the final pre-starter and starter diets were conducted at the commercial lab Dolfos®(Piotrków Trybunalski, Poland). Both the pre-starter and starter diets were designed to be iso-proteic and iso-energetic across treatments despite the presence of some deviation in the final products as shown in Table S1.

2.3. Recordings of Feed Intake and Body Weight

Piglets were ear-tagged at birth, and body weights of individual piglets were recorded 10 days before weaning (18 days of age), at weaning (28 days of age, and 14 days post-weaning (41 days of age). Thereafter, body weight was measured in each treatment on pen-basis until the piglets exited the experiment at 92 days of age. Feed consumption in each experiment was recorded daily on pen-basis throughout the experiment.

2.4. Piglets' Health Assessment and Fecal Sampling

Piglets were examined daily by farm staff for signs of diarrhea or signs of any other ailments including mortality and if possible, its causes, and any changes observed were recorded. Piglets with poor health conditions due to, among others, diarrhea were removed from the pen and experiment to be treated with antibiotics elsewhere. No antibiotics treatment was allowed for any piglet remaining in the experiment. Fecal samples were collected from each pen and analyzed for lactic acid content at the veterinary diagnostic laboratory, ALAB Weterynaria (Warsaw, Poland). A completion rate was calculated in percentage for each treatment based on the number of piglets remaining in a pen at the end of the experiment and the number of piglets assigned to pens at weaning.

2.5. Blood and Intestinal Samplings from Subgroups of Piglets

Fourteen days after weaning (41 days of age), six piglets were randomly selected from each treatment or pen, and transported to a commercial slaughter facility. Piglets were sacrificed by exsanguination following captive bolt stunning. Whole blood samples were collected during exsanguination from the jugular vein into ethylene diamine tetra acetic acid (EDTA) tubes for hematological analysis and plain vacutainer tubes for analysis of serum immunoglobulins and interleukin-6. Whole blood samples from EDTA tube were stored at +4 °C and the blood samples in plain tubes were allowed to clot to separate the serum and the serum samples were then stored at –20 °C until analyses the following day.

The abdominal wall was opened, and the cardiac sphincter, pyloroduodenal, ileocecal and anorectal junctions were ligated, and the gut segments were cut and taken out. Tissue sections of approximately 2 cm length were sampled from the middle of the small intestine and the apex of the ascending colon. Sampled tissue sections were gently rinsed for intestinal contents with saline and fixed in 10% buffered formalin.

2.6. Intestinal Morphology, Haematology and Serum Immunoglobulins

Histological analyses of mid-jejunum and colonic tissue were conducted at the commercial laboratory ALAB Weterynaria (Warsaw, Poland) using a standard light microscope Olympus BX41 and CellSens software (Olympus Corporation, Tokyo, Japan). All microscopic evaluations were performed in a blinded fashion. Initially, tissue samples fixed in 10% buffered formalin were dehydrated using graded ethanol and xylene baths and embedded in paraffin wax. Tissue sections of 3–4 μm were mounted on the microscopic slide and stained with hematoxylin and eosin (HE). General histopathological examinations were evaluated at a magnification of 10×, 40× and 100× (objective lens) and 10× (eyepiece), and photographic documentation was made. For the mid-jejunum the following were examined: the mucosa and submucosa, intestinal crypts, mucosal and submucosal blood vessels, enterocyte height and brush border integrity as well as the presence of goblet cells. The morphometric analysis delivered the following quantitative results: mid-jejunum villus height (JVH) at 10× magnification, mid-jejunum crypt depth (JCD) and colonic crypt depth (CCD) both at 10× magnification, and mid-jejunum enterocyte height (JEH) at 40× magnification. Two types of scoring analyses were performed: for intraepithelial lymphocytes infiltration (IEL) and stromal lymphocytes infiltration (SL). The following scale was used in both scoring evaluations: 0—normal; 1—slight; 2—moderate; 3—severe.

Hematological analysis of whole blood samples and serum immunoglobulin analysis were conducted at the ALAB Weterynaria (Warsaw, Poland). Analysis of red blood cell counts (RBC), hematocrit value (HCT), hemoglobin content (Hb), white blood cell counts (WBC) and differential leucocytes count were carried out using a Sysmex XT 2000i analyzer (Sysmex Corporation, Kobe, Japan). Commercially available swine-specific enzyme-linked immunosorbent assay (ELISA) kits (Cusabio®, Houston, TX, USA) were employed to quantify the concentrations of serum immunoglobulins such as immunoglobulin G (IgG, Catalog Number: CSB-E06804p), immunoglobulin A (IgA, Catalog Number: CSB-E13234p) and immunoglobulin M (IgM, Catalog Number: CSB-E06805p) as well as the pro-inflammatory cytokine interleukin-6 (IL-6, Catalog Number: CSB-E06786p) according to the manufacturer protocols.

2.7. Data Analyses

Data for growth performance, hematology and serum immunoglobulins, as well as a dose response effect of FRM on gut morphometry were analyzed using Linear Mixed Models of R version 3.6.0 [26]. Dietary treatment effects were evaluated using the following statistical model, and differences between individual treatments were assessed by using Tukey's Honest Significance Difference Test:

$$Yijkl = \mu + T_i + S_j + (TxS)_{ij} + (W + W^2)_k + P_l + \varepsilon_{ijkl} \tag{1}$$

where, $Yijkl$ is the response variable, μ is the overall mean, T_i is the effect of dietary treatment (i = 1, 2, 3, … , 9), S_j is the effect of sex of piglets (j = male or female), $(TxS)_{ij}$ is the treatment and sex interaction, $(W + W^2)_k$ is the linear and quadratic effects, respectively, of body weight at the start of the experiment (18 days post-partum) (k = 1,2,3, … , 50), P_l is the random effects of individual piglets and ε_{ijkl} is the residual.

Linear and quadratic relations for dietary inclusion level of FRM (FRM dose as % of dietary DM) and effects on gut morphometric parameters were analyzed using the following model:

$$Yijklm = \mu + T_i + S_j + (TxS)_{ij} + (W + W^2)_k + (D^2)_l + P_m + \varepsilon_{ijklm} \tag{2}$$

where, $Yijklm$ is the response variable, μ is the overall mean, T_i is the effect of dietary treatments (I = 1, 2, 3, … ., 9), S_j is the effect of sex of the piglet (j = male or female), $(TxS)_{ij}$ is the treatment and sex interaction, $(W + W^2)_k$ is the linear or quadratic effects, respectively, of body weight at the start of the experiment (18 days post-partum) (k = 1,2,3, … , 50), $(D^2)_l$ is the quadratic effects of the doses of the treatments, P_m is the random effect of individual piglets and ε_{ijklm} is the residual. The normality of the data was tested using normal quantile-quantile (Q-Q) plots. All non-normal data were log or box-cox (only in case of gut morphometry) transformed before statistical analysis. The best-fitting model was recruited based on Akaike information criterion (AIC) value. All results were expressed as least-square means (LSM) and standard error of the mean (SEM), and significance was considered at $p < 0.05$.

3. Results

3.1. Performance of Piglets

As shown in Table 1, piglets supplemented with ZnO (PC) showed no difference in average daily gain (ADG) and BW except for a significant increment ($p < 0.05$) during the 10 days before weaning (18 to 27 days of age) and at 27 days of age, respectively, compared to the NC piglets. FRM sustained similar ADG during 18–27 and 18–41 days of age, and BW at 27 and 41 days of age as the PC at all doses, and significantly increased ($p < 0.05$) ADG during 28–41 days of age when supplemented at 8%. ADG was increased during 18–27 days of age, and demonstrated no discernable pattern (despite significant increment at 8% FRM) during 28–41 days of age, and showed no difference during 18–41 days of age with increasing doses of FRM. Inclusion of AN on top of 10% FRM at either dose did not affect ADG compared to the 10% FRM. Results for BW and ADG after 41 days of age (42–92 days of age) were pen-based single observation per treatment, and hence, could only provide an inconclusive rough estimation of growth (Table S2). Based on this estimation, FRM resulted in numerically higher ADG at all doses compared to the controls (NC and PC) during 42–64 days of age and similar ADG to PC during 65–92 days of age when fed at 10% dietary DM. FRM resulted in the highest ADG during both periods (42–64 and 65–92 days of age) when supplemented at 10% dietary DM. In this study, average daily feed intake and feed conversion ratio were not provided, as daily feed intake was measured on pen-basis only (single observation per treatment), and hence, it was impossible to quantify the exact amount of feed consumed by each piglet housed in a group.

3.2. Incidence of Diarrhea and Fecal Lactic Acid Concentration

Due to the limited number of observations per treatment, statistical analyses were not applied to data from diarrhea and fecal lactic acid concentrations. The 2500 ppm ZnO (PC) resulted in a numerically lower incidence and higher completion rates than NC. At any of the doses supplemented, FRM without or with AN resulted in a higher completion rate than the NC but achieved similar completion rates to that of PC (Table S3). Dietary lactic acid concentration was increased with increasing doses of FRM (despite some fluctuation at 15% supplementation) demonstrating 3–10 and 4–12 fold increases compared to the NC and PC, respectively. It also showed 5–6 fold and 6–7 folds increment compared to the NC and PC, respectively, with inclusion of AN on top of 10% FRM (Figure S1A). Supplementation with FRM also resulted in lactic acid concentrations in the feces of a 10–20-fold increase compared to the NC and a 5–10-fold increase compared to the PC. The fecal lactic acid concentration was markedly increased with an increasing dose of FRM (except for 15% FRM) but was reduced when AN was added (Figure S1B).

3.3. Gut Tissue Morphometry and Immune Cell Infiltration

The effects of dietary supplementation on intestinal morphologies are shown in Figure 1, and the morphological features of intestinal tissues obtained from different dietary treatments are illustrated in Figure 2. Increasing doses of FRM resulted in quadratic effects on jejunum enterocyte height, villus height and crypt depth as well as colon crypt depth with maximum effects being observed at 8% FRM for colon and jejunal crypt depths, and 10% FRM for jejunal enterocyte and villus heights (Figure 1A) in weaned piglets 14 days after weaning. Inclusion of increasing doses of AN in 10% FRM counteracted all of these changes in intestinal histomorphometric features (Figure 1B). These intestinal histomorphometric features were not affected by the supplementation of ZnO compared to the NC.

Table 1. Effects of dietary addition of increasing doses of *Lactobacillus* pre-fermented rapeseed meal (FRM) or a combination of 10% FRM with increasing doses of AN on growth performance in piglets weaned at 28 days of age.

| Parameters | Treatments | | | | | | | | | SEM | *p*-Value | |
| | Controls | | FRM (%) | | | | | AN [1] (%) | | | | |
	NC	PC	8	10	12	15	25	0.6	1.0		TG	IBW
BW at d 18, kg	5.83 ± 0.88	5.28 ± 0.78	5.62 ± 0.83	4.92 ± 0.87	4.92 ± 0.83	4.92 ± 0.91	5.22 ± 0.75	5.22 ± 0.99	5.21 ± 1.10			
Days 18–27												
BW at d 27, kg	6.67 [a]	7.22 [bcd]	6.99 [ab]	6.95 [ac]	7.22 [bcd]	7.48 [bd]	7.50 [d]	7.29 [cd]	7.36 [bd]	0.092	***	***
ADG, g/d	143 [a]	198 [bcd]	175 [ab]	181 [bc]	198 [bcd]	224 [d]	226 [d]	205 [bcd]	212 [cd]	8.1	***	***
Days 28–41												
BW at d 41, kg	7.78 [a]	8.34 [ab]	8.96 [bc]	8.31 [ab]	9.18 [c]	8.79 [bc]	9.01 [bc]	8.91 [bc]	8.96 [bc]	0.19	***	***
ADG, g/d	108 [ab]	100 [a]	153 [b]	106 [ab]	148 [ab]	99 [a]	116 [ab]	124 [ab]	123 [ab]	12.2	**	NS
Days 18–41												
ADG, g/d	110 [a]	142 [ab]	162 [b]	141 [ab]	164 [b]	155 [b]	164 [b]	159 [b]	162 [b]	8.30	***	***

[a,b,c,d] Different letters within same row indicate significant difference ($p < 0.05$); *** $p \leq 0.001$; ** $p \leq 0.01$; * $p \leq 0.05$; NS = non-significant; ADG = average daily gain; IBW = initial body weight at 18 days of age; TG = treatment group; NC = negative control (basal diet without additives); PC = positive control (basal diet supplemented with 2500 ppm ZnO); AN = *Ascophyllum nodosum*; FRM = pre-fermented rapeseed meal (EP100i); [1] AN was included on top of 10% FRM at 0.6% or 1.0% giving 10% FRM +0.6% AN or 10% FRM + 1.0% AN overall supplements.

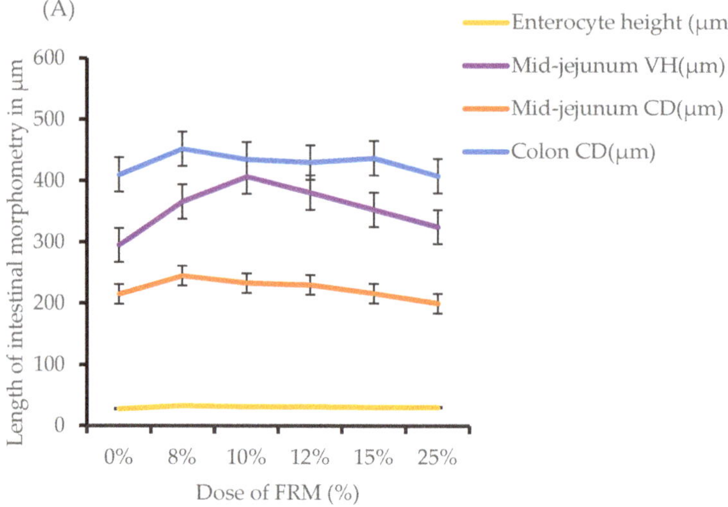

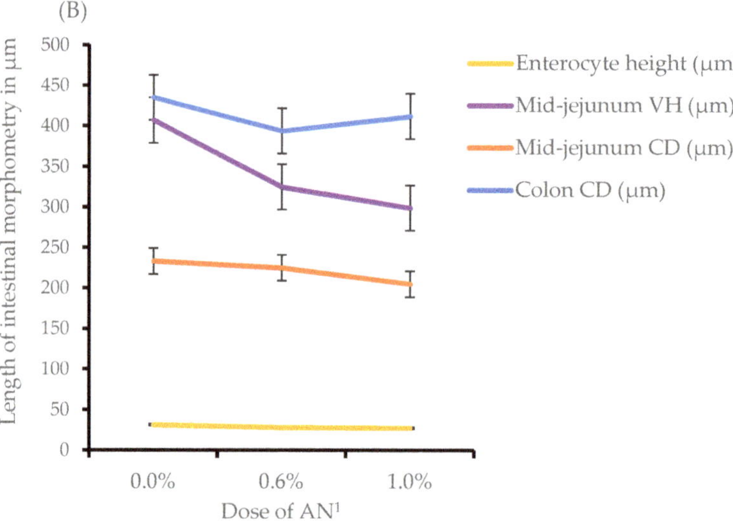

Figure 1. Effect of increasing doses of prefermented rapeseed meal (FRM) (**A**) or combination of 10% FRM with increasing doses of *Ascophyllum nodosum* (AN) (**B**) on enterocyte height, colon crypt depth, mid-jejunum crypt depth and villi in weaned piglets sacrificed at 14 days post-weaning. CD = crypt depth; VH = villus height; [1] AN was included on top of 10% FRM at 0.0%, 0.6% or 1.0% giving 10% FRM +0.6% AN or 10% FRM + 1.0% AN overall supplements.

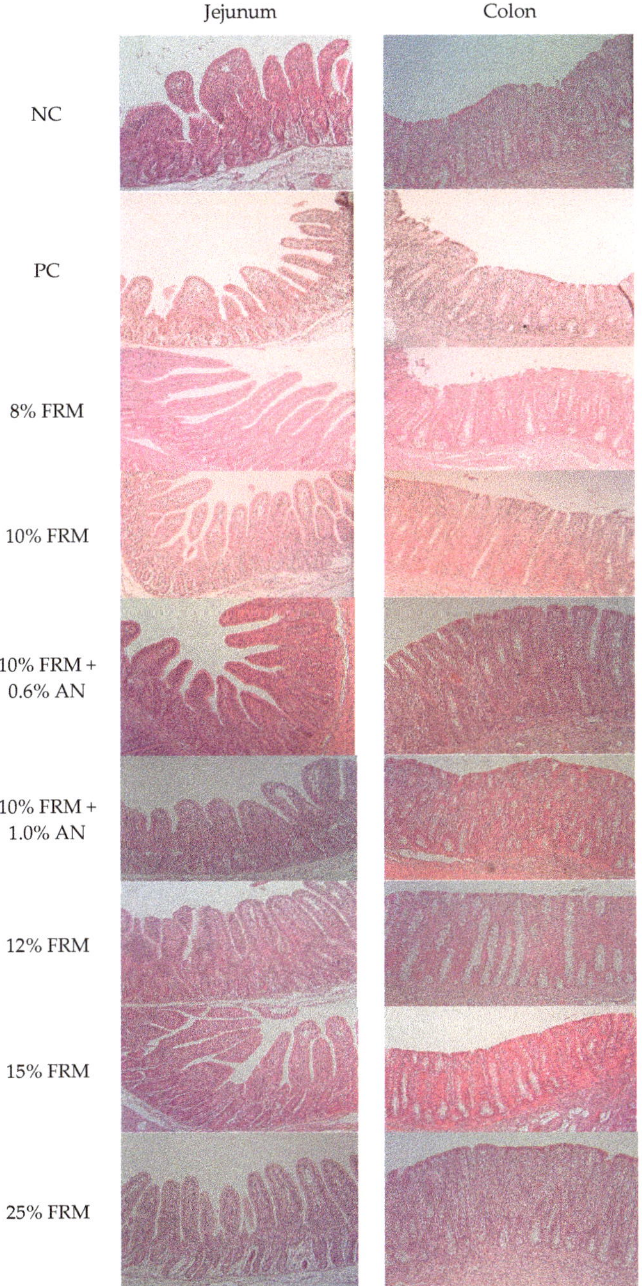

Figure 2. Effects of a pharmacological dose of ZnO, increasing doses of pre-fermented rapeseed meal (FRM) or combination of 10% FRM with increasing doses of *Ascophyllum nodosum* (AN) on the histomorphological features in the jejunal and colonic tissues in piglets 14 days after weaning. NC = negative control (basal diet without additives); PC = positive control (basal diet supplemented with 2500 ppm ZnO). Histomicrographs were taken at 10× magnification.

Supplementation of FRM at any dose or its 10% combination with any dose of AN resulted in numerically lower gut-associated lymphoid tissue (GALT) counts, and reduced immune cell infiltrations of the jejunal and colonic mucosa compared to the NC and PC piglets, which had similar levels of infiltration (Figure S2).

Assessment of histopathological changes revealed weak to strong stromal and intraepithelial immune cell infiltrations of the small intestine in both NC and PC piglets. A few piglets demonstrated disruption of the epithelial barrier and moderately visible brush borders in areas of significant intraepithelial lymphocyte infiltration in both controls. There were negligible immune cell infiltrations of the small intestine and colon at any doses of FRM compared to NC and PC, and infiltrations were limited only to the stroma in both intestinal segments. Inclusion of AN did not have significant impacts except for a slight increment in colonic stromal immune cell infiltration.

3.4. Hematological and Immunological Parameters

As shown in Table 2, FRM increased erythrocyte indices, WBC and lymphocyte (LYM) counts ($p < 0.05$), but reduced granulocyte (GRA) counts (except for 25% FRM) ($p < 0.05$) at any dose administered compared to NC and PC piglets. Inclusion of AN had no additional impact on hematological parameters except for a significant increment of RBC counts when 1.0% AN was supplemented. There were no differences in hematological values between PC and NC piglets except for higher WBC in NC ($p < 0.05$).

As shown in Table 3, PC piglets had higher plasma IgG, IgA and IgM titers, but not IL-6 level, compared to NC piglets. The highest immunoglobulins and IL-6 titers were reached when weaner diets contained 8% or 10% FRM. Inclusion of AN significantly enhanced IgG titers but significantly reduced that of IL-6 compared to 10% FRM. Inclusion of AN on top of 10% FRM increased IgG level but decreased blood levels of IgA, IgM and IL-6.

Table 2. Effects of a pharmacological dose of ZnO, increasing doses of *Lactobacillus* pre-fermented rapeseed meal (FRM) and combination of 10% FRM combination with increasing doses of *Ascophyllum nodosum* (AN) on hematological parameters in weaned piglets slaughtered 14 days post-weaning.

| Parameters | Treatments | | | | | | | | | SEM |
| | Controls | | FRM (%) | | | | | AN [1] (%) | | |
	NC	PC	8	10	12	15	25	0.6	1.0	
RBC indices										
RBC, $10^6/\mu L$	5.27 [a]	5.46 [a]	6.87 [bc]	6.53 [b]	6.87 [bc]	6.68 [b]	7.68 [c]	6.88 [bc]	7.67 [c]	0.223
Hb, g/dL	8.48 [a]	8.29 [a]	11.12 [b]	11.33 [bc]	11.12 [b]	11.85 [bc]	12.37 [c]	11.67 [bc]	11.33 [bc]	2.60
HCT, %	30.62 [a]	32.83 [a]	37.54 [b]	37.76 [b]	37.54 [b]	39.65 [b]	40.21 [b]	37.81 [b]	38.04 [b]	0.891
Total differential WBC counts										
WBC, $10^3/\mu L$	34.54 [d]	15.21 [a]	20.9 [bc]	19.60 [ab]	18.60 [ab]	22.20 [bc]	26.00 [c]	20.60 [bc]	20.30 [bc]	1.055
LYM, %	53.4 [ab]	51.4 [a]	62.7 [bc]	62.9 [bc]	69.5 [c]	69.3 [c]	62.5 [bc]	62.9 [bc]	64.8 [c]	2.43
GRA, %	45.7 [c]	45.0 [bc]	34.7 [a]	35.0 [a]	27.1 [a]	27.6 [a]	35.5 [ab]	34.5 [a]	33.9 [a]	2.451
MID, %	3.32	3.42	2.53	1.85	2.57	1.78	2.13	2.23	1.72	0.486

[a,b,c,d] Different letters within same row indicate significant difference ($p < 0.05$); initial body weight (IBW) has significant linear effect on Hb and quadratic effect on HCT and LYM %; NC = negative control (basal diet without additives); PC = positive control (basal diet supplemented with 2500 ppm ZnO); IBW = initial body weight; WBC = white blood cell count; LYM = lymphocyte count; GRA = the number and percentage of granulocytes; MID = mid-range absolute count (indicate non-classified cells); RBC = red blood cells count; HCT = hematocrite; Hb = hemoglobin concentration; [1] AN was included on top of 10% FRM at 0.6% or 1.0% giving 10% FRM +0.6% AN or 10% FRM + 1.0% AN overall supplements.

Table 3. Effects of a pharmacological dose of ZnO, increasing doses of *Lactobacillus* pre-fermented rapeseed meal (FRM) and a combination of 10% FRM with increasing doses of *Ascophyllum nodosum* (AN) on immunological parameters in weaned piglets slaughtered 14 days post-weaning.

| Parameters | Treatments | | | | | | | | | SEM | *p*-Value |
| | Controls | | FRM (%) | | | | | AN [1] (%) | | | TG |
	NC	PC	8	10	12	15	25	0.6	1.0		
IgG, mg/ml	2.11 [ab]	2.39 [cd]	3.63 [f]	2.22 [bc]	2.57 [de]	1.94 [a]	2.62 [de]	2.57 [d]	2.82 [e]	0.062	<0.0001
IgA, mg/ml	0.099 [bc]	0.136 [d]	0.107 [c]	0.170 [e]	0.077 [ab]	0.080 [ab]	0.066 [a]	0.137 [d]	0.081 [ab]	0.005	<0.0001
IgM, mg/ml	0.395 [bc]	0.545 [d]	0.427 [c]	0.679 [e]	0.309 [ab]	0.318 [ab]	0.263 [a]	0.547 [d]	0.323 [ab]	0.022	<0.0001
IL-6, pg/ml	208.6 [ab]	214.6 [ab]	370.9 [e]	260.0 [c]	195.5 [a]	226.8 [b]	285.4 [d]	224.8 [b]	214.9 [ab]	5.4	<0.0001

[a,b,c,d] Different letters within same row indicate significant difference ($p < 0.05$); live body weight (LBW) at slaughter has no effect any of the immunological parameters; NC = negative control (basal diet without additives); PC = positive control (basal diet supplemented with 2500 ppm ZnO); LBW = live body weight; TG = Treatment group; [1] AN was included on top of 10% FRM at 0.6% or 1.0% giving 10% FRM +0.6% AN or 10% FRM + 1.0% AN overall supplements.

4. Discussion

In the current study, FRM demonstrated the same desirable effects on ADG, and intestinal development and barrier function as we have previously observed and optimum effects were observed when 8% or 10% FRM was included in the weaner diets [14]. Similar improvements in ADG upon dietary addition of FRM in broiler chicken [27] and fermented soybean meal in piglets [28,29] were reported in other studies. This beneficial effect could most likely be explained by the fact that supplementation of fermented feed may provide beneficial probiotic bacteria, such as lactic acid bacteria, and also prebiotic products produced during the pre-fermentation process [30,31]. Rapeseed meal has a relatively high fiber and anti-nutritional contents compared to soybean, which can potentially decrease growth performance in monogastric species [32]. However, the negative effects of these components may be removed by pre-fermentation of rapeseed meal so that nutrient digestibility of the product could be increased [33], while anti-nutritional factors such as tannin, phytate and glucosinolate could be degraded during the pre-fermentation process [13,31].

In this study, AN did not have any impacts on ADG at any of the doses provided on top of the FRM. This is in contrast to Turner et al. [34], who reported a quadratic effect on ADG in piglets with increasing dietary levels of AN, with a maximum effect being achieved at 1% AN in dietary DM. However, Michiels et al. [24] have also found that feeding of intact AN to weaned piglets did not affect ADG and FCR, and supplementation of increased doses of AN to grower-finisher pigs has been reported to have a directly negative effect on ADG without affecting FCR and feed intake [22]. The lack of effects of intact seaweeds on piglet performance in the current study may be due to the presence of a large amount of anti-nutritional factors, such as phlorotannins in the intact macroalgae [35], and potentially also due to antagonistic actions of bioactive components in the algae. In some studies, purified extracts of special carbohydrates from macroalgae, such as laminarin and fucoidan, have been shown to improve ADG and overall performance of the animal when fed alone, but not when fed in combination [20,36]. Furthermore, the antimicrobial activities of AN, as confirmed by a drastic reduction in the concentration of fecal lactic acids compared to FRM, NC and PC piglets, may reduce hindgut fermentation and thus the availability of SCFAs and microbial vitamins that are essential for growth and also intestinal health (Figure S1B).

Inclusion of FRM at 8% or 10% of dietary DM in weaner diets in the current study resulted in improved intestinal morphometric indices (JVH, JEH, JCD and CCD), intestinal mucosal integrity and barrier functions consistent with observed improvements in ADG. Long villus height and intact mucosal integrity are essential for nutrient digestion and absorption as well as prevention of invasion of intestinal tissues by pathogenic microbes. Similar findings were reported when fermented rapeseed was fed to broilers where intestinal villus height and villus-to-crypt ratio showed significant improvements [13,37,38]. Feeding of fermented soybean to weaned piglets [39] and Japanese quail [40] also demonstrated significant improvement in villus height but reduced crypt depth along the intestine. However, some studies showed that feeding of lactobacilli fermented liquid feed (feed to water ratio of 1:3) significantly reduced JVH in weaned piglets [41]. The increase in villi development could be due to reduced anti-nutrition factors by pre-fermentation of the feed product [13,42] or increased fermentation products, mostly SCFAs, such as lactate, acetate and butyrate [43]. Stresses during weaning can increase oxidative stresses, which may lead to damage of intestinal tissues during weaning [44], but the generation of metabolites with antioxidant potential during fermentation could also be responsible for the healthy mucosal integrity and barrier functions observed in the current study [45].

The desirable effects of FRM on gut histomorphometry and GALT were counteracted by the addition of AN on top of 10% FRM. A similar result was reported where intestinal villus height was reduced in response to feeding of weaned piglets with a seaweed extract from *Laminaria digitata* [21]. In contrast, some studies reported villus height and villus-to-crypt ratio were improved when weaned piglets were supplemented with the seaweed extracts such as laminarin and fucoidan separately but these morphometric parameters were not affected when these extracts were supplemented in combination [46,47].

Although a high percentage of piglets completed the experiment in response to feeding of FRM (highest at 8% FRM) or a combination with AN (highest at 0.6% AN), piglets were not protected from diarrhea in the current study. We speculate that this diarrhea could be linked to osmotic pressure triggered by possibly high intestinal lactic acid concentrations as confirmed by detection of high fecal lactic acid concentration amidst intact intestinal mucosal integrity and barrier functions. In contrast, fermented feed such as soybean has been reported to prevent diarrhea in piglets weaned at 28 days of age [28,48]. Studies also revealed that the inclusion of fermented feeds in the diets of monogastric animals reduced the counts of enteric pathogens such as *Salmonella typhimurium* and *E. coli*, which are known etiologic agents for enteric infections and diarrhea [28,49,50]. Regarding the macroalgae AN, studies are limited and available ones reported inconsistent findings, where feeding of AN to growing–finisher pigs reduced coliform counts [22], while no effect on gut microbial profile was reported when AN was fed to weaned piglets [24]. However, others found that feeding of purified seaweeds extracts such as laminarin and fucoidan reduced enteropathogens such as *E. coli* which are the causative agent of infectious diarrhea during weaning [18,51].

RBC indices are essential markers for efficient oxygen transport to tissues and nutrient utilization for energy release. Although all of the RBC indices investigated were in the normal ranges for all of the treatments in the current study [52], they were increased in piglets fed FRM without or with AN compared to the NC and PC. Bhattarai and Nielsen [53] reported that a positive association exists between growth and hematological indices following weaning. Thus, it could be speculated that the increase in RBC indices demonstrated in the present study might be associated with increased growth and hence metabolic demands in pigs fed with FRM or its combination with AN.

In the current study, we found out that WBC indices were generally within normal ranges compared to reference values [52], and feeding of FRM and PC reduced WBC counts in weaned piglets in contrast to the NC. Conversely, levels of serum immunoglobulins (IgG, IgM, IgA) and IL-6 were increased in response to lower doses (8% or 10% FRM), while the inclusion of AN enhanced IgG but reduced the rests. A study showed that feeding of fermented products, such as the soybean meal, to weaned piglets significantly improved mucosal or systemic immunoglobulins [38,39,54,55]. The discrepancy between increased immunoglobulins and WBC counts could be related to the preferential activation of the adaptive immune responses, which culminate in the increment of immunoglobulins. Serum immunoglobulins may also increase in response to probiotic bacterial such as lactic acid bacteria and also bioactive components present in FRM [31,38]. The weaning period is usually marked by an ineffective immune response due to the immaturity of the immune system and also depression by weaning stresses. However, appropriate immune responses during this period could be very essential to prevent infections by enteric pathogens, and hence post-weaning diarrhea.

5. Conclusions

FRM improved piglet performance (ADG) and intestinal histomorphometric indices demonstrating optimum effects at 8% or 10% inclusion, but further inclusion of the brown macroalgae *Ascophyllum nodosum* (AN) did not show any additional benefits. Thus, FRM could be a potential alternative for medicinal ZnO when supplemented at $\leq$ 10% dietary DM. However, a comprehensive dose-response study is deemed necessary to assure optimum inclusion rate for desirable effects on performance and gut health.

Supplementary Materials: The following are available online at http://www.mdpi.com/2076-2615/10/4/559/s1, Figure S1: Effect pharmacological doses of ZnO, increasing doses of FRM or combination of 10% FRM with AN on dietary (Panel A) and fecal (PaneL B) lactic acid concentrations, Figure S2: Effects of various doses of FRM with or without AN on the density of GALT (Panel A), and jejunal (Panel B) and colonic (Panel C) immune cell infiltrations in piglets (LSM $\pm$ SEM) slaughtered two weeks post-weaning, Table S1: Ingredients and chemical compositions of the pe-starter and starter diets fed to piglets from 10 days before weaning until they exited the experiment at 92 days of age, Table S2: Effects of a pharmacological dose of ZnO, increasing doses of FRM and combination of 10% FRM with increasing doses of AN on growth performance in piglets weaned at 28 days of age, Table S3: Effects of a pharmacological dose of ZnO, increasing doses of FRM and a combination of 10% FRM with increasing doses of *Ascophyllum nodossum* on a number of cases of post-weaning diarrhea. The percentage of piglets that exited the weaner units are also shown.

Author Contributions: Conceptualization: M.O.N.; methodology: M.O.N. and G.D.S.; formal analysis: G.D.S. and R.D.; investigation: M.O.N. and G.D.S.; data curation: G.D.S.; writing—original draft preparation: G.D.S.; writing—review and editing: M.O.N., P.T.-H., S.K., E.V.-B.-P., R.D. and G.D.S.; supervision: M.O.N.; project administration: M.O.N.; funding acquisition: Fermentationexperts, M.O.N. All authors have read and agreed to the published version of the manuscript.

Funding: This research was funded by the Innovation Fund Denmark to Fermentationexperts and University of Copenhagen under grant agreement number 5157-00003A (MAB4—MacroAlgae Biorefinery for Value-Added Products).

Acknowledgments: The Authors would like to acknowledge Robert Chachaj, Feedstar Sp z.o.o., Poland, for help with coordination of experimental activities at the Chotycze Farm in Poland, and Fermentation Experts A/S, Bække, Denmark, for supply of the pre-fermented rapeseed meal and *Ascophyllum nodossum* feed ingredients.

Conflicts of Interest: The authors declare no conflict of interest.

References

1. Pluske, J.R.; Hampson, D.J.; Williams, I.H. Factors influencing the structure and function of the small intestine in the weaned pig: A review. *Livest. Prod. Sci.* **1997**, *51*, 215–236. [CrossRef]
2. Lallès, J.P.; Boudry, G.; Favier, C.; Le Floc'h, N.; Luron, I.; Montagne, L.; Oswald, I.P.; Pié, S.; Piel, C.; Sève, B. Gut function and dysfunction in young pigs: Physiology. *Anim. Res.* **2004**, *53*, 301–316. [CrossRef]
3. Campbell, J.M.; Crenshaw, J.D.; Polo, J. The biological stress of early weaned piglets. *J. Anim. Sci. Biotechnol.* **2013**, *4*, 1–19. [CrossRef] [PubMed]
4. Jensen, J.; Larsen, M.M.; Bak, J. National monitoring study in Denmark finds increased and critical levels of copper and zinc in arable soils fertilized with pig slurry. *Environ. Pollut.* **2016**, *214*, 334–340. [CrossRef] [PubMed]
5. Aarestrup, F.M.; Jensen, V.F.; Emborg, H.D.; Jacobsen, E.; Wegener, H.C. Changes in the use of antimicrobials and the effects on productivity of swine farms in Denmark. *Am. J. Vet. Res.* **2010**, *71*, 726–733. [CrossRef]
6. Cavaco, L.M.; Hasman, H.; Aarestrup, F.M.; Wagenaar, J.A.; Graveland, H.; Veldman, K.; Mevius, D.; Fetsch, A.; Tenhagen, B.A.; Porrero, M.C.; et al. Zinc resistance of Staphylococcus aureus of animal origin is strongly associated with methicillin resistance. *Vet. Microbiol.* **2011**, *150*, 344–348. [CrossRef]
7. EMA. Questions and answers on veterinary medicinal products containing zinc oxide to be administered orally to food-producing species. In *Outcome of a Referral Procedure under Article 35 of Directive 2001/82/EC (EMEA/V/A/118)—EMA/394961/2017*; European Medicines Agency: London, UK, 2017.
8. Xu, B.; Li, Z.; Wang, C.; Fu, J.; Zhang, Y.; Wang, Y.; Lu, Z. Effects of fermented feed supplementation on pig growth performance: A meta-analysis. *Anim. Feed Sci. Technol.* **2020**, *259*, 114315. [CrossRef]
9. Van Winsen, R.L.; Urlings, B.A.; Lipman, L.J.; Snijders, J.M.; Keuzenkamp, D.; Verheijden, J.H.; Van Knapen, F. Effect of fermented feed on the microbial population of the gastrointestinal tracts of pigs. *Appl. Environ. Microbiol.* **2001**, *67*, 3071–3076. [CrossRef]
10. Canibe, N.; Jensen, B.B. Fermented liquid feed—Microbial and nutritional aspects and impact on enteric diseases in pigs. *Anim. Feed Sci. Technol.* **2012**, *173*, 17–40. [CrossRef]
11. Hong, T.T.T.; Thuy, T.T.; Passoth, V.; Lindberg, J.E. Gut ecology, feed digestion and performance in weaned piglets fed liquid diets. *Livest. Sci.* **2009**, *125*, 232–237. [CrossRef]
12. Demecková, V.; Kelly, D.; Coutts, A.G.P.; Brooks, P.H.; Campbell, A. The effect of fermented liquid feeding on the faecal microbiology and colostrum quality of farrowing sows. *Int. J. Food Microbiol.* **2002**, *79*, 85–97. [CrossRef]
13. Hu, Y.; Wang, Y.; Li, A.; Wang, Z.; Zhang, X.; Yun, T.; Qiu, L.; Yin, Y. Effects of fermented rapeseed meal on antioxidant functions, serum biochemical parameters and intestinal morphology in broilers. *Food Agr. Immunol.* **2016**, *27*, 182–193. [CrossRef]
14. Satessa, G.D.; Tamez-Hidalgo, P.; Hui, Y.; Cieplak, T.; Krych, L.; Kjærulff, S.; Brunsgaard, G.; Nielsen, D.S.; Nielsen, M.O. Impact of Dietary Supplementation of Lactic Acid Bacteria Fermented Rapeseed with or without Macroalgae on Performance and Health of Piglets Following Omission of Medicinal Zinc from Weaner Diets. *Animals* **2020**, *10*, 137. [CrossRef] [PubMed]

15. Øverland, M.; Mydland, L.T.; Skrede, A. Marine macroalgae as sources of protein and bioactive compounds in feed for monogastric animals. *J. Sci. Food Agr.* **2019**, *99*, 13–24. [CrossRef] [PubMed]

16. Kadam, S.; O'Donnell, C.; Rai, D.; Hossain, M.; Burgess, C.; Walsh, D.; Tiwari, B. Laminarin from Irish brown seaweeds Ascophyllum nodosum and Laminaria hyperborea: Ultrasound assisted extraction, characterization and bioactivity. *Mar. Drugs* **2015**, *13*, 4270–4280. [CrossRef] [PubMed]

17. Jun, J.Y.; Jung, M.J.; Jeong, I.H.; Yamazaki, K.; Kawai, Y.; Kim, B.M. Antimicrobial and antibiofilm activities of sulfated polysaccharides from marine algae against dental plaque bacteria. *Mar. Drugs* **2018**, *16*, 301. [CrossRef]

18. O'Shea, C.J.; O'Doherty, J.V.; Callanan, J.J.; Doyle, D.; Thornton, K.; Sweeney, T. The effect of algal polysaccharides laminarin and fucoidan on colonic pathology, cytokine gene expression and Enterobacteriaceae in a dextran sodium sulfate-challenged porcine model. *J. Nutr. Sci.* **2016**, *5*, e15. [CrossRef]

19. Moroney, N.; O'Grady, M.; Lordan, S.; Stanton, C.; Kerry, J. Seaweed polysaccharides (laminarin and fucoidan) as functional ingredients in pork meat: An evaluation of anti-oxidative potential, thermal stability and bioaccessibility. *Mar. Drugs* **2015**, *13*, 2447–2464. [CrossRef]

20. McDonnell, P.; Figat, S.; O'doherty, J.V. The effect of dietary laminarin and fucoidan in the diet of the weanling piglet on performance, selected faecal microbial populations and volatile fatty acid concentrations. *Animal* **2010**, *4*, 579–585. [CrossRef]

21. Reilly, P.; O'doherty, J.V.; Pierce, K.M.; Callan, J.J.; O'sullivan, J.T.; Sweeney, T. The effects of seaweed extract inclusion on gut morphology, selected intestinal microbiota, nutrient digestibility, volatile fatty acid concentrations and the immune status of the weaned pig. *Animal* **2008**, *2*, 1465–1473. [CrossRef]

22. Gardiner, G.E.; Campbell, A.J.; O'Doherty, J.V.; Pierce, E.; Lynch, P.B.; Leonard, F.C.; Stanton, C.; Ross, R.P.; Lawlor, P.G. Effect of Ascophyllum nodosum extract on growth performance, digestibility, carcass characteristics and selected intestinal microflora populations of grower–finisher pigs. *Anim. Feed Sci. Technol.* **2008**, *141*, 259–273. [CrossRef]

23. Dierick, N.; Ovyn, A.; De Smet, S. Effect of feeding intact brown seaweed Ascophyllum nodosum on some digestive parameters and on iodine content in edible tissues in pigs. *J. Sci. Food Agr* **2009**, *89*, 584–594. [CrossRef]

24. Michiels, J.; Skrivanova, E.; Missotten, J.; Ovyn, A.; Mrazek, J.; De Smet, S.; Dierick, N. Intact brown seaweed (*Ascophyllum nodosum*) in diets of weaned piglets: Effects on performance, gut bacteria and morphology and plasma oxidative status. *J. Anim. Physiol. Anim. Nutr.* **2012**, *96*, 1101–1111. [CrossRef] [PubMed]

25. Tybirk, P.E.R. *Nutrient Recommendations for Pigs in Denmark*, 17th ed.; SEGES-VSP Danis Pig Research Center: Copenhague, Denmark, 2015.

26. R Core Team. *R: A Language and Environment for Statistical Computing*; R Foundation for Statistical Computing: Vienna, Austria, 2019; Available online: https://www.R-project.org/ (accessed on 21 January 2020).

27. Ashayerizadeh, A.; Dastar, B.; Shargh, M.S.; Mahoonak, A.S.; Zerehdaran, S. Effects of feeding fermented rapeseed meal on growth performance, gastrointestinal microflora population, blood metabolites, meat quality, and lipid metabolism in broiler chickens. *Livest. Sci.* **2018**, *216*, 183–190. [CrossRef]

28. Yuan, L.; Chang, J.; Yin, Q.; Lu, M.; Di, Y.; Wang, P.; Wang, Z.; Wang, E.; Lu, F. Fermented soybean meal improves the growth performance, nutrient digestibility, and microbial flora in piglets. *Anim. Nutr.* **2017**, *3*, 19–24. [CrossRef] [PubMed]

29. Zhang, Y.; Shi, C.; Wang, C.; Lu, Z.; Wang, F.; Feng, J.; Wang, Y. Effect of soybean meal fermented with Bacillus subtilis BS12 on growth performance and small intestinal immune status of piglets. *Food Agr. Immunol.* **2018**, *29*, 133–146. [CrossRef]

30. Shi, C.; He, J.; Yu, J.; Yu, B.; Huang, Z.; Mao, X.; Zheng, P.; Chen, D. Solid state fermentation of rapeseed cake with Aspergillus niger for degrading glucosinolates and upgrading nutritional value. *J. Anim. Sci. Biotechnol.* **2015**, *6*, 13. [CrossRef]

31. Drażbo, A.; Ognik, K.; Zaworska, A.; Ferenc, K.; Jankowski, J. The effect of raw and fermented rapeseed cake on the metabolic parameters, immune status, and intestinal morphology of turkeys. *Poult. Sci.* **2018**, *97*, 3910–3920. [CrossRef]

32. Zhu, L.P.; Wang, J.P.; Ding, X.M.; Bai, S.P.; Zeng, Q.F.; Su, Z.W.; Xuan, Y.; Applegate, T.J.; Zhang, K.Y. The effects of varieties and levels of rapeseed expeller cake on egg production performance, egg quality, nutrient digestibility, and duodenum morphology in laying hens. *Poult. Sci.* **2019**, *98*, 4942–4953. [CrossRef]

33. Koo, B.; Kim, J.W.; Nyachoti, C.M. Nutrient and energy digestibility, and microbial metabolites in weaned pigs fed diets containing Lactobacillus–fermented wheat. *Anim. Feed Sci. Technol.* **2018**, *241*, 27–37. [CrossRef]

34. Turner, J.L.; Dritz, S.S.; Higgins, J.J.; Minton, J.E. Effects of *Ascophyllum nodosum* extract on growth performance and immune function of young pigs challenged with *Salmonella typhimurium*. *J. Anim. Sci.* **2002**, *80*, 1947–1953. [CrossRef] [PubMed]

35. Huang, Q.; Liu, X.; Zhao, G.; Hu, T.; Wang, Y. Potential and challenges of tannins as an alternative to in-feed antibiotics for farm animal production. *Anim. Nutr.* **2018**, *4*, 137–150. [CrossRef] [PubMed]

36. Heim, G.; Walsh, A.M.; Sweeney, T.; Doyle, D.N.; O'shea, C.J.; Ryan, M.T.; O'doherty, J.V. Effect of seaweed-derived laminarin and fucoidan and zinc oxide on gut morphology, nutrient transporters, nutrient digestibility, growth performance and selected microbial populations in weaned pigs. *Br. J. Nutr.* **2014**, *111*, 1577–1585. [CrossRef]

37. Chiang, G.; Lu, W.Q.; Piao, X.S.; Hu, J.K.; Gong, L.M.; Thacker, P.A. Effects of feeding solid-state fermented rapeseed meal on performance, nutrient digestibility, intestinal ecology and intestinal morphology of broiler chickens. *Asian-Australas. J. Anim. Sci.* **2009**, *23*, 263–271. [CrossRef]

38. Xu, F.Z.; Zeng, X.G.; Ding, X.L. Effects of replacing soybean meal with fermented rapeseed meal on performance, serum biochemical variables and intestinal morphology of broilers. *Asian-Australas. J. Anim. Sci.* **2012**, *25*, 1734. [CrossRef]

39. Zhu, J.; Gao, M.; Zhang, R.; Sun, Z.; Wang, C.; Yang, F.; Huang, T.; Qu, S.; Zhao, L.; Li, Y.; et al. Effects of soybean meal fermented by L. plantarum, B. subtilis and S. cerevisieae on growth, immune function and intestinal morphology in weaned piglets. *Microb. Cell Fact.* **2017**, *16*, 191. [CrossRef]

40. Jazi, V.; Ashayerizadeh, A.; Toghyani, M.; Shabani, A.; Tellez, G.; Toghyani, M. Fermented soybean meal exhibits probiotic properties when included in Japanese quail diet in replacement of soybean meal. *Poult. Sci.* **2018**, *97*, 2113–2122. [CrossRef]

41. Missotten, J.A.M.; Michiels, J.; Willems, W.; Ovyn, A.; De Smet, S.; Dierick, N.A. Effect of fermented liquid feed on morpho-histological parameters in the piglet gut. *Livest. Sci.* **2010**, *134*, 155–157. [CrossRef]

42. Feng, J.; Liu, X.; Xu, Z.R.; Wang, Y.Z.; Liu, J.X. Effects of fermented soybean meal on digestive enzyme activities and intestinal morphology in broilers. *Poult. Sci.* **2007**, *86*, 1149–1154. [CrossRef]

43. Diao, H.; Jiao, A.R.; Yu, B.; Mao, X.B.; Chen, D.W. Gastric infusion of short-chain fatty acids can improve intestinal barrier function in weaned piglets. *Genes Nutr.* **2019**, *14*, 4. [CrossRef]

44. Liu, Y.; Xiong, X.; Tan, B.E.; Song, M.; Ji, P.; Kim, K.; Yin, Y. Nutritional intervention for the intestinal development and health of weaned pigs. *Front. Vet. Sci.* **2019**, *6*, 46. [CrossRef]

45. He, R.; Ju, X.; Yuan, J.; Wang, L.; Girgih, A.T.; Aluko, R.E. Antioxidant activities of rapeseed peptides produced by solid state fermentation. *Food Res. Int.* **2012**, *49*, 432–438. [CrossRef]

46. Walsh, A.M.; Sweeney, T.; O'Shea, C.J.; Doyle, D.N.; O'Doherty, J.V. Effects of supplementing dietary laminarin and fucoidan on intestinal morphology and the immune gene expression in the weaned pig. *J. Anim. Sci.* **2012**, *90*, 284–286. [CrossRef] [PubMed]

47. Walsh, A.M.; Sweeney, T.; O'shea, C.J.; Doyle, D.N.; O'doherty, J.V. Effect of dietary laminarin and fucoidan on selected microbiota, intestinal morphology and immune status of the newly weaned pig. *Br. J. Nutr.* **2013**, *110*, 1630–1638. [CrossRef]

48. Kiers, J.L.; Meijer, J.C.; Nout, M.J.R.; Rombouts, F.M.; Nabuurs, M.J.A.; Van der Meulen, J. Effect of fermented soya beans on diarrhoea and feed efficiency in weaned piglets. *J. Appl. Microbiol.* **2003**, *95*, 545–552. [CrossRef]

49. Ashayerizadeh, A.; Dastar, B.; Shargh, M.S.; Mahoonak, A.S.; Zerehdaran, S. Fermented rapeseed meal is effective in controlling *Salmonella enterica* serovar *Typhimurium* infection and improving growth performance in broiler chicks. *Vet. Microbiol.* **2017**, *201*, 93–102. [CrossRef]

50. Hu, J.; Lu, W.; Wang, C.; Zhu, R.; Qiao, J. Characteristics of solid-state fermented feed and its effects on performance and nutrient digestibility in growing-finishing pigs. *Asian-Australas. J. Anim. Sci.* **2008**, *21*, 1635–1641. [CrossRef]

51. Lynch, M.B.; Sweeney, T.; Callan, J.J.; O'Sullivan, J.T.; O'Doherty, J.V. The effect of dietary Laminaria-derived laminarin and fucoidan on nutrient digestibility, nitrogen utilisation, intestinal microflora and volatile fatty acid concentration in pigs. *J. Sci. Food Agr.* **2010**, *90*, 430–437. [CrossRef]

52. Ventrella, D.; Dondi, F.; Barone, F.; Serafini, F.; Elmi, A.; Giunti, M.; Romagnoli, N.; Forni, M.; Bacci, M.L. The biomedical piglet: Establishing reference intervals for haematology and clinical chemistry parameters of two age groups with and without iron supplementation. *BMC Vet. Res.* **2016**, *13*, 23. [CrossRef]

53. Bhattarai, S.; Nielsen, J.P. Association between hematological status at weaning and weight gain post-weaning in piglets. *Livest. Sci.* **2015**, *182*, 64–68. [CrossRef]

54. Tang, J.W.; Sun, H.; Yao, X.H.; Wu, Y.F.; Wang, X.; Feng, J. Effects of replacement of soybean meal by fermented cottonseed meal on growth performance, serum biochemical parameters and immune function of yellow-feathered broilers. *Asian-Australas. J. Anim. Sci.* **2012**, *25*, 393. [CrossRef] [PubMed]

55. Mizumachi, K.; Aoki, R.; Ohmori, H.; Saeki, M.; Kawashima, T. Effect of fermented liquid diet prepared with Lactobacillus plantarum LQ80 on the immune response in weaning pigs. *Animal* **2009**, *3*, 670–676. [CrossRef] [PubMed]

Article

Influence of the Fermented Feed and Vaccination and Their Interaction on Parameters of Large White/Norwegian Landrace Piglets

Laurynas Vadopalas [1], Sarunas Badaras [1], Modestas Ruzauskas [2,3], Vita Lele [1,4], Vytaute Starkute [1,4], Paulina Zavistanaviciute [1,4], Egle Zokaityte [1,4], Vadims Bartkevics [5,6], Dovile Klupsaite [1], Erika Mozuriene [1,4], Agila Dauksiene [1,2], Sonata Sidlauskiene [1], Romas Gruzauskas [7] and Elena Bartkiene [1,4,*]

[1] Institute of Animal Rearing Technologies, Lithuanian University of Health Sciences, Tilžės g. 18, LT-47181 Kaunas, Lithuania; laurynas.vadopalas@lsmuni.lt (L.V.); sarunas.badaras@gmail.com (S.B.); vita.lele@lsmuni.lt (V.L.); vytaute.starkute@lsmuni.lt (V.S.); paulina.zavistanaviciute@lsmuni.lt (P.Z.); egle.zokaityte@lsmuni.lt (E.Z.); dovile.klupsaite@lsmuni.lt (D.K.); erika.mozuriene@lsmuni.lt (E.M.); agila.dauksiene@lsmuni.lt (A.D.); sonata.sidlauskiene@lsmuni.lt (S.S.)
[2] Department of Physiology and Anatomy, Lithuanian University of Health Sciences, Tilžės g. 18, LT-47181 Kaunas, Lithuania; modestas.ruzauskas@lsmuni.lt
[3] Microbiology and Virology Institute, Lithuanian University of Health Sciences, Tilžės g. 18, LT-47181 Kaunas, Lithuania
[4] Department of Food Safety and Quality, Lithuanian University of Health Sciences, Tilžės g. 18, LT-47181 Kaunas, Lithuania
[5] University of Latvia Faculty of Chemistry, Jelgavas iela 1, LV-1004 Riga, Latvia; vadims.bartkevics@bior.gov.lv
[6] Institute of Food Safety, Animal Health and Environment BIOR, Lejupesiela 3, LV-1076 Riga, Latvia
[7] Department of Food Science and Technology, Kaunas University of Technology, Radvilenu str. 19, LT-50254 Kaunas, Lithuania; romas.gruzauskas@ktu.lt
* Correspondence: elena.bartkiene@lsmuni.lt; Tel.: +370-601-35837

Received: 2 June 2020; Accepted: 13 July 2020; Published: 15 July 2020

Simple Summary: Farm animals are constantly exposed to pathogenic microorganisms, which can lead to bacterial, as well as secondary infections caused by viruses. For this reason, vaccination is used. However, many discussions have been published about the effectiveness of this prophylactic measure. The aim of this study was to evaluate the influence of fermented feed on non-vaccinated (NV) and vaccinated with Circovac porcine circovirus type 2 vaccine piglets' blood parameters, gut microbial composition, growth performance and ammonia emission. The 36-day experiment was conducted using 25-day-old Large White/Norwegian Landrace piglets, which were randomly divided into four groups each comprising 100 piglets (NV piglets fed with soya meal, vaccinated piglets fed with soya meal, NV piglets fed with fermented rapeseed meal and vaccinated piglets fed with fermented rapeseed meal). The results revealed that vaccination, as a separate factor, did not significantly influence piglets' blood parameters. Finally, rapeseed meal fermented with the selected lactic acid bacteria strains can be used instead of expensive imported soya, because the fermented feed did not cause undesirable changes in piglets' health and growth performance. Furthermore, the process is more sustainable.

Abstract: The aim of this study was to evaluate the influence of fermented with a newly isolated lactic acid bacteria (LAB) strains combination (*Lactobacillus plantarum* LUHS122, *Lactobacillus casei* LUHS210, *Lactobacillus farraginis* LUHS206, *Pediococcus acidilactici* LUHS29, *Lactobacillus plantarum* LUHS135 and *Lactobacillus uvarum* LUHS245) feed on non-vaccinated (NV) and vaccinated with Circovac porcine circovirus type 2 vaccine (QI09AA07, CEVA-PHYLAXIA Co. Ltd. Szállás u. 5. 1107 Budapest, Hungary) piglets' blood parameters, gut microbial composition, growth performance and ammonia emission. The 36-day experiment was conducted using 25-day-old Large White/Norwegian

Landrace (LW/NL) piglets, which were randomly divided into four groups with 100 piglets each: S_{nonV}—non-vaccinated piglets fed with control group compound feed; S_V—vaccinated piglets fed with control group compound feed; RF_{nonV}—non-vaccinated piglets fed with fermented compound feed; RF_V—vaccinated piglets fed with fermented compound feed. Samples from 10 animals per group were collected at the beginning and end of the experiment. Metagenomic analysis showed that fermentation had a positive impact on the *Lactobacillus* prevalence during the post-weaning period of pigs, and vaccination had no negative impact on microbial communities. Although a higher amount of *Lactobacillus* was detected in vaccinated, compared with non-vaccinated groups. At the end of experiment, there was a significantly higher LAB count in the faeces of both vaccinated compared to non-vaccinated groups (26.6% for S_V and 17.2% for RF_V), with the highest LAB count in the S_V group. At the end of experiment, the S_V faeces also had the highest total bacteria count (TBC). The RF_V group had a 13.2% increase in total enterobacteria count (TEC) at the end of experiment, and the S_V group showed a 31.2% higher yeast/mould (Y/M) count. There were no significant differences in the average daily gain (ADG) among the groups; however, there were significant differences in the feed conversion ratios (FCR) between several groups: S_V vs. S_{nonV} (11.5% lower in the S_V group), RF_V vs. RF_{nonV} (10.2% lower in the RFnonV group) and S_V vs. RF_V (21.6% lower in the S_V group). Furthermore, there was a significant, very strong positive correlation between FCR and TEC in piglets' faeces (R = 0.919, p = 0.041). The lowest ammonia emission was in RF_V group section (58.2, 23.8, and 47.33% lower compared with the S_{nonV}, S_V and RF_{nonV} groups, respectively). Notably, there was lower ammonia emission in vaccinated groups (45.2% lower in S_V vs. S_{nonV} and 47.33% lower in RF_V vs. RF_{nonV}). There was also a significant, very strong positive correlation between ammonia emission and Y/M count in piglets' faeces at the end of the experiment (R = 0.974; p = 0.013). Vaccination as a separate factor did not significantly influence piglets' blood parameters. Overall, by changing from an extruded soya to cheaper rapeseed meal and applying the fermentation model with the selected LAB combination, it is possible to feed piglets without any undesirable changes in health and growth performance in a more sustainable manner. However, to evaluate the influence of vaccination and its interaction with other parameters (feed, piglets' age, breed, etc.) on piglets' parameters, additional studies should be performed and methods should be standardised to ensure the results may be compared.

Keywords: piglets; fermentation; antimicrobial properties; vaccination; microbiota; ammonia emission

---

1. Introduction

The challenges associated with livestock production become greater each year. This phenomenon is associated with population growth, as well as growing needs for animal-based products. Farms have become bigger, and animals are always exposed to opportunistic and pathogenic microorganisms. Most pathogen infections have a multifunctional influence on the host tissues and cells, and the response to the infection rather than the infection itself results in most of the damage and a general disturbance in physiological processes [1–5]. One consequence of infections on pig production is multisystemic wasting syndrome after weaning. Indeed, it is crucial to ensure piglets health at the earliest stage, because type 2 porcine circovirus (PCV2) leads to multisystemic wasting syndrome after weaning, and co-infection with other pathogens can be lethal [6,7]. Some published findings have indicated the potential participation of PPV2 in clinical diseases. PCV2 can occur as a coinfection with other pathogens in animals afflicted with respiratory disease [8], as well as in the lungs of animals with inflammatory lesions [9]. PCV2 infection leads to metabolic and neurological disorders; the latter cause multisystemic wasting syndrome [10–12]. To improve animal keeping conditions, understanding the interactions between desirable microorganisms and pathogens and reduce the risk of secondary infections is critical to develop preventive measures against various diseases. In this study, we hypothesised that, for vaccinated piglets, incorporating antimicrobial strains for feed fermentation can be a very attractive feeding option. Finally, incorporation and biomodification of local feedstock with antimicrobial lactic acid bacteria (LAB) strains can lead to higher value feedstock for farm animal nutrition and improve their conditions. This use can also increase the sustainability of pig farming and reduce dependence on imported soya stock. In this case, natural antimicrobial treatment, e.g., feed fermentation with antimicrobial LAB strains, is also desirable because the genetic diversity of PCV2 continuously increases, and novel subtypes are still emerging [13–15]. PCV2-related diseases are considered multifactorial in the swine industry [7], and they can promote co-infection with other viruses or bacteria [8,9,16]. In addition, co-infection of different viruses complicates the infection status and makes it more difficult to prevent and control the diseases [17]. Vaccines are used to prevent viral infections and antibiotics are used to treat bacterial diseases, but there is a lack of complex tools for prevention of diseases. Furthermore, viruses and bacteria can mutate very quickly, a phenomenon that remains a huge challenge for the livestock industry. Animals are always exposed to the environment, and thus changes in the microorganisms present in digestive tract should be studied to understand the mechanism of action, which can be associated with vaccination, as well as with the natural antimicrobials, as the microorganisms in the gut strongly influence the immune system and animals' health.

The aim of this study was to evaluate the influence of fermented with a newly isolated LAB strains combination (*Lactobacillus plantarum* LUHS122, *Lactobacillus casei* LUHS210, *Lactobacillus farraginis* LUHS206, *Pediococcus acidilactici* LUHS29, *Lactobacillus plantarum* LUHS135 and *Lactobacillus uvarum* LUHS245) feed on non-vaccinated (NV) and vaccinated with Circovac porcine circovirus type 2 vaccine (QI09AA07, CEVA-PHYLAXIA Co. Ltd. Szállás u. 5. 1107 Budapest, Hungary) piglets' blood parameters, gut microbial composition, growth performance and ammonia emission. This LAB starter combination, with antimicrobial characteristics, was used to ferment local feed stock (rapeseed meal), and the influence of changing from an extruded soya to biomodified rapeseed meal on the parameters of non-vaccinated piglets and those vaccinated with the Circovac PCV2 vaccine was evaluated.

2. Materials and Methods

2.1. LAB Strains Used for Feed Fermentation, Feed Fermentation and Fermented Feed Parameters

The *L. plantarum* LUHS122, *L. casei* LUHS210, *L. farraginis* LUHS206, *P. acidilactici* LUHS29, *L. plantarum* LUHS135 and *L. uvarum* LUHS245 strains were obtained from the Lithuanian University of Health Sciences collection (Kaunas, Lithuania). Our previous studies showed that the above-mentioned strains inhibit various pathogenic and opportunistic microorganisms and are suitable for fermentation of various cereal substrates [18–22], as well for rapeseed meal fermentation [23]. The rapeseed meal, water and LAB strains (equal parts of each strain by volume) suspension (3% from dry matter of feed mass, v/m), containing 8.9 $\log_{10}$ colony forming units CFU/mL, was fermented at 30 ± 2 °C for 12 h. The fermentation process involves wheat and rapeseed meal (1:1 ratio). The crude protein content of the rapeseed meal was 33.58%. Fermented feed was composed of 25% fermented rapeseed meal and 25% fermented wheat. This combination contained 23.24% crude protein. The compound feed for all groups contained 19% crude protein, crude fibre—3.15%, crude oil—6.51%, lysine—1.45%, methionine—0.55%, tryptophan—0.26%, threonine—0.93%, Ca—0.90%, total P—0.59% and Na—0.20%. The entire fermented wheat/rapeseed meal mass was divided in two parts: 70% (by mass) was used for piglet feeding, and 30% (by mass) was used as a starter for additional feed fermentation cycles (Figure 1). The detailed fermented feed technology and characteristics are described by Vadopalas et al. [23].

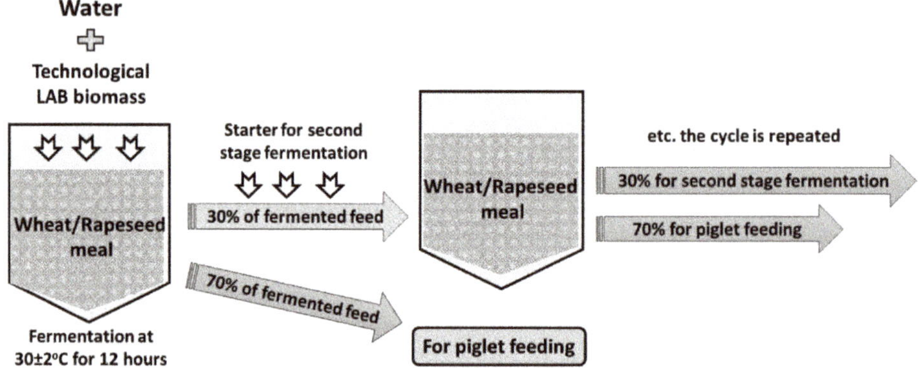

Figure 1. Scheme of fermented feed preparation. The technological lactic acid bacteria (LAB) biomass consisted of *Lactobacillus plantarum* LUHS122, *Lactobacillus casei* LUHS210, *Lactobacillus farraginis* LUHS206, *Pediococcus acidilactici* LUHS29, *Lactobacillus plantarum* LUHS135 and *Lactobacillus uvarum* LUHS245.

2.2. Animals and Housing

All animal procedures were conducted according to the European Union (EU) Directive of the European Parliament and Council, from 22 September, 2010 [24], on the protection of animals used for scientific purposes and the 'Requirements for the Keeping, Maintenance and Use of Animals Intended for Science and Education Purposes', approved by the order of the Lithuanian Director of the State Food and Veterinary Service [25]. The study was conducted at a pig farm in the Klaipeda district (Kontvainiai, Lithuania) and at the Institute of Animal Rearing Technologies, Lithuanian University of Health Sciences (Kaunas, Lithuania). A 36-day experiment was conducted using 25-day-old Large White/Norwegian Landrace (LW/NL) piglets with an initial body weight of 6.9–7.0 kg. Four groups were formed, each with 100 piglets: S_{nonV}—non-vaccinated piglets fed with control group compound feed; S_V—vaccinated piglets fed with control group compound feed; RF_{nonV}—non-vaccinated piglets fed with fermented compound feed; RF_V—vaccinated piglets fed with fermented compound feed. The weaner piglets were kept at the same conditions in a section with two climate zones. The first had a heated concrete floor (36 °C) and roof on it; the second had plastic piglet floors and optimum ventilated air and temperature for the active period. Drinking water and compound liquid feed were available ad libitum throughout the trial. Antibiotic treatment was not applied.

2.3. Experimental Design and Diets

Samples from 10 animals per group were collected at the beginning and end of experiment. Two groups (S_V and RF_V) were vaccinated by intramuscular injection with Circovac PCV2 vaccine (inactivated) (QI09AA07, CEVA-PHYLAXIA Co. Ltd. Szállás u. 5. 1107 Budapest, Hungary), which is used to protect pigs against PCV2. The influence of two dietary treatments on vaccinated and non-vaccinated piglets' parameters were compared. Fermented feed, comprising 500 g/kg of total feed, was included in the diet of treated group beginning at day 25 of life until day 61. Piglets' growth performance was evaluated by testing all 100 piglets in each group. The basal feed was formulated according to the nutritional requirements prescribed in the Nutrient Requirements of Swine [26]. The feed composition and nutritional value are shown in Table 1. The nutritional value of compound feed was determined according to the analytic methods described by Association of Official Analytical Chemists (AOAC) [27].

Table 1. Diet composition (dry matter 87 %).

Ingredients (%)	Control Group	Treated Group
Barley	38.45	33.25
Rapeseed meal	-	25.00
Wheat	32.12	25.00
Soya beans (extruded)	9.30	-
Potato protein	5.00	2.00
Soybean protein concentrate	2.00	-
Whey powder	5.80	5.80
Sunflower oil	2.72	4.51
Limestone	1.48	1.1
NaCl	0.38	0.35
Monocalcium phosphate	0.33	0.41
L-Lysine sulphate	0.87	1.1
DL-Methionine	0.25	0.16
Acidal NC (formic and acetic acids)	0.30	0.30
[1] Vitamins and trace elements (premix)	1.00	1.00
Nutritional value		
ME swine (MJ/kg)	13.86	13.95
Crude protein (%)	19.00	19.00
Crude fat (%)	6.51	6.51
Crude fibre (%)	3.15	5.14
Lysine (%)	1.45	1.45
Methionine (%)	0.55	0.55
Threonine (%)	0.93	0.94
Tryptophan (%)	0.26	0.25
Methionine + Cystine (%)	0.87	0.88
Ca (%)	0.90	0.90
Total P (%)	0.59	0.62
Available P (%)	0.37	0.38
Na (%)	0.20	0.21

ME—metabolisable energy. [1] Composition of premix per 1 kg of feed: Vitamin A—18,180 IU; vitamin D3—2040 IU; vitamin E—161 mg kg^{-1}; vitamin K3—5.03 mg; thiamine—3.64 mg; riboflavin—9.16 mg; choline chloride—404 mg; pyridoxine—4.60 mg; vitamin B12—0.05 mg; niacin—41 mg; pantothenic acid—22.85 mg; folic acid—1.85 mg; biotin—0.21 mg; Fe—152 mg; Cu—100 mg; Zn—91 mg; Mn—80 mg; I—0.81 mg; Co—0.53 mg; Se—0.30 mg.

2.4. Metagenomics and Microbial Profiling Analysis

Before the experiment, faeces from 25-day old piglets from each group (10 piglets per group) were collected. The same procedure, using 10 piglets per group, was performed at the end of the experiment (day 61 of the piglets' life). Faecal samples from each of the group were pooled making 4 samples before and 4 samples after the experiment. Therefore, 8 samples representing 4 tested groups (before and after the experiment) were sequenced for microbial profiling. Specimens were kept in DNA/RNA Shield diluted 1:10 (R1100-250, Zymo Research, Irvine, CA USA) at −80 °C before DNA extraction. DNA was extracted with a faecal DNA MiniPrep kit (D6010, Zymo Research, Irvine, CA USA). Library preparation, metagenomic sequencing and taxonomic characterisation of reads was performed as previously described [28]. ZymoBIOMICS Microbial Community Standard (D6300, Zymo Research, Irvine, CA USA) was used as a microbiome profiling quality control. The results of taxonomic classification were visualised using the interactive online platform https://genome-explorer.com.

2.5. Microbiological Analysis of Faecal Samples

Piglets' faecal samples were collected before and after the experiment, stored in vials (+4 °C) with a transport medium (Faecal Enteric Plus, Oxoid, Basingstoke, UK) and analysed on the same day. Evaluation of the microbiological parameters (LAB, total viable bacteria count [TVC], total enterobacteria count [TEC], and yeast and mould [Y/M] counts) was performed according to methods described by Zavistanaviciute et al. [29].

2.6. Blood Analysis

Piglets were bled from the jugular vein into vacuum blood tubes (BD Vacutainer, Wokingham, UK) before morning feeding. Tubes with clot activator were used for biochemical examination. Blood biochemical variables were evaluated before and after the experiment (on days 25 and 61 of the piglets' life). The parameters included aspartate aminotransferase (AST), alanine aminotransferase (ALT), cholesterol (Chol), high density lipoprotein cholesterol (HDL-C), low-density lipoprotein cholesterol (LDL-C), triglycerides (TG), total protein (TP), albumin (ALB), phosphorus (IP), Mg, K, Na, triiodothyronine (T3), thyroxine (T4), immunoglobulin G (IgG), vitamin B12, albumin (ALB), Fe, glucose (GLU), Ca, creatinine analysed by the Jaffe method (CREA), alkaline phosphatase (AP), thyroid-stimulating hormone (TSH), total bilirubin and urea. Blood parameters were analysed with an automatic biochemistry analyser in the accredited laboratory 'Anteja' (Klaipeda, Lithuania).

2.7. Evaluation of Piglets' Growth Performance and Cases of Mortality and Diarrhoea

Group body weight gain (BWG) was recorded on days 25, 32, 39, 46, 53, and 61 of age using an electronic weighing system (model type: IT1000, SysTec GmbH, Bergheim, Germany) and average daily gain (ADG) was calculated. The feed conversion ratio (FCR) was calculated from feed intake (87% of dry matter) and body weight gain (BWG). Feed consumption was measured using a WEDA (Dammann & Westerkamp GmbH, Goldenstedt, Germany) automated feeding system that has an electronic flowmeter and weighing system. Cases of mortality and diarrhoea were recorded in all tested groups throughout the experiment.

2.8. Analysis of Ammonia Emission

Analysis of ammonia emission was conducted according to the Environmental Protection Document LAND 88-2009 method, approved by the Nr. D1-862 order (31 December, 2009) of the Lithuanian Minister of Environment [30]. Ammonia concentration in the air was analysed by the accredited laboratory "Labtesta" (Kretinga, Lithuania). Air samples were taken on the first and the last day of experiment in four tested farm sectors: S_{nonV}, S_V, RF_{nonV}, and RF_V.

2.9. Statistical Analysis

Data were subjected to multivariate analysis of variance (ANOVA) using the statistical package SPSS for Windows (Ver.15.0, SPSS, Chicago, IL, USA). Baseline measurements were used as covariates to consider the experimental conditions. The mean values were compared using Duncan's multiple range test with a significance level defined at $p \leq 0.05$. In order to evaluate the influence of three different factors (piglets age related differences, the use of fermented feed, vaccination) and their interaction on piglets' parameters, data were subjected to three-way ANOVA and the post hoc Tukey honest significant difference (HSD) test. In the tables, the results are presented as mean values (n = 10). Differences in bacterial genera between the groups at the end of the experiment were assessed using the Z-Test Calculator for two Population Proportions (Social Science Statistics, socscistatistics.com, 2019). The results were considered statistically significant at $p \leq 0.05$.

3. Results and Discussion

3.1. Microbial Profiles of Pig Faeces

The number of bacterial reads in pig faeces before and after the experiment was quite similar: it varied between 35,000 and 40,000 reads among the groups. The number of species with a prevalence of at least 0.01% from the total amount of bacteria in different groups before the experiment was also similar: it varied from 400 in the RF_{nonV} group to 473 in the S_V group. The number of species with the same prevalence rate at the end of experiment varied from 340 to 387. Although the number of bacterial reads and bacterial composition among the groups was similar, we detected some distinct differences, particularly in the number of lactobacilli.

Before experiment, two bacterial genera—*Prevotella* and *Lactobacillus*, had the highest prevalence rates in all groups of pigs; they accounted for more than 40% of the total amount of bacterial composition (Figure 2). The prevalence of *Prevotella* varied from 21.8% to 38.2%, while *Lactobacillus* prevalence ranged from 19.9% to 29.7%. The lowest *Lactobacillus* prevalence was in the RF_V group, while the highest was in S_{nonV} group. In all groups, the most prevalent species among *Lactobacillus* was *Lactobacillus amylovorus* (Tables S1–S4)). The other most prevalent genera included *Barnesiella*, *Clostridium*, *Blautia*, *Faecalibacterium*, *Roseburia* and *Eubacterium*, which ranged from 1.2% (*Eubacterium*) up to 3.4% (*Barnesiella*) in the faeces. Overall, the microbial profiles were similar among all the groups before the experiment (Figure 2), however, it should be mentioned that *Bifidobacterium* was detected only in two groups—RF_{nonV} and RF_V with the prevalence of 4.5% and 1.7% respectively.

At the end of experiment, the microbial profiles had changed depending on the pig group (Figure 3). Overall, the main genera remained similar as before experiment, but there were obvious differences in the *Lactobacillus* prevalence. The lowest prevalence occurred in faeces from the S_{nonV} group, while the highest was in the RF_V group ($p \leq 0.05$). In the latter group, lactobacilli were the most abundant bacteria: their prevalence reached 47.9% of the total bacteria, while in the rest of the groups the highest prevalence was *Prevotella*. When comparing groups with fermented vs. non-fermented feed, pigs that received fermented feed had a notably higher *Lactobacillus* prevalence compared to pigs fed non-fermented feed. The pigs from non-fermented feed groups had higher *Clostridium* and *Terrisporobacter* prevalence ($p \leq 0.05$), while the prevalence of other bacterial genera was low overall in all groups.

When comparing non-vaccinated with vaccinated groups, the main differences were also associated with the *Lactobacillus* prevalence: it was higher in vaccinated compared to non-vaccinated groups ($p \leq 0.05$). At the same time, the amount of *Prevotella* was higher in non-vaccinated groups, although the differences between RF_{nonV} and S_{nonV} groups were small (31.4% vs. 30.2%). The species prevalence and variety at the end of the experiment are presented in Tables S5–S8).

The microbial communities in the pig gut perform a variety of beneficial functions [31]. Previous studies of the gut microbial community have illustrated how populations of constituents are shaped by environmental exposure to microbes, diet, immunological pressures, host genetics and ecological forces within the ecosystem itself [32–34]. Understanding which microbial populations are influenced by fermented feed provides insight into how dietary changes in pigs shape the gut microbiome. Although there have been studies about microbial communities within the gut of healthy pigs, the microbial composition can differ depending on breed, age, place, feed, hygiene conditions and other factors [31,35,36]. This study demonstrated changes in microbial profiles using feed prepared by different technologies within similar groups of animals and the same farm. While the exact mechanism(s) by which vaccination can influence microbial changes within the gut of mammals remains unknown, this study suggests that vaccination has a certain influence, particularly on the number of *Lactobacillus* spp.

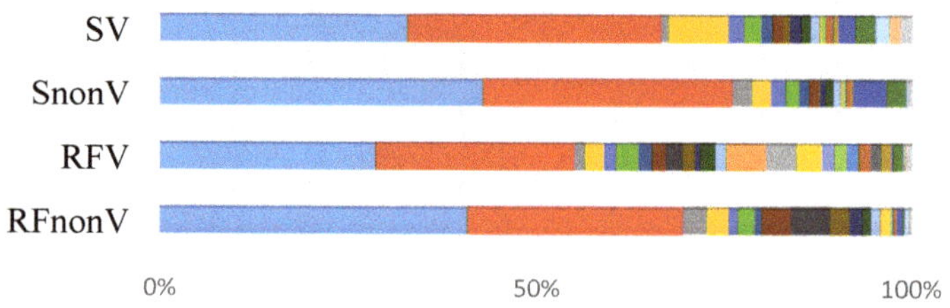

	RFnonV	RFV	SnonV	SV
■ Prevotella	34.6	21.8	38.2	28.1
■ Lactobacillus	24.3	19.9	29.7	28.8
▦ Clostridium	2.7	1.1	2.4	0.9
■ Barnesiella	2.5	1.9	2.2	6.8
■ Blautia	1.1	1.2	1.8	1.9
■ Faecalibacterium	1.9	2.3	1.6	1.9
■ Eubacterium	0.8	1.4	1.2	1.3
■ Roseburia	3.3	1.4	1.2	1.2
▦ Bifidobacterium	4.5	1.7	0	0
■ Megasphaera	2.2	1.2	0.1	0.8
■ Anaerovibrio	1.5	0.6	0.8	1.3
■ Ruminococcus	1	1.5	0.9	1.1
▦ Alloprevotella	1	1.1	0.8	1
■ Collinsella	0.2	4.1	0.1	0.1
▦ Olsenella	0	3.1	0.1	0
■ Selenomonas	1	2.6	0.1	0.1
■ Oligosphaera	0	1.3	0	0
▦ Oscillibacter	0.2	1.3	0.3	0.5
■ Denitrobacterium	0.1	1.2	0.1	0
■ Intestinimonas	0.3	1.2	0.5	0.8
▦ Succinivibrio	0	1.1	0	0.2
■ Oscillospira	0.1	1	0.3	0.5
■ Terrisporobacter	0.7	0.2	4.1	2
■ Gemmiger	0.2	1	2.3	2.3
▦ Catenibacterium	0.5	0.2	0.3	1.6
■ Anaerobium	0.1	0.2	0.1	1.3
▦ Escherichia	0.2	0.5	0.3	1.3

Figure 2. Microbial profiles (prevalence, %) at the genus level in pig faeces before the experiment.

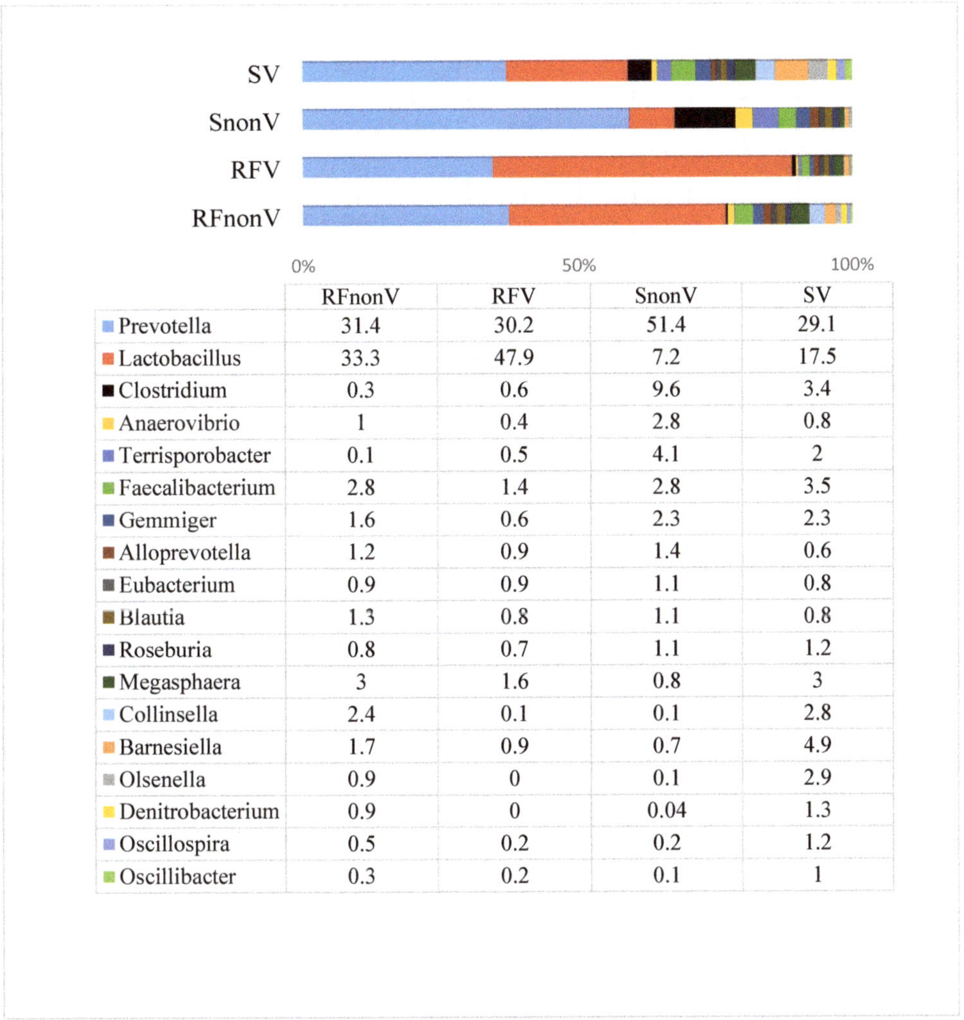

	RFnonV	RFV	SnonV	SV
Prevotella	31.4	30.2	51.4	29.1
Lactobacillus	33.3	47.9	7.2	17.5
Clostridium	0.3	0.6	9.6	3.4
Anaerovibrio	1	0.4	2.8	0.8
Terrisporobacter	0.1	0.5	4.1	2
Faecalibacterium	2.8	1.4	2.8	3.5
Gemmiger	1.6	0.6	2.3	2.3
Alloprevotella	1.2	0.9	1.4	0.6
Eubacterium	0.9	0.9	1.1	0.8
Blautia	1.3	0.8	1.1	0.8
Roseburia	0.8	0.7	1.1	1.2
Megasphaera	3	1.6	0.8	3
Collinsella	2.4	0.1	0.1	2.8
Barnesiella	1.7	0.9	0.7	4.9
Olsenella	0.9	0	0.1	2.9
Denitrobacterium	0.9	0	0.04	1.3
Oscillospira	0.5	0.2	0.2	1.2
Oscillibacter	0.3	0.2	0.1	1

Figure 3. Microbial profiles (prevalence, %) at the genus level in pig faeces after experiment.

Before the experiment, the microbial composition in all groups was very similar, with the highest prevalence for *Prevotella* and *Lactobacillus*, both of which accounted for 56% of the bacterial count. These genera include abundant microorganisms in young weaned healthy pigs [37]. *Prevotella* spp. are usually dominant in pigs' gut and gradually increase in number with age [38,39]. *Prevotella* spp. are key microbial members of the gastrointestinal tracts of adult animals; they are crucial for the degradation of starch and plant polysaccharides but also have a strong capacity for mucoprotein catabolism [40,41]. *Lactobacillus* spp. are common in both the proximal and distal regions of the porcine digestive tract; they colonise soon after birth [42]. This genus influences intestinal physiology, regulates the immune system and balances the intestinal ecology of the host [43]. In addition, *Lactobacillus* spp. have been known to metabolise carbohydrates, including oligosaccharides and starch, which are fermented in the large intestine to short chain fatty acids by lactobacilli for subsequent utilisation by the pigs [44]. Previous studies have indicated that stress greatly affects the gastrointestinal microbiota: it decreases total *Lactobacillus* populations and thus provides an opportunity for pathogen overgrowth [45]. Such stress usually occurs during the weaning period. According to the literature, when compared to diarrhoeic piglets, the gut microbiota of healthy piglets has a higher abundance of *Prevotellaceae*, *Lachnospiraceae*, *Ruminococcaceae* and *Lactobacillaceae* [46]. These data suggest that the gut microbial composition may be used as a biomarker to predict the health status of piglets. The present study demonstrated the positive impact of fermented feed on the porcine microbial composition compared with a conventional feeding regimen in healthy pigs. At the end of the experiment (day 61 of the piglets' life), the *Lactobacillus* prevalence decreased on average 2.4 fold in the groups fed non-fermented feed, whereas the prevalence increased on average 1.8 fold in the groups fed fermented feed. Hence, providing fermented feed prevents the decrease in *Lactobacillus* prevalence after weaning. This effect may help prevent digestive disorders associated with microbial changes because this period is one of the most critical regarding different infections [46]. It is known that during this age, *Lactobacillus* spp. in pigs normally decrease in number [37]. The *Lactobacillus* prevalence was also higher in vaccinated compared with non-vaccinated groups irrespective of the feed type. However, the data do not indicate whether vaccination against PCV2 leads to an increase in *Lactobacillus* prevalence; more experiments are required in this area. Moreover, different types of vaccines and larger experiments should be performed to better understand the possible influence of vaccination on the microbial communities. According to this study, it may be assumed that at the very least vaccination against PCV2 does not negatively affect the microbial composition within the gastrointestinal tract of pigs.

3.2. LAB, TVC, TEC and M/Y Counts in Piglets' Faeces

Microbiological parameters of the piglet faeces (from 25- and 61-day-old piglets) are shown in Table 2 (Table S9—Differences between microbiological parameters of the piglets' faeces between all the tested groups). In most groups (except the S_V group), the LAB count was significantly lower at the end compared with the beginning of the experiment (20.1% lower in the S_{nonV} group; 37.9% lower in the RF_{nonV} group; and 25.3% lower in the RF_V). The LAB count at the end of experiment was significantly higher in both vaccinated compared with non-vaccinated groups (26.6% for S_V and 17.2% for RF_V). The LAB count was 26.3% higher in S_V compared to the RF_V group at the end of the experiment. In a previous study, the faeces of pigs receiving a diet with fermented feed contained significantly fewer total bacteria and fungi, as well as coliform bacteria (including Escherichia coli) and anaerobic *Clostridium perfringens* counts. This phenomenon is due in part to the reduction in pH, an increase in the amount of lactic acid and other volatile fatty acids in the intestinal contents and a reduction in the number of *Enterobacteriaceae* [47,48]. From birth until weaning and then during the post-weaning period, the gut microbiota is dynamic and undergoes major compositional changes that are driven by age, exposure to microbes, environmental conditions and diet [49]. Many authors have described the great impact of early-life events in mammals, and particularly in pigs, on their future health. These experiences shape immune system development through changes in the pattern of microbial intestinal colonisation [50,51]. Colonisation is initiated at birth and is shaped by consumption of the sow's milk, which provides nutritional advantages to the LAB population, building a milk-oriented microbiome that includes *Bacteroidaceae* and *Lactobacillaceae*. This composition rapidly changes after weaning when a (largely) plant-based diet is introduced [35]. The rapidly varying microbiome of young piglets seems to increase in microbial diversity and richness in the suckling phase and gradually stabilises post-weaning [35,52,53]. Our results are in agreement with Bian et al. [54], namely that members of the *Lactobacillaceae* became predominant on days 7, 14 and 28, but had a lower relative abundance again on day 49.

When comparing the TVC in piglets' faeces, in all but the S_V group it was significantly lower at the end of experiment compared with the beginning (9.5% reduction in the S_{nonV} group; 23.6% reduction in the RF_{nonV} group; 3.4% reduction in the RF_V group). TVC was significantly higher in the vaccinated piglets compared to the non-vaccinated piglet faeces at the end of the experiment (22.2% for S_V and 13.8% for RF_V). Between the vaccinated groups, TVC was 10.0% higher in the S_V compared to the RF_V group. During weaning, there is gut microbiota dysbiosis, including a loss of microbial diversity. Studies have noted significant reductions in total bacterial number, including coliform bacteria and anaerobic *C. perfringens*, as well as an increase in *Acetivibrio*, *Dialister*, *Oribacterium*, *Prevotella* and *Proteobacteriaceae*, including *E. coli* [48,49,55,56]. A decrease in *Lactobacillus* spp. can lead to decrease in TVC. Zimmerman et al. [57] demonstrated that *Lactobacillus*, *Bifidobacterium* and *Lactobacillus* spp. have a positive immunomodulatory effect on vaccines in animals.

Table 2. Microbiological parameters in faeces from 25- and 61-day-old piglets.

Microbiological Parameters ($\log_{10}$ CFU/g)	Day	Treatments			
		S_{nonV}	S_V	RF_{nonV}	RF_V
LAB	Baseline	7.78 ± 0.03 [A;a]	8.22 ± 0.04 [A;b]	8.33 ± 0.05 [A;a]	8.35 ± 0.06 [A;a]
	61	6.22 ± 0.04 [B;a]	8.47 ± 0.02 [B;b]	5.17 ± 0.06 [B;a]	6.24 ± 0.07 [B;b]
TVC	Baseline	7.08 ± 0.02 [A;a]	8.25 ± 0.05 [A;b]	8.38 ± 0.07 [A;b]	7.68 ± 0.07 [A;a]
	61	6.41 ± 0.04. [B;a]	8.24 ± 0.03. [A;b]	6.40 ± 0.03. [B;a]	7.42 ± 0.08 [B;b]
TEC	Baseline	7.22 ± 0.06 [A;a]	7.79 ± 0.06 [A;b]	7.40 ± 0.08 [A;a]	7.58 ± 0.04 [A;b]
	61	6.36 ± 0.02 [B;b]	6.18 ± 0.05 [B;a]	6.87 ± 0.07 [B;a]	7.12 ± 0.03 [B;b]
Y/M	Baseline	6.73 ± 0.04 [A;a]	7.71 ± 0.08 [A;b]	6.21 ± 0.04 [A;a]	6.63 ± 0.04 [A;b]
	61	6.36 ± 0.08 [B;b]	4.31 ± 0.06 [B;a]	5.72±0.05 [B;a]	6.26±0.04 [B;b]

Tests of between-subject effects: influence of the analysed factors and their interactions on microbiological parameters in piglets' faeces

Dependent Variable	Treatment Duration	Fermented Feed	Vaccination	Treatment Duration × Fermented Feed	Treatment Duration × Vaccination	Fermented Feed × Vaccination	Treatment Duration × Fermented Feed × Vaccination
			Significance of the differences between groups (p)				
LAB	0.0001	0.0001	0.0001	0.0001	0.0001	0.0001	0.0001
TBC	0.0001	0.0001	0.0001	0.0001	0.0001	0.246	0.0001
TVC	0.0001	0.0001	0.001	0.0001	0.021	0.0001	0.0001
Y/M	0.0001	0.0001	0.0001	0.0001	0.0001	0.0001	0.0001

S_{nonV}—non-vaccinated piglets fed with control group compound feed; S_V—vaccinated piglets fed with control group compound feed; RF_{nonV}—non-vaccinated piglets fed with fermented compound feed; RF_V—vaccinated piglets fed with fermented compound feed. CFU—colony forming units; LAB—lactic acid bacteria; TVC—total viable bacteria count; TEC—total enterobacteria count; Y/M—yeast/mould count. Measurements were made at baseline—day 25, before the start of the feeding experiment—and on day 61, the end of the experiment. [A,B] Different uppercase letters indicate significant treatment duration-related differences ($p \leq 0.05$). [a,b] Different lowercase letters indicate differences among treatments, comparing S_{nonV} and S_V; RF_{nonV} and RF_V ($p \leq 0.05$). Data are presented as mean values ± standard error (n = 10 per group).

In all the groups, the TEC and Y/M counts were lower in piglets' faeces at the end compared with the beginning of the experiment (11.9 and 5.5%, respectively, for the S_{nonV} group; 26.1 and 44.1%, respectively, for the S_V group; 7.2 and 7.9%, respectively, for the RF_{nonV} group; 6.1 and 5.6%, respectively, for the RF_V group). Both TEC and Y/M were significantly lower in the S_V compared to S_{nonV} group faeces at the end of experiment, but the opposite was true for the RF_V compared to the RF_{nonV} group. At the end of the experiment, the TEC and Y/M counts were higher in the RF_V compared to the S_V group (13.2 and 31.2%, respectively). Our results are in agreement with Dowarah et al. [58], who showed that fermented feed improved the number of beneficial microbes (LAB and bifidobacteria) and reduced *E. coli* and clostridia count, the pathogens responsible for diarrhoea, in faces. Arfken et al. [59] investigated the mutualistic relationship between yeast in the piglet gastrointestinal (GI) tract and *Lactobacillus* count. There was a significant reduction in the total number of fungi compared with the control group [48].

The ANOVA results indicated that there were significant treatment duration, fermented feed and vaccination main effects, as well as significant treatment duration × fermented feed, treatment duration × vaccination, fermented feed × vaccination and treatment duration × fermented feed × vaccination interactions on microbiological parameters of the pig faeces. Indeed, only the fermented feed × vaccination interaction was not significant for TVC in pig faeces.

Finally, microbiological parameters of the piglets' faeces were varied and, in most of the cases, influenced by the analysed factors and their interaction. Considering that genera and species as well as the interaction between strains can lead changes of piglets' growth performance, correlation between the above-mentioned parameters were calculated.

3.3. Piglet Blood Parameters

The piglets' blood parameters are shown in Table 3 (Table S10—Differences between blood parameters of the piglets' between all the tested groups). At the end of experiment, there were significantly lower ALT, T3, T4, IP, Mg, K, Ca and vitamin B12 concentrations in the S_V compared with the S_{nonV} group. However, in the S_V group, Chol, HDL-C, TP and GLU concentrations were significantly higher compared with the SnonV group. At the end of experiment, there were significantly lower ALT, HDL-C, LDL-C, TP, ALB and Fe blood concentrations in the RF_V compared to RF_{nonV} group. However, in the RF_V group blood, TG and total bilirubin concentrations were significantly higher compared with the RF_{nonV} group.

When comparing blood parameters between the vaccinated groups (S_V and RF_V) at the end of experiment, ALT, TG, IgG, Mg, K, Fe and total bilirubin were significantly higher in the RF_V group. By contrast, HDL-C, LDL-C, TP, ALB, T3, T4 and vitamin B12 were significantly lower in the RF_V compared with the S_V group.

In general, blood biochemical indices reflect comprehensive functions of piglet's nutritional metabolism, health and welfare [60]. The IgG level reflects immune status, plays important roles in the humoral immune response and controls bacterial infections in the body; it can also function to control diarrhoeal infections by binding multiple pathogenic antigens [46]. Lu et al. [59] reported that IgG and immunoglobulin M (IgM) concentrations were significantly greater and serum ALT and AST concentrations were decreased in pigs that received fermented feed ($p \leq 0.05$). The activities of the hepatic enzymes ALT and AST are key indices that reflect the acute injury of liver. The ALT blood concentration increases when the liver cell membrane is damaged. The AST content significantly increases when the liver mitochondrial membrane is impaired [60]. The blood TP concentration reflects the relationship between protein absorption in vivo and humoral immunity [60]. The hydrolysis products of glucosinolates are known to depress iodine metabolism in the thyroid gland and inhibit the synthesis of thyroid hormones T3 and T4. When these compounds, especially thiocyanates, interfere with iodine uptake, hypothyroidism and thyroid gland enlargement may ensue [46]. However, in this study the utilised rapeseed meal contained very low concentrations of erucic acid and glucosinolates (7.5–11.2 µmol/g). Hence, the influence of the above-mentioned compounds on T3 and T4 should be minimal. The fermentation process can reduce dietary anti-nutrients (phytates, glucosinolates and trypsin inhibitors) and improve the absorption and use of nutrients, including amino acids and minerals (e.g., P, Ca, Zn, Cu and Fe) [48,61]. These changes can lead to haematological, biochemical and mineral profile variations in the blood [61,62]. However, it was published that the TP concentration was increased in the blood of piglets fed fermented feed [63]. In opposition to above-mentioned findings, Min [64] reported that the TP blood concentration was not affected by the fermented feed diets. In serum samples from piglets fed with fermented feed, there were higher ALP, TP, ALB and GLU concentrations [62,65,66]. Liu et al. [15] published that piglets fed fermented feed had lower levels of serum IgG, but there was no difference in the immunoglobulin IgA and IgM serum levels. Furthermore, the serum glucose level was decreased in piglets fed fermented feed; however, TP, Chol, TG, LDL-C, ALT, AST, Ca, P and Mg were not different among the treatments [67]. Satessa et al. [68] found that fermented rapeseed meal reduced the glucose, HDL-C and LDL-C concentrations, but TP, P, AST, ALT concentrations were increased, and significant changes in blood IgG were not established. According to Czech et al. [61], the blood concentrations of Chol and TG AST in piglets fed with fermented rapeseed meal was reduced, and there was better availability of minerals.

Table 3. Piglets' blood parameters.

Blood Parameters	Day	Treatments			
		S$_{nonV}$	S$_V$	RF$_{nonV}$	RF$_V$
Aspartate aminotransferase (AST), U/L	Baseline	29.4 [A;a]	42.67 [A;a]	51.4 [A;a]	61.8 [A;a]
	61	34.0 [B;a]	41.0. [A;a]	44.0. [A;a]	42.4. [A;a]
Alanine aminotransferase (ALT), U/L	Baseline	48.4 [A;a]	43.67 [A;a]	53.2 [A;b]	60.8 [A;a]
	61	76.2. [B;a]	69.6 [B;b]	87.0 [B;a]	80.8 [A;b]
Cholesterol (Chol), mmol/L	Baseline	1.64 [A;a]	1.64 [A;a]	1.88 [A;a]	2.04 [A;a]
	61	2.06. [B;a]	2.26. [B;b]	2.34. [A;a]	2.25. [B;a]
High-density lipoprotein cholesterol (HDL-C), mmol/L	Baseline	0.714 [A;a]	0.744 [A;b]	0.898 [A;a]	0.846 [A;a]
	61	0.840 [B;a]	0.944 [B;b]	1.028 [B;a]	0.872 [A;b]
Low-density lipoprotein cholesterol (LDL-C), mmol/L	Baseline	0.758 [A;a]	0.726 [A;a]	0.814 [A;a]	0.976 [A; a]
	61	0.980 [B;a]	1.102 [B;a]	1.032 [A;a]	0.860 [A; b]
Triglycerides (TG), mmol/L	Baseline	0.360 [A;a]	0.372 [A;a]	0.366 [A;a]	0.478 [A;a]
	61	0.466 [B;a]	0.466 [A;a]	0.620 [B;a]	0.650 [B;b]
Total protein (TP), g/L	Baseline	46.2 [A;a]	44.9 [A;a]	44.2 [A;a]	45.6 [A;a]
	61	51.8 [B;a]	56.3 [B;b]	52.8 [B;a]	49.5 [B;b]
Albumin (ALB), g/L	Baseline	30.0 [A;a]	29.0 [A;a]	32.6 [A;a]	32.8 [A;a]
	61	35.8 [A;a]	37.4 [B;a]	36.2 [A;a]	33.4 [B;b]
Immunoglobulin IgG, g/L	Baseline	2.65 [A;a]	2.85 [A;a]	2.35 [A;a]	2.96 [A;a]
	61	3.74 [B;a]	3.44 [A;a]	3.05 [B;a]	5.45 [A;b]
Triiodothyronine (T3), nmol/L	Baseline	1.21 [A;a]	1.14 [A;b]	1.29 [A;a]	1.22 [A;a]
	61	2.14 [B;a]	2.02 [B;b]	1.59 [A;a]	1.49 [A;a]
Thyroxine (T4), μ d/L	Baseline	4.50 [A;a]	4.82 [A;a]	3.50 [A;a]	5.22 [A;a]
	61	4.84 [A;a]	3.84 [A;b]	2.92 [B;a]	2.90 [A;a]
Glucose (GLU), nmol/L	Baseline	5.84 [A;a]	6.10 [A;a]	6.12 [A;a]	5.66 [A;a]
	61	5.74 [A;a]	6.22 [A;b]	6.08 [A;a]	6.18 [B;a]
Phosphorus (IP), mmol/L	Baseline	2.94 [A;a]	2.87 [A;a]	2.61 [A;a]	2.76 [A;a]
	61	3.50 [B;a]	3.39 [A;b]	3.28 [B;a]	3.39 [B;a]
Magnesium (Mg), mmol/L	Baseline	1.02 [A;a]	0.928 [A;b]	0.996 [A;a]	0.932 [A;b]
	61	1.07 [A;a]	0.936 [A;b]	0.960 [A;a]	0.966 [A;b]
Potassium (K)	Baseline	4.96 [A;a]	5.66 [A;b]	4.65 [A;a]	5.69 [A;b]
	61	5.81 [B;a]	5.34 [B;b]	4.96 [A;a]	4.74 [B;a]
Sodium (Na)	Baseline	143.4 [A;a]	141.0 [A;a]	144.0 [A;a]	142.0 [A;b]
	61	147.2 [A;a]	147.2 [B; a]	146.6 [B;a]	144.4 [A;a]
Iron (Fe), μmol/L	Baseline	23.6 [A;a]	23.8 [A;a]	31.5 [A;a]	23.5 [A;b]
	61	28.1 [A;a]	26.9 [B;a]	47.1 [A;a]	51.4 [B;a]
Calcium (Ca), nmol/L	Baseline	2.60 [A;a]	2.57 [A;a]	2.71 [A;a]	2.55 [A;a]
	61	2.87 [A;a]	2.78 [B;b]	2.79 [A;a]	2.78 [B;a]
Vitamin B12, pmol/L	Baseline	142.2 [A;a]	93.8 [A;b]	78.2 [A;a]	131.0 [A;b]
	61	214.6 [A;a]	122.2 [B;b]	94.2 [A; a]	98.2 [B;a]
Creatinine (CREA), μmol/L	Baseline	64.2 [A;a]	57.4 [A;a]	78.8 [A;a]	63.4 [A;a]
	61	57.4 [A;a]	54.8 [A;a]	48.2 [B;a]	49.4 [A;a]
Alkaline phosphatase (AP), U/L	Baseline	336.2 [A;a]	285.0 [A;a]	408.6 [A;a]	318.0 [A;a]
	61	263.6 [A;a]	220.4 [A;a]	242.6 [B;a]	245.4 [A;a]
Urea, mmol/L	Baseline	2.36 [A;a]	3.26 [A;a]	2.64 [A;a]	2.58 [A;a]
	61	2.02 [A;a]	2.38 [A;a]	3.19 [B;a]	2.63 [B;a]
Thyroid-stimulating hormone (TSH)	Baseline	0.0200 [A;a]	0.0190 [A;a]	0.0208 [A;a]	0.0236 [A;a]
	61	0.0208 [A;a]	0.0228 [A;a]	0.0230 [A;a]	0.0260 [B;a]
Total bilirubin (pmol/L)	Baseline	1.88 [A;a]	1.99 [A;a]	1.99 [A;a]	1.99 [A;a]
	61	1.98 [A;a]	1.99 [A;a]	1.99 [A;a]	2.65 [A;b]

S$_{nonV}$—non-vaccinated piglets fed with control group compound feed; S$_V$—vaccinated piglets fed with control group compound feed; RF$_{nonV}$—non-vaccinated piglets fed with fermented compound feed; RF$_V$—vaccinated piglets fed with fermented compound feed. Measurements were made at baseline—day 25, before the start of the feeding experiment—and on day 61, the end of the experiment. [A,B] Different uppercase letters indicate significant treatment duration-related differences ($p \leq 0.05$). [a,b] Different lowercase letters indicate differences among treatments: S$_{nonV}$ and S$_V$; RF$_{nonV}$ and RF$_V$ ($p \leq 0.05$). Data are presented as mean values (n = 10 per group).

The influence of the analysed factors and their interaction on piglets' blood parameters is shown in Table 4. Vaccination as a separate factor did not significantly influence piglets' blood parameters. However, the treatment duration × fermented feed interaction was significant for T3 ($p = 0.038$), T4 ($p = 0.041$) and Fe ($p = 0.008$) concentrations in piglets' blood. There was a significant treatment duration × vaccination interaction on T4 ($p = 0.008$) and K ($p = 0.007$) concentrations in piglets' blood. Finally, there was a significant fermented feed × vaccination interaction for T4 ($p = 0.033$), vitamin B12 ($p = 0.003$) and urea ($p = 0.014$) concentrations in piglets' blood. All three factors were significant with regard to the TP content in piglets' blood ($p = 0.041$).

Table 4. Tests of between-subject effects: influence of the analysed factors and their interaction on piglets' blood parameters.

Dependent Variable	TD	FF	V	TD × FF	TD × V	FF × V	TD × FF × V
			Significance (p)				
AST, U/L	0.279	**0.025**	0.191	0.182	0.404	0.598	0.791
ALT, U/L	**0.0001**	**0.017**	0.557	0.997	0.358	0.453	0.481
Chol, mmol/L	**0.004**	0.088	0.608	0.467	0.907	0.809	0.396
HDL-C, mmol/L	**0.069**	0.123	0.768	0.501	0.905	0.185	0.482
LDL-C, mmol/L	**0.045**	0.723	0.807	0.143	0.584	0.760	0.149
TG, mmol/L	**0.002**	**0.018**	0.382	0.206	0.591	0.459	0.688
TP, g/L	**0.0001**	0.155	0.773	0.345	0.811	0.310	**0.041**
ALB, g/L	**0.002**	0.581	0.693	0.061	0.937	0.529	0.276
IgG, g/L	0.112	0.699	0.329	0.612	0.660	0.299	0.442
T3, nmol/L	**0.001**	0.120	0.522	0.038	0.909	0.994	0.949
T4, µ d/L	**0.003**	**0.004**	0.331	0.041	**0.008**	**0.033**	0.685
GLU, nmol/L	0.633	0.716	0.660	0.458	0.300	0.633	0.745
IP, mmol/L	**0.001**	0.179	0.878	0.658	0.851	0.349	0.993
Mg, mmol/L	0.675	0.487	**0.060**	0.655	0.867	0.250	0.422
K	0.885	**0.041**	0.203	0.153	**0.007**	0.462	0.902
Na	**0.0001**	0.548	**0.039**	0.108	0.464	0.548	0.388
Fe, µmol/L	**0.001**	**0.001**	0.694	**0.008**	0.370	0.830	0.267
Ca, nmol/L	**0.001**	0.985	0.176	0.379	0.657	0.816	0.313
B12, pmol/L	0.158	**0.008**	0.158	0.055	0.121	**0.003**	0.934
CREA, µmol/L	0.069	0.831	0.406	0.222	0.463	0.864	0.660
AP, U/L	**0.027**	0.490	0.256	0.521	0.521	0.966	0.588
Urea, mmol/L	0.164	0.797	0.899	0.286	0.075	**0.014**	0.456
TSH	0.604	0.523	0.736	0.854	0.951	0.854	0.691
Total bilirubin (pmol/L)	0.338	0.280	0.280	0.290	0.290	0.349	0.338

AST—Aspartate aminotransferase, ALT—Alanine aminotransferase, Chol—Cholesterol, HDL-C—High-density lipoprotein cholesterol, LDL-C—Low-density lipoprotein cholesterol, TG—Triglycerides, TP—Total protein, ALB—Albumin, T3—Triiodothyronine, T4—Thyroxine, GLU—Glucose, IP—Phosphorus, Mg—Magnesium, K—Potassium, Na—Sodium, Fe—Iron, Ca—Calcium, CREA—Creatinine, AP—Alkaline phosphatase, TSH—Thyroid-stimulating hormone. TD—treatment duration, FF—fermented feed, V—vaccination. Bold values indicate significant differences ($p \leq 0.05$).

3.4. Piglets' Growth Performance

The piglets' average FCR, ADG and the influence of analysed factors and their interaction on piglets' growth performance parameters from day 25 to 61 are shown in Table 5. For ADG, there were no significant differences among the groups. In addition, the analysed factors as well as their interactions did not exert significant effects on piglets' ADG. However, there were significant differences in FCR between S_V and S_{nonV} (11.5% lower in the SV group), between RF_V and RF_{nonV} (10.2% lower in the RF_{nonV} group) and between S_V and RF_V (21.6% lower in the S_V group). There was a significant, very strong positive correlation between FCR and TEC in piglets' faeces ($R = 0.919, p = 0.041$). There were no significant correlations between the other analysed faecal microbiological parameters (LAB, TBC and Y/M count) and FCR or ADG.

Mortality and diarrhoea cases were similar in all groups throughout the experiment. The mortality of piglets in the non-vaccinated groups (S_{nonV} and RF_{nonV}) was 2%, while mortality in the vaccinated groups (S_V and RF_V) was 2% and 3%, respectively. There was more intense diarrhoea in the S_{nonV} group at day 31 and 36. There were several diarrhoea cases in the S_V, RF_{nonV} and RF_V groups from day 28 to 46.

Table 5. Influence of the analysed factors and their interactions on piglets' growth performance.

Piglets Growth Performance	Average from day 25 to 61	Treatments			
		S_{nonV}	S_V	RF_{nonV}	RF_V
FCR		1.56	1.38	1.58	1.76
ADG		0.395	0.392	0.397	0.399

Tests of between-subjects effects: influence of analysed factors and their interactions on FCR and ADG

Dependent Variable	Fermented Feed	Vaccination	Fermented Feed × Vaccination
		Significance (p)	
FCR	0.154	1.000	0.194
ADG	0.962	0.996	0.979

Differences among treatments ($p < 0.05$)

S_V vs. S_{nonV}	RF_V vs. RF_{nonV}	S_V vs. RF_V	S_{nonV} vs. RF_{nonV}	S_{nonV} vs. RF_V	RF_{nonV} vs. S_V
		ADG			
0.901	0.783	0.55	0.943	0.91	0.286
		FCR			
0.001	0.002	0.0001	0.773	0.91	0.074

S_{nonV}—non-vaccinated piglets fed with control group compound feed; S_V—vaccinated piglets fed with control group compound feed; RF_{nonV}—non-vaccinated piglets fed with fermented compound feed; RF_V—vaccinated piglets fed with fermented compound feed. FCR—*feed conversion ratio*; ADG—average daily gain. Measurements were made at baseline—on day 25, before the start of the feeding experiment—and on day 61, the end of the experiment. Bold values indicate significant differences ($p \leq 0.05$). Data are presented as mean values ($n = 100$ per group).

The available published information about the influence of vaccination on piglets' growth performance is varied. According to Oliver-Ferrando et al. [69], the optimal time for PCV2 vaccination is at either 3 or 6 weeks of age. In our study, piglets were vaccinated on day 25 of life. Duivon et al. [70] published that the added PCV2 valence in the vaccination protocol helps counter the negative impact of subclinical PCV2 infection on growth. Jeong et al. [71] indicated that ADG showed improvement in vaccinated compared with non-vaccinated animals. Da Silva et al. [72] compared the ADG between vaccinated and non-vaccinated animals from weaning to finishing for four commercial vaccine products. All products significantly increased ADG values. However, according to Woźniak et al. [73] data from four farms where vaccination was used, the results were similar to those from non-vaccinated farms. The authors suggested revising the vaccination protocol. The above-mentioned findings can be explained by a major PCV2 genotype shift from the predominant PCV2 genotype 2b towards 2d [74]. While the commercial vaccines that were first introduced to the US in 2006 have been highly effective in reducing clinical signs and improving production, recent studies have indicated a declining level of PCV2 prevalence and viraemia in the field. Hence, the efficiency of current vaccines against new and emerging strains, as well as new vaccine development, is crucial [75]. Vaccination against PCV2 is an important co-working agent that confers unintended benefits in the protection against the other agents [76]. Notably, the breed information is frequently omitted from experimental publications of vaccine studies, as well as environmental conditions (temperature, stocking density, etc.). These factors are very important for the further standardisation across studies before logical comparisons can be made among studies [77].

While we found differences in the ADG between different groups, the influence of vaccination on different feed groups was the opposite. Specifically, the FCR was lower in the S_V compared to the S_{nonV} group. However, the effect was the opposite in the rapeseed meal groups: FCR was lower in the R_{nonV} compared to the RF_V group. The lowest FCR (1.38) occurred in the S_V group. Notably, soya-based feed is much more expensive compared with the feed composed of local rapeseed meal. Overall, the data support the use of fermented local feedstock for piglet feeding because it provides similar growth parameters as imported soya.

3.5. Influence of Analysed Factors on Ammonia Emission

Table 6 presents the influence of the analysed factors on ammonia emission. At the end of experiment, the RF_V group had the lowest ammonia emission: 58.2, 23.8 and 47.33% lower than the S_{nonV}, S_V, and RF_{nonV} groups, respectively. There was lower ammonia emission in vaccinated groups at the end of the experiment: 45.2% lower in the S_V compared to S_{nonV} group and 47.33% lower in the RF_V compared to the RF_{nonV} group. ANOVA revealed a significant effect for all the analysed factors and their interaction on ammonia emission. Furthermore, there was a significant, very strong positive correlation between ammonia emission and Y/M count in piglets' faeces at the end of experiment ($R = 0.974$; $p = 0.013$).

Table 6. Influence of the analysed factors on ammonia emission.

Ammonia Emission	Day	Treatments			
		S_{nonV}	S_V	RF_{nonV}	RF_V
	Baseline	0.837 [A;a]	1.172 [A;b]	1.584 [A;a]	1.252 [A;b]
	61	1.938 [B;a]	1.063 [A;b]	1.538 [A;a]	0.810 [B;b]

Tests of between-subject effects: influence of the analysed factors and their interactions on ammonia emission

Treatment Duration	Fermented Feed	Vaccination	Treatment Duration × Fermented Feed	Treatment Duration × Vaccination	Fermented Feed × Vaccination	Treatment Duration × Fermented Feed × Vaccination
			Significance (*p*)			
0.0001	0.003	0.0001	0.0001	0.0001	0.0001	0.0001

S_{nonV}—non-vaccinated piglets fed with soya meal; S_V—vaccinated piglets fed with soya meal; RF_{nonV}—non-vaccinated piglets fed with fermented rapeseed meal; RF_V—vaccinated piglets fed with fermented rapeseed meal. Measurements were done at baseline, on day 25 before the start of the feeding experiment, and on day 61, the end of the experiment. [A,B] Different uppercase letters indicate significant treatment duration-related differences ($p \leq 0.05$). [a,b] Different lowercase letters indicate differences among treatments ($p \leq 0.05$). Bold values indicate significant differences ($p \leq 0.05$). Data are presented as the mean values ($n = 3$).

There has been no information published about the influence of vaccination on ammonia emission; most of the published studies have focussed on dietary aspects. Yi et al. [37] and Lee et al. [78] reported that protected organic acids used for weanling pigs reduce faecal ammonia. Nguyen et al. [79] concluded that diets with probiotics mixture decreased pigs' ammonia emission. According to Wang et al. [51], diets with fermented feed ingredients tended to decrease ammonia emissions, but this effect was not statistically significant. According to Bindas et al. [80], ammonia in the faeces of experimental animals fed with fermented feed was significantly lower compared with the control animals. Reducing ammonia emissions by changing the dietary composition is considered economical and feasible because it can improve nitrogen utilisation and, consequently, reduce ammonia emissions [53]. Lactobacilli-based feed fermentation decreases the emissions of total organic carbon and ammonium [67]. Finally, many factors can contribute to ammonia emission on pig farms. Our data indicate that vaccination as a factor should be considered, as well as microbiological parameters of the piglets' faeces. Indeed, we found a very strong correlation between ammonia emission and Y/M count in piglets' faeces.

4. Conclusions

Fermented feed had a positive impact on the amount of *Lactobacillus* during the post-weaning period of pigs. In addition, vaccination had no negative impact on microbial communities. There was a higher *Lactobacillus* prevalence in vaccinated compared with non-vaccinated groups, as well as a higher viable LAB count in faeces from both vaccinated groups (S_V and RF_V) at the end of experiment. There were no significant differences in ADG between piglets' groups. There were differences in FCR: 11.5% lower in the S_V compared with the S_{nonV} group, 10.2% lower in the RF_{nonV} compared with the RF_V group, and 21.6% lower in the S_V compared with the RF_V group. There were a significant, very strong positive correlations between FCR and TEC in piglets' faeces ($R = 0.919$, $p = 0.041$) and between ammonia emission and Y/M count in piglets' faeces at the end of experiment ($R = 0.974$, $p = 0.013$). The lowest ammonia emission was in the RF_V group section. Vaccination as a separate factor did not significantly influence piglets' blood parameters. Overall, by changing from an extruded soya to cheaper rapeseed meal and applying the fermentation model with the selected LAB combination, it is possible to feed piglets without any undesirable changes in health and growth performance, as well as in a more sustainable manner. However, to evaluate the influence of vaccination and its interaction with other parameters (feed, piglet age, breed, etc.) on piglets' parameters, additional studies should be performed and methods should be standardised so that the results may be compared.

Supplementary Materials: The following are available online at http://www.mdpi.com/2076-2615/10/7/1201/s1, Table S1: Species S_{nonV} group before experiment; Table S2: Species S_V group before experiment Species S_V group before experiment; Table S3: Species RF_{nonV} group before experiment; Table S4: Species RFV group before experiment; Table S5: Species S_{nonV} group after experiment; Table S6: Species SV group after experiment; Table S7: Species RF_{nonV} group after experiment; Table S8: Species RFV group after experiment; Table S9: Differences between microbiological parameters of the piglets' faeces between all the tested groups; Table S10: Differences between blood parameters of the piglets between all the tested groups.

Author Contributions: Conceptualisation, methodology, E.B., R.G., L.V., S.B. and M.R.; investigation, L.V., S.B., S.S., A.D., V.S., P.Z., E.Z. and V.L; writing—original draft preparation, L.V., D.K., P.Z., V.S., E.M. and V.L.; writing—review and editing, E.B., V.B. and R.G.; visualisation, P.Z., V.S., V.L., D.K. and E.M.; supervision, E.B. All authors have read and agreed to the published version of the manuscript.

Funding: This research received no external funding.

Acknowledgments: The authors gratefully acknowledge the EUREKA Network Project E!13309 "SUSFEETECH" (Nr. 01.2.2-MITA-K-702-05-0001) and COST Action CA18101 'SOURDOugh biotechnology network towards novel, healthier and sustainable food and bIoproCesseS'.

Conflicts of Interest: The authors declare no conflict of interest.

References

1. Tischer, I.; Mields, W.; Wolff, D.; Vagt, M.; Griem, W. Studies on epidemiology and pathogenicity of porcine circovirus. *Arch. Virol.* **1986**, *91*, 271–276. [CrossRef]

2. Ellis, J.; Hassard, L.; Clark, E.; Harding, J.; Allan, G.; Willson, P.; Strokappe, J.; Martin, K.; McNeilly, F.; Meehan, B.; et al. Isolation of circovirus from lesions of pigs with postweaning multisystemic wasting syndrome. *Can. Vet. J.* **1998**, *39*, 44–51.

3. Argilés, J.M.; Busquets, S.; Toledo, M.; López-Soriano, F.J. The role of cytokines in cancer cachexia. *Curr. Opin. Support Palliat. Care* **2009**, *3*, 263–268. [CrossRef] [PubMed]

4. Ottenhoff, T.H.M. The knowns and unknowns of the immunopathogenesis of tuberculosis. *Int. J. Tuberc. Lung Dis.* **2012**, *16*, 1424–1432. [CrossRef] [PubMed]

5. Szabo, G.; Csak, T. Inflammasomes in liver diseases. *J. Hepatol.* **2012**, *57*, 642–654. [CrossRef]

6. Ramamoorthy, S.; Meng, X.-J. Porcine circoviruses: A minuscule yet mammoth paradox. *Anim. Health Res. Rev.* **2009**, *10*, 1–20. [CrossRef]

7. Gillespie, J.; Opriessnig, T.; Meng, X.J.; Pelzer, K.; Buechner-Maxwell, V. Porcine circovirus type 2 and porcine circovirus-associated disease. *J. Vet. Intern. Med.* **2009**, *23*, 1151–1163. [CrossRef]

8. Cságola, A.; Zádori, Z.; Mészáros, I.; Tuboly, T. Detection of porcine parvovirus 2 (*Ungulate Tetraparvovirus* 3) Specific antibodies and examination of the serological profile of an infected swine herd. *PLoS ONE* **2016**, *11*, e0151036. [CrossRef]

9. Novosel, D.; Cadar, D.; Tuboly, T.; Jungic, A.; Stadejek, T.; Ait-Ali, T.; Cságola, A. Investigating porcine parvoviruses genogroup 2 infection using in situ polymerase chain reaction. *BMC Vet. Res.* **2018**, *14*, 163. [CrossRef] [PubMed]

10. Thomson, J.; Smith, B.; Alan, G.; McNeilly, F.; McVicar, C. PDNS, PMWS and porcine circovirus type 2 in Scotland. *Vet. Rec.* **2000**, *146*, 651–652.

11. Dán, Á.; Molnár, T.; Biksi, I.; Glávits, R.; Shaheim, M.; Harrach, B. Characterisation of Hungarian porcine circovirus 2 genomes associated with PMWS and PDNS cases. *Acta Vet. Hung.* **2003**, *51*, 551–562. [CrossRef]

12. An, D.; Roh, I.; Song, D.; Park, C.; Park, B. Phylogenetic characterization of porcine circovirus type 2 in PMWS and PDNS Korean pigs between 1999 and 2006. *Virus Res.* **2007**, *129*, 115–122. [CrossRef] [PubMed]

13. Zheng, G.; Lu, Q.; Wang, F.; Xing, G.; Feng, H.; Jin, Q.; Guo, Z.; Teng, M.; Hao, H.; Li, D.; et al. Phylogenetic analysis of porcine circovirus type 2 (PCV2) between 2015 and 2018 in Henan Province, China. *BMC Vet. Res.* **2020**, *16*, 6. [CrossRef] [PubMed]

14. Xiao, C.-T.; Halbur, P.G.; Opriessnig, T. Global molecular genetic analysis of porcine circovirus type 2 (PCV2) sequences confirms the presence of four main PCV2 genotypes and reveals a rapid increase of PCV2d. *J. Gen. Virol.* **2015**, *96*, 1830–1841. [CrossRef]

15. Liu, J.; Wei, C.; Dai, A.; Lin, Z.; Fan, K.; Fan, J.; Liu, J.; Luo, M.; Yang, X. Detection of PCV2e strains in Southeast China. *PeerJ* **2018**, *6*, e4476. [CrossRef]

16. Opriessnig, T.; Xiao, C.-T.; Gerber, P.F.; Halbur, P.G. Identification of recently described porcine parvoviruses in archived North American samples from 1996 and association with porcine circovirus associated disease. *Vet. Microbiol.* **2014**, *173*, 9–16. [CrossRef]

17. Chen, N.; Huang, Y.; Ye, M.; Li, S.; Xiao, Y.; Cui, B.; Zhu, J. Co-infection status of classical swine fever virus (CSFV), porcine reproductive and respiratory syndrome virus (PRRSV) and porcine circoviruses (PCV2 and PCV3) in eight regions of China from 2016 to 2018. *Infect. Genet. Evol.* **2019**, *68*, 127–135. [CrossRef] [PubMed]

18. Bartkiene, E.; Bartkevics, V.; Krungleviciute, V.; Pugajeva, I.; Zadeike, D.; Juodeikiene, G. Lactic acid bacteria combinations for wheat sourdough preparation and their influence on wheat bread quality and acrylamide formation. *J. Food Sci.* **2017**, *82*, 2371–2378. [CrossRef]

19. Bartkiene, E.; Bartkevics, V.; Lele, V.; Pugajeva, I.; Zavistanaviciute, P.; Mickiene, R.; Zadeike, D.; Juodeikiene, G. A concept of mould spoilage prevention and acrylamide reduction in wheat bread: Application of lactobacilli in combination with a cranberry coating. *Food Control* **2018**, *91*, 284–293. [CrossRef]

20. Bartkiene, E.; Sakiene, V.; Bartkevics, V.; Juodeikiene, G.; Lele, V.; Wiacek, C.; Braun, P.G. Modulation of the nutritional value of lupine wholemeal and protein isolates using submerged and solid-state fermentation with *Pediococcus pentosaceus* strains. *Int. J. Food Sci. Technol.* **2018**, *53*, 1896–1905. [CrossRef]

21. Bartkiene, E.; Sakiene, V.; Lele, V.; Bartkevics, V.; Rusko, J.; Wiacek, C.; Ruzauskas, M.; Braun, P.G.; Matusevicius, P.; Zdunczyk, Z.; et al. Perspectives of lupine wholemeal protein and protein isolates biodegradation. *Int. J. Food Sci. Technol.* **2019**, *54*, 1989–2001. [CrossRef]

22. Bartkiene, E.; Mozuriene, E.; Lele, V.; Zokaityte, E.; Gruzauskas, R.; Jakobsone, I.; Juodeikiene, G.; Ruibys, R.; Bartkevics, V. Changes of bioactive compounds in barley industry by-products during submerged and solid state fermentation with antimicrobial *Pediococcus acidilactici* strain LUHS29. *Food Sci. Nutr.* **2020**, *8*, 340–350. [CrossRef] [PubMed]

23. Vadopalas, L.; Ruzauskas, M.; Lele, V.; Starkute, V.; Zavistanaviciute, P.; Zokaityte, E.; Bartkevics, V.; Badaras, S.; Klupsaite, D.; Mozuriene, E.; et al. Pigs' feed fermentation model with antimicrobial lactic acid bacteria strains combination by changing extruded soya to biomodified local feed stock. *Animals* **2020**, *10*, 783. [CrossRef] [PubMed]

24. European Union Law. European Parliament Directive 2010/63/EU of the European Parliament and of the Council of 22 September 2010 on the Protection of Animals Used for Scientific Purposes Text with EEA Relevance. Available online: https://eur-lex.europa.eu/legal-content/EN/TXT/?uri=celex%3A32010L0063 (accessed on 13 July 2020).

25. Lithuanian Director of the State Food and Veterinary Service. B1-866 Dėl Mokslo ir mokymo tikslais naudojamų gyvūnų laikymo, priežiūros ir naudojimo reikalavimų patvirtinimo. Available online: https://e-seimas.lrs.lt/portal/legalAct/lt/T

AD/TAIS.437081?positionInSearchResults=0&searchModelUUID=64a1f51f-6356-4b60-a21e-3a67ac3b1107 (accessed on 10 March 2020).

26. National Research Council. *Nutrient Requirements of Swine: Eleventh Revised Edition*; The National Academies Press: Washington, DC, USA, 2012.
27. AOAC. *Official Methods of Analysis*, 21st ed.; AOAC International: Rockland, MD, USA, 2019.
28. Merkeviciene, L.; Ruzauskaite, N.; Klimiene, I.; Siugzdiniene, R.; Dailidaviciene, J.; Virgailis, M.; Mockeliunas, R.; Ruzauskas, M. Microbiome and antimicrobial resistance genes in microbiota of cloacal samples from European herring gulls (*Larus Argentatus*). *J. Vet. Res.* **2017**, *4*, 27–35. [CrossRef]
29. Zavistanaviciute, P.; Poskiene, I.; Lele, V.; Antanaitis, R.; Kantautaite, J.; Bartkiene, E. The influence of the newly isolated Lactobacillus plantarum LUHS135 and *Lactobacillus paracasei* LUHS244 strains on blood and faeces parameters in endurance horses. *Pol. J. Vet. Sci.* **2019**, *22*, 513–521. [CrossRef]
30. The Lithuanian Minister of Environment. D1-862. Dėl Lietuvos Respublikos Aplinkos Apsaugos Normatyvinio Dokumento LAND 88-2009 "Amoniako Koncentracijos Nustatymas Aplinkos Ore Spektrometriniu Metodu" Patvirtinimo. Available online: https://e-seimas.lrs.lt/portal/legalAct/lt/TAD/TAIS.363222 (accessed on 13 July 2020).
31. Pajarillo, E.A.B.; Chae, J.P.; Kim, H.B.; Kim, I.H.; Kang, D.-K. Barcoded pyrosequencing-based metagenomic analysis of the faecal microbiome of three purebred pig lines after cohabitation. *Appl. Microbiol. Biotechnol.* **2015**, *99*, 5647–5656. [CrossRef]
32. Round, J.L.; Mazmanian, S.K. The gut microbiota shapes intestinal immune responses during health and disease. *Nat. Rev. Immunol.* **2009**, *9*, 313–323. [CrossRef]
33. Benson, A.K.; Kelly, S.A.; Legge, R.; Ma, F.; Low, S.J.; Kim, J.; Zhang, M.; Oh, P.L.; Nehrenberg, D.; Hua, K.; et al. Individuality in gut microbiota composition is a complex polygenic trait shaped by multiple environmental and host genetic factors. *Proc. Natl. Acad. Sci. USA* **2010**, *107*, 18933–18938. [CrossRef]
34. Leamy, L.J.; Kelly, S.A.; Nietfeldt, J.; Legge, R.M.; Ma, F.; Hua, K.; Sinha, R.; Peterson, D.A.; Walter, J.; Benson, A.K.; et al. Host genetics and diet, but not immunoglobulin A expression, converge to shape compositional features of the gut microbiome in an advanced intercross population of mice. *Genome Biol.* **2014**, *15*, 552. [CrossRef]
35. Frese, S.A.; Parker, K.; Calvert, C.C.; Mills, D.A. Diet shapes the gut microbiome of pigs during nursing and weaning. *Microbiome* **2015**, *3*, 28. [CrossRef] [PubMed]
36. Le Floc'h, N.; Knudsen, C.; Gidenne, T.; Montagne, L.; Merlot, E. Impact of feed restriction on health, digestion and faecal microbiota of growing pigs housed in good or poor hygiene conditions. *Animal* **2014**, *8*, 1632–1642. [CrossRef] [PubMed]
37. Yang, Q.; Huang, X.; Wang, P.; Yan, Z.; Sun, W.; Zhao, S. Longitudinal development of the gut microbiota in healthy and diarrheic piglets induced by age-related dietary changes. *Microbiologyopen* **2019**, *8*, e923. [CrossRef] [PubMed]
38. Lamendella, R.; Domingo, J.W.; Ghosh, S.; Martinson, J.; Oerther, D.B. Comparative fecal metagenomics unveils unique functional capacity of the swine gut. *BMC Microbiol.* **2011**, *15*, 103. [CrossRef] [PubMed]
39. Pajarillo, E.A.B.; Chae, J.-P.; Balolong, M.P.; Kim, H.B.; Kang, D.-K. Assessment of fecal bacterial diversity among healthy piglets during the weaning transition. *J. Gen. Appl. Microbiol.* **2014**, *60*, 140–146. [CrossRef]
40. Ivarsson, E.; Roos, S.; Liu, H.Y.; Lindberg, J.E. Fermentable non-starch polysaccharides increase the abundance of *Bacteroides-Prevotella-Porphyromonas* in ileal microbial community of growing pigs. *Animal* **2014**, *8*, 1777–1787. [CrossRef] [PubMed]
41. Rho, J.; Wright, D.P.; Christie, D.L.; Clinch, K.; Furneaux, R.H.; Roberton, A.M. A novel mechanism for desulfation of mucin: Identification and cloning of a mucin-desulfating glycosidase (*sulfoglycosidase*) from Prevotella strain RS2. *J. Bacteriol.* **2005**, *187*, 1543–1551. [CrossRef] [PubMed]
42. Delia, E.; Tafaj, M.; Männer, K. Efficiency of Probiotics in Farm Animals. Available online: https://www.intechopen.com/books/probiotic-in-animals/efficiency-of-probiotics-in-farm-animals (accessed on 17 May 2020).
43. Valeriano, V.D.V.; Balolong, M.P.; Kang, D.-K. Probiotic roles of Lactobacillus sp. in swine: Insights from gut microbiota. *J. Appl. Microbiol.* **2017**, *122*, 554–567. [CrossRef]
44. Gänzle, M.G.; Follador, R. Metabolism of oligosaccharides and starch in lactobacilli: A review. *Front. Microbiol.* **2012**, *3*, 340. [CrossRef]

45. Konstantinov, S.R.; Awati, A.A.; Williams, B.A.; Miller, B.G.; Jones, P.; Stokes, C.R.; Akkermans, A.D.L.; Smidt, H.; de Vos, W.M. Post-natal development of the porcine microbiota composition and activities. *Environ. Microbiol.* **2006**, *8*, 1191–1199. [CrossRef]

46. Dou, S.; Gadonna-Widehem, P.; Rome, V.; Hamoudi, D.; Rhazi, L.; Lakhal, L.; Larcher, T.; Bahi-Jaber, N.; Pinon-Quintana, A.; Guyonvarch, A.; et al. Characterisation of early-life fecal microbiota in susceptible and healthy pigs to post-weaning diarrhoea. *PLoS ONE* **2017**, *12*, e0169851. [CrossRef]

47. Canibe, N.; Jensen, B.B. Fermented liquid feed—Microbial and nutritional aspects and impact on enteric diseases in pigs. *Anim. Feed Sci. Technol.* **2012**, *173*, 17–40. [CrossRef]

48. Grela, E.R.; Czech, A.; Kiesz, M.; Wlazło, Ł.; Nowakowicz-Dębek, B. A fermented rapeseed meal additive: Effects on production performance, nutrient digestibility, colostrum immunoglobulin content and microbial flora in sows. *Anim. Nutr.* **2019**, *5*, 373–379. [CrossRef] [PubMed]

49. Mach, N.; Berri, M.; Estellé, J.; Levenez, F.; Lemonnier, G.; Denis, C.; Leplat, J.-J.; Chevaleyre, C.; Billon, Y.; Doré, J.; et al. Early-life establishment of the swine gut microbiome and impact on host phenotypes. *Environ. Microbiol. Rep.* **2015**, *7*, 554–569. [CrossRef]

50. Shin, D.; Chang, S.Y.; Bogere, P.; Won, K.; Choi, J.-Y.; Choi, Y.-J.; Lee, H.K.; Hur, J.; Park, B.-Y.; Kim, Y.; et al. Beneficial roles of probiotics on the modulation of gut microbiota and immune response in pigs. *PLoS ONE* **2019**, *14*, e0220843. [CrossRef]

51. Wang, L.; Zhu, L.; Qin, S. Gut microbiota modulation on intestinal mucosal adaptive immunity. *J. Immunol. Res.* **2019**, *2019*, 4735040. [CrossRef] [PubMed]

52. Slifierz, M.J.; Friendship, R.; Weese, J.S. Zinc oxide therapy increases prevalence and persistence of methicillin-resistant Staphylococcus aureus in pigs: A randomized controlled trial. *Zoonoses Public Health* **2015**, *62*, 301–308. [CrossRef]

53. Chen, L.; Xu, Y.; Chen, X.; Fang, C.; Zhao, L.; Chen, F. The maturing development of gut microbiota in commercial piglets during the weaning transition. *Front. Microbiol.* **2017**, *8*, 1688. [CrossRef]

54. Bian, G.; Ma, S.; Zhu, Z.; Su, Y.; Zoetendal, E.G.; Mackie, R.; Liu, J.; Mu, C.; Huang, R.; Smidt, H.; et al. Age, introduction of solid feed and weaning are more important determinants of gut bacterial succession in piglets than breed and nursing mother as revealed by a reciprocal cross-fostering model. *Environ. Microbiol.* **2016**, *18*, 1566–1577. [CrossRef]

55. Levesque, C.L.; Hooda, S.; Swanson, K.S.; de Lange, K. Alterations in ileal mucosa bacteria related to diet complexity and growth performance in young pigs. *PLoS ONE* **2014**, *9*, e108472. [CrossRef]

56. Zhu, L.; Liao, R.; Tu, W.; Lu, Y.; Cai, X. Pyrodextrin enhances intestinal function through changing the intestinal microbiota composition and metabolism in early weaned piglets. *Appl. Microbiol. Biotechnol.* **2020**, *104*, 4141–4154. [CrossRef]

57. Zimmermann, P.; Curtis, N. The influence of probiotics on vaccine responses—A systematic review. *Vaccine* **2018**, *36*, 207–213. [CrossRef] [PubMed]

58. Dowarah, R.; Verma, A.K.; Agarwal, N.; Singh, P.; Singh, B.R. Selection and characterization of probiotic lactic acid bacteria and its impact on growth, nutrient digestibility, health and antioxidant status in weaned piglets. *PLoS ONE* **2018**, *13*, e0192978. [CrossRef] [PubMed]

59. Arfken, A.M.; Frey, J.F.; Ramsay, T.G.; Summers, K.L. Yeasts of burden: Exploring the mycobiome–bacteriome of the piglet GI tract. *Front. Microbiol.* **2019**, *10*, 2286. [CrossRef] [PubMed]

60. Xu, X.; Li, L.-M.; Li, B.; Guo, W.-J.; Ding, X.-L.; Xu, F.-Z. Effect of fermented biogas residue on growth performance, serum biochemical parameters, and meat quality in pigs. *Asian Austral. J. Anim. Sci.* **2017**, *30*, 1464–1470. [CrossRef]

61. Czech, A.; Grela, E.; Kiesz, M.; Kłys, S. Biochemical and haematological blood parameters of sows and piglets fed a diet with a dried fermented rapeseed meal. *Ann. Anim. Sci.* **2020**, *20*, 535–550. [CrossRef]

62. Shi, C.; He, J.; Wang, J.; Yu, J.; Yu, B.; Mao, X.; Zheng, P.; Huang, Z.; Chen, D. Effects of Aspergillus niger fermented rapeseed meal on nutrient digestibility, growth performance and serum parameters in growing pigs. *Anim. Sci. J.* **2016**, *87*, 557–563. [CrossRef]

63. Cho, J.; Min, B.; Chen, Y.; Yoo, J.S.; Wang, Q. Evaluation of FSP (fermented soy protein) to replace soybean meal in weaned pigs: Growth performance, blood urea nitrogen and total protein concentrations in serum and nutrient digestibility. *Asian Austral. J. Anim. Sci.* **2007**, *20*, 1874–1879. [CrossRef]

64. Min, B. Effects of diet complexity and fermented soy protein on growth performance and apparent ileal amino acid digestibility in weanling pigs. *Asian Austral. J. Anim. Sci.* **2010**, *23*, 1496–1502. [CrossRef]

65. Zhu, J.; Gao, M.; Zhang, R.; Sun, Z.; Wang, C.; Yang, F.; Huang, T.; Qu, S.; Zhao, L.; Li, Y.; et al. Effects of soybean meal fermented by *L. plantarum, B. subtilis* and *S. cerevisieae* on growth, immune function and intestinal morphology in weaned piglets. *Microb. Cell Fact.* **2017**, *16*, 191. [CrossRef]

66. Liu, X.; Feng, J.; Xu, Z. The effects of fermented soybean meal on growth performance and immune characteristics in weaned piglets. *Turk. J. Vet. Anim. Sci.* **2007**, *31*, 341–345.

67. Jeong, Y.D.; Ko, H.S.; Hosseindoust, A.; Choi, Y.H.; Chae, B.J.; Yu, D.J.; Cho, E.S.; Cho, K.H.; Shim, S.M.; Ra, C.S.; et al. Lactobacillus-based fermentation product and lactose level in the feed for weanling pigs: Effects on intestinal morphology, microbiota, gas emission, and targeted intestinal coliforms. *Livest. Sci.* **2019**, *227*, 90–96. [CrossRef]

68. Satessa, G.D.; Tamez-Hidalgo, P.; Hui, Y.; Cieplak, T.; Krych, L.; Kjærulff, S.; Brunsgaard, G.; Nielsen, D.S.; Nielsen, M.O. Impact of dietary supplementation of lactic acid bacteria fermented rapeseed with or without macroalgae on performance and health of piglets following omission of medicinal zinc from weaner diets. *Animals* **2020**, *10*, 137. [CrossRef] [PubMed]

69. Oliver-Ferrando, S.; Segalés, J.; López-Soria, S.; Callén, A.; Merdy, O.; Joisel, F.; Sibila, M. Evaluation of natural porcine circovirus type 2 (PCV2) subclinical infection and seroconversion dynamics in piglets vaccinated at different ages. *Vet. Res.* **2016**, *47*, 121. [CrossRef]

70. Duivon, D.; Corrégé, I.; Hémonic, A.; Rigaut, M.; Roudaut, D.; Jolie, R. Field evaluation of piglet vaccination with a Mycoplasma hyopneumoniae bacterin as compared to a ready-to-use product including porcine circovirus 2 and M. hyopneumoniae in a conventional French farrow-to-finish farm. *Porc. Health Manag.* **2018**, *4*, 4. [CrossRef] [PubMed]

71. Jeong, J.; Park, C.; Choi, K.; Chae, C. Comparison of three commercial one-dose porcine circovirus type 2 (PCV2) vaccines in a herd with concurrent circulation of PCV2b and mutant PCV2b. *Vet. Microbiol.* **2015**, *177*, 43–52. [CrossRef] [PubMed]

72. da Silva, N.; Carriquiry, A.; O'Neill, K.; Opriessnig, T.; O'Connor, A.M. Mixed treatment comparison meta-analysis of porcine circovirus type 2 (PCV2) vaccines used in piglets. *Prev. Vet. Med.* **2014**, *117*, 413–424. [CrossRef]

73. Woźniak, A.; Miłek, D.; Matyba, P.; Stadejek, T. Real-time PCR detection patterns of porcine circovirus type 2 (PCV2) in Polish farms with different statuses of vaccination against PCV2. *Viruses* **2019**, *11*, 1135. [CrossRef]

74. Opriessnig, T.; Xiao, C.-T.; Halbur, P.G.; Gerber, P.F.; Matzinger, S.R.; Meng, X.-J. A commercial porcine circovirus (PCV) type 2a-based vaccine reduces PCV2d viremia and shedding and prevents PCV2d transmission to naïve pigs under experimental conditions. *Vaccine* **2017**, *35*, 248–254. [CrossRef]

75. Afghah, Z.; Webb, B.; Meng, X.-J.; Ramamoorthy, S. Ten years of PCV2 vaccines and vaccination: Is eradication a possibility? *Vet. Microbiol.* **2017**, *206*, 21–28. [CrossRef]

76. Chae, C. Porcine respiratory disease complex: Interaction of vaccination and porcine circovirus type 2, porcine reproductive and respiratory syndrome virus, and Mycoplasma hyopneumoniae. *Vet. J.* **2016**, *212*, 1–6. [CrossRef]

77. Patterson, R.; Nevel, A.; Diaz, A.V.; Martineau, H.M.; Demmers, T.; Browne, C.; Mavrommatis, B.; Werling, D. Exposure to environmental stressors result in increased viral load and further reduction of production parameters in pigs experimentally infected with PCV2b. *Vet. Microbiol.* **2015**, *177*, 261–269. [CrossRef] [PubMed]

78. Lee, D.J.; Yang, Y.; Jung, H.I.; Nguyen, D.H.; Kim, I.H. Effect of a protected blend of organic acids and medium-chain fatty acids on growth performance, nutrient digestibility, blood profiles, meat quality, faecal microflora, and faecal gas emission in finishing pigs. *Can. J. Anim. Sci.* **2019**, *99*, 448–455.

79. Nguyen, D.H.; Nyachoti, C.; Kim, I.H. Evaluation of effect of probiotics mixture supplementation on growth performance, nutrient digestibility, faecal bacterial enumeration, and noxious gas emission in weaning pigs. *Ital. J. Anim. Sci.* **2019**, *18*, 466–473. [CrossRef]

80. Bindas, L.; Bujnak, L.; Maskaľová, I.; Mihok, T.; Lacková, P.; Nad, P. Effects of feeding low protein diets on serum and faeces parameters in weaned piglets. *Folia Vet.* **2019**, *63*, 37–44. [CrossRef]

Review

The Effects of Fungal Feed Additives in Animals: A Review

Wen Yang Chuang [1], Yun Chen Hsieh [1] and Tzu-Tai Lee [1,2,*

[1] Department of Animal Science, National Chung Hsing University, Taichung 402, Taiwan;
xssaazxssaaz@yahoo.com.tw (W.Y.C.); richard840909@gmail.com (Y.C.H.)
[2] The iEGG and Animal Biotechnology Center, National Chung Hsing University, Taichung 402, Taiwan
* Correspondence: ttlee@dragon.nchu.edu.tw; Tel.: +886-4-22840366; Fax: +886-4-22860265

Received: 5 April 2020; Accepted: 5 May 2020; Published: 6 May 2020

Simple Summary: Fungal probiotics and ferments have potential as feed additives, but their use has long been ignored. The main goal of this review article is to report on the potential benefits and hazards of fungal feed additives. Previous research indicates that fungal feed additives enhance antioxidant capacity and decrease the inflammatory response in animals through polysaccharides, triterpenes, polyphenols, ergosterol, and adenosine. Accordingly, fungal feed additives could further enhance growth performance and animal health and could be of functional use.

Abstract: As probiotics, fungi enhance animal health and are suitable animal feed additives. In addition to brewing fungi, there are also edible and medicinal fungi. Common fungi utilized in feeding programs include *Saccharomyces cerevisiae*, *Aspergillus oryzae*, *Pleurotus* spp., *Antrodia cinnamomea*, and *Cordyceps militaris*. These fungi are rich in glucans, polysaccharides, polyphenols, triterpenes, ergosterol, adenosine, and laccases. These functional components play important roles in antioxidant, anti-inflammatory, anti-obesity, and immune system regulation. As such, fungal feed additives could be of potential use when breeding livestock. In previous studies, fungal feed additives enhanced body weight and egg production in poultry and improved the feed conversion rate. Several mycotoxins can be produced by hazardous fungi but fortunately, the cell walls constituents and enzymes of fungal probiotics can also act to decrease the toxicity of mycotoxins. Overall, fungal feed additives are of value, but their safety and usage must be studied further, including cost-benefit economic analyses.

Keywords: fungi; probiotic; feed additive; mushroom waste compost

1. Introduction

Fungi, as ancient eukaryotes, have been present on Earth for at least 2.4 billion years [1], and include 120,000 different species [2]. Fungi have traditionally been classified according to their morphology but can now be identified by DNA. The existence of fungi is inseparable from human history and development. Fungi can be classified as either edible or fermenting, based on their usage. The former includes common fungi such as *Pleurotus eryngii*, *Flammulina velutipes*, *P. ostreatus*, and mushrooms; the latter includes probiotics such as *Aspergillus oryzae*. Currently, although studies have shown that *Pennisetum* can be used as a culture medium for edible fungi, most are still planted in space packages filled with wood chips [3,4]. Mature fungi are rich in minerals, vitamins, and probiotics and, due to their powerful antioxidative functions, are increasingly attractive to consumers [5]. However, due to the widespread cultivation of mushrooms, increasing attention has been paid to the treatment of mushroom waste compost, the residual culture medium [4,6].

Saccharomyces cerevisiae and *A. oryzae* have been integrated into human life since ancient times. The former is related mainly to alcohol production, while the latter is involved in the fermentation process of soy sauce. Chuang et al. [7] and Khempaka et al. [8] both pointed out that, following fermentation by *A. oryzae* or *Saccharomyces cerevisiae*, neutral detergent fibers decrease and hemicellulose and crude protein content increase in the fermented product. In addition, our research shows that *S. cerevisiae* and *A. oryzae* can be used as feed additives and assist in reducing the inflammatory response in animals [7].

In addition to probiotics and edible mushrooms, the earliest discovered antibiotic—penicillin—is also produced by fungi (*Penicillium*). With the discovery and use of antibiotics, animals can grow rapidly and be free of many diseases. However, the abuse of antibiotics in recent years has caused many microorganisms to develop resistance, therefore reducing antibiotics in the diet has become increasingly important. Since 2006, the European Union has banned antibiotic use in animal feed (Commission Implementing Regulation (EU) 1831/2003 and 1463/2004, European Parliament) and other countries are also heading in this direction. In order to reduce the use of antibiotics, beneficial fungi whose ferments do not produce antibiotics or mycotoxins have been investigated, due to their increased antioxidant capacity [9].

Mycotoxins are fungal toxins often found in improperly preserved feed whose negative effects on animals include decreased growth performance and intestinal and liver damage [10,11]. However, other fungal probiotics could reduce damage by encasing mycotoxins in their cell walls or even degrading them [7,12,13].

Although most of the discussion on probiotics has focused on bacteria, fungi have considerable potential as probiotics and have been ignored for too long. Over the past five years, our research team has investigated the effects of fungal feed supplements on poultry health [4,7,9,14–21]. Although there have been many studies on the use of fungi and their ferments as feed supplements and their effect on animal health, most review articles have focused on the addition of probiotics or the effects of mycotoxins. This review aims instead to explore the use of fungal feed additives and their potential risks.

2. Edible Fungi and Their Potential Uses

There are more than 2000 known species of edible fungi worldwide, several of which have been commercially cultivated, including medicinal fungi. Common medicinal fungi include *Cordyceps* spp. and *Antrodia* spp., while edible fungi include *Pleurotus* spp., *Lentinula* spp., *Agaricus* spp., and *Flammulina* spp. *Cordyceps militaris*, *Pleurotus eryngii*, *Pleurotus sajor-caju*, and *Flammulina velutipes* are 4 fungi renowned for their nutritional benefits to animals.

Pleurotus eryngii, *Pleurotus sajor-caju* and *Flammulina velutipes* are the most commonly cultured mushrooms in Taiwan (Taiwan Agricultural Research Institute, Council of Agriculture, Executive Yuan). According to the estimates of the Taiwan Agricultural Research Institute, the annual production volume of fresh mushrooms is 140,000 tons, and the output value exceeds 10 billion new Taiwan dollars. It is therefore important to exploit any added value of mushroom usage. Although wild *Cordyceps sinensis* has many functional components and strong biological activity, it takes a long time to grow under strict conditions in order to form fruiting bodies and its yield is insufficient for human needs. Furthermore, *Cordyceps sinensis* cannot be cultivated artificially and can only be harvested once. In contrast, *Cordyceps militaris* only takes one to three months to mature and therefore has greater potential for development and use. In addition, *Cordyceps militaris* can be cultivated in media and has functional components similar to *Cordyceps sinensis*, garnering it greater attention [22]. *Cordyceps militaris* is rich in a variety of bioactive compounds, such as cordycepin, polysaccharides, ergosterol, and mannitol [23]. Some studies have also extracted polysaccharides with antioxidative or immunomodulatory activities from *Cordyceps militaris* [24–26].

Antrodia cinnamomea, an endemic species in Taiwan, has been used for centuries due to its high antioxidative and anti-inflammatory capacities [27,28]. The major functional components in *A. cinnamomea* are polysaccharides, triterpenes, sterols, benzenoids, benzoquinone derivatives, and maleic acid. [29]. These components could enhance antioxidative capacities and reduce the damage from tumors [30]. According to Lee et al. [20], the addition of 0.1%, 0.2%, or 0.4% *A. cinnamomea* powder not only decreased the coliform count and increased lactobacilli in broiler intestines, but it also enhanced the antioxidative capacity in serum. As a result, *A. cinnamomea* was also considered as a potential feed additive in chicken.

Pleurotus eryngii contains many functional ingredients, such as polysaccharides and peptides, is high in fiber, and has high anti-inflammatory and antioxidative capacities [31]. Widely used by humans as an important edible mushroom [32], the whole plant can be processed and is used widely as a health food [32]. *Pleurotus sajor-caju*, a mushroom similar to *Pleurotus eryngii*, is low in calories with very low lipid and starch levels but is rich in protein, fiber, minerals, and vitamins [5]. *Pleurotus sajor-caju* has a variety of biologically active compounds, including polysaccharides, phenols, terpenes, and sterols. [5]. It also has antiviral, antibacterial, antifungal, antioxidative, and anti-inflammatory activities, and is therefore widely used in traditional medicine and nutritional research [33].

In addition to its delightful aroma and taste, *Flammulina velutipes* has a variety of pharmacological properties [34]. It is also well known for its curative properties in liver disease and gastroenteric ulcers [35]. Yang et al. [36] indicated that the extractable polysaccharides in *Flammulina velutipes* have high antioxidative capacities, while the mushroom itself has anti-cancer, antibacterial, and immunomodulatory properties [36]. Based on these characteristics, *Flammulina velutipes* is very promising for further research. However, no matter the species, one must keep in mind that when the mature mushroom is removed, the waste compost remains, potentially causing new environmental problems [4]. Previous research shows that mushroom byproducts have a similar nutritional composition as edible parts [37,38]. As such, based on its low cost, rich nutritional composition, large supply, and numerous biologically active functions, it has potential as a feed additive.

3. Hazardous Fungi Species

Mycotoxins are secondary metabolites produced by some pathogenic fungi. Mycotoxins can affect animal health and are one of the main detriments to growth [10]. Unfortunately, corn and soybeans, the main ingredients in animal feed, are prone to mold when improperly stored and may accumulate mycotoxins [11]. Common mycotoxins include aflatoxins, citrinin, ochratoxin, T-2 toxin, and vomitoxin [39]. Mycotoxins reduce growth performance in animals and are also nephrotoxic, hepatotoxic, and neurotoxic [10,40]. At present, the harm from mycotoxins can be reduced by mold adsorbents, mycotoxin-degrading enzymes, and probiotic degradation [12,40]. Interestingly, although produced by pathogenic fungi, mycotoxins can be treated with yeast cell walls or enzymes produced by edible fungi [12,40]. The main components in yeast cell walls, glucan, and mannan, are mycotoxin-adsorbing substances, while laccase and oxidase produced by *Pleurotus eryngii* reduce the toxicity of mycotoxins [12,13]. The main concern regarding the use of mushroom waste compost is whether its culture medium is conducive to mold growth and could produce mycotoxins. Fortunately, according to Loi et al. [12] and Haque et al. [13], the mushroom mycelium and enzymes contained in waste compost are not conducive to mycotoxin production. In addition, under proper storage conditions (including rapid drying and crushing), the risk of mycotoxins in mushroom waste compost can be further minimized.

4. The Functional Components of Fungi

During fungus cultivation, several enzymes assist the fungi in degrading and utilizing medium nutrients [7]. During this process, many fungal metabolites, secondary metabolites, and synthetics are produced [5,41] (Figure 1). These substances include fungal cell walls, polysaccharides, ergosterol, adenosine, and triterpenes. [5]. In addition to anti-inflammatory, antimicrobial, anti-fatigue, and anti-malarial effects, these functional substances can also protect the lungs and liver and enhance immunity and antioxidant capacity [41].

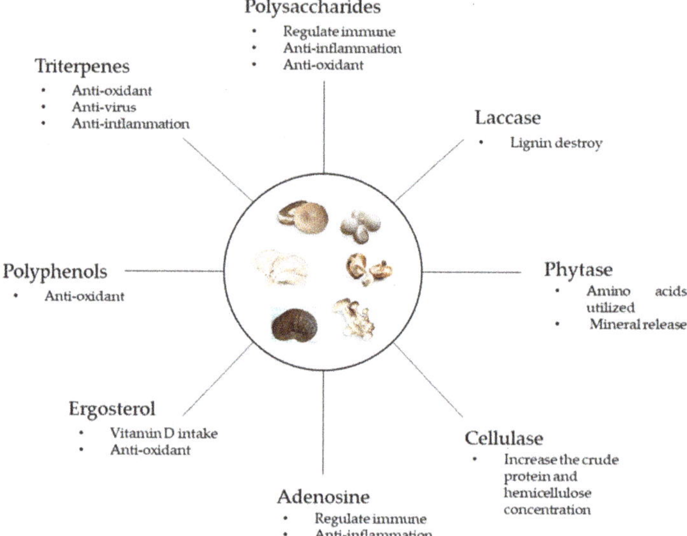

Figure 1. Fungal compounds and their potential functions. Mushroom images were adapted from Du et al. [42], and the photograph of *Ganoderma* was taken by the author.

4.1. Triterpenes

Triterpenes are functional sterol metabolites commonly found in mushrooms, that are mainly present in lanostane [5,43]. However, although Chuang et al. [4] reported that triterpenes can be found in mushroom waste compost, most research has focused on triterpenes in medicinal fungi. Mushroom triterpenes can inhibit α-glucosidase activity [44] and thereby reduce blood sugar concentration in animals [45]. Triterpenoids present in *Ganoderma lucidum* can reduce LPS/ D-Galactosamine-induced liver damage by reducing tumor necrosis factor-alpha (TNF-α) and interleukin-6 (IL-6) expression [46]. These triterpenes can also inhibit TLR4-MyD88 (Toll-like receptor 4-myeloid differentiation primary response 88) activity and therefore decrease NF-κB expression [46,47]. In addition, Choi et al. [48] indicated that the anostane triterpene content in *G. lucidum* enhanced heme oxygenase-1 expression and decreased the inflammatory response in RAW264.7 cells. Novak [49] reported on the antibacterial activity of diterpenes in *P. mutilis*, while Ma et al. [43] described the functionality of other terpenoids in mushrooms, including antioxidant and anti-virus capacities.

4.2. Polyphenols and Flavonoids

Polyphenols are functional substances commonly found in nature, especially plants and fungi [50,51], with flavonoids being among the most effective [50]. Gil-Ramirez et al. [52] indicated there were no flavonoids in mushrooms, due to the absence of flavonoid absorption and synthesis enzymes. This claim was challenged four years later by Mohanta [53], who pointed to evidence that a large number of fungi contain genes and enzymes related to flavonoid synthesis.

According to Gil-Ramirez et al. [52], it comes down to four points: (1) the traditional method of measuring flavonoid content via absorbance is wrong; (2) flavonoids in the medium cannot be absorbed by the mushrooms; (3) mushrooms do not have genes or enzymes related to flavonoid synthesis; and (4) something went wrong in about 9% of the reports that used HPLC analysis to determine flavonoid content in mushrooms. However, although Gil-Ramirez et al. [52] provided sufficient evidence to prove the first and second points, the third point was challenged by Mohanta [53]. In addition, there is no evidence for the fourth point, only the authors' speculations. For these reasons, we agree that we should avoid "considering fungi as plants" but the components of fungi should not be ignored. Therefore, without further scientific proof to the contrary, we assume fungi contain flavonoids.

The feed additives used with animals are usually agriculture byproducts or ferments. It was confirmed that fermented products, including waste compost and fungi-fermented products, contain high amounts of flavonoids, which come from the plant-based medium [4]. The phenolic components in fungi play important roles in the antioxidant capacities of animals [51,54–56]. It is well known that phenols can chelate free radicals and decrease oxidative stress in animal cells [57,58]. In vivo, phenols enhance antioxidant systems, such as Nrf-2 and glutamate-cysteine ligase catalytic (GCLC) expression, and therefore improve lipid metabolism and meat quality [4,20,59,60].

4.3. Ergosterol

First discovered in yeast, ergosterol is commonly found in fungi and has anti-inflammatory effects [5]. Kuo et al. [61] found that ergosterol inhibits the expression of RAW264.7 NF–κB induced by lipopolysaccharides (LPS) and reduces the inflammatory response of TNF-α and other substances (IC50 = 24.5 μg/mL). Ergosterol inhibits the performance of cyclooxygenase-2 (COX-2) but has no effect on nitric oxide production [61]. Not only can ergosterol reduce the inflammatory response in animals, but it is also converted into vitamin D following exposure to ultraviolet light [62]. Additional vitamin D intake increases serum levels in animals [63] and may therefore enhance antioxidant and fat metabolism capabilities.

4.4. Adenosine

Adenosine, an end product from the breakdown of ATP, is common in fungi and animal cells [64,65]. It regulates energy utilization in animal cells and reduces the inflammatory response. Adnosine binds to G proteins (A1, A2A, A2B, and A3) to regulate the immune capacity and inhibit the production of harmful enzymes and proteins [64,66]. When combined with different G protein receptors, adenosine improves the growth of macrophages (A1), suppresses the inflammatory response by inhibiting Th1 activities and promoting Th2 activities (A2A), upregulates Th2 activities (A2B), and suppresses immunity (A3) [66]. In *Cordyceps militaris*, there is another special adenosine, 3′-deoxyadenosine, also known as cordycepin [65]. Cordycepin upregulates pro-apoptosis genes such as P53, BCL2-associated X protein, caspase-3, and caspase-9, while downregulating antioxidant gene expression. Apoptosis is triggered by the destruction of mitochondria in the tumor cell, thereby inhibiting the growth of cancer cells in the brain [67,68].

4.5. Fungal Cell Wall and Polysaccharides

The fungal cell wall consists mainly of β-glucan, mannoprotein, and chitin [69]. Of these, the first two are very effective at coating LPS and mycotoxins and can reduce their damage [69–71]. In addition, β-glucan, mannoprotein, and chitin can regulate immunity, increase antioxidant capacity, and reduce the inflammatory response [70,72]. A common fungal polysaccharide is yeast cell walls.

Yeast cell walls are dominated by mannan-oligosaccharides and β-glucans, with less chitin [70]. Mannan-oligosaccharides and β-glucans reduce the harm of ochratoxin A to broilers and regulate their immunity [70]. β-glucans also reduce the inflammatory response and the harm from LPS [73,74]. The concentration of polysaccharides in *Cordyceps*, as with other functional metabolites, is significantly affected by medium composition (moisture, geographical environment, and climate) [37,75]. In general, *Cordyceps militaris* contains about 3–8% of dry matter and is stored in fruiting bodies and mycelium in solid and liquid media [37,75]. *Cordyceps* polysaccharides enhance immune cell phagocytosis, improve humoral and cell immunity, enhance the activity of macrophages, monocytes, and lymphocytes, and reduce oxidative stress by enhancing the activity of SOD and Gpx. In addition, crude extracts of *Cordyceps militaris* can reduce the amount of lipid oxidation [75].

Polysaccharides derived from *Flammulina velutipes* decrease the pH value in mice intestines and increase both the amount of short-chain fatty acids in mice intestines and serum immunoglobin [34].

In addition to acting as a kind of prebiotic, polysaccharides in edible mushrooms are also well known for their antioxidant and anti-inflammatory capacities [76–81]. In mice, the polysaccharides in *L. edodes* and *Phallus atrovolvatus* decrease TNF-α, IL-6, IL-1β, blood urea nitrogen, and uric acid levels in blood serum while increasing antioxidant enzymes such as SOD, CAT, and Gpx in the kidney [77,79]. Accordingly, due to the antioxidant and anti-inflammatory capacities of their polysaccharides, mushrooms have potential as a feed additive which improves animal health.

4.6. Enzymes

Fungi produce multiple enzymes during cultivation, which can be divided into three categories: peroxidases, carbohydrases, and phytases. Peroxidases assist fungi in facing the challenge of toxic phenols and increase their environmental competitiveness [80]. Laccase, one of the most well-known peroxidases in mushrooms, degrades lignin in the culture medium [81]. By degrading lignin, fungi effectively use nutrients that are difficult for other microorganisms to consume [82]. Based on this, the fermentation of plant-based feed supplements can reduce lignin content and increase the concentration of hemicellulose [7,8]. Fang et al. [83] further indicated that fungi could transform fiber into volatile fatty acids as carbohydrate enzymes such as laccase efficiently destroyed the structure of lignin.

Once laccase destroys the structure of lignin, cellulase, and hemicellulase further degrade the fiber in plant-based ingredients. As such, fungi reduce the concentration of difficult-to-digest fiber and produce more hemicellulose-based prebiotics in the fermented medium [7,8]. In previous research, cellulose enzyme groups produced by fungi included cellulase, β-glucanase, xylanase, and β-glucosidase [7,84]. In addition to degrading fiber in the culture medium and providing a carbon source for fungal growth, these enzymes increase crude protein and hemicellulose concentrations in plant-based ingredients following fungal fermentation thanks to the destruction of cellulose and condensed nutrients [7,8]. These additional nutrients can be further utilized by animals, suggesting that fungi-fermented plant-based ingredients have potential as feed additives [4,7].

Another common fungal enzyme is phytase. Phytase degrades phytate in plant materials and increases the animal's use of amino acids and minerals, especially phosphorus [85]. In general, phytase is found in *Aspergillus niger* and *Escherichia coli* [86]. However, Wanzenbock et al. [87] indicated that the phytate content in wheat bran decreased after *P. eryngii* fermentation, which reveals potential phytase activities in *P. eryngii*. Phytase significantly increases the use of feed nutrients by animals and reduces the excretion of inorganic phosphorus [88]. Interestingly, co-fermentation of phytase with fungal microorganisms enhances the efficiency of phytase by increasing the release of phytate from plant-based ingredients [7]. This indirectly reduces the harm to the environment when raising livestock [7,88].

5. The Potential Use of Fungal Feed Additives

T.T. Lee's research team has discussed in detail the effects of fungal feed additives on poultry health. Fungal species investigated include *S. cerevisiae*, *A. oryzae*, *A. cinnamomea*, *Trichoderma pseudokoningii*, *Cordyceps militaris*, *Pleurotus ostreatus*, *Pleurotus eryngii*, and *Aureobasidium pullulans*. Additives were added directly to the diet or a portion of an ingredient was replaced with either pure probiotic powder, the fermented product, or mushroom waste compost. Fungal feed additives may enhance animal production performance; the results are listed in Table 1.

Table 1. The effects of fungal feed additives on animal growth [1].

Animal Type	Treatment	Rearing Period	Body (Egg) Weight	Feed Conversion Rate	References
Broilers	0.5% *A. pullulans* fermentation [2]	22–35 d 1–35 d	+12% * +8% *	−4% −3%	[14]
Hendrix laying hens	2% *C. militaris* wastes residue	5–8 weeks 9–12 weeks 0–12 weeks	+4% * +3% * +7% *	−10% * −5% * −6% *	[15]
White Roman geese	5% mushroom waste compost	5–8 weeks 9–12 weeks	+2% −1%	0% +31%	[16]
Male broilers	0.1% *T. pseudokoningii* powder	1–21 d 22–35 d	+10% * +6%	−4% −3%	[18]
Broilers	10% *P. eryngii* mushroom compost fermented wheat bran	1–21 d 22–35 d 1–35 d	+2% +3% +2%	−8% 0% −3%	[6]
Male broilers	0.1% *A. cinnamomea*	1–21 d 22–35 d 1–35 d	+14% * +10% * +11% *	−1% −5% −7%	[20]
Male broilers	10% *A. cinnamomea* fermented wheat bran	1–21 d 22–35 d 1–35 d	+7% * −1% −2%	−5% * +1% +2%	[21]
Male broilers	0.5% mushroom waste compost	1–21 d 22–35 d 1–35 d	+3% +5% +4%	−6% * +5% −7%	[4]

[1] All percentage changes calculated by the difference between the treatment group and control group. [2] 75% *Aureobasidium pullulans* fermented soybean hulls in combination with 25% *Pleurotus eryngii* stalk residue * Significant difference between control and treatment group ($p < 0.05$).

In addition to improving production performance, fungal feed additives could further enhance animal health (Table 2). Both *S. cerevisiae* and *A. oryzae* are ancient probiotics used in the brewing industry. However, they could also enhance broiler health as feed additives. Chuang et al. [7] indicated that the product from co-fermenting 0.1% *S. cerevisiae* or *A. oryzae* with phytase could reduce the inflammatory response in broilers by decreasing the number of *Clostridium perfringens* in the ileum and suppressing inflammation-related mRNA expression, including NF-κB and iNOS. Furthermore, the *A. oryzae* and phytase co-fermented product enhanced villus height in the jejunum. According to the research of Teng et al. [19], villus height and lactic acid concentration in the ileum increased significantly after supplementation with 10% *S. cerevisiae* wheat bran. Previous research revealed that *A. oryzae* contains multiple enzymes, especially cellulase and hemicellulase, which improves the fiber type in feed and enhances nutrient absorption [8,89,90]. A specific kind of yeast, *Aureobasidium pullulans* increases the hemicellulose concentration of soybean hulls and *Pleurotus eryngii* stalk residue by about 1.5 times following fermentation [60]. *Aureobasidium pullulans* fermented soybean hulls and *Pleurotus eryngii* stalk residue could also decrease the ammonia nitrogen concentration in the cecum [14]. Furthermore, Lai et al. [14] indicated that the *Aureobasidium pullulans* ferment contained high amounts of xylanase and mannanase, and these enzymes proportionally increased relative to the days of fermentation.

Table 2. The effects of fungal feed additives on animal health.

Animal Type	Treatment	Functional Components	Functions	References
Broilers	0.5% *A. pullulans* ferment [1]	_[2]	Increased SOD and CAT activities, ileum villus height and lactic acid bacteria number in cecum	[14]
Hendrix laying hens	2% *C. militaris* waste residue	Cordycepin, cordycepic acid, crude polysaccharides, flavonoid, adenosine, and crude triterpenoid	Increased egg mass and eggshell strength; improved feed conversion rate throughout the entire experimental period; decreased cholesterol content in egg yolk	[15]
White Roman geese	5% mushroom waste compost	-	Increased SOD activities and decreased MDA content in serum; improved flavor, color, and acceptability on sensory evaluation	[16]
Male broilers	10% *T. pseudokoningii* fermented wheat bran	Cellulase, xylanase, and reducing sugar	Decreased coliform count and increased villus height in ileum	[17]
Male broilers	0.1% *T. pseudokoningii* powder	Phenols, xylanase, and cellulase	Increased SOD and CAT activities, jejunum villus height, and lactic acid bacteria levels in cecum	[18]
Broilers	10% *S. cerevisiae* fermented wheat bran	-	Increased villus height and lactic acid content in ileum	[19]
Male broilers	0.4% *A. cinnamomea* addition	Crude triterpenoids, crude polysaccharides, and phenols	Enhanced Nrf-2, GCLC, SOD, CAT and HO-1 mRNA expression and decreased NF-κB and IL-1β mRNA expression	[20]
Male broilers	10% *A. cinnamomea* fermented wheat bran	Crude triterpenoids, crude polysaccharides, and phenols	Enhanced SOD and CAT activities and decreased total cholesterol and low-density lipoprotein content in serum	[21]
Male broilers	0.1% *S. cerevisiae* or *A. oryzae* powder	Xylanase, protease, cellulase, and ß-glucanase	Decreased the number of *Clostridium perfringens* in ileum and suppressed inflammation-related mRNA expression	[7]
Male broilers	0.5% mushroom waste compost	Crude triterpenes, phenols, flavonoids, gallocatechin, and epigallocatechin	Increased antioxidant capacity, adipolysis, and gut barrier expression	[4]
Female broilers	1% *C. militaris* waste residue	Polysaccharides, triterpenes, phenols, and flavonoids	Enhanced antioxidant-related mRNA expression	[9]

[1] 75% *Aureobasidium pullulans* fermented soybean hulls in combination with 25% *Pleurotus eryngii* stalk residue;
[2] The symbol "-" indicates the author did not mention the functional components.

Among fungal microbes, there is one special species: *Trichoderma pseudokoningii*. Although *T. pseudokoningii* is not a legal feed additive, its high potential in the poultry industry was reported by both Lin et al. [18] and Chu et al. [17]. Supplementation with 10% *T. pseudokoningii* fermented wheat bran enhanced the villus height, thereby increasing the villus height:crypt depth ratio, and the coliform count in the ileum decreased [17]. The addition of 0.1%, 0.2%, or 0.4% *T. pseudokoningii* powder improved animal growth performance in the first 21 days and improved the feed conversion ratio [18]. Furthermore, supplementation with *T. pseudokoningii* powder increased lactate concentration and lactic acid bacteria levels in the cecum [18].

C. militaris and *A. cinnamomea* are both well-known traditional Chinese medicines, due to their antioxidative, anti-inflammatory, and antitumor capacities [9,20]. Lee et al. [20,21] reported on the effects of *A. cinnamomea* powder and fermentation (fermented wheat bran) supplements on antioxidant, anti-inflammation, and fat metabolism in broilers. *A. cinnamomea* powder decreased the coliform count and increased lactobacilli in an in vitro test [20]. Supplementation with 0.1%, 0.2%, or 0.4% *A. cinnamomea* powder enhanced SOD and CAT activities in mRNA as well as protein levels in

21- and 35-day-old animals. The expression of inflammation-related mRNA, such as NF-κB and IL-1β, also decreased [20]. Furthermore, the addition of *A. cinnamomea*-fermented wheat bran to the broiler diet enhanced SOD and CAT activities and reduced total cholesterol and low-density lipoprotein levels in broiler serum [21]. Due to its ease of cultivation, *C. militaris* is more popular than *C. sinensis*. However, as with other types of mushrooms, its waste residue is rich in mycelium and other functional compounds, suggesting it has potential as a feed additive. Wang et al. [15] indicated that supplementation with 0.5%, 1%, or 2% *C. militaris* waste residue increased egg mass and eggshell strength and improved the feed conversion rate throughout the entire experimental period. The cholesterol content in egg yolk also decreased after supplementation with 1% and 2 % *C. militaris* waste residue [15]. Furthermore, Hsieh et al. [9] revealed that *C. militaris* waste residue was about 9 times higher in polysaccharide content than *Pleurotus eryngii*, *Pleurotus sajor-caju*, or *Fammulina velutipes* waste residues. As such, supplementation with *C. militaris* waste residue could significantly increase antioxidant capacity via activated Keap-1 and Nrf-2 mRNA expression [9].

Another common fungal feed additive is mushroom waste compost. As the residue left after cultivation, mushroom waste compost is high in mycelium and wood fiber, which enhances animal antioxidants and adipolysis [4]. Although the real effect of mushroom waste compost on adipose metabolism is unclear, it appears to increase KCTD15 and adiponectin expression, thereby enhancing adipolysis-related mRNA expression [4]. Interestingly, in recent years the anti-obesity capacity of vitamin D, which is converted from ergosterol, has attracted attention [62]. The addition of vitamin D could enhance antioxidant capacity, decrease the total cholesterol and triglyceride levels in serum, and further enhance the peroxisome proliferator-activated receptor γ (PPARγ) and perilipin-2 expression to enhance adipolysis [91–93]. Bindhu and Arunave [94] also indicated that *Pleurotus ostreatus* and its bioactive anthraquinone could enhance the PPARγ and CCAAT enhancer-binding protein α (CEBPα), thereby decreasing adipose accumulation. This suggests that the anti-obesity capacity of mushrooms may be due to ergosterol and anthraquinone intake.

Several papers point out that mushroom waste compost could enhance antioxidant capacity [4,6,16]. By enhancing Nrf-2 and GCLC mRNA expression, mushroom waste compost could upregulate the antioxidant capacity of poultry and further decrease MDA concentration [4]. Based on its high fiber content, the gut barrier would also be improved by supplementation with mushroom waste compost [4]. Interestingly, Wang et al. [6] indicated that a 10% supplementation of wheat bran fermented by the remaining mycelium in *P. eryngii* mushroom compost could also enhance CAT activities. Overall, whether they're used in traditional Chinese medicine or are common edible species, fungi play an important role in upregulating animal health and could protect against stress. Based on the effects of the different fungal feed additives on growth performance and animal health, supplementation with 0.1% *A. cinnamomea*, 0.5% mushroom waste compost, or 2% *C. militaris* waste residue are recommended [4,15,20]. Furthermore, in addition to avoiding illegal or harmful fungi, it is not recommended to add more than 5% of any fungal feed additive since better results were achieved with a small amount (<2%) (Table 1).

6. Future Perspectives

In general, probiotic discussions have focused on bacterial microorganisms and the potential use of fungi has long been ignored. Through this review, we have tried our best to fully explain the potential uses of fungi. Fungi can enhance antioxidant capacity and fat metabolism in animals and maintain intestinal health. However, pathogenic fungi should be avoided, even if probiotic fungi can offset their negative effects. The most suitable amount depends on the species and type of fungi. Overall, the value of fungi as probiotics or functional feed additives has been highlighted in this review.

Author Contributions: Conceptualization, W.Y.C. and T.-T.L.; supervision, T.-T.L.; investigation, W.Y.C. and T.-T.L.; writing—original draft preparation, W.Y.C., Y.C.H. and T.-T.L.; writing—review and editing, W.Y.C. and T.-T.L. All authors have read and agreed to the published version of the manuscript.

Funding: This research received no external funding.

Acknowledgments: The authors thank the Ministry of Science and Technology (MOST 107-2313-B-005-037-MY2) and the iEGG and Animal Biotechnology Center from The Feature Areas Research Center Program within the framework of the Higher Education Sprout Project by the Ministry of Education (MOE) in Taiwan for supporting this study.

Conflicts of Interest: The authors declare no conflict of interest.

Abbreviations

TNF-α	Tumor necrosis factor alpha
Gpx	Glutathione peroxidase
SOD	Superoxidase dismutase
MDA	Malondialdehyde
TLR4	Toll-like receptor 4
MyD88	Myeloid differentiation primary response 88
COX-2	Cyclooxygenase-2
CAT	Catalase
NFκB	Nuclear factor kappa B p 65
iNOS	Inducible nitric oxide synthases
IFN-γ	Interferon-γ
IL-1ß	Interleukin-1ß
Nrf-2	Nuclear factor erythroid 2–related factor 2
GCLC	Glutamate-cysteine ligase catalytic
KCTD-15	Potassium channel tetramerization domain-containing 15
CEBPα	CCAAT-enhancer-binding proteins-alpha
CPT-1	Carnitine palmitoyltransferase I
PPAR-γ	Peroxisome proliferator-activated receptor gamma

References

1. Bengtson, S.; Rasmussen, B.; Ivarsson, M.; Muhling, J.; Broman, C.; Marone, F.; Stampanoni, M.; Bekker, A. Fungus-like mycelial fossils in 2.4-billion-year-old vesicular basalt. *Nat. Ecol. Evol.* **2017**, *1*, 141. [CrossRef] [PubMed]

2. Mueller, G.M.; Schmit, J.P. Fungal biodiversity: What do we know? What can we predict? *Biodivers. Conserv.* **2006**, *16*, 1–5. [CrossRef]

3. Maleko, D.; Mwilawa, A.; Msalya, G.; Pasape, L.; Mtei, K. Forage growth, yield and nutritional characteristics of four varieties of napier grass (*Pennisetum purpureum Schumach*) in the west Usambara highlands. *Afr. Crop Sci. J.* **2019**, *6*, e00214. [CrossRef]

4. Chuang, W.Y.; Liu, C.L.; Tsai, C.F.; Lin, W.C.; Chang, S.C.; Shih, H.D.; Shy, Y.M.; Lee, T.T. Evaluation of waste mushroom compost as a feed supplement and its effects on the fat metabolism and anti-oxidant capacity of broilers. *Animals* **2020**, *10*, 445. [CrossRef] [PubMed]

5. Finimundy, T.C.; Barros, L.; Calhelha, R.C.; Alves, M.J.; Prieto, M.A.; Abreu, R.M.V.; Dilon, A.J.P.; Henriques, J.A.P.; Roesch, M.; Ferreira, I.C.F.R. Multifunctions of *pleurotus sajor-caju* (fr.) singer: A highly nutritious food and a source for bioactive compounds. *Food Chem.* **2018**, *245*, 150–158. [CrossRef] [PubMed]

6. Wang, C.C.; Lin, L.J.; Chao, Y.P.; Chiang, C.J.; Lee, M.T.; Chang, S.C.; Yu, B.; Lee, T.T. Anti-oxidant molecular targets of wheat bran fermented by white rot fungi and its potential modulation of antioxidative status in broiler chickens. *Br. Poult. Sci.* **2017**, *58*, 262–271. [CrossRef] [PubMed]

7. Chuang, W.Y.; Lin, W.C.; Hsieh, Y.C.; Huang, C.M.; Chang, S.C.; Lee, T.T. Evaluation of the combined use of *Saccharomyces Cerevisiae* and *Aspergillus Oryzae* with phytase fermentation products on growth, inflammatory, and intestinal morphology in broilers. *Animals* **2019**, *9*, E1051. [CrossRef]

8. Khempaka, S.; Thongkratok, R.; Okrathok, S.; Molee, W. An evaluation of cassava pulp feedstuff fermented with *A. oryzae*, on growth performance, nutrient digestibility and carcass quality of broilers. *J. Poult. Sci.* **2013**, *51*, 71–79. [CrossRef]

9. Hsieh, Y.C.; Lin, W.C.; Chuang, W.Y.; Chen, M.H.; Chang, S.C.; Lee, T.T. Effects of mushroom waster medium and stalk residues on the growth performance and oxidative status in broilers. *Asian-Australas. J. Anim. Sci.* **2020**. [CrossRef]

10. Maciorowski, K.G.; Herrera, P.; Jones, F.T.; Pillai, S.D.; Ricke, S. C Effects on poultry and livestock of feed contamination with bacteria and fungi. *Anim. Feed Sci. Technol.* **2007**, *133*, 109–136. [CrossRef]

11. Munkvold, G.P.; Arias, S.; Taschl, I.; Gruber-Dorninger, C. Chapter 9—Mycotoxins in corn: Occurrence, impacts, and management. In *Corn*; Elsevier Inc.: Amsterdam, The Netherlands, 2019; pp. 235–287. [CrossRef]

12. Loi, M.; Fanelli, F.; Cimmarusti, M.T.; Mirabelli, V.; Haidukowski, M.; Logrieco, A.F.; Caliandro, R.; Mule, G. In vitro single and combined mycotoxins degradation by Ery4 laccase from *Pleurotus eryngii* and redox mediators. *Food Control* **2018**, *90*, 401–406. [CrossRef]

13. Haque, M.A.; Wang, Y.; Shen, Z.; Li, X.; Saleemi, M.K.; He, C. Mycotoxin contamination and control strategy in human, domestic animal and poultry: A review. *Microb. Pathog.* **2020**, *142*, 104095. [CrossRef] [PubMed]

14. Lai, L.P.; Lee, M.T.; Chen, C.S.; Yu, B.; Lee, T.T. Effects of co-fermented *Pleurotus eryngii* stalk residues and soybean hulls by *Aureobasidium pullulans* on performance and intestinal morphology in broiler chickens. *Poult. Sci.* **2015**, *94*, 2959–2969. [CrossRef] [PubMed]

15. Wang, C.L.; Chiang, C.J.; Chao, Y.P.; Yu, B.; Lee, T.T. Effect of *Cordyceps Militaris* waster medium on production performance, egg traits and egg yolk cholesterol of laying hens. *J. Poult. Sci.* **2015**, *52*, 188–196. [CrossRef]

16. Chang, S.C.; Lin, M.J.; Chao, Y.P.; Chiang, C.J.; Jea, Y.S.; Lee, T.T. Effects of spent mushroom compost meal on growth performance and meat characteristics of grower geese. *R. Bras. Zootec.* **2016**, *45*, 281–287. [CrossRef]

17. Chu, Y.T.; Lo, C.T.; Chang, S.C.; Lee, T.T. Effects of *Trichoderma* fermented wheat bran on growth performance, intestinal morphology and histological findings in broiler chickens. *Ital. J. Anim. Sci.* **2017**, *16*, 82–92. [CrossRef]

18. Lin, W.C.; Lee, M.T.; Lo, C.T.; Chang, S.C.; Lee, T.T. Effects of dietary supplementation of *Trichoderma pseudokoningii* fermented enzyme powder on growth performance, intestinal morphology, microflora and serum anti-oxidantive status in broiler chickens. *Ital. J. Anim. Sci.* **2017**, *17*, 153–164.

19. Teng, P.Y.; Chang, C.L.; Huang, C.M.; Chang, S.C.; Lee, T.T. Effects of solid-state fermented wheat bran by *Bacillus amyloliquefaciens* and *Saccharomyces cerevisiae* on growth performance and intestinal microbiota in broiler chickens. *Ital. J. Poult. Sci.* **2017**, *54*, 134–141. [CrossRef]

20. Lee, M.T.; Lin, W.C.; Wang, S.Y.; Lin, L.J.; Yu, B.; Lee, T.T. Evaluation of potential anti-oxidant and anti-inflammatory effects of *Antrodia cinnamomea* powder and the underlying molecular mechanisms via Nrf2- and NF-κB-dominated pathways in broiler chickens. *Poult. Sci.* **2018**, *97*, 2419–2434. [CrossRef]

21. Lee, M.T.; Lin, W.C.; Lin, L.J.; Wang, S.Y.; Chang, S.C.; Lee, T.T. Effects of dietary *Antrodia cinnamomea* fermented product supplementation on antioxidation, anti-inflammation, and lipid metabolism in broiler chickens. *Asian-Australas. J. Anim. Sci.* **2019**. In press. [CrossRef]

22. Wu, L.; Sun, H.; Hao, Y.; Zheng, X.; Song, Q.; Dai, S.; Zhu, Z. Chemical structure and inhibition on α-glucosidase of the polysaccharides from *Cordyceps militaris* with different developmental stages. *Int. J. Biol. Macromol.* **2020**, *148*, 722–736. [CrossRef] [PubMed]

23. Ng, T.B.; Wang, H.X. Pharmacological actions of cordyceps, a prized folk medicine. *J. Pharm. Pharmacol.* **2005**, *57*, 1509–1519. [CrossRef] [PubMed]

24. Chen, X.; Wu, G.; Huang, Z. Structural analysis and anti-oxidant activities of polysaccharides from cultured *Cordyceps militaris*. *Int. J. Biol. Macromol.* **2013**, *58*, 18–22. [CrossRef] [PubMed]

25. Wang, M.; Meng, X.Y.; Yang, R.L.; Qin, T.; Wang, X.Y.; Zhang, K.Y.; Fei, C.Z.; Li, Y.; Hu, Y.L.; Xue, F.Q. *Cordyceps militaris* polysaccharides can enhance the immunity and antioxidation activity in immunosuppressed mice. *Carbohydr. Polym.* **2012**, *89*, 461–466. [CrossRef]

26. Won, S.Y.; Park, E.H. Anti-inflammatory and related pharmacological activities of cultured mycelia and fruiting bodies of cordyceps militaris. *J. Ethnopharmacol.* **2005**, *96*, 555–561. [CrossRef]

27. Wang, H.C.; Chu, F.H.; Chien, S.C.; Liao, J.W.; Hsieh, H.W.; Li, W.H.; Lin, C.C.; Shaw, J.F.; Kuo, Y.H.; Wang, S.Y. Establishment of the metabolite profile for an *Antrodia cinnamomea* health food product and investigation of its chemoprevention activity. *J. Agric. Food Chem.* **2013**, *61*, 8556–8564. [CrossRef]

28. Yang, F.C.; Yang, Y.H.; Lu, H.C. Enhanced anti-oxidant and antitumor activities of *Antrodia cinnamomea* cultured with cereal substrates in solid state fermentation. *Biochem. Eng. J.* **2013**, *78*, 108–113. [CrossRef]

29. Lee, I.H.; Huang, R.L.; Chen, C.T.; Chen, H.C.; Hsu, W.C.; Lu, M.K. Antrodia camphorate polysaccharides exhibit anti-hepatitis B virus effects. *FEMS Microbiol. Lett.* **2002**, *209*, 63–67. [CrossRef]

30. Wu, H.; Pan, C.L.; Yao, Y.C.; Chang, S.S.; Li, S.L.; Wu, T.F. Proteomic analysis of the effect of Antrodia camphorate extract on human lung cancer A549 cell. *Proteomics* **2006**, *6*, 826–835. [CrossRef]

31. Li, S.; Shah, N.P. Anti-inflammatory and anti-proliferative activities of natural and sulphonated polysaccharides from *Pleurotus eryngii*. *J. Funct. Foods* **2016**, *23*, 80–86. [CrossRef]

32. Zhang, C.; Li, S.; Zhang, J.; Hu, C.; Che, G.; Zhou, M.; Jia, L. anti-oxidant and hepatoprotective activities of intracellular polysaccharide from *Pleurotus eryngii* si-04. *Int. J. Biol. Macromol.* **2016**, *91*, 568–577. [CrossRef] [PubMed]

33. Duru, M.E.; Cayan, G.T. Biologically active terpenoids from mushroom origin: A review. *Rec. Nat. Prod.* **2015**, *9*, 456–483.

34. Zhao, R.; Hu, Q.; Ma, G.; Su, A.; Xie, M.; Li, X.; Chen, G.; Zhao, L. Effects of *Flammulina velutipes* polysaccharide on immune response and intestinal microbiota in mice. *J. Funct. Foods* **2019**, *56*, 255–264. [CrossRef]

35. Tang, C.; Hoo, P.C.; Tan, L.T.; Pusparajah, P.; Khan, T.M.; Lee, L.H.; Goh, B.H.; Chan, K.G. Golden Needle Mushroom: A culinary medicine with evidenced-based biological activities and health promoting properties. *Front. Pharmacol.* **2016**, *7*, 474–501. [CrossRef]

36. Yang, W.; Yu, J.; Zhao, L.; Ma, N.; Fang, Y.; Pei, F.; Mariga, M.; Hu, Q. Polysaccharides from *Flammulina velutipes* improve scopolamine-induced impairment of learning and memory of rats. *J. Funct. Foods* **2015**, *18*, 411–422. [CrossRef]

37. Liu, L.; Chen, Y.; Luo, Q.; Xu, N.; Zhou, M.; Gao, B.; Wang, C.; Shi, Y. Fermenting liquid vinegar with higher taste, flavor and healthy value by using discarded *Cordyceps militaris* solid culture medium. *LWT-Food Sci. Technol.* **2018**, *98*, 654–660. [CrossRef]

38. Mahfuz, S.; Song, H.; Liu, Z.; Liu, X.; Diao, Z.; Ren, G.; Guo, Z.; Cui, Y. Effect of golden needle mushroom (*Flammulina velutipes*) stem waste on laying performance, calcium utilization, immune response and serum immunity at early phase of production. *Asian-Australas. J. Anim. Sci.* **2018**, *31*, 705–711. [CrossRef]

39. Whitlow, L.W.; Hagler, W.M.J. Mycotoxins in Dairy Cattle: Occurrence, Toxicity, Prevention and Treatment. *Proc. Southwest Nutr. Conf.* **2005**, 124–138.

40. Peng, W.X.; Marchal, J.L.M.; vanderPoel, A.F.B. Strategies to prevent and reduce mycotoxins for compound feed manufacturing. *Anim. Feed Sci. Technol.* **2018**, *237*, 129–153. [CrossRef]

41. Das, S.K.; Masuda, M.; Sakurai, A.; Sakakibara, M. Medicinal uses of the mushroom *Cordyceps militaris*: Current state and prospects. *Fitote* **2010**, 961–968. [CrossRef]

42. Du, B.; Zhu, F.; Xu, B. An insight into the anti-inflammatory properties of edible and medicinal mushrooms. *J. Funct. Foods* **2018**, *47*, 334–342. [CrossRef]

43. Ma, G.; Yang, W.; Zhao, L.; Pei, F.; Fang, D.; Hu, Q. A critical review on the health promoting effects of mushrooms nutraceuticals. *Food Sci. Hum. Wellness* **2018**, *7*, 125–133. [CrossRef]

44. Chen, X.Q.; Zhao, J.; Chen, L.X.; Wang, S.F.; Wang, Y.; Li, S.P. Lanostane triterpenes from the mushroom Ganoderma resinaceum and their inhibitory activities against a-glucosidase. *Phytochemistry* **2018**, *149*, 103–115. [CrossRef]

45. Ali, K.M.; Chatterjee, K.; De, D.; Jana, K.; Bera, T.K.; Ghosh, D. Inhibitory effect of hydro-methanolic extract of seed of *Holarrhena antidysenterica* on alpha-glucosidase activity and postprandial blood glucose level in normoglycemic rat. *J. Ethnopharmacol.* **2011**, *135*, 194–196. [CrossRef] [PubMed]

46. Hu, Z.; Du, R.; Xiu, L.; Bian, Z.; Ma, C.; Sato, N.; Hattori, M.; Zhang, H.; Liang, Y.; Yu, S.; et al. Protective effect of triterpenes of *Ganoderma lucidum* on lipopolysaccharide induced inflammatory responses and acute liver injury. *Cytokine* **2020**, *127*, 154917. [CrossRef] [PubMed]

47. Dudhgaonkar, S.; Thyagarajan, A.; Sliva, D. Suppression of the inflammatory response by triterpenes isolated from the mushroom *Ganoderma lucidum*. *Int. Immunopharmacol.* **2009**, *9*, 1272–1280. [CrossRef]

48. Choi, S.; Nguyen, V.T.; Tae, N.; Lee, S.; Ryoo, S.; Min, B.S.; Lee, J.H. Anti-inflammatory and heme oxygenase-1 inducing activities of lanostane triterpenes isolated from mushroom *Ganoderma lucidum* in RAW264.7 cells. *Toxicol. Appl. Pharmacol.* **2014**, *280*, 434–442. [CrossRef]

49. Novak, R. Are pleuromutilin antibiotics finally fit for human use? *Ann. N.Y. Acad. Sci.* **2011**, *1241*, 71–81. [CrossRef]

50. Chiang, Y.M.; Chuang, D.Y.; Wang, S.Y.; Guo, Y.H.; Tsai, P.W.; Shyur, L.F. Metabolite profiling and chemopreventive bioactivity of plant extracts from *Bidens Pilosa*. *J. Ethnopharmacol.* **2004**, *95*, 409–419. [CrossRef]

51. Palacios, I.; Lozano, M.; Moro, C.; Arrigo, M.D.; Rostagno, M.A.; Martinez, J.A.; GarciaLafuente, A.; Guillamon, E.; Villares, A. Anti-oxidant properties of phenolic compounds occurring in edible mushrooms. *Food Chem.* **2011**, *128*, 674–678. [CrossRef]

52. Gil-Ramírez, A.; Pavo-Caballero, C.; Baeza, E.; Baenas, N.; Garcia-Viguera, C.; Marín, F.R.; Soler-Rivas, C. Mushrooms do not contain flavonoids. *J. Funct. Food* **2016**, *25*, 1–13. [CrossRef]

53. Mohanta, T.K. Fungi contain genes associated with flavonoid biosynthesis pathway. *J. Funct. Foods* **2020**, *68*, 103910. [CrossRef]

54. Vaz, J.A.; Barros, L.; Martins, A.; Morais, J.S.; Vasconcelos, M.H.; Ferreira, I.C.F.R. Phenolic profile of seventeen Portuguese wild mushrooms. *Food Sci. Technol.* **2011**, *44*, 343–346. [CrossRef]

55. 56Gasecka, M.; Mleczek, M.; Siwulski, M.; Niedzielski, P. Phenolic composition and anti-oxidant properties of *Pleurotus ostreatus* and *Pleurotus eryngii* enriched with selenium and zinc. *Eur. Food Res. Technol.* **2016**, *242*, 723–732. [CrossRef]

56. Sun, Y.; Zhang, M.; Fang, Z. Efficient physical extraction of active constituents from edible fungi and their potential bioactivities: A review. *Trends Food Sci. Technol.* **2019**. In press. [CrossRef]

57. Yan, J.; Zhao, Y.; Suo, S.; Liu, Y.; Zhao, B. Green tea catechins ameliorate adipose insulin resistance by improving oxidative stress. *Free Radic. Biol. Med.* **2012**, *52*, 1648–1657. [CrossRef]

58. Guo, H.; Saravanakumar, K.; Wang, M.H. Total phenolic, flavonoid contents and free radical scavenging capacity of extracts from tubers of *Stachys affinis*. *Biocatal. Agric. Biotechnol.* **2018**, *15*, 235–239. [CrossRef]

59. Starčević, K.; Krstulović, L.; Brozić, D.; Maurić, M.; Stojević, Z.; Mikulec, Ž.; Bajić, M.; Mašek, T. Production performance, meat composition and oxidative susceptibility in broiler chicken fed with different phenolic compounds. *J. Sci. Food Agric.* **2015**, *95*, 1172–1178. [CrossRef]

60. Lee, M.T.; Lai, L.P.; Lin, W.C.; Ciou, J.Y.; Chang, S.C.; Yu, B.; Lee, T.T. Improving nutrition utilization and meat Quality of broiler chickens through solid-state fermentation of agricultural by-products by *Aureobasidium Pullulans*. *Rev. Bras. Cienc. Avic.* **2017**, *19*, 645–654. [CrossRef]

61. Kuo, C.F.; Hsieh, C.H.; Lin, W.Y. Proteomic response of LAP-activated RAW 264.7 macrophages to the anti-inflammatory property of fungal ergosterol. *Food Chem.* **2011**, *126*, 207–212. [CrossRef]

62. Papoutsis, K.; Grasso, S.; Menon, A.; Brunton, N.P.; Lyng, J.G.; Jacquier, J.C.; Bhuyan, D.J. Recovery of ergosterol and vitamin D2 from mushroom waste - Potential valorization by food and pharmaceutical industries. *Trends Food Sci. Technol.* **2020**, *99*, 351–366. [CrossRef]

63. Baur, A.C.; Kühn, J.; Brandsch, C.; Hirche, F.; Stangl, G.I. Intake of ergosterol increases the vitamin D concentrations in serum and liver of mice. *J. Steroid Biochem.* **2019**, *194*, 105435. [CrossRef] [PubMed]

64. Caiazzo, E.; Maione, F.; Morello, S.; Lapucci, A.; Paccosi, S.; Steckel, B.; Lavecchia, A.; Parenti, A.; Iuvone, T.; Schrader, J.; et al. Adenosine signaling mediates the anti-inflammatory effects of the COX-2 inhibitor nimesulide. *Biochem. Pharmacol.* **2016**, *112*, 72–81. [CrossRef]

65. Chiu, C.P.; Hwang, T.L.; Chan, Y.; El-Shazly, M.; Wug, T.Y.; Lo, I.W.; Hsu, Y.M.; Lai, K.H.; Houh, M.F.; Yuani, S.S.; et al. Research and development of *Cordyceps* in Taiwan. *Food Sci. Hum. Wellness* **2016**, *5*, 177–185. [CrossRef]

66. Haskó, G.; Antonioli, L.; Cronstein, B.N. Adenosine metabolism, immunity and joint health. *Biochem. Pharmacol.* **2018**, *151*, 307–313. [CrossRef]

67. Nakamura, K.; Shinozuka, K.; Yoshikawa, N. Anticancer and antimetastatic effects of cordycepin, an active component of *Cordyceps sinensis*. *J. Pharma. Sci.* **2014**, *127*, 53–56. [CrossRef]

68. Chaicharoenaudomrung, N.; Jaroonwitchawan, T.; Noisa, P. *Cordycepin* induces apoptotic cell death of human brain cancer through the modulation of autophagy. *Toxicol. In Vitro* **2018**, *46*, 113–121. [CrossRef]

69. Rodríguez, A.; Cuesta, A.; Ortuño, J.; Esteban, M.A.; Meseguer, J. Immunostimulant properties of a cell wall-modified whole *Saccharomyces cerevisiae* strain administered by diet to seabream (*Sparus aurata* L.). *Vet. Immunol. Immunopathol.* **2003**, *96*, 183–192. [CrossRef]

70. Awaad, M.; Atta, A.M.; El-Ghany, W.A.; Elmenawey, M.; Ahmed, A.; Hassan, A.A.; Nada, A.; Abdelaleem, G.A.; Kawkab, A.A. Effect of a specific combination of mannan-oligosaccharides and β-glucans extracted from yeast cell wall on the health status and growth performance of ochratoxicated broiler chickens. *J. Anim. Sci.* **2011**, *7*, 82–96.

71. Al-Khalaifah, H.S. Benefits of probiotics and/or prebiotics for antibiotic-reduced poultry. *Poult. Sci.* **2018**, *11*, 3807–3815. [CrossRef]

72. Carrasco-González, J.A.; Serna-Saldívar, S.O.; Gutierrez-Uribe, J.A. Mycochemical changes induced by selenium enrichment in *P. ostreatus* fruiting bodies. *J. Agric. Food Chem.* **2017**, *65*, 4074–4082. [CrossRef] [PubMed]

73. Penney, J.; Lu, Y.; Pan, B.; Feng, Y.; Walk, C.; Li, J. Pure yeast beta-glucan and two types of yeast cell wall extracts enhance cell migration in porcine intestine model. *J. Funct. Foods* **2019**, *59*, 129–137. [CrossRef]

74. Peng, Q.H.; Cheng, L.; Kang, K.; Tian, G.; Mohammad, A.M.; Xue, B.; Wang, L.Z.; Zou, H.W.; Mathew, G.G.; Wang, Z.S. Effects of yeast and yeast cell wall polysaccharides supplementation on beef cattle growth performance, rumen microbial populations and lipopolysaccharides production. *J. Integr. Agric.* **2020**, *19*, 810–819. [CrossRef]

75. Nie, S.; Cui, S.; Xie, M. Bioactive polysaccharides from *Cordyceps sinensis*: Isolation, structure features and bioactivities. *Bioact. Carbohydr. Diet. Fibre.* **2017**, *1*, 38–52. [CrossRef]

76. Zhang, L.; Hu, Y.; Duan, X.; Tang, T.; Shen, Y.; Hu, B.; Liu, A.; Chen, H.; Li, C.; Liu, Y. Characterization and anti-oxidant activities of polysaccharides from thirteen boletus mushrooms. *Int. J. Biol. Macromol.* **2018**, *113*, 1–7. [CrossRef] [PubMed]

77. Song, X.; Ren, Z.; Wang, X.; Jia, L.; Zhang, C. Anti-oxidant, anti-inflammatory and renoprotective effects of acidichydrolytic polysaccharides by spent mushroom compost (*Lentinula edodes*) on LPS-induced kidney injury. *Int. J. Biol. Macromol.* **2019**. [CrossRef]

78. Yan, J.; Zhu, L.; Qu, Y.; Qu, X.; Mu, M.; Zhang, M.; Muneer, G.; Zhou, Y.; Sun, L. Analyses of active anti-oxidant polysaccharides from four edible mushrooms. *Int. J. Biol. Macromol.* **2019**, *123*, 945–956. [CrossRef]

79. Chaiyama, V.; Keawsompong, S.; LeBlanc, J.G.; deLeBlanc, A.M.; Chatel, J.M.; Chanput, W. Action modes of the immune modulating activities of crude mushroom polysaccharide from *Phallus atrovolvatus*. *Bioact. Carbohydr. Diet. Fibre.* **2020**, *23*, 100216. [CrossRef]

80. Bettin, F.; Cousseau, F.; Martins, K.; Boff, N.A.; Zaccaria, S.; daSilveira, M.M.; Dillon, A.J.P. Phenol removal by laccases and other phenol oxidases of *Pleurotus sajor-caju* PS-2001 in submerged cultivations and aqueous mixtures. *J. Environ. Manag.* **2019**, *236*, 581–590. [CrossRef]

81. Junior, J.A.; Vieira, Y.A.; Cruz, I.A.; Vilar, D.S.; Aguiar, M.M.; Torres, N.H.; Bharagava, R.N.; Lima, Á.S.; deSouza, R.L.; Ferreira, L.F.R. Sequential degradation of raw vinasse by a laccase enzyme producing fungus *Pleurotus sajor-caju* and its ATPS purification. *Biotechnol. Rep.* **2020**, *25*, e00411. [CrossRef]

82. Abdel-Hamid, A.M.; Solbiati, J.O.; Cann, I.K.O. Chapter One—Insights into lignin degradation and its potential industrial applications. *Adv. Appl. Microbiol.* **2013**, *82*, 1–28. [PubMed]

83. Fang, W.; Zhang, P.; Zhang, X.; Zhu, X.; van Lier, J.B.; Spanjers, H. White rot fungi pretreatment to advance volatile fatty acid production from solid-state fermentation of solid digestate: Efficiency and mechanisms. *Energy* **2018**, *162*, 534–541. [CrossRef]

84. Hu, T.; Wang, X.; Zhen, L.; Gu, J.; Zhang, K.; Wang, Q.; Ma, J.; Peng, H.; Lei, L.; Zhao, W. Effects of inoculating with lignocellulose-degrading consortium on cellulose degrading genes and fungal community during co-composting of spent mushroom substrate with swine manure. *Bioresour. Technol.* **2019**, *291*, 121876. [CrossRef] [PubMed]

85. Cowieson, A.J.; Ruckebusch, J.P.; Sorbara, J.O.B.; Wilson, J.W.; Guggenbuhl, P.; Roos, F.F. A systematic view on the effect of phytase on ileal amino acid digestibility in broilers. *Anim. Feed Sci. Technol.* **2017**, *225*, 182–194. [CrossRef]

86. Menezes-Blackburn, D.; Jorquera, M.; Gianfreda, L.; Rao, M.; Greiner, R.; Garrido, E.; Mora, M.L. Activity stabilization of *Aspergillus niger* and *Escherichia coli* phytases immobilized on allophanic synthetic compounds and montmorillonite nanoclays. *Bioresour. Technol.* **2011**, *102*, 9360–9367. [CrossRef]

87. Wanzenbock, E.; Apprich, S.; Tirpanalan, O.; Zitz, U.; Kracher, D.; Schedle, K.; Kneifel, W. Wheat bran biodegradation by edible *Pleurotus* fungi e A sustainable perspective for food and feed. *Food Sci. Technol.* **2017**, *86*, 123–131. [CrossRef]

88. Hamdi, M.; Perez, J.F.; L´etourneau-Montminy, M.P.; Franco-Rossell'o, R.; Aligue, R.; Sol'a-Oriol, D. The effects of microbisal phytase and dietary calcium and phosphorus levels on the productive performance and bone mineralization of broilers. *Anim. Feed Sci. Technol.* **2018**, *243*, 41–51. [CrossRef]

89. Ndrepepa, G. Uric acid and cardiovascular disease. *Eur. Caediol.* **2018**, *484*, 150–163. [CrossRef]

90. Sandhya, C.; Alagarsamy, S.; Szakacs, G.; Pandey, A. Comparative evaluation of neutral protease production by *Aspergillus oryzae* in submerged and solid-state fermentation. *Process. Biochem.* **2005**, *40*, 2689–2694. [CrossRef]

91. Li, J.; Mihalcioiu, M.; Li, L.; Zakikhani, M.; Camirand, A.; Kremer, R. Vitamin D prevents lipid accumulation in murine muscle through regulation of PPARγ and perilipin-2 expression. *J. Steroid Biochem.* **2018**, *177*, 116–124. [CrossRef]

92. Barzegari, M.; Sarbakhsh, P.; Mobasseri, M.; Noshad, H.; Esfandiari, A.; Khodadadi, B.; Gargari, B.P. The effects of vitamin D supplementation on lipid profiles and oxidative indices among diabetic nephropathy patients with marginal vitamin D status. *Diabetes Metab. Syndr.* **2019**, *13*, 542–547. [CrossRef] [PubMed]

93. Asbaghi, O.; Kashkooli, S.; Choghakhori, R.; Hasanvand, A.; Abbasnezhad, A. Effect of calcium and vitamin D co-supplementation on lipid profile of overweight/obese subjects: A systematic review and meta-analysis of the randomized clinical trials. *Obes. Med.* **2019**, *15*, 100124. [CrossRef]

94. Bindhu, J.; Arunava, D. An edible fungi *Pleurotus ostreatus* inhibits adipogenesis via suppressing expression of PPAR γ and C/EBP α in 3T3-L1 cells: In vitro validation of gene knock out of RNAs in PPAR γ using CRISPR spcas9. *Biomed. Pharmacother.* **2019**, *116*, 109030.

Article

In Vivo Toxicity and In Vitro Solubility Assessment of Pre-Treated Struvite as a Potential Alternative Phosphorus Source in Animal Feed

Soomin Shim [1,†], Seunggun Won [2,†], Arif Reza [1,3,†], Seungsoo Kim [1], Sungil Ahn [1], Baedong Jung [4], Byungil Yoon [4] and Changsix Ra [1,*]

[1]   Department of Animal Industry Convergence, College of Animal Life Sciences, Kangwon National University, Chuncheon 24341, Korea; vibrato7@naver.com (S.S.); reza.arif@kangwon.ac.kr (A.R.); seung_su89@nate.com (S.K.); sorzente@naver.com (S.A.)

[2]   Department of Animal Resources, College of Life and Environmental Science, Daegu University, Gyeongsan 38453, Korea; swon@daegu.ac.kr

[3]   Department of Environmental Science, College of Agricultural Sciences, IUBAT—International University of Business Agriculture and Technology, Dhaka 1230, Bangladesh

[4]   Department of Veterinary Science, Kangwon National University, Chuncheon 24341, Korea; bdjung@kangwon.ac.kr (B.J.); byoon@kangwon.ac.kr (B.Y.)

*   Correspondence: changsix@kangwon.ac.kr; Tel.: +82-33-250-8618

†   These authors contributed equally to this work.

Received: 13 August 2019; Accepted: 7 October 2019; Published: 11 October 2019

Simple Summary: Phosphorus (P) is an indispensable element needed for the growth and development of all living organisms. Phosphorus is mined primarily from non-renewable phosphate rock reserves. The use of P as fertilizer and feed additives continues to rise with the increasing global population. We therefore need to find an alternative and renewable, as well as sustainable, source of P to fulfill the demands of agriculture and livestock. Phosphorus recovered as struvite from livestock wastewater could be a sustainable alternative to commercial P sources, while its application has only been limited to arable lands as fertilizer. This study elucidates that struvite can also be used as an alternative P source in animal feed other than fertilizer through proper pre-treatment. Phosphorus recovery from livestock wastewater and its reutilization as an animal feed ingredient would therefore be a good strategy to substitute commercial P sources and ensure societal sustainability.

Abstract: Apart from using as fertilizer for plants, the application of struvite may be expanded to animal feed industries through proper pre-treatment. This study aimed to investigate the safety and efficacy of using pre-treated struvite (microwave irradiated struvite (MS) and incinerated struvite (IS)) in animal feeds. For safety assessment, an in vivo toxicity experiment using thirty female Sprague Dawley rats (average body weight (BW) of 200 ± 10 g) was conducted. The rats were randomly divided into five groups, including a control. Based on the BW, MS and IS were applied daily by oral administration with 1 and 10 mg kg^{-1}-BW (MS1 and MS10; IS1 and IS10) using dimethyl sulfoxide (DMSO) as a vehicle. A series of jar tests were conducted for four hours to check the solubility of the MS and IS at different pH (pH 2, 4, 5, 6 and 7) and compared to a commercial P source (monocalcium phosphate, MCP, control). The toxicity experiment results showed no significant differences among the treatments in BW and organ (liver, kidney, heart, and lung) weight of rats ($p > 0.05$). There were no adverse effects on blood parameters and the histopathological examination showed no inflammation in the organ tissues in MS and IS treated groups compared to the control. In an in vitro solubility test, no significant difference was observed in ortho-phosphate (O-P) solubility from the MCP and MS at pH 2 and 4 ($p > 0.05$), while O-P solubility from MS at pH 5 to 7 was higher than MCP and found to be significantly different ($p < 0.05$). O-P solubility from IS was the lowest among the treatments and significantly different from MCP and MS in all the experiments ($p < 0.05$). The results of this study not only suggest that the struvite pre-treated as MS could be a potential alternative source of P in animal feed but also motivate further studies with more stringent designs to better examine the potential of struvite application in diverse fields.

Keywords: phosphorus; struvite; swine wastewater; pre-treatment; toxicity; solubility; animal feed

---

1. Introduction

Phosphorus (P) is a key element for animal and plant growth as well as an important nutrient for future food security. It has been estimated that the present reserve of phosphate rock is 70,000 million tons as P_2O_5, of which 40 million tons of phosphate rock is mined every year [1,2]. Studies reported that the resource may be exhausted within the next 90–300 years, with an expected annual increase in P consumption of 2% [1,3]. Phosphate rock was recently categorized as one of the 20 critical raw materials in Europe since P utilization has been known as a one-way flow and it is classified as a non-renewable resource. In the world, it is mainly three countries (Morocco, China, and the United States), which take the lead in intensive production for phosphate rock [4]. In addition, due to non-replaceability and high economic importance, there might be a political and social tension built over the phosphate rock reserves and the world could move from an oil-based to a phosphate-based economy, as the aforementioned three countries control more than 85% of the known global phosphorus reserves [5].

The majority (80 to 90%) of the P mined from phosphate rock reserves has been used in the agricultural sector for fertilizer and animal feed production [6]. Whereas, of the total phosphate fertilizer applied in the arable lands, 57% can be lost in soil leaching, erosion, and runoff [7,8]. Moreover, livestock wastewater contains relatively high strengths of P concentration, which leads to eutrophication [9], resulting from excessive P contents inevitably given to achieve high growth performance. Every year 7 million tons of P may be released to the environment through animal manure and excreta [10]. P recovery and recycling from highly polluted livestock wastewater could therefore be a sustainable alternative to commercial P sources.

For P recovery from different wastewater, various physicochemical techniques have been introduced. Among those methods, struvite precipitation has been widely studied with its very straightforward regime and is considered as an eco-friendly and sustainable process in terms of nutrient recovery and recycling from nutrient-rich waste streams.

Along with economic development, swine farming practices in Korea are becoming more intensive. Intensive swine farming practices result in an increase in swine wastewater generation. Swine wastewater treatment practice in Korea is quite different compared to European and North American countries. The Korean government is currently operating more than 200 centralized swine wastewater treatment plants with a treatment capacity of more than 20,000 tonnes per day [11,12]. The bulk recovery of P in the form struvite from the centralized treatment plants might therefore be economical and sustainable. Moreover, the strategy of recovering and recycling P from swine wastewater and its reutilization as an animal feed ingredient can help protect the environment by reducing nutrient contents in discharged effluents as well as initiating a P circular economy.

Currently, the application of struvite has only been limited to arable lands as fertilizer and reported to be non-toxic to plants [13]. Moreover, P bioavailability from struvite has shown similar performance as the chemical fertilizers in crop trials [14]. Furthermore, to our knowledge, none of the studies focused on exploring the feasibility of struvite application in animal feed as an alternative P source. Swine wastewater generally contains high amounts of ortho-phosphate (O-P) and ammonium nitrogen (NH_4-N). NH_4-N is highly toxic for the animals and its accumulation in cells can damage the tissues [15]. It is therefore necessary to pre-treat the struvite recovered from swine wastewater for its further use in animal feed. We have already successfully tested the feasibility of reutilizing P recovered from swine wastewater as a dietary supplement in juvenile far eastern catfish (*Silurus asotus*) by comparing with a commercial P source [16,17]. In this study, we suggest manufacturing options for struvite recovered from swine wastewater to ensure biological safety and present the possibility of struvite application as an alternative P source in animal feed rations, with the results of an in vivo toxicity test on rats and an in vitro solubility test.

2. Materials and Methods

2.1. Struvite Production from Swine Wastewater and Pre-Treatment

Swine wastewater was collected from the storage tank of a gestation pig barn located in Chuncheon, South Korea. After collection, the solid/liquid separation was done using centrifugation at 3000 rpm for 10 min and the liquid part was used for struvite production. The reactor had an effective volume of 20 L, including a compartment of about 2 L as a reaction zone, and the rest was composed of a settling zone and recovery zone. The influent with the flow rate of 0.11 L min^{-1} (160 L d) was introduced from the top and passed through the compartment with holes at the bottom. Crystals formed in the reaction zone were precipitated and the flux flew up to the outlet, which resulted in a hydraulic retention time (HRT) of 3 h in the entire P recovery process. The precipitates were collected from the bottom where the valve was equipped. $MgCl_2$ in the liquid phase was provided into the reactor to meet the appropriate Mg to P ratio (~1.3) in the influent, which was found as an optimal condition in our previous research [18]. An air diffuser was set at the bottom of the compartment in the reaction zone for CO_2 stripping and, thus, a pH of over 8.5 was maintained. An X-ray diffraction (XRD) analysis (X'Pert PRO MPD, PANalytical B.V., Netherlands) was conducted to give the precipitates an identity, where the peaks indicate that the precipitates recovered were identified as struvite. The supernatant was put back into the reactor after collecting precipitates from the bottom of the struvite reactor. The settled materials were spread on the tray and stored in a dry-oven at 20 °C for 2 days, then the moisture content of pre-dried sample was measured at 105 °C following the standard methods [19], which resulted in a moisture content of 35%, but water parts in struvite evaporated during the second drying at 105 °C. To remove all possible toxicity, recovered precipitates were pre-treated in a microwave (80 Hz g^{-1} for 5 min) and a muffle furnace at 550 °C for over 30 min after drying [3,17]. The heavy metal contents in the pre-treated struvites were determined using inductively coupled plasma atomic emission spectroscopy (ICP-AES) (Optima 7300 DV, PerkinElmer, Waltham, MA, USA).

2.2. In Vivo Toxicity Assessment in Rats

An in vivo toxicity experiment using rats was conducted to evaluate the safety of using pretreated struvite (microwave irradiated struvite (MS); incinerated struvite (IS)) as an alternative to commercial P sources in animal feeds. The protocol of the experiment was approved and the rats were cared for according to the guidelines of the Institutional Animal Care and Use Committee of Kangwon National University, Chuncheon, South Korea (ACUC # KW 130620-1). For the toxicity experiment, 30 female Sprague Dawley rats (average body weight (BW) of 200 ± 10 g) were procured from Bio Genomic Inc. (Charles River Technology, Gapyung-Gun, Korea) and were allotted to each of five groups, including a control, in a randomized complete block design, resulting in six rats per treatment group. The sample size was calculated using the resource equation method [20]. All groups of rats were given ad libitum access to water and conventional laboratory diets (5L79, Labdiet, St. Louis, MO, USA) to maintain their normal physiology. The amount of crude protein, crude fat, crude fiber, ash, Ca, and P was 18, 5, 5, 8, 0.85, and 0.70%, respectively, in the supplied lab diet. To observe the toxic effects of test materials, subacute in vivo toxicity tests were conducted for 28 days [21]. Based on the observations regarding the highest dose of ammonia without effect reported by Tuleka et al., pre-treated struvite doses of 1 and 10 mg kg^{-1} BW were designated to evaluate the repeated dose toxicity [22]. MS and IS of 1 and 10 mg kg^{-1} BW were dissolved in 1% dimethyl sulfoxide (DMSO) as a vehicle, which were represented by MS1, MS10, IS1, and IS10, respectively, in 4 treatment groups. Nutrient contents of the treatments are shown in Table 1. The prepared testing materials were given daily by oral administration at once. A control group was only provided with the carrier (1% DMSO). During the experimental period, the cycle of light/dark was maintained at 12 h/12 h, respectively, and the temperature of the experimental cages was maintained at 20–25 °C. The feed intake and body weight were recorded regularly. Rats were monitored for any abnormalities in activities, behavior, and general health status, and fed with respective experimental diets.

Table 1. Nutrient contents in the treatments.

Group	Treatment	Nutrient Content (mg kg^{-1})			
		P	**N**	**Ca**	**Mg**
I	DMSO (Control) [a]	-	-	-	-
II	IS1 [b]	0.26	0.00003	0.10	0.12
III	IS10 [c]	2.60	0.0003	1.03	1.15
IV	MS1 [d]	0.22	0.04	0.09	0.10
V	MS10 [e]	2.21	0.37	0.88	0.98

[a] DMSO = dimethyl sulfoxide; [b] IS1 = rats fed with IS at 1 mg kg^{-1} BW; [c] IS10 = rats fed with IS at 10 mg kg^{-1} BW; [d] MS1 = rats fed with MS at 1 mg kg^{-1} BW; [e] MS10 = rats fed with MS at 10 mg kg^{-1} BW.

At the end of the experimental feeding (28 d), the final BW was measured and all the rats were euthanized and sacrificed. Blood samples of 5 mL were collected from all rats by heart puncture using the sterilized needles and syringes in a disposable vacutainer tube without any anticoagulant and stored at 4 °C for 2 h. After centrifugation (3000 rpm for 10 min at 4 °C), serum samples were separated and stored at −20 °C in Eppendorf tubes. Internal organs (liver, kidney, lung, and heart) were removed and weighed from all rats after they were sacrificed. For histopathological assay of the internal organs, the tissue samples of heart, liver, kidney, and lungs were collected and fixed in 10% neutral buffered formalin, embedded in paraffin wax, and cut into 5 μm-thin sections. For routine examination, all samples were stained with hematoxylin and eosin (H&E) and observed under an optical microscope (×100 or ×400) [23]. The blood tests were conducted at the end of the experimental feeding; i.e., after 24 h from the last dose of testing materials. Assay kits for the analysis of serum biochemicals were obtained from Chemon Inc., South Korea. Serum biochemical parameters including alanine aminotransferase (ALT), aspartate aminotransferase (AST), blood urea nitrogen (BUN), creatinine (CRE), inorganic P (P_i), and calcium (Ca) were examined to assess the effect of the treatments on the internal organs [24,25]. The selected serum biochemical parameters were analyzed using an automated blood biochemical analyzer (Model AU400, Olympus, Tokyo, Japan) and an automated electrolyte analyzer (Model M744 (Na$^+$/K$^+$/Cl$^-$) analyzer, Siemens, Deerfield, MA, USA).

2.3. In Vitro Solubility Test

A series of jar tests were conducted for four hours at 25 °C to check the solubility of the test materials by simulating the gastrointestinal pH conditions (pH 2, 4, 5, 6, and 7). The required amount of test materials was calculated based on the total P content of the materials. A total of 0.5 g (particle size < 100μm) of each of the test materials were weighed exactly in a 1L beaker. The beakers were filled with de-ionized water. The pH of the solution was monitored constantly and the necessary pH adjustment was done with 1M HCl throughout the experiment. All the experiments were performed in triplicate. Considering the variation of gastrointestinal transit time among animals, immediate release of P at low pH, and the availability for absorption, 2 mL of sample was taken per treatment at 15, 30, 60, 90, 120, 150, 180, 210, and 240 min intervals, diluted with 18 mL de-ionized water, centrifuged at 3000 rpm for 10 min, and O-P content was determined using an auto-analyzer (Quik Chem 8500, LACHAT, Loveland, CO, USA).

2.4. Statistical Analysis

Statistical analysis was conducted using GraphPad Prism (version 5.03, 2009). Data obtained from the in vivo toxicity test were analyzed with the one-way analysis of variance (ANOVA) test, considering treatments as a fixed factor. Multiple comparisons were performed using Tukey's HSD tests. In vitro solubility of O-P was analyzed by comparing the O-P concentrations at the beginning with those at the end of the experiment (240 min) using an ANOVA with O-P concentrations at 9 time points as a repeated measure. A *p*-value of < 0.05 was designated to determine statistical significance.

3. Results and Discussion

3.1. P Recovery and Manufacturing

Using a lab-scale reactor (160 L d^{-1}), an O-P removal efficiency of 93.1% was observed and the precipitate production rate and yield of 8.7 ± 0.6 kg struvite m$^{-3}_{reactor}$.d^{-1} and 9.8 ± 0.5 g struvite g^{-1}O-P$_{input}$, respectively, were achieved. The recovered precipitates were identified by XRD analysis, which indicated the recovered material was struvite (Figure 1a). The precipitates contained a P content of 21.6%, which was higher over the theoretical P of 12.7% in struvite since the P in the form of organic compounds might be present in the precipitates.

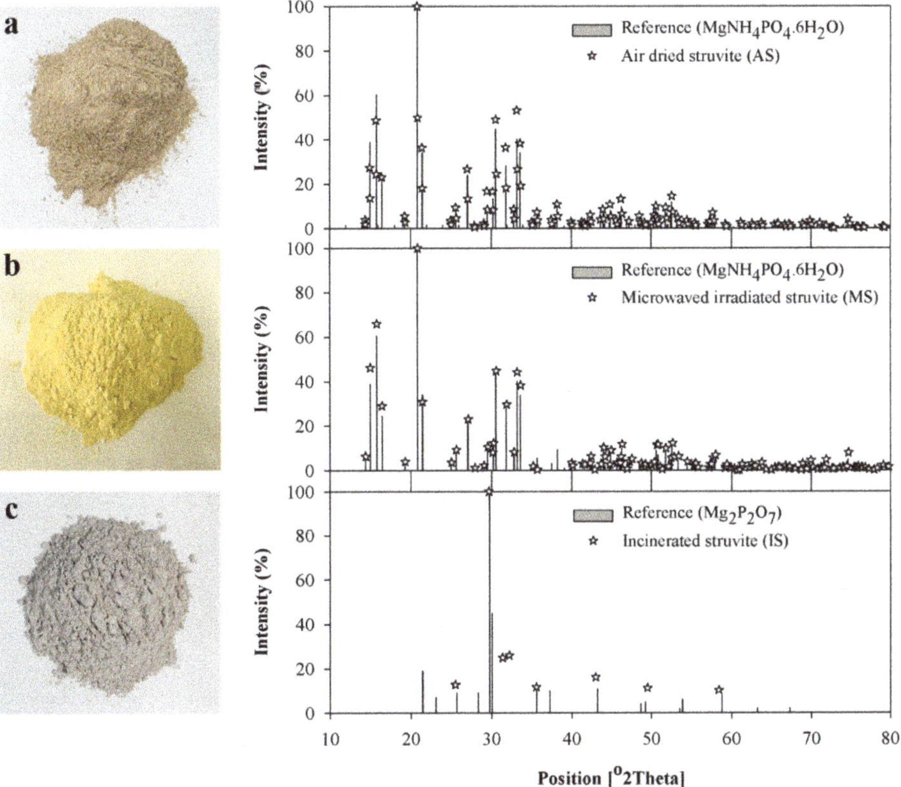

Figure 1. Pictures of recovered and pre-treated struvites with the identification by XRD analyses. (**a**) Air-dried at 20 °C for 2 days, (**b**) microwave irradiated at 80 Hz g^{-1} for 5 min, and (**c**) incinerated at 550 °C for 30 min (* Asterisks in each graph indicate reference materials from XRD analyses).

Different from the struvite recovered from municipal and industrial wastewater, struvite recovered from livestock wastewater is relatively safer for its application to agricultural land [13] because all the ingredients in precipitated struvite from livestock wastewater originate from animal feedstock and include relatively lower amounts of harmful metals. Instead of phosphate rock mined from P reserves, recovered P via struvite precipitation contains lower heavy metal content and needs lesser efforts to purify for target use, such as for fertilizer and chemical industries [26]. During this study, the trace heavy metals in the recovered struvite were rarely found or not detected and satisfied the highest standard limits of feedstock for animal growth as suggested by the Korean Regulatory Authority (Table 2) [27]. In addition to the use in fertilizer sector, struvite as a way of P recycling can therefore be applied in the animal feed industry like phosphate rock. However, it is very important to ensure that the heavy metal contents in the recovered struvite are within the safe limits prior to use as animal feed additive.

Table 2. Chemical composition of recovered and pre-treated struvites and the highest standard limits of heavy metals in feedstock for animals, based on the Korean Regulatory Authority [27].

Parameters	Test Materials			Highest Standard Limits
	AS [c]	MS [b]	IS [a]	
P (g kg^{-1} DM)	216	221	260	
Ca (g kg^{-1} DM)	85	88	103	
N (g kg^{-1} DM)	50	3713	0.03	
Mg (g kg^{-1} DM)	95	98	115	
K (mg kg^{-1} DM)	3567.6		4407	-
Zn (mg kg^{-1} DM)	ND [d]		ND [d]	-
Ni (mg kg^{-1} DM)	ND [d]		ND [d]	-
Cu (mg kg^{-1} DM)	15.7		19.4	-
Cd (mg kg^{-1} DM)	ND [d]	NA [e]	ND [d]	1.0
Pb (mg kg^{-1} DM)	0.0001		0.0001	10.0
As (mg kg^{-1} DM)	0.0012		0.0015	2.0
Cr (mg kg^{-1} DM)	ND [d]		ND [d]	100.0
Hg (mg kg^{-1} DM)	ND [d]		ND [d]	0.4
Se (mg kg^{-1} DM)	ND [d]		ND [d]	2.0

[a] AS = air-dried struvite; [b] MS = microwave irradiated struvite; [c] IS = incinerated struvite; [d] ND = not detected; [e] NA = not analyzed.

The theoretical N content is only around 5.8% in struvite, with NH_4-N being the main N species, and further N content could be decreased through air drying (Table 2). However, N content in air-dried struvite was still higher than the typical N content in animal feeds (1.9 to 4.8%) [28]. Therefore, additional treatment is required to reduce the N content before its application in feed production. To ascertain cost-effective treatments for the reduction of the N content and the removal of organic compounds, microwave irradiation and incineration treatments were selected according to our previous study [3]. Both treatments were conducted to remove mainly NH_4-N from struvite. Microwave irradiation treatment did not show much influence on its morphology and it was still determined as struvite with XRD analysis (Figure 1b), but incineration under 550 °C denatured struvite to magnesium pyrophosphate ($Mg_2P_2O_7$) by removing ammonium and hydrates, which was confirmed by XRD analysis as well (Figure 1c). Ammonium and hydrate contained in struvite started to be released as a gas phase at 55 °C and completed at 250 °C due to the thermal decomposition [29]. The N concentration of 5.0% in air-dried struvite was decreased 3.713 and 0.003% through microwave irradiation and incineration treatments, respectively (Table 2). Moreover, after pre-treatment of recovered struvite with the decrease in moisture content, an increase in the P, Ca, and Mg concentrations in the pre-treated materials were observed (Table 2). Hence, if a specific amount of struvite is added to the animal feed, considering the dietary P requirement, the concentration of other minerals present in the struvite might be higher than the requirement. Further toxicological studies are therefore needed to confirm the safety and efficacy of pre-treated struvites for their use in animal feed, although pre-treatments reduce the N content in struvite.

3.2. Toxicity Test in Rats

Prior to applying struvite recovered swine wastewater as a P additive in the animal feed ration, it is necessary to conduct the toxicity test to ensure biological safety. We therefore pre-treated struvite recovered from swine wastewater as MS and IS. After 28 d of feeding MS and IS, along with control feed, to the rats, the average total BW gain of nearly 121.4 ± 20.3 g was observed in the treated groups, including the control group. However, no statistical differences were found in BW, average daily feed intake, and the internal organs (liver, kidney, heart, and lung) among all the experimental groups fed with different doses of MS and IS, including the control ($p > 0.05$) (Figure 2a).

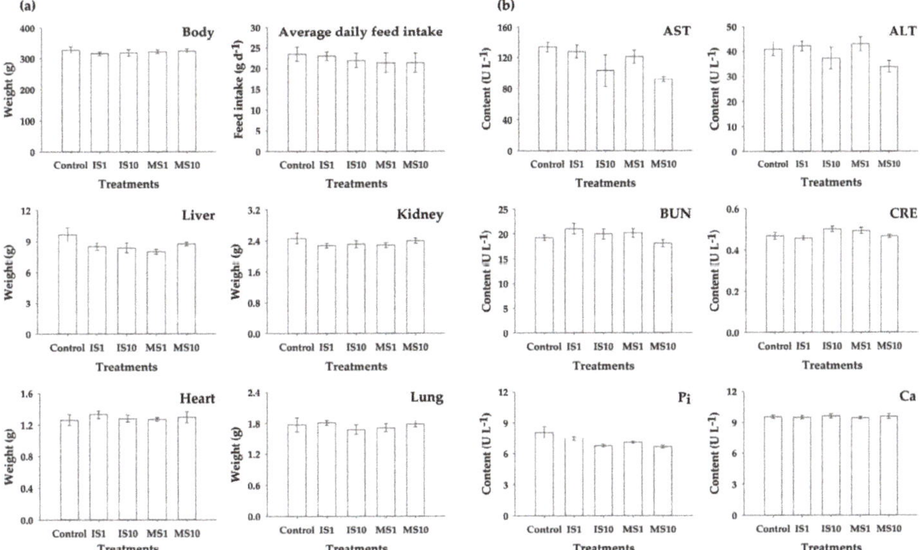

Figure 2. Comparison of (**a**) body and internal organ weights and (**b**) blood metabolites of rats treated with control, IS, and MS groups (Control (only 1% DMSO)); IS1 and IS10, rats fed with IS at 1 and 10 mg kg^{-1} BW; MS1 and MS10, rats fed with MS at 1 and 10mg kg^{-1} BW; Error bar: Standard error of means in each group (n = 6).

In the administration with only the vehicle (control), the blood metabolites levels in rats were found to be normal (Figure 2b). No significant alteration in AST and ALT levels between IS1, IS10, MS1, MS10 (1 and 10 mg kg^{-1} BW), and the control groups were observed. The BUN, CRE, Pi, and Ca levels (mg dL^{-1}) of blood were found within the normal range and no statistical differences ($p > 0.05$) were observed in MS and IS treated groups compared to the control.

The histopathological analyses of the treatment groups applied 10 mg kg^{-1} BW of MS and IS displayed the normal shape, size, and color of the internal organs. Tissues from each organ (liver, kidney, lung, and heart) were found to be normal and no mark of inflammation, necrosis, edema, and/or any other specific pathological symptoms were noticed at such a higher dose of struvite supplementation (Figure 3).

Struvite contains an ammonium ion, which is thought to be toxic for cellular components for animals. Ammonia and ammonium ions are toxic to animal cells and especially influence the cell membrane [15,30] through interrupting the enzymatic reactions and intracellular pH changes, which may lead to the disturbance of proton gradients and the inhibition of endocytosis and exocytosis [31].

A higher dose of ammonium compound (300 ppm ammonium perfluorooctanoate) in rats caused an elevated liver weight, an increase in the incidence of diffuse hepatocellular hypertrophy, portal mononuclear cell infiltration, mild hepatocellular vacuolation [32], and some reproductive abnormalities in male and female rats were also found with a higher dose of ammonium compounds [33]. The present study did not find any abnormalities or lesions in the internal structures of the lungs, liver, kidney, and heart. It might be stated that, although struvite contains ammonium, it would not be harmful to animals after microwave irradiation treatment or incineration, thus ensuring animal welfare.

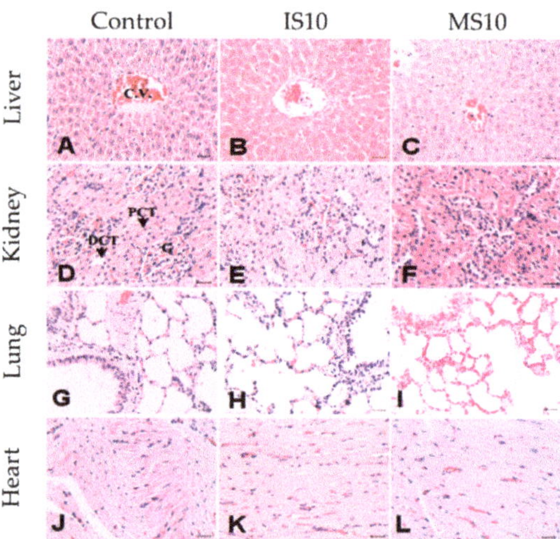

Figure 3. Comparison of histopathological examination of the internal organs of rats treated with control, IS, and MS groups (Control (only 1% DMSO); IS10, rats fed with IS at 10 mg kg^{-1} BW; MS10, rats fed with MS at 10 mg kg^{-1} BW.

The homeostatic condition of P, Ca, and Mg in the body is important as they control regular cellular activities such as energy metabolism and cell signaling [34]. Numerous cellular and tissue injuries including impaired bone mineralization, increased cell death, impaired cell signaling, vascular calcification, pre-mature aging, renal dysfunction, increases tumorigenesis, impaired fertility, skeletal development, etc., are associated with excessive retention of the above-mentioned macronutrients in the body [35–40]. In this study, no abnormalities in internal organ weights and serum biochemicals were observed. Therefore, the results obtained from growth performance, blood biochemical report, and histopathological studies of the internal organs indicated that the administration of MS and IS as a P additive retrieved from swine wastewater had no toxic effects on rats.

3.3. In Vitro Solubility Test

Variations in O-P concentration of the test materials at different pH levels (pH 2, 4, and 5) during the in vitro test are shown in Figure 4. All the experiments started with a constant concentration of the test materials. At pH 2, O-P concentrations of MCP, MS, and IS varied from 110.3 to 194.5 mg L^{-1}, 164.4 to 172.9 mg L^{-1}, and 131.4 to 138.2 mg L^{-1}, respectively. At the end of the experiment, 98.5%, 83.4%, and 69.1% of the total P in MCP, MS, and IS, respectively, was dissolved at pH 2 (Table 3). At pH 4, O-P concentrations of MCP, MS, and IS ranged from 111.3 to 158.8 mg L^{-1}, 143.8 to 158.1 mg L^{-1}, and 124.5 to 143.4 mg L^{-1}, respectively, and it was found that at the end of the experiment 79.7%, 77.2%, and 70.4% of the total P in MCP, MS, and IS, respectively, was dissolved. At pH 5, O-P concentrations of MCP, MS, and IM were varied from 109.7 to 119 mg L^{-1}, 146.1 to 159.2 mg L^{-1}, and 119.9 to 135.4 mg L^{-1}, respectively, and, at the end of the experiment, showed solubilities of 57.8%, 78.6%, and 67.6%, respectively. At pH 6, O-P concentrations of MCP, MS, and IS were ranged from 116.6

to 123.1 mg L^{-1}, 142.1 to 154.3 mg L^{-1}, and 104.5 to 128.6 mg L^{-1}, respectively and, at the end of the experiment, it was found that 61.6%, 76.2%, and 64.3% of the total P in MCP, MS, and IS, respectively, was dissolved. At pH 7, O-P concentrations of MCP, MS, and IS were varied from 95.4 to 117.1 mg L^{-1}, 135.9 to 152.5 mg L^{-1}, and 90.1 to 112.9 mg L^{-1}, respectively, and, at the end of the experiment, showed solubility of 58.5%, 76.2%, and 56.4%, respectively. Although the solubility of O-P from MCP was higher than MS at pH 2 and 4, no significant difference was observed ($p > 0.05$). However, the solubility of O-P from MS increased with the increase in pH and was found to be significantly different from MCP ($p < 0.05$). The in vitro solubility of O-P from IS was the lowest among the treatments and significantly different from MCP and MS in all the experiments ($p < 0.05$).

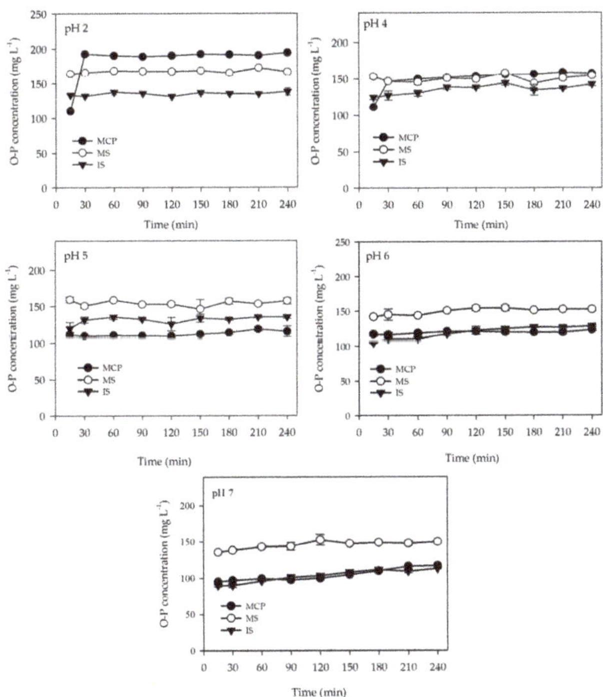

Figure 4. Variations in O-P concentration in an in vitro solubility test of each test materials over time.

Table 3. P solubility from test materials at pH 2, 4, 5, 6, and 7 after four hours of incubation.

P Sources	P Solubility (%)				
	pH 2	**pH 4**	**pH 5**	**pH 6**	**pH 7**
MCP *	98.5 ± 0.7 [a]	79.7 ± 0.1 [a]	57.8 ± 3.7 [a]	61.6 ± 0.5 [a]	58.5 ± 0.5 [a]
MS	83.4 ± 0.6 [a]	77.2 ± 1.3 [a]	78.6 ± 2.2 [b]	76.2 ± 1.0 [b]	74.8 ± 0.4 [b]
IS	69.1 ± 2.6 [b]	70.4 ± 0.3 [b]	67.6 ± 1.0 [c]	64.3 ± 0.6 [c]	56.4 ± 0.2 [c]

* MCP = Monocalcium phosphate; [a,b,c] Different superscript in the same column indicates statistical differences among the different treatment groups.

Animals are generally fed for higher performance, either in terms of reproduction or growth. Phosphorus plays a vital role in maintaining normal muscle growth, egg formation, and animal reproduction [41]. It also plays an important role in regulating acid-base and osmotic balance, energy and amino acid metabolism, and protein synthesis. Due to limited resources and high cost, nutritionists are trying to formulate diets considering the safety margins as well as dietary P requirements [42].

P solubility is one of the deterministic factors that influences the amount of supplement required for the animals. Factors including transit and retention time, pH, and particle size may control the P solubility in the gastrointestinal tract [43]. Moreover, P from inorganic sources are more soluble in the acid medium than the P from plant materials [44]. The amount of P available for absorption is controlled by its sustained solubility in the gastrointestinal tract in animals [45]. Primarily, P is absorbed in the small intestine within the pH range of 2 to 7 [46]. Studies reported that around 35% of the P is absorbed in the duodenum (pH 2 to 6.4), 5% in the jejunum (pH 6), and 40% in the ileum (pH 7) [47,48]. Struvite, an alternative P source has been extensively used as fertilizers in agriculture [13]; however, it is rarely tested as the P source in animal diets. As struvite is readily soluble under acidic conditions, the in vitro P solubility study was therefore conducted using pre-treated struvite as MS and IS by simulating the conditions in the gastrointestinal tract and compared with MCP. Among the test materials, P solubility from MS and MCP was not significantly different at a low pH. Moreover, due to an increase in P solubility from MS at comparatively higher pH (pH 5, 6, and 7), it can be said that, between the pre-treated struvites, MS could be better absorbed in the gastrointestinal tract. The solubility experiment results provide preliminary support of using MS as a potential alternative P source in animal diets. However, in vitro solubility testing has some limitations in extrapolating conditions in animals and further in vivo experiments are required for a better understanding of P metabolism.

4. Conclusions

In this study, neither any histopathological conditions in the internal organs nor any growth inhibition was found in rats fed with pre-treated struvites (MS and IS). The solubility test showed no significant difference in P solubility from MS and MCP at pH 2 and 4, while P solubility from MS increased at pH 5 to 7 and was found to be significantly different compared to MCP and IS. Considering resource recovery and recycling, efficacy as well as societal sustainability, struvite pre-treated as MS has the potential to be a sustainable alternative source of supplemental P in animal diets. Moreover, the results of this study motivate further studies with more stringent designs to better examine the potential of struvite application in diverse fields. Although commercialization of struvite may not be straightforward and is presumably dependent on circumstances such as legislation and national policies and economic feasibility in each country, P recycling is inevitable for both environmental protection and resource security in the world.

Author Contributions: Conceptualization, S.W., B.J., B.Y., and C.R.; methodology, S.W., B.J., B.Y., and C.R.; software, A.R.; formal analysis, S.S., S.K., and S.A.; investigation, S.S., S.K., and S.A.; resources, C.R.; data curation, S.S. and A.R.; writing—original draft preparation, S.S., S.W., and A.R.; writing—review and editing, A.R., S.W., B.J., B.Y., and C.R.; supervision, S.W. and C.R.; project administration, C.R.; funding acquisition, C.R.

Funding: This research was funded by the Rural Development Administration of Korea, grant number PJ011623 and partially supported by Kangwon National University.

References

1. Phosphate Rock Statistics and Information. Available online: http://https://www.usgs.gov/centers/nmic/phosphate-rock-statistics-and-information (accessed on 25 July 2019).
2. Steen, I. Phosphorus availability in the 21st Century: Management of a non-renewable resource. *Phosphorus Potassium* **1998**, *217*, 25–31.
3. Reza, A.; Shim, S.; Kim, S.; Ahmed, N.; Won, S.; Ra, C. Nutrient Leaching Loss of Pre-Treated Struvite and Its Application in Sudan Grass Cultivation as an Eco-Friendly and Sustainable Fertilizer Source. *Sustainability* **2019**, *11*, 4204. [CrossRef]
4. 20 Critical Raw Materials—Major Challenge for EU Industry. Available online: https://ec.europa.eu/growth/content/20-critical-raw-materials-major-challenge-eu-industry-0_mt (accessed on 20 July 2019).

5.	Gilbert, N. Environment: The disappearing nutrient. *Nature* **2009**, *461*, 716–718. [CrossRef] [PubMed]

6.	Van Kauwenbergh, S.J.; Stewart, M.; Mikkelsen, R. World reserves of phosphate rock—A dynamic and unfolding story. *Better Crop.* **2013**, *97*, 18–20.

7.	Elser, J.; Bennett, E. Phosphorus cycle: A broken biogeochemical cycle. *Nature* **2011**, *478*, 29–31. [CrossRef] [PubMed]

8.	Reza, A.; Eum, J.; Jung, S.; Choi, Y.; Jang, C.; Kim, K.; Owen, J.S.; Kim, B. Phosphorus Budget for a Forested-Agricultural Watershed in Korea. *Water* **2019**, *11*, 4. [CrossRef]

9.	Khan, F.A.; Ansari, A.A. Eutrophication: An ecological vision. *Bot. Rev.* **2005**, *71*, 449–482. [CrossRef]

10.	Cordell, D.; Drangert, J.O.; White, S. The story of phosphorus: Global food security and food for thought. *Glob. Environ. Chang.* **2009**, *19*, 292–305. [CrossRef]

11.	Rural Development Administration (RDA). *Centralized Livestock Wastewater Treatment Plan 2019*; Rural Development Administration: Jeonju, Korea, 2019.

12.	Ministry of Environment (MoE). *Establishment and Operation of Centralized Livestock Wastewater Treatment Plant*; Ministry of Environment: Sejong, Korea, 2018.

13.	Ahmed, N.; Shim, S.; Won, S.; Ra, C. Struvite recovered from various types of wastewaters: Characteristics, soil leaching behaviour, and plant growth. *Land Degrad. Dev.* **2018**, *29*. [CrossRef]

14.	Horta, C. Bioavailability of phosphorus from composts and struvite in acid soils. *Rev. Bras. Eng. Agric. Ambient.* **2017**, *21*, 459–464. [CrossRef]

15.	Martinelle, K.; Haggstrom, L. Mechanisms of ammonia and ammonium ion toxicity in animal cells: Transport across cell membranes. *J. Biotechnol.* **1993**, *30*, 339–350. [CrossRef]

16.	Yoon, T.H.; Lee, D.H.; Won, S.G.; Ra, C.S.; Kim, J.D. Effects of dietary supplementation of magnesium hydrogen phosphate (MgHPO4) as an alternative phosphorus source on growth and feed utilization of juvenile far eastern catfish (*Silurus asotus*). *Asian Austral. J. Anim.* **2014**, *27*, 1141–1149. [CrossRef]

17.	Yoon, T.H.; Lee, D.H.; Won, S.G.; Ra, C.S.; Kim, J.D. Optimal incorporation level of dietary alternative phosphate (MgHPO4) and requirement for phosphorus in juvenile far eastern catfish (*Silurus asotus*). *Asian Austral. J. Anim.* **2015**, *28*, 111–119. [CrossRef] [PubMed]

18.	Rahman, M.M.; Liu, Y.; Kwag, J.H.; Ra, C. Recovery of struvite from animal wastewater and its nutrient leaching loss in soil. *J. Hazard. Mater.* **2011**, *186*, 2026–2030. [CrossRef] [PubMed]

19.	Eaton, A.D. *Standard Methods for the Examination of Water and Wastewater*, 21st ed.; American Public Health Association, American Water Works Association, Water Environment Federation: Washington, DC, USA, 2005; ISBN 978-08-7553-047-5.

20.	Charan, J.; Kantharia, N.D. How to calculate sample size in animal studies? *J. Pharmacol. Pharmacother.* **2013**, *4*, 303. [CrossRef] [PubMed]

21.	Park, E.; Lee, M.Y.; Seo, C.S.; Yoo, S.R.; Jeon, W.Y.; Shin, H.K. Acute and subacute toxicity of an ethanolic extract of Melandrii Herba in Crl: CD sprague dawley rats and cytotoxicity of the extract in vitro. *BMC Complement. Altern. Med.* **2016**, *16*, 370.

22.	Tualeka, A.R.; Faradisha, J.; Maharja, R. Determination of No-Observed-Adverse-Effect Level Ammonia in White Mice Through CD4 Expression. *Dose-Response* **2018**, *16*, 1–9. [CrossRef] [PubMed]

23.	Karahan, N.; Işler, M.; Koyu, A.; Karahan, A.G.; BaşyığıtKiliç, G.; Cırış, I.M.; Sütçü, R.; Onaran, I.; Cam, H.; Keskın, M. Effects of probiotics on methionine choline deficient diet-induced steatohepatitis in rats. *Turk. J. Gastroenterol.* **2012**, *23*, 110–121. [CrossRef] [PubMed]

24.	Pierre, P.J.; Sequeira, M.K.; Corcoran, C.A.; Blevins, M.W.; Gee, M.; Laudenslager, M.L.; Bennett, A.J. Hematological and serum biochemical indices in healthy bonnet macaques (Macaca radiata). *J. Med. Primatol.* **2011**, *40*, 287–293. [CrossRef]

25.	Ahmed, Z.; Ahmed, U.; Walayat, S.; Ren, J.; Martin, D.K.; Moole, H.; Koppe, S.; Yong, S.; Dhillon, S. Liver function tests in identifying patients with liver disease. *Clin. Exp. Gastroenterol.* **2018**, *11*, 301–307. [CrossRef]

26.	Liu, Y.H.; Kumar, S.; Kwag, J.H.; Ra, C.S. Magnesium ammonium phosphate formation, recovery and its application as valuable resources: A review. *J. Chem. Technol. Biotechnol.* **2013**, *88*, 181–189. [CrossRef]

27.	Control of Livestock and Fish Feed Act. Available online: http://www.law.go.kr/LSW/eng/engLsSc.do?menuId=2§ion=lawNm&query=feed&x=0&y=0#liBgcolor0 (accessed on 3 October 2019).

28.	Protein Nutrition Requirements of Farmed Livestock and Dietary Supply. Available online: http://www.fao.org/3/y5019e/y5019e06.htm (accessed on 2 October 2019).

29. Bhuiyan, M.I.H.; Mavinic, D.S.; Koch, F.A. Thermal decomposition of struvite and its phase transition. *Chemosphere* **2008**, *70*, 1347–1356. [CrossRef] [PubMed]

30. Ozturk, S.S.; Riley, M.R.; Palsson, B.O. Effects of ammonia and lactate on hybridoma growth, metabolism, and antibody production. *Biotechnol. Bioeng.* **1992**, *39*, 418–431. [CrossRef] [PubMed]

31. Docherty, P.A.; Snider, M.D. Effects of hypertonic and sodium-free medium on transport of a membrane glycoprotein along the secretory pathway in cultured mammalian cells. *J. Cell Physiol.* **1991**, *146*, 34–42. [CrossRef]

32. Butenhoff, J.L.; Kennedy, J.L., Jr.; Chang, S.C.; Olsen, G.W. Chronic dietary toxicity and carcinogenicity study with ammonium perfluorooctanoate in Sprague–Dawley rats. *Toxicol.* **2012**, *298*, 1–13. [CrossRef]

33. Koizumi, M.H.; Fujii, S.; Ono, A.; Hirose, A.; Imai, T.; Ogawa, K.; Ema, M.; Nishikawa, A. Evaluation of the reproductive and developmental toxicity of aluminium ammonium sulfate in a two-generation study in rats. *Food Chem. Toxicol.* **2011**, *49*, 1948–1959.

34. Razzaque, M.S. Phosphate toxicity: New insights into an old problem. *Clin. Sci.* **2011**, *120*, 91–97. [CrossRef]

35. Ohnishi, M.; Nakatani, T.; Lanske, B.; Razzaque, M.S. In vivo genetic evidence for suppressing vascular and soft-tissue calcification through the reduction of serum phosphate levels, even in the presence of high serum calcium and 1,25-dihydroxyvitamin D levels. *Circ.-Cardiovasc. Genet.* **2009**, *2*, 583–590. [CrossRef]

36. Kuro-o, M. Klotho as a regulator of fibroblast growth factor signaling and phosphate/calcium metabolism. *Curr. Opin. Nephrol. Hypertens.* **2006**, *15*, 437–441. [CrossRef]

37. Kuro-o, M. A potential link between phosphate and aging: Lessons from Klotho-deficient mice. *Mech. Ageing. Dev.* **2010**, *131*, 270–275. [CrossRef]

38. Jin, H.; Xu, C.X.; Lim, H.T.; Park, S.J.; Shin, J.Y.; Chung, Y.S.; Park, S.C.; Chang, S.H.; Youn, H.J.; Lee, K.H.; et al. High dietary inorganic phosphate increases lung tumorigenesis and alters Akt signaling. *Am. J. Respir. Crit. Care Med.* **2009**, *179*, 59–68. [CrossRef]

39. Amato, D.; Maravilla, A.; Montoya, C.; Gaja, O.; Revilla, C.; Guerra, R.; Paniagua, R. Acute effects of soft drink intake on calcium and phosphate metabolism in immature and adult rats. *Rev. Invest. Clin.* **1998**, *50*, 185–189. [PubMed]

40. Toba, Y.; Kajita, Y.; Masuyama, R.; Takada, Y.; Suzuki, K.; Aoe, S. Dietary magnesium supplementation affects bone metabolism and dynamic strength of bone in ovariectomized rats. *J. Nutr.* **2000**, *130*, 216–220. [CrossRef]

41. Zhang, B.; Wang, C.; Wei, Z.H.; Sun, H.Z.; Xu, G.Z.; Liu, J.X.; Liu, H.Y. The effects of dietary phosphorus on the growth performance and phosphorus excretion of dairy heifers. *Asian Australas. J. Anim. Sci.* **2016**, *29*, 960–964. [CrossRef] [PubMed]

42. Li, X.; Zhang, D.; Yang, T.Y.; Bryden, W.L. Phosphorus Bioavailability: A Key Aspect for Conserving this Critical Animal Feed Resource with Reference to Broiler Nutrition. *Agriculture* **2016**, *6*, 25. [CrossRef]

43. Walk, C.L.; Bedford, M.R.; McElroy, A.P. Influence of diet, phytase, and incubation time on calcium and phosphorus solubility in the gastric and small intestinal phase of an in vitro digestion assay. *J. Anim. Sci.* **2012**, *90*, 3120–3125. [CrossRef] [PubMed]

44. Lineva, A.; Kirchner, R.; Kienzle, E.; Kamphues, J.; Dobenecker, B. A pilot study on in vitro solubility of phosphorus from mineral sources, feed ingredients and compound feed for pigs, poultry, dogs and cats. *J. Anim. Physio. Anim. Nutr.* **2019**, *103*, 317–323. [CrossRef]

45. Breves, G.; Schröder, B. Comparative aspects of gastrointestinal phosphorus metabolism. *Nutr. Res. Rev.* **1991**, *4*, 125–140. [CrossRef]

46. Kiela, P.R.; Ghishan, F.K. Recent advances in the renal–skeletal–gut axis that controls phosphate homeostasis. *Lab. Invest.* **2009**, *89*, 7–14. [CrossRef]

47. Phosphorus Binders: Relative Potency of Available Agents: Phosphorus Absorption. Available online: https://www.medscape.org/viewarticle/506489_2 (accessed on 4 October 2019).

48. Wagner, A.L. Phytase Impacts Various Non-Starch Polysaccharidase Activities on Distillers Dried Grains with Solubles. Doctoral dissertation, Virginia Tech, Blacksburg, VA, USA, 2008.

Article

Evaluation of Extrusion Temperatures, Pelleting Parameters, and Vitamin Forms on Vitamin Stability in Feed

Pan Yang, Huakai Wang, Min Zhu and Yongxi Ma *

State Key Laboratory of Animal Nutrition, College of Animal Science and Technology, China Agricultural University, Beijing 100193, China; ypan23@163.com (P.Y.); huakaiwhk@cau.edu.cn (H.W.); s20173040486@cau.edu.cn (M.Z.)
* Correspondence: mayongxi@cau.edu.cn; Tel.: +86-1381-139-8538

Received: 23 April 2020; Accepted: 18 May 2020; Published: 20 May 2020

Simple Summary: Since African Swine Fever is a pandemic in China, the Chinese feed mills implemented specific thermal processing to inactivate the virus. Farmers and animal nutritionists gradually focus on the destructive effects of feed processing on substances, e.g., vitamins, because vitamins are labile nutrients that are sensitive to the chemical and physical factors during thermal processing. The objectives of this study were to determine the effects of vitamin forms, extrusion temperature, and pelleting parameters on vitamin stability, and to determine which vitamins are destroyed by thermal processing. The deleterious impact of feed processing is of practical relevance to vitamin stability. The majority of B complex vitamins have great stability in feed processing, but the stability of fat-soluble vitamins was negatively affected by feed processing. In addition, microencapsulated vitamins had greater stability compared to non-microencapsulated vitamins. Based on the current results, decreasing the strength of feed processing and choice of suitable forms of the vitamin could be recommended in feed production.

Abstract: Two experiments were conducted to determine the stability of microencapsulated and non-microencapsulated forms of vitamins in diets during extrusion and pelleting. We investigated the recovery of vitamins in swine diets after extrusion at 100 °C, 140 °C, or 180 °C. Next, two diets were conditioned at 65 °C (low temperature; LT) or 85 °C (high temperature; HT), and pellets were formed using a 2.5 × 15.0 mm (low length-to-diameter ratio; LR) or 2.5 × 20.0 mm (high length-to-diameter ratio; HR) die. The extrusion temperature had a significant effect on the recovery of vitamins E, B_1, B_2, B_3, and B_5 in the diets. The diet extruded at 100 °C had higher B_1, B_2, B_3, and B_5 vitamin recoveries than diets extruded at 140 °C and 180 °C. Microencapsulated vitamins A and K_3 had greater stability than non-microencapsulated vitamins A and K_3 at 100 °C and 140 °C extrusion. In the diet extruded at 180 °C, microencapsulated vitamins A, D_3, and K_3 had higher recoveries than non-microencapsulated vitamins A, D_3, and K_3. The recovery of vitamin K_3 in diets after LTLR (low temperature + low length-to-diameter ratio) or HTLR (high temperature + low length-to-diameter ratio) pelleting was greater ($p < 0.05$) than after LTHR (low temperature + high length-to-diameter ratio) and HTHR (high temperature + high length-to-diameter ratio) pelleting. Our results clearly show that low extrusion temperature and low pellet temperature, and a low length-to-diameter ratio (L:D ratio) for pellet mill die are recommended for pig feed. Moreover, microencapsulated vitamins had greater stability compared to non-microencapsulated vitamins.

Keywords: extrusion; pelleting; feed; vitamin stability; vitamin form

1. Introduction

Feed processing (sieving, grinding, extrusion, and pelleting) can lead to increased nutrient availability for an animal [1–3]. Specifically, pelleting has been shown to increase the average daily gain while decreasing the average daily feed intake for pigs [4]. It is widely accepted that pelleting improves average daily gain and feed efficiency in modern pig genotypes by 4%–8%, and the improvements are due to enhanced palatability, reduced waste, and the potential for improved nutrient utilization due to heat treatment of the ingredients [1–3]. Although thermal processes have variable effects on mycotoxins [5], the implementation of thermal processing can reduce microbial contaminations and biological hazards (i.e., the potential porcine epidemic diarrhea virus, *salmonella*) in feedstuffs [6,7]. Also, high-temperature extrusion is used frequently to improve feed hygiene and nutrient availability, as well as reduce anti-nutritional factors in feed [6,8]. Pelleting is the most prevalent processing in feed production. It has been demonstrated that pelleting increases feed quality, hygiene, and also reduces feed-carried pathogens [3,6,9]. Although extrusion and pelleting processes can improve feed hygiene, and the availability of some nutrients, the potential negative effects of these conditions should also be considered. Potential adverse effects include changes in lipid oxidation, protein denaturation, and cross-linking, starch gelatinization and dextrinization, degradation of vitamins and denaturation of enzymes, browning, and flavor formation [10,11]. Previous studies indicated that the stability of some vitamins could be as low as 50% when feeds are subjected to high-temperature processing [4,6,10].

Microencapsulated vitamins are currently available as a vitamin source, where the chemical structure of these vitamins is protected by microencapsulation [12]. It is believed that vitamin stability is much higher for microencapsulated vitamins post-feed processing because the protective wall materials of the microcapsules resist the heat, shear stress, and long periods of oxygen exposure during high-temperature processing. However, no information is available regarding the effect of microencapsulation on vitamin stability under different extrusion temperatures and pelleting parameters. Therefore, the objective of this study was to determine the effect of microencapsulation and extrusion or pelleting on the stability of commercially available vitamins in feed and to determine which vitamins are destroyed by thermal processing.

2. Materials and Methods

This study was conducted at the State Key Laboratory of Animal Nutrition at the China Agricultural University (Beijing, China). The extrusion of diets was conducted at the Institute of Food Science and Technology Chinese Academy of Agricultural Sciences (Beijing, China), and pelleting was completed at the China Agricultural University Feed Mill Educational Unit (Beijing, China). Approval from the Animal Care and Use Committee was not obtained for these experiments, because no animals were used.

2.1. Processing Parameters and Diets

Two experimental diets based on corn-soybean meal were used in this research. Diets were formulated to meet amino acid, vitamin, and mineral requirements for pigs according to the Nutrient Requirements of Swine [13]. Diet composition is shown in Table 1. The non-microencapsulated (NM) diet was formulated using non-microencapsulated vitamins, whereas the microencapsulated (M) diet was formulated with hydrogenated fat microcapsules containing microencapsulated vitamins, which have a coating structure and are made by the spray-drying technique. The processing parameters of the extrusion and pelleting for the NM and M diets are presented in Table 2.

Table 1. Ingredient composition and calculated nutrients of the experimental diet (%, as-fed basis).

Ingredient	Diets [1]	
	NM	M
Corn	68.81	68.81
Soybean meal	24.00	24.00
Monocalcium phosphate	1.60	1.60
Limestone	0.80	0.80
Salt	0.30	0.30
Soybean oil	2.75	2.75
L-lysine HCl	0.53	0.53
DL-methionine	0.13	0.13
L-threonine	0.24	0.24
Tryptophan	0.04	0.04
NMVP [2]	0.30	-
MVP [3]	-	0.30
Trace mineral premix [4]	0.50	0.50
Total	100.00	100.00
Calculated values		
SID Lysine	1.23	1.23
SID Methionine	0.36	0.36
SID Threonine	0.74	0.74
SID Tryptophan	0.20	0.20
ME, kcal/kg	3383.56	3383.56
CP	17.1	17.1
Ca	0.72	0.72
P	0.60	0.60
Available P	0.41	0.41

[1] NM, non-microencapsulated; M, microencapsulated. [2] NMVP = Non-microencapsulated vitamin premixes. NMVP provided the following per kilogram of feed: Vitamin A, 13,500 IU; vitamin D_3, 3,000 IU; vitamin E, 30 mg; vitamin K_3, 3 mg; vitamin B_{12}, 24 µg; riboflavin, 6 mg; pantothenic acid, 18 mg; niacin, 30 mg; choline chloride, 400 mg; folacin, 0.12 mg; thiamine, 1.5 mg; pyridoxine, 3 mg; biotin 0.03 mg. [3] MVP = Microencapsulated vitamin premixes. MVP provided the following per kilogram of feed: Vitamin A, 13,500 IU; vitamin D_3, 3,000 IU; vitamin E, 30 mg; vitamin K_3, 3 mg; vitamin B_{12}, 24 µg; riboflavin, 6 mg; pantothenic acid, 18 mg; niacin, 30 mg; choline chloride, 400 mg; folacin, 0.12 mg; thiamine, 1.5 mg; pyridoxine, 3 mg; biotin 0.03 mg. [4] Trace-mineral premixes provided per kilogram of diet: 80 mg Fe as iron sulfate, 60 mg Cu as copper sulfate, 17.5 mg Mn as manganese oxide, 65 mg Zn as zinc oxide, 0.3 mg I as ethylenediamine dihydroiodide, and 0.2 mg Se as sodium selenite.

Table 2. Processing parameters of experimental diets.

Parameters	Extrusion			Pelleting			
	100 °C	140 °C	180 °C	LTLR	LTHR	HTLR	HTHR
Machine type	DSE-25 [1]	DSE-25	DSE-25	MUZL 180 [2]	MUZL 180	MUZL 180	MUZL 180
Feeding rate, rpm	40	40	40	30	30	30	30
Screw speed, rpm	160	160	160	-	-	-	-
Condition time, s	-	-	-	60	60	60	60
Temperature intake, °C	100	140	180	65	65	85	85
Barrel pressure, MPa	5	2	1	ND	ND	ND	ND
Stream pressure, MPa	-	-	-	0.3	0.3	0.3	0.3
Diameter, mm	25	25	25	2.5	2.5	2.5	2.5
L:D ratio	20:1	20:1	20:1	6:1	8:1	6:1	8:1
Moisture content, %	25	25	25	12	12	12	12

ND = Not detected; L:D ratio, pellet mill dies length-to-diameter ratio; LTLR, low temperature + low L:D ratio; LTHR, low temperature + high L:D ratio; HTLR, high temperature + low L:D ratio; HTHR, high temperature + high L:D ratio. [1] DSE-25, extruder, Brabender Technologie GmbH & Co. KG, Germany. [2] MUZL 180, pellet mill, FAMSUN, China.

2.2. Characterization of Non-microencapsulated and Microencapsulated Vitamins

Non-microencapsulated and microencapsulated vitamins were obtained from the Wellroad Animal Health Co. Ltd., China. High-resolution scanning electron microscopy (SEM) was used to observe the physical characteristics of non-microencapsulated and microencapsulated vitamins. The particle size distribution was measured with a series of 13 selected US standard sieves (Nos. 6, 8, 12, 16, 20, 30, 40, 50, 70, 100, 140, 200, and 270) and a pan, fitted into a sieve shaker. The sieving procedure was carried out according to a standard method [14]. In brief, 100 g of sample was sieved by shaking for 10 min. The mass of the material retained on each sieve, as well as that on the pan, was weighed and recorded. All measurements were performed in duplicate. The mass frequency (%) for material retained on each sieve size was calculated and plotted against each particle size category. The geometric diameter average (dgw) and geometric standard deviation (Sgw) were also calculated for each sieving replicate, based on the formula described in the ASAE Standards (2008) [14].

2.3. Extrusion Processing

In experiment 1 (Exp. 1), a 2 × 3 factorial design was used to evaluate the microencapsulated and non-microencapsulated vitamin forms and three extruder temperatures (100 °C, 140 °C, or 180 °C) on vitamin stability in the feed. The corn and soybean were ground to a mean particle size of 600 μm with a FAMSUN hammer mill (SFSP56 × 40C, FAMSUN, China). Vitamins from each product were initially mixed with 150 kg of corn. This premixing was completed to ensure proper mixing of the vitamins throughout the subsequent 400-kg batches of complete feed used for processing. Unprocessed samples were used to measure initial vitamin concentrations in the diet. The diets were extruded in a double screw annular gap extruder (DSE-25, Brabender Technologies GmbH & Co. KG, Germany) and extruded at 100 °C, 140 °C, or 180 °C. The screw diameter was 25 mm, and the pellet mill die length-to-diameter ratio (L:D ratio) was 20:1. These temperatures were selected because they are common parameters attainable in most commercial mills, and previous research exists for comparison.

The extrusion was repeated three times in three days, resulting in three replicates/temperature. At the beginning of the extrusion, flush feed containing no vitamins was used to warm the extruder to the first extrusion temperature. When the extruder barrel temperature was stable, the NM diet was added to the hopper and delivered into the extruder barrel by the feeder. After the NM diet was extruded at 100 °C and removed from the barrel, 10 kg of flush diet was added to the hopper to flush the system. While the extruder was maintained at 100 °C, the M diet was added to the hopper for extrusion. When both diets had been extruded at 100 °C, the flush feed was added to the hopper, and the temperature was increased to 140 °C. We ensured the temperature was stable before adding experimental diets. The NM and M diets were processed through the extruder (extruded at 140 °C or 180 °C) using the same procedures. The extruder was shut off between each run. These extrusion processes were repeated on days 2 and 3 to create replications.

2.4. Pelleting Processing

In experiment 2 (Exp. 2), a 2 × 4 factorial design was arranged with two diets (supplemented microencapsulated or non-microencapsulated vitamins) and four pellet parameters: two condition temperatures (65 °C vs. 85 °C) and two length-to-diameter ratios (L:D ratio, 6:1 vs. 8:1). The pelleting parameters were similarly selected because they are attainable variables in most commercial pellet mills, and previous research exists for comparison. The LTLR refers to low temperature + low L:D ratio, using a 15-mm-thick die with a 2.5 mm diameter hole at 65 °C. The LTHR refers to low temperature + low L:D ratio, using a 20-mm-thick die with a 2.5 mm diameter hole at 65 °C. The HTLR refers to a high temperature + low L:D ratio, using a 15-mm-thick die with a 2.5 mm diameter hole at 85 °C. The HTHR refers to a high temperature + low L:D ratio, using a 20-mm-thick die with a 2.5 mm diameter hole at 85 °C. Unprocessed feeds (mash) were prepared using the same procedure in Section 2.3. Steam conditioning (65 °C or 85 °C) of homogenized mixtures (NM diets or M diet) was carried out using a double-shaft steam conditioner (FAMSUN, SBTZ 10, China). Conditioning ended when the target temperature was achieved. After conditioning, the complete feed was pelleted in a pellet mill (FAMSUN MUZL 180, China) equipped with 15.0- or 20.0-mm-thick die having 2.5 mm openings.

The pelleting process was conducted on four consecutive days to create four replications. Flush feed containing no vitamins was used to warm the pellet mill equipped with 15.0-mm-thick die at the initial conditioning temperature. After the NM diet was pelleted (LTLR), 20 kg of flush diet was added to hopper to flush the system. While the pellet mill was still at 65 °C conditioning temperature, the M diet was added to the hopper. When the two diets had been processed at 65 °C using the 15.0-mm-thick die with 2.5 mm openings, the flush feed was added to the hopper, and the temperature was increased to 85 °C. We ensured the mill temperature was stable before proceeding with additional experimental diets. When the pelleting of LTLR and HTLR was complete, the pellet mill was shut off, and the die was changed to 20.0-mm-thick with 2.5 mm openings. Flush feed containing no vitamin was used to warm the pellet mill at 65 °C. After 65 °C pelleting (LTHR) for the NM diet was finished, 20 kg of flush diet was added to hopper to flush the system. While the pellet mill was still at 65 °C, the M diet was added to the hopper. When the two diets had been processed at 65 °C, the flush feed was added to the hopper, and the temperature was increased to 85 °C. The pellet mill was completely shut off when all runs were finished. These pelleting processes were repeated on days 2–4 to create replications.

The extruded and pelleted diets were dried in a double pass and cross-flow dryer. Samples of the extruded and pelleted diets were collected and stored for several hours at room temperature (22 °C) until a stable temperature was achieved. Samples were ground in Retsch ZM200 Mill (Retsch GMBH, Haan, Germany) and stored at -20 °C until analysis. The proximate and vitamin analyses were completed on all samples.

2.5. Chemical Analyses of Ingredients

Proximate analysis and analysis of vitamins in all feed ingredients were conducted at the Ministry of Agriculture and Rural Affairs Feed Potency and Safety Supervision and Testing Center located at the China Agricultural University, Beijing, China. Dry matter (DM) content of diets was determined by drying 5 g of sample in a forced-air oven (model GZX-9140 MBE; Boxun Company, Shanghai, China) at 105 °C to a constant weight (method 934.01) [15]. All diet samples were analyzed for crude protein (CP) (method 990.03) [15] using a Kjeldahl analyzer (Foss KjeltecTM 2100, Foss Kemao Inc., Beijing, China). Ether extract (EE) content in diet samples was determined using the petroleum ether extraction method (Method 920.39) [15] and an automated analyzer (Ankom XT15 Extractor; Ankom Technology, Macedon, NY, USA). Crude fiber (CF; Method 978.10) and ash (Method 942.05) were analyzed for all samples [15]. The nutrient composition of raw and thermally processed diets (Table 3) was confirmed. Diet nutrient concentrations were similar between raw and processed diets, except for the ether extract. In addition, samples were analyzed for vitamins A, D_3, E, K_3, B_1, B_2, B_3, B_5, and B_6 content. Standards used were retinyl esters, cholecalciferol, α-tocopherol acetate, menadione, thiamine, riboflavin, pyridoxine, niacin, and pantothenic acid (Fluka, Sigma–Aldrich, Steinheim, Germany).

Table 3. Analyzed nutrient concentration in manufacturing productions (%, as-fed basis) [1].

Items [2]	NM Diet [2]									M Diet [3]								
	Extrusion				Pelleting					Extrusion				Pelleting				
	Mash	Temperature, °C			Mash	65 °C		85 °C		Mash	Temperature, °C			Mash	65 °C		85 °C	
		100	140	180		6:01	8:01	6:01	8:01		100	140	180		6:01	8:01	6:01	8:01
DM	88.24	88.70	88.74	88.05	88.51	88.45	88.72	88.07	88.23	88.15	88.55	88.23	88.67	88.47	88.81	88.49	88.03	88.20
CP	17.13	17.19	17.38	17.03	16.96	17.11	17.36	17.09	17.28	17.16	16.93	17.38	17.52	16.93	17.39	16.97	17.26	17.20
CF	4.25	4.38	3.99	4.11	4.03	4.06	4.21	3.95	4.02	4.27	3.93	4.29	4.20	4.17	4.09	3.98	4.14	4.08
EE	6.41	2.63	2.77	2.90	6.37	6.41	6.58	6.49	6.41	6.32	2.79	2.98	3.17	6.41	6.22	6.79	6.82	6.45
Ash	4.72	4.31	4.40	4.59	4.99	4.61	4.73	4.51	4.61	4.83	4.80	4.53	4.27	4.81	4.85	4.57	4.6	4.49

[1] Analyses were completed in duplicate according to the Association of Official Agricultural Chemists (AOAC) International official methods 934.01 (moisture), 984.13 (crude protein), 978.10 (crude fiber), 942.05 (Ash) and 920.39 (EE). [2] The NM diet was formulated using non-microencapsulated vitamins. [3] The M Diet was formulated with microencapsulated vitamins.

Vitamins A (VA) and E (VE) were determined by method 2012.10 [16]. In brief, feed samples were dissolved in 2% papain solution. Samples were placed in a 37 °C ± 2 °C water bath, and methanol was added to each sample tube for extraction. Vitamins A and E in this extract were analyzed by HPLC (Agilent 1200 Series; Agilent Technologies Inc., Santa Clara, CA, USA). For the extraction of vitamin D_3 (VD_3) from the feed, method 992.26 [16] was used. In brief, 5 g of sample was transferred to a centrifuge tube, and 15 mL anhydrous ethanol, 400 mg ascorbic acid, and 7.5 g KOH were added. The tube was placed in a 75 °C water bath. After saponification, the tube contents were extracted using ethyl ether/petroleum ether extraction. The supernatant was transferred to a clean tube and evaporated to dryness under nitrogen. Analysis of VD_3 was completed using HPLC, followed by UV detection at 254 nm. For the determination of vitamin K_3 (VK_3), 5 g of feed was extracted with trichloromethane. The extract (1 mL) was transferred to an HPLC vial for direct injection. The injection volume was 10 µL, and UV absorbance was measured at 251 nm [17]. For the extraction of water-soluble vitamins (vitamins B_1, B_2, B_3, B_5, and B_6) from diets, the procedure published by Chen et al. [18] was used. A total of 5 g of feed sample was extracted with 10 mM phosphate buffer and sonication in the dark. The supernatant was filtered through 0.1 µm polytetrafluoroethylene syringe filters. The extracted sample was analyzed using a 250 × 4.5 mm, 5 µm, Eclipse Plus C18 column (Agilent Technologies Inc., USA) on an Agilent liquid chromatography-tandem mass spectrometer (electrospray ionization source). Methods for analyzing vitamin concentration in samples were validated for repeatability, between-day precision, long-term precision, limits of quantitation, and linearity (data not shown) by the staff of the Ministry of Agriculture and Rural Affairs Feed Efficacy and Safety Evaluation Center in Beijing, China. The expected and analyzed vitamin contents found in diets NM and M are shown in Table 4. Although there is no published, accepted standard for vitamin recovery in animal feed, the analysis showed the experimental diets were within 10% of their formulated targets, which is consistent with the acceptable analytical variation. True recovery of vitamins through the feed processing methods was calculated using the following equation [19]: %True recovery = nutrient content per gram of processed feed × feed weight (gram) after processing/ (nutrient content per gram of raw feed × feed weight (gram) before processing) × 100.

Table 4. Vitamin concentration in unmanufactured mash diet (as-fed basis) [1].

Item [2]	NM Diet			M Diet		
	Calculated [3]	Analyzed	Ratio [4]	Calculated	Analyzed	Ratio
Vitamin A, IU/kg	13,500	13,660.08	101.19	13,500	13,784.01	102.10
Vitamin D_3, IU/kg	3000	3084.27	102.81	3000	3115.35	103.85
Vitamin E, mg/kg	30	32.12	107.09	30	31.10	103.68
Vitamin K_3, mg/kg	3	3.12	104.10	3	3.19	106.20
Vitamin B_1, mg/kg	3	3.05	101.57	3	3.02	100.80
Vitamin B_2, mg/kg	6	6.12	102.05	6	6.11	101.80
Vitamin B_3, mg/kg	30	30.41	101.38	30	30.86	102.87
Vitamin B_5, mg/kg	18	18.42	102.31	18	18.30	101.68
Vitamin B_6, mg/kg	3	3.10	103.43	3	3.15	105.00

[1] The NM diet was formulated using non-microencapsulated vitamins, whereas M diet was formulated with microencapsulated vitamins. [2] Values represent means of replicate samples each analyzed in duplicate (method 2012.10 for vitamin A (VA) and vitamin E (VE) analysis; AOAC 2012, method 992.26 for vitamin D_3 (VD_3) analysis; AOAC 2012, GB/T 18872-2017 for vitamin K_3 (VK_3) analysis; National standard 2017, the method for water-soluble vitamins analysis; Chen et al. [18], Ministry of Agriculture and Rural Affairs Feed Efficacy and Safety Evaluation Center, Beijing, CN). [3] Calculated values were determined from manufacturers guaranteed minimum. [4] Analyzed to calculated ratio.

2.6. Vitamin Stability Ranking

For each vitamin form, the stability was ranked based on the percent of each vitamin retained and a previously reported method [20]. The vitamin with the lowest stability at any of the three extrusion temperatures (100 °C, 140 °C, and 180 °C) or four pelleting parameters (low temperature + low L:D ratio, low temperature + high L:D ratio, high temperature + low L:D ratio, and high temperature + high L:D ratio) was ranked number one. Alternatively, the vitamin with the highest stability in any of those processing parameters was ranked number nine. Based on each vitamin stability rank order at different extrusion temperatures (100 °C, 140 °C, and 180 °C), the average vitamin stability rank order was calculated to represent the overall vitamin stability rank for extrusion. Also, based on each vitamin stability rank order at different pelleting parameters (low temperature + low L:D ratio, low temperature + high L:D ratio, high temperature + low L:D ratio, and high temperature + high L:D ratio), the average vitamin stability rank order was calculated to represent the overall vitamin stability rank for pelleting. Based on each vitamin's overall stability rank order for extrusion and pelleting, the average vitamin stability rank order was calculated to represent the overall vitamin stability rank for feed processing.

2.7. Statistical Analysis

The normality of data was verified using the UNIVARIATE procedure of SAS (SAS Institute, Cary, NC, USA). The BOXPLOT procedure of SAS was used to check for outliers. Data were analyzed using the Proc Mixed procedure of SAS. The data from the extrusion and pelleting experiments were analyzed as completely random designs in 2 × 3 and 2 × 4 factorial arrangements, respectively, for the vitamin premix form and feed processing. The vitamin premix form, feed processing, and their interaction served as fixed variables. Only the main effects were discussed for responses in which the interaction was not significant, whereas contrasts were discussed where a significant interaction was detected. The LSMEANS statement was used to calculate treatment means. Significantly different means were identified using Tukey's test. The results were considered significant at $p < 0.05$.

3. Results

As shown in Figure 1, the physical analysis revealed that the average geometric diameter (dgw) of microencapsulated (M) vitamins was higher than non-microencapsulated (NM) vitamins. Regarding the geometric standard deviation (Sgw), microencapsulated vitamins were superior to their non-microencapsulated forms.

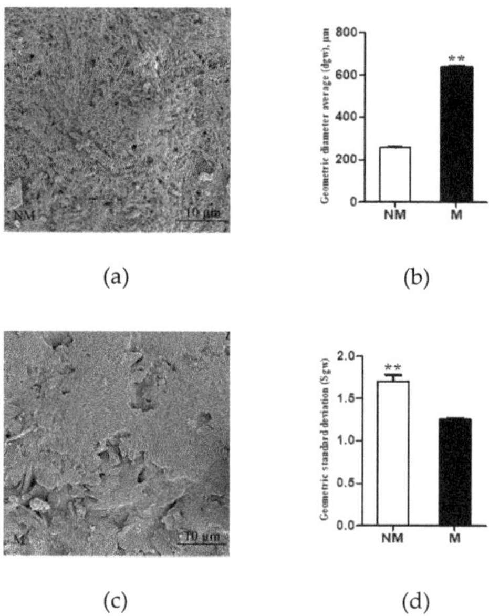

(a) (b)

(c) (d)

Figure 1. (**a**) The physical characteristics of non-microencapsulated vitamin form high-resolution scanning electron microscopy. (**b**) The geometric diameter average (dgw) of NM (non-microencapsulated) and M (microencapsulated) vitamins. (**c**) The physical characteristics of microencapsulated vitamin form high-resolution scanning electron microscopy. (**d**) The geometric standard deviation (Sgw) of NM (non-microencapsulated) and M (microencapsulated) vitamins. ** represent $p < 0.01$.

There was significant form × temperature interaction for vitamin A (VA), vitamin D_3 (VD_3), and vitamin K_3 (VK_3) recoveries (Table 5). At 100 °C and 140 °C extrusion temperatures, the diet containing the microencapsulated vitamins had higher recoveries of VA and VK_3 than the diet containing the non-microencapsulated vitamins. At 180 °C extrusion temperature, the diet containing microencapsulated vitamins had higher VA, VD_3, and VK_3 recoveries than the diet containing non-microencapsulated vitamins ($p < 0.05$). The NM Diet, which contained the non-microencapsulated vitamins, had higher VA and VD_3 recoveries at 100 °C extrusion than at 140 °C and 180 °C extrusion ($p < 0.05$), and the NM diet had higher VA and VD_3 recoveries at 140 °C extrusion than at 180 °C extrusion ($p < 0.05$). However, the NM diet had lower VK_3 recovery at 100 °C and 140 °C extrusion than at 180 °C extrusion ($p < 0.05$). The 100 °C extrusion conditions resulted in higher VA, VD_3, and VK_3 recoveries in diets containing the microencapsulated vitamins than at 140 °C and 180 °C extrusion ($p < 0.05$). In addition, the M diet had higher VA, VD_3, and VK_3 recoveries at 140 °C extrusion than at 180 °C extrusion ($p < 0.05$).

Table 5. Effects of extruded temperature (Temp.) and vitamin forms (non-microencapsulated or microencapsulated) on the percentage of vitamins in diets (Experiment 1) [1].

Form	Temp.	VA	VD_3	VE	VK_3	VB_1	VB_2	VB_3	VB_5	VB_6
	100 °C	46.34	73.94	47.34	7.17	94.91	98.71	102.59	99.19	74.35
NM	140 °C	34.75	53.14	42.23	6.12	79.24	87.77	89.75	84.32	78.12
	180 °C	30.76	40.26	45.17	11.64	78.31	90.16	87.18	85.51	71.82
	100 °C	56.41	75.73	49.17	48.04	100.71	105.16	104.81	100.83	101.88
M	140 °C	40.25	56.86	43.09	35.20	83.81	91.39	90.59	87.47	92.46
	180 °C	41.80	60.74	46.41	38.55	83.60	90.76	90.20	84.41	89.25
SEM		0.74	3.40	1.10	1.81	0.68	1.28	1.14	1.75	3.16
Main effects										
Form	NM	37.28	55.78	44.91	8.31	84.15 [y]	92.21 [y]	93.18 [y]	89.67	74.76 [y]
	M	46.15	64.44	46.22	40.60	89.37 [x]	95.77 [x]	95.20 [x]	90.90	94.53 [x]
Temp.	100 °C	51.37	74.83	48.26 [a]	27.60	97.81 [a]	101.93 [a]	103.70 [a]	100.00 [a]	88.11
	140 °C	37.50	55.00	42.66 [b]	20.66	81.52 [b]	90.46 [b]	90.17 [b]	85.90 [b]	85.29
	180 °C	36.28	50.50	45.79 [a]	25.10	80.96 [b]	89.58 [b]	88.69 [b]	84.96 [b]	80.54
p-value										
Form		<0.001	0.015	0.154	<0.001	<0.001	0.002	0.038	0.397	<0.001
Temp.		<0.001	<0.001	< 0.001	0.002	<0.001	<0.001	<0.001	<0.001	0.069
Form × temp.		<0.001	0.035	0.905	0.001	0.667	0.09	0.110	0.478	0.110
Significant *p*-values for contrasts										
NM 100 °C vs. M 100 °C		<0.001	-	-	<0.001	-	-	-	-	-
NM 140 °C vs. M 140 °C		<0.001	-	-	<0.001	-	-	-	-	-
NM 180 °C vs. M 180 °C		<0.001	<0.001	-	<0.001	-	-	-	-	-
NM 100 °C vs. NM 140 °C		<0.001	<0.001	-	-	-	-	-	-	-
NM. 100 °C vs. NM 180 °C		<0.001	<0.001	-	-	-	-	-	-	-
NM 140 °C vs. NM 18 0 °C		<0.001	0.013	-	0.039	-	-	-	-	-
M 100 °C vs. M 140 °C		<0.001	0.011	-	<0.001	-	-	-	-	-
M 100 °C vs. M. 180 °C		<0.001	0.037	-	<0.001	-	-	-	-	-
M 140 °C vs. M 180 °C		-	-	-	<0.001	-	-	-	-	-

[1] Main effects are shown for responses in which the interaction was not significant, whereas contrasts are shown where a significant interaction was detected. VA; vitamin A, VD_3; vitamin D_3, VE; vitamin E, VK_3; vitamin K_3, VB_1; vitamin B_1, VB_2; vitamin B_2, VB_3; vitamin B_3, VB_5; vitamin B_5, VB_6; vitamin B_6. [x,y] Means within a column that lack a common superscript differ ($p < 0.05$). [a,b,c] Means within a column that lack a common superscript differ ($p < 0.05$). NM, non-microencapsulated; M, microencapsulated. The changes in vitamin concentration during extrusion are presented in Table S1.

There was no significant form × temperature interaction for vitamin E, vitamin B_1, vitamin B_2, vitamin B_3, and vitamin B_6 (see Table 5). Vitamin forms had a significant effect on vitamin B_1, vitamin B_2, vitamin B_3, and vitamin B_6 recoveries. The degradation of vitamins B_1, B_2, B_3, and B_6 in the microencapsulated form was less ($p < 0.05$) than the non-microencapsulated form. Additionally, the extrusion temperature had a significant effect on vitamin E, vitamin B_1, vitamin B_2, vitamin B_3, and vitamin B_5 recoveries. No significant differences were detected among vitamin B_1, vitamin B_2, vitamin B_3, and vitamin B_5 recoveries in diets at 140 °C or 180 °C extrusion. At the 100 °C extrusion temperature, diets had higher recoveries of vitamin B_1, vitamin B_2, vitamin B_3, and vitamin B_5 than at 140 °C and 180 °C extrusions. Interestingly, there was no significant difference in the recovery of vitamin E at 100 °C and 180 °C extrusion, but diets at 140 °C extrusion had the lowest vitamin E recovery.

3.1. Effects of Pelleting on the Stability of Vitamins in Exp. 2

The recovery of vitamins after pelleting is shown in Table 6. There was no significant form × processing interaction on vitamin recovery. Moreover, vitamin form and processing had no significant effect on vitamin D_3, vitamin E, vitamin B_2, vitamin B_3, vitamin B_5, and vitamin B_6 recoveries. Microencapsulation had a significant effect on the recoveries of vitamin A, vitamin K_3, and vitamin B_1. The recoveries of microencapsulated vitamin A, vitamin K_3, and vitamin B_1 were higher ($p < 0.05$) than their non-microencapsulated forms. Processing had a significant effect on vitamin K_3 recovery. The recovery of vitamin K_3 in LTLR (low temperature + low L:D ratio) and HTLR (high temperature + low L:D ratio) was greater than in LTHR (low temperature + high L:D ratio) and HTHR (high temperature + high L:D ratio) ($p < 0.05$).

Table 6. Effects of pelleting parameters and vitamin forms (non-microencapsulated or microencapsulated) on the percentage of vitamins in diets (Experiment 2) [1].

Form	Processing	VA	VD_3	VE	VK_3	VB_1	VB_2	VB_3	VB_5	VB_6
NM	LTLR	94.29	90.03	97.38	31.74	78.58	96.27	94.19	97.77	91.01
	LTHR	93.25	84.42	95.15	27.74	76.05	94.69	92.92	95.54	96.92
	HTLR	92.57	87.81	97.54	29.69	82.21	96.48	96.80	97.22	96.61
	HTHR	88.57	88.71	95.76	23.49	81.52	96.78	96.13	95.30	96.56
M	LTLR	98.11	89.90	97.61	38.43	90.04	96.03	94.54	97.06	90.70
	LTHR	96.12	91.62	97.28	32.91	89.59	96.27	94.13	98.61	89.69
	HTLR	96.48	88.87	96.18	39.44	85.60	98.00	95.75	95.96	94.97
	HTHR	95.42	87.53	95.95	30.95	92.39	96.66	94.65	95.96	95.16
SEM		1.76	4.64	2.43	1.32	1.73	0.73	1.15	2.07	2.69
Main effects										
Form	NM	92.17 [y]	87.74	96.46	28.16 [y]	79.59 [y]	96.05	95.01	96.46	95.27
	M	96.53 [x]	89.48	96.75	35.43 [x]	89.41 [x]	96.74	94.77	96.90	92.63
Processing	LTLR	96.20	89.97	97.50	35.08 [a]	84.31	96.15	94.37	97.41	90.85
	LTHR	94.69	88.02	96.86	30.33 [b]	82.82	95.48	93.52	97.07	93.30
	HTLR	94.53	88.34	96.21	34.57 [a]	83.91	97.24	96.28	96.59	95.79
	HTHR	92.00	88.12	95.85	27.22 [b]	86.96	96.72	95.39	95.63	95.86
p-value										
Form		0.001	0.599	0.864	< 0.001	< 0.001	0.231	0.778	0.766	0.592
Processing		0.135	0.972	0.911	< 0.001	0.164	0.169	0.111	0.839	0.105
Form × processing		0.697	0.806	0.914	0.386	0.135	0.495	0.672	0.729	0.440

[1] NM, non-microencapsulated; M, microencapsulated; L:D ratio, pellet mill die length-to-diameter ratio; LTLR, low temperature + low L:D ratio; LTHR, low temperature + high L:D ratio; HTLR, high temperature + low L:D ratio; HTHR, high temperature + high L:D ratio. VA; vitamin A, VD_3; vitamin D_3, VE; vitamin E, VK_3; vitamin K_3, VB_1; vitamin B_1, VB_2; vitamin B_2, VB_3; vitamin B_3, VB_5; vitamin B_5, VB_6; vitamin B_6. [x,y] Means within a column that lack a common superscript differ ($p < 0.05$). [a,b] Means within a column that lack a common superscript differ ($p < 0.05$). The changes in vitamin concentration during pelleting are presented in Table S2.

3.2. Ranking of Vitamins

Vitamin stability was ranked to identify those vitamins that may be at greater risk of degradation during feed processing and could be used as indicators of vitamin quality in mixed diets. The vitamins were ranked to evaluate vitamin stability during pelleting and extrusion. Vitamin rankings are shown in Tables 7 and 8. Non-microencapsulated vitamin K_3 exhibited the greatest loss at all the three extrusion temperatures (100 °C, 140 °C, and 180 °C) and the four pelleting processes (LTLR, low temperature + low L:D ratio; LTHR, low temperature + high L:D ratio; HTLR, high temperature + low L:D ratio; HTHR, high temperature + high L:D ratio). Thus, vitamin K_3 was ranked 1, whereas non-microencapsulated vitamin B_3 was ranked 9 for all extrusion temperatures (Table 7). The non-microencapsulated vitamin B_5 had the greatest stability in LTLR pelleting and was ranked 9. The non-microencapsulated vitamin B_6 had the highest stability in LTHR pelleting and was ranked 9. Non-microencapsulated vitamin B_2 had the greatest concentration in HTLR and HTHR pelleting and ranked 9 for these pelleting conditions. With respect to the microencapsulated forms, vitamin K_3 exhibited the greatest loss at all three extrusion temperatures and four pelleting processes, and was consistently ranked 1. Conversely, vitamin B_2 showed the greatest stability at 100 °C and 180 °C extrusion and HTLR and HTHR pelleting processes. It was ranked 9 (Table 8). Vitamin B_6 at 140 °C extrusion had the highest stability and was ranked 9. Vitamin A in LTLR pelleting exhibited the greatest stability and was ranked 9. Vitamin B_5 in LTHR pelleting showed the lowest degradation and was ranked 9. Based on the vitamin ranking tables we used, vitamin A, vitamin D_3, vitamin K_3, and vitamin B_1 are the most easily degraded vitamins during feed processing, regardless of vitamin form.

Table 7. Comparison and ranking of activity loss of non-microencapsulated vitamins in extrusion and pelleting [1].

Item	Extrusion			Overall Rank in Extrusion	Pelleting [2]				Overall Rank in Pelleting	Overall Rank
	100 °C	140 °C	180 °C		LTLR	LTHR	HTLR	HTHR		
VA	2	2	2	2	6	5	4	3	4	2
VD_3	4	4	4	4	3	3	3	4	3	3
VE	3	3	3	3	8	7	8	6	7	5
VK_3	1	1	1	1	1	1	1	1	1	1
VB_1	6	6	6	6	2	2	2	2	2	4
VB_2	7	8	8	8	7	6	9	9	8	8
VB_3	9	9	9	9	5	4	6	7	5	7
VB_5	8	7	7	7	9	8	7	5	7	7
VB_6	5	5	5	5	4	9	5	8	6	6

[1] Based on the percent of commercial vitamin loss and vitamin recovery ranked method reported by Shurson et al. [20]. Vitamins that exhibited the highest % loss in three extrusions or four pelleting were, respectively, ranked 1, whereas vitamins with the lowest activity loss in this processing were separately ranked 9. Overall, the rank in the extrusion is according to mean of the three ranks, and the overall rank in pelleting is according to the mean of the four ranks. Overall, the rank is according to mean of the ranks in the extrusion and pelleting. VA; vitamin A, VD_3; vitamin D_3, VE; vitamin E, VK_3; vitamin K_3, VB_1; vitamin B_1, VB_2; vitamin B_2, VB_3; vitamin B_3, VB_5; vitamin B_5, VB_6; vitamin B_6. [2] L:D ratio, pellet mill die length-to-diameter ratio; LTLR, low temperature + low L:D ratio; LTHR, low temperature + high L:D ratio; HTLR, high temperature + low L:D ratio; HTHR, high temperature + high L:D ratio.

Table 8. Comparison and ranking of activity loss of microencapsulated vitamins in extrusion and pelleting [1].

Item	Extrusion			Overall Rank in Extrusion	Pelleting [2]				Overall Rank in Pelleting	Overall Rank
	100 °C	140 °C	180 °C		LTLR	LTHR	HTLR	HTHR		
VA	3	2	2	2	9	6	8	6	6	3
VD$_3$	4	4	4	4	3	4	3	2	3	2
VE	2	3	3	3	8	8	7	7	7	4
VK$_3$	1	1	1	1	1	1	1	1	1	1
VB$_1$	5	5	5	5	2	2	2	3	2	2
VB$_2$	9	8	9	8	6	7	9	9	9	8
VB$_3$	8	7	8	7	5	5	5	4	5	6
VB$_5$	6	6	6	6	7	9	6	8	8	7
VB$_6$	7	9	7	7	4	3	4	5	4	5

[1] Based on the percent of commercial vitamin loss and vitamin recovery ranked method reported by Shurson et al. [20]. Vitamins that exhibited the highest % loss in three extrusions or four pelleting were, respectively, ranked 1, whereas vitamins with the lowest activity loss in this processing were separately ranked 9. Overall, the rank in the extrusion is according to mean of the three ranks, and the overall rank in pelleting is according to the mean of the four ranks. Overall, the rank is according to mean of the ranks in the extrusion and pelleting. VA; vitamin A, VD$_3$; vitamin D$_3$, VE; vitamin E, VK$_3$; vitamin K$_3$, VB$_1$; vitamin B$_1$, VB$_2$; vitamin B$_2$, VB$_3$; vitamin B$_3$, VB$_5$; vitamin B$_5$, VB$_6$; vitamin B$_6$. [2] L:D ratio, pellet mill die length-to-diameter ratio; LTLR, low temperature + low L:D ratio; LTHR, low temperature + high L:D ratio; HTLR, high temperature + low L:D ratio; HTHR, high temperature + high L:D ratio.

4. Discussion

There were no differences in the analyzed nutrients except the ether extract between the mash and thermally processed treatments. A decrease in the recovery of ether extract after expanding has been reported in several studies [21,22], which was explained by the hydrothermal degradation of fat and by the formation of amylose-lipid complexes, which could not be extracted by ether. Vitamin supplementation of swine diets became essential when pig production moved into complete confinement housing. In confinement systems, pigs have no access to vitamin-rich pasture and are housed on slatted concrete floors, which limit the consumption of vitamins found in feces. Consequently, supplemental vitamins now play a vital role in meeting the vitamin requirements of pigs. Vitamin A, vitamin E, vitamin B$_2$, vitamin B$_3$, and vitamin B$_5$ are commonly marginal or deficient in corn-soybean diets [11]. Compared to vitamin B$_2$, vitamin B$_3$, and vitamin B$_5$, vitamins A and E appeared to be readily degraded under the varied processing conditions used in the study; thus, there is a high risk of insufficient vitamins A and E in manufactured diets.

The geometric diameter average (dgw) and geometric standard deviation (Sgw) are the average particle size of a sample and particle size variation, respectively. Microencapsulated vitamins have larger particle size and greater uniformity than their non-microencapsulated forms. Vitamin uniformity is critical because it will be further diluted hundred-fold in feeds; the difference in particle size between non-microencapsulated vitamins and feed constituents can give rise to the segregation of micro-ingredients. Also, high-resolution scanning electron microscopy revealed that the microcapsule wall material covered the vitamin crystal, protecting the vitamin in the feed matrix during processing.

4.1. Effects of Extrusion on Vitamin Stability

Extrusion and pelleting have a positive impact on feed hygiene and nutrient availability. The extruded temperature under normal commercial mill conditions is above 90 °C. This high temperature is used frequently to produce more hygienic compounded feeds, specifically to reduce contamination with *Salmonella* and *Escherichia coli* [5,8]. The current data suggest that extrusion temperatures above 100 °C negatively affect vitamin recovery in mixed feed and are consistent with previous studies that heat processing of feed reduces vitamin stability [4,5]. The stability of vitamins depends on various conditions, such as temperature, oxidation, abrasion, and moisture. Our results revealed that feed processing (extrusion and pelleting) could reduce vitamin stability. A similar result has been previously observed from Charlton and Ewing [23] and Riaz et al. [24]. They concluded that higher extruded temperatures resulted in a low recovery of vitamins. Furthermore, it has been reported that the extrusion process has a negative impact on the degradation of vitamins A and E [25,26]. An investigation of high-heat cooking of foodstuffs resulted in a low recovery of vitamin D_3 at 39%–45% [27], which corresponds well with our results. The chemical structures of vitamin A and vitamin D_3 (the hydroxy group and double bonds) make them susceptible to oxidation during the feed processing. In addition, mash feed absorbs water vapor and heat during extrusion or pelleting, which decreases the stability of vitamin A and vitamin D_3. This results in further destruction of vitamins A and D_3 due to heightened oxidation reactions. This result is in agreement with a study by Tiwari and Cummins [28], who found that high-temperature, fast extrusion cooking decreased the stability of fat-soluble vitamins. Zielinski et al. [29] observed a significant decrease in vitamin E content of buckwheat during extrusion. Similarly, we observed that extrusion could cause a significant decrease in vitamin A, vitamin D_3, vitamin E, and vitamin K_3 contents of extruded diets.

We observed that water-soluble vitamins were stable during extrusion, but the recovery of vitamins B_1 and B_6 was lower than other B-complex vitamins. This result corresponds with several previous studies. Gadient and Fenster [30] reported that vitamin B_2 in extruded feed was not affected by extrusion temperatures up to 150 °C. Furthermore, Li et al. [31] investigated the stability of B-complex vitamins in feeds extruded at 160 °C. They reported that vitamin B_2, vitamin B_3, and vitamin B_5 had recovery rates of 100%, 96.3%, and 100%, respectively, but the recovery rates of vitamin B_1 and vitamin B_6 were 65.1% and 70.3%, respectively [31]. Marchetti et al. [32] reported the recovery rates of vitamin B_1, vitamin B_2, vitamin B_3, vitamin B_5, and vitamin B_6 after extrusion at 96 °C as 87.89%, 86.05%, 92.17%, 85.84%, and 66.50%, respectively. The methylene bridge connecting the pyrimidine and thiazole moieties of vitamin B_1 can easily be broken down during feed processing [10]. The thiazole moiety is less stable than the pyrimidine moiety and is easily cleaved by hydrolysis. Pyridoxine (vitamin B_6) is commonly used for feed because pyridoxine is more stable than either pyridoxal or pyridoxamine [33], but pyridoxine is sensitive to light, particularly in neutral and alkaline solutions [11]. Major changes were observed in the levels of important water-soluble vitamins, including vitamins B_1 and B_6, which are regarded as the most heat-sensitive [34]. For B group vitamins, the thermostabilities of vitamins B_1 and B_6 were poor, and vitamin B_1 showed the greatest reductions during processing. Vitamins B_2, B_3, and B_5 are stable upon heating and not affected by processes such as hot air convection, infrared, high-pressure steam, or microwave during cooking [33]. Thermochemistry of B vitamins shows the molecules of vitamins B_2, B_3, or B_5 are tightly bound and easily form hydrogen bonds, which improves their thermostability.

In addition, there are some extrusion parameters that increase vitamin destruction, e.g., screw rpm, moisture, energy input, etc. [34]. Generally, the recovery of the vitamin in extrusion decreases with increasing screw speed, increasing pressure in the barrel, decreasing moisture, and increasing energy input [24,26,34]. In the present study, pressure in the barrel at low-temperature treatment was greater than at high-temperature treatment, which may contribute to huge vitamins A, E, and K_3 losses of more than 50% during extrusion. However, no serious vitamin degradation occurred in other vitamins. The reason may be explained that those vitamins are more sensitive to temperature than pressure. During extrusion, very complex reactions were ongoing; therefore, it is not easy to study the influence of each factor. On the other hand, vitamins differ greatly in chemical structure, and available form, their stability during extrusion also varies [24,34]. The degradation of most vitamins could be avoided by reducing the temperature and pressure in the extruder.

4.2. Effects of Pelleting on the Stability of Vitamins

Pelleting of pig feed has been practiced for decades; the specific energy input in the pellet mill has increased due to harder pellets requested in the global market and used for cereal by-production [2,3,6]. Prolonged conditioning and double pelleting increase the aggressiveness of the pelleting process [35], which could compromise bioactive substances [36] and the stability of vitamins in diets [32,37].

After extrusion, the recovery of vitamins was generally lower than after pelleting. This observation is in line with Marchetti et al. [32], where the extrusion process involves much higher energy input, pressure, and temperatures than pelleting. Conditioning temperatures in pelleting from 65 °C to 85 °C and an L:D ratio of 6:1 to 8:1 are typical commercial mill conditions, depending on the type of diet. The current data suggest that conditioning temperatures above 65 °C and an increased L:D ratio can negatively affect the vitamin recovery of feed products. The resistance of vitamins A, D_3, and E recoveries to increased temperature during pelleting is consistent with other reports. Jones [38] reported that the vitamin A concentration in feed was reduced by 6.5% by pelleting at 80 °C, and Cutlip et al. [39] observed that vitamin A lost 6.7% of its biological activity when pelleted at 93.3 °C. Furthermore, the concentrations of vitamins A and E decreased to 6.62% and 4.83% of their initial values in the samples of pelleted feed (60 °C, L:D ratio 6:1) [25,26], and pelleting temperatures of up to 88 °C did not affect the stability of vitamin D_3 [40]. Modern vitamin production gives vitamin A and vitamin D_3 greater protection against moisture, heat, and pressure during pelleting [11,26]. Commercial forms of vitamins A and D_3 exist in the matrix as a cross-linked beadlet generally composed of gelatin [10]. The cross-linked beadlet also reduces the effect of shear force or pressure during pelleting. The antioxidant capabilities of the tocopherols, due to the free phenolic hydroxy group, compromise the stability of vitamin E acetate. Esterification with acetic acid eliminates its antioxidative nature, thereby improving stability in pelleting [10,11,33]. We observed that the recoveries of vitamin A, vitamin D_3, and vitamin E in extruded diets were lower than pelleted diets, which was likely due to higher abrasion, heat, and pressure in extrusion than pelleting. Interestingly, the loss of vitamin K_3 was over 50% in the present study, which is higher than the results of Marchetti et al. [32]. This result may be caused by prolonged conditioning times in our experiment, which increases the leaching of vitamin K_3 and prolongs oxidation-reduction reactions. Additionally, the VK_3 source used in the present study was menadione sodium bisulfite (MSB), which is most commonly used in the feed industry [11,12]. However, MSB has a high solubility in water, which led to increased leaching during feed processing. MSB also has limited stability to light, heat, humidity, and pressure [11,12,20], in which the molecule is destroyed.

Although there was no significant difference, the recovery of B complex vitamins at 85 °C was slightly higher than that at 65 °C. In addition, the recovery of water-soluble vitamins was slightly higher at low pressure (L:D ratio 6:1) than at high pressure (L:D ratio 8:1). The reason was due to the degradation of B vitamins that can easily occur at 65 °C, and this is consistent with Lewis et al. [40]. Kimura et al. [41] compared various methods of cooking pork, and found that vitamin B_1 loss was the highest during boiling, followed by steaming, parching, and frying. This was explained by the leaching of vitamin B_1 into water, due to its water-soluble nature. Vitamins B_3 and B_6 destruction occur during the precooking process, when they may leach into water but processing alone is not expected to destroy the vitamin [42]. However, cooking beef products to 57 °C internal temperature has demonstrated a loss of both vitamins B_2 and B_3 [43]. Vitamin B_5 is also sensitive to cooking in water. Stability factors for vitamin B_5 in legumes (33%–76%) were significantly influenced by the pre-soaking method and cooking times [42]. Those experiments were conducted using food materials and may not correspond to water-soluble vitamin degradation in feed nutrition, but the degradation of B vitamins can occur at low temperatures. To the best of our knowledge, there is limited information about vitamin stability data during the pelleting of swine feed. Due to the lack of water-soluble vitamin stability data during the pelleting of swine feed, data presented herein, and from other reports, on the stability of vitamins could be useful in modifying vitamin premixes.

4.3. Effects of Microencapsulation on Vitamin Stability During Extrusion and Pelleting

Microencapsulation is a process by which substances are coated with a continuous film of polymeric material [33]. The substance, if sensitive to oxygen, moisture, or light, can be stabilized by microencapsulation [12]. In the current study, the stability of microencapsulated vitamin A was higher than that of non-microencapsulated vitamin A regardless of the extruded temperature. The reason is likely because vitamin A contains four double bonds and one hydroxyl group, which are unstable under high temperature [11,26,34,37]. Microencapsulation can give crystalline vitamin A greater protection against oxidation reactions because the polymeric film provides effective protection of core material during processing [11,12]. Cutlip et al. [39] reported that diets containing vitamin A (in the form of retinyl acetate in a gelatin-lactose coating with antioxidants) exhibited only slight oxidative damage when pelleted at 93.3 °C. Extrusion and pelleting may cause a detrimental loss of vitamin D, but it depends on the heating process [10,11,24,34]. Our results show that the recovery rate of microencapsulated vitamin D_3 was higher during extrusion at 180 °C, compared with non-microencapsulated vitamin D_3. Microencapsulation can reduce the reactivity and incompatibility of compounds with the outside environment, enhancing their stability in conditions of heat, moisture, oxygen, among others [12,33]. Thus, the stability of vitamin D_3 during feed processing can be greatly improved. In addition, the stability of the microencapsulated vitamin E was slightly higher than the non-microencapsulated vitamin E. During microencapsulation of vitamin E; it is esterified with acid, which provides protection and improves stability [10,11]. Microencapsulation significantly reduced the loss of vitamin K_3 during extrusion and pelleting. The result was consistent with Marchetti et al. [32] who reported that the recovery of menadione from coated forms after extrusion or pelleting was significantly higher than non-coated forms.

There were no significant differences in the recoveries of vitamin B_2, vitamin B_3, and vitamin B_5 when comparing the two forms of vitamins. The result is consistent with Riaz et al. [24], who reported that vitamin B_2, vitamin B_3, and vitamin B_5 were stable during thermal processing. The recoveries of microencapsulated vitamins B_1 and B_6 were higher than their non-microencapsulated forms when extruded at 100 °C, 140 °C, and 180 °C. This is likely because vitamin B_1 is highly unstable upon heating and will actively participate in the Maillard reaction during heat treatments [44]. Additionally, vitamin B_6 rapidly degrades with increasing temperature. Microencapsulation likely protected vitamin B_1 against the Maillard reactions caused by heat and pressure during extrusion and pelleting [44]. Microencapsulation could also minimize contact with carbonyl groups of reducing sugars. Increased thermal stability of microencapsulation was also observed by Chatterjee et al. [45].

To our knowledge, the present study is the first to establish vitamin ranking for the stability of various vitamins following extrusion and pelleting. This ranking could be used to identify vitamins that are vulnerable, and producers can analysis those vitamins for the feed manufacturing quality control program. Based on our vitamin ranking, vitamin A, vitamin D_3, vitamin K_3, and vitamin B_1 are the top four sensitive vitamins under feed processing, regardless of vitamin form, and they could be used as indicators to determine overall dietary vitamin quality after feed manufacturing. We suggest monitoring the content of these vitamins in the manufacturing of feed.

5. Conclusions

The deleterious impact of feed processing, particularly extrusion, is of practical relevance to vitamin stability and feed quality. Our results clearly show a low extrusion temperature is recommended for pig feed. Based on the current results, we generally suggest that a reduction of pelleting strength, low temperature, and low L:D ratio may recommend for compound vitamin mixes used in the production of pig feed. Our research found that the majority of B complex vitamins have great stability in feed processing, but recovery of fat-soluble vitamins (vitamin A, vitamin D_3, vitamin E, and vitamin K_3), vitamin B_1, and vitamin B_6 was negatively affected by feed processing. Our results clearly indicate improved stability of microencapsulated vitamins, particularly vitamin A, vitamin D_3, vitamin K_3, vitamin B_1, and vitamin B_6.

Supplementary Materials: The following are available online at http://www.mdpi.com/2076-2615/10/5/894/s1, Table S1: Effects of extruded temperature (Temp.) and vitamin forms (non-microencapsulated or microencapsulated) on concentrations of vitamin in diets (Dry matter basis, Experiment 1), Table S2: Effects of pelleting parameters and vitamin forms (non-microencapsulated or microencapsulated) on the concentration of vitamins in diets (Dry matter basis, Experiment 2).

Author Contributions: Conceptualization, P.Y. and Y.M.; methodology, P.Y. and M.Z.; validation, Y.M.; formal analysis, P.Y.; investigation, P.Y.; resources, P.Y. and H.W.; data curation, P.Y. and H.W.; writing—original draft preparation, P.Y.; writing—review and editing, Y.M.; project administration, Y.M.; funding acquisition, Y.M. All authors have read and agreed to the published version of the manuscript.

Funding: This study was financially supported by the China Agricultural University Research Funding (201705510410056) and the Ministry of Agriculture and Rural Affairs Funding (21178259).

Acknowledgments: The authors express their appreciation to the Institute of Food Science and Technology Chinese Academy of Agricultural Sciences, and China Agricultural University Feed Mill Educational Unit assistance with extrusion and pelleting throughout the study. We thank Accdon (www.accdon.com) for its linguistic assistance during the preparation of this manuscript.

Conflicts of Interest: The authors declare no conflict of interest.

References

1. Lundblad, K.K.; Issa, S.; Hancock, J.D.; Behnke, K.C.; McKinney, L.J.; Alavi, S.; Prestløkken, E.; Fledderus, J.; Sørensen, M. Effects of Steam Conditioning at Low and High Temperature, Expander Conditioning and Extruder Processing Prior to Pelleting on Growth Performance and Nutrient Digestibility in Nursery Pigs and Broiler Chickens. *Anim. Feed Sci. Technol.* **2011**, *169*, 208–217. [CrossRef]

2. De Vries, S.; Pustjens, A.M.; Schols, H.A.; Hendriks, W.H.; Gerrits, W.J.J. Improving Digestive Utilization of Fiber-Rich Feedstuffs in Pigs and Poultry by Processing and Enzyme Technologies: A Review. *Anim. Feed Sci. Technol.* **2012**, *178*, 123–138. [CrossRef]

3. Vukmirović, Đ.; Čolović, R.; Rakita, S.; Brlek, T.; Đuragić, O.; Solà-Oriol, D. Importance of Feed Structure (Particle Size) and Feed Form (Mash vs. Pellets) in Pig Nutrition—A Review. *Anim. Feed Sci. Technol.* **2017**, *233*, 133–144. [CrossRef]

4. Abdollahi, M.R.; Ravindran, V.; Svihus, B. Pelleting of broiler diets: An overview with emphasis on pellet quality and nutritional value. *Anim. Feed Sci. Tech.* **2013**, *179*, 1–23. [CrossRef]

5. Čolović, R.; Puvača, N.; Cheli, F.; Avantaggiato, G.; Greco, D.; Đuragić, O.; Kos, J.; Pinotti, L. Decontamination of Mycotoxin-Contaminated Feedstuffs and Compound Feed. *Toxins* **2019**, *11*, 617. [CrossRef]

6. Boroojeni, F.G.; Svihus, B.; von Reichenbach, H.G.; Zentek, J. The Effects of Hydrothermal Processing on Feed Hygiene, Nutrient Availability, Intestinal Microbiota and Morphology in Poultry—A Review. *Anim. Feed Sci. Technol.* **2016**, *220*, 187–215. [CrossRef]

7. Cochrane, R.A.; Schumacher, L.L.; Dritz, S.S.; Woodworth, J.C.; Huss, A.R.; Stark, C.R.; DeRouchey, J.M.; Tokach, M.D.; Goodband, R.D.; Bia, J.; et al. Effect Of Pelleting on Survival of Porcine Epidemic Diarrhea Virus—Contaminated Feed. *J. Anim. Sci.* **2017**, *95*, 1170–1178. [CrossRef]

8. Puvača, N.; Stanaćev, V.; Glamočić, D.; Lević, J.; Stanaćev, V.; Milić, D.; Vukelić, N.; Ljubojević, D. Application of the Process of Extrusion and Micronisation and their Influence on Nutritive Value of Feedstuffs. In Conference Proceeding, International Conference on BioScience: Biotechnology and Biodiversity–Step in the Future-The Forth Joint UNS-PSU Conference, Novi Sad, Serbia, 18–20 June 2012; pp. 197–202.

9. Vukmirović, D.; Fišteš, A.; Lević, J.; Čolović, R.; Rakić, D.; Brlek, T.; Banjac, V. Possibilities for Preservation of Coarse Particles in Pelleting Process to Improve Feed Quality Characteristics. *J. Anim. Physiol. Anim. Nutr.* **2017**, *101*, 857–867. [CrossRef]

10. Coelho, M. Vitamin Stability in Premixes and Feeds A Practical Approach in Ruminant Diets. In Proceedings of the 13th Annual Florida Ruminant Nutrition Symposium, Gainesville, FL, USA, 10–11 January 2002; pp. 127–145.

11. McDowell, L. *Vitamins in Animal Nutrition: Comparative Aspects to Human Nutrition*, 2nd ed.; Academic Press Inc.: San Diego, CA, USA, 2000; pp. 1–812.

12. Yang, P.; Wang, H.; Zhu, M.; Ma, Y. Effects of Choline Chloride, Copper Sulfate and Zinc Oxide on Long-Term Stabilization of Microencapsulated Vitamins in Premixes for Weanling Piglets. *Animals* **2019**, *9*, 122. [CrossRef]

13. National Research Council. *Nutrient Requirements of Swine*, 11th ed.; National Research Council of the National Academies: Washington, DC, USA, 2012.

14. Method of Determining and Expressing Fineness of Feed Materials by Sieving. Available online: http://citeseerx.ist.psu.edu/viewdoc/download?doi=10.1.1.214.2429&rep=rep1&type=pdf (accessed on 8 April 2020).

15. AOAC. *Official Methods of Analysis of AOAC International*, 18th ed.; Rev. 2; Hortwitz, W., Latimer, G.W., Jr., Eds.; AOAC Int.: Gaithersburg, MD, USA, 2007.

16. AOAC. *Official Methods of Analysis of AOAC International*, 19th ed.; Rev. 2; Hortwitz, W., Latimer, G.W., Jr., Eds.; AOAC Int.: Gaithersburg, MD, USA, 2012.

17. *National Standard Determination of Vitamin K_3 in Feeds- High Performance Liquid Chromatography*; General Administration of Quality Supervision, Inspection and Quarantine of the People's Republic of China: Beijing, China, 2017; Available online: https://www.gb-standards.com/GB_standard/GB-T%2018872-2017.html (accessed on 20 March 2020).

18. Chen, P.; Atkinson, R.; Wolf, W.R. Single-Laboratory Validation of a High-Performance Liquid Chromatographic-Diode Array Detector-Fluorescence Detector/Mass Spectrometric Method for Simultaneous Determination of Water-Soluble Vitamins in Multivitamin Dietary Tablets. *J. AOAC Int.* **2009**, *92*, 680–687.

19. Murphy, E.W.; Criner, P.E.; Gray, B.C. Comparisons of Methods for Calculating Retentions of Nutrients in Cooked Foods. *J. Agric. Food Chem.* **1975**, *23*, 1153–1157. [CrossRef]

20. Shurson, G.C.; Salzer, T.M.; Koehler, D.D.; Whitney, M.H. Effect of metal specific amino acid complexes and inorganic trace minerals on vitamin stability in premixes. *Anim. Feed Sci. Technol.* **2011**, *163*, 200–206. [CrossRef]

21. Mercier, C.; Charbonniere, R.; Grebaut, J.; De la Gueriviere, J.F. Formation of Amylose-Lipid Complexes by Twin-Screw Extrusion Cooking of Manioc Starch. *Cereal Chem.* **1980**, *57*, 4–9.

22. Vranjes, M.V.; Pfirter, H.P.; Wenk, C. Influence of Processing on Dietary Enzyme Effect and Nutritive Value of Diets for Laying Hens. *Can. J. Anim. Sci.* **1996**, *75*, 453–460. [CrossRef]

23. Charlton, S.J.; Ewing, W.N. *The Vitamins Directory: Your Easy to Use Guide to Animal Nutrition*; Context Products Ltd.: Leicestershire, UK, 2007.

24. Riaz, M.N.; Asif, M.; Ali, R. Stability of Vitamins during Extrusion. *Crit. Rev. Food Sci. Nutr.* **2009**, *49*, 361–368. [CrossRef]

25. Kostadinović, L.M.; Teodosin, S.J.; Spasevski, N.J.; Đuragić, O.M.; Banjac, V.V.; Vukmirović, Đ.M.; Sredanović, S.A. Effect of Pelleting and Expanding Processes on Stability of Vitamin E in Animal Feeds. *Food Feed Res.* **2013**, *40*, 109–114.

26. Kostadinović, L.; Teodosin, S.; Lević, J.; Čolović, R.; Banjac, V.; Vukmirović, Đ.; Sredanović, S. Effect of Pelleting and Expanding Processes on Vitamin A Stability in Animal Feeds. *J. Process. Energy Agric.* **2014**, *18*, 44–46.

27. Jakobsen, J.; Knuthsen, P. Stability of Vitamin D in Foodstuffs during Cooking. *Food Chem.* **2014**, *148*, 170–175. [CrossRef]

28. Tiwari, U.; Cummins, E. Nutritional Importance and Effect of Processing on Tocols in Cereals. *Trends Food Sci. Tech.* **2009**, *20*, 511–520. [CrossRef]

29. Zielinski, H.; Michalska, A.; Piskula, M.K.; Kozlowska, H. Antioxidants in Thermally Treated Buckwheat Groats. *Mol. Nutr. Food Res.* **2006**, *50*, 824–832. [CrossRef]

30. Gadient, M.; Fenster, R. Stability of Ascorbic Acid and Other Vitamins in Extruded Fish Feeds. *Aquaculture* **1994**, *124*, 207–211. [CrossRef]

31. Li, M.H.; Rushing, J.B.; Robinson, E.H. Stability of B-complex Vitamins in Extruded Catfish Feeds. *J. Appl. Aquac.* **1996**, *6*, 67–71. [CrossRef]

32. Marchetti, M.; Tossani, N.; Marchetti, S.; Bauce, G. Stability of Crystalline and Coated Vitamins during Manufacture and Storage of Fish Feeds. *Aquac. Nutr.* **1999**, *5*, 115–120. [CrossRef]

33. Combs, G.F., Jr. *The Vitamins—Fundamental Aspects in Nutrition and Health*, 4th ed.; Academic press: Cambridge, MA, USA, 2016; p. 1570.

34. Killeit, U. Vitamin Retention in Extrusion Cooking. *Food Chem.* **1994**, *49*, 149–155. [CrossRef]

35. Vukmirović, Đ.; Ivanov, D.; Čolović, R.; Kokić, B.; Lević, J.; Đuragić, O.; & Sredanović, S. Effect of Steam Conditioning on Physical Properties of Pellets and Energy Consumption in Pelleting Process. *J. Process. Energy Agri.* **2010**, *14*, 106–108.

36. Spasevski, N.; Kokić, B.; Bliznikas, S.; Švirmickas, G.; Vukmirović, Đ.; Čolović, R.; Lević, J. Effect of Different Thickness of Die on the Stability of Amino Acids in Pelleting Pig Feed. In Proceedings of the 2nd Workshop Feed-to-Food FP7 REGPOT-3. XIV International Symposium Feed Technology, Proceedings, Novi Sad, Serbia, 19–21 October 2010; pp. 447–453.

37. Athar, N.; Hardacre, A.; Taylor, G.; Clark, S.; Harding, R.; McLaughlin, J. Vitamin Retention in Extruded Food Products. *J. Food Compos. Anal.* **2006**, *19*, 379–383. [CrossRef]

38. Jones, F.T. Effect of Pelleting on Vitamin A Assay Levels of Poultry Feed. *Poult. Sci.* **1986**, *65*, 1421–1422. [CrossRef]

39. Cutlip, S.E.; Hott, J.M.; Buchanan, N.P.; Rack, A.L.; Latshaw, J.D.; Moritz, J.S. The Effect of Steam-Conditioning Practices on Pellet Quality and Growing Broiler Nutritional Value. *J. Appl. Poult. Res.* **2008**, *17*, 249–261. [CrossRef]

40. Lewis, L.L.; Stark, C.R.; Fahrenholz, A.C.; Bergstrom, J.R.; Jones, C.K. Evaluation of Conditioning Time and Temperature on Gelatinized Starch and Vitamin Retention in a Pelleted Swine Diet. *J. Anim. Sci.* **2015**, *93*, 615–619. [CrossRef]

41. Kimura, M.; Itokawa, Y.; Fujiwara, M. Cooking Losses of Thiamin in Food and its Nutritional Significance. *J. Nutr. Sci. Vitaminol.* **1990**, *36*, 17–24. [CrossRef]

42. Lešková, E.; Kubíková, J.; Kováčiková, E.; Košická, M.; Porubská, J.; Holčíková, K. Vitamin Losses: Retention during Heat Treatment and Continual Changes Expressed by Mathematical Models. *J. Food Compos. Anal.* **2006**, *19*, 252–276. [CrossRef]

43. Dawson, K.R.; Unklesbay, N.F.; Hedrick, H.B. HPLC Determination of Riboflavin, Niacin and Thiamin in Beef, Pork, and Lamb after Alternate Heat-Processing Methods. *J. Agric. Food Chem.* **1988**, *36*, 1176–1179. [CrossRef]

44. Mahan, L.; Escott, S.; Stump, S. *Krause's Food Nutrition and Diet Therapy*, 14th ed.; Elsevier Health Sciences: London, UK, 2016; pp. 1–1152.

45. Chatterjee, N.S.; Anandan, R.; Navitha, M.; Asha, K.K.; Kumar, K.A.; Mathew, S.; Ravishankar, C.N. Development of Thiamine and Pyridoxine Loaded Ferulic Acid-Grafted Chitosan Microspheres for Dietary Supplementation. *J. Food Sci. Technol.* **2016**, *53*, 551–560. [CrossRef]

Article

Former Foodstuff Products in *Tenebrio Molitor* Rearing: Effects on Growth, Chemical Composition, Microbiological Load, and Antioxidant Status

Simone Mancini [1,*], Filippo Fratini [1,2], Barbara Turchi [1], Simona Mattioli [3], Alessandro Dal Bosco [3], Tiziano Tuccinardi [4], Sanjin Nozic [1,4] and Gisella Paci [1,2]

[1] Department of Veterinary Sciences, University of Pisa, Viale delle Piagge 2, 56124 Pisa, Italy
[2] Interdepartmental Research Center "Nutraceuticals and Food for Health", University of Pisa, Via del Borghetto 80, 56124 Pisa, Italy
[3] Department of Agricultural, Food and Environmental Science, University of Perugia, Borgo XX Giugno 74, 06100 Perugia, Italy
[4] Department of Pharmacy, University of Pisa, Via Bonanno 6, 56124 Pisa, Italy
* Correspondence: simone.mancini@for.unipi.it or simafo@gmail.com; Tel.: +0039-050-2216803

Received: 28 June 2019; Accepted: 23 July 2019; Published: 25 July 2019

Simple Summary: Insects represent a possible alternative nutrient source for food and feed purposes. Insects could be reared on a feed basis alternative to conventional ones of animal origin and could help to face the rising demand of proteins. Mealworm could be reared directly on former foodstuff products allowing to reduce waste materials and enhance profits in several sectors. This study demonstrates that *Tenebrio molitor* rearing can be done on leftovers and by-products with proficient outcomes and high-quality final products. However, rearing substrates must be carefully selected in order to maximize the outcomes in relation to the prefixed goals.

Abstract: *Tenebrio molitor* (mealworm) larvae represent one of the most interesting edible insects and could be reared on alternative feeds, such as former foodstuff products (FFPs). In the present work, five different FFPs (brewery spent grains, bread and cookie leftovers, and mixes of brewer's spent grain or bread with cookies) were employed as feeding substrates. Larvae's growth performances, chemical composition, microbial loads, and antioxidant status were determined. Chemical compositions of the substrates affected all the tested parameters. Brewery spent grains-fed larvae showed a faster growth period and higher crude protein and carbohydrate contents. The use of cookies as a single substrate or their addition to spent grains or bread increased the lipids contents, while growth was delayed. Microbial loads were partially affected by the fed diet. The antioxidant status of larvae showed different concentrations of tocopherols isoforms (δ, γ, α) in relation to the diet; however, no differences were detected in relation to the global antioxidant capacity (2,2-azinobis-(3 ethylbenzothiazoline-6-sulfonic acid), ABTS reducing activity; 1,1-diphenyl-2-pircydrazyl, DPPH radical scavenging activity; ferric reducing ability, FRAP). Results point out a high plasticity of mealworm larvae and the potential to tailor the final outcomes in relation to the substrate employed. Mealworms could be practically reared on FFPs to produce food-feed with high nutrient values.

Keywords: mealworm; edible insects; animal protein; by-product; entomophagy

---

1. Introduction

In the light of the increasing world population and the demand of proteins from the developing countries, insects could represent one of the more suitable answers [1]. Insects, compared to conventional production animals, have a lower environmental impact (mostly linked to greenhouse gas production, use of water, and use of arable land); moreover, due to their higher reproductive capacity, nutritional quality, and feed conversion efficiency, they could be taken into account for both food and feed purposes.

Despite researchers' interests in edible insects, western consumers seem to be cautious to practice entomophagy, mainly in relation to disgust related to western socio-cultural ideas of these animals [2,3]. During the last years, several research studies were carried out to determine which items could be positively accepted by consumers in order to introduce edible insects on the market and increase their use both as food and feed. The main drivers which induced an increase of familiarity and a decrease in the awareness in relation to entomophagy have been identified with the visual presence of insects (visible/invisible, powder/whole), the familiarity with the products containing insects, the correct information about nutritional values, and possible positive effects of edible insects from sustainability and environmental perspectives [4–6]. With respect to these last perspectives, the rearing conditions could play a major role, as insects can be reared on sustainable feeds, such as waste or by-products that do not meet the nutritional values needed for the rearing of other farmed animals [7]. Former foodstuff products (FFPs) are defined by the European Commission in the Commission Regulation (EU) No 68/2013 of the catalogue of feed materials as "foodstuffs other than catering reflux, which are manufactured in full compliance with EU food law but are no longer intended for human consumption for practical and logistical reasons or due to problems in manufacturing or packaging which are unlikely to cause any health risks when used as feed". Therefore, in this category, both leftovers from the food industry mainly composed of bakery products, such as bread, and leftovers composed of confectionery products, such as cookies, are listed. The use of FFPs as feed leads to a decrease in waste material and pollution and represents an interesting opportunity in terms of the circular economy [8]. Thus, FFPs represent a valid and cheap source of energy, and they could be used in insect feeding concerning their essential nutrients [9].

Based on the above-mentioned considerations, the use of FFPs in the rearing of edible insects seems a good compromise to reduce food waste materials, produce valuable new food/feed products, and have a positive impact on the consumers' perception. However, it has to be mentioned that, even in the insect production sector, there are some limitations for the employment of certain feeds as the diet composition could affect the insect's development rate and body nutrient composition [7,10].

The mealworm (*Tenebrio molitor* L. 1758; Coleoptera Tenebrionidae) represents one of the most studied edible insects, both as food and feed. The mealworm is a holometabolic insect (complete metamorphosis) and probably originated in the Mediterranean region, but it is nowadays cosmopolitan in its distribution. Due to its low rearing request, the mealworm is one of the promising candidates for an industrial scale production. Indeed, in the last years, several farms and food-feed companies worldwide have started to employ this insect. Nutritional composition of mealworm larvae revealed a relatively high content of protein, 50% on a dry-matter basis [11,12], and lipids, about 30%–34% on a dry-matter basis [12–14], with a good composition in amino acids (good source of the essential amino acids), vitamins (i.e., vitamin E, vitamin B12, niacin, riboflavin, pantothenic acid, and biotin), and minerals (P, K, Mg, Zn, and Mn) [15].

Few research studies were carried out on the rearing conditions and on the effects of different substrates on the development and chemical-biological characteristics of mealworm larvae [7,10,13,14,16,17]. Most of the cited studies highlighted a plasticity of *T. molitor* in relation to the substrate, with variations of both development times and nutritional values of the larvae. However, researchers employed mixes of several ingredients as substrates in their studies, since the main focus was to determine the effects on combinations of high/low amounts of proteins, lipids, or carbohydrates. In this scenario, it is very difficult to assess if mealworms could be reared on FFPs directly gathered from the producers (one single type of FFP or a mix of a few). Indeed, as insect rearing does not request specific features in geographical or natural environmental conditions (as carried out in indoor controlled conditions), it is possible to speculate that new farms could be located near the substrate suppliers, and feed manipulation will be kept at a minimum in order to maintain low costs.

Hence, the main aim of this research study was to evaluate the use of three FFPs in *T. molitor* rearing in order to produce larvae for food/feed purposes. In particular, larval development, chemical characteristics, and the microbiological and antioxidant status were determined in mealworm larvae fed with brewery spent grains, bread and cookie leftovers, and on mixes of brewer's spent grain or bread with cookies (mixed to increase brewer's spent grain or bread diets in lipid content).

2. Materials and Methods

2.1. Diet Preparation

The mealworms were fed with five different diets: brewery spent grains (SG), bread (B), cookies (C), 50% brewer's spent grain and 50% cookies (SG-C), and 50% bread and 50% cookies (B-C). Spent grains were directly collected from a local brewery and immediately frozen at −20 °C. Bread and cookies were collected from a market shop as daily remains (bread) or soon to expire products (cookies). Excessive humidity of spent grains, bread, and cookies was removed in an oven at 90 °C until the product was completely dry. Spent grains were previously thawed for 18 h at 4 °C. The proximate composition of SG, B, and C is reported in Table 1. The three substrates were then finely ground, and the five diets were formulated.

Table 1. Proximate composition of brewery spent grains (SG), bread (B) and cookies (C).

Item	Unit	SG	B	C
Dry matter (DM)	%	94.81	97.09	99.99
Ether extract	%	3.29	0.31	10.44
Crude protein [1]	%	17.98	11.15	6.55
Ash	%	3.43	1.88	0.70
Carbohydrates	%	70.10	83.75	82.29
Ether extract	% on DM	3.47	0.32	10.45
Crude protein [1]	% on DM	18.97	11.49	6.55
Ash	% on DM	3.62	1.93	0.70
Carbohydrates	% on DM	73.94	86.26	82.30

[1] Conversion factors of 5.83 for SG [18], of 5.70 for B, and 6.25 for C [19] (protocol numbers 950.36 and 935.39, respectively).

2.2. Insect Rearing and Growth Performances

The mealworms were reared in plastic containers ($39 \times 28 \times 14$ cm) at the Department of Veterinary Sciences (University of Pisa, Pisa, Italy) under a laboratory scale production. The temperature was maintained at 25 ± 1 °C with 50%–60% of relative humidity. In the first stage of the trial, 15 boxes were employed (three per diet), in which adult beetles (one to two weeks old) were placed for one week to deposit eggs. Then beetles were removed, and the larvae were left to grow in the substrate. Experimental diets were added weekly if needed (*ad libitum*, weighed before adding), and in order to provide moisture, potatoes slices were deposited once a week. Boxes were visually evaluated three times per week to check larvae health and to remove dead ones.

The mealworms were harvested when the first pupa was observed or after one year of rearing time (in order to maintain a profitable low cost). The development time was calculated between the first day of the experiment and the day of harvesting. Every week, representative samples of larvae (100 per box) were weighed to quantify the growth performance. After harvesting, the larvae were starved for 24 h (apart from samples for microbiological analyses performed with and without starving, see below) in sterile plastic containers with plastic web on the base in order to separate feces and to avoid fecal contact.

The feed conversion ratio (FCR) and efficiency of conversion of ingested food (ECI) were calculated assuming that all the provided feed was consumed as reported by Oonincx et al. [7]. Furthermore, the nitrogen conversion efficiency (N-ECI) was also calculated as reported by Oonincx et al. [7].

2.3. Proximate Composition of Feed and Larvae

Proximate composition analyses were performed in triplicate for each sample. Larvae were ground in a blender before being analyzed.

The dry matter content was determined by dehydration in a drying oven at 105 °C until constant weight. The lipids content was quantified by the Soxhlet method using petroleum ether as a solvent. The ash content was determined by incineration in a muffle furnace at 550 °C. The crude protein content was determined by the Kjeldahl method, two protein-to-nitrogen conversion factors were used: 6.25 as normally calculated for meat samples and 4.76 as suggested by Janssen et al. [20] for insects. Carbohydrates were calculated as: 100 − crude protein − lipid content − moisture − ash.

2.4. Microbiological Analyses

For microbiological analyses both un-starved and starved larvae were employed in order to quantify the effect of fasting. Substrates were also analyzed, and microbiological amounts lower than the detection limits were reported.

Ten grams of larvae were weighed in sterile stomacher bags, and then 90 mL of sterile saline solution was added. The mixture was homogenized for 60 s in a stomacher (Stomacher® 400 Circulator, VWR International Sr, Milan, Italy). Ten-fold serial dilution series were performed and plated on different media.

Plate count agar (PCA) was employed for the quantification of the total viable aerobic count (incubated at 30 °C for 72 h) and aerobic bacterial endospores after heating the 10:1 dilution at 80 °C for 10 min and performing ten-fold serial dilutions (incubated at 30 °C for 72 h); violet red bile glucose agar (VRBGA) was used to enumerate Enterobacteriaceae (incubation at 37 °C for 24 h); Tryptone Bile X-Glucuronide medium (TBX) was employed to quantify *Escherichia coli* (42 °C for 24 h); Baird Parker medium (BP) was employed for the enumeration of *staphylococci* both presumptive coagulase-positive and -negative (incubated at 37 °C for 24–48 h); yeast extract, dextrose, chloramphenicol medium (YEDC) was used for yeast and mold counts (incubation at 25 °C for 120 h); de Man–Rogosa–Sharpe agar (MRS) was employed to count lactic acid bacteria (37 °C for 48 h in anaerobic conditions); Bacillus Cereus MYP agar was used for the evaluation of the presence of presumptive *Bacillus cereus* (37 °C for 24 h). Absence of *Listeria monocytogenes* and *Salmonella* spp. in 25 g was assessed according to ISO 11290 and ISO 6579, respectively.

All culture media and supplements were purchased from ThermoFisher Scientific, (Milan, Italy) except for Bacillus Cereus MYP agar, which was purchased from Biolife (Milan, Italy).

Microbial counts were expressed in log colony-forming unit (CFU)/g as mean of three replicates (when one or more samples showed values below the detection threshold, 1.0 log CFU/g, the detection threshold divided by two, 0.5 log CFU/g, was used to calculate the mean and standard deviation as reported by Stoops et al. [21]).

2.5. Antioxidant Status

Antioxidant activity of the larvae was quantified as reported by Mancini et al. [22] for meat samples with minor modifications. Three grams of larvae were homogenized in 10 mL ethanol, followed by a centrifugation and filtration (Whatman number 4 filter paper) steps. The filtrate was used to measure 2,2-azinobis-(3 ethylbenzothiazoline-6-sulfonic acid) (ABTS) reducing activity, 1,1-diphenyl-2-pircydrazyl (DPPH) radical scavenging activity, and ferric reducing ability (FRAP). The results of antioxidant capacity were expressed as mmol of Trolox equivalent per kilogram of sample (calibration curves obtained with Trolox at 0–2000 µM, final concentrations).

The tocopherol (δ-, γ-, α-isoforms) contents of the samples were quantified by a high performance liquid chromatography (HPLC) system, according to Zaspel and Csallany [23]. About 3 g of larvae were saponified in 60 g/100 mL KOH in ethanol in a thermostat bath at 70 °C for 30 min. Then the content was sonicated and extracted twice with 15 mL of n-hexane/ethyl acetate (9:1, v/v). After collecting the upper phase, the samples were nitrogen dried and then reconstituted in 200 µL of acetonitrile. Fifty 50 µL was injected into the HPLC system (Perkin Elmer series 200, equipped with an autosampler system, model AS 950–10, Tokyo, Japan) on a Synergy Hydro-RP column (4 µm, 4.6 × 100 mm; Phenomenex, Bologna, Italy). An Fluorometric detector (FD) (model, FP-1525; Jasco, Tokyo, Japan; excitation and emission wavelengths of 295 and 328 nm, respectively) were used to identify the different isoforms. External calibration curves were used to quantify isoforms by increasing amounts of pure tocopherols in ethanol.

2.6. Statistical Analysis

The data obtained from proximate composition, microbial determinations, antioxidant capacity, and antioxidant compounds were statistically analyzed using a one-way ANOVA with regards to the different diets (SG, B, C, SG-C, and B-C). The effect of starvation (starved and un-starved) was tested against each different microbiological quantification by a Student's *t*-test. Statistical significance was set at 0.05 and differences were assessed using Tukey's test. Linear regression and second-order polynomial quadratic equation were performed to evaluate the effect of each diet on the growth performance. Principal components analysis (PCA) was performed on proximate composition (protein conversion factor of 4.76), microbial loads (fasted samples), growth performance (g weight gain per day), antioxidant capacity (ABTS, DPPH, and FRAP), and antioxidant compounds (tocopherol δ-, γ-, α-isoforms); all the data were mean centered and scaled to a unit standard deviation before analysis. R free statistical software was used [24].

3. Results and Discussion

3.1. Growth Performances and Proximate Compositions

Proximate compositions of the larvae reared on the different substrates are reported in Table 2.

Table 2. Proximate compositions of the larvae reared on the different substrates.

Item	Unit	Rearing Substrates					RMSE	*p*-Value
		SG	B	C	SG-C	B-C		
Dry matter (DM)	%	33.33 [c]	32.62 [c]	35.55 [b]	37.53 [a]	35.33 [b]	0.733	<0.001
Ether extract	%	6.46 [d]	14.82 [b]	17.77 [a]	11.77 [c]	17.48 [a]	1.189	<0.001
Crude protein [1]	%	17.36 [a]	14.09 [ab]	13.33 [b]	17.65 [a]	14.07 [ab]	1.392	0.020
Crude protein [2]	%	13.22 [a]	10.73 [ab]	10.15 [b]	13.44 [a]	10.72 [ab]	1.392	0.020
Ash	%	1.10	0.98	0.95	1.07	1.01	0.097	0.371
Carbohydrates [1]	%	8.40 [a]	2.73 [b]	3.50 [b]	7.05 [a]	2.76 [b]	1.620	<0.001
Carbohydrates [2]	%	12.54 [a]	6.09 [b]	6.72 [b]	11.26 [a]	6.12 [b]	1.523	<0.001
Ether extract	% on DM	19.38 [c]	45.43 [a]	50.00 [a]	31.35 [b]	49.48 [a]	3.389	<0.001
Crude protein [1]	% on DM	51.34 [a]	42.28 [ab]	37.31 [b]	46.40 [ab]	40.96 [ab]	3.898	0.022
Crude protein [2]	% on DM	39.10 [a]	32.20 [ab]	28.41 [b]	35.33 [ab]	31.20 [ab]	3.898	0.022
Ash	% on DM	3.36	3.05	2.69	2.88	2.81	0.268	0.084
Carbohydrates [1]	% on DM	25.15 [a]	8.28 [b]	9.83 [b]	18.76 [a]	7.81 [b]	4.523	<0.001
Carbohydrates [2]	% on DM	37.58 [a]	18.58 [c]	18.77 [c]	29.97 [b]	17.31 [c]	4.123	<0.001

SG: brewery spent grains; B: bread; C: cookies; SG-C: 50% spent grains, 50% cookies; B-C: 50% bread, 50% cookies.
[1] Conversion factors of 6.25 for crude protein. [2] Conversion factors of 4.76 for crude protein. Means in the same row with no common superscripts (a–d) differ significantly (*p* < 0.05).

Dry matter, ether extract, and crude protein contents were affected by the diet. Dry matter of the SG-C larvae was the highest, followed by that of the C and B-C diets. From these data, it would seem that the presence of the cookies induced a decrease of humidity in the larval body, as the larvae fed with spent grains and bread alone (SG and B, respectively) showed the lowest values of dry matter. Cookies also played a major role in the lipid content of the larvae. Diets C and B-C showed the highest contents, followed by B. Spent grains contained low amounts of lipids (3.29%), and consequently larvae fed with SG and SG-C showed lower values compared to those fed other diets.

The highest crude protein contents were shown by diets which included spent grains (SG and SG-C) with minor differences. This value was lower in diets with bread (B and B-C). In this case, the low amount of crude proteins of the cookies (6.55%) negatively affected the chemical composition of the larvae with a consequent lowest body protein content.

As the dry matter analyses showed to be significantly affected by the diet, the proximate composition was also reported as % of dry matter (Table 2). Expressing the data as % of dry matter, the ether extract content showed that B, C, and B-C had the highest lipid contents, followed by SG-C and SG, respectively. Thus, the effect of the cookie fat content was confirmed but the bread diet (B) also played a key role.

Crude protein contents, as % of dry matter, showed only minor differences in the statistical analyses if compared to the expression as % of fresh sample. The conversion factors of 6.25 or 4.76, as well as the expression as % of fresh sample or % of dry matter, drastically affected the final numerical expression; nevertheless, as no standardization is currently used in the insect research field, we choose to report all the data expressions.

Carbohydrates were higher in larvae fed SG and SG-C diets than those fed the other diets. Interestingly, SG had the lowest amount of carbohydrates within the employed substrates. Particularly, moisture and lipids content seemed to affect the calculations of carbohydrates in larvae reared in SG and SG-C.

Chemical compositions of mealworm confirm the data already published [7,10,20] and in particular that body composition showed a considerable plasticity in relation to the diet [7,10]. Notably, in our trial, mealworms fed cookies (C) showed a high amount of lipids (50% on DM base) mostly related to the detriment of the protein content. This data could be interesting as edible insects are almost exclusively recommended as a protein source even though they could contain a high amount of fats.

The larvae weights are reported in Figure 1; larvae fed SG and SG-C reached the harvesting day approximately after 5 to 6.5 months. After one year of rearing, larvae fed diets with lower content of protein, C, B, and B-C did not reach the pupae stage; therefore, in order to not exceed in the rearing time, their growth was stopped. Larvae fed diets C, B, and B-C reached the final weights of 87, 95, and 112.5 mg, respectively. In linear regression and second-order polynomial quadratic equations (Figure 1), the coefficient of determination indicates that the models explain a high percentage of the variability (R2 between 0.94 and 0.99). Between linear regression and second-order polynomial quadratic equation, the second seems to better fit the larval growth rate.

The FCR (calculated on fresh basis), the ECI, and the N-ECI were severely affected by the dietary treatments. Dietary efficiency of mealworm larvae was strictly related to the chemical composition of the substrate; indeed, larvae fed SG and SG-C showed the lowest values of FCR, 2.22 and 2.76 respectively, confirming the extremely positive potential of this species to convert feed into body weight. On the other hand, larvae fed B, C, and the mix B-C showed FCR values about 8.86, 7.31, and 4.02, which are similar to those of conventional production animals (4.0 for pork and 8.8 for cereal beef; [9]). The variability in FCR values was related to the protein contents of the diets; indeed, SG and SG-C showed the highest amounts of protein. A large variability of the FCR value in relation to the protein content was also reported by Oonincx et al. [7] and van Broekhoven et al. [10] in mealworms fed different diets (from 4.1 to 19.1 and from 2.62 to 6.05, respectively).

The ECI values were higher for SG and SG-C diets (15.85 and 13.94, respectively) than for B, C, and B-C ones (3.79, 4.87, and 8.92, respectively). These data are comparable with those reported by Oonincx et al. [7] and van Broekhoven et al. [10]. As reported before for FCR, the ECI values were affected by the protein content of the diets, which confirms that mealworm efficiency could be modelled through the rearing substrate's chemical composition.

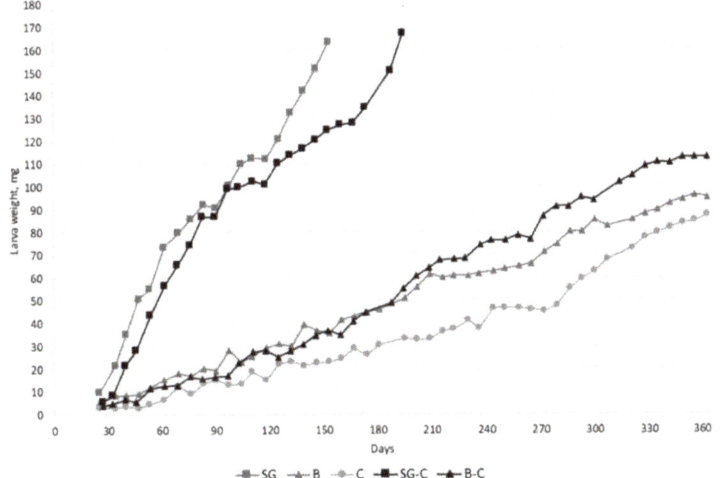

Rearing substrate	Linear regression	R^2	Second order polynomial quadratic equation	R^2
SG	y = 1.0665x - 4.0790	0.972	y = -0.0024x² + 1.4989x - 19.763	0.978
B	y = 0.2774x - 3.0232	0.993	y = 0.00001x² + 0.2734x - 2.7408	0.993
C	y = 0.2434x - 10.839	0.952	y = 0.0005x² + 0.0449x + 3.3146	0.985
SG-C	y = 0.8480x - 0.5992	0.943	y = -0.0033x² + 1.5648x - 31.4230	0.973
B-C	y = 0.3583x - 13.1800	0.989	y = 0.0002x² + 0.2820x - 7.7445	0.991

Figure 1. *Tenebrio molitor* larval growth in relation to the diet (SG: brewery spent grains; B: bread; C: cookies; SG-C: 50% spent grains, 50% cookies; B-C: 50% bread, 50% cookies).

Nitrogen was more efficiently converted than the other diet components in almost all diets, as in the SG, SG-C, C, and B-C diets, the N-ECIs were higher than the ECI (59.11, 71.75, 13.83, and 14.05, respectively). The diet exclusively composed of bread (B) showed the lowest value of N-ECI (2.77), which is lower than the ECI value. This means that mealworm fed only bread did not convert N of the substrate efficiently into body mass [7]. Indeed, larvae fed only cookies (C) reached approximately the same amount of crude protein content (Table 2) even though the substrate showed to contain half the protein (6.55% in C vs. 11.15% in B, Table 1). The high N-ECI showed by SG and SG-C larvae represents a starting point to reach a relevant benefit of insects over conventional production animals [25].

3.2. Microbiological Analyses

Microbial determinations of starved and un-starved larvae are reported in Table 3.

In the un-starved larvae, staphylococci, yeast-molds, and bacterial endospores were significantly different in relation to the diet. Staphylococci and yeast-molds loads were higher in larvae fed B, C, and B-C, followed by SG-C and SG. Minor differences in the staphylococci amount was highlighted between SG and SG-C in relation to the presence of the cookies. Bacterial endospores were absent in SG, and their presence was related to B and C diets. Indeed, feeding the larvae with a mix of SG and C induced an increase of these bacteria. The highest number of bacterial endospores was detected in the B-C diet, followed by B and C substrates.

Microbial analyses highlighted the total absence of *Escherichia coli* and *Bacillus cereus*, as well as the absence in 25 g of *Listeria monocytogenes* and *Salmonella* spp. Other research studies reported the absence of these bacteria in edible insects reared for human consumption [26–29].

Table 3. Microbiological analyses of the larvae (un-starved and starved) reared on the different substrates.

Item	Rearing Substrates					RMSE	*p*-Value
	SG	B	C	SG-C	B-C		
No starvation							
Total viable aerobic counts	7.08	7.63	6.46	6.84	7.77	0.776	0.266
Enterobacteriaceae	6.31	6.44	6.33	5.87	6.30	0.502	0.688
Staphylococci	3.85 [c]	5.96 [a]	5.29 [ab]	4.38 [b]	5.19 [ab]	0.579	0.005
Yeast and molds	3.30 [b]	5.60 [a]	6.31 [a]	3.13 [b]	5.34 [ab]	0.944	0.001
Lactic acid bacteria	6.20	5.18	5.57	5.44	5.31	0.685	0.343
Bacterial endospores	0.00 [d]	4.44 [ab]	3.44 [b]	2.05 [c]	5.32 [a]	0.240	<0.001
Starvation, 24 h							
Total viable aerobic counts	7.04	7.63	6.41	7.17	6.91	0.689	0.295
Enterobacteriaceae	6.36	6.01	4.96	5.52	5.55	0.744	0.211
Staphylococci	4.87	4.20	3.92	4.40	3.73	0.705	0.370
Yeast and molds	2.56 [b]	4.45 [a]	5.11 [a]	3.54 [ab]	4.44 [a]	0.561	0.002
Lactic acid bacteria	6.10	5.91	4.91	5.73	5.13	0.666	0.168
Bacterial endospores	0.00 [c]	3.62 [a]	2.44 [b]	2.36 [b]	3.60 [a]	0.243	0.038
Effect of starvation, *p*-Value							
Total viable aerobic counts	0.954	0.988	0.915	0.658	0.187		
Enterobacteriaceae	0.913	0.364	0.006	0.639	0.285		
Staphylococci	0.115	0.031	0.012	0.968	0.039		
Yeast and molds	0.506	0.014	< 0.001	0.598	0.021		
Lactic acid bacteria	0.835	0.271	0.250	0.653	0.629		
Bacterial endospores	-	0.120	0.003	0.312	0.009		

SG: brewery spent grains; B: bread; C: cookies; SG-C: 50% spent grains, 50% cookies; B-C: 50% bread, 50% cookies. Data were reported as log CFU/g. Means in the same row with no common superscripts (a–d) differ significantly ($p < 0.05$).

In starved larvae, diets affected only yeast-molds and bacterial endospores, showing a similar trend of that reported for the un-starved larvae with minor differences. Starvation was not effective in larvae fed SG and SG-C diets, while in larvae fed B, C, and B-C it partially affected the microbial flora. Total viable aerobic counts and lactic acid bacteria were not affected by the starvation. Staphylococci and yeast-molds amounts were significantly decreased in B, C, and B-C fed larvae. Starvation in larvae fed C induced a significant decrease in the Enterobacteriaceae amount, as well as in bacterial endospores of larvae fed C and B-C.

Studying larvae fed wheat bran supplemented with carrots, Wynants et al. [30] reported that fasting for 24 or 48 h, both with and without fecal contact of the larvae, did not significantly affect total viable aerobic bacteria, Enterobacteriaceae, aerobic bacterial endospores, psychrotrophic aerobic bacteria, and yeast and mold amounts.

Contrarily, Mancini et al. [31] reported that a starvation treatment for 24 h affected the bacterial endospores amount in mealworm larvae reared on wheat bran supplemented with potato slices as a water source without influencing the other microorganisms.

Moreover, starvation also resulted effective in the reduction of *Salmonella enterica* and *Listeria monocytogenes* in mealworm larvae reared in artificially contaminated substrates [32,33]. Thus, the effect of the starvation process seems to vary in relation to the diet as well as the response of different tested bacteria.

3.3. Antioxidant Status

Antioxidant capacity of mealworm larvae is reported in Table 4.

Table 4. Antioxidant capacity (ABTS, DPPH, and FRAP) and antioxidant compounds (tocopherols) of *Tenebrio molitor* larvae reared on different substrates.

Item	Rearing Substrates					RMSE	*p*-Value
	SG	B	C	SG-C	B-C		
Antioxidant Capacity							
ABTS	1.73	2.17	2.45	2.01	1.70	0.407	0.415
DPPH	0.35	0.30	0.34	0.30	0.28	0.042	0.504
FRAP	0.75	0.80	1.04	0.86	0.75	0.139	0.332
Antioxidant Compounds							
δ-tocopherol	0.12 [c]	0.13 [c]	0.18 [c]	0.24 [b]	0.33 [a]	0.018	<0.001
γ-tocopherol	0.07 [b]	0.02 [b]	0.15 [a]	0.17 [a]	0.04 [b]	0.017	<0.001
α-tocopherol	0.48 [c]	0.35 [c]	4.95 [a]	2.86 [b]	4.59 [a]	0.055	<0.001
Total tocopherols	0.79 [c]	0.51 [c]	5.28 [a]	3.16 [b]	4.96 [ab]	0.601	<0.001

SG: brewery spent grains; B: bread; C: cookies; SG-C: 50% spent grains, 50% cookies; B-C: 50% bread, 50% cookies. ABTS, DPPH and FRAP expressed as mmol of Trolox equivalent per kilogram of sample. Tocopherols and carotenes expressed as mg per kilogram of samples. Means in the same row with no common superscripts (a–c) differ significantly ($p < 0.05$).

No significant differences were found among the experimental diets. However, the C diet seems to slightly improve the antioxidant capacity of the larvae with respect to the other groups (as highlighted by ABTS and FRAP values, even if no statistical significances were determined). Such a trend is justified also by the significantly higher concentration of tocopherols (mainly due to the α-isoform amount) found in the cookies-supplemented groups (C, SG-C, and B-C larvae). Nevertheless, vitamin E being a fat-soluble vitamin, it is not surprising that its concentration was higher in the mealworms with a higher fat content (ether extract values). To the best of our knowledge, this is the first evaluation of the antioxidant capacity of *T. molitor* larvae.

The antioxidant capacity of mealworms could be very interesting in order to enhance dietary antioxidant ingestion; however, as no thermal process was applied to the samples (cooking or drying), enzymes present in the larvae body may have altered the final outcomes. Few published data reported the antioxidant concentration of insects: Finke [15] summarized that the vitamin E contents ranged from 3.3–24.0 mg/kg for mealworms, 5.3–9.1 mg/kg for superworms (*Zophobas morio*), and 8.6–69.2 mg/kg for waxworms (*Galleria mellonella*). In the present study, the total vitamin E concentration of mealworms ranged from 0.51 to 5.28 mg/kg, close to the lower values reported in the literature. However, a higher concentration (>10 mg/kg) of vitamin E is reported to be common in the wild-caught insects with respect to the reared ones [34].

3.4. Principal Component Analysis

A principal component analysis of proximate composition, microbial loads, growth performance, antioxidant capacity, and antioxidant compounds was performed in order to detect the principal components that better describe the modifications highlighted (Figure 2).

Eigenvalues, eigenvectors, and cumulative % of the first three principal components are reported in Table 5. The first two principal components (PC1: 41.07% and PC2: 17.91%) well differentiate the samples in relation to the diet.

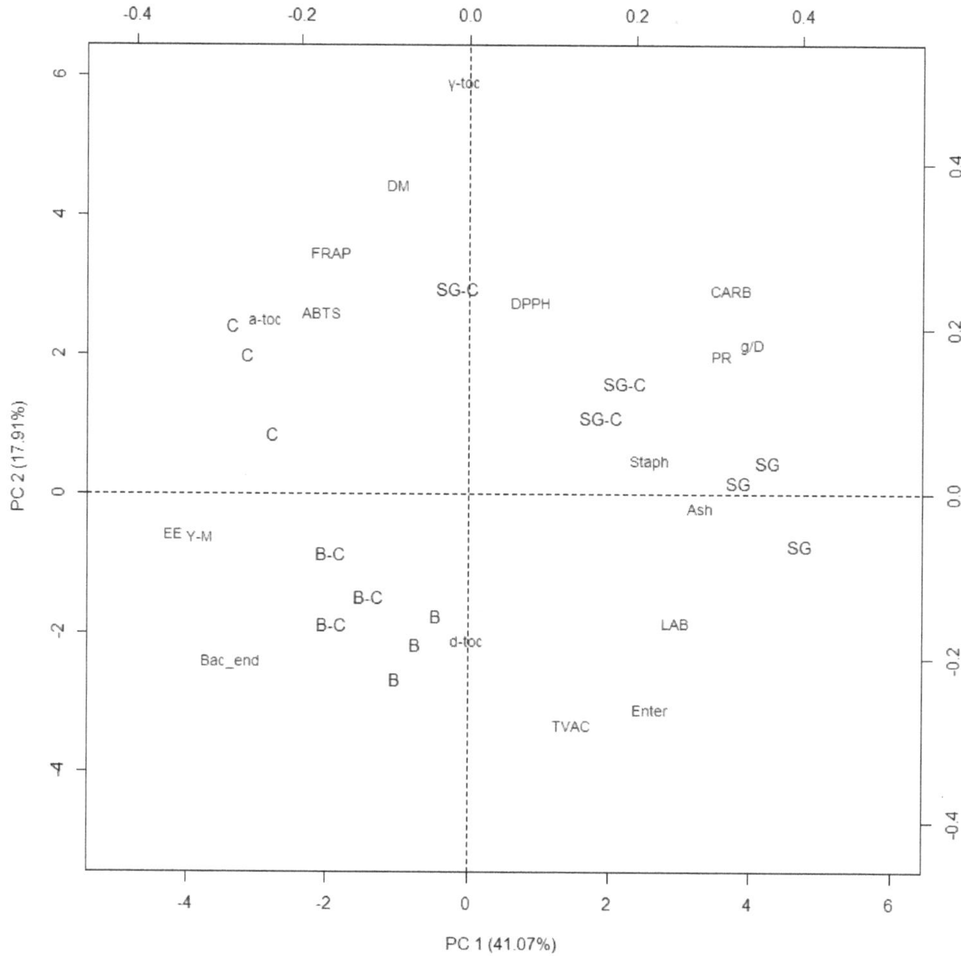

Figure 2. Biplot of the principal component (PC) analysis of mealworm larvae fed different diets (SG: brewery spent grains; B: bread; C: cookies; SG-C: 50% spent grains, 50% cookies; B-C: 50% bread, 50% cookies).

Positives PC1 eigenvectors collocate crude proteins on the right side of the biplot, near to SG and SG-C samples, as well as the growth performance. This relation highlights that diets rich in proteins (SG and SG-C) could promote growth and increase the protein content of the larvae. On the other hand, negative PC1 eigenvectors collocated ether extract content on the left side of the biplot, in contraposition with the protein content and the growth performances. The presence of the cookies in B-C and SG-C diets induced a shift on the left side of the plotted samples and then evidence a worsening of the growth performances and a higher deposition of fats.

Almost all the parameters that showed a statistical significance were plotted in the upper right or the lower left squares (both eigenvectors, positives or negatives) except for dry matter and tocopherol γ- and α-isoforms, which were plotted in the upper left square highlighting that another undetermined effect plays a role. As PC2 differentiates the antioxidant capacity and the tocopherol γ- and α-isoforms in the upper part of the plot and the microbial loads in the lower part, it could possible to hypothesize that a higher content of antioxidants played a role against microbial loads.

The observations reported above suggest that PC1 and PC2 were both a part of the diet effect related to the chemical composition of the substrates.

A residual percentage of variance (13.01%) was expressed also by PC3; this principal component was mostly related to (as absolute values) the antioxidant capacity and tocopherols δ- and α-isoforms.

Table 5. Eigenvalues and eigenvectors of the first three principal components (PC) of principal components analysis performed on proximate composition, microbial loads, growth performance, antioxidant capacity, and antioxidant compounds of mealworm larvae fed experimental diets.

Item	PC		
	PC 1	**PC 2**	**PC 3**
Eigenvalues	7.392	3.223	2.341
Eigenvectors			
Dry matter	−0.085	0.377	−0.269
Ether extract	−0.356	−0.049	−0.052
Crude protein	0.304	0.170	−0.133
Ash	0.279	−0.016	−0.204
Carbohydrates	0.314	0.250	−0.020
Weight gain (g/D)	0.340	0.182	−0.093
Total viable aerobic counts	0.127	−0.279	0.198
Enterobacteriaceae	0.220	−0.260	0.117
Staphylococci	0.219	0.042	0.156
Yeast and molds	−0.321	−0.053	0.117
Lactic acid bacteria	0.250	−0.154	0.221
Bacterial endospores	−0.284	−0.205	−0.041
ABTS	−0.176	0.220	0.451
DPPH	0.075	0.233	0.389
FRAP	−0.165	0.294	0.312
δ-tocopherol	−0.002	−0.180	−0.389
γ-tocopherol	−0.010	0.499	−0.064
α-tocopherol	−0.244	0.212	−0.329
Cumulative %	41.07	58.97	71.98

4. Conclusions

In conclusion, FFPs could be employed as feed in *T. molitor* rearing with effects on the chemical composition of the larvae and growth performance. Results point out a high plasticity of mealworm larvae and the potential to tailor the final outcomes. Single FFPs or mixes of multiple FFPs could be employed in relation to the requests and benefit–cost criteria determined by the producers. In general, it seems that feeds with high protein contents determine a better performance in rearing time, at the expense of the lipid content of the larvae.

Author Contributions: Conceptualization, S.M. (Simone Mancini) and F.F.; Data curation, S.M. (Simone Mancini); Formal analysis, S.M. (Simone Mancini), F.F., S.M. (Simona Mattioli) and S.N.; Investigation, S.M. (Simone Mancini), F.F., S.M. (Simona Mattioli) and S.N.; Methodology, S.M. (Simone Mancini), F.F. and S.M. (Simona Mattioli); Supervision, G.P.; Writing–original draft, S.M. (Simone Mancini), F.F. and S.M. (Simona Mattioli); Writing–review & editing, S.M. (Simone Mancini), F.F., B.T., S.M. (Simona Mattioli), A.D.B., T.T. and G.P.

Funding: This research received no external funding.

Conflicts of Interest: The authors declare no conflict of interest.

References

1. Van Huis, A.; Van Itterbeeck, J.; Klunder, H.; Mertens, E.; Halloran, A.; Muir, G.; Vantomme, P. *Edible Insects: Future Prospects for Food and Feed Security*; Food and Agriculture Organization of the United Nations (FAO): Rome, Italy, 2013; Volume 171, ISBN 978-92-5-107595-1.

2. Anankware, J.P.; Fening, K.; Obeng-Ofori, D. Insects as food and feed: A review. *Int. J. Agric. Res. Rev.* **2015**, *3*, 143–151.

3. Verbeke, W. Profiling consumers who are ready to adopt insects as a meat substitute in a Western society. *Food Qual. Prefer.* **2015**, *39*, 147–155. [CrossRef]

4. Tan, H.S.G.; van Berg, E.V.; Stieger, M. The influence of product preparation, familiarity and individual traits on the consumer acceptance of insects as food. *Food Qual. Prefer.* **2016**, *52*, 222–231. [CrossRef]

5. Hartmann, C.; Ruby, M.B.; Schmidt, P.; Siegrist, M. Brave, health-conscious, and environmentally friendly: Positive impressions of insect food product consumers. *Food Qual. Prefer.* **2018**, *68*, 64–71. [CrossRef]

6. Mancini, S.; Moruzzo, R.; Riccioli, F.; Paci, G. European consumers' readiness to adopt insects as food. A review. *Food Res. Int.* **2019**, *122*, 661–678. [CrossRef] [PubMed]

7. Oonincx, D.G.A.B.; van Broekhoven, S.; van Huis, A.; van Loon, J.J.A. Feed Conversion, Survival and Development, and Composition of Four Insect Species on Diets Composed of Food By-Products. *PLoS One* **2015**, *10*, e0144601. [CrossRef]

8. Pinotti, L.; Giromini, C.; Ottoboni, M.; Tretola, M.; Marchis, D. Review: Insects and former foodstuffs for upgrading food waste biomasses/streams to feed ingredients for farm animals. *Animal* **2019**, *13*, 1365–1375. [CrossRef] [PubMed]

9. Wilkinson, J.M. Re-defining efficiency of feed use by livestock. *Animal* **2011**, *5*, 1014–1022. [CrossRef]

10. Van Broekhoven, S.; Oonincx, D.G.A.B.; van Huis, A.; van Loon, J.J.A. Growth performance and feed conversion efficiency of three edible mealworm species (Coleoptera: Tenebrionidae) on diets composed of organic by-products. *J. Insect Physiol.* **2015**, *73*, 1–10. [CrossRef]

11. Azagoh, C.; Ducept, F.; Garcia, R.; Rakotozafy, L.; Cuvelier, M.-E.; Keller, S.; Lewandowski, R.; Mezdour, S. Extraction and physicochemical characterization of Tenebrio molitor proteins. *Food Res. Int.* **2016**, *88*, 24–31. [CrossRef]

12. Osimani, A.; Garofalo, C.; Milanović, V.; Taccari, M.; Cardinali, F.; Aquilanti, L.; Pasquini, M.; Mozzon, M.; Raffaelli, N.; Ruschioni, S.; et al. Insight into the proximate composition and microbial diversity of edible insects marketed in the European Union. *Eur. Food Res. Technol.* **2017**, *243*, 1157–1171. [CrossRef]

13. Dreassi, E.; Cito, A.; Zanfini, A.; Materozzi, L.; Botta, M.; Francardi, V. Dietary fatty acids influence the growth and fatty acid composition of the yellow mealworm Tenebrio molitor (Coleoptera: Tenebrionidae). *Lipids* **2017**, *52*, 285–294. [CrossRef] [PubMed]

14. Fasel, N.J.; Mene-Saffrane, L.; Ruczynski, I.; Komar, E.; Christe, P. Diet induced modifications of fatty-acid composition in mealworm larvae (Tenebrio molitor). *J. Food Res.* **2017**, *6*, 22. [CrossRef]

15. Finke, M.D. Complete nutrient content of four species of commercially available feeder insects fed enhanced diets during growth. *Zoo Biol.* **2015**, *34*, 554–564. [CrossRef] [PubMed]

16. Ramos-Elorduy, J.; González, E.A.; Hernández, A.R.; Pino, J.M. Use of Tenebrio molito (Coleoptera: Tenebrionidae) to Recycle Organic Wastes and as Feed for Broiler Chickens. *J. Econ. Entomol.* **2002**, *95*, 214–220. [CrossRef] [PubMed]

17. Oonincx, D.G.A.B.; de Boer, I.J.M. Environmental impact of the production of mealworms as a protein source for humans A Life Cycle Assessment. *PLoS One* **2012**, *7*, e51145. [CrossRef] [PubMed]

18. Merrill, A.L.; Watt, B.K. Energy value of food: Basis and derivation. *USDA. United States Dep. Agric.* **1973**.

19. Association of Official Analytical Chemists (AOAC). *Official Methods of Analysis*, 19th ed.; Association of Official Analytical Chemists: Washington, DC, USA, 2012; ISBN 9783642312410.

20. Janssen, R.H.; Vincken, J.-P.; van den Broek, L.A.M.; Fogliano, V.; Lakemond, C.M.M. Nitrogen-to-protein conversion factors for three edible insects: Tenebrio molitor, Alphitobius diaperinus, and Hermetia illucens. *J. Agric. Food Chem.* **2017**, *65*, 2275–2278. [CrossRef]

21. Stoops, J.; Vandeweyer, D.; Crauwels, S.; Verreth, C.; Boeckx, H.; Van Der Borght, M.; Claes, J.; Lievens, B.; Van Campenhout, L. Minced meat-like products from mealworm larvae (Tenebrio molitor and Alphitobius diaperinus): microbial dynamics during production and storage. *Innov. Food Sci. Emerg. Technol.* **2017**, *41*, 1–9. [CrossRef]

22. Mancini, S.; Preziuso, G.; Dal Bosco, A.; Roscini, V.; Szendro, Z.; Fratini, F.; Paci, G.; Szendrő, Z.; Fratini, F.; Paci, G. Effect of turmeric powder (Curcuma longa L.) and ascorbic acid on physical characteristics and oxidative status of fresh and stored rabbit burgers. *Meat Sci.* **2015**, *110*, 93–100. [CrossRef]

23. Zaspel, B.J.; Csallany, A.S. Determination of alpha-tocopherol in tissues and plasma by high-performance liquid chromatography. *Anal. Biochem.* **1983**, *130*, 146–150. [CrossRef]

24. R Core Team. *R: A Language and Environment for Statistical Computing*; R Foundation for Statistical Computing: Vienna, Austria. Available online: https://www.r-project.org/ (accessed on 24 July 2019).

25. Herrero, M.; Thornton, P.K. Livestock and global change: Emerging issues for sustainable food systems. *Proc. Natl. Acad. Sci.* **2013**, *110*, 20878–20881. [CrossRef] [PubMed]

26. Grabowski, N.T.; Klein, G. Microbiology of processed edible insect products – Results of a preliminary survey. *Int. J. Food Microbiol.* **2017**, *243*, 103–107. [CrossRef] [PubMed]

27. Grabowski, N.T.; Klein, G. Microbiological analysis of raw edible insects. *J. Insects as Food Feed* **2017**, *3*, 7–14. [CrossRef]

28. Wynants, E.; Crauwels, S.; Verreth, C.; Gianotten, N.; Lievens, B.; Claes, J.; Van Campenhout, L. Microbial dynamics during production of lesser mealworms (Alphitobius diaperinus) for human consumption at industrial scale. *Food Microbiol.* **2018**, *70*, 181–191. [CrossRef] [PubMed]

29. Vandeweyer, D.; Crauwels, S.; Lievens, B.; Van Campenhout, L. Microbial counts of mealworm larvae (Tenebrio molitor) and crickets (Acheta domesticus and Gryllodes sigillatus) from different rearing companies and different production batches. *Int. J. Food Microbiol.* **2017**, *242*, 13–18. [CrossRef] [PubMed]

30. Wynants, E.; Crauwels, S.; Lievens, B.; Luca, S.; Claes, J.; Borremans, A.; Bruyninckx, L.; Van Campenhout, L. Effect of post-harvest starvation and rinsing on the microbial numbers and the bacterial community composition of mealworm larvae (Tenebrio molitor). *Innov. Food Sci. Emerg. Technol.* **2017**, *42*, 8–15. [CrossRef]

31. Mancini, S.; Fratini, F.; Tuccinardi, T.; Turchi, B.; Nuvoloni, R.; Paci, G. Effects of different blanching treatments on microbiological profile and quality of the mealworm (Tenebrio molitor). *J. Insects as Food Feed* **2019**, *5*, 225–234. [CrossRef]

32. Wynants, E.; Frooninckx, L.; Van Miert, S.; Geeraerd, A.; Claes, J.; Van Campenhout, L. Risks related to the presence of Salmonella sp. during rearing of mealworms (Tenebrio molitor) for food or feed: Survival in the substrate and transmission to the larvae. *Food Control.* **2019**, *100*, 227–234. [CrossRef]

33. Mancini, S.; Paci, G.; Ciardelli, V.; Turchi, B.; Pedonese, F.; Fratini, F. Listeria monocytogenes contamination of Tenebrio molitor larvae rearing substrate: Preliminary evaluations. *Food Microbiol.* **2019**, *83*, 104–108. [CrossRef]

34. Arnold, K.E.; Ramsay, S.L.; Henderson, L.; Larcombe, S.D. Seasonal variation in diet quality: antioxidants, invertebrates and blue tits Cyanistes caeruleus. *Biol. J. Linn. Soc.* **2010**, *99*, 708–717. [CrossRef]

MDPI

St. Alban-Anlage 66

4052 Basel

Switzerland

Tel. +41 61 683 77 34

Fax +41 61 302 89 18

www.mdpi.com

Animals Editorial Office

E-mail: animals@mdpi.com

www.mdpi.com/journal/animals